CAMBRIDGE
International Examinations

BIOLOGY IN CONTEXT

for Cambridge International AS & A Level

Second edition

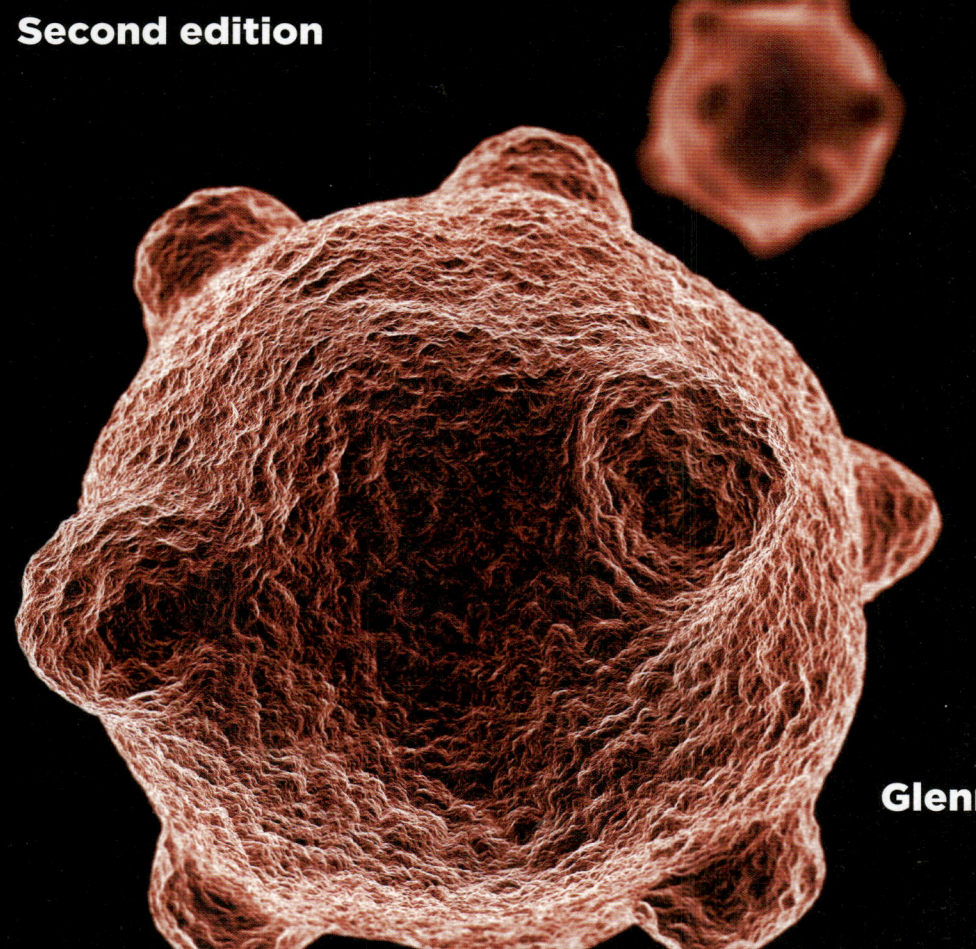

Glenn and Susan Toole

Stephanie Fowler

OXFORD
UNIVERSITY PRESS

Great Clarendon Street, Oxford, OX2 6DP, United Kingdom

Oxford University Press is a department of the University of Oxford.
It furthers the University's objective of excellence in research, scholarship, and education by publishing worldwide. Oxford is a registered trade mark of Oxford University Press in the UK and in certain other countries

British Library Cataloguing in Publication Data
Data available

978-0-19-835478-9

1 3 5 7 9 10 8 6 4 2

MIX
Paper from responsible sources
FSC® C007785
www.fsc.org

Paper used in the production of this book is a natural, recyclable product made from wood grown in sustainable forests. The manufacturing process conforms to the environmental regulations of the country of origin.

Typeset by GreenGate Publishing Services, Tonbridge, Kent

Printed in the UK by Bell and Bain Ltd, Glasgow

Acknowledgements

The publisher would like to thank Cambridge International Examinations for their kind permission to reproduce past paper questions. Cambridge International Examinations bears no responsibility for the example answers to questions taken from its past question papers which are contained in this publication.

Unless otherwise indicated, the questions, example answers, marks awarded, and comments that appear in this title were written by the authors. In examination, the way marks are awarded to answers like these may be different.

The publisher would like to thank the following for permissions to use their photographs:

Cover image: © gecko753 / iStock; p3: DR JEREMY BURGESS/SCIENCE PHOTO LIBRARY; p4: ANDREW LAMBERT PHOTOGRAPHY/SCIENCE PHOTO LIBRARY; p5: DR GOPAL MURTI/SCIENCE PHOTO LIBRARY; p5: CLAUDE NURIDSANY & MARIE PERENNOU/SCIENCE PHOTO LIBRARY; p6: © MARKA / Alamy; p7: DR GOPAL MURTI/ SCIENCE PHOTO LIBRARY; p7: DR.JEREMY BURGESS/SCIENCE PHOTO LIBRARY; p9: DR GOPAL MURTI/SCIENCE PHOTO LIBRARY; p12: © PHOTOTAKE Inc. / Alamy; p12: DR KARI LOUNATMAA/SCIENCE PHOTO LIBRARY; p14: MEDIMAGE/SCIENCE PHOTO LIBRARY; p14: Don W. Fawcett/SCIENCE PHOTO LIBRARY; p15: SCIENCE PHOTO LIBRARY; p16: DON W. FAWCETT/SCIENCE PHOTO LIBRARY; p17: DON W. FAWCETT/ SCIENCE PHOTO LIBRARY; p18: POWER AND SYRED/SCIENCE PHOTO LIBRARY; p38: mjf99/Shutterstock; p39: © Corbis Flirt / Alamy; p43: J.C. REVY, ISM/SCIENCE PHOTO LIBRARY; p45: © imagebroker / Alamy; p65: LOOK AT SCIENCES/SCIENCE PHOTO LIBRARY; p64: POWER AND SYRED/SCIENCE PHOTO LIBRARY; p66: PR. G GIMENEZ-MARTIN/SCIENCE PHOTO LIBRARY; p66: PR. G GIMENEZ-MARTIN/SCIENCE PHOTO LIBRARY; p66: PR. G GIMENEZ-MARTIN/SCIENCE PHOTO LIBRARY; p66: PR. G GIMENEZ-MARTIN/SCIENCE PHOTO LIBRARY; p69: VERONIQUE LEPLAT/SCIENCE PHOTO LIBRARY; p68: DR YORGOS NIKAS/SCIENCE PHOTO LIBRARY; p69: STEVE GSCHMEISSNER/SCIENCE PHOTO LIBRARY; p70: HERVE CONGE, ISM /SCIENCE PHOTO LIBRARY; p80: WILL & DENI MCINTYRE/SCIENCE PHOTO LIBRARY; p88 (top): © travelstock44 / Alamy; p88 (bottom): © D. Hurst / Alamy; p92 (top): DR KEITH WHEELER/SCIENCE PHOTO LIBRARY; p92 (bottom): DR KEITH WHEELER/SCIENCE PHOTO LIBRARY; p93 (top): BIOPHOTO ASSOCIATES/SCIENCE PHOTO LIBRARY; p93 (middle): GARRY DELONG/SCIENCE PHOTO LIBRARY; p93 (bottom): DR KEITH WHEELER/SCIENCE PHOTO LIBRARY; p94 (top): CLAUDE NURIDSANY & MARIE PERENNOU/SCIENCE PHOTO LIBRARY; p94 (bottom): DR DAVID FURNESS, KEELE UNIVERSITY/SCIENCE PHOTO LIBRARY; p99: PETER BOND, EM CENTRE, UNIVERSITY OF PLYMOUTH/SCIENCE PHOTO LIBRARY; p100: DR JEREMY BURGESS/SCIENCE PHOTO LIBRARY; p102 (top): DR KEITH WHEELER/SCIENCE PHOTO LIBRARY; p102 (middle): © Steve Hamblin / Alamy; p102 (bottom): Authors; p103: JOHN CLEGG/ SCIENCE PHOTO LIBRARY; p104 (top): DR KEITH WHEELER/SCIENCE PHOTO LIBRARY; p104 (bottom): J.C. REVY, ISM/SCIENCE PHOTO LIBRARY; p112: Dr. Gladden Willis/Visuals Unlimited, Inc.; p113: THOMAS DEERINCK, NCMIR/SCIENCE PHOTO LIBRARY; p114: NATIONAL CANCER INSTITUTE/SCIENCE PHOTO LIBRARY; p115: STEVE GSCHMEISSNER/SCIENCE PHOTO LIBRARY; p119: FRANCIS LEROY, BIOCOSMOS/SCIENCE PHOTO LIBRARY; p122 (top): © Jake Norton / Alamy; p122 (bottom): © Lou Linwei / Alamy; p124: DR KEITH WHEELER/SCIENCE PHOTO LIBRARY; p130: michaeljung/Shutterstock; p131: Maridav; p133: ZEPHYR/SCIENCE PHOTO LIBRARY; p135 (top): PASIEKA/SCIENCE PHOTO LIBRARY; p135 (bottom): PROF. P. MOTTA/DEPT. OF ANATOMY/UNIVERSITY "LA SAPIENZA", ROME/SCIENCE PHOTO LIBRARY; p136: Dr. John Sasner/UNH/Visuals Unlimited, Inc.; p139: © Dennis Cox / Alamy; p138: © Robert Neumann / Alamy; p140 (top): JAMES STEVESON/ SCIENCE PHOTO LIBRARY; p140 (second image from top: TISSUEPIX/SCIENCE PHOTO LIBRARY; p140 (third image from top): BIOPHOTO ASSOCIATES/SCIENCE PHOTO LIBRARY; p140 (bottom): MANFRED KAGE/SCIENCE PHOTO LIBRARY.; p143 (left): PROFESSOR P.M. MOTTA, G. MACCHIARELLI, S.A NOTTOLA/SCIENCE PHOTO LIBRARY; p143 (right): PROFESSOR P.M. MOTTA, G. MACCHIARELLI, S.A NOTTOLA/ SCIENCE PHOTO LIBRARY; p145: ARTHUR GLAUBERMAN/SCIENCE PHOTO LIBRARY; p146: © Homer W Sykes / Alamy; p149: Mike Orlov/Shutterstock; p151: STEVE GSCHMEISSNER/SCIENCE PHOTO LIBRARY; p152: © Andrew McConnell / Alamy; p153: © Medical-on-Line / Alamy; p154: PASIEKA/SCIENCE PHOTO LIBRARY; p155: SCIENCE SOURCE/SCIENCE PHOTO LIBRARY; p156: Konstantin Sutyagin/ Shutterstock; 158 (top): CNRI/SCIENCE PHOTO LIBRARY; 158 (bottom): © Julio Etchart / Alamy; 160 (top): CNRI/SCIENCE PHOTO LIBRARY; 160 (bottom): SINCLAIR STAMMERS/SCIENCE PHOTO LIBRARY; p162: PASIEKA/SCIENCE PHOTO LIBRARY; p164: Dr. Hans Gelderblom/Visuals Unlimited, Inc.; p166: © Medical-on-Line / Alamy; p167: DAVID SCHARF/SCIENCE PHOTO LIBRARY; p168: © Jenny Matthews / Alamy; p169: JOHN DURHAM/SCIENCE PHOTO LIBRARY; p172: BIOLOGY MEDIA/SCIENCE PHOTO LIBRARY; p174: STEVE GSCHMEISSNER/SCIENCE PHOTO LIBRARY; p176: DR JEREMY BURGESS/SCIENCE PHOTO LIBRARY; p176: anyaivanova/Shutterstock; p177: © eye35 / Alamy; p182: Jaimie Duplass/Shutterstock; p184: © PHOTOTAKE Inc. / Alamy; p185: ROBERT LONGUEHAYE, NIBSC/SCIENCE PHOTO LIBRARY; p195: Credit: CNRI/SCIENCE PHOTO LIBRARY; p201 (left): © blickwinkel / Alamy; p201 (right): © Eric Gevaert / Alamy; p202: TONY CAMACHO/SCIENCE PHOTO LIBRARY; p204 (left): Bob Denelzen/Shutterstock; p204 (right): mchin/Shutterstock; p207: DR.JEREMY BURGESS/SCIENCE PHOTO LIBRARY; p217: BILL BARKSDALE/AGSTOCKUSA/SCIENCE PHOTO LIBRARY; p218 (left): DR KENNETH R. MILLER/SCIENCE PHOTO LIBRARY; p220 (top): © Mitsuaki Iwago/Minden Pictures/Corbis; p220 (middle): © imagebroker / Alamy; p220 (bottom): © Mike Goldwater / Alamy; p224 (top): ASTRID & HANNS-FRIEDER MICHLER/SCIENCE PHOTO LIBRARY; p224 (bottom): Dr. John D. Cunningham/Visuals Unlimited, Inc.; p225: PROF. P. MOTTA/DEPT. OF ANATOMY/ UNIVERSITY "LA SAPIENZA", ROME/SCIENCE PHOTO LIBRARY; p227: THOMAS DEERINCK, NCMIR/SCIENCE PHOTO LIBRARY; p233: STEVE GSCHMEISSNER/ SCIENCE PHOTO LIBRARY; p234: J.C. REVY, ISM/SCIENCE PHOTO LIBRARY; p237: SATURN STILLS/SCIENCE PHOTO LIBRARY; p239: DR JEREMY BURGESS/SCIENCE PHOTO LIBRARY; p240 (left): RALPH HUTCHINGS, VISUALS UNLIMITED /SCIENCE PHOTO LIBRARY; p240 (right): STEVE GSCHMEISSNER/SCIENCE PHOTO LIBRARY; p243: STEVE GSCHMEISSNER/SCIENCE PHOTO LIBRARY; p247: MARTYN F. CHILLMAID/SCIENCE PHOTO LIBRARY; p253: STEVE GSCHMEISSNER/SCIENCE PHOTO LIBRARY; p254: CNRI/SCIENCE PHOTO LIBRARY; p259 (top): BIOLOGY MEDIA/ SCIENCE PHOTO LIBRARY; p259 (bottom): ASTRID andamp;amp; HANNS-FRIEDER MICHLER/SCIENCE PHOTO LIBRARY; p265: DR YORGOS NIKAS/SCIENCE PHOTO LIBRARY; p266: SATURN STILLS/SCIENCE PHOTO LIBRARY; p267: mage Point Fr/ Shutterstock; p268: Kitty Bern/Shutterstock; p274 (top): SCIENCE PICTURES LTD/ SCIENCE PHOTO LIBRARY; p274 (bottom): ADRIAN T SUMNER/SCIENCE PHOTO LIBRARY; p275 (top): ADRIAN T SUMNER/SCIENCE PHOTO LIBRARY; p275 (bottom): © PHOTOTAKE Inc. / Alamy; p288: SCIENCE PHOTO LIBRARY; p291 (left): CreativeNature R.Zwerver; p291 (middle): Vasiliy Koval; p291 (right): Vasiliy Koval; p292 (top): © FLPA / Alamy; p292 (bottom right): © Jan Kupracz / Alamy; p292 (bottom left): © Vasiliy Vishnevskiy / Alamy; p299: DR GOPAL MURTI/SCIENCE PHOTO LIBRARY; p306 (top): chromatos/Shutterstock; p306 (middle): bozulek/ Shutterstock; p306 (bottom): outdoorsman/Shutterstock; p313 (top): MICHAEL W. TWEEDIE/SCIENCE PHOTO LIBRARY; p313 (bottom): MICHAEL W. TWEEDIE/ SCIENCE PHOTO LIBRARY; p315: JEANNE WHITE/SCIENCE PHOTO LIBRARY; p320: EYE OF SCIENCE/SCIENCE PHOTO LIBRARY; p324 (top): Menno Schaefer/ Shutterstock; p324 (bottom): © imageBROKER / Alamy; p325: PETER MENZEL/ SCIENCE PHOTO LIBRARY; p327: © Pierre BRYE / Alamy; p329 (top): © Top-Pics TBK / Alamy; p329 (bottom): © Juniors Bildarchiv GmbH / Alamy; p330: © Universal Images Group Limited / Alamy; p333 (top): © Glyn Ryland / Alamy; p333 (bottom): © John T.L / Alamy; p334 (top): CNRI/SCIENCE PHOTO LIBRARY; p334 (bottom): Wim van Egmond/Visuals Unlimited, Inc.; p335 (left): VERONIQUE LEPLAT/SCIENCE PHOTO LIBRARY; p335 (right): BOB GIBBONS/SCIENCE PHOTO LIBRARY; p336 (top): © Dennis Frates / Alamy; p336 (bottom): SIMON FRASER/SCIENCE PHOTO LIBRARY; p337: © Geoffrey Morgan / Alamy; p338: © Nature Picture Library / Alamy; p340: © Frans Lanting Studio / Alamy; p344: DR JEREMY BURGESS/SCIENCE PHOTO LIBRARY; p347: MARTYN F. CHILLMAID/SCIENCE PHOTO LIBRARY; p352: WAYNE LAWLER/SCIENCE PHOTO LIBRARY; p353 (top): © Ivy Close Images / Alamy; p353 (bottom): JULIE DERMANSKY/SCIENCE PHOTO LIBRARY; p354 (left): © Jonathan Dorey - China / Alamy; p354 (right): © blickwinkel / Alamy; p355: MATTHEW OLDFIELD/SCIENCE PHOTO LIBRARY; p356: PATRICK LANDMANN/SCIENCE PHOTO LIBRARY; p362: PHILIPPE PSAILA/SCIENCE PHOTO LIBRARY; p365: DR GOPAL MURTI/SCIENCE PHOTO LIBRARY; p367: PASCAL GOETGHELUCK/SCIENCE PHOTO LIBRARY; p369: IAN HOOTON/SCIENCE PHOTO LIBRARY; p370: SCOTT SINKLIER/AGSTOCKUSA/ SCIENCE PHOTO LIBRARY; p372 (top): © Martin Shields / Alamy; p372 (bottom): JAMES KING-HOLMES/SCIENCE PHOTO LIBRARY; p373: Alila Medical Media; p381: MAURO FERMARIELLO/SCIENCE PHOTO LIBRARY; p382: © Wayne HUTCHINSON / Alamy; p383: © Nigel Cattlin / Alamy.

OxNav acknowledgements: S. Fowler and C. Tranter

Artwork by GreenGate Publishing Services and OUP.

Contents

A section on practical skills can be found on the accompanying CD. The relevant sections in the book are highlighted using conical flask icons.

 CD icons indicate additional assessment material on the accompanying CD.

Introduction

Biology in Context aims to make your study of biology successful and interesting.

It has been written specifically to meet the requirements of Supplementary Level (AS Level) and Advanced Level (A Level) Biology, and is particularly suited to the Cambridge International syllabus. The content also covers the requirements of the Caribbean Advanced Proficiency Examination (CAPE)®.

The book is divided into 20 chapters:

Chapters 1–11 cover AS Level.

Chapters 12–20 cover A Level.

New ideas are presented in the book in a careful step-by-step manner to allow you to develop a firm understanding of concepts and ideas. Biology at this level will require you to describe and explain facts and processes in detail and with accuracy. The course is also about developing skills so that you can apply what you have learned.

The more basic topics are covered in the early chapters of the book and each chapter contains related information. However, the topics do not have to be studied in strict numerical order, but can be followed in any sequence which suits you or your teacher.

New biological discoveries are made each day, which add to our knowledge and understanding of the world and make it a safer and healthier place to live. Biology in Context explores these discoveries in a way that not only provides the facts, but also considers the moral, ethical and economic implications which they present.

The layout of the book is designed to cover information in a clear way that is easy to access. Its features include:

- **Double page spreads** to divide the material into manageable portions by covering a single topic within the pages on view.
- **Full colour diagrams** to illustrate points made in the text. Labels and annotations are included so that these diagrams aid your understanding and improve clarity.
- **Colour photographs** to improve your understanding further and add realism to the information and ideas within the text.
- **Extensive use of bullet points** to produce lists of information which you will find easy to follow. They permit you to quickly reference information. They often have key introductory words that make your learning, and hence your revision, easier.
- **Frequent cross referencing** to link different topics and provide you with a fully integrated understanding of biology as a whole.
- **Accessible language** to improve your comprehension and understanding.
- **Bold type** to emphasise key terms in the text which you will need to be able to define and understand. They also allow for quick reference when you are looking through the book. This should make your revision more effective.
- **Purple type** to highlight biological words that are not defined within a specific topic, but which can be found in the glossary. This enables you to easily access a full explanation of important biological terms used in the text.
- **Extension material** helps to widen your horizons and stimulate an interest in broader aspects of biology. Much of the 'Extension material' includes content that is required for the Caribbean Advanced Proficiency Examination (CAPE)®.

- **Hints in boxes** to give you useful advice about certain aspects of a topic and so aid your learning. These hints may be memory aids, helpful information, useful tips on answering examination questions, or warnings about common errors made by students.
- **Comprehensive glossary** to provide you with a clear and concise definition of over 250 biological terms used throughout the book.
- **Summary tests** to provide a quick check on how well you have learnt and understood the factual content of each topic. Answers are provided at the back of the book, so that the accurately completed test gives you a concise summary of the information in each topic.
- **Cambridge International Examination questions** to give you practice of the type of questions you can expect in the final examination. This will allow you to check your progress. Arranged at the end of each chapter, these questions test the full range of skills expected at AS and A Level, including application of knowledge, understanding, analysis, synthesis and evaluation. The questions cover mostly material in the same chapter but may sometimes include information from other chapters, largely earlier ones.
- **Suggested answers to examination questions** which indicate the possible type of responses that are likely to bring you credit in examinations. These answers are more than just a mark scheme. They often include explanations of the answers to aid your learning and understanding. The answers are not exhaustive and there may be acceptable alternatives.

Cambridge International Examinations bears no responsibility for the example answers to questions taken from its past questions papers which are contained in this publication.

Key concepts in Biology

In many textbooks, including Biology in Context, the subject of biology is broken down into small parts to help you understand its often complex elements. In reality, however, biology has just a few underlying principles that run through all aspects of the subject. These are known as **key concepts**. In learning each of the individual components of biology, it is possible to lose sight of these key concepts, or even to ignore them altogether. To do so would be a mistake, as an appreciation of the key concepts not only unifies biology but also aids your understanding of the subject.

Key concepts are essential principles, theories and ideas that help you to develop a deeper comprehension of biology and to make relevant links between different topics. They are, in effect, the foundations upon which the whole subject is based. An awareness of key concepts allows you to see biology, not as a set of isolated topics, but rather an interrelated and coherent whole. Once you have mastered the key concepts you will be able to use them to help you solve problems and to understand related biological material that is completely new to you.

There are a number of possible ways to divide up the key concepts in Biology. For the purposes of the Cambridge International AS and A Level Biology syllabus the six key concepts are:

- **Cells as units of life**
  A cell is the basic unit of life and all organisms are composed of one or more cells. There are two fundamental types of cell: prokaryotic and eukaryotic.

- **Biochemical processes**
  Cells are dynamic: biochemistry and molecular biology help to explain how and why cells function as they do.

- **DNA, the molecule of heredity**
  Cells contain the molecule of heredity, DNA. Heredity is based on the inheritance of genes.

- **Natural selection**

  Natural selection is the major mechanism to explain the theory of evolution.

- **Organisms in their environment**

  All organisms interact with their biotic and abiotic environment.

- **Observation and experiment**

  The different fields of biology are intertwined and cannot be studied in isolation; observation and enquiry, experimentation and fieldwork are fundamental to biology.

As these key concepts are the fundamental ideas upon which the subject of biology is based, they can be found, in some form or other, within every chapter of this book. To illustrate this, the table opposite provides one example of each key concept for each of the 19 chapters that relate to the Cambridge International AS and A Level Biology syllabus.

The table is for illustration purposes only. The content does not need to be learned, as key concepts will not be assessed as such. They will, however, help you to understand the syllabus as a whole and to make useful connections between different aspects of biology. They are something to keep at the back of your mind – a framework upon which to hang the various elements of the subject.

Each entry in the table is just a single example of that key concept in the particular chapter. There are many more possible examples, and those chosen have no particular significance.

We trust that you will enjoy using this book and find it interesting and informative. We hope that it will build upon the knowledge and skills that you have already acquired and so stimulate a further interest in biology that encourages you to pursue your study beyond this level. Above all, we hope it will contribute to your success in the AS and A Level examinations.

Glenn and Susan Toole

Chapter in Biology in Context	Cells as the units of life	Biochemical processes	Key concept			Observation and experiment
			DNA, the molecule of heredity	Natural selection	Organisms in their environment	
1 Cell structure	All organisms are composed of cells	Biochemical processes are carried out by cell organelles e.g. the Golgi body forms glycoproteins	DNA is contained within the nucleus and instructs cell organelles e.g. ribosomes in protein synthesis	Natural selection has determined the structure of cells and their organelles e.g. muscle cells with many mitochondria	Environment affects cells e.g. phagocytes are attracted to chemicals produced by non-self cells	Microscopy is used to investigate cell structure
2 Biological molecules	Cells are made up of biological molecules such as carbohydrates, lipids and proteins	Biochemical processes convert molecules into one another e.g. hydrolysis of starch to glucose	DNA is a complex biological molecule made up of a long chain of nucleotides	Natural selection has determined the structure of molecules for certain functions e.g. collagen for strength in tendons	Environment changes structure of molecules e.g. temperature can change the shape of globular proteins	Food tests are used to identify reducing sugars, starch, lipids and proteins
3 Enzymes	Cells produce enzymes to catalyse reactions inside and outside of them e.g. amylase to breakdown starch.	Biochemical processes are controlled by enzymes e.g. lipase controls the breakdown of lipids to fatty acids and glycerol	The replication of DNA involves enzymes e.g. DNA polymerase	Natural selection has determined which enzymes any individual cell produces e.g. cells lining the small intestine produce lactase	Environmental changes affect enzymes e.g. temperatures above 60°C denature many enzymes	Experiments can be carried out to determine the effect of temperature and pH on enzyme action
4 Cell membranes and transport	Cell membranes control the movement of molecules in and out of cells e.g. co-transport of glucose by protein carriers	Active transport is a biochemical process involved in the movement of ions e.g. potassium ions across the cell membrane	DNA is contained within the nucleus by the plasma membrane that surrounds it	The complex structure of cell membranes has evolved through natural selection	Cell membranes are the interface between cells and their environment and control the interchange of molecules between the two.	Osmosis can be observed in experiments with potato sticks, beetroot and onion epidermis
5 Cell and nuclear division	Diploid cells divide by mitosis to give diploid cells and by meiosis to give haploid cells	During interphase of the cell cycle biochemical processes produce new cell organelles	DNA replication takes place during the S phase of interphase in the cell cycle	The process of meiosis produces variety in offspring which is essential to natural selection	Environmental factors, such as certain chemicals and radiation, can affect mitosis and lead to cancer	Mitosis can be observed through a microscope in stained preparations of cells
6 Nucleic acids and protein synthesis	Cells contain the organelles required for DNA replication and for protein synthesis e.g. ribosomes,	Protein synthesis involves a variety of biochemical processes using enzymes like DNA helicase and RNA polymerase	DNA codes for the transcription of mRNA that is translated into specific polypeptides	It is variation in DNA that determines the range of characteristics in organisms from which the best adapted are selected	Environmental factors can determine which forms of DNA are best suited to survival e.g. selection of the sickle cell trait in malarial regions.	The experiments of Meselsohn and Stahl showed that DNA replication was semi-conservative
7 Transport in plants	Cells called sieve tubes are specially adapted for sugar transport and xylem vessels for water transport	Biochemical process are involved in the movement of sucrose into sieve tubes and the opening and closing of stomata	Lengths of DNA code for the structure of sieve tubes and xylem vessels	The special adaptations of sieve tubes and xylem vessels that make them suitable for transporting materials arose as the result of natural selection	The rate of transpiration is affected by environmental factors like humidity, light and air movement	A potometer can be used to measure water uptake by a plant under different environmental conditions
8 Transport in mammals	Many different cell types are involved in animal transport e.g. red blood cells, endothelial cells and cardiac muscle	The formation of oxy-haemoglobin and carbamino-haemoglobin are the result of biochemical processes	DNA determines the structure of haemoglobin – a molecule adapted to transporting oxygen in blood	Native highlanders adapted to the transport of oxygen at low atmospheric oxygen partial pressures are the result of natural selection	The loading of haemoglobin with oxygen changes under different environmental conditions e.g. it is greater at low CO_2 concentrations	The heart rate before and after exercise, can be calculated by taking the pulse

Chapter in Biology in Context	Cells as the units of life	Biochemical processes	DNA, the molecule of heredity	Natural selection	Organisms in their environment	Observation and experiment
			Key concept			
9 Gas exchange and smoking	Specialised cells are involved in gas exchange e.g. ciliated epithelial and goblet cells in lungs	Nicotine in tobacco smoke causes biochemical changes because it stimulates production of adrenaline	Carcinogens in tobacco smoke cause mutations in DNA that lead to lung cancer	Natural selection favours individuals living at high altitude if they have greater lung capacities	Environment conditions such as air pollution and smoking can cause COPD and so affect gas exchange in the lungs	Using a microscope, comparisons can be made of lung tissue of smokers and non-smokers
10 Infectious disease	Some prokaryotic cells cause disease e.g.the bacterium *Vibrio cholerae* causes cholera	Pathogens often cause disease by interfering with the normal biochemical processes of host cells e.g. HIV takes over the processes of T helper cells	HIV prevents the host cell DNA determining the proteins to be made and uses the cells to make the proteins it requires instead	Natural selection of mutant bacteria that produce penicillinase has lead to antibiotic resistance	Environmental conditions determine the geographical distribution of vectors and so the diseases they spread e.g. *Anopheles* mosquitoes, and hence malaria, occurs in tropical/sub tropical areas	Antibiotic discs placed on bacterial lawns grown on agar can identify bacterial resistance
11 Immunity	White blood cells protect organisms from disease and produce antibodies which provide immunity from further infection	T killer cells produce proteins called perforins which make holes in cell surface membranes	DNA codes for the structure of the different types of white blood cells involved in the immune process	Immunity is a factor in selection. Organisms with immunity to a disease are more likely to survive to breed in the presence of that disease than those without	It is a response to the environment, e.g. exposure to a pathogen, that stimulates an immune response	Monoclonal antibodies immobilized on a dipstick can be used in the detection of substances in urine
12 Energy and respiration	Cells possess mitochondria which are essential for cellular respiration	The respiratory processes glycolysis and Krebs cycle are examples of biochemical pathways	The enzymes of glycolysis and Krebs cycle are coded for by DNA	Anaerobic organisms have evolved, through natural selection, respiratory pathways that enable them to release energy in the absence of molecular oxygen	The rate of cellular respiration is influenced by environmental conditions e.g. in most organisms it slows in low temperatures, cyanide can halt it altogether	A respirometer can be used to measure the rate of respiration in different environmental conditions
13 Photosynthesis	Photosynthesis takes place in palisade cells which possess the cell organelles, chloroplasts	The light independent reaction (Calvin cycle) of photosynthesis is a biochemical process	Chloroplasts contain their own DNA, separate from that in the nucleus	It is thought that chloroplasts evolved from a symbiotic relationship between a eukaryotic cell and a photosynthetic prokaryotic one	The rate of photosynthesis is determined by environmental factors such as light intensity and carbon dioxide concentration	The rate of photosynthesis under differing conditions can be measured using a photosynthometer
14 Homeostasis	Homeostatic processes are important in maintaining a constant internal environment in cells	The homeostatic control of blood sugar involves the biochemical inter-conversion of glucose and glycogen	Variations in certain genes (DNA) increase the likelihood of individuals developing diabetes	Organisms that can maintain a constant internal environment have a selective advantage as they are more independent of the external environment	Homeostasis allows organisms a degree of independence from changes to the external environment	Biosensors such as dipsticks can be used to analyse the level of glucose in sample material
15 Control and coordination	Neurones are highly specialised cells adapted to carry nerve impulses	The nerve impulse is the result of chemical changes across the plasma membrane of the axon	Huntington's disease is a degenerative disorder of the nervous system due to a single gene mutation.	Natural selection has led to the evolution of a nervous system and a variety of sense organs adapted to the predominant stimuli received by an organism	The nervous system allows organisms to respond to environmental stimuli	Experiments can be carried out to investigate the responses of plants to stimuli such as light and gravity

Chapter in Biology in Context	Key concept					
	Cells as the units of life	Biochemical processes	DNA, the molecule of heredity	Natural selection	Organisms in their environment	Observation and experiment
16 Inherited change	Meiosis is the means by which cells reduce the number of chromosomes from the diploid to haploid number, usually in the formation of gametes	Biochemical processes such as the production of lactase are controlled by groups of genes e.g. the lac operon	A gene is a length of DNA that codes for a particular polypeptide	Mutations of genes produce new varieties of offspring that, through the process of natural selection, may evolve into new species	The phenotype of an organism is the result of interaction between the genotype of an organism and its environment	Genetic experiments can be carried out to illustrate Mendel's Laws e.g. using fruit flies
17 Selection and evolution	Variation results from the random fusion of egg and sperm cells	Kangaroo rats survive water shortage by adapting their biochemistry to oxidise fat instead of protein and to produce concentrated urine	Changes to the nucleotides in DNA give rise to mutations and lead to increased variety in organisms	Variation and overproduction of offspring are important factors in natural selection	The environment affects the expression of genes e.g. the effect of altitude on the size of genetically identical *Potentilla grandulosa* plants	Observation of features of organisms show how they have become adapted to different conditions e.g. xeromorphic features in plants
18 Biodiversity and conservation	The classification of organisms into the three domains (Bacteria, Archaea and Eukarya) is by cell type	The nitrogen cycle is an example of the chemical changes involved in the cycling of an element	Protecting endangered species is a means of preventing certain genes from being lost forever	Natural classification is based on the evolutionary relationships between organisms	Ecology is the study of the inter-relationships between organisms and their environment	Fieldwork involving a variety of techniques is essential to the study of ecology
19 Genetic technology	Plasmids from bacterial cells are the vectors used in gene technology	Reverse transcriptase and restriction endonucleases are important enzymes in the biochemical process of gene technology	Genetic technology entails the combining of DNA from two different organisms	Genetic technology is a form of artificial selection rather than natural selection and raises ethical issues about reducing variety.	Crop improvement through genetic modification can enable plants to better survive hostile environments e.g. tolerate dry conditions	Gel electrophoresis can be used to separate out proteins or DNA fragments

Cell structure

The microscope in cell studies

On these pages you will learn to:

- Calculate the linear magnifications of drawings, photomicrographs and electron micrographs
- Explain and distinguish between resolution and magnification, with reference to light microscopy

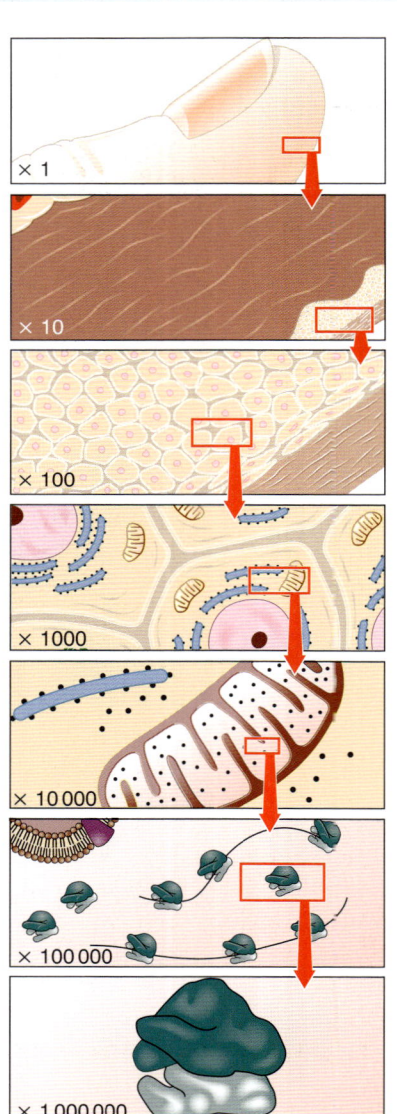

Figure 1 *The effect of progressive magnification of a portion of human skin*

× 1
× 10
× 100
× 1000
× 10 000
× 100 000
× 1 000 000

Have you ever wished that you could see just that little bit better? How annoying it is to be not quite able to see the numbers on the doors of houses when you are looking for a particular address, or to pick out a friend in a large crowd. Early scientists must have felt much the same as they struggled to make out the details of the objects they were studying. Imagine their joy at the development of the first glass lenses and then the compound light microscope – a whole new world was revealed to them.

In time, however, frustration returned as they tried to discover the sub-cellular detail (detail within the cell) that was beyond the limit of the light microscope. Their curiosity was satisfied by the development of the electron microscope in 1933 – an instrument that is now capable of allowing us to see images of objects that can be up to 500 000 times larger than their actual size, whereas the best light microscope achieves only 1500 times.

Thanks to microscopes, we now take it for granted that living organisms are composed of cells. This knowledge came, not as the result of a single piece of research, but rather from a series of discoveries made over more than two centuries.

The cell theory

The cell is the basic unit of life and current cell theory states that:

- All living organisms are made up of one cell (**unicellular**) or more cells (**multicellular**).
- Within cells biochemical (metabolic) reactions take place.
- New cells arise from existing ones.
- Cells possess the genetic material of an organism which is passed from parent cells to daughter cells.
- A cell is the smallest unit of an organism capable of surviving independently.

Cells are necessary because the chemical reactions of vital processes, such as respiration and photosynthesis, require molecules to come into contact with one another. Keeping the required molecules within the boundary of a cell surface membrane, rather than letting them disperse freely throughout the organism, ensures that they react more effectively. In addition, each cell surface membrane can control, to some extent, which molecules to allow in and which to exclude. In this way different cells can carry out different functions.

Microscopy

Microscopes are instruments that magnify the image of an object in some way. A simple convex glass lens can act as a magnifying glass but such lenses work more effectively if they are in a compound light microscope. The relatively long wavelength of light rays means that a light microscope can only distinguish between two objects if they are 0.2 µm, or further, apart. This restriction can be overcome by using beams of electrons rather than beams of light. With their shorter wavelengths, the beam of electrons in the electron microscope can distinguish two objects as close together as 0.1 nm.

However good a microscope is, it will only be effective if the material to be viewed under it is properly prepared. This often involves the material being stained in some way to make the parts more easily visible.

Magnification

The material put under the microscope can be termed the **object**. The appearance of this material when viewed under the microscope is known as the **image**. The magnification of an object is how many times bigger the image is when compared to the object.

$$\text{magnification} = \frac{\text{size of image}}{\text{size of object}}$$

You may be asked to calculate the size of an object when you know the size of the image and the magnification. In this case:

$$\text{size of object} = \frac{\text{size of image}}{\text{magnification}}$$

The important thing to remember when calculating the magnification is to ensure that the units of length (Table 1) are the same for both the object and the image.

Imagine, for example, that you know an object is actually 100 nm in length and you are asked how much it is magnified in a photograph. You should first measure the object in the photograph. Suppose it is 10 mm long. The magnification is:

$$\frac{\text{size of image}}{\text{size of object}} = \frac{10\,\text{mm}}{100\,\text{nm}}$$

Now convert the measurements to the same units – normally the smallest – which in this case is nanometres. In 10 millimetres there are 10 000 (10 × 1000) micrometres and in 10 000 micrometres there are 10 000 000 (10 000 × 1000) nanometres. Therefore the magnification is:

$$\frac{\text{size of image}}{\text{size of object}} = \frac{10\,000\,000\,\text{nm}}{100\,\text{nm}} = \frac{100\,000}{1}$$

$$= \times 100\,000$$

Resolution

The **resolution**, or **resolving power**, of a microscope is the minimum distance apart that two objects can be in order for them to appear as separate items. Whatever the type of microscope, the resolution depends on the wavelength or form of radiation used and can be estimated as approximately half the wavelength of the radiation. So in a light microscope it is about 0.2 μm – any two objects which are 0.2 μm or more apart will be seen separately, but any objects closer than 0.2 μm will appear as a single item. In other words, higher resolution means greater clarity. The higher the resolution, the more clear and detailed is the image produced.

Increasing the magnification will increase the size of an image, but does not always increase the resolution. Every microscope has a limit of resolution. Up to this point increasing the magnification **will** reveal more detail but beyond this point increasing the magnification **will not** – the object, while appearing larger, will just be more blurred.

Table 1 Units of length

Unit	Symbol	Equivalent in metres
kilometre	km	10^3
metre	m	1
millimetre	mm	10^{-3}
micrometre	μm	10^{-6}
nanometre	nm	10^{-9}

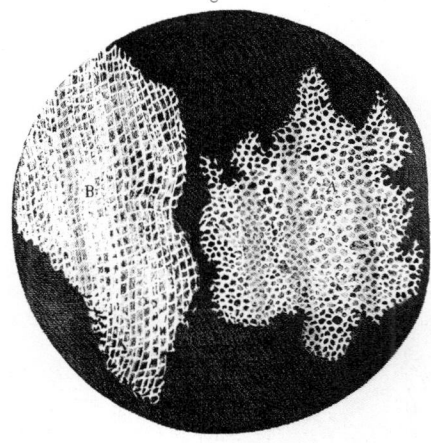

Figure 2 Cork cells as drawn by the English physicist Robert Hooke in 1665

SUMMARY TEST 1.1

The magnification of an object is how many times bigger the image is when compared to the object. If an object is 10 μm long and its image under the microscope is 1 mm long, then the object has been magnified **(1)** times. If an object is measured as 100 μm in diameter and it has been magnified 1000 times, the actual size of the object is **(2)** mm. The resolution of a microscope is the minimum distance apart that two objects can be in order for them to appear as **(3)** items. A light microscope has a resolution of around 0.2 **(4)** [units], whereas an **(5)** microscope has a much better resolution. The greater the resolution, the more **(6)** the image will appear.

The light microscope

On these pages you will learn to:

- Compare the structure of typical animal and plant cells as seen using the light microscope

In its simplest form, the light microscope consists of a single lens which operates as a magnifying glass. More effective is the **compound light microscope** (Figure 2), which has three systems of lenses.

- **The condenser lenses** are located beneath the microscope stage and can be adjusted in height to ensure that light is focused on the specimen being examined. This allows the resolving power of the microscope to be used to its full effect.
- **The objective lenses** produce an initial magnified image of the specimen.
- **The eyepiece lenses** further magnify the image produced by the objective lenses.

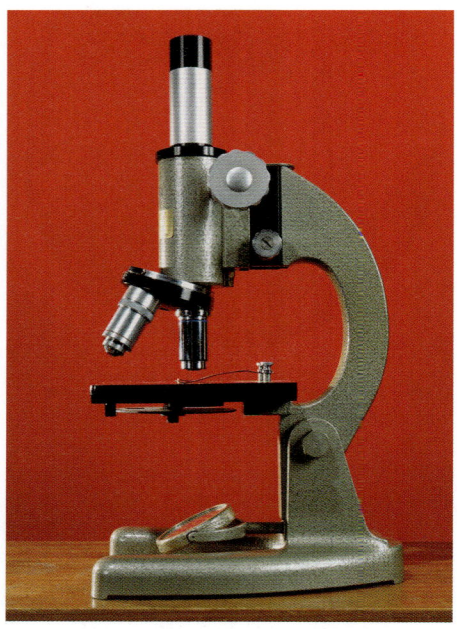

Figure 1 *A compound light microscope*

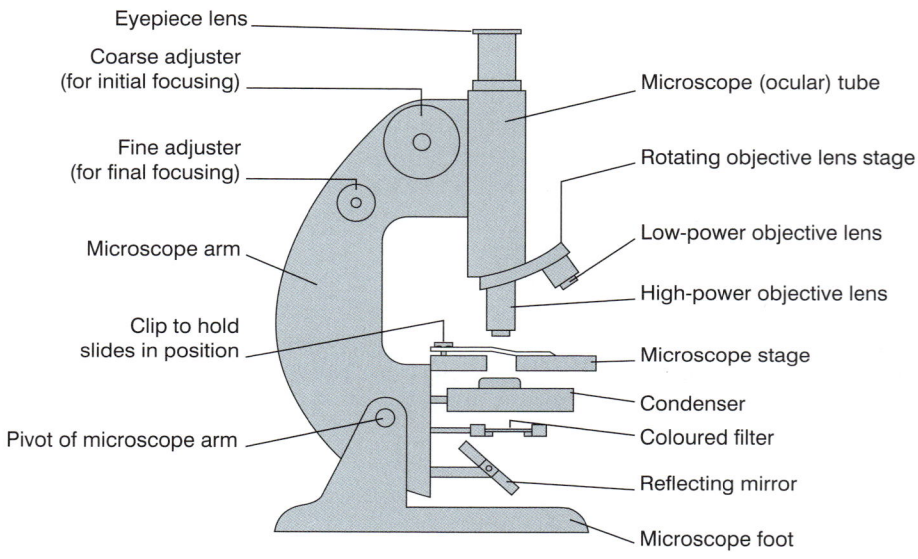

Eyepiece lens

Coarse adjuster (for initial focusing)

Fine adjuster (for final focusing)

Microscope arm

Clip to hold slides in position

Pivot of microscope arm

Microscope (ocular) tube

Rotating objective lens stage

Low-power objective lens

High-power objective lens

Microscope stage

Condenser

Coloured filter

Reflecting mirror

Microscope foot

Figure 2 *A compound light microscope*

REMEMBER

The compound light microscope is the type of microscope that you will use when studying the structure of plant and animal cells.

An animal cell as seen under a light microscope

When viewed under a light microscope, a typical animal cell is made up of:

- **Cell surface membrane** – a phospholipid bilayer containing proteins that controls the movement of materials in and out of the cell (Topic 4.1).
- **Cytoplasm** – watery material with a jelly-like consistency.
- **Nucleus** – made up of one or more nucleoli (singular: nucleolus) and thread-like chromatin (Topic 1.6) in a fluid nucleoplasm.
- **Mitochondria** – often rod-shaped structures within which the reactions of aerobic respiration occur (Topic 1.6).
- **Centrioles** – a pair of hollow cylinders involved in cell division (Topic 1.8).
- **Glycogen granules** – small particles found in some cells.

The structure of a typical animal cell as seen under a light microscope is illustrated in Figure 3.

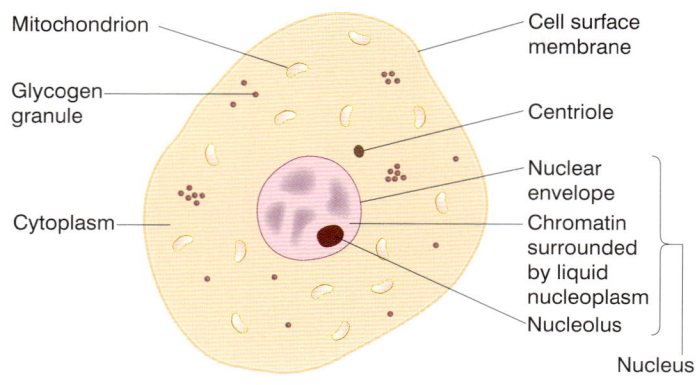

Figure 3 *Generalised animal cell as seen under a light microscope*

A plant cell as seen under a light microscope

A typical plant cell seen under a light microscope comprises:

- **Cell wall** – a tough wall containing cellulose fibres that surrounds the cell.
- **Cell surface membrane** – a phospholipid bilayer containing proteins, that controls movement of materials in and out of the cell (Topic 4.1).
- **Cytoplasm** – watery material with a jelly-like consistency.
- **Nucleus** – made up of one or more nucleoli (singular: nucleolus) and thread-like chromatin (Topic 1.6) in a fluid nucleoplasm.
- **Vacuole** – a large permanent structure that contains a solution of mineral salts and sugars known as **cell sap**.
- **Tonoplast** – a membrane that surrounds the vacuole.
- **Mitochondria** – often rod-shaped structures involved in aerobic respiration (Topic 1.6).
- **Starch grains** – spherical bodies that store carbohydrate.

The structure of a typical plant cell as seen under a light microscope is illustrated in Figure 6.

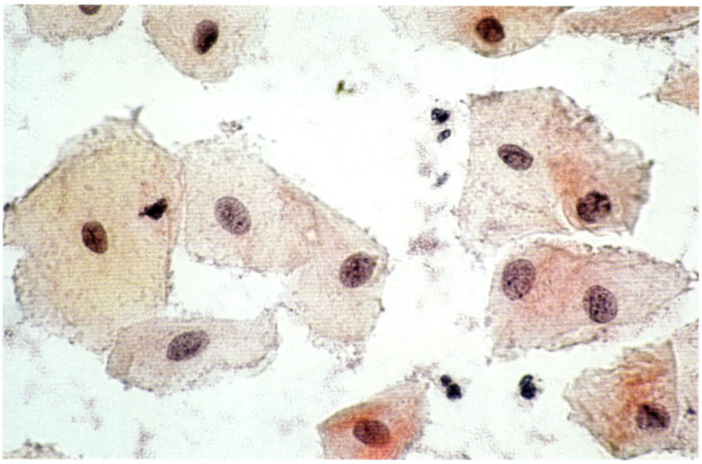

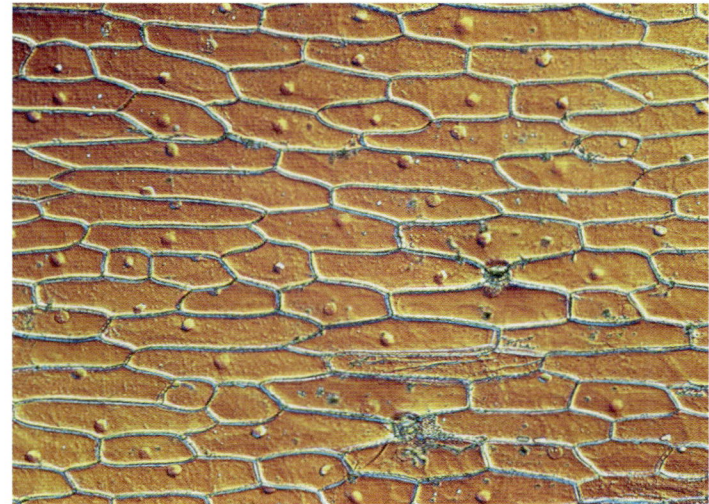

Figure 4 *A photomicrograph of animal cells (epithelial cells from the human mouth) as seen under a light microscope*

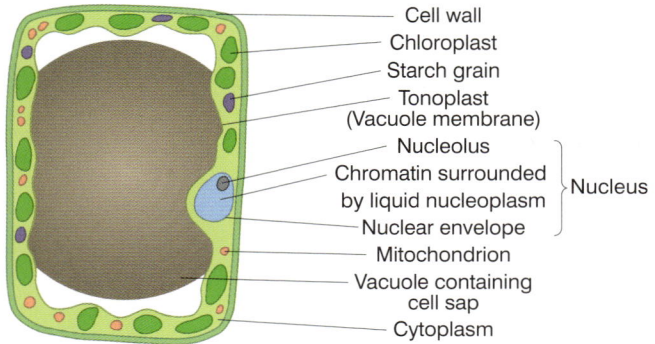

Figure 5 *A photomicrograph of plant cells (onion epidermal cells) as seen under a light microscope*

SUMMARY TEST 1.2

Viewed under a light microscope, an animal cell is seen to comprise a cell surface membrane made up of protein and **(1)**, cytoplasm, a nucleus with thread-like **(2)**, rod-shaped structures involved in respiration called **(3)**, centrioles and some cells have **(4)** granules which store carbohydrate; in plant cells these stores are seen as **(5)** grains. Plant cells are also surrounded by a **(6)**, which is made of the tough material called **(7)**. In the centre of a plant cell is a vacuole which contains a solution of salts and sugars called **(8)**. This vacuole is surrounded by a membrane known as the **(9)**.

Figure 6 *Generalised plant cell as seen under a light microscope*

1.3 The electron microscope

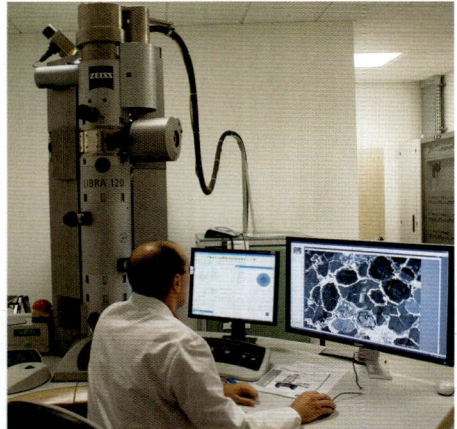

Figure 1 *An electron microscope*

Development of the electron microscope

In the 19th century scientists fully exploited the use of the light microscope but were increasingly aware of its limitations due to its poor resolution. This poor resolution is the result of the relatively long wavelength of light and so the search for a form of electromagnetic radiation with a shorter wavelength began. X-rays, in theory, should increase magnification but cannot easily be focused. In the 1930s, however, a microscope was developed that used a beam of **electrons**. This new microscope had two main advantages:

- The electron beam had a very short wavelength and the microscope could therefore resolve objects well – it had a high resolution.
- As electrons are negatively charged the beam could be focused using electromagnets (Figure 2).

The first electron microscopes could only resolve objects that were 100 nm apart – just twice the resolution of a light microscope. The best modern electron microscopes, by contrast, can resolve objects that are just 0.1 nm apart – 2000 times better than a light microscope. A comparison of the advantages and disadvantages of the light and electron microscopes is given in Table 1.

Table 1 *Comparison of advantages and disadvantages of the light and electron microscopes*

Light microscope	Electron microscope
Advantages	*Disadvantages*
Small and portable – can be used almost anywhere	Very large and must be operated in special rooms
Living and dead material can be observed	Only dead material can be observed
Preparation of material is relatively quick and simple, requiring only a little expertise	Preparation of material is lengthy and requires considerable expertise and sometimes complex equipment
Material rarely distorted by preparation	Preparation of material may distort it
Natural colour of the material can be observed	All images are in black and white but colour can be artificially added
Cheap to purchase and operate	Expensive to purchase and operate
Disadvantages	*Advantages*
Magnifies objects up to 1500× only	Magnifies objects more than 500 000×
Low resolution	High resolution
The depth of field is restricted	It is possible to investigate a greater depth of field

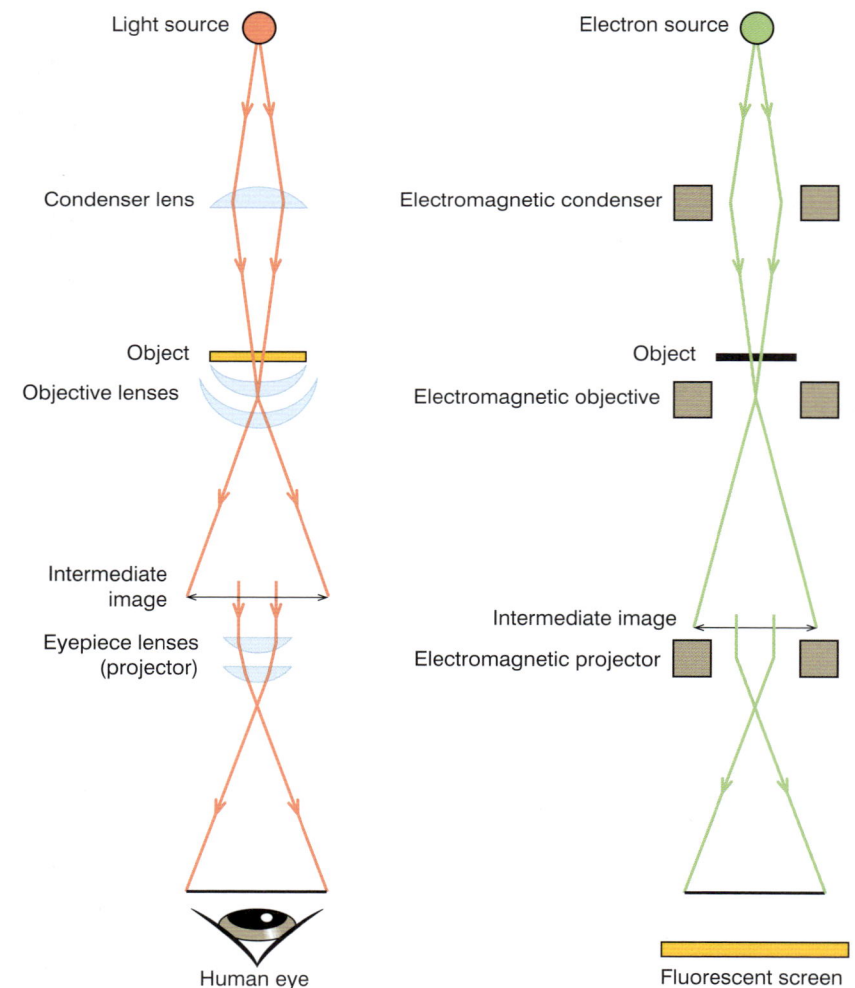

Figure 2 *Comparison of radiation pathways in light and electron microscopes*

Because electrons are absorbed or deflected by the molecules in air, a near-vacuum has to be created within the chamber of an electron microscope in order for it to work effectively. There are two types of electron microscope:

- the transmission electron microscope (TEM)
- the scanning electron microscope (SEM).

The transmission electron microscope (TEM)

The **transmission electron microscope (TEM)** consists of an electron gun that produces a beam of electrons by heating a tungsten filament. The beam is then 'focused' onto the specimen by means of the electromagnets which make up the condenser. The greater the electrical current applied to these electromagnets, the more the beam is deflected. After passing through the condenser, the image is enlarged by passing it through objective and projector lenses. The human eye cannot detect electrons and so the image is made visible by directing the electron beam onto a fluorescent screen. In a TEM, the beam passes through a thin section of the specimen. Parts of this specimen absorb electrons and therefore appear dark. Other parts of the specimen allow the electrons to pass through and hit the screen, which fluoresces, and so appears bright. The image produced on the screen can be photographed to give a **transmission electron micrograph**.

The scanning electron microscope (SEM)

One disadvantage of the TEM is that specimens must be extremely thin to allow electrons to penetrate. The result is therefore a flat, two-dimensional image. Basically similar to a TEM, the **scanning electron microscope (SEM)** overcomes this problem by directing a beam of electrons onto the surface of the specimen from above, rather than penetrating it from below. The specimen must first be dried and coated with a metal (to produce secondary electrons). The beam is then passed back and forth across a portion of the specimen in a regular pattern. The electrons are scattered by the specimen and the pattern of this scattering depends on the contours of the specimen surface. By computer analysis of the pattern of scattered electrons and secondary electrons produced, a three-dimensional image can be built up. The basic SEM has a lower resolution than a TEM, around 10 nm, but still 20 times better than a light microscope. The photographed image is known as a **scanning electron micrograph**.

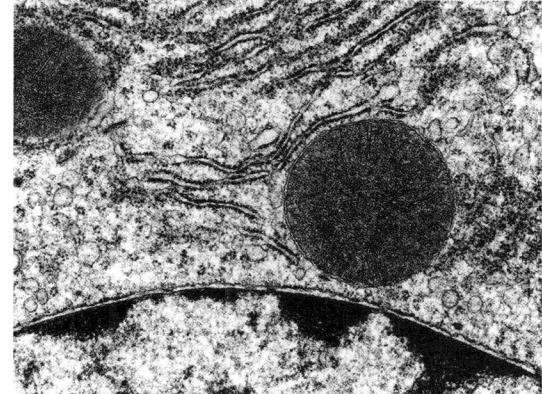

Figure 3 *Part of an animal cell as seen with a transmission electron microscope (×16 000)*

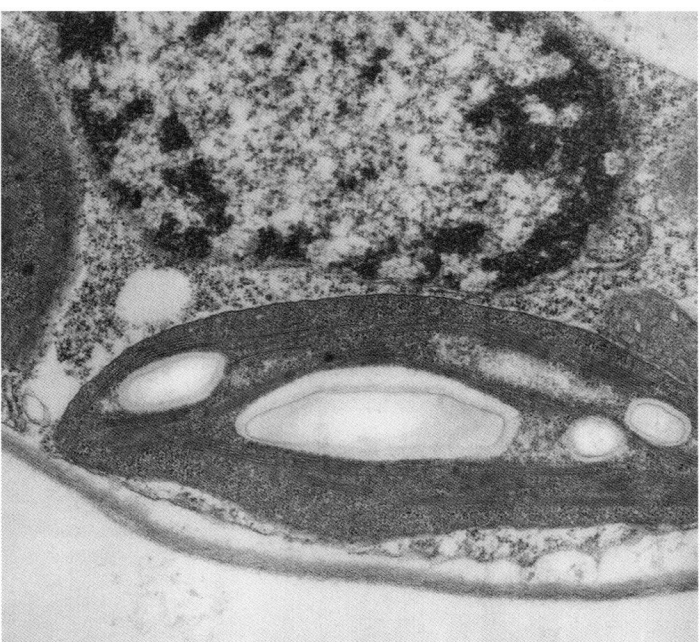

Figure 4 *Part of a plant cell as seen with a transmission electron microscope (×10 000)*

SUMMARY TEST 1.3

An electron microscope can resolve objects that are as little as **(1)** apart in distance. They use a beam of electrons which can be focused using **(2)**. To allow the electrons to move across the chamber of the electron microscope, the **(3)** within it must be largely removed. The type where a beam of electrons passes through a thin section of the specimen is called a **(4)** electron microscope. If a beam of electrons is passed backwards and forwards over the surface of the specimen, it is called a **(5)** electron microscope.

1.4   Microscopic measurements and calculations

On these pages you will learn to:

- Calculate the linear magnifications of drawings, photomicrographs and electron micrographs
- Use an eyepiece graticule and stage micrometer scale to measure cells and be familiar with units (millimetre, micrometre, nanometre) used in cell studies
- Calculate actual sizes of specimens from drawings, photomicrographs and electron micrographs

Measuring cells

When using a light microscope, we can measure the size of objects using an **eyepiece graticule**. The graticule is a glass disc, which is placed in the eyepiece of a microscope. On the glass disc is etched a scale. This scale is typically 10 mm long and is divided into 100 sub-divisions as shown in Figure 1. The scale is visible when looking down the eyepiece of the microscope.

The scale on the eyepiece graticule cannot be used directly to measure the size of objects under a microscope's objective lens because each objective lens will magnify to a different degree. The graticule must first be calibrated for a particular objective lens. Once calibrated for a certain objective lens, the graticule can remain in position for future use, provided the same objective lens is used.

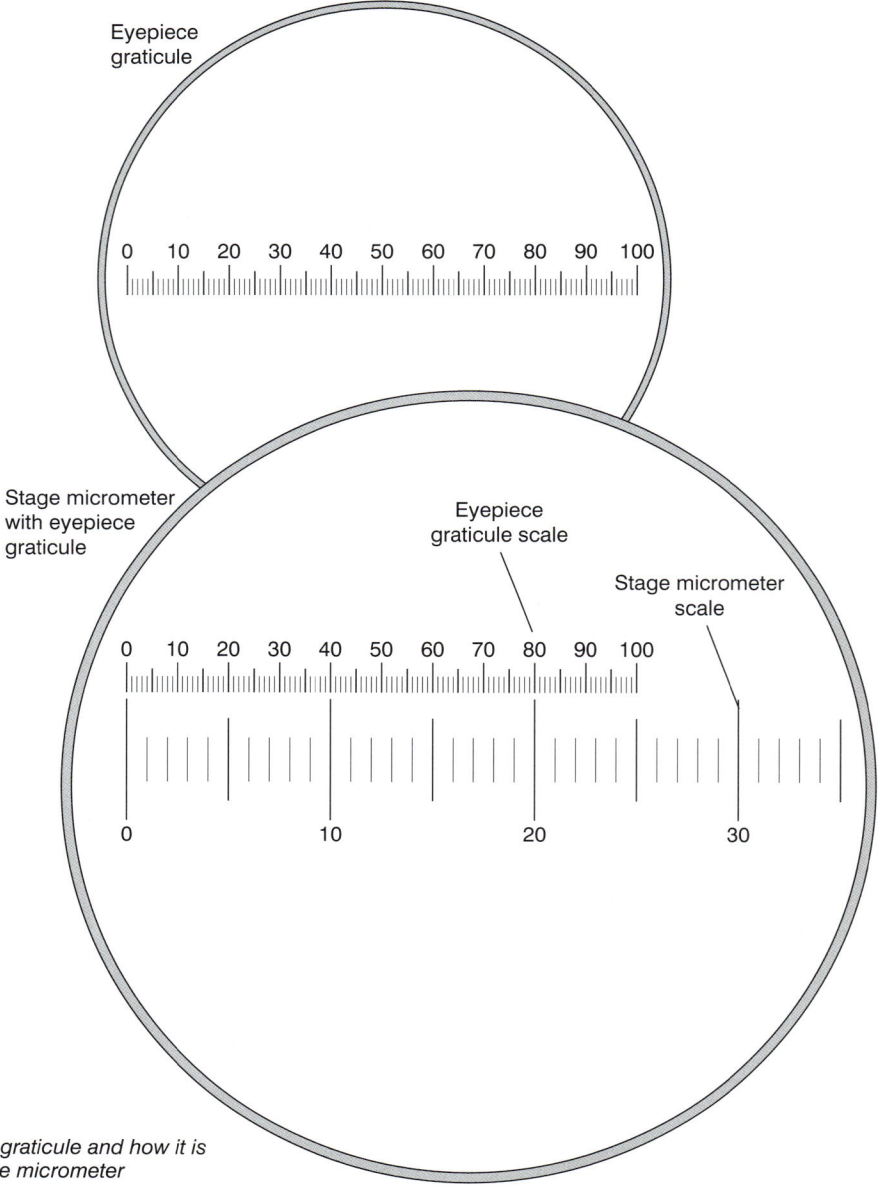

Figure 1 *An eyepiece graticule and how it is calibrated using a stage micrometer*

It is therefore sensible to record the results of the calibration for a particular objective lens and to leave this attached to the microscope. This will save you having to recalibrate each time you want to measure the size of the object being viewed under the microscope.

Calibrating the eyepiece graticule

To calibrate an eyepiece graticule you need to use a special microscope slide called a **stage micrometer**. This slide also has a scale etched onto it. Usually the scale is 2 mm long and its smallest sub-divisions are 0.01 mm (10 µm).

When the eyepiece graticule scale and the stage micrometer scales are lined up as shown in Figure 1, it is possible to calculate the length of the divisions on the eyepiece graticule. For example, you can see in Figure 1:

- 10 units on the micrometer scale are equivalent to 40 units on the graticule scale.
- Therefore one unit on the micrometer scale equals 4 units on the graticule scale.
- As each unit on the micrometer scale equals 10 µm, each unit on the graticule equals 10 ÷ 4 = 2.5 µm.

Parallax error occurs due to the apparent difference in position of an object when viewed from two different lines of sight. To avoid parallax error when reading a scale and pointer, the observer's eye and pointer must be in a perpendicular line to the plane of the scale.

Calculating linear magnifications of drawings and photographs

You may be asked to calculate the magnification of a drawing or photograph of an object under a microscope. We have looked at how the calculation is made. Now let us try an example using an actual electron micrograph.

Look at Figure 2, which is a transmission micrograph showing part of an animal cell. On this electron micrograph a red line X–Y is marked. This line represents a length of 1.0 µm of the actual cell. Using this information we can calculate the magnification of the electron micrograph as follows:

- If we measure the length of X–Y as drawn on the photograph we find it is 25 mm long.
- As the line represents 1.0 µm in the cell, we need to convert our measurement in the electron micrograph to micrometres (µm) as well. (25 mm = 25 000 µm)
- If 25 000 µm on the photograph is equivalent to 1 µm in the cell, then the magnification must be 25 000 ÷ 1.0 = 25 000 times.

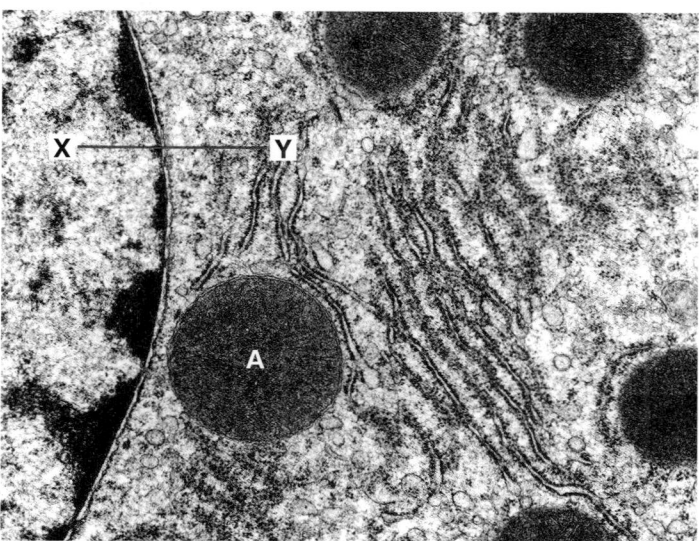

Figure 2 *Part of an animal cell seen with a transmission electron microscope (TEM)*

Calculating actual sizes of specimens from drawings and photographs

We have seen how the size of an object can be calculated when we know the magnification and the image size. Let us again see how this is done using an electron micrograph. In Figure 2, suppose we are told that the magnification is 5000×. We can now calculate the diameter of the mitochondrion labelled. This mitochondrion is labelled A and appears circular as it is a cross section through the rod-shaped structure. To calculate its actual size:

- Measure the diameter of mitochondrion A. As it is not truly spherical we need to calculate the mean of a number of different diameters. The mean is 23 mm.
- The size of the image is 23 mm (23 000 µm).
- The actual size of the mitochondrion equals the size of the image ÷ magnification or 23 000 µm ÷ 25 000 = 0.92 µm.

SUMMARY TEST 1.4

To measure the size of an object under a **(1)** microscope we can use an **(2)** graticule and a **(3)** micrometer. Before we can use the graticule to measure the size of objects it must first be **(4)**. A micrograph of a cell from an electron microscope is magnified 5000 times. On the micrograph the nucleus measures 100 mm in diameter. The actual size of the nucleus is therefore **(5)** µm. A chloroplast that is 5 µm in diameter measures 15 mm in a drawing made of a plant cell as seen under a microscope. The magnification of this drawing is therefore **(6)** times.

1.5 Cell structure 💿

On these pages you will learn to:

• Describe drawings of typical plant and animal cells as seen with the electron microscope

Each cell of an organism can be regarded as a metabolic compartment designed to perform a particular function. Depending on that function, each cell type has an internal structure that suits it for the job it does. To illustrate the range of cell structure, the plant cell and animal cell shown in this topic are 'generalised'. They represent a combination of many different types of cell rather than any cell in particular – indeed, no cell actually displays **all** the features shown here.

The animal cell

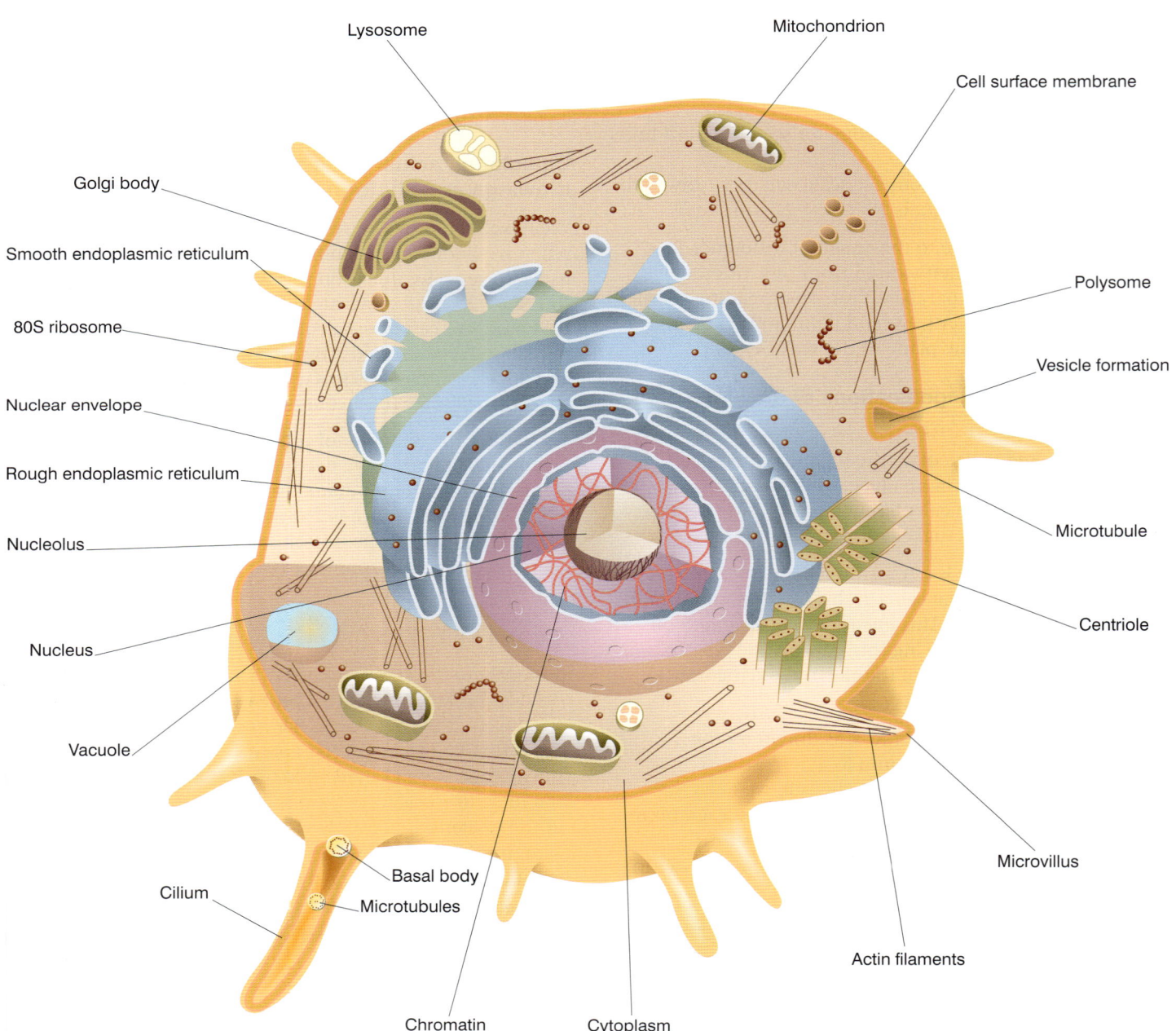

Lysosome
Mitochondrion
Cell surface membrane
Golgi body
Polysome
Smooth endoplasmic reticulum
Vesicle formation
80S ribosome
Nuclear envelope
Rough endoplasmic reticulum
Microtubule
Nucleolus
Centriole
Nucleus
Vacuole
Microvillus
Cilium
Basal body
Microtubules
Actin filaments
Chromatin
Cytoplasm

Figure 1 *A generalised animal cell*

The plant cell

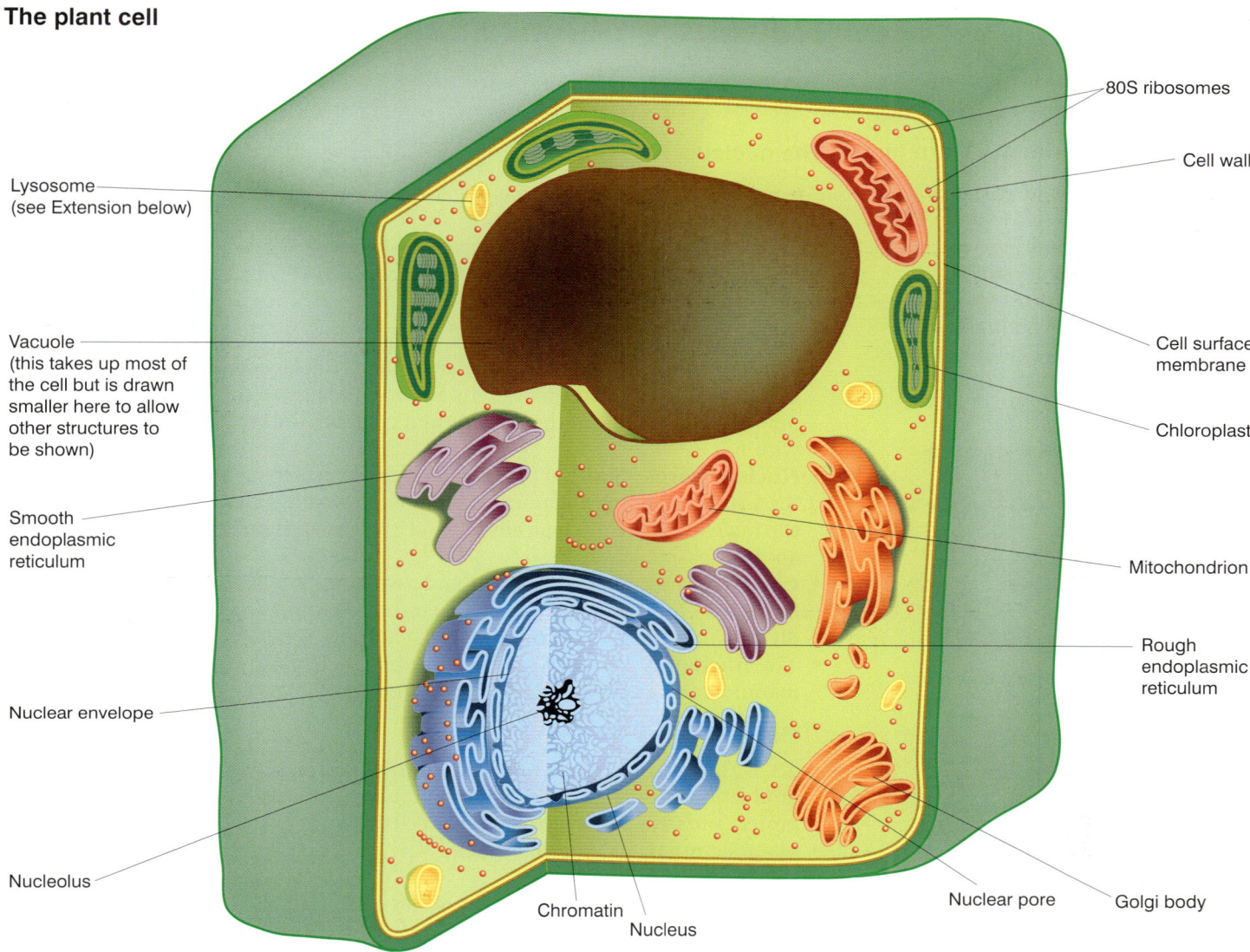

Lysosome
(see Extension below)

Vacuole
(this takes up most of
the cell but is drawn
smaller here to allow
other structures to
be shown)

Smooth
endoplasmic
reticulum

Nuclear envelope

Nucleolus

Chromatin

Nucleus

80S ribosomes

Cell wall

Cell surface
membrane

Chloroplast

Mitochondrion

Rough
endoplasmic
reticulum

Nuclear pore

Golgi body

Figure 2 *A generalised plant cell*

Table 1 *Differences between the structure of plant and animal cells*

Plant cells	Animal cells
Tough cellulose cell wall present (in addition to the cell surface membrane)	Cell wall absent – only a cell surface membrane surrounds the cell
Pits and **plasmodesmata** present in the cell wall	No cell wall and therefore no pits or plasmodesmata
Chloroplasts present in large numbers in cells carrying out photosynthesis	Chloroplasts absent
Mature cells normally have a large single, central permanent vacuole filled with cell sap which is surrounded by a vacuolar membrane (tonoplast)	Temporary vacuoles if present, are small and scattered throughout the cell
Cytoplasm normally confined to a thin layer at the edge of the cell	Cytoplasm present throughout the cell
Nucleus at the edge of the cell	Nucleus anywhere in the cell, but often central
Centrioles absent in higher plants	Centrioles present
Cilia and flagella absent in higher plants	Cilia or flagella may be present
Starch grains used for carbohydrate storage	Glycogen granules used for carbohydrate storage

Extension

Do plant cells have lysosomes?

There is a debate about whether plant cells have lysosomes, as these structures are normally associated with animal cells. As some vacuoles in plant cells have hydrolytic enzymes and perform lysosomal activities, some botanists think that plant cells have lysosomes. This view is not universally accepted, as the 'lysosomes' in plants do not carry out all the usual activities of the lysosomes found in animal cells.

Nucleus, chloroplast and mitochondria

On these pages you will learn to:

- Recognise the following cell structures and outline their functions: nucleus, nuclear envelope and nucleolus, mitochondria and chloroplasts
- State that ATP is produced in mitochondria and chloroplasts and outline its role in cells

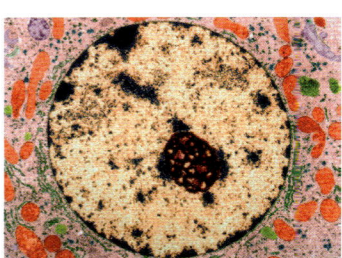

Figure 1 Colourised transmission electron micrograph of liver cell nucleus

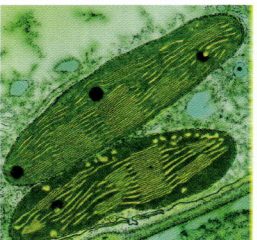

Figure 2 Colourised transmission electron micrograph of chloroplasts showing thylakoids (yellow) stacked to form grana. The black circles are oil droplets (× 10 000)

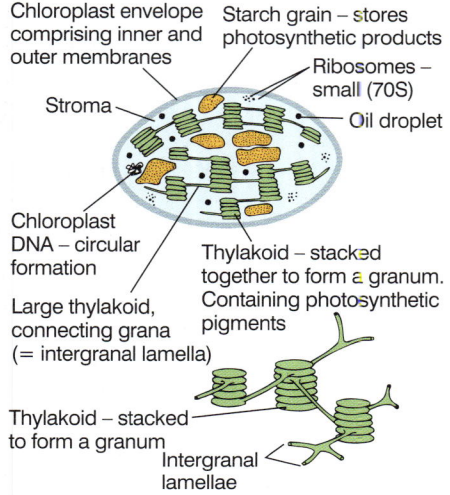

Chloroplast envelope comprising inner and outer membranes

Starch grain – stores photosynthetic products

Ribosomes – small (70S)

Stroma

Oil droplet

Chloroplast DNA – circular formation

Thylakoid – stacked together to form a granum. Containing photosynthetic pigments

Large thylakoid, connecting grana (= intergranal lamella)

Thylakoid – stacked to form a granum

Intergranal lamellae

Figure 3 Structure of chloroplasts

The nucleus

The nucleus is the most prominent feature of a **eukaryotic cell** when viewed under the microscope. It may vary in shape, size and position from cell to cell. It contains the genetic (hereditary) material of the cell, DNA, and it controls the activities of the cell. Usually spherical and between 10 and 20 μm in diameter, the nucleus has a number of parts:

- **The nuclear envelope** is a double membrane which surrounds the nucleus. Its outer membrane is continuous with the endoplasmic reticulum of the cell and often has ribosomes on its surface. It controls the entry and exit of materials in and out of the nucleus and contains the reactions taking place within it.
- **Nuclear pores** allow the passage of large molecules such as messenger RNA (mRNA) out of the nucleus but are too small to allow DNA to leave the nucleus. There are typically around 3000 in each nucleus, each being 40–100 nm in diameter.
- **Chromatin** is composed of DNA and associated proteins. This is the diffuse form that chromosomes take up when the cell is not dividing. When the cell divides, the chromatin condenses into chromosomes.
- **The nucleolus** is a small spherical region (there may be more than one) that manufactures ribosomal RNA (rRNA) and assembles the ribosomes.

The functions of the nucleus are to:

- act as the control centre of the cell, controlling cell activities through the production of mRNA and tRNA and protein synthesis
- protect the DNA from the rest of the cell
- manufacture rRNA and ribosomes.

Chloroplasts

Chloroplasts are found in eukaryotic cells which photosynthesise. They are flat discs, usually 3–10 μm in diameter and 1 μm thick (Figure 3), and are made up of a number of parts:

- **The chloroplast envelope** is a double membrane. It controls the entry and exit of substances in and out of the chloroplast.
- **The stroma** is a colourless, gelatinous matrix which contains the enzymes necessary for the light independent stage of photosynthesis (Topic 13.5). A small circular piece of DNA, 70S ribosomes and oil droplets (lipid globules) are also found in the stroma.
- **The grana** are structures that look like a stack of coins. There are typically 50 grana in a chloroplast, and each is made up of up to 100 stacked, flattened sacs called **thylakoids**, or **lamellae** (Figure 3). It is to the thylakoids that the chlorophyll molecules are attached. The grana carry out the light dependent stage of photosynthesis in which ATP is produced (Topic 13.4). ATP is used as the intermediate energy source of a cell (Topic 12.2). The role of ATP in cells is to provide the source of energy for building large molecules from small ones, for muscle contraction, active transport and the secretion of cell products. It is also has a role in making some chemicals react more easily.
- **Starch grains** act as temporary stores of the carbohydrate that is produced during photosynthesis.

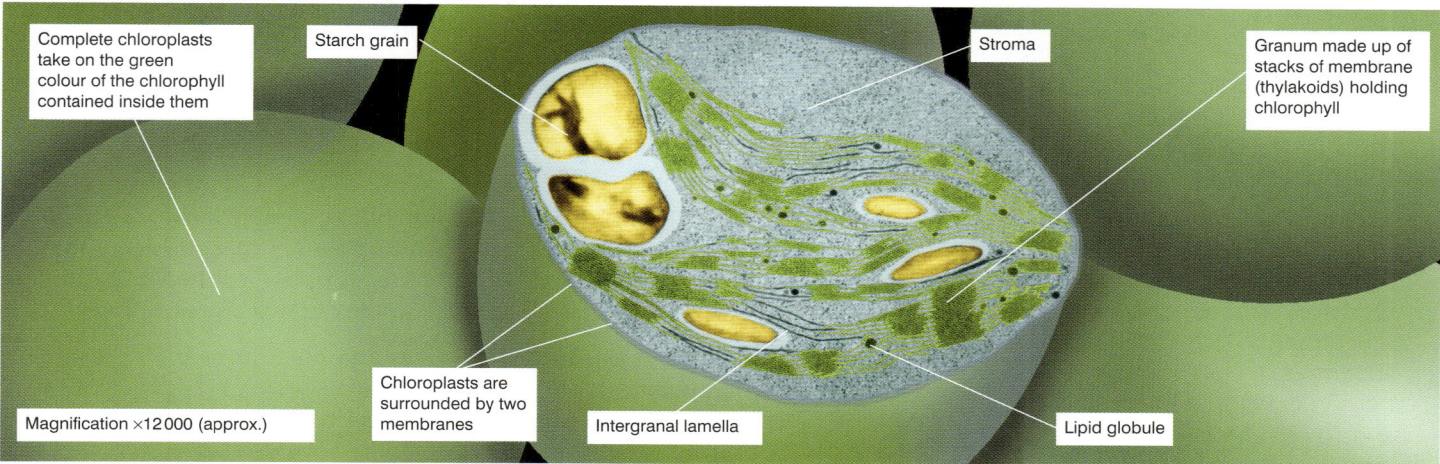

Figure 4 *Chloroplast structure*

The mitochondrion

Present in all but a few eukaryotic cells, mitochondria (Figure 6) are rod-shaped and 1–7 μm in length and 0.5–1.0 μm in diameter. They are made up of a number of parts:

- **A double membrane** around the organelle that controls the entry and exit of substances. The inner of the two membranes is folded to form extensions known as cristae.
- **Cristae** are infoldings of the inner membrane, some of which extend across the whole width of the mitochondrion. As they contain enzymes and other molecules involved in aerobic respiration, they provide a large surface area for a stage of aerobic respiration known as oxidative phosphorylation, which results in the synthesis of ATP molecules (Topic 12.5).
- **The matrix** makes up the remainder of the mitochondrion. It is a semi-rigid material containing protein, lipids, 70S ribosomes and a small circular piece of DNA that allows them to control the production of some of their own proteins. Many enzymes involved in **Krebs cycle**, another stage of aerobic respiration, are found in the matrix.

Functions of mitochondria

Mitochondria act as the sites for the Krebs cycle and oxidative phosphorylation stages of respiration. They are therefore responsible for the synthesis of ATP molecules from carbohydrates. ATP is the universal energy currency of the cell that is used as a molecule of energy transfer. Because of this, the number of mitochondria, their size and the number of cristae all increase in cells that have a high level of metabolic activity and therefore need a good supply of ATP. Such cells include those of the muscles and the liver.

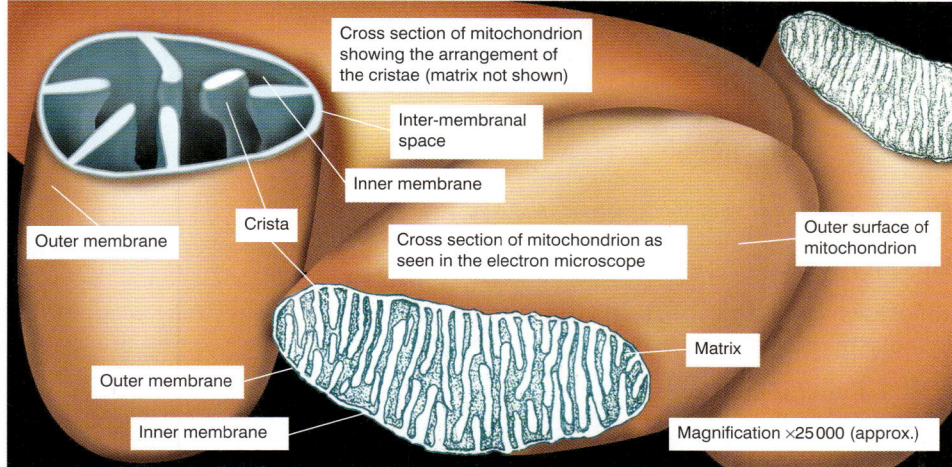

Figure 6 *Mitochondria*

On these pages you will learn to:

- Recognise the following cell structures and outline their functions: rough endoplasmic reticulum, smooth endoplasmic reticulum, Golgi body (Golgi apparatus or Golgi complex), ribosomes (80S in the cytoplasm and 70S in chloroplasts and mitochondria) and lysosomes

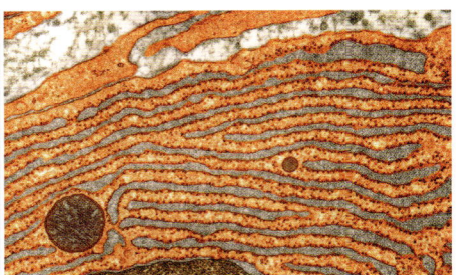

Figure 1 *Colourised transmission electron micrograph of rough endoplasmic reticulum (orange) (×12 000)*

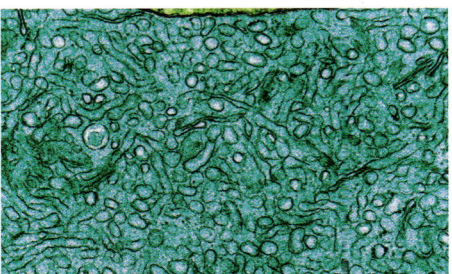

Figure 2 *Colourised transmission electron micrograph of smooth endoplasmic reticulum*

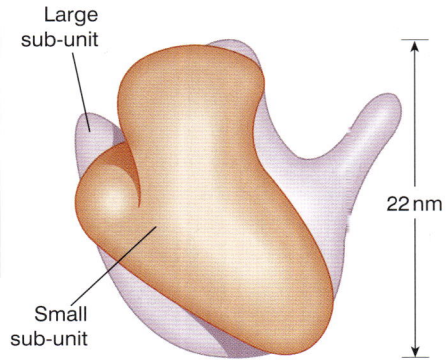

Figure 4 *Structure of an 80S ribosome*

Large sub-unit

Small sub-unit

22 nm

Endoplasmic reticulum

The endoplasmic reticulum (ER) is an elaborate, three-dimensional system of sheet-like membranes spreading through the cytoplasm of cells. It is continuous with the outer nuclear membrane. The membranes enclose a network of tubules and flattened sacs called cisternae (Figure 3). There are two types of ER:

- **Rough endoplasmic reticulum (RER)** has ribosomes present on the outer surfaces of the membranes.
- **Smooth endoplasmic reticulum (SER)** lacks ribosomes on its surface and is often more tubular in appearance.

Functions of endoplasmic reticulum

- Provides a large surface area for the synthesis of proteins (RER).
- Provides a pathway for the transport of materials, especially proteins, throughout the cell (RER).
- Synthesises, stores and transports lipids, including steroids such as cholesterol (SER).
- Synthesises glycogen in some cells, e.g. liver cells.
- Forms transport vesicles, small membrane-bound sacs that transport synthesised materials to other destinations in the cell, such as the Golgi body (RER and SER).

It follows that cells that need to manufacture and store large quantities of carbohydrates, proteins and lipids have a very extensive ER. Such cells include liver and secretory cells.

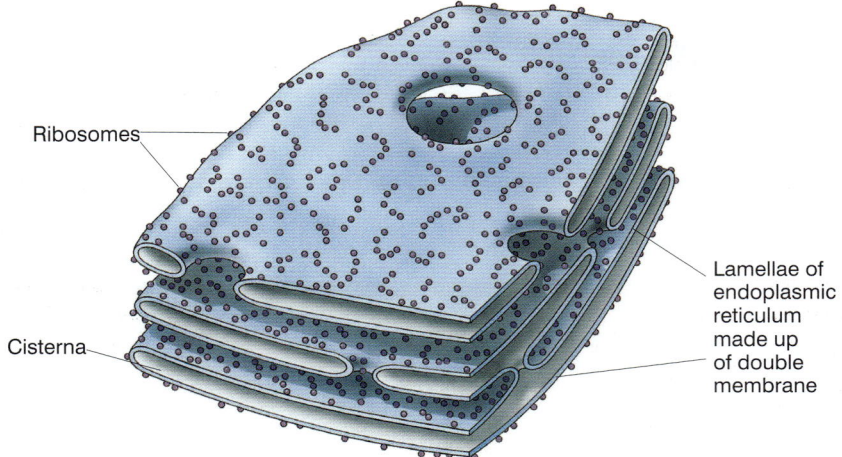

Ribosomes

Cisterna

Lamellae of endoplasmic reticulum made up of double membrane

Figure 3 *Structure of rough endoplasmic reticulum*

Ribosomes

Ribosomes are small spherical structures found in all cells. They may occur in the cytoplasm or be associated with the RER. There are two types:

- **80S**, found in **eukaryotic cells**, is around 22 nm in diameter.
- **70S**, found in **prokaryotic cells** and in the mitochondria and chloroplasts of eukaryotic cells (Topic 1.10), is slightly smaller (17 nm).

Each ribosome has two sub-units – one large and one small (Figure 4) – each of which contains ribosomal RNA and protein. Despite their small size, they occur in such vast numbers that they can account for up to 25% of the dry mass of a cell. Ribosomes are important in protein synthesis (Topics 6.6 and 6.7).

Golgi body

The **Golgi body (Golgi apparatus)** occurs in almost all eukaryotic cells and is similar to SER in structure, except that it is more compact. It consists of a stack of membranes which make up flattened sacs, or **cisternae**, and associated hollow **vesicles** (Figure 5). The proteins and lipids produced by the ER are passed through the Golgi body in strict sequence. The Golgi modifies these proteins, often adding non-protein components such as carbohydrate to them. It also 'labels' them, allowing them to be accurately sorted and sent to their correct destinations. Once sorted, the modified proteins and lipids are transported in vesicles which are regularly pinched off from the ends of the Golgi cisternae (Figure 6). Some of these vesicles move to the cell surface, where they fuse with the membrane and release their contents to the outside.

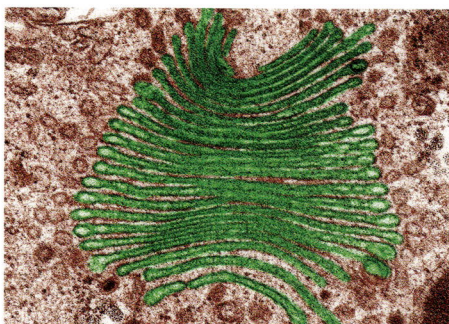

Figure 5 *Colourised transmission electron micrograph of Golgi body (green) (× 6000)*

Functions of the Golgi body

In general, the Golgi body acts as the cell's post office, receiving, sorting, processing and delivering proteins and lipids. More specifically, it:

- adds carbohydrates to proteins to form glycoproteins
- produces secretory enzymes such as those secreted by the pancreas
- secretes carbohydrates such as those used in making cell walls in plants
- transports, modifies and stores lipids
- forms primary lysosomes.

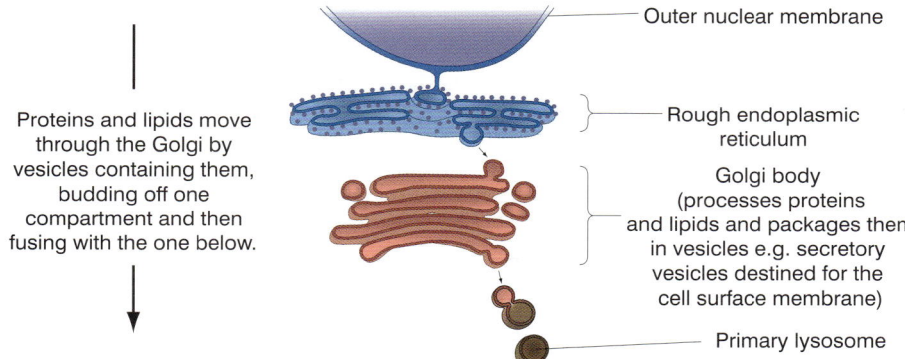

Proteins and lipids move through the Golgi by vesicles containing them, budding off one compartment and then fusing with the one below.

Outer nuclear membrane

Rough endoplasmic reticulum

Golgi body (processes proteins and lipids and packages them in vesicles e.g. secretory vesicles destined for the cell surface membrane)

Primary lysosome

Figure 6 *The Golgi body and its relationship to the nucleus, ER and lysosomes (the exact details of how this functions is not yet known)*

Lysosomes

Lysosomes are bound by a single membrane and are formed when the vesicles produced by the Golgi body include within them enzymes such as proteases and lipases. Up to 50 such enzymes may be contained in a single lysosome. Up to 1.0 μm in diameter, lysosomes isolate these potentially harmful enzymes from the rest of the cell, before releasing them, either to the outside or into a **phagocytic** vesicle within the cell (Figure 7).

Functions of lysosomes

Lysosomes are used to destroy foreign material inside or outside the cell. More particularly, they:

- break down material ingested by phagocytic cells such as macrophages and neutrophils (Chapter 11)
- digest worn out organelles **(autophagy)** so that the useful chemicals of which they are made can be re-used
- completely break down cells after they have died **(autolysis)**
- Some lysosomes have a role in releasing hydrolytic enzymes to the outside of the cell to destroy material around the cell.

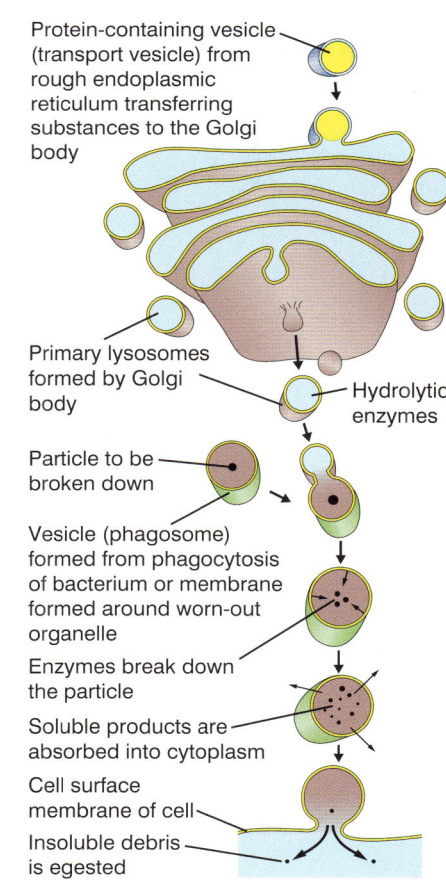

Protein-containing vesicle (transport vesicle) from rough endoplasmic reticulum transferring substances to the Golgi body

Primary lysosomes formed by Golgi body

Hydrolytic enzymes

Particle to be broken down

Vesicle (phagosome) formed from phagocytosis of bacterium or membrane formed around worn-out organelle

Enzymes break down the particle

Soluble products are absorbed into cytoplasm

Cell surface membrane of cell

Insoluble debris is egested

Figure 7 *The formation and functioning of a lysosome*

REMEMBER

When you look at a group of animal
cells under a light microscope you
cannot see the cell surface membrane
because it is too thin to be observed.
What you see is the boundary between
cells.

Cell surface membrane

The cell surface membrane is the boundary between the cell cytoplasm and the
environment. It controls the movement of substances into and out of the cell,
permanently excluding some and permanently containing others. Others may
cross the membrane on one occasion and be prevented from doing so at another
time. The cell surface membrane is therefore said to be **partially permeable**. The
structure of the cell surface membrane is described and illustrated in Topic 4.1.

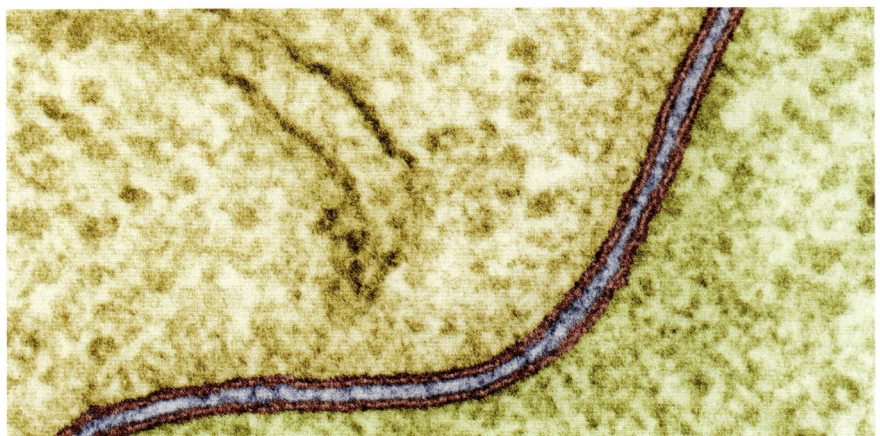

Figure 1 *Colourised transmission electron micrograph of the cell surface membranes of two
adjacent cells separated by intercellular fluid (blue)*

Functions of the cell surface membrane

The cell surface membrane defines the cell and encloses the cell contents. It:

- controls movement of substances in and out of the cell
- is important in cell recognition
- has receptor sites which bind specific hormones and neurotransmitters
- in animal cells, may be folded to form microvilli to provide a larger surface area
  for the absorption of substances
- has components that attach to one another and so help cells to form tissues.

Cell wall

Characteristic of all plant cells, the cell wall consists of cellulose fibres containing
cellulose microfibrils which contain the polysaccharide cellulose (Topic 2.4),
embedded in a matrix. Cellulose microfibrils have considerable strength and so
contribute to the overall strength of the cell wall.
Cell walls have the following features:

- They consist of a number of polysaccharides, such as cellulose. In addition to
  cellulose, they contain other polysaccharides such as hemicellulose and pectin.

- There is a thin layer, called the **middle lamella**, which marks the boundary between adjacent cell walls and cements adjacent cells together.

The functions of the cellulose cell wall are to:

- provide mechanical strength in order to prevent the cell bursting (cell lysis) under the pressure created by the osmotic entry of water (Topic 4.4)
- give mechanical strength to the plant as a whole
- allow water to pass along it and so contribute to the movement of water through the plant.

Small thin strands called **plasmodesmata** are present within cell walls. These connect the cytoplasm of adjacent cells and allow substances to pass between them without having to pass through the cell wall or cell surface membrane. They form part of the symplast pathway (Topic 7.5) by which water moves through plants. Mineral ions also pass between cells via the plasmodesmata.

Large permanent vacuole

Plant cells have a large permanent vacuole that is surrounded by a membrane called the **tonoplast** and contains a fluid known as cell sap.
The vacuole stores water, ions, sugars and pigments, pushes chloroplasts to the edge of the cell and gives turgidity to the cell to help support the plant.

Centrioles and microtubules

Centrioles are found in almost all animal cells, as well as in the cells of certain algae and fungi, but not in plant cells of higher plants. They are hollow cylinders 0.5 μm in length and 0.2 μm in diameter made up of nine sets of three microtubules (Figure 2). Microtubules are composed of a globular protein called tubulin. There are two centrioles in a cell and they lie at right angles to one another near to the nucleus.

Functions of centrioles and microtubules
Centrioles and microtubules have several functions.

- The centrioles have a role in organising microtubules to form the spindle fibres during nuclear division (see Section 5.2).
- Basal bodies (see Figure 1, Section 1.5) are modified centrioles that organise microtubules to form cilia and flagella.
- Microtubules form part of the cytoskeleton, the internal 'skeleton' of the cell that provides support and gives shape to the cell.
- Microtubules serve as a scaffold for the movement and positioning of organelles within the cell, including transport vesicles.
- Microtubules are involved in movement of the cell, for example in phagocytosis.

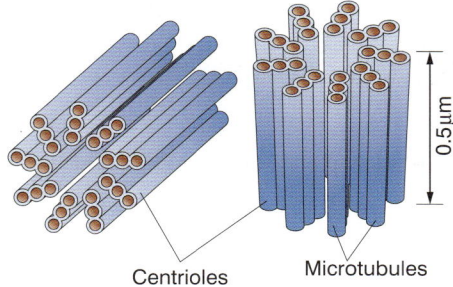

Figure 2 Centrioles

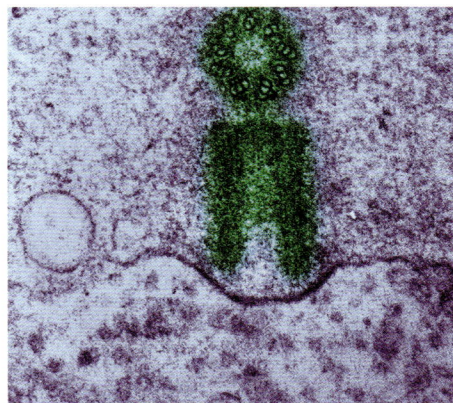

Figure 3 Colourised transmission electron micrograph of two centrioles in transverse (top) and longitudinal (bottom) section

SUMMARY TEST 1.8

The cell surface membrane of a typical cell is made up of 45% phospholipid and 45% (**1**); the remaining 10% comprises glycolipids, (**2**) and (**3**). In animal cells there is a total of (**4**) centrioles. These are important in (**5**), where they form (**6**) fibres that help position and move (**7**) during the process.

1.9 Prokaryotic cells

Viruses

Viruses are not prokaryotic cells; in fact, they are not cells at all. They are made up of a protein coat that surrounds a nucleic acid – either DNA or RNA. They can only replicate within living cells using the biochemical processes of these cells. Outside cells they are inert and are therefore considered to be on the border between living and non-living. As viruses do not fit the key concept of cells as the unit of life, they are often considered as non-living. As they take over living cells to reproduce, all viruses cause harm to some degree. Many are the cause of human diseases such as HIV/AIDS, influenza and colds.

REMEMBER

'Naked DNA' refers to a lack of protein associated with DNA, while 'free DNA' refers to DNA that is not confined to the nucleus.

Although cells come in a bewildering variety of size, shape and function, they nevertheless fall into two basic groups:

- **Prokaryotic cells** ('pro' = before, 'karyote' = nucleus) have no nucleus or nuclear membranes.
- **Eukaryotic cells** ('eu' = true, 'karyote' = nucleus) have a nucleus bounded by nuclear membranes.

Other differences between prokaryotic and eukaryotic cells are listed in Table 1.

Table 1 *Comparison of prokaryotic and eukaryotic cells*

Prokaryotic cells	Eukaryotic cells
No true nucleus or nuclear envelope	Distinct nucleus, with a nuclear envelope
No nucleolus	Nucleolus is present
DNA is not associated with proteins (naked DNA)	DNA is associated with histone proteins
Some DNA may be in the form of circular strands called plasmids	There are no plasmids and DNA is linear
No double membrane-bounded organelles	Double membrane-bounded organelles such as mitochondria are present
No chloroplasts, only bacterial chlorophyll associated with cell surface membranes in some bacteria	Chloroplasts present in plants and algae
Ribosomes are smaller (70S)	Ribosomes are larger (80S)
Flagella (if present) lack internal 9+2 microtubule arrangement	Flagella, where present, have a 9+2 internal microtubule arrangement
No endoplasmic reticulum or associated Golgi body and lysosomes	Endoplasmic reticulum present along with Golgi body and lysosomes
Cell wall made of peptidoglycan (murein)	Where present, cell wall is made mostly of cellulose or chitin

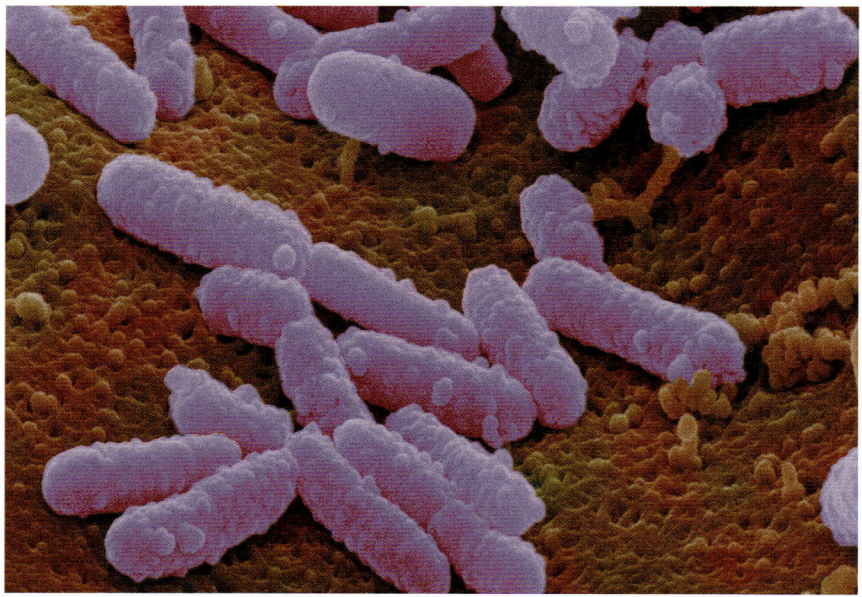

Figure 1 *Colourised scanning electron micrograph of a colony of the rod-shaped bacterium – Escherichia coli*

Structure of a bacterial cell

Bacteria occur in every habitat in the world; they are versatile, adaptable and successful. Much of their success is a result of their small size, normally in the range 0.1–5 μm in length. Their cellular structure is relatively simple (Figure 2). All bacteria possess a cell wall which is made up of **peptidoglycan (murein)** – a polysaccharide cross-linked by peptide molecules. Around this wall, many bacteria further protect themselves by secreting a **capsule** of mucilaginous slime. Hair-like structures, made of protein and called **pili**, extend through the cell wall in some species. These enable the bacteria to stick to one another or to other surfaces. **Flagella** occur in certain types of bacteria. These lack microtubules and so do not beat, which they do in eukaryotic cells. Their rigid corkscrew shape and rotating base, however, cause bacteria to spin through fluids.

Within the cytoplasm of bacterial cells are scattered **ribosomes** (70S). These are smaller than those of eukaryotic cells (80S), but nevertheless serve the same function in protein synthesis. **Glycogen granules** and **oil droplets** are used for storage. Infoldings of the cell surface membrane can occur in some bacteria. These provide a large surface area for the attachment of respiratory enzymes. In photosynthetic bacteria there is bacterial chlorophyll associated with cell surface membranes that is essential to photosynthesis. The genetic material in bacteria is usually circular (double-stranded) DNA. Separate from this, and not necessary for growth and metabolism, are smaller circular pieces of DNA called **plasmids**. These can replicate themselves independently and may contain genes that give the bacterium an advantage, for example, resistance to chemicals such as **antibiotics**. Plasmids are used extensively as vectors (carriers of genetic information) in **genetic engineering**.

Table 2 *Roles of structures found in a bacterial cell*

Cell structure	Role
Cell wall	Physical barrier which protects against mechanical damage and lysis
Cell surface membrane	Acts as a partially permeable layer which controls the entry and exit of chemicals
Circular DNA	Possesses the genetic information for the replication of bacterial cells
Plasmids	Possess genes which aid the survival of bacteria in adverse conditions, e.g. produce enzymes which break down antibiotics
Ribosomes	Site of protein synthesis (70S)

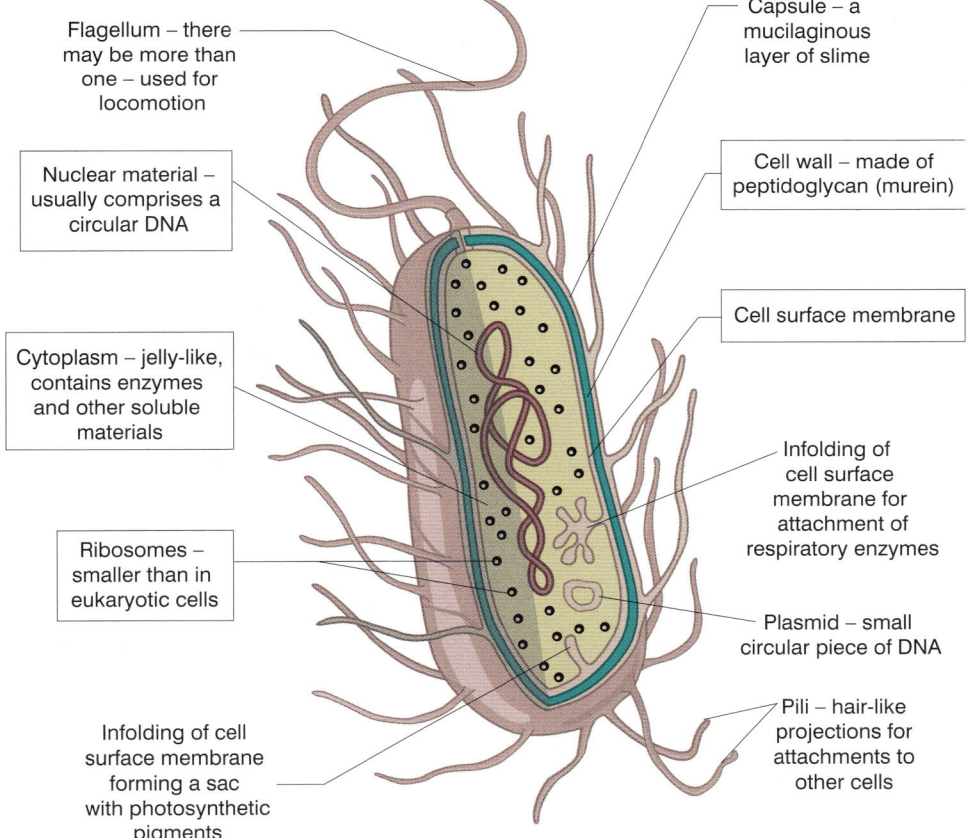

Flagellum – there may be more than one – used for locomotion

Nuclear material – usually comprises a circular DNA

Cytoplasm – jelly-like, contains enzymes and other soluble materials

Ribosomes – smaller than in eukaryotic cells

Infolding of cell surface membrane forming a sac with photosynthetic pigments

Capsule – a mucilaginous layer of slime

Cell wall – made of peptidoglycan (murein)

Cell surface membrane

Infolding of cell surface membrane for attachment of respiratory enzymes

Plasmid – small circular piece of DNA

Pili – hair-like projections for attachments to other cells

Figure 2 *Structure of a generalised bacterial cell. Structures labelled in boxes occur in all bacteria whereas the others occur only in certain species*

SUMMARY TEST 1.10

Prokaryotic cells lack a distinct **(1)** and include the group of organisms called **(2)**. Their DNA is **(3)** in shape and their ribosomes are smaller than in **(4)** cells and are known as **(5)** ribosomes. The cell wall of bacteria is made of **(6)**.

Examination and Practice Questions

1 Examination Questions

1 Figure 1 is an electron micrograph of part of an animal cell. A centriole is labelled.

Figure 1

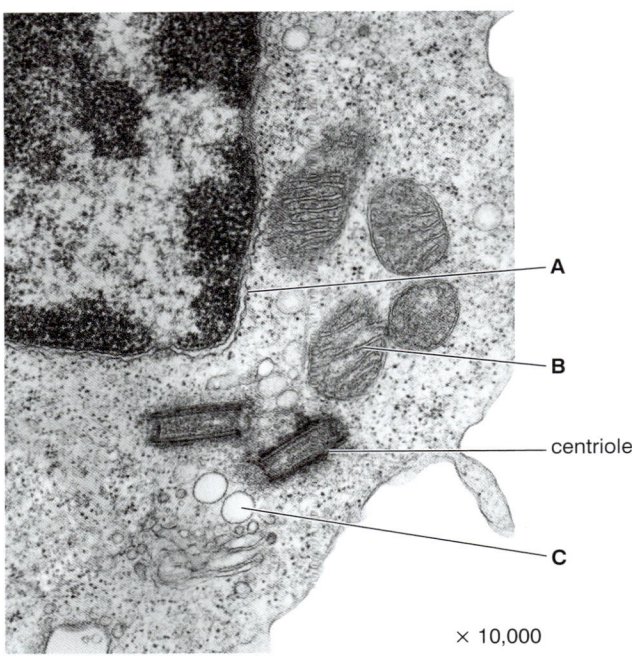

× 10,000

a Name the structures labelled **A** to **C**. *(3 marks)*

b Describe the roles of centrioles in animal cells. *(3 marks)*

c Explain why it is possible to see the internal membranes of a cell in electron micrographs, such as Figure 1, but it is not possible to see them when using the light microscope. *(3 marks)*

d A student investigated the effect of temperature on beetroot tissue. Beetroot cells contain a dark red pigment known as betalain, which is stored inside their vacuoles.

The student
 • cut the beetroot tissue into cubes of the same size
 • washed the cubes thoroughly in distilled water
 • placed the same number of cubes into distilled water at seven different temperatures.

After 30 minutes, samples of the water were removed and placed in a colorimeter to measure the transmission of light. The lower the percentage transmission the more betalain is present in the water.

The results are shown in Figure 2.

Figure 2

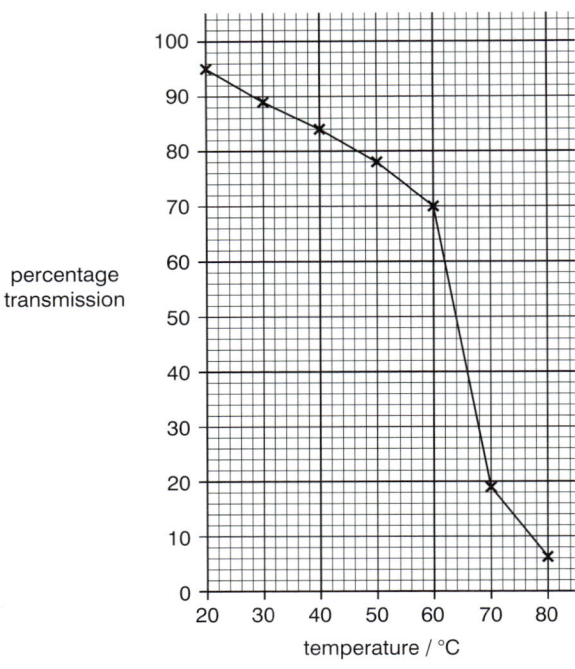

percentage transmission

temperature / °C

Using the information in Figure 2,
i describe the student's results; *(3 marks)*
ii explain the effect of increasing temperature on the beetroot tissue. (You may need to read Chapter 4 to help you answer this question.) *(3 marks)*
(Total 15 marks)

Cambridge International AS and A Level Biology 9700 Paper 2 Q2 June 2008

2 *Candida albicans* is a yeast-like fungus that lives in human lungs. It is the causative agent of one of the opportunistic infections that may develop during HIV/AIDS.

C. albicans is eukaryotic. Figure 3 shows its structure.

Figure 3

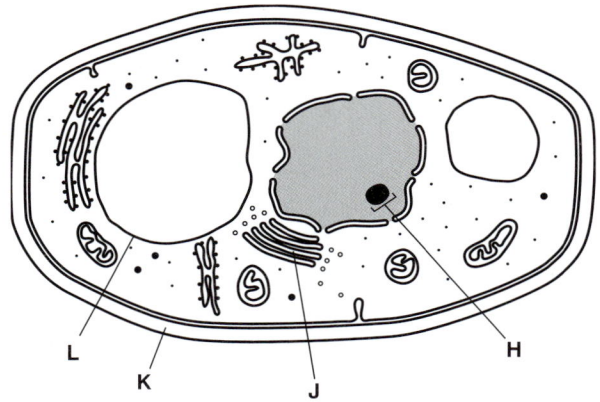

a i Name **H** to **L**.   *(4 marks)*

  ii State two ways in which the **structure** of a
       prokaryotic cell differs from that shown in Figure 3.
                                                    (2 marks)

b *C. albicans* uses a transport protein, TMP1, to absorb
   sugar molecules from the inside of the mouth.
   TMP1 is encoded by a gene within the nucleus
   and is produced when sugars are present in the
   surroundings.

   Explain how the structures within the cell shown
   in Figure 3 are involved with the production of
   functioning TMP1.   *(4 marks)*
                                               (Total 10 marks)

Cambridge International AS and A Level Biology 9700 Paper 22 Q5 November 2009

1 Practice Questions

3 The table below lists some features of cells. For the letter
   in each box, write down one of the following: 'present' if
   the feature always occurs, 'absent' if it never occurs and
   'sometimes' if it occurs in some cells but not others.

Feature	Prokaryotic cell	Eukaryotic cell
nuclear envelope	A	B
cell wall	C	D
flagellum	E	F
ribosomes	G	H
plasmid	I	J
cell surface membrane	K	L
mitochondria	M	N

4 a Distinguish between magnification and resolution.

  b An organelle that is 5 μm in diameter appears under
      a microscope to have a diameter of 1 mm. How many
      times has the organelle been magnified?

  c A ribosome is 25 nm in diameter. If viewed under an
      electron microscope that magnified it 400 000 times,
      what would the diameter of the ribosome appear to be
      in millimetres?

  d At a magnification of 12 000 times a structure appears
      to be 6 mm long. What is its actual length?

  e Why is the electron microscope able to resolve objects
      better than the light microscope?

  f Why do specimens have to be kept in a near-vacuum
      in order to be viewed effectively using an electron
      microscope?

  g Of the following list of biological structures: plant cell
      (100 μm); DNA molecule (2 nm); virus (100 nm); actin
      molecule (3.5 nm) and a bacterium (1 μm) which ones

can, in theory, be resolved by:
  i a light microscope
  ii a transmission electron microscope
  iii a scanning electron microscope?

h In practice, the theoretical resolution of an electron
   microscope cannot always be achieved. Why not?

5 a Ribosomes are important in which process?

  b In each of the following, name the organelle being
      referred to.
      i Possesses structures called cristae; **(ii)** Contains
         chromatin; **(iii)** Synthesises glycoproteins;
         (iv) Digests worn out organelles.

  c The following list names a type of cell and a brief
      description of its role. Suggest in each case two
      organelles that might be numerous and/or well
      developed in that cell.
      i A sperm cell swims a considerable distance and when
         it reaches an egg releases enzymes to digest a path.
      ii A type of white blood cell engulfs and digests
         foreign material.
      iii Cells lining a kidney tubule reabsorb soluble
         substances against a concentration gradient.
      iv Liver cells manufacture proteins and lipids at a
         rapid rate.

6 The drawing in the figure has been made from an
   electron micrograph of a cell lining the human intestine.

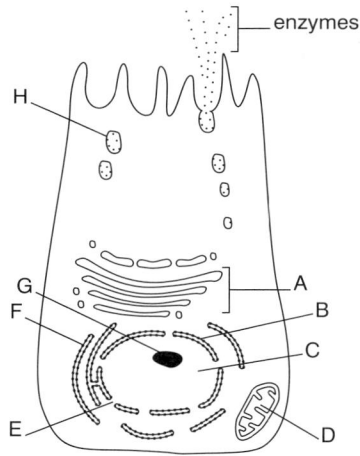

  a Give the names of the structures labelled **A–H**.

  b Give the letter of the structure that best fits the
      following descriptions
      i Produces ATP molecules
      ii Manufactures ribosomal RNA
      iii Possesses structures called cristae
      iv Produce the enzymes shown being released from
         the cell.

2.1

Introduction to biological molecules

On these pages you will learn to:

- Understand atoms, isotopes, oxidation, reduction and bonding
- Define the terms 'monomer', 'polymer', 'macromolecule', 'monosaccharide', 'disaccharide' and 'polysaccharide'

Biological molecules are particular groups of chemicals that are found in living organisms. A good understanding of biological molecules is necessary for **molecular biology** and for **biochemistry**. All molecules, whether biological or not, are made up of units called atoms.

Atoms

Atoms are the smallest units of a chemical element that can exist independently. An atom comprises a nucleus that contains particles called protons and neutrons (the hydrogen atom is the only exception as it has no neutrons). Tiny particles called electrons orbit the nucleus of the atom. The main features of these sub-atomic particles are:

- **Neutrons** – occur in the nucleus of an atom and have the same mass as protons but no electrical charge.
- **Protons** – occur in the nucleus of an atom and have the same mass as neutrons but have a positive charge.
- **Electrons** – orbit in fixed shells around the nucleus but a long way from it. They have such a small mass that their contribution to the overall mass of the atom is negligible. They are, however, negatively charged and their number determines the chemical properties of an atom.

In an atom the number of protons and electrons is the same and therefore there is no overall charge. The **mass number** of an atom is the total number of protons and neutrons in a given atom. The atomic structure of three biological elements is given in Figure 1.

Oxidation, reduction and the formation of ions

If an atom loses or receives an electron it becomes an ion.

- The loss of an electron is called **oxidation** and leads to the formation of a positive ion, e.g. the loss of an electron from a hydrogen atom produces a positively charged hydrogen ion, written as H^+.
- The receiving of an electron is called **reduction** and leads to the formation of a negative ion, e.g. if a chlorine atom receives an additional electron it becomes a negatively charged chloride ion, written as Cl^-.

More than one electron may be lost or received, e.g. the loss of two electrons from a calcium atom forms the calcium ion, Ca^{2+}. Ions may be made up of more than one type of atom, e.g. a sulfate ion is formed when one sulfur atom and four oxygen atoms receive two electrons and form the sulfate ion, SO_4^{2-}.

Isotopes

While the number of protons in an element always remains the same, the number of neutrons can vary. The different types of the atom so produced are called **isotopes**. Isotopes of any one element have the same chemical properties but differ in mass. Each type is therefore recognised by its different mass number. Isotopes, especially radioactive ones, are very useful in biology for tracing the route of certain elements in biological processes and for dating fossils.

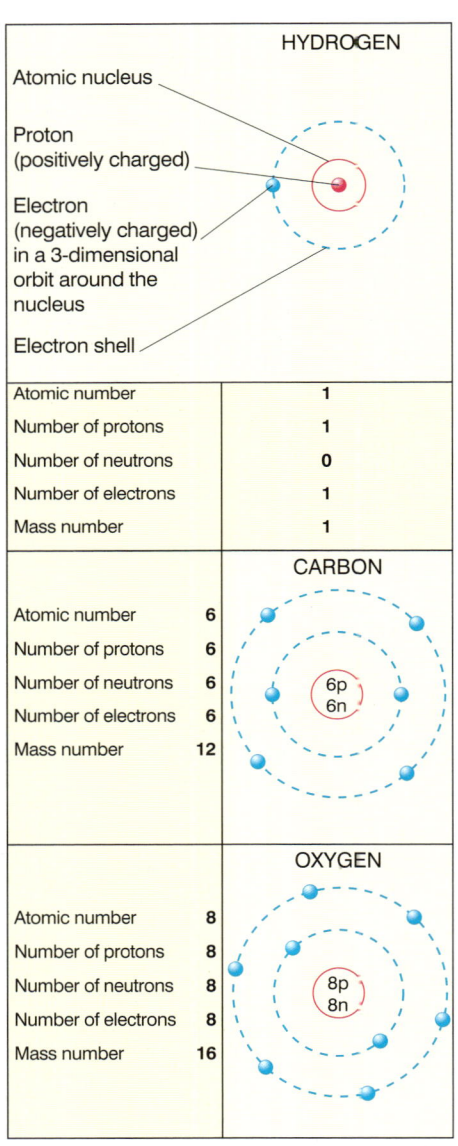

Figure 1 *Atomic structure of three commonly occurring biological elements*

HYDROGEN	
Atomic number	1
Number of protons	1
Number of neutrons	0
Number of electrons	1
Mass number	1

CARBON	
Atomic number	6
Number of protons	6
Number of neutrons	6
Number of electrons	6
Mass number	12

OXYGEN	
Atomic number	8
Number of protons	8
Number of neutrons	8
Number of electrons	8
Mass number	16

Bonding and the formation of molecules

Atoms may combine with each other in a number of ways:

- **Covalent bonding** – atoms share a pair of electrons in their outer shells. As a result the outer shell of both atoms is filled and a more stable compound, called a **molecule**, is formed.
- **Ionic bonding** – ions with opposite charges attract one another. This electrostatic attraction is known as an ionic bond. For example the positively charged sodium Na^+ and negatively charged chloride Cl^- form an ionic bond to make sodium chloride. Ionic bonds are weaker than covalent bonds.
- **Hydrogen bonding** – occurs when a weak attractive force occurs between the hydrogen of an electronegative atom of one molecule and an electronegative atom of a different molecule. It is due to hydrogen bonding that water molecules tend to stick together.

Formation of macromolecules

Certain molecules can be linked together to form long chains of similar sub-units. These chains form large molecules called **macromolecules**. The sub-units are known as **monomers**. The chains of monomer sub-units are called **polymers** and the process by which they are formed is therefore called **polymerisation**. The monomers of a polymer are usually based on carbon. Many, such as polythene and polyesters, are industrially produced. Others, like polysaccharides, polypeptides and polynucleotides, are made naturally by living organisms. The basic sub-unit of a polysaccharide is a monosaccharide or single sugar (Topic 2.2), e.g. glucose. Polynucleotides are formed from mononucleotide sub-units (Topic 6.1). Polypeptides are formed by linking together peptides which have amino acids as their basic sub-unit (Topic 2.6).

Condensation and hydrolysis reactions

In the formation of polymers by polymerisation in organisms, each time a new sub-unit is attached a molecule of water is formed. Reactions that produce water in this way are termed **condensation reactions**. Therefore the formation of a polypeptide from amino acids and that of the polysaccharide starch from the monosaccharide glucose are both condensation reactions.

Polymers can be broken down through the addition of water. Water molecules break down the bonds that link the sub-units of a polymer, thereby splitting the molecule into its constituent parts. This type of reaction is called **hydrolysis** ('hydro' = water; 'lysis' = splitting). Thus polypeptides can be hydrolysed into amino acids, and starch can be hydrolysed into glucose. Figure 2 summarises atomic and molecular organisation.

Metabolism

All the chemical processes that take place in living organisms are collectively called metabolism. Metabolism can be divided into two parts:

- **Anabolism** – an energy-requiring process in which small molecules are combined to make larger ones. The condensation reactions that build polymers from basic sub-units, e.g. polypeptides formed from amino acids, are examples of anabolic reactions.
- **Catabolism** – chemical reactions involving the release of energy in the breakdown of larger molecules into smaller ones. The hydrolysis reactions that split polymers into their basic sub-units, e.g. polypeptides being split into amino acids, are examples of catabolic reactions.

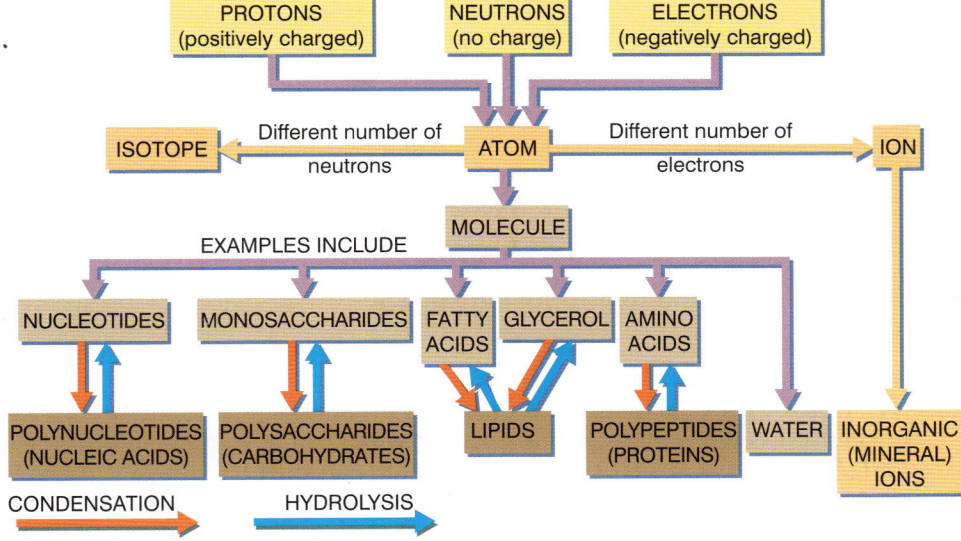

Figure 2 *Summary of atomic and molecular organisation*

REMEMBER

To avoid confusing the two terms oxidation and reduction simply look at the first vowel in each of the key words in the table below:

Process	Oxidation	Reduction
Change of electrons in atoms	L**o**ses	R**e**ceives
Atom becomes	**O**xidised	R**e**duced
Ion produced	P**o**sitive	N**e**gative
First vowel	**O**	**e**

A useful acronym is OIL RIG – Oxidation Is Loss, Reduction Is Gain.

Carbohydrates – monosaccharides and disaccharides

On these pages you will learn to:

- Carry out tests for reducing sugars and non-reducing sugars to identify the contents of solutions
- Carry out a semi-quantitative Benedict's test on a reducing sugar using dilution, standardising the test and using the results to estimate the concentration
- Describe the ring forms of α-glucose and β-glucose
- Describe the formation of a glycosidic bond by condensation, with reference both to polysaccharides and to disaccharides, including sucrose
- Describe the breakage of glycosidic bonds in polysaccharides and disaccharides by hydrolysis, with reference to the non-reducing sugar test

As the word suggests, carbohydrates are carbon molecules (carbo) combined with water (hydrate); their general formula is $C_x(H_2O)_y$. Like many organic molecules, they are made up of individual units called **monomers**, which can be combined to form larger units called **polymers**. In carbohydrates the basic unit is a sugar, or **saccharide**. A single unit is called a **monosaccharide**, pairs of which can be combined to form a **disaccharide**. Monosaccharides are usually combined in much larger numbers to form **polysaccharides** (see Topics 2.3 and 2.4).

Monosaccharides

Monosaccharides are sweet-tasting, soluble substances that have the general formula $(CH_2O)_n$. While 'n' can be any number from 3 to 7, the three most common groups of monosaccharides are shown in Table 1.

Structure of monosaccharides

Perhaps the best-known monosaccharide is **glucose**. This molecule is a hexose (6-carbon) sugar and has the formula $C_6H_{12}O_6$. However, the atoms of carbon, hydrogen and oxygen can be arranged in many different ways. Although the molecular arrangement is often shown as a straight chain for convenience, the atoms actually form a ring which can take a number of forms, as shown in Figure 1. Different molecular structures are given different names, e.g. glucose, fructose and galactose, and further differences are shown by a letter before the molecule's name, e.g. α-glucose, β-glucose. You will see from Figure 1 that the hydroxyl group —OH on carbon atom 1 is at the bottom of the ring in α-glucose and at the top of the ring in β-glucose. Although some of these differences are small, they often give the resulting molecules very different properties.

Disaccharides

When combined in pairs, monosaccharides form a **disaccharide**. As Table 2 shows, the two monosaccharides that combine can be the same or different. When they join, a molecule of water is removed and the reaction is therefore called a **condensation reaction**. The bond that is formed is called a **glycosidic bond**. In the formation of sucrose this bond is between carbon atom 1 of α-glucose and carbon atom 2 of β-fructose. It is known as a 1,2 glycosidic bond. Figure 2 illustrates the formation and breaking of a 1,2 glycosidic bond.

Table 1 Types of monosaccharide

Formula	Name	Examples
$C_3H_6O_3$ (n = 3)	Triose	Glyceraldehyde
$C_5H_{10}O_5$ (n = 5)	Pentose	Ribose Deoxyribose
$C_6H_{12}O_6$ (n = 6)	Hexose	Glucose, Fructose, Galactose

Table 2 Types of disaccharide

glucose	+	glucose	=	maltose
glucose	+	fructose	=	sucrose
glucose	+	galactose	=	lactose

When water is added to a disaccharide under suitable conditions, it breaks the glycosidic bond into its constituent monosaccharides. This is called **hydrolysis** (breakdown by water). The breakdown is very slow, however, unless it is catalysed by the appropriate enzyme.

Tests for reducing and non-reducing sugars

All monosaccharides and some disaccharides (e.g. maltose) are reducing sugars. The test for a reducing sugar is known as the **Benedict's test**. When a reducing sugar is heated with an alkaline solution of copper(II) sulfate (Benedict's reagent) it forms an insoluble precipitate of copper(I) oxide. The colour of the precipitate changes from green through yellow, orange and brown to deep red, depending on the quantity of reducing sugar present (see Table 3). The disaccharide sucrose is a non-reducing sugar. There is no direct test for a non-reducing sugar, but they can be identified by first hydrolysing them with a dilute

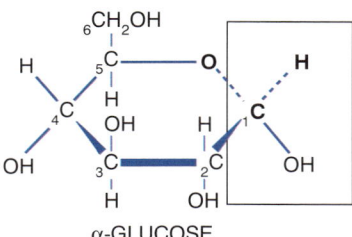

α-GLUCOSE

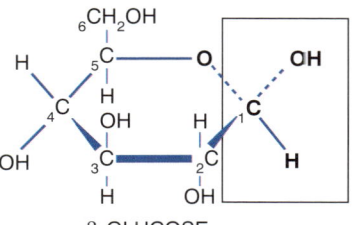

β-GLUCOSE

Figure 1 Molecular arrangement of α-glucose and β-glucose

(a) Formation of glycosidic bond by removal of water (condensation reaction)

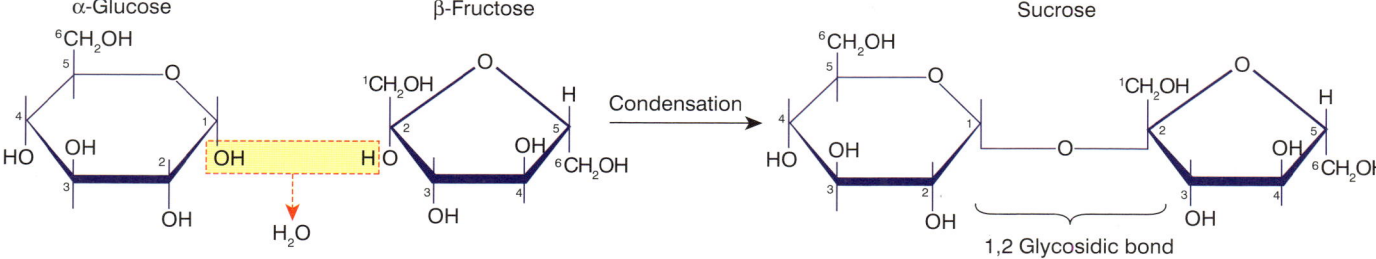

(b) Breaking of glycosidic bond by addition of water (hydrolysis reaction)

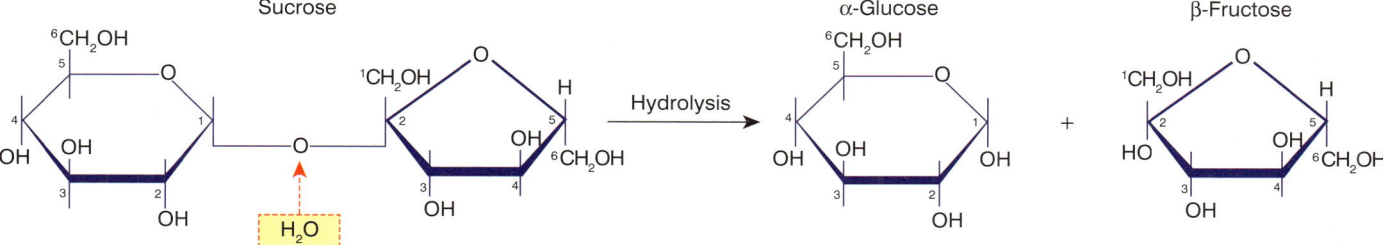

Figure 2 *Formation and breaking of glycosidic bond (some carbon and hydrogen atoms have been omitted for simplicity)*

acid and then detecting the resulting reducing sugars by the Benedict's test. The process is as follows.

- Heat a sample with Benedict's reagent in a water bath. If there is no change (solution remains blue), then no reducing sugar is present.
- Heat the sample in a water bath for five minutes with dilute hydrochloric acid to hydrolyse the non-reducing sugar, then neutralise with sodium hydrogencarbonate and allow to cool.
- Re-test the resulting solution by heating in a water bath with Benedict's reagent, which will now turn yellow/brown/red due to the reducing sugars made from hydrolysis of the non-reducing sugar.

Roles of monosaccharides and disaccharides

Monosaccharides and disaccharides function as respiratory substrates that are broken down to produce **ATP**. They are particularly useful because they have a large number of C—H groups and these can be easily oxidised, yielding a lot of energy.

REMEMBER

Condensation is the **giving out** of water in reactions while hydrolysis is the **taking in** of water to split molecules in reactions.

Semi-quantitative nature of the Benedict's test

Table 3 shows the relationship between the concentration of reducing sugar and the colour of the solution and precipitate formed during the Benedict's test. The differences in colour mean that the Benedict's test is **semi-quantitative**, i.e. it can be used to estimate the approximate concentration of reducing sugar in a sample. First a range of colour standards is produced by preparing a series of glucose solutions of known concentration. To an equal volume of each is added an excess of Benedict's reagent and they are then heated for the same length of time before being cooled to room temperature. An

equal volume of an unknown sample is then treated in the same way and the colour compared with that of the colour standards. As shown in the table, samples that turn red contain more reducing sugar than those that turn yellow.

A further extension of this experiment would be to carry out the reducing sugar test and then to filter the suspensions. The precipitate can then be dried and weighed. Alternatively, the filtrate can be placed in a colorimeter – the more intense the blue colour the less concentrated the reducing sugar. The greater the mass of precipitate, the more reducing sugar is present.

Table 3 *The Benedict's test*

Concentration of reducing sugar	Colour of solution and precipitate
None	Blue
Very low	Green
Low	Yellow
Medium	Brown
High	Red

2.3　Carbohydrates – starch and glycogen

On these pages you will learn to:

- Carry out the iodine in potassium iodide solution test for starch to identify the contents of solutions
- Describe the molecular structure of polysaccharides including starch (amylose and amylopectin) and glycogen and relate these structures to their functions in living organisms

Starch and glycogen are examples of **polysaccharides**. Polysaccharides are polymers, formed from combining together many monosaccharide units. The monosaccharides are joined by glycosidic bonds that are formed by **condensation reactions**. The resulting chain may vary in length and be branched and folded in various ways. All these features affect the properties of the polysaccharide that is formed. As polysaccharides are very large molecules (**macromolecules**), they are insoluble – a feature which suits them for storage. When they are **hydrolysed**, polysaccharides break down into monosaccharides or disaccharides. Some polysaccharides, such as cellulose (Topic 2.4), are not used for storage, but give structural support to plant cells.

Starch

Starch is a polysaccharide found in many parts of a plant in the form of small granules, or grains, e.g. starch grains in chloroplasts. Especially large amounts occur in seeds and storage organs such as potato tubers. It forms an important component of food and is the major energy source in most diets. Apart from the starch produced for eating, it is extracted from plants across the world for other purposes. These include wallpaper pastes, paper coatings, textiles, paints, cosmetics and medicines. Starch is a mixture of two substances – amylose and amylopectin.

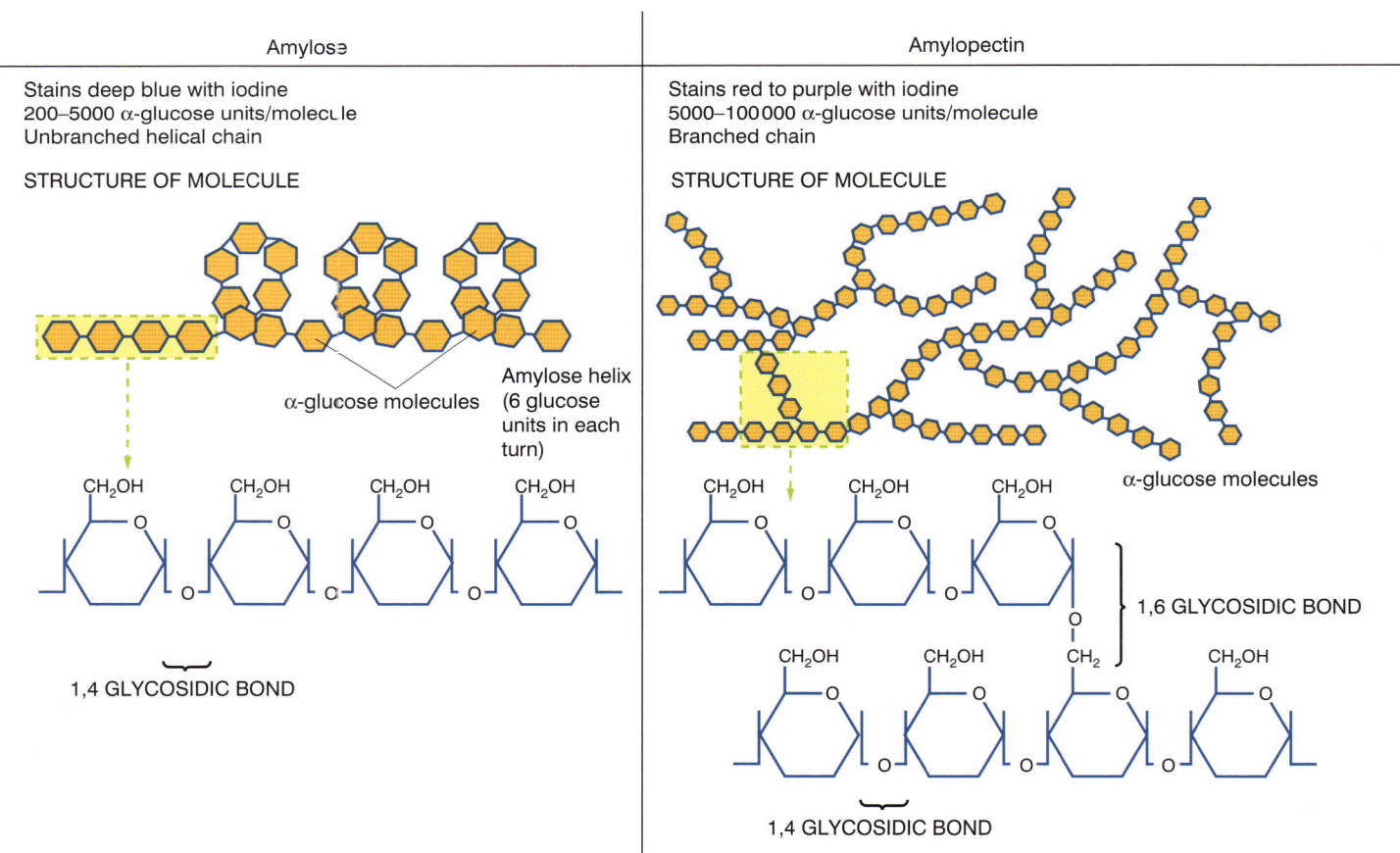

Amylose	Amylopectin
Stains deep blue with iodine 200–5000 α-glucose units/molecule Unbranched helical chain	Stains red to purple with iodine 5000–100 000 α-glucose units/molecule Branched chain

Figure 1 *Comparison of amylose and amylopectin*

- **Amylose** is composed of between 200 and 5000 α-glucose units, which are joined in a straight chain by 1,4 glycosidic bonds. This chain is then wound into a tight helix which makes the molecule more compact and therefore it can be stored more efficiently as it takes up less space.
- **Amylopectin** is made up of between 5000 and 100 000 α-glucose units joined to each other by 1,4 and 1,6 glycosidic bonds.

A comparison of amylose and amylopectin is given in Figure 1. About 80% of starch is amylopectin and the remaining 20% is amylose. These relative proportions can vary slightly depending on the source of the starch. The main role of starch is for energy storage, something it is especially suited for because:

- It is insoluble and therefore does not have any **osmotic** effects within cells, i.e. it does not affect the water potential of the cell.
- Being insoluble, it does not diffuse out of cells.
- Amylose is a helical molecule and is compact, so a lot of it can be stored in a small space.
- Amylopectin is branched and so has many free ends that amylase, the enzyme that catalyses the hydrolysis of starch, can work on simultaneously, meaning that glucose monomers are rapidly released.
- When hydrolysed it forms glucose, which is both easily transported and readily used in respiration, to provide ATP.

Starch is never found in animal cells. Instead a similar polysaccharide, called **glycogen**, serves the same role.

Glycogen

Glycogen is very similar in structure to amylopectin but has shorter chains and is more highly branched. It is the major carbohydrate storage product of animals, in which it is stored as small granules mainly in the muscles and the liver. Its structure suits it for storage for the same reasons as those given for starch except that it is more highly branched than starch and so has more ends that can be simultaneously acted on by enzymes. It is therefore more rapidly broken down, which is important to animals, which are more active than plants.

Test for starch

Starch is easily detected by its ability to turn the iodine in potassium iodide solution from an orange-yellow colour to blue-black. The colouration is due to the iodine molecules becoming fixed in the centre of the helix of each starch molecule (Figure 2). It is important that this test is carried out at room temperature (or below), as high temperatures cause the starch helix to unwind, releasing the iodine, which then returns to its usual yellow colour.

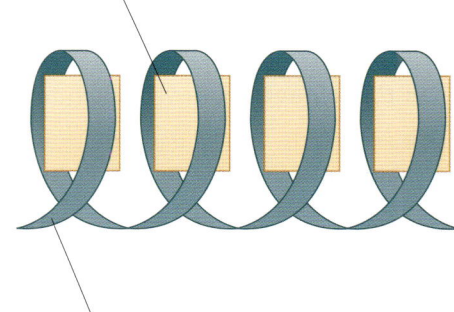

Iodine molecule in the centre of the amylose helix

Amylose helix formed by α-glucose molecules (6 per turn of helix). The dimensions of the centre are just sufficient to fit iodine molecules within it

Figure 2 *Amylose – iodine staining reaction*

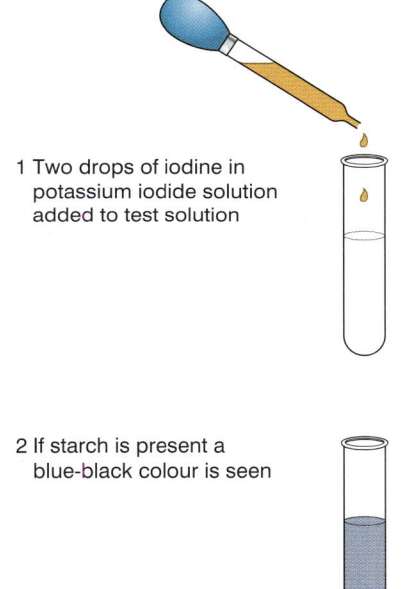

1 Two drops of iodine in potassium iodide solution added to test solution

2 If starch is present a blue-black colour is seen

Figure 3 *Test for starch*

SUMMARY TEST 2.3

From the following list of carbohydrates choose one or more which most closely fit each of the statements 1–9 below. Each carbohydrate may be used once, more than once or not at all.
- α-glucose
- β-glucose
- amylose
- amylopectin
- glycogen

1 Stains deep blue with iodine
2 The storage carbohydrate in animal cells
3 Found in plants and possesses 1,6 glycosidic bonds
4 Monosaccharide found in starch
5 Possesses 1,4 glycosidic bonds
6 Makes up 80% of starch
7 Can be hydrolysed
8 Branched molecule formed by condensation
9 Easily diffuses in and out of cells

2.4

Carbohydrates – cellulose

On these pages you will learn to:

- Describe the molecular structure of polysaccharides, including cellulose

Cellulose is a polysaccharide that makes up around 50% of all organic carbon and is therefore the most abundant organic molecule on Earth.

Structure of cellulose

Cellulose differs from starch and glycogen in one major respect – it is made of monomers of β-glucose rather than α-glucose. This seemingly small variation produces fundamental differences in the structure and function of this polysaccharide. The main reason for this is that, in the β-glucose units, the positions of the —H group and the —OH group on carbon atom 1 are reversed (Figure 1). In β-glucose the —OH group is above, rather than below, the ring. This means that, in order for 1,4 glycosidic links to form, each β-glucose molecule must be rotated by 180° compared to its neighbour. The result is that carbon atom 6 (the one forming part of the —CH₂OH group) on each β-glucose molecule alternates between being above and below the chain (Figure 1). Rather than forming a coiled chain like starch, cellulose has straight, unbranched chains. These run parallel to one another, allowing hydrogen bonds (Topic 2.1) to form cross-linkages between adjacent chains. While each individual hydrogen bond adds very little to the strength of the molecule, the sheer overall number of them makes a considerable contribution to strengthening cellulose and making it the valuable structural material it is. The arrangement of β-glucose chains in a cellulose molecule is shown in Figure 2.

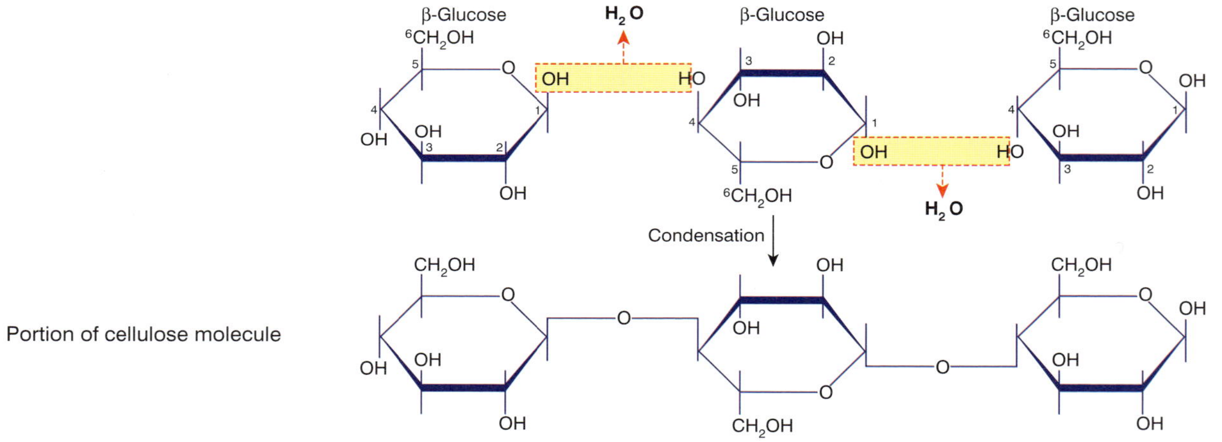

Figure 1 Formation of 1,4 glycosidic bonds between three β-glucose molecules (some carbon and hydrogen atoms have been omitted for simplicity)

Being composed of fewer β-glucose units, the chain, unlike that of starch, has adjacent glucose molecules rotated by 180°. This allows hydrogen bonds to be formed between the hydroxyl (—OH) groups on adjacent parallel chains which help to give cellulose its structural stability.

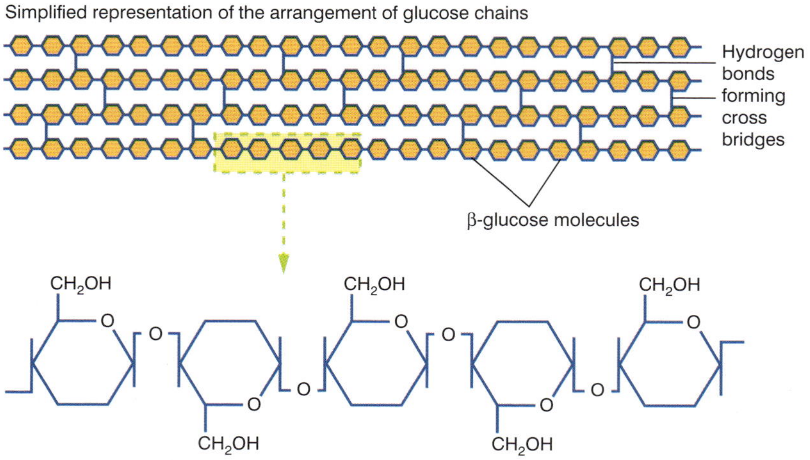

Figure 2 Structure of the cellulose molecule

The cellulose molecules are grouped together to form microfibrils which, in turn, are arranged in parallel groups called fibres.

Functions of cellulose

Cellulose frequently makes up between 20% and 50% of plant cell walls. It is, however, not part of the living cell but rather a non-living covering that encases the **protoplast** within. The cellulose cell wall is therefore freely permeable, allowing materials to access the cell surface membrane. It also allows the movement of water along the cell walls of adjacent cells (Topic 7.5). Cellulose performs a mainly structural role by providing rigidity to the plant cell. It also prevents the cell from bursting as water enters it by **osmosis**. It does this by the cellulose cell wall exerting an inward pressure that stops any further influx of water. As a result, living plant cells are turgid and push against one another, making herbaceous parts of the plant semi-rigid. This is especially important in maintaining stems and leaves in a turgid state so that they can provide the maximum surface area for photosynthesis.

The molecular stability of the cellulose molecule means it is well suited to giving structural support. However, it is also difficult to digest. It is therefore not a useful food for animals, which rarely produce cellulose-digesting enzymes. Some animals get round this by forming **mutualistic** relationships with cellulose-digesting microorganisms in their intestines. The structural strength of cellulose has been made use of by humans. Cotton and rayon used in fabrics are largely cellulose. Cellophane used in packaging and celluloid used in photographic films are also derived from cellulose. Paper is perhaps the best-known cellulose product.

Comparison of cellulose and other carbohydrates

Table 1 compares cellulose with the other polysaccharides, amylose, amylopectin and glycogen.

Table 1 Comparison of the polysaccharides amylose, amylopectin, glycogen and cellulose

Characteristic	Amylose	Amylopectin	Glycogen	Cellulose
Found in	Plants	Plants	Animals and fungi	Plants
Found as	Grains	Grains	Tiny granules	Fibres
Function	Energy store	Energy store	Energy store	Structural support
Basic monomer unit	α-glucose	α-glucose	α-glucose	β-glucose
Type of bond between monomer units	1,4 glycosidic	1,4 and 1,6 glycosidic	1,4 and 1,6 glycosidic	1,4 glycosidic
Type of chain	Unbranched and helical (coiled)	Branched, but less highly branched than glycogen	Short and highly branched	Long, unbranched straight chains with no coiling

SUMMARY TEST 2.4

Cellulose is made up of (1) monomers joined together by (2) links. It forms straight, unbranched chains that run parallel to each other and are cross-linked by (3). These cellulose molecules are then grouped together to form (4), which in turn are grouped into fibres. Cellulose performs a (5) function in plants by giving a plant cell rigidity. It also prevents cells from (6) when water enters by (7). In this way it keeps herbaceous parts of plants (8) so that they provide the maximum surface area for (9).

On these pages you will learn to:

- Carry out the emulsion test for lipids to identify the contents of solutions
- Describe the molecular structure of a triglyceride with reference to the formation of ester bonds and relate the structure of triglycerides to their functions in living organisms
- Describe the structure of a phospholipid and relate the structure of phospholipids to their functions in living organisms

Lipids make up a varied and diverse group of substances that share the following characteristics:

- They contain carbon, hydrogen and oxygen.
- The proportion of oxygen to carbon and hydrogen is smaller than in carbohydrates.
- They are insoluble in water.
- They are soluble in organic solvents such as alcohols.

The main groups of lipids are **triglycerides (fats and oils)** and **phospholipids**. Other forms include waxes, steroids and cholesterol.

Triglycerides (fats and oils)

There is no fundamental chemical difference between a fat and an oil. Fats are solid at room temperature (10–20 °C), whereas oils are liquid. Triglycerides are so called because they have three (tri) fatty acids combined with glycerol (glyceride). Each fatty acid forms an **ester bond** with glycerol in a **condensation reaction** (Figure 1). Hydrolysis of a triglyceride therefore produces glycerol and three fatty acids.

The three fatty acids may all be the same, thereby forming a simple triglyceride, or they may be different, in which case a mixed triglyceride is produced. In either case it is a condensation reaction.

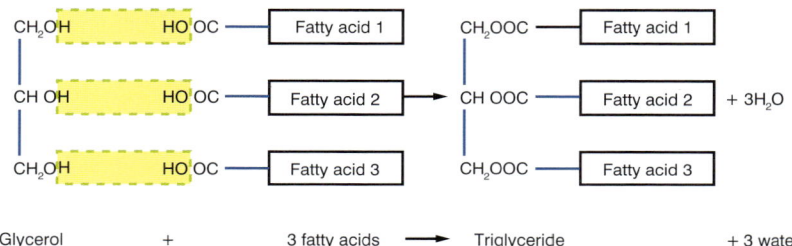

Figure 1 *Formation of a triglyceride*

Fatty acids

As the glycerol molecule in all triglycerides is the same, the differences in the properties of different fats and oils come from variations in the fatty acids. There are over 70 fatty acids and all have a carboxyl (—COOH) group with a hydrocarbon chain attached. This chain may possess no double bonds and is then described as **saturated**, because all the carbon atoms are linked to the maximum possible number of hydrogen atoms, i.e. they are saturated with hydrogen atoms. If there is a single double bond, it is **mono-unsaturated**; if more than one double bond is present, it is **polyunsaturated**.

REMEMBER

Some phospholipids contain nitrogen, for example cholines attached to the phosphate.

Phospholipids

Phospholipids are similar to lipids except that one of the fatty acid molecules is replaced by a phosphate molecule (Figure 2). Whereas fatty acid molecules repel water (are **hydrophobic**), phosphate molecules attract water (are **hydrophilic**).

Figure 2 *Structure of a phospholipid*

Phospholipids are important components of cell surface membranes. Both the inside of a cell and the environment outside are watery, and the phospholipids in cell surface membranes form a double layer, with the hydrophilic heads of the molecules pointing into either the watery environment outside the membrane or the watery medium inside the cell. The hydrophobic tails point into the middle of the membrane to form a hydrophobic core (Figure 3). This **bilayer** arrangement makes cell surface membranes fluid and easily crossed by lipid-soluble substances.

The structure of triglycerides related to their functions

- Triglycerides have a high ratio of energy-storing carbon–hydrogen bonds to carbon atoms and are therefore an excellent source of energy. They therefore supply many hydrogens for the reduction of NAD (Topic 12.5).
- Triglycerides have low mass to volume ratio making them good storage molecules because much energy can be stored in a small volume. This is especially beneficial to animals, as it reduces the mass they have to carry as they move around.
- Being large, non polar molecules, triglycerides are insoluble in water. As a result their storage does not affect the **water potential** of cells.
- As they have a high ratio of hydrogen to oxygen atoms, triglycerides release water when oxidised and therefore provide an important source of water, especially for organisms living in dry deserts.

The structure of phospholipids related to their functions

- Phospholipids are polar molecules, having a hydrophilic phosphate 'head' and a hydrophobic 'tail' of two fatty acids. This means that in an aqueous environment, phospholipid molecules form a bilayer of cell surface membranes. As a result, a hydrophobic barrier is formed between the inside and outside of a cell.
- Double bonds in the hydrocarbon chains/fatty acid tails form a kink and therefore there is more room between molecules and also less hydrophobic interaction. This gives greater fluidity to the membrane.
- Help to hold some peripheral proteins at the surface of the cell surface membrane.
- Phospholipid structure allows them to form glycolipids by combining with carbohydrates within the cell surface membrane. These glycolipids act as antigens and so are important in cell recognition.

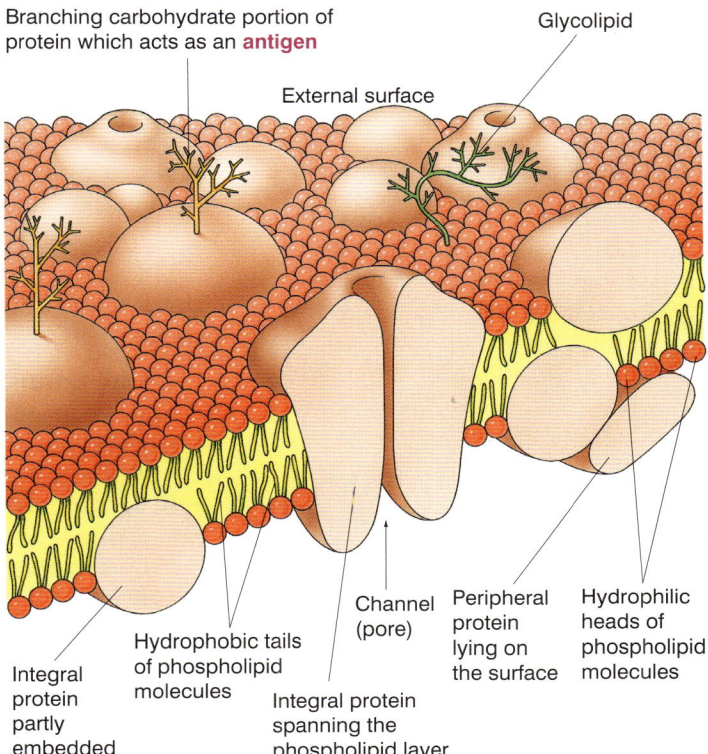

Branching carbohydrate portion of protein which acts as an **antigen**

Glycolipid

External surface

Channel (pore)

Peripheral protein lying on the surface

Hydrophilic heads of phospholipid molecules

Hydrophobic tails of phospholipid molecules

Integral protein partly embedded

Integral protein spanning the phospholipid layer

Figure 3 *Arrangement of the phospholipid bilayer in the cell surface membrane*

Test for lipids

The test for lipids is known as the **emulsion test** and is carried out as follows.

- Take a completely dry and grease-free test tube.
- Add about 2 cm³ of the sample being tested and add 5 cm³ of ethanol.
- Shake the tube thoroughly to dissolve any lipid in the sample.
- Add 5 cm³ of water and shake gently.
- A cloudy-white colour indicates the presence of a lipid.
- As a control, repeat the procedures using water instead of the sample; the final solution should remain clear.

The cloudy colour is due to any lipid in the sample being finely dispersed in the water to form an emulsion. Light passing through this emulsion is refracted as it passes from oil droplets to water droplets, making it appear cloudy.

SUMMARY TEST 2.5

Fats and oils make up a group of lipids called **(1)** which when hydrolysed form **(2)** and fatty acids. A fatty acid with more than one double bond is called **(3)**. In a phospholipid the number of fatty acids is **(4)**; these are called **(5)** because they repel water. Triglycerides have a high ratio of carbon to **(6)** atoms and so are an excellent source of **(7)**.

2.6 Amino acids and polypeptides

On these pages you will learn to:

- Describe the structure of an amino acid and the formation and breakage of a peptide bond
- Explain the meaning of the terms 'primary structure' and 'secondary structure' of proteins

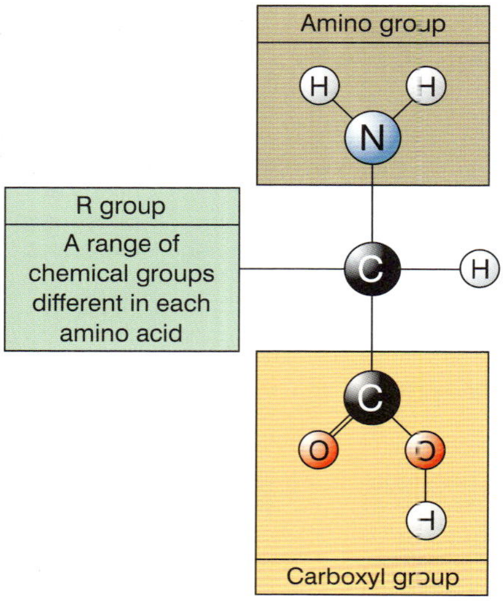

Figure 1 *General structure of an amino acid*

Amino acids are the basic monomer units which combine to make up proteins. Around 100 amino acids have been identified, of which 20 occur naturally in proteins.

Structure of an amino acid

Every amino acid has a central carbon atom to which are attached four different chemical groups:

- **amino group** (—NH_2) – a basic group from which part of the name amino acid is derived
- **carboxyl group** (—COOH) – an acid group which gives the amino acid the rest of its name
- **hydrogen atom** (—H)
- **R group** – a variety of different chemical groups ranging from a single hydrogen atom, as in glycine, to a double ring structure, as in tyrosine. Each amino acid has a different R group.

The general structure of an amino acid is shown in Figure 1.

As the carboxyl group is acidic and the amino group is basic, an amino acid is both an acid and a base – it is said to be **amphoteric**. Amphoteric compounds act as **buffer** solutions in that they resist the tendency to alter their pH, despite the addition of acids or bases. This property is important in cells because it helps them maintain the stable pH that is necessary for the efficient functioning of enzymes.

Formation and breakage of a peptide bond

In a similar way that monosaccharide monomers combine to form disaccharides (Topic 2.2), so amino acid monomers can combine to form a **dipeptide**. The process is essentially the same – namely the removal of a water molecule in a condensation reaction (Topic 2.1). The water is made by combining an —OH from the carboxyl group of one amino acid with an —H from the amino group of another amino acid. The two amino acids then become linked by a new **covalent bond** between the carbon atom of one amino acid and the nitrogen atom of the other. The formation of a peptide bond is illustrated in Figure 2. In the same way as a glycosidic bond of a disaccharide can be broken by the addition of water (hydrolysis), so the peptide bond of a dipeptide can also be broken by hydrolysis (Topic 2.1), to give its two constituent amino acids.

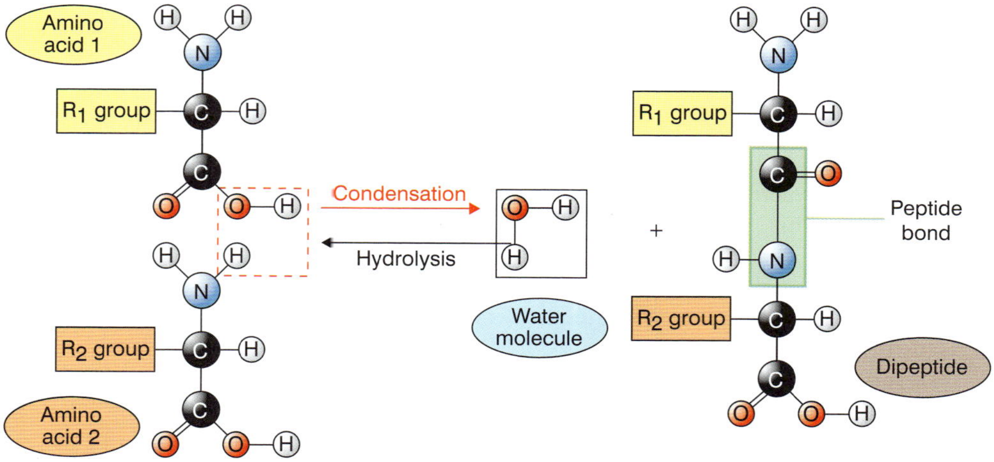

Figure 2 *Formation of a peptide bond*

EXTENSION

Plants can manufacture all the amino acids they need by combining nitrates, which they absorb from the soil, with various carbohydrates that they make during photosynthesis. Animals, however, need to obtain eight amino acids from their food. These are called **essential amino acids**. The remaining twelve amino acids, called **non-essential amino acids**, can be manufactured from the eight essential ones if they are not provided in adequate amounts in the diet.

The primary structure of proteins – polypeptides

Through a series of condensation reactions, many amino acid monomers can be joined together in a process called **polymerisation**. The resulting chain of many hundreds of amino acids is called a **polypeptide**. The sequence of amino acids in a polypeptide chain forms the **primary structure** of any protein.

The secondary structure of proteins

The linked amino acids that make up a polypeptide possess both —NH and —C=O groups on either side of every peptide bond. Both these groups are polar, i.e. their **electrons** are unevenly distributed. As a result, the hydrogen of the —NH group has an overall positive charge while the O of the —C=O group has an overall negative charge. These two groups therefore readily form hydrogen bonds (Topic 2.1). This causes sections of the polypeptide chain to form one of two regular three-dimensional shapes that have a repeating pattern:

* **α-helix** – the polypeptide chain is coiled into a spiral shape
* **β-pleated sheet** – the polypeptide chains are linked in parallel flat sheets.

Figure 3 illustrates these two basic types of protein structure. Some areas of the protein will not have the regular arrangement shown by alpha helices and beta pleated sheets and form what is termed a random coil. Different proteins have differences in the extent of the alpha helices and beta-pleated sheets formed.

SUMMARY TEST 2.6

Amino acids always contain an acid carboxyl group, which has the chemical formula **(1)**, as well as a basic group called an **(2)** group. Amino acids therefore have both acidic and basic properties and are said to be **(3)**. Any two amino acids can combine in a **(4)** reaction to form a **(5)** bond between them. Many amino acids can combine to form a polypeptide chain which can become altered in shape due to **(6)** bonds formed between certain groups. These secondary shapes may be spiral, in which case they are called **(7)**, or flatter, in which case they are called **(8)**.

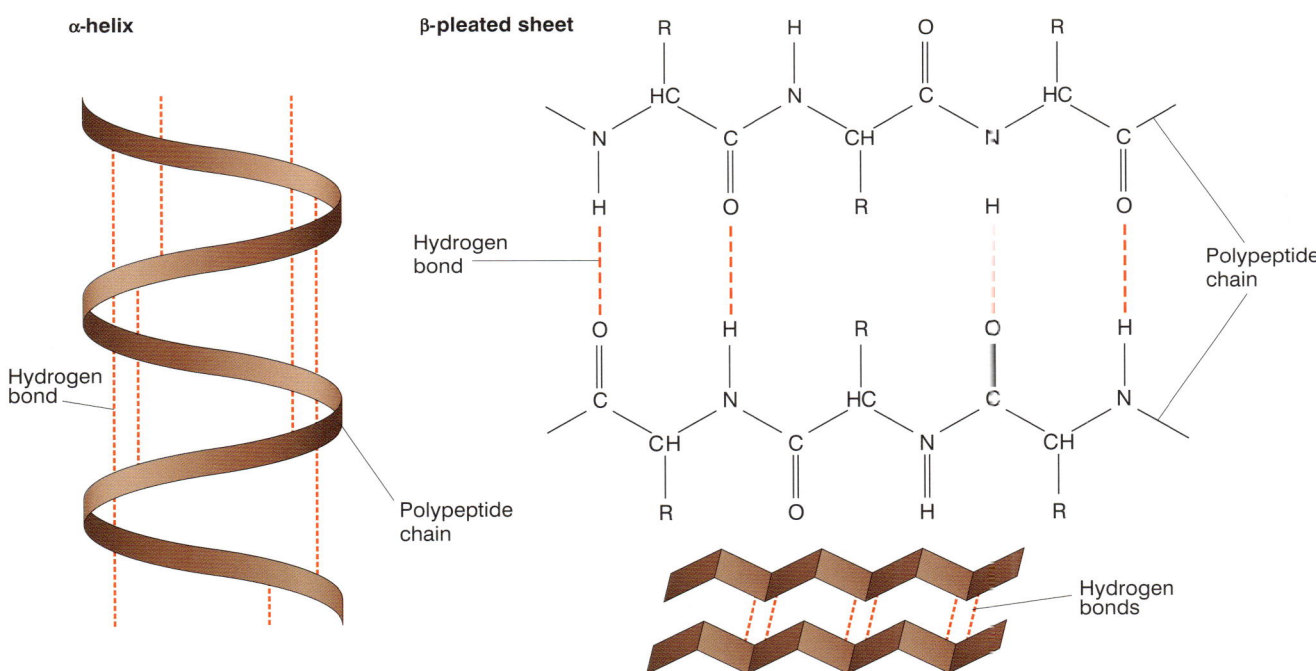

Figure 3 Structure of the α-helix and the β-pleated sheet

2.7 Protein structure

On these pages you will learn to:

- Carry out the biuret test for proteins to identify the contents of solutions
- Explain the meaning of the terms 'tertiary structure' and 'quaternary structure' of proteins and describe the types of bonding (hydrogen, ionic, disulfide and hydrophobic interactions) that hold these molecules in shape

Proteins are large molecules with relative molecular masses ranging from many thousands up to 40 million. While the types of carbohydrates and lipids of all organisms are relatively few and very similar, their proteins are numerous (estimated to be up to two million types in humans) and differ from species to species. The shape of any one type of protein molecule differs from that of other proteins. As the amino acids of which they are made differ only in their R groups, it is these that determine the shape, and therefore the functions, of a protein. We have already seen in Topic 2.6 how the primary and secondary shapes of a protein are determined. We shall now see how these structures are moulded into the tertiary and quaternary structures that make up a protein's final shape, or **configuration**.

Tertiary structure of proteins

The polypeptide chain, which may already have sections of secondary structure (alpha helixes and beta-pleated sheets), undergoes further folding and coiling to give the complex, and often unique, three-dimensional structure of each protein (Figure 2). This is known as the tertiary structure and is the result of four possible types of bonds that can arise between the R-groups (side chains) of each amino acid. The interactions between R-groups, in order of relative strength, are:

- **Disulfide bridges** – found between sulfur atoms in the molecules of the amino acid cysteine. They are **covalent bonds** and, as such, form very strong links which make the tertiary protein structure very stable.
- **Ionic bonds** – occur between any carboxyl and amino groups that have not been involved in forming peptide bonds. These groups ionise to give —NH_3^+ and —COO^- groups, which then form electrostatic bonds due to their mutual attraction. These bonds are weaker than disulfide bridges and can be broken by changes in pH.
- **Hydrogen bonds** – result from the attraction between the electronegative oxygen atoms on the —CO groups and the electropositive H atoms on either the —OH or —NH groups. Although they are individually weaker than ionic bonds, their large number makes them an important factor in maintaining the tertiary structure of a protein. Note that they are not the same hydrogen bonds as with secondary structure – these are hydrogen bonds between the R-groups.
- **Hydrophobic interactions** – due to certain non-polar R groups in amino acids that have side groups which repel water. As a result they may fold or twist the polypeptide chain as they take up a position towards the centre of the protein, further away from the watery medium outside.

Figure 1 illustrates how each of these bonds is formed.

Quaternary structure of proteins

A protein with quaternary structure is composed of more than one polypeptide chain. There may also be non-protein (prosthetic) groups associated with the molecules (Figure 2). Examples of quaternary structure are illustrated by the protein haemoglobin (Topic 2.8).

(a) **Disulfide bridges** – covalent bond between R groups of cysteine amino acids

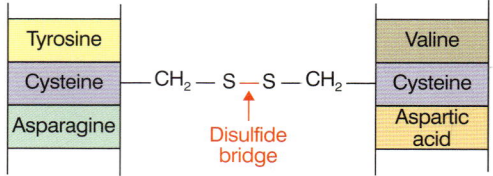

(b) **Ionic bonds** – between NH_3^+ and COO^- ions on basic amino acids such as asparagine and acid ones such as aspartic acid

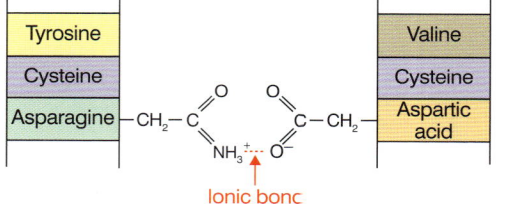

(c) **Hydrophobic interactions** – between non-polar R groups such as those on the amino acids tyrosine and valine

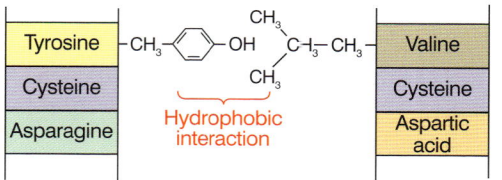

(d) **Hydrogen bonds** – between electronegative oxygen atoms on CO groups and electropositive H atoms on NH groups

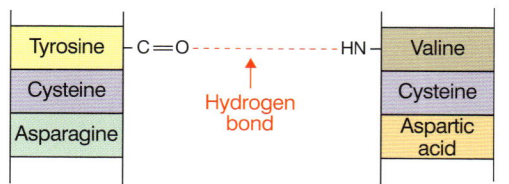

Figure 1 Types of bond that determine the shape of a protein

(a) The primary structure of a protein is the sequence of amino acids found in its polypeptide chains. This sequence determines its properties and shape. Following the discovery of the amino acid sequence of the hormone insulin by Frederick Sanger in 1954, the primary structure of many other proteins is now known.

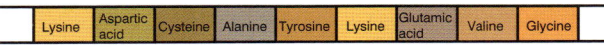

(b) The secondary structure is the shape which the polypeptide chain forms as a result of bonding between the hydrogens of the amine group and the oxygens of the carboxyl group. This may be spiral or a beta-pleated sheet.

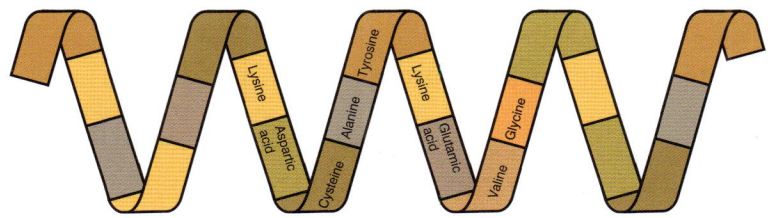

(c) The tertiary structure is due to the coiling and folding of the polypeptide chain into a specific three-dimensional structure. All four types of interaction between R-groups contribute to the maintenance of the tertiary structure.

(d) The quaternary structure arises from the combination of a number of different polypeptide chains and associated non-protein (prosthetic) groups into a large, complex protein molecule.

The tertiary structure is held in shape by all four types of interaction

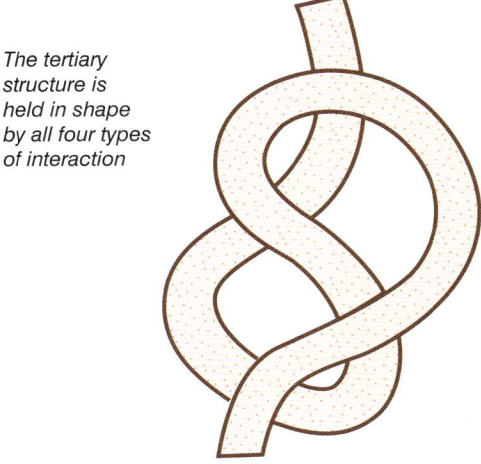

Polypeptide chain 1

Polypeptide chain 2

Prosthetic group

Polypeptide chain 3

Figure 2 Structure of proteins

Test for proteins

The most reliable protein test is the **biuret test**, which detects peptide links. It is performed as follows.

- To a sample of the solution thought to contain protein, add an equal volume of prepared biuret solution.
- Add a few drops of very dilute (0.05%) copper(II) sulfate solution and mix gently.

- A purple colouration indicates the presence of peptide bonds and hence a protein. If no protein is present, the solution remains blue.
- A control should be carried out by performing the above stages, but with water replacing the sample under test.

SUMMARY TEST 2.7

The primary structure of proteins is determined by the sequence of **(1)** which make up the **(2)** chain. The secondary structure results from coiling or folding of the chain due to **(3)** formed between the —NH of one amino acid and the **(4)** group of another **(5)**. Four types of bond between R-groups cause further twisting and folding of the chain. The first of these bonds arises between **(6)** atoms in cysteine molecules and is called **(7)**. The second type is called **(8)** and results from electrostatic forces between carboxyl and R-groups of amino acids. Thirdly there are forces due to amino acid R-groups which repel water and these are called **(9)**. The fourth type of bond is the hydrogen bond. The quaternary structure of proteins results from a number of chains combining, sometimes also incorporating non-protein groups known as **(10)** groups.

2.8 Fibrous and globular proteins

On these pages you will learn to:

- Describe the molecular structure of haemoglobin as an example of a globular protein, and cf collagen as an example of a fibrous protein, and relate these structures to their functions

Proteins perform many different roles in living organisms. In one form or another they are essential for the efficient functioning of every characteristic of life. Their roles depend on their molecular configurations, which are of two basic types.

- **Fibrous proteins**, such as collagen and keratin, have structural functions.
- **Globular proteins**, such as enzymes, haemoglobin and insulin, carry out metabolic functions.

Table 1 lists the differences between fibrous and globular proteins.

The fibrous protein collagen

Fibrous proteins form long chains which run parallel to one another. These chains are linked by cross-bridges and so form very stable molecules. One example is **collagen**, a protein found in tissues requiring physical strength, e.g. tendons, walls of blood vessels, bone and the fibres that hold teeth in place. Collagen is extremely strong and stable, being able to withstand immense pulling forces without stretching. At the same time it is flexible, so that, while the collagen in a tendon transmits the pull of a muscle to the bone without stretching, it can still bend around a joint as it flexes during movement. The ability of collagen to do this is the result of the following features of its structure.

- Its primary structure is largely a repeat of the amino acid sequence glycine–proline–alanine, which forms an unbranched polypeptide chain.
- The collagen molecule is made up of three such polypeptide chains wound in a triple helix that is held together by hydrogen bonds.
- As every third amino acid is the relatively small and compact glycine molecule, the triple helix produced is very tightly wound. Larger amino acids would have led to a more loosely wound, and therefore less strong, triple helix.
- The triple-stranded molecules run parallel to others, forming even stronger units called fibrils, with fibrils forming collagen fibres.
- The collagen molecules in the fibres are held together by cross-linkages formed by covalent bonds between lysine amino acids of adjacent fibres. This adds greater strength and stability to the structure.
- The points where one collagen molecule ends and the next begins are spread throughout the structure. If they were all joined together in the same region this would be a weak point and therefore prone to breaking under tension.

The structure of collagen is illustrated in Figure 1.

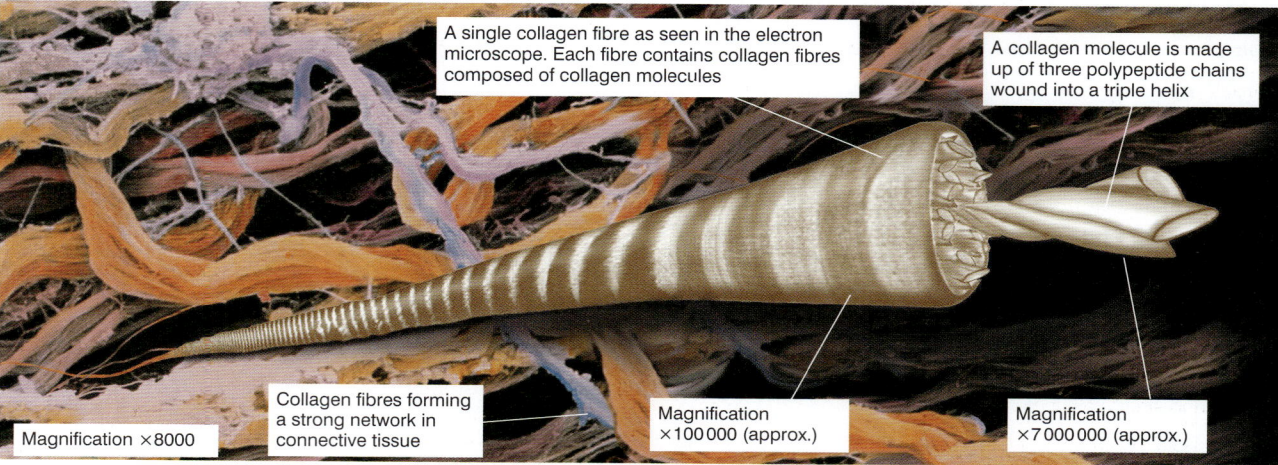

A single collagen fibre as seen in the electron microscope. Each fibre contains collagen fibres composed of collagen molecules

A collagen molecule is made up of three polypeptide chains wound into a triple helix

Collagen fibres forming a strong network in connective tissue

Magnification ×8000

Magnification ×100 000 (approx.)

Magnification ×7 000 000 (approx.)

Figure 1 *Fine structure of the fibrous protein collagen*

The globular protein haemoglobin

The sequence of amino acids in globular proteins is far more varied, and they form a compact spherical-shaped structure, unlike fibrous proteins. If the polypeptide chains of a fibrous protein are thought of as string twisted into a rope, then a globular protein is like the same string rolled into a ball. One example of a globular protein is haemoglobin (Figure 2). This is an oxygen-carrying respiratory pigment found in most animal groups. Its ability to transport oxygen is related to its structural features, which in adult humans, include the following:

- It has a quaternary structure of four polypeptide chains.
- Two of these chains, called α-globin polypeptides, are identical and each consists of 141 amino acids.
- The other two chains, called β-globin polypeptides, also form an identical pair and each consists of 146 amino acids.
- Each polypeptide chain is folded into a compact shape and all four are linked together to form an almost spherical haemoglobin molecule.
- Hydrophobic interactions (Topic 2.7) of groups within the haemoglobin molecule help to maintain its precise shape – an important factor in its ability to carry oxygen.
- Hydrophilic groups in the molecule tend to orient themselves to point outwards. This enables haemoglobin to be soluble and mix more readily with a watery medium (the cytoplasm of the red blood cell).
- Associated with each polypeptide is a haem group – which contains a ferrous (Fe^{2+}) ion. Non-protein groups such as this are called **prosthetic groups** and they from an important and integral part of the protein molecule.
- Each Fe^{2+} ion can combine with a single oxygen molecule (O_2), making a total of four O_2 molecules that can be carried by a single haemoglobin molecule in humans. (Other types of haemoglobin have different numbers of haem groups and so carry different numbers of O_2 molecules.)

When haemoglobin combines with oxygen it forms a molecule called **oxyhaemoglobin** and changes colour from purple to bright red. The structure of a haemoglobin molecule is shown in Figure 2.

Globular proteins such as haemoglobin and enzymes have a very specific shape. Even slight changes to their structure can make such molecules far less efficient at carrying out their functions. In the case of haemoglobin a slight alteration in shape as a result of the condition sickle cell anaemia, makes it far less able to transport oxygen.

Table 2 lists some other important proteins and their functions.

Table 1 *Comparison of fibrous and globular proteins*

Fibrous proteins	Globular proteins
Repetitive regular sequences of amino acids	Irregular amino acid sequences
Actual sequences may vary slightly between two examples of the same protein	Sequence highly specific and never varies between two examples of the same protein
Polypeptide chains form long parallel strands	Polypeptide chains folded into a spherical shape
Length of chain may vary in two examples of the same protein	Length always identical in two examples of the same protein
Stable structure	Relatively unstable structure
Insoluble	Water soluble – forms **colloidal** suspensions
Support and structural functions	Metabolic functions
Examples include collagen and keratin	Examples include all enzymes, some hormones (e.g. insulin), antibodies and haemoglobin

EXTENSION

Table 2 *Protein functions*

Protein	Function of protein
Trypsin and pepsin	Digestion of proteins/polypeptides
Myoglobin	Stores oxygen in muscle
Actin and myosin	Needed for contraction of muscle
Antibodies	Defend against bacterial invasion
Gluten	Storage protein in seeds
Histone	Gives structural support to **chromosomes**

Four polypeptide chains make up the haemoglobin molecule. Each molecule contains 574 amino acids

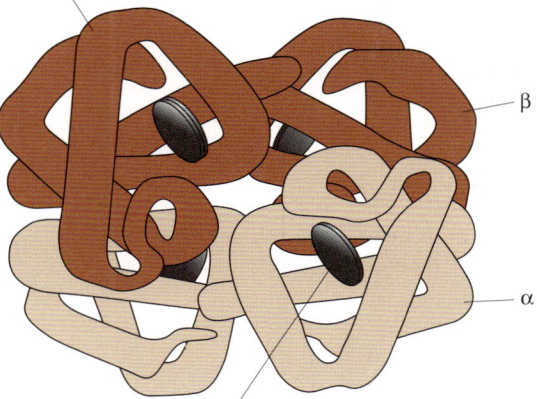

β

α

Each chain is attached to a haem group that can combine with oxygen

Figure 2 *Quaternary structure of a haemoglobin molecule*

SUMMARY TEST 2.8

Proteins are of two basic types: fibrous proteins such as collagen and **(1)** proteins such as the respiratory pigment haemoglobin. Proteins like actin and myosin have a structural function and are therefore examples of a **(2)** protein. Collagen is a fibrous protein with the repeating amino acid sequence of **(3)**. It is found in structures such as **(4),** which attach muscle to bone, where its properties of **(5)**, and **(6)** suit it to its role. A single haemoglobin molecule is made up of polypeptides, which total **(7)** in number. Each polypeptide contains a **(8)** group that contains a single **(9)** ion, to which can be attached a single oxygen molecule.

Water and its functions

On these pages you will learn to:

- Explain how hydrogen bonding occurs between water molecules and relate the properties of water to its roles in living organisms (limited to solvent action, specific heat capacity and latent heat of vaporisation)

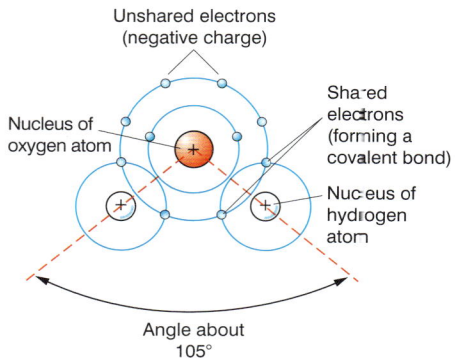

Figure 1 A water molecule

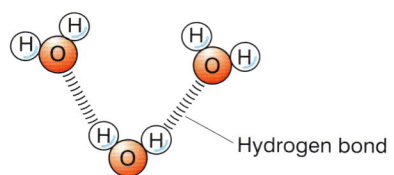

Figure 2 Water molecules showing hydrogen bonding

Although water is the most abundant liquid on Earth, it is certainly no ordinary molecule. Its unusual properties are due to its dipolar nature and the subsequent hydrogen bonding that this allows.

The dipolar water molecule

A water molecule is made up of two atoms of hydrogen and one of oxygen as shown in Figure 1. The atoms form a triangular shape. Although the molecule has no overall charge, the distribution of negatively charged electrons is uneven because the oxygen atom pulls them away from the hydrogen atoms. The oxygen atom therefore has a slight negative charge, while the hydrogen atoms have a slight positive one. In other words, the water molecule has both positive and negative poles and is therefore described as **dipolar**.

Water and hydrogen bonding

Different poles attract, and therefore the positive pole of one water molecule will be attracted to the negative pole of another water molecule. The attractive force between these opposite charges is called a **hydrogen bond** (Figure 2). Although each bond is fairly weak (about one-tenth as strong as a **covalent bond**), together they form important forces that cause the water molecules to stick together, giving water its unusual properties.

Specific heat capacity

As water molecules stick together (cohesion), it takes more energy (heat) to separate them than would be needed if they did not bond to one another. For this reason the boiling point of water is higher than expected. Without its hydrogen bonding, water would be a gas (water vapour) at the temperatures commonly found on Earth and life as we know it would not exist. For the same reason, it takes more energy to heat a given mass of water, i.e. water has a high **specific heat capacity**. Water therefore acts as a buffer against sudden temperature variations, making the aquatic environment a temperature-stable one.

EXTENSION

Cohesion and surface tension in water

The tendency of molecules to stick together is known as **cohesion**. With its hydrogen bonding, water has large cohesive forces and these allow it to be pulled up through a tube, such as a **xylem vessel** in plants. In the same way, water molecules at the surface of a body of water tend to be pulled back into the body of water rather than escaping from it. This force is called **surface tension** and means that the water surface acts like a skin and is strong enough to support small organisms such as pond skaters (Figure 3).

The density of water

Most substances are at their least dense when a gas and at their most dense when a solid, with the liquid phase having an intermediate density. Water is different. Water is actually less dense in the form of ice than when it is a liquid. This property is crucial to the survival of aquatic organisms as it means that ponds, lakes, etc. freeze from the top down rather than from the bottom up. The ice formed at the top then acts as an insulating layer that delays the freezing of the water beneath it. Large bodies of water almost never freeze completely, allowing their inhabitants to survive.

Figure 3 Due to surface tension, pond skaters walk on water

Latent heat of vaporisation

Hydrogen bonding between water molecules means that it requires a lot of energy to evaporate one gram of water. The energy is called the **latent heat of vaporisation**. Evaporation of water such as sweat in mammals is therefore a very effective means of cooling because body heat is used to evaporate the water.

Solvent action

The dipolar nature of the water molecule means that other polar molecules and ions readily dissolve in water. As a result, water is a very good solvent and a wide range of substances dissolve in it. This property means that water is used for:

- transport, e.g. sugars in blood and phloem
- removal of wastes, e.g. ammonia, urea
- secretions, e.g. digestive juices, tears
- the environment in which enzyme reactions take place.

Figure 4 *Evaporation of water during sweating helps to maintain body temperature*

EXTENSION
The importance of water to living organisms

Water is the main constituent of all organisms – up to 98% of a jellyfish is water and mammals are typically 65% water. Water is also where life on Earth arose and it is the environment in which many species still live. It is important for other reasons too. In this book you will meet examples of the importance of water to living organisms. These include the following.

Water in metabolism
- Water is used to break down many complex molecules by **hydrolysis**, e.g. proteins to amino acids.
- Chemical reactions take place in an aqueous medium.
- Water is a major raw material in photosynthesis.

Water as a solvent
The importance of water as a solvent was discussed above.

Water giving support
Water is not easily compressed and therefore is used in:

- the hydrostatic skeleton of animals such as earthworms
- the amniotic fluid to support the fetus
- creating the turgor pressure of leaf cells that contribute to the support of leaves.

Other important features of water
- Its evaporation cools organisms and allows them to control their temperature.
- In plants, it is involved in cell elongation and expansion.

Water as an environment for living organisms
- Water acts as a buffer against sudden temperature variations, making the aquatic environment a temperature-stable one.
- Large bodies of water almost never freeze completely, allowing living organisms to survive.
- Water is transparent and therefore aquatic plants can photosynthesise.
- Water is a dense medium and so provides support for organisms which therefore require less supporting tissue than on land.

SUMMARY TEST 2.9

A water molecule is said to be **(1)** because it has a positive and a negative pole as a result of the uneven distribution of **(2)** within it. This creates attractive forces called **(3)** between water molecules, causing them to stick together. This stickiness of water means that it takes more energy to heat a given mass of water. Water therefore has a high **(4)**. Hydrogen bonding between water molecules means that it requires a lot of energy to evaporate one gram of water. This energy is called the **(5)**.

2 Examination Questions

1 Polysaccharides, such as glycogen, amylopectin and amylase, are formed by polymerisation of glucose. Figure 1 shows part of a glycogen molecule.

Figure 1

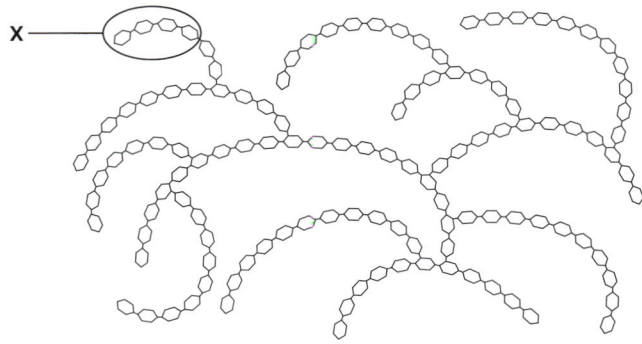

a With reference to Figure 1,

  i describe how the **structure** of glycogen differs from the structure of amylose; *(2 marks)*

  ii describe the advantages for organisms in storing polysaccharides, such as glycogen, rather than storing glucose. *(3 marks)*

b Glycogen may be broken down to form glucose.

Figure 2 shows region **X** from the glycogen molecule in Figure 1 in more detail.

Figure 2

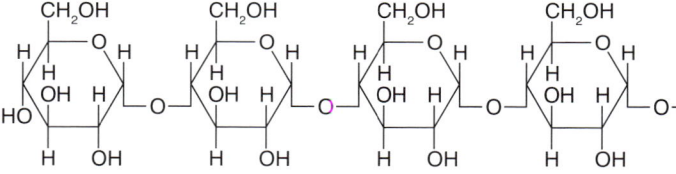

Draw an annotated diagram to explain how a glucose molecule is formed from the free end of the glycogen molecule shown in Figure 2. *(3 marks)*

(Total 8 marks)

Cambridge International AS and A Level Biology 9700 Paper 2 Q2 November 2008

2 a Figure 3 shows the breakdown of a molecule of sucrose.

  i Name the bond indicated by **T**. *(1 mark)*

  ii State the name given to this type of reaction in which water is involved. *(1 mark)*

  iii State two roles of water **within plant cells** other than taking part in breakdown reactions. *(2 marks)*

Figure 3

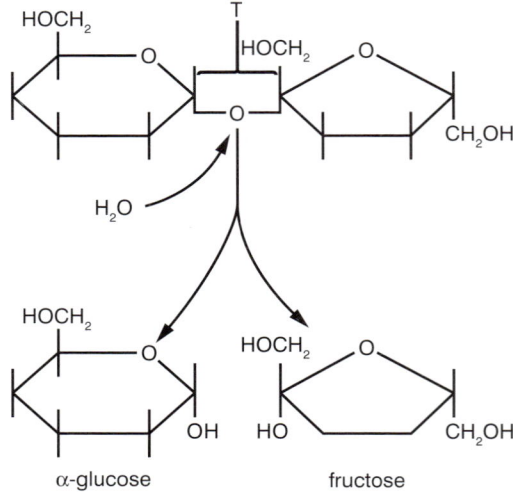

α-glucose   fructose

b Enzymes are globular proteins.

State what is meant by the term *globular*. *(2 marks)*

c The reaction shown in Figure 3 is catalysed by the enzyme sucrase. Figure 4 shows an enzyme-catalysed reaction.

Figure 4

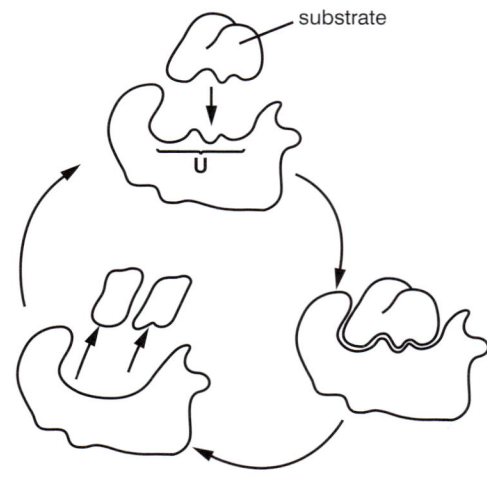

substrate

  i Name the part of the enzyme labelled **U**. *(1 mark)*

  ii With reference to Figure 4, explain the mode of action of enzymes. *(4 marks)*

(Total 11 marks)

Cambridge International AS and A Level Biology 9700 Paper 21 Q1 June 2010

2 Practice Questions

3 The figure below represents a phospholipid molecule.

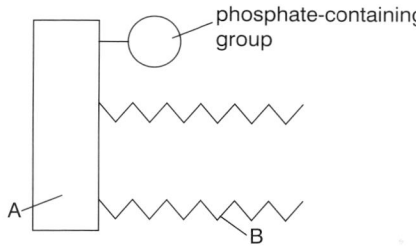

phosphate-containing group

A

B

a Give the names of the structures labelled **A** and **B**.

b State how the structures **A** and **B** differ in the way they react to water.

c Which chemical elements are found in fats?

d What is meant by a 'saturated' fatty acid?

e A 200 g portion of chips (French fries) from a restaurant was found to contain 19.2 g of fat. The same mass of chips from a frozen oven-ready portion was found to contain 11.6 g of fat. When broken down, fat releases 38 kJ g^{-1}. How much more energy is released from a portion of chips from a restaurant compared to an oven-ready portion? Show your working.

4 The figure below represents a polypeptide made up of seven amino acids, **A–G**.

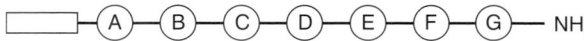

— A — B — C — D — E — F — G — NH$_2$

a What is the chemical formula of the group represented by the box?

b Name the type of bond that links one amino acid to another.

c What is the type of reaction that links amino acids together called?

d Name the test that is used to test for proteins.

e Protein molecules are held together by a combination of the following: peptide bonds, hydrogen bonds, ionic bonds, disulfide bridges, hydrophobic interactions. Which one or more of these bonds:
   i maintain the primary structure of a protein
   ii maintain the secondary structure of a protein
   iii maintain the tertiary structure of a protein
   iv are individually the two strongest?

5 The figure below shows eight biological molecules A–H.

For each of the following give the letter(s) of one or more molecules that fits the description in each case.

a A triglyceride

b A disaccharide

c Can be polymerised to make a protein

d An inorganic molecule

e Has hydrophilic and hydrophobic portions

f An amino acid that can form disulfide bridges

g Is insoluble in water and stores energy in organisms

h A reducing sugar

3

3.1

Enzymes

Enzyme structure and mode of action

On these pages you will learn to:

- Explain that enzymes are globular proteins that catalyse metabolic reactions
- State that enzymes function inside cells (intracellular enzymes) and outside cells (extracellular enzymes)
- Explain the mode of action of enzymes in terms of an active site, enzyme/substrate complex, lowering of activation energy and enzyme specificity (the lock and key hypothesis and the induced fit hypothesis should be included)

Enzymes are globular proteins which catalyse metabolic reactions. A catalyst alters the rate of a chemical reaction without itself undergoing permanent change. It can therefore be used repeatedly and so is effective in tiny amounts. Enzymes do not make a reaction happen; they simply speed up ones which already occur, sometimes by a factor of many millions.

Enzyme structure

As globular proteins, enzymes have a specific three-dimensional shape which is determined by their sequence of amino acids (Topic 2.7). Despite their large overall size, enzyme molecules only have a small region that is functional. This is known as the **active site**. Only a few amino acids of the enzyme molecule make up this active site. The active site forms a hollow depression within the much larger enzyme molecule. The substrate molecule is held within the active site by bonds that temporarily form between the R groups of the amino acids of the active site and groups on the substrate molecule. This structure is known as the **enzyme–substrate complex** (Figure 1).

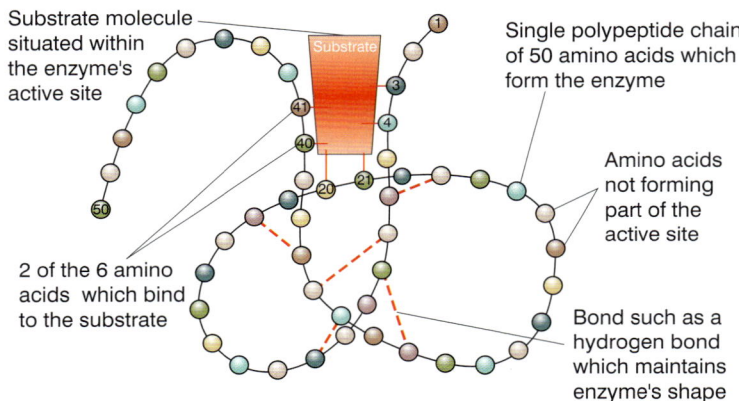

Figure 1 *Enzyme–substrate complex showing the six out of 50 enzyme amino acids that form the active site*

Enzymes and activation energy

Consider a typical chemical reaction:

$$\text{sucrose} + \text{water} \rightarrow \text{glucose} + \text{fructose}$$
$$\text{(substrates)} \qquad\qquad \text{(products)}$$

For such a reaction to occur naturally, the energy of the products (glucose + fructose) must be less than that of the substrates (sucrose + water). Such reactions, however, need an initial boost of energy to get them kick-started. This is known as the **activation energy**. It can be likened to a small stone lying on a hillside. If initial energy is applied to the stone, e.g. by giving it a push, then it will begin to move. It will move downhill, rather than uphill, because this lowers its potential energy (it is a fundamental law of thermodynamics that materials will naturally tend towards a state of low energy). Once set in motion, however, the stone gathers its own momentum and reaches the bottom with no further input of energy. This comparison shows how an initial input of energy (activation energy) can cause a reaction to continue on its own. In other words, there is an energy hill, or barrier, which must be overcome before the reaction can proceed. What enzymes do is to lower this activation energy level so that the reaction can happen more easily (Figure 2). For example, they allow many reactions to take place at a lower temperature than normal. As a result, some metabolic processes occur rapidly at the human body temperature of 37 °C, which is relatively cool in terms of chemical reactions. These would take place too slowly to sustain life as we know it were they to take place without enzymes.

How enzymes work

In one sense, enzymes work in the same way as a key operates a lock: each key has a very specific shape which, on the whole, fits and operates only one lock. In the same way, a substrate will only fit the active site of one particular enzyme. Enzymes are therefore **specific** in the reactions that they catalyse. The shape of the substrate (key) exactly fits the active site of the enzyme (lock). This is known as the **lock and key hypothesis** and explains, in a simple way, what happens (Figure 4). In practice, for most enzymes, the process is slightly different. Rather than being a rigid lock, the enzyme actually changes its form slightly to fit the shape of the substrate. In other words, it is flexible and moulds itself around the substrate, just as a glove moulds itself to the shape of someone's hand. The enzyme has a certain basic shape, just as a glove has, but this becomes slightly different as it alters in the presence of the substrate. As it alters its shape, the enzyme puts a strain on the substrate molecule and thereby lowers its activation energy. This whole process is called the **induced fit hypothesis** of enzyme action.

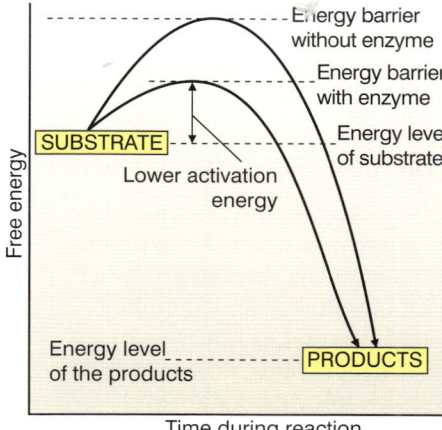

Figure 2 *How enzymes lower activation energy*

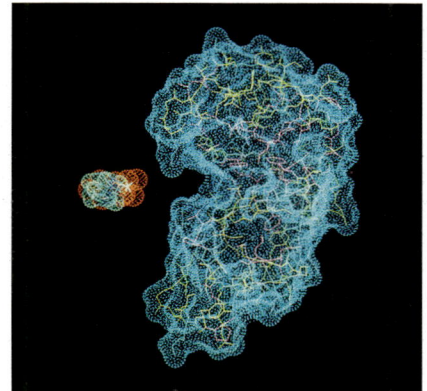

Figure 3 *The ribonuclease A enzyme and its substrate close to the enzyme's active site*

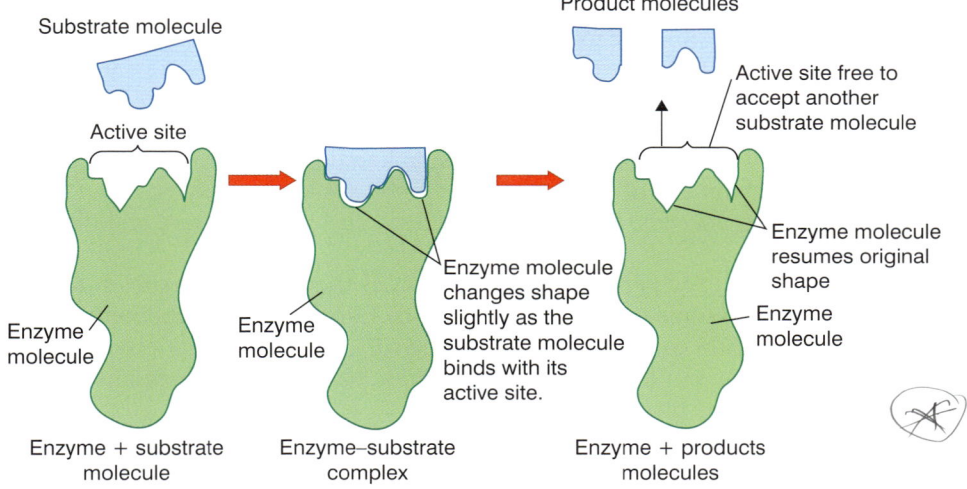

Figure 4 *Induced fit mechanism of enzyme action*

SUMMARY TEST 3.1

Enzymes act as biological (**1**). They are (**2**) proteins that have a specific shape within which there is a functional portion known as the (**3**). Enzymes lower the (**4**) of a reaction, allowing it to proceed at a lower temperature than it would normally. In an enzyme-controlled reaction, the general term for the substance on which the enzyme acts is (**5**) and the substances formed at the end of the reaction are known as the (**6**). The enzyme molecule and the substance it acts on fit together very precisely, giving rise to the name (**7**) hypothesis of enzyme action. In practice, the enzyme changes shape slightly and so moulds itself to the shape of the substance it acts on. This is called the (**8**) hypothesis of enzyme action.

Effect of temperature and pH on enzyme action

On these pages you will learn to:

- Investigate the progress of an enzyme-catalysed reaction by measuring rates of formation of products (for example, using catalase) or rates of disappearance of substrate (for example, using amylase)
- Investigate and explain the effects of the following factors on the rate of enzyme-catalysed reactions: temperature, pH (using buffer solutions)

Before considering how pH and temperature affect enzymes, it is worth bearing in mind that, for an enzyme to work, it must:

- come into physical contact with its substrate
- have an **active site** which fits the substrate.

Almost all factors that influence the rate at which an enzyme works do so by affecting one or both of the above two circumstances. In order to investigate how enzymes are affected by various factors we need to be able to measure the reactions they catalyse.

Measuring enzyme-catalysed reactions

To measure the progress of an enzyme-catalysed reaction we usually measure its time-course, i.e. how long it takes for a particular event to run its course. The two 'events' most frequently measured are:

- **the formation of products** of the reaction, e.g. the volume of oxygen produced when catalase acts on hydrogen peroxide (Figure 1)
- **the disappearance of the substrate**, e.g. the reduction in concentration of starch when it is acted upon by amylase (Figure 2).

Although the graphs in Figures 2 and 3 differ, the explanation for their shapes is the same:

- At first there is a lot of substrate (hydrogen peroxide / starch) but no product (water and oxygen / maltose).
- It is very easy for substrate molecules to come into contact with the empty active sites on the enzyme molecules.
- All enzyme active sites are filled and the substrate is rapidly broken down into its products.
- The amount of substrate decreases as it is broken down, resulting in an increase in the amount of product.
- As the reaction proceeds, there is less and less substrate and more and more product.
- The product produced per unit time decreases as there are fewer substrate molecules and so some active sites may not be filled at any one moment.
- The rate of reaction continues to slow as the substrate concentration decreases.
- The graphs flatten out because all the substrate has been used up and so no new product can be produced.

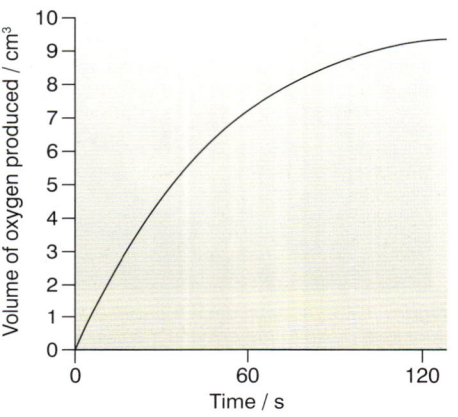

Figure 1 *Measurement of the formation of oxygen due to the action of catalase on hydrogen peroxide*

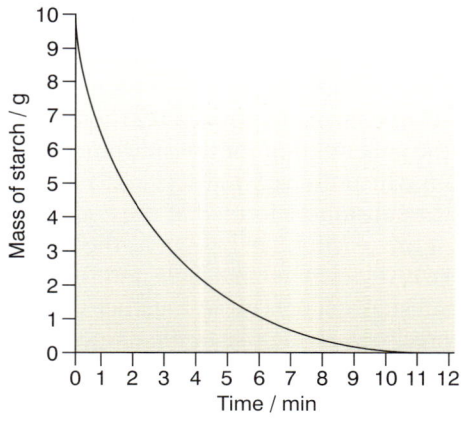

Figure 2 *Measurement of the disappearance of starch due to the action of amylase*

Effect of temperature on enzymes

A rise in temperature increases the **kinetic energy** of molecules, which therefore move around more rapidly and collide with one another more often. In an enzyme-catalysed reaction, this means that the enzyme and substrate molecules come together more often in a given time, so that the rate of reaction is increased. Shown on a graph, this gives a rising curve. However, the temperature rise also increases the energy of the atoms that make up the enzyme molecule. Its atoms begin to vibrate and cause bonds to break, with weaker bonds such as hydrogen bonds, breaking first. Gradually, the shape of the active sites is changed. At first, the substrate fits less easily into the active site slowing the rate of reaction. For many human enzymes this may begin at temperatures of around 45 °C. At some point, usually around 60 °C, the tertiary structure of the enzyme and shape of the active site is so changed that it stops working altogether. It is said to be **denatured**. Shown on a graph, the rate of this reaction follows a falling curve. The actual effect of temperature on the rate of an enzyme reaction is a combination of these two factors, increased kinetic energy of molecules and denaturation of the enzyme (Figure 4). The optimum working temperature differs from enzyme to enzyme. Some work best at around 10 °C, while others continue to work well at 80 °C. Each enzyme in the human body has a different optimum working temperature. Our body temperatures have, however, evolved to be 37 °C because:

- Although higher body temperatures would increase the metabolic rate slightly, the advantages are offset by the additional energy (food) that would be needed to maintain the higher temperature.
- Proteins, other than enzymes, may be denatured at higher temperatures.
- At higher temperatures, any further rise in temperature, e.g. during illness, might denature the enzymes.

Denaturing enzymes at high temperatures is used to prevent the spoilage (breakdown) by various enzymes found in food materials. This is the basis for heating food before canning or bottling it and for blanching vegetables before freezing.

An investigation involving temperature and the enzyme catalase is discussed in Making decisions in investigations *in the Practical skills section on the accompanying CD.*

Effect of pH on enzymes

The pH of a solution is a measure of its hydrogen ion concentration. Each enzyme has an optimum pH, i.e. a pH at which it works fastest (Figure 5). This is because the exact arrangement of the active site of an enzyme is partly fixed by hydrogen and ionic bonds between —NH$_2$ and —COOH groups of the polypeptides that make up the enzyme. Changes in pH can affect this bonding, causing changes of shape in the active site. As a result, the substrate can no longer become attached to the active site and the enzyme–substrate complex cannot be formed. In a similar way to a rise in temperature, this reduces the effectiveness of an enzyme and eventually causes it to stop working altogether, i.e. it becomes denatured. This is why foods can be preserved in vinegar: the low pH denatures the enzymes that would otherwise cause the food to break down. Solutions, known as **buffer solutions**, can be used to prevent fluctuations in pH.

Figure 3 *Bacteria (red) growing in this hot spring in New Zealand are not killed, and their enzymes are not denatured, at temperatures in excess of 80°C*

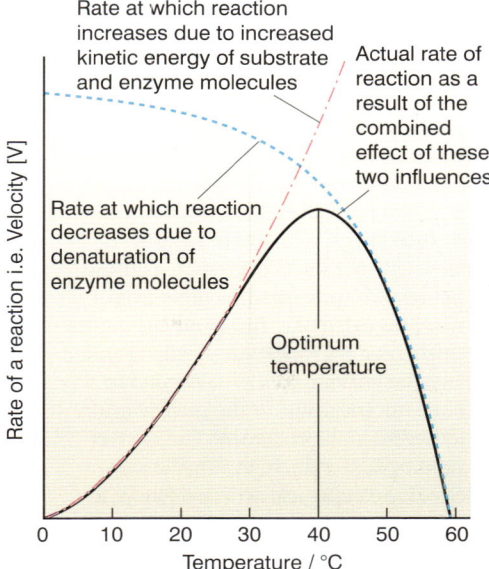

Figure 4 *Effect of temperature on the rate of an enzyme-controlled reaction*

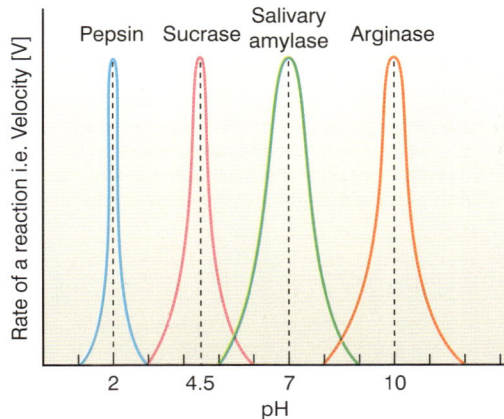

Figure 5 *Effect of pH on the rate of an enzyme-controlled reaction*

SUMMARY TEST 3.2

We can measure the progress of an enzyme-catalysed reaction by measuring its **(1)**. This is usually done by measuring either the **(2)** of the substrate or the formation of the **(3)**. For example, in the case of the enzyme amylase, we could either measure the rate at which **(4)** is produced or the rate at which **(5)** is used up. If the temperature is increased, the rate of enzyme action will **(6)** up to a point at which its molecular structure is disrupted and the shape of its **(7)** is altered so that the substrate no longer fits it. At this point the enzyme is said to be **(8)**. Many human enzymes have an optimum working temperature of **(9)**. Enzymes also have an optimum pH at which they operate. Some, like pepsin, work fastest at a pH of **(10)** while others, such as **(11)**, function fastest in neutral conditions.

On these pages you will learn to:

- Investigate and explain the effects of the following factors on the rate of enzyme-catalysed reactions: enzyme concentration, substrate concentration
- Explain that the maximum rate of reaction (V_{max}) is used to derive the Michaelis–Menten constant (K_m), which is used to compare the affinity of different enzymes for their substrates

EXTENSION

Non-protein biological catalysts

It used to be thought that all biological catalysts were enzymes and were therefore made of protein. We now know that some reactions in cells are catalysed by RNA molecules, also known as **ribozymes**. Some ribozymes work on other RNA molecules, such as those that cut out unwanted sections from **messenger RNA**. This could answer the 'chicken or egg' question – which came first, the enzyme (protein) needed to make nucleic acids, or the nucleic acids, needed to make enzymes? The answer could be RNA which, in this sense at least, is both nucleic acid and 'enzyme'.

In addition to external factors such as temperature and pH, the substrate and enzyme concentrations affect the rate of enzyme-catalysed reactions. An enzyme reaction is always most rapid at first because the enzyme and substrate molecules can freely collide with one another. As the reaction proceeds substrate molecules decrease in concentration and there are fewer successful collisions between enzyme and substrate molecules. The rate of reaction therefore slows.

Effect of enzyme concentration on the rate of reaction

Once an **active site** on an enzyme has acted on its substrate, it is free to repeat the procedure on another substrate molecule. This means that enzymes are not used up in the reaction and therefore work efficiently at very low concentrations. In some cases, a single enzyme molecule can act on millions of substrate molecules in one minute.

As long as there is an excess of substrate, an increase in the amount of enzyme leads to a proportionate increase in the rate of reaction. A graph of the rate of reaction against enzyme concentration will initially show a proportionate increase (straight line). This is because there is more substrate than the enzyme's active sites can cope with. If we therefore increase the enzyme concentration, more substrate will be acted upon and the rate of reaction will increase. If, however, the substrate is limiting, i.e. there is not sufficient to supply all the enzyme's active sites at one time, then any increase in enzyme concentration will have no effect on the rate of reaction. The rate of reaction will therefore stabilise at a constant level, i.e. the graph will level off. This is because the available substrate is already being used as rapidly as it can be by the existing enzyme molecules. These events are summarised in Figure 1.

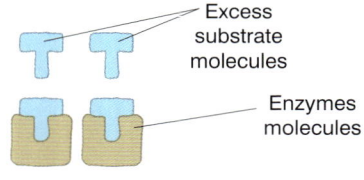

Low enzyme concentration

There are too few enzyme molecules to allow all substrate molecules to find an active site at one time. In this example the rate of reaction is therefore only half the maximum possible for the number of substrate molecules available.

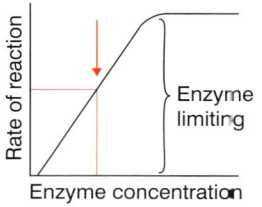

Intermediate enzyme concentration

With twice as many enzyme molecules available, all the substrate molecules can occupy an active site at the same time. The rate of reaction has doubled to its maximum because all active site are filled.

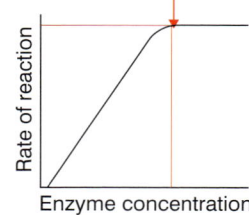

High enzyme concentration

An increase in the number of enzyme molecules present has no effect as there will not be enough substrate to fill all the available active sites. There is no increase in the rate of reaction.

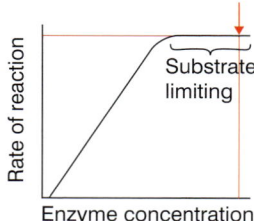

Figure 1 *Effect of enzyme concentration on the rate of enzyme action*

Effects of substrate concentration on the rate of enzyme action

If the concentration of enzyme is fixed at a constant level and substrate concentration is increased, the rate of reaction increases in proportion to the increase in substrate concentration. If a higher concentration of substrate is used the active sites become fully occupied. They are said to be fully saturated at the point where they are all working as fast as they can. The rate of reaction is at its maximum (V_{max}). After that, the addition of more substrate will have no effect on the rate of reaction. In other words, when the substrate is in excess the rate of reaction levels off. A summary of the effect of substrate concentration on the rate of enzyme action is given in Figure 2.

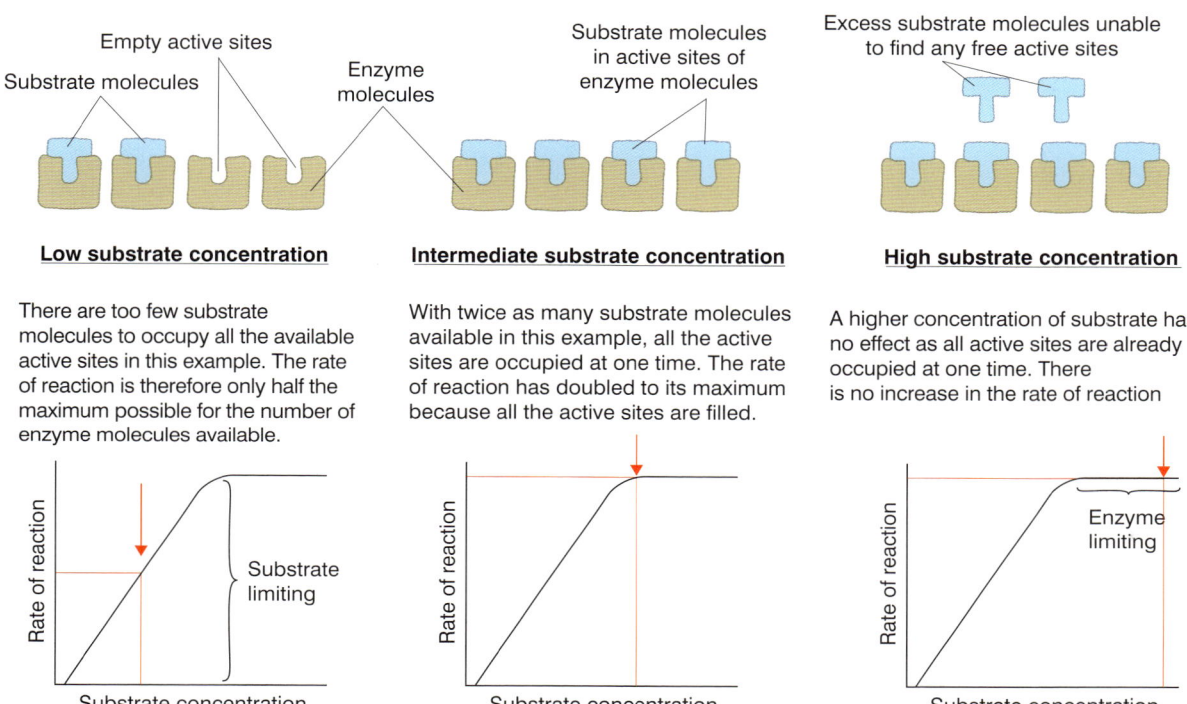

Figure 2 *Effect of substrate concentration on the rate of enzyme action*

Michaelis–Menten constant

The Michaelis–Menten constant is the substrate concentration needed for an enzyme reaction to proceed at half of its maximum rate (Figure 3). The constant is the same for any one enzyme but varies for different enzymes. It gives a measure of how easily an enzyme reacts with its substrate. In other words, it measures the affinity of an enzyme for its substrate. A high Michaelis–Menten constant shows that an enzyme has a low affinity for its substrate, whereas a low constant shows that it has a high affinity for its substrate. The Michaelis–Menten constant can therefore be used to compare the efficiency of different enzymes for their substrates.

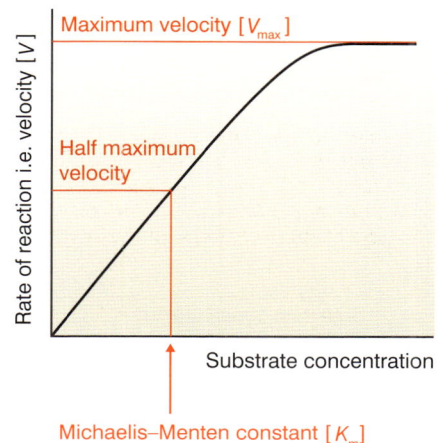

Figure 3

SUMMARY TEST 3.3

Enzymes work fastest at the start of a process and this is called the **(1)**. As the enzyme concentration increases, the rate of reaction **(2)**, provided there is excess substrate. The graph may later 'tail off' if the concentration of substrate is limited because not all the **(3)** of the enzyme molecules are filled. If the substrate concentration of an enzyme-controlled reaction is halved then the rate of reaction will be **(4)**, but when the substrate is in excess the rate of reaction will **(5)**.

On these pages you will learn to:

- Explain the effects of reversible inhibitors, both competitive and non-competitive, on the rate of enzyme activity
- Investigate and explain the effects of the following factor on the rate of enzyme-catalysed reactions: inhibitor concentration
- Investigate and explain the effect of immobilising an enzyme in alginate on its activity as compared with its activity when free in solution

Enzyme inhibitors are substances that directly or indirectly interfere with the functioning of the **active site** of an enzyme and so reduce its activity. Most inhibitors only make temporary attachments to the active site. These are called **reversible inhibitors** and are of two types:

- **competitive** (active site directed) – inhibitor binds to the active site of the enzyme
- **non-competitive** (non-active site directed) – inhibitor binds to the enzyme at a position other than the active site.

Competitive (active site directed) inhibitors

Competitive inhibitors have a molecular shape that is complementary to that of the substrate, which allows them to occupy the active site of an enzyme. They therefore compete with the substrate for the available active sites (Figure 1). It is the difference between the concentration of the inhibitor and the concentration of the substrate that determines the effect this has on enzyme activity: if the substrate concentration is increased, the effect of the inhibitor is reduced. The inhibitor is not permanently bound to the active site and so, when it leaves, another molecule can take its place. This could be a substrate or inhibitor molecule, depending on how much of each type is present. Sooner or later, all the substrate molecules will find an active site, but the greater the concentration of inhibitor, the longer this will take. Examples of competitive inhibitors include malonate, which inhibits succinic dehydrogenase in the Krebs cycle (stage of aerobic respiration).

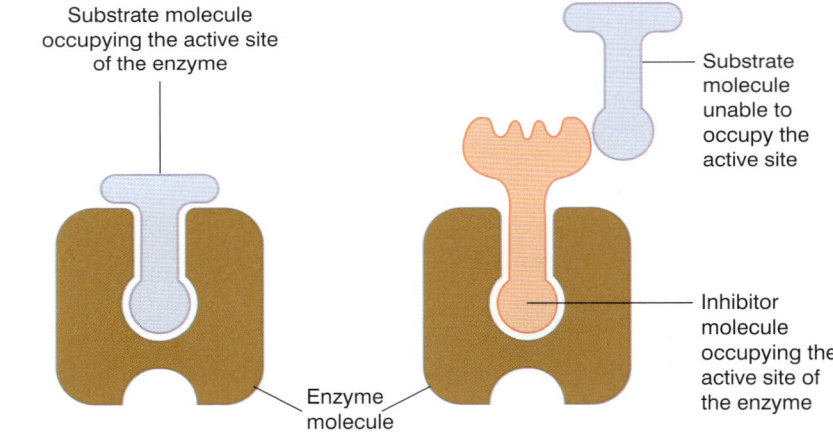

Figure 1 Competitive inhibition

Non-competitive (non-active site directed) inhibitors

Non-competitive inhibitors attach themselves to the enzyme at a binding site which is not the active site. This is known as the **allosteric site** (allosteric = 'at another place'). Upon attaching to the enzyme, the inhibitor alters the shape of the enzyme's active site in such a way that substrate molecules can no longer occupy it, and so the enzyme cannot function (Figure 3). As the substrate and the inhibitor are not competing for the same site, an increase in substrate concentration does not decrease the effect of the inhibitor (Figure 2).

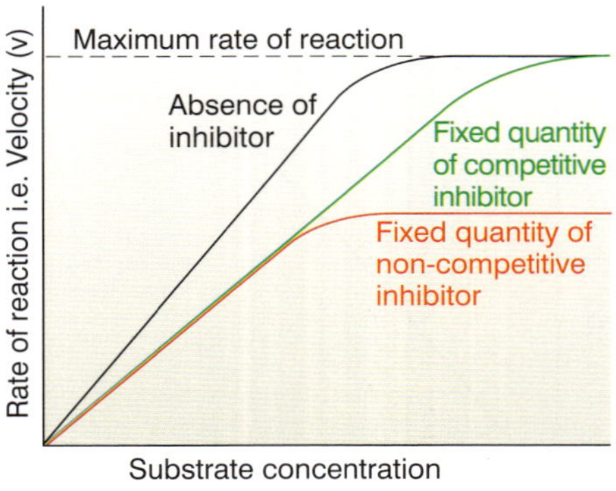

Figure 2 Comparison of competitive and non-competitive inhibition on the rate of an enzyme-controlled reaction at different substrate concentrations

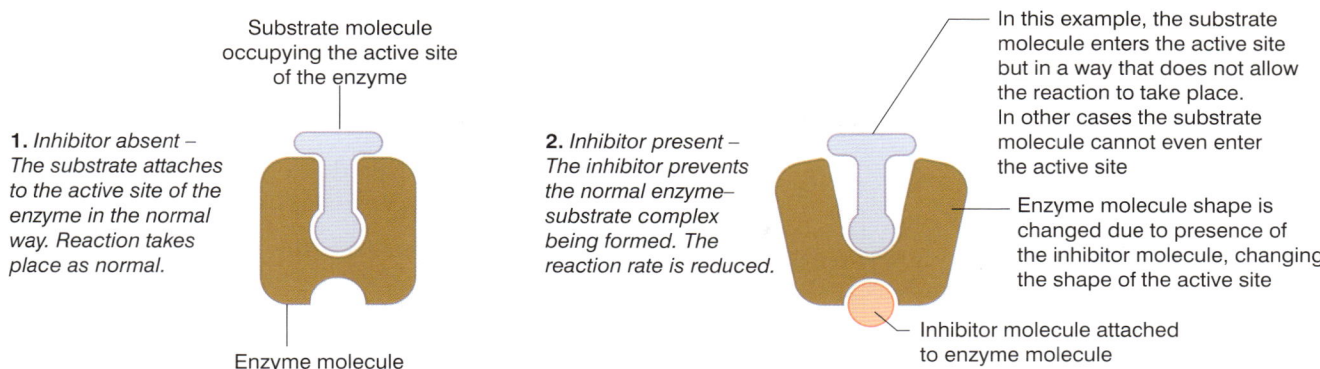

1. *Inhibitor absent – The substrate attaches to the active site of the enzyme in the normal way. Reaction takes place as normal.*

Substrate molecule occupying the active site of the enzyme

2. *Inhibitor present – The inhibitor prevents the normal enzyme–substrate complex being formed. The reaction rate is reduced.*

Enzyme molecule

In this example, the substrate molecule enters the active site but in a way that does not allow the reaction to take place. In other cases the substrate molecule cannot even enter the active site

Enzyme molecule shape is changed due to presence of the inhibitor molecule, changing the shape of the active site

Inhibitor molecule attached to enzyme molecule

Figure 3 *Non-competitive inhibition*

Immobilising enzymes

One method of immobilising enzymes is to trap them in small alginate beads. This is carried out in a laboratory as follows.

- The enzyme to be immobilised is mixed with a solution of sodium alginate.
- Tiny droplets of this mixture are added to a solution of calcium chloride, one at a time using a syringe.
- A reaction takes place between the sodium alginate–enzyme mixture and the calcium, causing calcium ions to replace sodium ions.
- As a result of this reaction, jelly-like beads are formed in which the enzyme is trapped.

To catalyse reactions, the beads are packed into a long column in a vessel and the substrate is poured in at the top. As the substrate enters the beads, it is converted to the product by the enzymes immobilised in the beads. Although the process cannot continue indefinitely, as impurities accumulate, it can proceed for a considerable time without renewing the enzyme.

Enzymes are used in a wide range of industrial processes, including the production of foods, agrochemicals and drugs. In many cases the enzyme and substrate are mixed together to form a product. This is then extracted and the rest of the mixture, including the enzyme, is discarded. Enzymes are costly to produce so this is both wasteful and expensive. As enzymes are not used up in reactions, keeping them for future use clearly has a cost benefit. One way of doing this is to immobilise the enzymes so that they can be retained rather than discarded.

An example of the commercial use of immobilised enzymes is in the manufacture of lactose-free milk for people with lactose intolerance. In this case the immobilised enzyme is β-galactosidase. Milk contains the sugar lactose which is the cause of the discomfort experienced by lactose intolerant individuals. To produce lactose-free milk, the milk is passed over the immobilised β-galactosidase which catalyses the **hydrolysis** of lactose to glucose and galactose. Figure 4 illustrates the process.

Advantages of using immobilised enzymes or cells

There are a number of advantages to immobilising enzymes or cells.

- With enzyme immobilisation, the enzyme can be used repeatedly as it is not lost in the process making it more economic, especially where the enzyme is expensive.
- Enzymes are vulnerable to changes in temperature and pH. The beads in which they are trapped can buffer them against these changes.
- The immobilised enzymes and cells, being held in place, cannot contaminate the substance being made leading to a purer product.
- With whole cell immobilisation, a number of enzymes can act together at the same time in a single process.

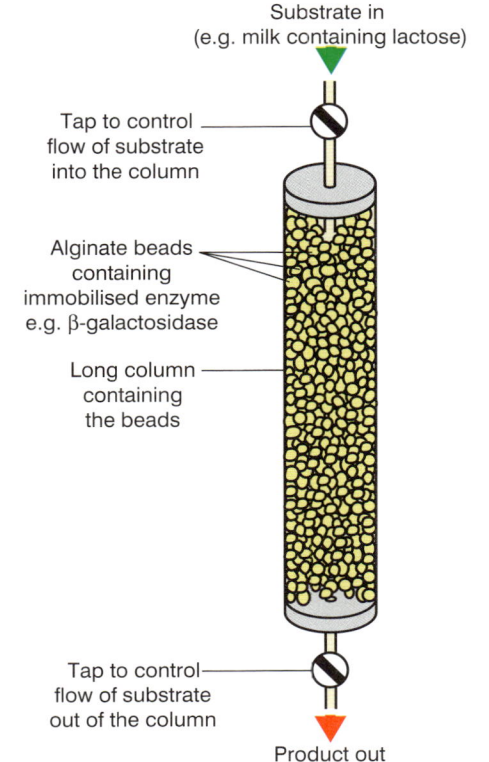

Substrate in (e.g. milk containing lactose)

Tap to control flow of substrate into the column

Alginate beads containing immobilised enzyme e.g. β-galactosidase

Long column containing the beads

Tap to control flow of substrate out of the column

Product out (e.g. milk with glucose and galactose, but no lactose)

Figure 4 *Using immobilised enzymes to produce lactose-free milk*

3 Examination Questions

1 The enzyme, catechol oxidase, causes a brown colour to develop when slices of many fruits, such as apples, are exposed to air.

The enzyme catalyses the following reaction :

$$catechol + oxygen \xrightarrow{\text{catechol oxidase}} quinone + water$$

Quinone is then immediately further oxidized in air to a brown-coloured substance. Catechol and quinone are colourless.

A student investigated the rate of this reaction under different conditions.

a State how the student could follow the progress of this reaction. *(1 mark)*

In the first investigation, the student measured the initial rate of the reaction in varying concentrations of catechol. The results are shown in Figure 1.

Figure 1

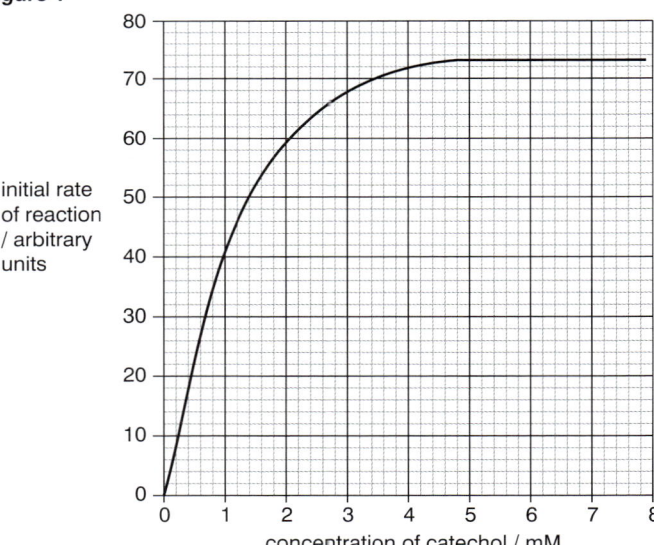

concentration of catechol / mM

b Explain the results shown in Figure 1. *(5 marks)*

c In a second investigation, the student repeated the experiment, but this time added a competitive inhibitor, para-hydroxybenzoic acid (PHBA), to each reaction mixture.
   i Sketch on Figure 1 the results that would be obtained for this second investigation. *(2 marks)*

ii Explain the effect that PHBA will have on the action of catechol oxidase. *(2 marks)*

d Lemon juice contains citric acid. Adding even a small amount of diluted lemon juice to apple slices slows the appearance of the brown colour.

Suggest an explanation for this observation. *(2 marks)*
(Total 12 marks)

Cambridge International AS and A Level Biology 9700 Paper 23 Q4 June 2013

2 Sucrase is the enzyme that catalyses the hydrolysis of sucrose. A student investigated the effect of substrate concentration on the activity of this enzyme.

Six test-tubes were set up each containing $10\,cm^3$ of different concentrations of sucrose solutions. The test-tubes were left in a water bath at $30\,°C$ for ten minutes.

After ten minutes, $5\,cm^3$ of a sucrase solution at $30\,°C$ was added to each test-tube and the reaction mixtures were stirred.

After a further five minutes, the temperature of the water bath was raised to above $85\,°C$ and the same volume of Benedict's solution added to each test-tube in turn. The student recorded the time when a green colour first became visible in each test-tube.

The concentrations used and the student's results are shown in Table 2.

Table 2

concentration of sucrose/$g\,dm^{-3}$	time taken for green colour to appear/s
5	278
10	145
15	95
20	75
50	47
100	45

a Explain why the temperature of the water bath was raised to above $85\,°C$. *(2 marks)*

b Copy the axes and sketch a graph to show the effect of substrate concentration on the **rate of hydrolysis** of sucrose by sucrase. *(2 marks)*

c With reference to the student's results, describe **and** explain the effect of increasing substrate concentration on the rate of hydrolysis of sucrose by sucrase. *(5 marks)*
(Total 9 marks)

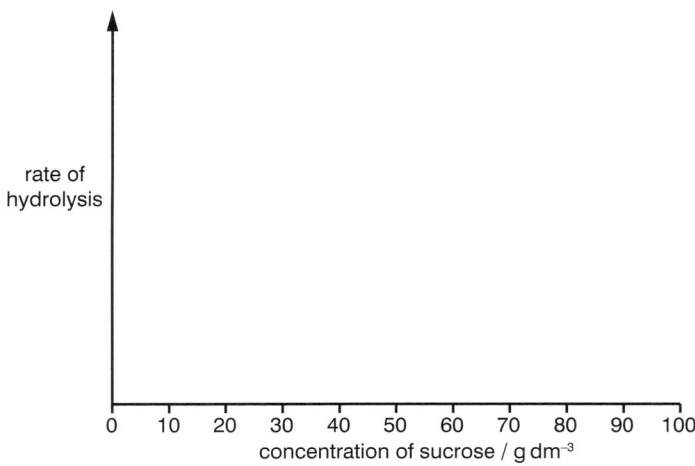

Cambridge International AS and A Level Biology 9700 Paper 22 Q2 November 2009

3 Practice Questions

3 Enzymes can function at a wide-range of temperatures. Shrimps that live in Arctic waters have enzymes that function best around 4 °C and are denatured at around 15 °C. By contrast, bacteria that live in hot springs have enzymes that function best at 95 °C and continue to operate effectively above 100 °C. These bacteria are called thermophilic (heat-loving) bacteria.

Enzyme X is produced by thermophilic bacteria and hydrolyses many proteins, including haemoglobin and egg albumin.

Enzyme Y is found in the stomach of young mammals, where it acts on a single soluble protein found in milk, causing it to coagulate (clot).

a i From the descriptions, comment on the differences in the specificity of the two enzymes.
ii Enzymes X and Y are each used for different commercial purposes. Suggest what this might be in each case.
iii Suggest a possible purpose of enzyme Y in the mammalian stomach.
iv Use the information about the two enzymes to suggest a possible difference in the type of bonding found in the tertiary structure of each. Explain your reasoning.

b An experiment was carried out with enzyme X in which the time taken for it to fully hydolyse 5 g of its protein substrate was measured at different temperatures. The following data were obtained:

Temperature/°C	Time for hydrolysis of protein/min	Rate of reaction $\frac{1}{time}$
15	5.8	
25	3.4	
35	1.7	
45	0.7	
55	0.6	
65	0.9	
75	7.1	

i Calculate the values for 1/time for each of the temperatures.
ii Plot a graph that shows the effect of temperature on the rate of reaction of enzyme X.
iii Measure the optimum temperature for the action of enzyme X.
iv Suggest how you might determine this optimum temperature more precisely.

4 a Explain why enzymes function less well at lower temperatures.

b Explain how high temperatures may prevent enzymes from functioning at all.

c Enzymes produced by microorganisms are responsible for spoiling food. Using this fact and your knowledge of enzymes suggest a reason in each case why the following procedures are carried out:
i Food is heated to a high temperature before being canned.
ii Some foods, such as onions, are preserved in vinegar.

5 The figure below represents an enzyme and its substrate. Of the other four molecules shown, one is a competitive inhibitor and one is a non-competitive inhibitor of the enzyme.

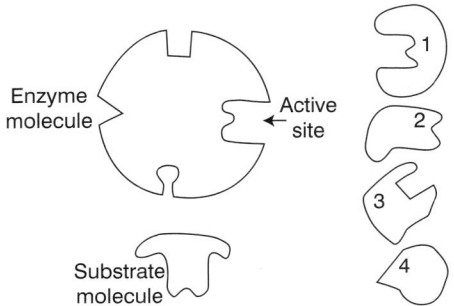

a What is the number of the molecule that is a competitive inhibitor?
b What is the number of the molecule that is a non-competitive inhibitor?

4.1

Structure of the cell surface membrane

On these pages you will learn to:

- Recognise the following cell structure and outline its functions: cell surface membrane
- Describe and explain the fluid mosaic model of membrane structure, including an outline of the roles of phospholipids, cholesterol, glycolipids, proteins and glycoproteins
- Outline the roles of cell surface membranes including references to carrier proteins, channel proteins, cell surface receptors and cell surface antigens
- Outline the process of cell signalling involving the release of chemicals that combine with cell surface receptors on target cells, leading to specific responses

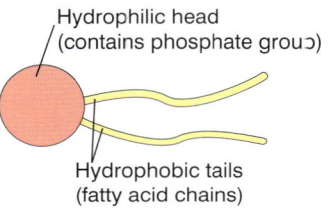

Hydrophilic head
(contains phosphate group)

Hydrophobic tails
(fatty acid chains)

Figure 1 *Structure of a phospholipid*

Cell signalling

Cell signalling is a means by which cells interact with their environment and with cells around them. One method involves the release of a chemical (signalling molecule) that has a complementary shape to a receptor (usually a glycoprotein or protein; sometimes a glycolipid) on the cell surface membrane of a target cell. The signalling molecule binds to its complementary receptor and so starts a chain of reactions within the target cell, leading to a specific response. Hormones are often used as signalling molecules, and specific examples of how they cause a response in target cells are given in Topic 14.7.

All membranes around and within cells (including those around and within cell organelles) have the same basic structure. The cell surface membrane is the membrane that surrounds cells. It is the boundary between the cell cytoplasm and the environment and, as such, controls the movement of substances in and out of the cell. Before we look at how the cell surface membrane achieves this we need first to look in more detail at the molecules that form its structure – phospholipids, proteins, cholesterol, glycolipids and glycoproteins. The arrangement of these molecules within the cell surface membrane is detailed in Figure 2.

Phospholipids

The molecular structure of a phospholipid is described in Topic 2.5 and shown in Figure 1. A phospholipid is made up of two parts:

- **a hydrophilic 'head'**, which is attracted to water but not to lipid
- **a hydrophobic 'tail'**, which is repelled by water but mixes readily with lipid.

This means that when phospholipid molecules are placed in water they take up positions that place the hydrophilic heads as close to the water as possible, and the hydrophobic tails as far away from the water as possible. The phospholipids in the cell surface membrane form a bilayer, which has:

- one layer of phospholipids with its hydrophilic heads pointing inwards (attracted by the water in the cell cytoplasm)
- the other layer of phospholipids with its hydrophilic heads pointing outwards (attracted by the watery environment which surrounds all cells)
- the hydrophobic tails of both phospholipid layers pointing into the centre of the membrane – repelled by the water on both sides and forming a hydrophobic core. The more unsaturated these tails, the more fluid is the membrane.

Lipid-soluble material moves through the membrane via the phospholipid bilayer.

Proteins

The proteins of the cell surface membrane are arranged more randomly than the regular pattern of phospholipids. They are associated with the phospholipid bilayer in two main ways:

- **Extrinsic (peripheral) proteins** occur on both surfaces of the bilayer but never extend completely across it. They may act to give mechanical support to the membrane or may have other functions, such as enzymes that have a role in cell signalling.
- **Intrinsic (integral) proteins** may have regions embedded in the bilayer from one side to the other. Some are **channel proteins**, which form open water-filled tubes to allow water-soluble ions to diffuse across the membrane. Some are **carrier proteins** that bind to molecules such as glucose and amino acids,

Table 1 *Functions of membranes within cells*

- Control the entry and exit of materials in discrete organelles such as mitochondria and chloroplasts
- Isolate organelles so that specific metabolic reactions can take place within them
- Provide an internal transport system, e.g. endoplasmic reticulum
- Isolate enzymes that might damage the cell, e.g. lysosomes
- Provide surfaces on which processes can occur, e.g. protein synthesis using ribosomes on rough endoplasmic reticulum

then change shape in order to move these molecules across the membrane. Other functions are shown in Table 2.

Cholesterol

Cholesterol molecules occur within the phospholipid bilayer. They have hydrophilic heads and hydrophobic tails. They add strength and stability to animal cell surface membranes – a necessary feature in the absence of a cell wall. The tails of cholesterol molecules, being hydrophobic, have an important role in preventing the passage of dissolved ions and polar molecules across the bilayer. They also interact with the fatty acid tails of the phospholipid molecules. They prevent close packing when temperatures decrease and prevent excessive movement when temperatures increase. Therefore they can be described as regulating the fluidity of membranes.

Glycolipids

Glycolipids are made up of a carbohydrate covalently bonded with a phospholipid. The carbohydrate portion extends from the phospholipid bilayer into the watery environment outside the cell, where it acts as a recognition site for specific chemicals, e.g. the human ABO blood system operates as a result of glycolipids on the cell surface membrane.

Glycoproteins

Carbohydrate chains are attached to many proteins on the outer surface of the cell membrane. These glycoproteins also act as recognition sites, more particularly for hormones and neurotransmitters.

The way in which all the various molecules above are combined into the structure of the cell surface membrane is shown in Figure 2. This arrangement is known as the **fluid-mosaic model** for the following reasons:

- **fluid** because the individual phospholipid and protein molecules can move relative to one another. This gives the membrane a flexible structure that is constantly changing in shape. Many protein molecules move within the bilayer.
- **mosaic** because the proteins that occur in the phospholipid bilayer vary in shape, size and pattern (scattered) in the same way as the stones or tiles in a mosaic.

Table 2 *Summary of functions of the components of the cell surface membrane*

Proteins
- provide structural support
- act as carrier proteins transporting glucose and amino acids
- function as enzymes
- form **channel proteins** for sodium, potassium, etc
- act as energy **transducers**
- form cell surface antigens to allow recognition
- help cells adhere together
- act as cell surface receptors, e.g. for hormones

Phospholipids
- (as part of the bilayer) allow lipid-soluble substances to enter and leave the cell
- (as part of the bilayer) prevent water-soluble substances entering and leaving the cell
- give the membrane fluidity

Cholesterol
- reduces lateral movement of phospholipids
- regulates membrane fluidity depending on temperature
- prevents passage of dissolved polar molecules and ions across the membrane

Glycolipids
- act as recognition sites, e.g. ABO blood system
- help maintain stability of the membrane
- help cells attach to one another and so form tissues

Glycoproteins
- act as recognition sites for hormones and neurotransmitters
- help cells attach to one another and so form tissues
- allow cells to recognise one another, e.g. lymphocytes can recognise an organism's own cells.

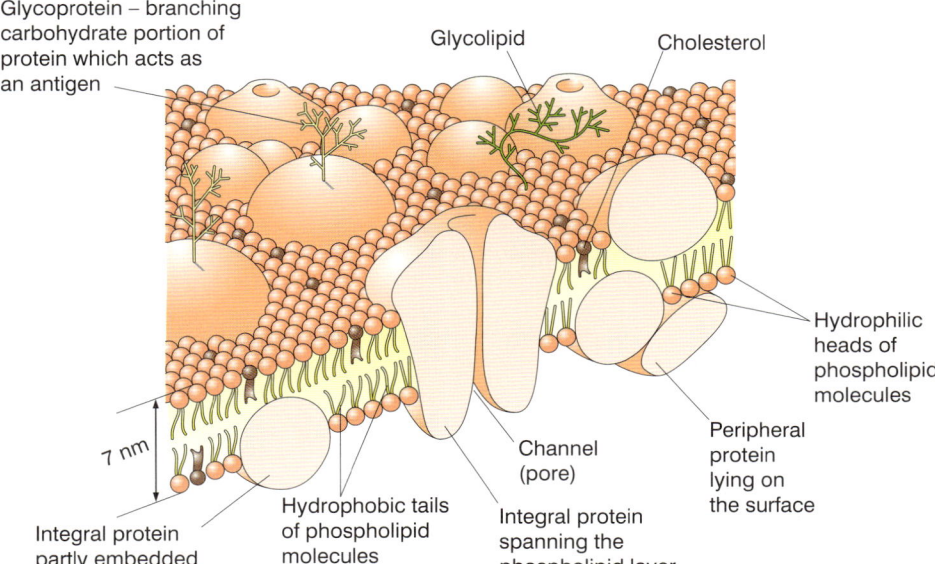

Glycoprotein – branching carbohydrate portion of protein which acts as an antigen

Glycolipid

Cholesterol

Hydrophilic heads of phospholipid molecules

7 nm

Integral protein partly embedded

Hydrophobic tails of phospholipid molecules

Channel (pore)

Integral protein spanning the phospholipid layer

Peripheral protein lying on the surface

Figure 2 *Structure of the cell surface membrane*

SUMMARY TEST 4.1

The cell surface (plasma) membrane is made up of five main types of molecules. Phospholipid molecules from a **(1)** in which their **(2)** heads point both inwards towards the cell cytoplasm and outwards towards the external environment. The **(3)** tails of the phospholipids point into the centre of the membrane. Within the phospholipid layer are both **(4)** proteins that span the complete membrane or are partly embedded in it and **(5)** proteins that occur on the membrane surface. The remaining types of molecules within a cell surface membrane are **(6)**, **(7)** and **(8)**.

The movement of material into and out of cells occurs in a number of ways, some of which require energy **(active transport)** and some of which do not **(passive transport)**. Diffusion is an example of passive transport.

Explanation of simple diffusion

As all movement requires energy, it is possibly confusing to describe diffusion as passive transport. What is meant by passive, in this sense, is that the energy comes from the natural, inbuilt motion **(kinetic energy)** of particles, rather than ATP from respiration. To help understand diffusion and other passive forms of transport it is necessary to understand that:

• all particles are constantly in motion due to the kinetic energy that they possess
• this motion is random, with no set pattern to the way the particles move around
• particles are constantly bouncing off one another as well as other objects, e.g. the sides of a vessel in which they are contained.

Given those facts, Figure 1 shows in a series of diagrams how particles concentrated together in part of a closed vessel will, of their own accord, distribute themselves evenly throughout the vessel, due to diffusion. Diffusion is therefore defined as **'the net movement of molecules or ions from a region where they are more highly concentrated to one where their concentration is lower'**.

1. *If 10 particles occupying the left-hand side of a closed vessel are in random motion, they will collide with each other and the sides of the vessel. Some particles from the left-hand side move to the right, but initially there are no available particles to move in the opposite direction, so the movement is in one direction only. There is a large concentration gradient and diffusion is rapid.*

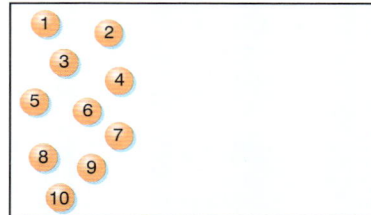

2. *After a short time the particles (still in random motion) have spread themselves more evenly. Particles can now move from right to left as well as from left to right. However, with a higher concentration of particles (7 particles) on the left than on the right (3 particles), there is a greater probability of a particle moving to the right than in the reverse direction There is a smaller concentration gradient and diffusion is slower.*

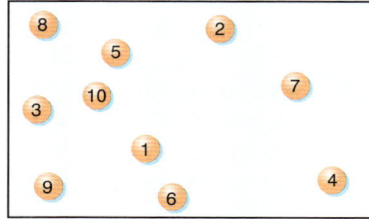

3. *Some time later, the particles will be evenly distributed throughout the vessel and the concentrations will be equal on each side. The system is in equilibrium. However, the particles are not static but remain in random motion. With equal concentrations on each side, the probability of a particle moving from left to right is equal to the probability of one moving in the opposite direction. There is no concentration gradient and no net diffusion.*

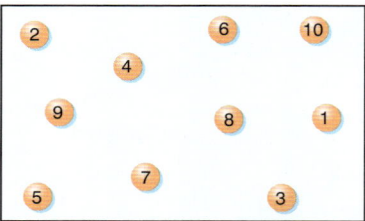

4. *At a later stage, the particles remain evenly distributed and will continue to do so. Although the number of particles on each side remains the same, individual particles are continuously changing position. This situation is called **dynamic equilibrium**.*

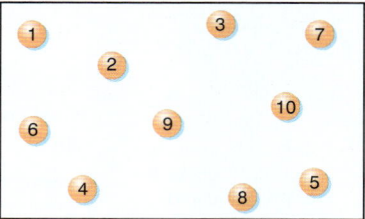

Figure 1 Diffusion

Rate of diffusion

A number of factors affect the rate at which molecules or ions diffuse. These include:

- **The concentration gradient** – the greater the difference in concentration between two regions of molecules or ions, the faster the rate of diffusion.
- **The area over which diffusion takes place** – the larger the area, the faster the rate of diffusion.
- **The distance over which diffusion occurs** – the shorter the distance, the faster the rate of diffusion.

The relationship between these three factors is expressed in **Fick's Law**, which states:

Diffusion is proportional to:

$$\frac{\text{surface area} \times \text{difference in concentration}}{\text{length of diffusion path}}$$

Although Fick's Law gives a good guide to the rate of diffusion, it is not wholly applicable to cells, because diffusion is also affected by:

- **the nature of the cell surface membrane** – its composition and number of carrier and channel proteins
- **the size and nature of the diffusing molecule** – for example:
  - small molecules diffuse faster than large ones
  - fat-soluble molecules such as glycerol diffuse faster than water-soluble ones like glucose.

Facilitated diffusion

Facilitated diffusion is a passive process relying only on the kinetic energy of the diffusing molecules. Like diffusion, it occurs along a concentration gradient, but it differs in that it occurs at specific points on the membrane where there are special protein molecules. These proteins form water-filled channels **(channel proteins)** across the membrane and therefore allow water-soluble ions to pass through. Such ions would usually diffuse only very slowly through the phospholipid bilayer of the membrane. The channels are selective, each opening only in the presence of a specific ion. When the particular ion is not present, the channel remains closed. In this way, some control is kept over the entry and exit of substances. An alternative form of facilitated diffusion involves **carrier proteins** which also span the membrane. When a particular molecule specific to the protein is present it binds with the protein at a specific binding point, causing it to change shape (conformational change) in such a way that the molecule is released to the inside of the membrane (Figure 2). Again, there is no use of energy from ATP, and the molecules move from a region where they are highly concentrated to one of lower concentration, using only the kinetic energy of the molecules themselves.

SUMMARY TEST 4.2

Diffusion is the net movement of molecules or ions from where they are in a **(1)** concentration to a region where their concentration is **(2)**. The energy for this movement comes from the **(3)** energy of the molecules themselves and the process is therefore said to be a **(4)** one. If the area over which diffusion takes place is made smaller, its rate becomes **(5)**. If the concentration gradient is reduced, the rate becomes **(6)** and if the distance over which diffusion takes place is made shorter, its rate becomes **(7)**. Facilitated diffusion, which is faster than diffusion, may involve molecules known as **(8)** that span the cell surface membrane.

Surface area to volume ratio

As a shape like a cube becomes larger, its volume increases at a greater rate than its surface area. In other words, the cube's surface area to volume ratio becomes smaller. This is illustrated with specific examples in Topic 7.2.

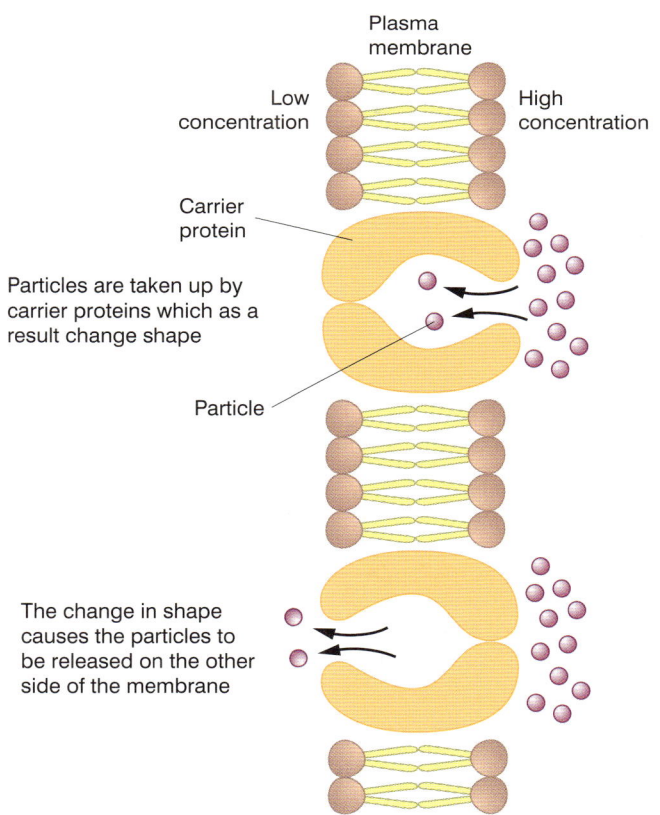

Figure 2 Facilitated diffusion involving carrier proteins

On these pages you will learn to:

- Describe and explain the process of osmosis

Osmosis is a special form of diffusion involving only water molecules. It is defined as **the passage of water from a region where it has a higher water potential to a region where it has a lower water potential through a partially permeable membrane.** Cell surface membranes and other membranes, such as those surrounding organelles, are **partially (selectively) permeable**, i.e. they are permeable to water molecules and certain solute molecules but not to many other molecules. A **solute** is any substance that is dissolved in a **solvent**, e.g. glucose in water. The solute and the solvent together form a **solution**.

Explanation of osmosis

Consider the hypothetical situation in Figure 1, in which a partially permeable membrane separates two solutions.

- The solution on the left has more free water molecules while the solution on the right has fewer free water molecules.
- Both the solute and water molecules are in random motion due to their **kinetic energy**.
- The partially permeable membrane, however, only allows water molecules across it and not solute molecules.
- The water molecules diffuse from the left-hand side, which has the higher water potential, to the right-hand side, which has the lower water potential, i.e. down a water potential gradient.
- At the point where the water potentials on either side of the membrane are equal, a dynamic equilibrium is established and there is no **net** movement of water.
- Solute molecules are unable to cross membranes because they are too large or they cannot cross the phospholipid bilayer and there are no specific carrier proteins to allow them through.

Water potential

Water potential is represented by the Greek letter psi (Ψ), and is measured in units of pressure, usually kilopascals (kPa). It is the pressure created by water

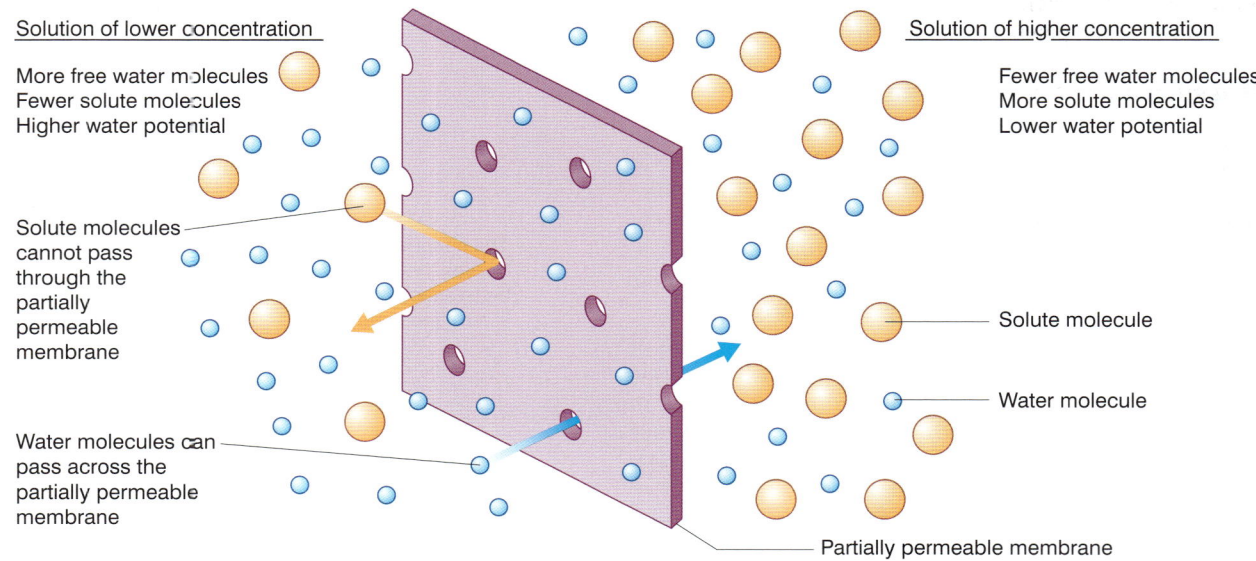

Solution of lower concentration

More free water molecules
Fewer solute molecules
Higher water potential

Solute molecules cannot pass through the partially permeable membrane

Water molecules can pass across the partially permeable membrane

Solution of higher concentration

Fewer free water molecules
More solute molecules
Lower water potential

Solute molecule

Water molecule

Partially permeable membrane

Figure 1 *Osmosis*

molecules. Under standard conditions of temperature and pressure (25 °C and 100 kPa), pure water is said to have a water potential of zero. It follows that:

- the addition of a solute to pure water will lower its water potential
- the water potential of a solution (water + solute) must always be less than zero, i.e. a negative value
- the more solute that is added (i.e. the more concentrated a solution), the lower (more negative) its water potential
- water will move by osmosis from a region of higher (less negative) water potential (e.g. −10 kPa) to one of lower (more negative) water potential (e.g. −20 kPa). An example of how water moves between cells of different water potentials is shown in Figure 2.

EXTENSION

Hypotonic, isotonic and hypertonic solutions

You may come across the following terms that are sometimes used when referring to animal cells:

- **hypotonic solution** – a solution with a higher concentration of water (higher water potential), i.e. the more dilute of the two solutions

- **hypertonic solution** – a solution with a lower concentration of water (lower water potential), i.e. the more concentrated of the two solutions

- **isotonic solutions** – solutions with the same concentration of water.

There is always net movement of water from the hypotonic solution to the hypertonic solution because there are more water molecules in the hypotonic solution. When the solutions are isotonic, the amount of water diffusing in one direction is exactly offset by that moving in the other direction, so there is no net movement of water.

EXTENSION

Solute potential

The presence of solute molecules in a solution lowers its water potential. The greater the concentration of solutes, the lower (more negative) is the water potential. This change in water potential as a consequence of the presence of solute molecules is called **solute potential** (Ψs). As the solute molecules always lower the water potential, the value of the solute potential is always negative.

Pressure potential

Take a situation such as that in Figure 1 where water is entering a solution by osmosis. If a pressure were applied to the solution on the right of Figure 1 then this would resist the entry of water. This pressure is known as the **pressure potential** and is given the symbol Ψp. Pressure potential is of importance in the water relations of plant cells as the cell wall can exert an influence.

The relationship between water potential, solute potential and pressure potential can be expressed in the equation;

$$\Psi = \Psi s + \Psi p$$

Key

x kPa water potential of cell
⟶ direction of water movment

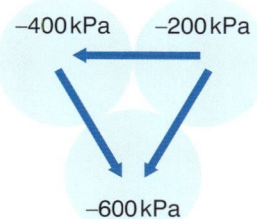

Water moves from higher water potential to lower water potential. The highest water potential is zero.

Figure 2 *Movement of water between cells down a water potential gradient*

4.4 Osmosis and cells

On these pages you will learn to:

- Explain the movement of water between cells and solutions with different water potentials and explain the different effects on plant and animal cells

In Topic 4.3 we looked at what osmosis is and why it occurs. We now need to turn our attention to how it affects living cells. Due to the structural differences between plant and animal cells, osmosis affects each of them in a different way.

Osmosis and animal cells

Animal cells, such as a red blood cell, contain a variety of solutes dissolved in their watery cytoplasm. If a red blood cell is placed in pure water, water will move in by osmosis because it has a lower **water potential**. Cell surface membranes are very thin (7 nm) and, although they are flexible, they cannot stretch to any great extent. The cell surface membrane will therefore break, bursting the cell and releasing its contents – an event known as **lysis**. To prevent this happening, animal cells are normally bathed in a liquid which has the same water potential as the cells. In our example, the blood plasma and red blood cells have the same water potential. If a red blood cell is placed in a solution with a lower water potential than its own, water leaves by osmosis and the cell shrinks, causing its shape to shrivel (Figure 1).

Water potential (Ψ) of external solution compared to cell solution	Higher (less negative)	Equal	Lower (more negative)
Net movement of water	Enters cell	Neither enters nor leaves	Leaves cell
State of cell	Swells and bursts	No change	Shrinks

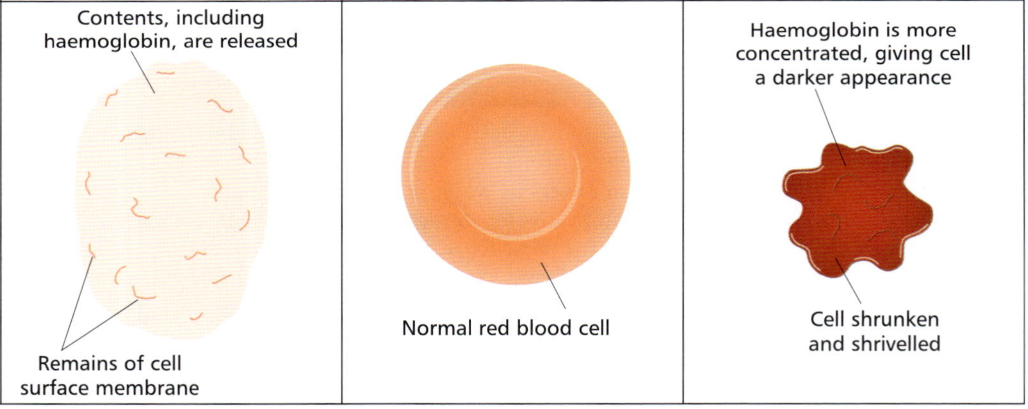

Contents, including haemoglobin, are released
Remains of cell surface membrane
Normal red blood cell
Haemoglobin is more concentrated, giving cell a darker appearance
Cell shrunken and shrivelled

Figure 1 *Summary of osmosis in an animal cell, e.g. a red blood cell*

REMEMBER

- Pure water has a water potential (Ψ) of zero.
- Solute makes the water potential lower (more negative).
- Water always moves from a higher (less negative) water potential to a lower (more negative) one.

Osmosis and plant cells

When studying osmosis in plant cells, the term **protoplast** is sometimes used. The **protoplast** is the plant cell without the cell wall, that is, the cell surface membrane and the cell contents, including the large permanent vacuole containing cell sap.

Like animal cells, plant cells also contain a variety of solutes, largely dissolved in the water of the large cell vacuole that each possesses. When placed in pure water, water moves in by osmosis because of their lower (more negative) water potential. Unlike animal cells, however, they are unable to control the composition of the fluid around their cells. Indeed, plant cells are normally permanently bathed in almost pure water, which is constantly absorbed from the plant's roots (Topic 7.5). So why don't the cells burst? The answer lies in the cellulose cell wall which surrounds every plant cell. If this wall, or something similar, had not evolved, plants would not exist as we know them today. How then does it work?

- The cellulose fibres of the cell wall have great tensile strength. The fibres are laid down in layers, with different layers at different angles to contribute to the overall strength and rigidity of the cell wall.
- Water entering a plant cell by osmosis enters the vacuole, which will increase in size and will cause the protoplast to swell and push against the cellulose cell wall.
- Because the cell wall is capable of only very limited extension, a pressure builds up on it that resists the further entry of water.
- Because it prevents more water entering, this pressure increases the water potential of the cell.
- In this situation, the protoplast of the cell is kept pushed against the cell wall and the cell is said to be turgid.

If the same plant cell is placed in a solution with a lower water potential than its own, water leaves by osmosis. The volume of the vacuole and hence the cell decreases. A stage is reached where the protoplast no longer presses on the cellulose cell wall. At this point the cell is said to be at **incipient plasmolysis**. Further loss of water will cause the cell to shrink further and the cell surface membrane begins to peel away from the cell wall. This condition is called **plasmolysis**. Figure 2 summarises osmosis in plant cells.

SUMMARY TEST 4.4

A red blood cell will burst if placed in a solution which has a (1) water potential than itself. To prevent this, animal cells such as red blood cells are normally bathed in a solution that has the same (2) as themselves. A plant cell placed in pure water takes in water by (3), causing the living contents, called the (4), to swell. This creates a pressure on the (5). This pressure prevents water entering and therefore makes the water potential (6). If a plant cell is placed in a solution with a lower water potential than itself, the cell contents shrink away from the cell wall, a process called (7).

Water potential (Ψ) of external solution compared to cell solution	Less negative (higher)	Equal	More negative (lower)
Net movement of water	Enters cell	Neither enters nor leaves cell	Leaves cell
Protoplast	Swells	No change	Shrinks
Condition of cell	Turgid	Incipient plasmolysis	Plasmolysed

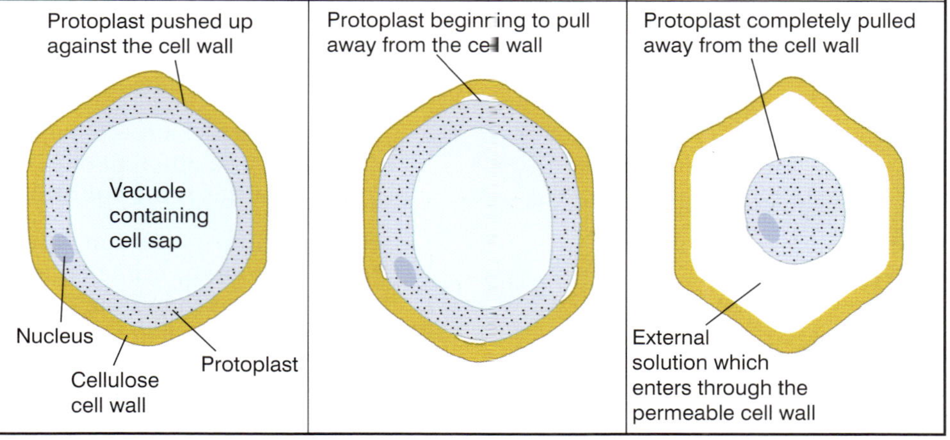

Figure 2 Summary of osmosis in a plant cell

Active transport, endocytosis and exocytosis

- Describe and explain the processes of active transport, endocytosis and exocytosis

REMEMBER

Active transport and facilitated diffusion both use carrier proteins but while facilitated diffusion occurs down a concentration gradient, active transport occurs against a concentration gradient. This means that facilitated diffusion does not require metabolic energy, while active transport does. The metabolic energy is provided in the form of ATP.

Active transport is the movement of molecules or ions into or out of a cell from a region of lower concentration to a region of higher concentration using energy and carrier molecules.

The main features of active transport are:

- Metabolic energy supplied by ATP is needed.
- Materials are moved against a concentration gradient (i.e. from a lower to a higher concentration).
- Carrier protein molecules which act as 'pumps' are involved.
- The protein molecules undergo a change in shape (conformational change).
- The process is very selective, with specific substances being transported.

Mechanism of active transport

Active transport uses ATP in two ways:

- ATP is used directly to move molecules
- by using a concentration gradient that has already been set up by direct active transport. This is also known as **co-transport**.

An example of direct active transport of an ion is described below:

- The carrier proteins span the cell surface membrane and accept the ions to be transported on one side of it.
- The ions bind to binding sites on the inside of the carrier protein.
- On the inside of the cell, ATP binds to the protein causing it to split into ADP and a phosphate molecule. As a result, the protein molecule changes shape and opens to the opposite side of the membrane.
- The ions are then released to the other side of the membrane.
- The phosphate molecule is released from the protein and recombines with the ADP to form ATP and can be used in the synthesis of ATP in aerobic respiration.
- This causes the protein to return to its original shape, ready for the process to be repeated.

These events are illustrated in Figure 1.

Occasionally, the ion is moved into a cell or organelle at the same time as a different one is being removed from it. One example of this is the **sodium–potassium pump**.

In the sodium–potassium pump, sodium ions are actively removed from the cell or organelle while potassium ions are actively taken in from the surroundings. This process is essential to a number of important processes in organisms including the creation of a nerve impulse.

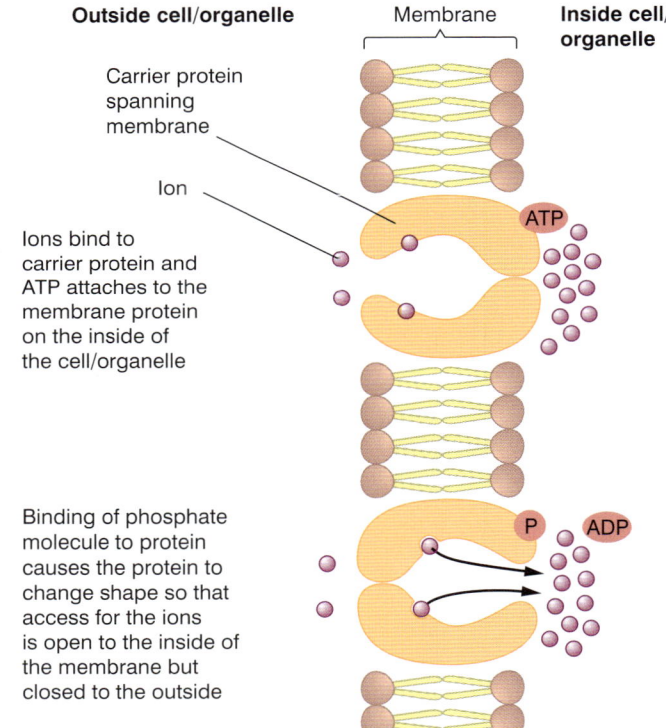

Outside cell/organelle   Membrane   **Inside cell/organelle**

Carrier protein spanning membrane

Ion

Ions bind to carrier protein and ATP attaches to the membrane protein on the inside of the cell/organelle

Binding of phosphate molecule to protein causes the protein to change shape so that access for the ions is open to the inside of the membrane but closed to the outside

Figure 1 *Active transport*

Requirements for active transport

Certain conditions are necessary if a cell is to carry out active transport effectively. These include:

- the presence of numerous mitochondria
- a ready supply of ATP
- a high respiratory rate.

Clearly, any factor that affects respiratory rate will affect active transport, and therefore higher temperatures up to an optimum or an increased supply of oxygen will increase the rate of active transport. Lower temperatures, less oxygen or the presence of respiratory inhibitors, such as cyanide, will slow the rate of active transport.

Table 1 *Comparison of different forms of transport in cells*

Process	Occurs against a concentration gradient	Needs ATP to supply energy	May use carrier molecules
Diffusion	No	No	No
Facilitated diffusion	No	No	Yes
Osmosis	No	No	No
Active transport	Yes	Yes	Yes

Materials may be transported, either as molecules or in bulk as larger particles, in a process called cytosis. There are two forms of cytosis, both of which require ATP (Figure 2):

- **endocytosis** – the bulk movement of material into the cell
- **exocytosis** – the bulk movement of the material out of the cell.

Endocytosis

Endocytosis is the bulk movement of material into a cell by active means. It takes two forms:

- Phagocytosis involves the invagination of the cell to form a cup-shaped depression in which large particles or even whole organisms are contained. The depression is then pinched off to the inside of the cell, forming a vesicle. In single-celled organisms such as *Amoeba* sp., it is used as a method of feeding. A few specialised cells in higher organisms also carry out phagocytosis. These cells are called phagocytes and include certain types of white blood cells that ingest harmful bacteria.
- Pinocytosis is very similar to phagocytosis, except that the vesicles formed are smaller. Pinocytosis is often used for the uptake of liquids.

Exocytosis

Exocytosis is the bulk movement of material out of the cell. It is the reverse of phagocytosis and pinocytosis. Vesicles budded off from a Golgi body within the cell fuse with the cell surface membrane and their contents are expelled into the medium outside. In higher organisms, exocytosis is used to release hormones, e.g. insulin, from the cells that manufacture them.

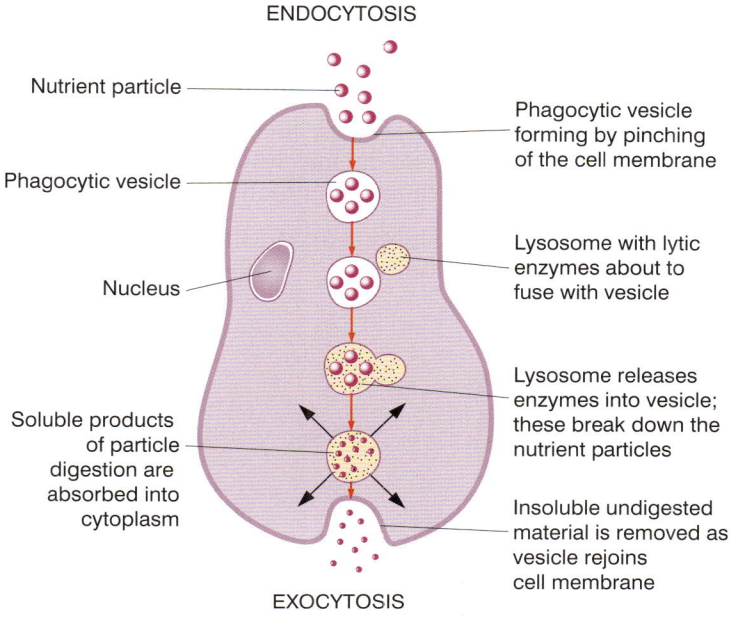

Figure 2 *Endocytosis and exocytosis*

SUMMARY TEST 4.5

Active transport occurs **(1)** a concentration gradient. It requires energy supplied by **(2)** and cells exhibiting active transport therefore have numerous **(3)** and a high **(4)** rate. The bulk movement of molecules and particles across the cell surface membrane is called **(5)**. Where this movement is into the cell, it is called **(6)**. Where this movement is out of the cell, it is called **(7)**. The invagination of a cell surface membrane to ingest particles and form a vesicle is known as **(8)**, whereas the invagination of the membrane to ingest liquids and form a vesicle is known as **(9)**.

4 Examination Questions

1 a Cell surface membranes are involved with the movement of substances into and out of cells.

Calcium pumps in cell surface membranes maintain a concentration of calcium ions inside the cytoplasm that is a thousand times lower than outside the cell.

Figure 1 shows the movement of calcium ions across a cell surface membrane.

Figure 1

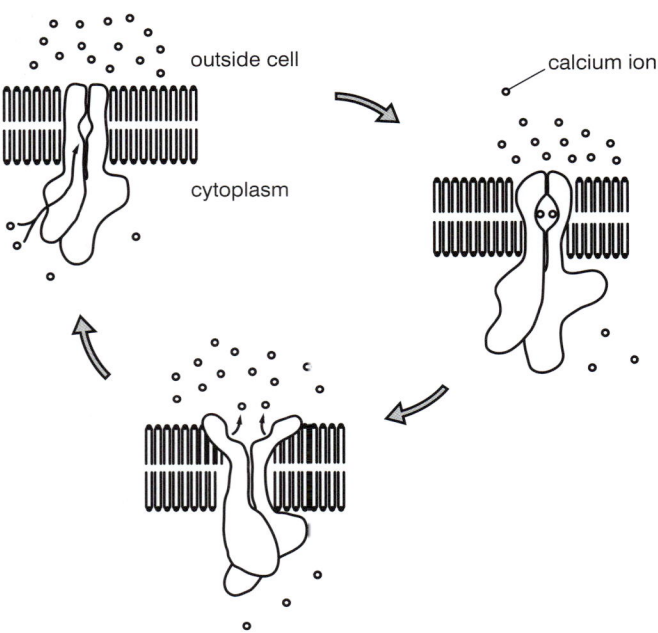

With reference to Figure 1,
  i explain why calcium ions do not pass through the phospholipid bilayer; *(2 marks)*
  ii name and describe the process by which calcium ions are moved across the membrane. *(3 marks)*

b Phagocytosis is the process by which bacteria are ingested by cells.

Describe the role of the cell surface membrane during phagocytosis. *(3 marks)*

c Phagocytic cells contain many lysosomes.

Describe the function of lysosomes in destroying ingested bacterial cells. *(4 marks)*

(Total 12 marks)

Cambridge International AS and A Level Biology 9700 Paper 22 Q1 November 2009

2 Figure 2 shows a diagram of a plasma (cell surface) membrane.

Figure 2

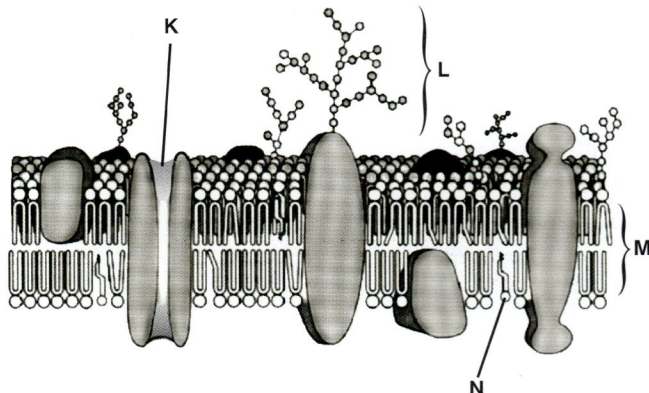

a Indicate, by putting a circle, around **one** of the following, the width of the membrane shown in Figure 1.

0.7 nm   7.0 nm   70 nm 7×10^{-5} m   700 μm   7.0 μm
(1 mark)

b Outline the functions of the following components of the plasma membrane: **K**, **L**, **M**, **N**. *(4 marks)*

c Some substances may cross plasma membranes by simple diffusion. Glucose, however, does not.

Explain why glucose cannot pass across membranes by simple diffusion. *(2 marks)*

d In an investigation, animal cells were exposed to different concentrations of glucose. The rate of uptake of glucose into the cells across the plasma membrane was determined for each concentration. Figure 3 shows the results.

Figure 3

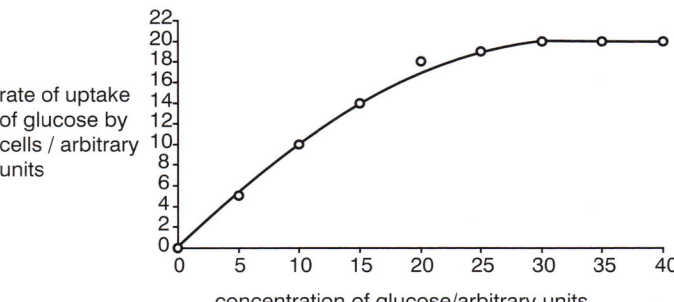

Using the information in Figure 3, explain how the results of the investigation support the idea that glucose enters cells by facilitated diffusion. *(2 marks)*

e State how active transport differs from facilitated diffusion. *(1 mark)*

(Total 10 marks)

Cambridge International AS and A Level Biology 9700 Paper 2 Q6 June 2007

3 One role of the cell surface membrane is to control the entry and exit of substances.

a Complete Table 1 to show the transport mechanisms across cell surface membranes and examples of materials transported.

Table 1

transport mechanism across cell surface membrane	example of material transported across membrane
active transport	sodium ions
	oxygen molecules
	bacteria
exocytosis	mucin (for mucus)
facilitated diffusion	
osmosis	

(2 marks)

b Each transport mechanism across cell surface membranes has a characteristic set of features.

In **each** of the boxes below, state one example of a transport mechanism that matches the pathway shown.

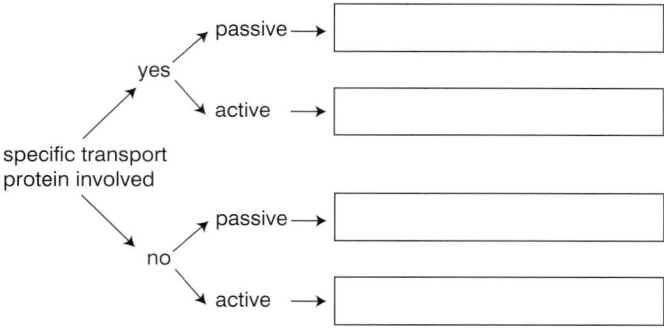

(4 marks)

(Total 6 marks)

Cambridge International AS and A Level Biology 9700 Paper 22 Q1 June 2012

4 Practice Questions

4 Starch in the diet is digested by the enzymes amylase and then maltase to form glucose. Glucose must be absorbed into the body so that it can be used by cells as a substrate for respiration. The glucose is absorbed from the exchange surface of the small intestine into the epithelial cells that line it. This absorption occurs partly by facilitated diffusion.

a Glucose molecules mostly diffuse into cells through the carrier proteins that span the phospholipid bilayer. Why do they not pass easily through the phospholipid bilayer?

b State **two** changes to the structure of cell surface membranes that would increase the rate at which glucose diffuses into a cell.

c The other molecule required by cells for respiration is oxygen. This diffuses into the blood through the epithelial layers of the alveoli and blood capillaries. By how much would each of the following changes increase or decrease the rate of diffusion of oxygen?
   i The surface area of the alveoli is doubled.
   ii The surface area of the alveoli is halved and the oxygen concentration gradient is doubled.
   iii The oxygen concentration gradient is halved and the total thickness of the epithelial layers is doubled.
   iv The oxygen concentration of the blood is halved and the carbon dioxide concentration of the alveoli is doubled.

5 a What is meant by a partially permeable membrane?

b Under standard conditions of pressure and temperature, what is the water potential of pure water?

c Four cells have the following water potentials:

Cell A $= -200\,kPa$ Cell C $= -100\,kPa$
Cell B $= -250\,kPa$ Cell D $= -150\,kPa$.

In what order could the cells be placed for water to pass from one cell to the next if arranged in a line?

6 a Explain why an animal cell placed in pure water bursts while a plant cell placed in pure water does not.

b Plant cells that have a water potential of $-600\,kPa$ are placed in solutions of different water potentials. State in each of the following cases whether, after 10 minutes, the cells would be turgid, plasmolysed or at incipient plasmolysis.

Solution A $= -400\,kPa$ Solution C $= -900\,kPa$
Solution B $= -600\,kPa$ Solution D $=$ pure water

c If an animal cell with a water potential of $-700\,kPa$ were placed in each of the solutions, in which solutions is it likely to burst?

7 a State one similarity and one difference between active transport and facilitated diffusion.

b The presence of many mitochondria is typical of cells that carry out active transport. Explain why this is so.

c In the making of urine, glucose is initially lost from the blood but is then reabsorbed back into it by cells in the kidneys. Explain why it is important that this reabsorption occurs by active transport rather than diffusion.

The mitotic cell cycle

Chromosomes and the cell cycle

The cells that make up organisms always arise from existing cells, by the process of division, which occurs in two main stages:

- **Nuclear division**, is the process by which the nucleus divides. There are two types of nuclear division:
  - **Mitosis**, results in two daughter nuclei having the same number of chromosomes as the parent nucleus. The nuclei formed are normally genetically identical to the parent one. If the parent nucleus has two complete sets of chromosome (diploid), then each daughter nucleus contains one set (haploid).
  - **Meiosis**, results in four daughter nuclei having half the number of chromosomes as the parent nucleus. The nuclei formed have a genetic composition different from the parent one.
- **Cytokinesis (cell division)** follows nuclear division, and is the process by which the whole cell divides.

The cell cycle

Cells do not divide continuously, but undergo a regular cycle of division separated by periods of cell growth. This is known as the **cell cycle** (Figure 3) and has three stages:

- **Interphase** – occupies most of the cell cycle, and is sometimes known as the resting phase, because no division takes place. In one sense, this could hardly be further from the truth, as interphase is a period of intense chemical activity, divided into three parts:
  - **First growth (G_1) phase**, when the proteins from which cell organelles are synthesised are produced.
  - **Synthesis (S) phase**, when DNA is replicated.
  - **Second growth (G_2) phase**, when organelles grow and divide and energy stores are increased.

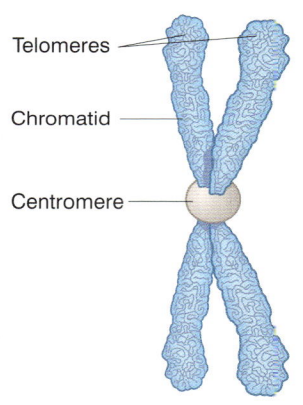

Figure 1 *Structure of a chromosome*

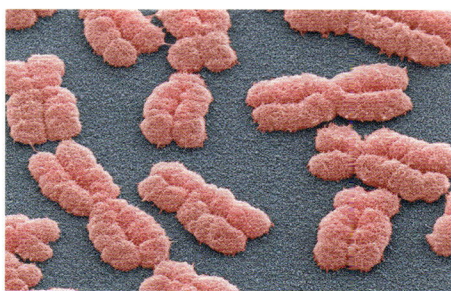

Figure 2 *Colourised scanning electron micrograph of a group of human chromosomes*

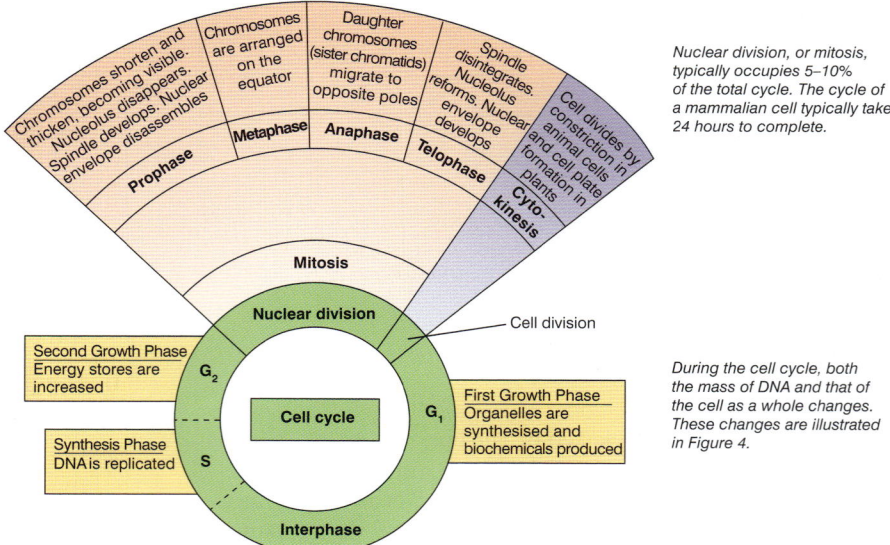

Nuclear division, or mitosis, typically occupies 5–10% of the total cycle. The cycle of a mammalian cell typically takes 24 hours to complete.

During the cell cycle, both the mass of DNA and that of the cell as a whole changes. These changes are illustrated in Figure 4.

Figure 3 *The cell cycle*

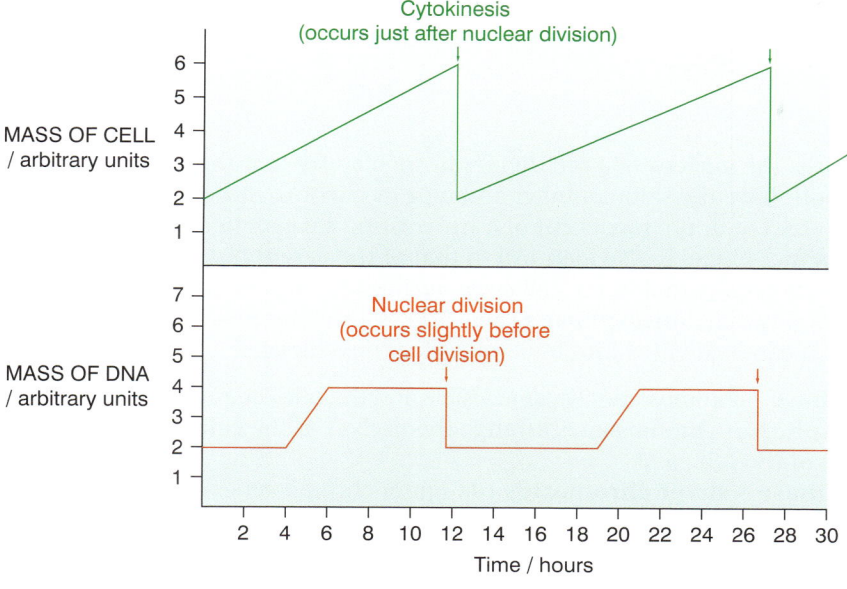

Figure 4 *Variations in the mass of a diploid cell and DNA within it during the cell cycle*

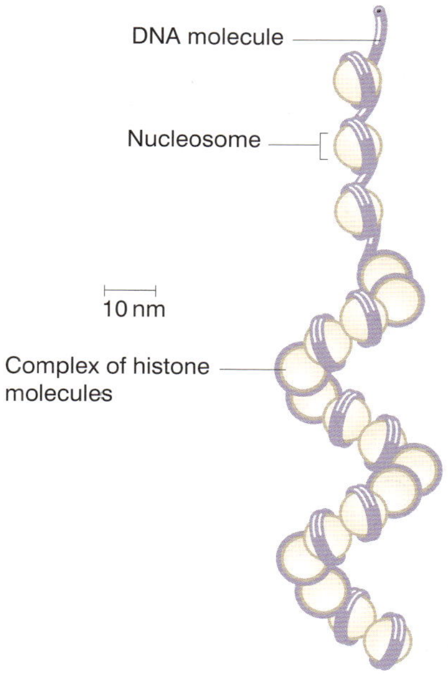

Figure 5 *Detailed structure of a chromosome*

- **Nuclear division**, when the nucleus divides into two (mitosis) or four (meiosis). In animal cells the centrioles (Topic 1.8), which lie at right angles to each other, are replicated to give two pairs.
- **Cytokinesis**, when the cell divides into two (mitosis) or four (meiosis).

Chromosome structure

Chromosomes have a characteristic shape, occur in pairs, and carry the hereditary (genetic) material of the cell. Chromosomes are only visible as discrete structures when a cell is dividing. The rest of the time, they consist of widely spread areas of darkly staining material called **chromatin**. When they are first visible, chromosomes appear as long, thin threads around 50 μm long. They are made up of two identical sister **chromatids**, joined at a point called the **centromere** (Figure 1).

Chromosomes are made up mainly of:

- **proteins** (70%), mostly in the form of **histones**
- **deoxyribonucleic acid** – DNA (15%).

To fit in, the considerable length of DNA found in each cell (around 2 metres in humans) is highly coiled and folded. This DNA is held in position by proteins called **histones**, which together form a complex known as **chromatin**. The chromatin has a beaded appearance due to the presence of **nucleosomes** (Figure 5). A nucleosome consists of a portion of DNA which is 146 base pairs in length and wrapped around eight histone molecules.

For convenience, and to make them easier to study, photographs of chromosomes are cut out and pasted into a logical format where they are arranged in their pairs and given numbers to identify them (Figure 6).

Telomeres and their significance

At the end of each chromatid there is a region, called the telomere, where a sequence of **nucleotide bases** is repeated many times. When cells divide, the enzymes that replicate DNA are unable to continue to work to the end of the chromosome. As a result some genetic material is missing from the end of the new chromosome. Telomeres act like disposable caps without which chromosomes would lose valuable genetic material and the cell would die. After repeated cell divisions, the telomeres are used up and the cell dies. The length of telomeres therefore determines the life span of cells.

Telomerase is an enzyme that adds bases to telomeres and so prevents them from shortening. Where cells divide repeatedly, e.g. cancer cells, there is insufficient telomerase to prevent this shortening. Cancer cells make more telomerase so the telomeres do not shorten and the cells do not die. Scientists are researching into whether telomerase could be used in humans to allow our cells to live longer and therefore increase our life span. One problem would be that the use of telomerase could increase the risk of cancer.

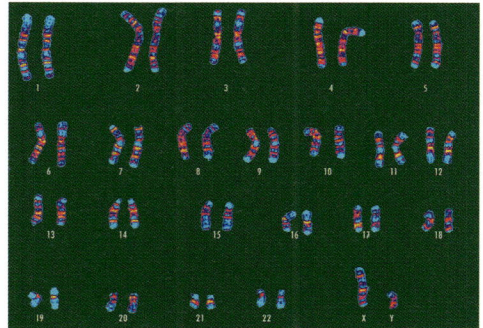

Figure 6 *Human male chromosomes*

On these pages you will learn to:

- Describe, with the aid of photomicrographs and diagrams, the behaviour of chromosomes in plant and animal cells during the mitotic cell cycle and the associated behaviour of the nuclear envelope, cell surface membrane and the spindle
- Observe and draw the mitotic stages visible in temporary root tip squash preparations

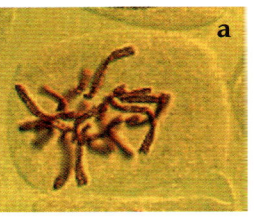

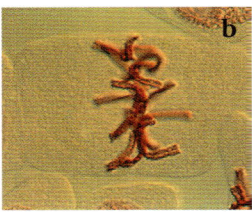

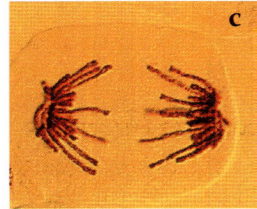

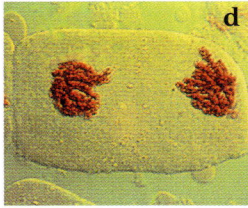

Figure 1 *The main stages of mitosis: a) prophase, b) metaphase, c) anaphase and d) telophase*

By mitosis, the nucleus of a cell divides in such a way that the two resulting nuclei both have the same number and type of **chromosomes** as the parent nucleus. Except in the rare event of a **mutation**, the genetic make-up of the two daughter nuclei is also identical to that of the parent nucleus. Before mitosis there is always a period in the cell cycle during which the cell is not dividing. This period is called **interphase** (Topic 5.1). Although mitosis is a continuous process, it can be divided into four stages for convenience:

- **Prophase** – chromosomes become visible and the nuclear envelope disassembles.
- **Metaphase** – chromosomes arrange themselves at the equator (metaphase plate) of the cell.
- **Anaphase** – **sister chromatids** (daughter chromosomes) move to opposite poles.
- **Telophase** – the nuclear envelope reforms.

The different stages are shown in Figure 1 and illustrated in Figure 2.

Prophase

In prophase, the chromosomes first become visible, to begin with as long thin threads, which later shorten and thicken. In animal cells the two pairs of centrioles separate and each pair moves to opposite poles. From each pair of centrioles, **microtubules** develop and form spindle fibres which span the cell from pole to pole. Collectively, these microtubules are called the **spindle apparatus**. Microtubules are made up of a globular protein called tubulin. As plant cells lack centrioles but do develop a spindle apparatus, centrioles are clearly not essential to microtubule formation. Plant cells have an area known as a microtubule organising centre (MTOC). The nucleolus disappears and the nuclear envelope breaks down, leaving the chromosomes free in the cytoplasm of the cell. These chromosomes become attached by their centromeres to microtubules of spindle fibres and are moved towards the equator of the cell towards the end of prophase.

Metaphase

During metaphase, the chromosomes are arranged at the spindle equator. They remain attached to microtubules at their centromere.

Anaphase

In anaphase, the centromeres divide into two and the microtubules joined to each contract, causing the sister chromatids that make up the chromosome to separate and move to opposite poles of the cell. The chromatids, which are now known as daughter chromosomes, move rapidly to their respective poles. The energy for the process is provided by mitochondria, which gather around the spindle fibres. If cells are treated with chemicals that destroy the spindle, the chromosomes remain at the equator, unable to reach the poles.

Telophase

In this stage, the daughter chromosomes reach their respective poles and become longer and thinner, finally disappearing altogether, leaving only widely spread **chromatin**. The spindle fibres disintegrate, and the nuclear envelope and nucleolus re-form.

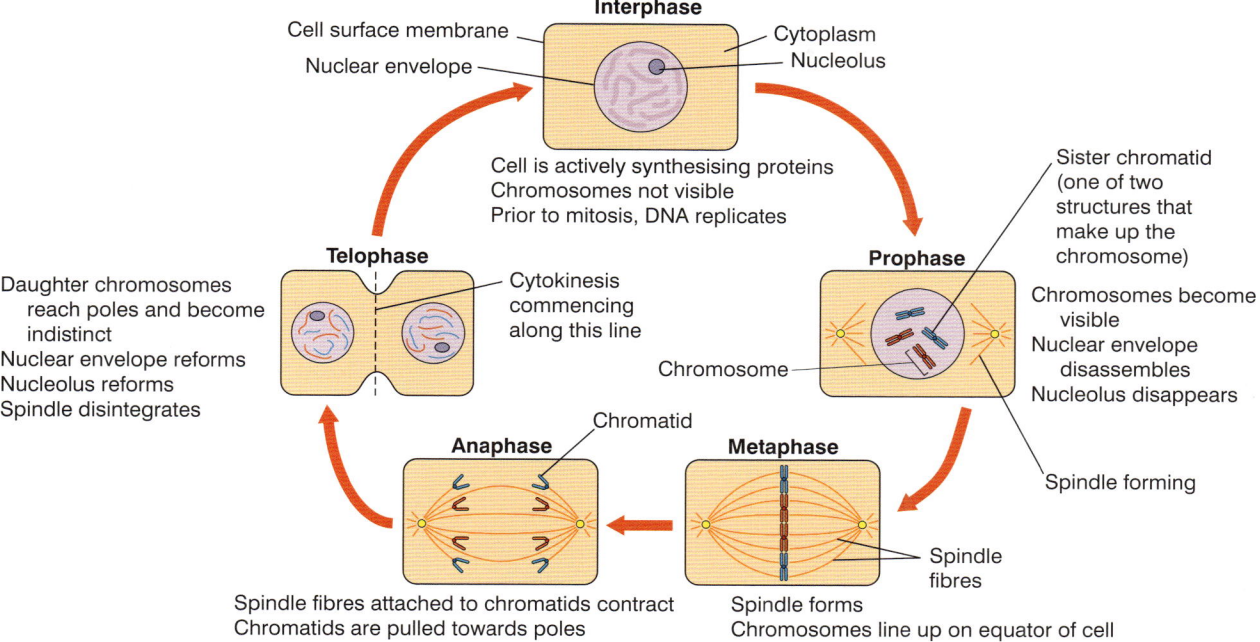

Figure 2 *Stages of mitosis in an animal cell*

Differences between mitosis in plant and animal cells

Centrioles are found in the cells of animals and some lower plants. However, they are absent from the cells of higher plants, although they still form spindles. Mitosis can still occur in animal cells that have had the centrioles removed, but not always with successful results.

In animal cells, cytokinesis occurs by the constriction of the centre of the parent cell from the outside inwards (see the telophase stage in Figure 2). In plant cells, division occurs by the growth of a cell plate across the equator of the parent cell from the centre outwards. Cellulose is laid down on this plate to form the cell wall.

In plants, mitosis occurs in a specialised tissue known as meristematic tissue. Plant meristems occur in the growing regions, for example in root and shoot tips and in the cambium of stems and roots (Topic 7.3). In animals, stem cells are able to divide by mitosis. Stem cells occur where there is a requirement for growth, tissue repair and cell replacement (Topic 5.3).

REMEMBER

The replication of DNA takes place during interphase before the nucleus and cell divide.

The Practical skills section on the accompanying CD covers temporary root tip squash preparations and discusses how to make high-powered drawings of cells in stages of mitosis.

EXTENSION

In animal cells, the pair of centrioles are replicated before the start of mitosis. The centrosome is the area containing the centrioles and associated proteins. During prophase, the centrosome divides so that each centriole pair moves to the opposite poles of the cell and the spindle apparatus forms. Animal cells that have the centrioles removed can still form a spindle, but not always with successful results. Plant cells do not have centrioles but do have microtubule organising centres (MTOCs) that perform the same role as the centrosome containing the centrioles in animal cells.

SUMMARY TEST 5.2

The stage when a cell is not dividing is called **(1)**. The first stage of mitosis is called **(2)**. During this stage in animal cells, each pair of cylindrical structures, called **(3)** move to the opposite **(4)** of the cell. Thin structures, called microtubules develop and form the spindle fibres that span the cell and together form the **(5)**. Towards the end of this stage, the **(6)** disassembles and the **(7)** disappears. During the second stage, called **(8)**, the chromosomes arrange themselves at the **(9)**. By the third stage, called **(10)**, the **(11)** of each chromosome divides into two and the microtubules attached to each pull the individual **(12)** to opposite ends of the cell. In the final stage, known as **(13)**, the nuclear envelope and nucleolus reform and chromosomes become longer and thinner to form chromatin.

The importance of mitosis

On these pages you will learn to:

- Explain the importance of mitosis in the production of genetically identical cells, growth, cell replacement, repair of tissues and asexual reproduction
- Outline the significance of mitosis in cell replacement and tissue repair by stem cells and state that uncontrolled cell division can result in the formation of a tumour and state that uncontrolled cell division can result in the formation of a tumour

We have seen in Topic 5.2 how cells divide by mitosis. Mitosis produces daughter cells that are genetically identical to the parent cells. Why, then, is it so essential to make exact copies of existing cells? The reason is that are certain processes in living organisms that require new cells to be identical. These are:

- Growth
- Cell replacement
- Repair of tissues
- Asexual reproduction.

Let us look at each in turn to see why mitosis is essential.

Growth

When two **haploid** cells (e.g. a sperm and an ovum) fuse together to form a diploid cell (e.g. a zygote), this diploid cell has all the genetic information needed to form the new organism. If the new organism is to resemble its parents, all the cells that grow from the original cell must possess this identical set of genetic information. Mitosis ensures that this is the case. The cell firstly divides to give a group of genetically identical cells like those shown in Figure 1. This is growth by increase in cell numbers. Although all cells have a complete set of genetic information, only part of it is expressed in any one cell. Depending on which part is expressed, cells change (differentiate) to give groups of specialised cells, e.g. muscle or epithelium in animals, xylem or phloem in plants. Cells formed by mitosis must all be genetically identical so that no genetic information is lost and the newly formed cells can become specialised with their particular structure and function.

Cell replacement

All cells have a limited life span and so die naturally at some stage. These cells need to be replaced if the organism is to continue to function normally. Red blood cells, for example, live for only around 120 days. If an organism is to continue to transport oxygen around its body, these red blood cells must be replaced. They are produced by stem cells in the bone marrow by the process of mitosis. Other examples include replacing epithelial cells lost from the skin surface and the cells of the uterus lining shed during menstruation. The renewal of cells by mitosis takes place constantly during an organism's lifetime.

Repair of tissues

If cells in a tissue are damaged or destroyed in some way, it is important that they are replaced as rapidly as possible. The replacement cells need to have an identical structure and function to the ones that have been damaged. If they were not exact copies the tissue would not function as effectively as before. An example is liver cells damaged by the hepatitis virus. Once again, mitosis is the means by which new cells replace damaged or dead ones.

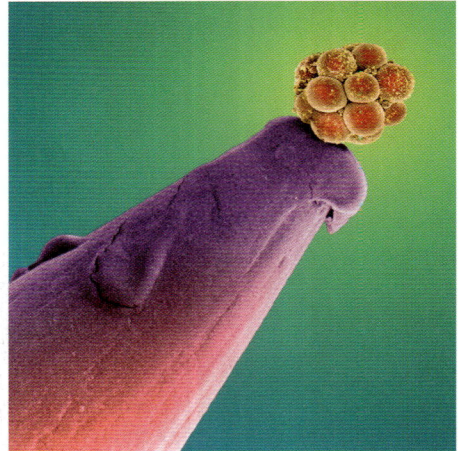

Figure 1 *Scanning electron micrograph of a 3-day-old human embryo at the 16 cell stage on the top of a pin. The cells are produced mitotically from the zygote and are genetically identical.*

Asexual reproduction

Mitosis is the means by which certain organisms carry out asexual reproduction. It produces offspring that are genetically identical (assuming no **mutations**) to their parents and to each other. This has one main advantage. Because they have been able to survive, grow and produce offspring, the parents must be well enough suited to the environmental conditions in which they currently live. By producing genetically identical offspring they can be sure that these too will thrive as long as conditions do not change.

Asexual reproduction is a relatively rapid form of reproduction (there is no delay while searching for a mate) and so large numbers can be quickly built up and the local area colonised. This is of particular advantage to plants, which can gain a competitive advantage, especially for light, by this means. Asexual reproduction by mitosis is therefore common in plants, where a variety of forms of vegetative propagation are used. An example is shown in Figure 2. In eukaryotic, unicellular organisms, mitosis **is** asexual reproduction. Mitosis is also the basis of natural and artificial **cloning**.

Stem cells and the significance of mitosis

Adult stem cells are undifferentiated dividing cells that occur in animal tissues. They are found, for example, in the inner lining of the small intestine, in the skin, in the lining of the gas exchange system and also in the bone marrow, which produces red and white blood cells. Under certain conditions, stem cells can develop into any other types of cell. In addition to the adult stem cells found in mature organisms, stem cells also occur at the earliest stage of the development of an embryo, before the cells have differentiated. These are called embryonic stem cells.

Stem cells need to be constantly replaced and this is done by mitosis because it produces genetically identical undifferentiated cells that can then develop into any type of tissue. This has significance because stem cells can be used to treat a variety of genetic disorders, such as the blood diseases thalassaemia and sickle cell anaemia. Research is also taking place to use stem cells to replace tissues that have been damaged by injury or disease, for example, in replacing heart tissue damaged by a heart attack.

Mitosis and tumour formation

Damage to the genes that regulate mitosis and the cell cycle can lead to uncontrolled cell division. As a consequence, a group of abnormal cells, called a **tumour**, develops and continues to expand in size as new cells are formed. Tumours can develop in any organ of the body, but are most commonly found in the lungs, prostate (male), breast and ovaries (female), large intestine, stomach, oesophagus and pancreas. Some tumours are cancerous (malignant) (Figure 3) while others are non-cancerous (benign). Uncontrolled cell division in plants may also lead to the formation of a tumour.

Figure 2 *This rhizome of a ginger plant is an organ of asexual reproduction. It grows by mitosis and allows the ginger plant to rapidly colonise a local area.*

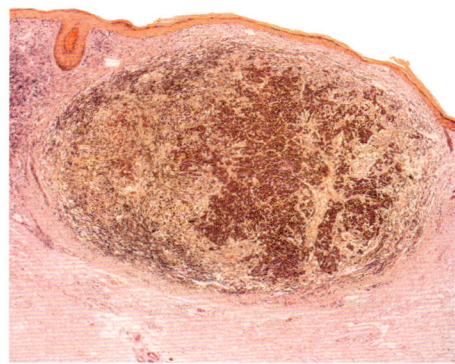

Figure 3 *This cancerous tissue in the skin is produced by the uncontrolled mitotic growth of skin cells.*

SUMMARY TEST 5.3

Mitosis produces cells that are genetically identical unless a **(1)** occurs. Mitosis is important in replacing cells that die naturally, especially **(2)** cells that only have a life span of 120 days. Other processes in which mitosis is important include **(3)**, **(4)** and **(5)**. Mitosis replaces undifferentiated dividing cells, called **(6)** that occur in adult animal tissues. Damage may occur to the genes that regulate mitosis and the cell cycle. This can lead to uncontrolled cell division and as a consequence, a group of abnormal cells, called a **(7)**, develops and continues to expand in size.

Examination and Practice Questions

5 Examination Questions

1 a Explain the importance of mitosis in multicellular organisms.   *(3 marks)*

A protein, mitosis-promoting factor (MPF), has been identified in cells. MPF is a globular protein made from two polypeptide chains.

b Place a tick (✓) in the box next to the type, or types, of protein structure shown by MPF.

primary ☐

secondary ☐

tertiary ☐

quaternary ☐   *(1 mark)*

The presence of MPF is known to cause prophase to start.

c Describe the changes that occur during prophase in an animal cell.   *(4 marks)*

d MPF normally begins to break down and stops functioning during anaphase.

Suggest the possible consequences of MPF not breaking down.   *(3 marks)*
                                              (Total 11 marks)

Cambridge International AS and A Level Biology 9700 Paper 23 Q5 June 2013

2 Muntjac are small deer found throughout Asia. Cells at the base of the epidermis in the skin continually divide by mitosis. Figure 1 shows the chromosomes from a skin cell of a female Indian muntjac deer at metaphase of mitosis.

Figure 1

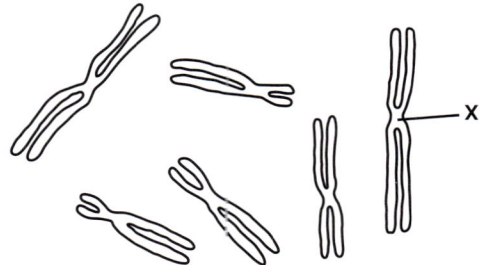

a i State the diploid chromosome number of the female Indian muntjac deer.   *(1 mark)*

ii Name **X** and state its role in mitosis.   *(2 marks)*

iii On a copy of Figure 1, **shade in** a pair of homologous chromosomes.   *(1 mark)*

iv Draw one of the chromosomes shown in Figure 2 as it would appear during **anaphase** of mitosis.
                                              (2 marks)

b Outline what happens to **a chromosome** between the end of anaphase and the start of the next mitosis.
                                              (3 marks)

c During the formation of eggs in the ovary of the female Indian muntjac deer, the chromosome number changes.

State what happens to the chromosome number and explain why this change is necessary.   *(2 marks)*
                                              (Total 11 marks)

Cambridge International AS and A Level Biology 9700 Paper 2 Q3 June 2007

3 Figure 2 shows a stage in the mitotic cell cycle in an animal cell .

Figure 2

a i Name the stage of mitosis shown in Figure 2.
                                              (1 mark)

ii State three features which are characteristic of the stage of mitosis shown in Figure 2.   *(3 marks)*

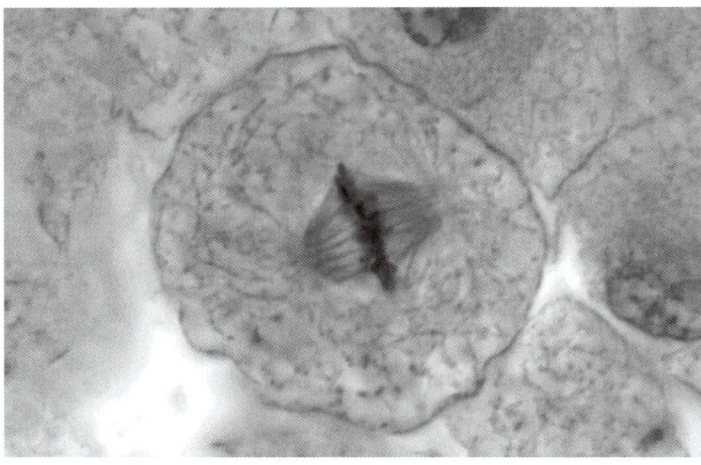

b Explain the importance of mitosis in organisms.
                                              (3 marks)

c In many multicellular organisms, such as mammals, the time taken for the mitotic cell cycle varies considerably between different tissues, but is very carefully controlled in each cell.

Suggest the importance of this control in mammals.
                                              (2 marks)
                                              (Total 9 marks)

Cambridge International AS and A Level Biology 9700 Paper 23 Q1 June 2011

5 Practice Questions

4 a State two ways in which the DNA of a prokaryotic cell differs from that of a eukaryotic cell.

  b What is the function of the protein found in chromosomes?

  c How is the considerable length of a DNA molecule compacted into a chromosome?

  d Suppose the total length of all the DNA in a single human muscle cell is 2.3 metres.
    i If all the DNA were distributed equally between the chromosomes, what would be the length of DNA in each one?
    ii What do you think the length of DNA is in a brain cell?

5 Mitosis is a continuous process. When mitosis is viewed under a microscope, the observer only gets a snapshot of the process at one moment in time. In this snapshot, the number of cells at each stage of mitosis is proportional to the time each cell spends undergoing that stage. Table 1 shows the number of cells at each stage of mitosis during one observation.

Table 1

Stage	Number of cells
Interphase	890
Prophase	73
Metaphase	20
Anaphase	9
Telophase	8

  a If one complete cycle takes 20 hours, how many minutes were spent in metaphase? Show your working.

  b In what percentage of cells would the chromosomes have been visible? Show your working.

6 Cancer is a group of diseases which results from uncontrolled growth and division of cells. As a consequence, a group of abnormal cells, called a tumour, develops and continues to expand in size.

The treatment of cancer often involves blocking some part of the cell cycle with drugs. In this way the cell cycle is disrupted and cell division, and hence cancer growth, ceases.

The problem with such drugs is that they also disrupt the cell cycle of normal cells. However, the drugs are more effective against rapidly dividing cells. As cancer cells have a particularly fast rate of division, they are damaged to a greater degree than normal cells.

The graph below shows the effect of a chemotherapy drug that kills dividing cells. It was given to a person with cancer once every three weeks starting at time 0. The graph plots the changes in the number of healthy cells and cancer cells in a tissue over the treatment period of 12 weeks.

Figure 3 *Changes in the number of healthy cells and cancer cells in a tissue during a chemotherapy treatment of 12 weeks*

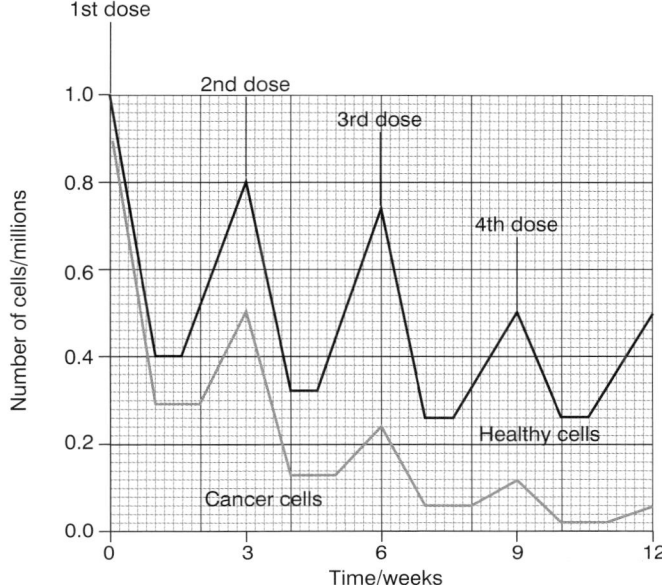

  a How many fewer healthy cells were there after three weeks compared to the start of the treatment?

  b What percentage of the original number of healthy cells were still present at 12 weeks?

  c How many times greater is the number of healthy cells compared to the number of cancer cells after 12 weeks?

  d Give a reason for the lower number of cancer cells compared to healthy cells at 12 weeks.

  e Describe two differences between the effect of the drug on cancer cells compared with healthy cells throughout the treatment.

  f Use the graph to explain why chemotherapy drugs have to be given a number of times if they are to be effective in treating cancer.

71

Nucleic acids and protein synthesis

Nucleotides and ribonucleic acid (RNA)

On these pages you will learn to:

- Describe the structure of nucleotides, including the phosphorylated nucleotide ATP
- Describe the structure of RNA

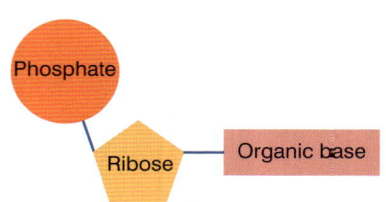

Figure 1 *Generalised structure of an RNA nucleotide*

NAME OF MOLECULE	REPRESENTATIVE SHAPE
Phosphate	
Pentose sugar	
Adenine (a purine)	 Adenine
Guanine (a purine)	 Guanine
Cytosine (a pyrimidine)	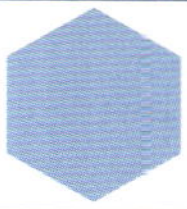 Cytosine
Thymine (a pyrimidine)	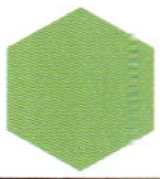 Thymine
Uracil (a pyrimidine)	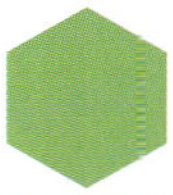 Uracil

Figure 2 *Molecules found in nucleotides*

Nucleotides are the basic units which make up a group of the most important chemicals in all organisms. These are the **nucleic acids**, of which the best known are:

- **ribonucleic acid (RNA)**
- **deoxyribonucleic acid (DNA)**.

Nucleotide structure

Individual nucleotides are made up of three components:

- **a pentose sugar**, of which there are two types: ribose and deoxyribose
- **a phosphate group**
- **a nitrogenous organic base**, of which five different forms are found in nucleic acids.

The five organic bases (Figure 2) are divided into two groups:

- **pyrimidines**, which are made up of a single six-sided ring, include **cytosine**, **thymine** and **uracil**
- **purines**, which are made up of a six-sided ring joined to a five-sided one. The two examples found in nucleic acids are **adenine** and **guanine**.

The pentose sugar, phosphate group and organic base are combined, as a result of **condensation reactions**, to give a **mononucleotide** (often shortened to nucleotide) (Figure 1). Two mononucleotides may, in turn, be combined as a result of a condensation reaction between the pentose sugar of one mononucleotide and the phosphate group of another. The strong covalent bond formed is called a **phosphodiester bond**. The new structure is called a **dinucleotide**. Continued linking of mononucleotides in this way forms a **polynucleotide**, such as RNA (Figure 3).

Ribonucleic acid (RNA) structure

Ribonucleic acid (RNA) is a polymer made up of repeating mononucleotide sub-units. It forms a single strand in which the pentose sugar is always **ribose** and the organic bases are adenine, guanine, cytosine and uracil (Figure 3). There are three types of RNA, all of which are important in protein synthesis:

- **ribosomal RNA (rRNA)**
- **transfer RNA (tRNA)**
- **messenger RNA (mRNA)**.

Ribosomal RNA (rRNA)

Ribosomal RNA (rRNA) is a large molecule that is complexed with proteins to form the subunits of ribosomes. It has a sequence of organic bases which is very similar in organisms within the same kingdom (Topic 18.1).

Transfer RNA (tRNA)

Transfer RNA (tRNA) is a relatively small molecule which is made up of around 80 nucleotides. It is manufactured by DNA and makes up 10–15% of the total

RNA in a cell. Although there are a number of types of tRNA, they are very similar, each having a single-stranded chain folded into a clover-leaf shape, with one end of the chain extending beyond the other. This extended chain always has the organic base sequence of cytosine–cytosine–adenine; this is the part of the tRNA molecule to which an amino acid can easily attach. There are at least 20 types of tRNA, each able to carry a different amino acid.

At the opposite end of the tRNA molecule is a sequence of three other organic bases, known as the **anticodon**. For each amino acid there is a different sequence of organic bases on the anticodon. During protein synthesis, this anticodon pairs with the complementary three organic bases that make up the triplet of bases on mRNA, known as the **codon**. The tRNA structure (Topic 6.7), with its end chain for attaching amino acids and its anticodon for pairing with the codon of the mRNA, is structually suited to its role of lining up amino acids on the mRNA template during protein synthesis.

Messenger RNA (mRNA)

Consisting of thousands of mononucleotides, mRNA is a long strand which is arranged in a single helix. The base sequence of an mRNA is determined by the sequence of bases in a length of DNA (gene) in a process called transcription. There is a great variety of different types of mRNA. Once formed, mRNA leaves the nucleus via pores in the nuclear envelope and enters the cytoplasm, where it associates with the ribosomes. There it acts as a template for protein synthesis. Its structure is suited to this function because it possesses the correct sequence of triplets that code for specific polypeptides. It is usually broken down quickly and normally exists only for as long as it is needed to manufacture a given protein.

Adenosine triphosphate (ATP)

ATP is a phosphorylated nucleotide made up of an adenine molecule, a ribose molecule and three phosphate molecules. It is the universal energy currency of all cells. The structure of an ATP molecule is shown in Figure 4, and details of its role in cells are given in Topic 12.2.

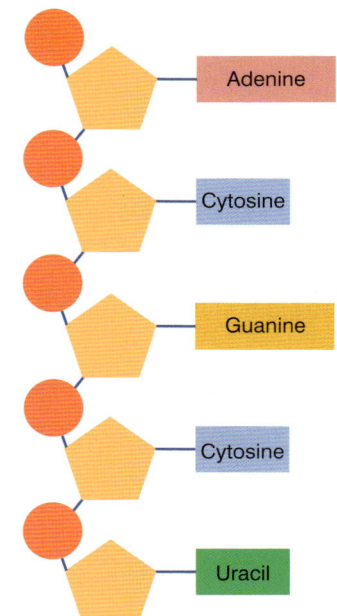

Figure 3 Section of an RNA polynucleotide

> **REMEMBER**
>
> Spelling can make a difference. 'Thymine' is a base in DNA. Thiamine is vitamin B_1.

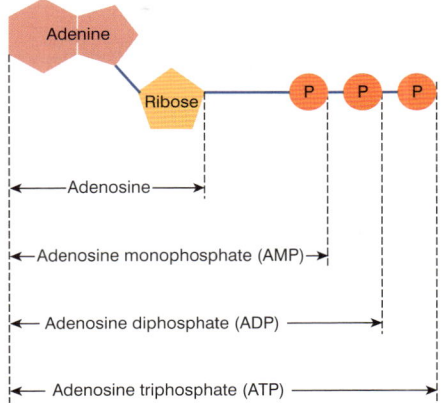

Figure 4 Structure of ATP

SUMMARY TEST 6.1

Nucleotides are organic compounds that contain the elements carbon, hydrogen, oxygen, **(1)** and **(2)**. A nucleotide contains a **(3)** sugar, which has **(4)** carbon atoms and has two forms: **(5)** and **(6)**. It also contains one of five organic bases, which fall into two groups. Those with a six-carbon ring only are called **(7)** and exist in three forms: thymine, **(8)** and **(9)**. Those of the second group, called **(10)**, have a six-sided ring joined to a **(11)**-sided ring; there are two such molecules: **(12)** and **(13)**. Ribonucleic acid, which never has the organic base **(14)**, exists in three forms. The form that has the same sequence of organic bases in all living organisms is called **(15)**; the form that has a sequence of three bases called an anticodon is **(16)**; and the remaining form, upon which proteins are formed, is **(17)**.

Deoxyribonucleic acid (DNA) structure

On these pages you will learn to:

- Describe the structure of DNA and explain the importance of base pairing and the different hydrogen bonding between bases

Deoxyribonucleic acid (DNA) is made up of two **nucleotide polymer** strands. In DNA, the pentose sugar is **deoxyribose** and the organic bases are adenine, guanine, cytosine and thymine. Each of the two polynucleotide strands is extremely long, and they are wound around one another to form a double helix. The differences between ribonucleic acid (RNA) and DNA are listed in Table 1.

DNA structure

In 1953, James Watson and Francis Crick worked out the structure of DNA following work by Rosalind Franklin on X-ray diffraction patterns of DNA. This opened the door for many of the major developments in biology. The basic structure of DNA is shown in Figure 1. For each strand of the DNA double helix, the deoxyribose sugars and phosphates alternate to form a strong sugar-phosphate backbone. The two strands are always the same distance apart from each other and are held together by hydrogen bonding between precise pairings of the organic bases. The organic bases are of two types: the purines (adenine and guanine) are longer molecules than the pyrimidines (cytosine and thymine). It follows that, if the two strands are the same distance apart, then the base-pairs must all be the same length and must always be made up of one purine and one pyrimidine. In fact, the pairings are even more precise than this:

- Adenine always pairs with thymine by means of two **hydrogen bonds**.
- Guanine always pairs with cytosine by means of three hydrogen bonds.

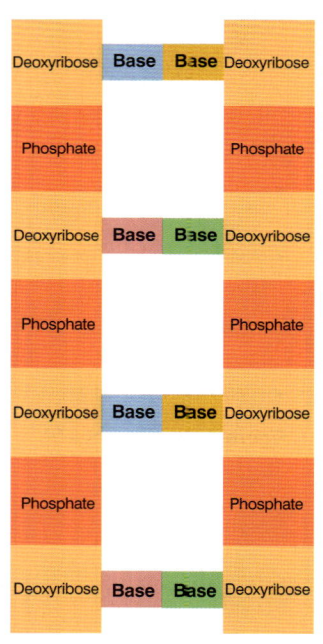

Simplified structure

Alternating phosphate and deoxyribose molecules make up the sugar-phophate backbones and pairs of organic bases hold the two together.

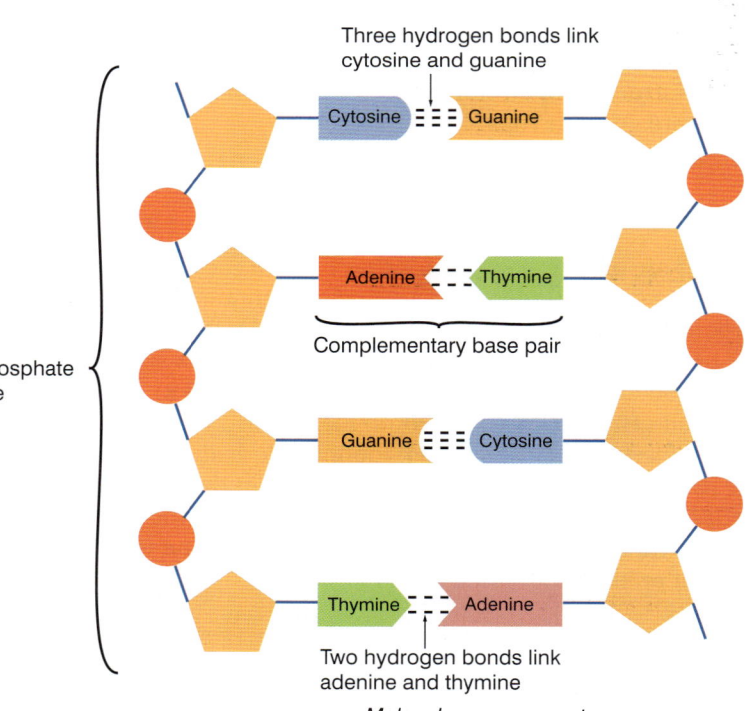

Molecular arrangement

Note the base pairings are always cytosine–guanine and adenine–thymine. This allows the two strands to be the same distance apart along the length of the molecule. Note that the sugar-phosphate backbones 'run' in the opposite direction to each other (i.e. are antiparallel).

Figure 1 *Basic structure of DNA*

It follows that, in DNA, the quantity of adenine is the same as the quantity of thymine, and similarly, the quantity of guanine is equal to the quantity of cytosine. However, the ratio of adenine and thymine to guanine and cytosine varies from species to species.

The structure of DNA is in the form of a **double helix** (Figure 3). The sugar-phosphate backbones run in the opposite direction to each other and are therefore said to be **antiparallel**. They form the overall structural backbone of the DNA molecule. For each complete turn of the double helix, there are 10 base pairs (Figure 2). In total, there are around 3.2 billion base pairs in the DNA of a typical mammalian cell. This vast number means that there is an almost infinite number of sequences of bases along the length of a DNA molecule and it is this variety that provides the genetic diversity of living organisms.

The DNA molecule is adapted to carry out its functions in a number of ways:

- It is very stable and can pass from generation to generation without change.
- Its two separate strands are, however, joined only with hydrogen bonds, allowing them to separate during replication (Topic 6.3) and form **messenger RNA** during protein synthesis (Topic 6.6).
- It can pass from generation to generation without change as it can be copied accurately (see Topic 6.3) and is a stable molecule.
- It is an extremely large molecule and it therefore carries an immense amount of genetic information in the sequence of nucleotides.
- By having the base pairs within the helical cylinder of the sugar-phosphate backbone, the genetic information is protected to some extent from being corrupted by outside chemical and physical forces.

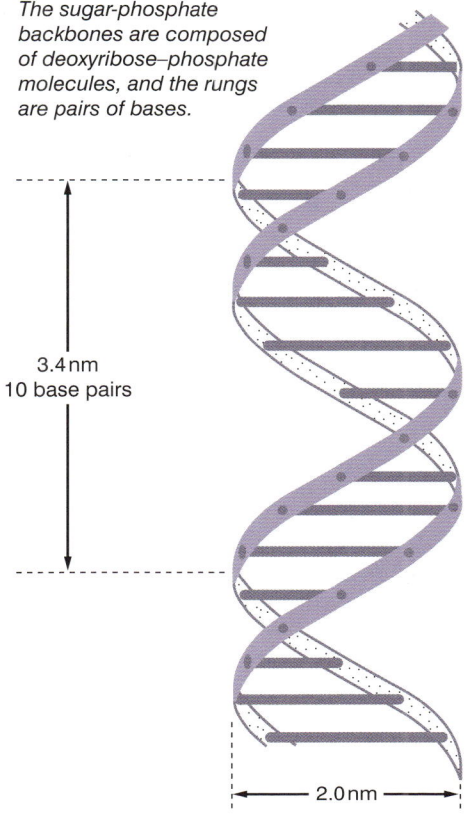

The sugar-phosphate backbones are composed of deoxyribose–phosphate molecules, and the rungs are pairs of bases.

3.4 nm
10 base pairs

2.0 nm

Figure 2 The DNA double helix structure

Table 1 *Differences between RNA and DNA*

RNA	DNA
Single polynucleotide strand (chain)	Double polynucleotide chain (double helix)
Smaller molecular mass (20 000–2 000 000)	Larger molecular mass (100 000–150 000 000)
Pentose sugar is ribose	Pentose sugar is deoxyribose
Organic bases present are adenine, guanine, cytosine and uracil	Organic bases present are adenine, guanine, cytosine and thymine
Ratio of adenine to uracil and the ratio of cytosine to guanine varies	Ratio of adenine to thymine and the ratio of cytosine to guanine is one
Manufactured in the nucleus but found throughout the cell	Found mostly in the nucleus with some present in mitochondria and chloroplasts
Amount varies from cell to cell (and within a cell according to metabolic acivity)	Amount is constant for all cells of a species (except gametes and spores)
Chemically less stable	Chemically very stable
Three basic forms: messenger, transfer and ribosomal RNA	Only one basic form, but with an almost infinite variety within that form

SUMMARY TEST 6.2

DNA is made up of a **(1)** sugar called **(2)**. This forms a structure, with the sugar and **(3)** groups forming the backbone and the organic bases making up the base pairs. These organic bases are always the same pairings: thymine is always paired with **(4)**, while guanine is always paired with **(5)**. The two backbones run in opposite directions, i.e. they are **(6)** and are twisted around one another to form a **(7)** shape.

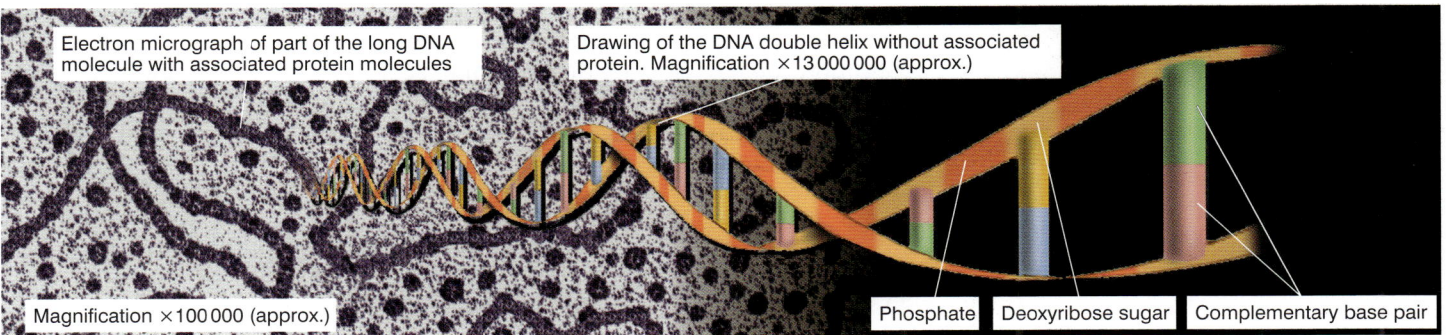

Electron micrograph of part of the long DNA molecule with associated protein molecules

Drawing of the DNA double helix without associated protein. Magnification ×13 000 000 (approx.)

Magnification ×100 000 (approx.)

Phosphate   Deoxyribose sugar   Complementary base pair

Figure 3 Deoxyribonucleic acid

6.3

DNA replication 💿

On these pages you will learn to:

• Describe the semi-conservative replication of DNA during interphase

REMEMBER

When smaller molecules are built up into larger ones, energy is required. When larger molcules are broken down into smaller ones, energy is released.

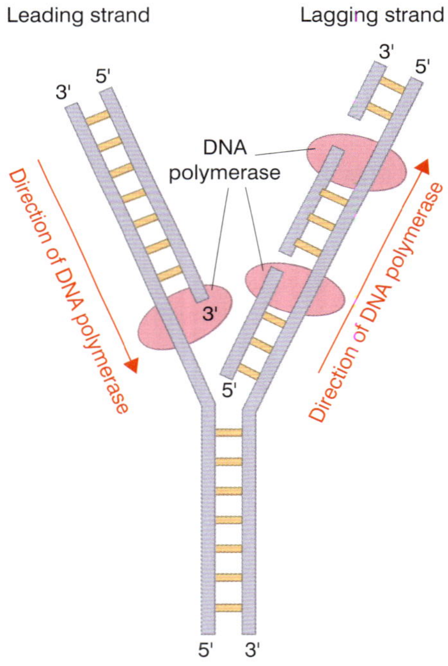

We know that DNA is the molecule of inheritance (the hereditary material) and that the genetic information held within DNA is passed from parent cell to daughter cell and from generation to generation. We have only to look at identical twins to see just how perfectly the 3.2 billion base pairs of DNA in the human genome can be copied. The process of DNA replication is clearly very precise. How then is it achieved?

Semi-conservative replication

The Watson–Crick model of DNA structure (Topic 6.2) allows for a logical explanation of how DNA produces exact copies of itself. Basically, the **hydrogen bonds** linking the base pairs of DNA break and the two strands of the double helix separate. Each exposed strand then acts as a template to which complementary free DNA **nucleotides** bind by complementary base pairing. These nucleotides have been activated by having two phosphate molecules added to them to make them more reactive. Adjacent activated nucleotides are joined together one nucleotide at a time to form the new strand. This results in two DNA molecules that are identical to each other and to the original DNA molecule (Figure 2). Each of the new DNA molecules contains one of the original parental DNA strands – and one newly synthesised strand. The process is therefore termed **semi-conservative replication**. It takes place during interphase in the cell cycle (Topic 5.1).

The process outlined so far is, in practice, a complex one involving a series of enzymes. It can be summarised as follows:

• **Opening up the DNA double helix** – an enzyme (DNA helicase) breaks the hydrogen bonds between the two parental strands of DNA, at a number of different points, called replication origins, which form **replication forks** (Figure 1).

• **Unwinding the DNA** – where the DNA is separated it unwinds. To prevent a tangled mess, one strand of the parental DNA is temporarily broken by an enzyme (topoisomerase) and then rejoined once unwinding has occurred.

• **Assembling the leading strand** – one strand of DNA is made by **DNA polymerase** in a continuous process that occurs in the same direction ($5' \rightarrow 3'$) as the replication fork is moving.

• **Assembling the lagging strand** – the other strand of the DNA is antiparallel (Topic 6.2) and runs in the $3' \rightarrow 5'$ direction. DNA polymerase can only work in the $5' \rightarrow 3'$ direction. Therefore short sections of complementary DNA are made simultaneously in the $5' \rightarrow 3'$ direction and these small sections are then linked together using the enzyme ligase.

• **Removing wrongly coded DNA** – errors that inevitably occur during replication could lead to a different amino acid being coded for and so a different primary structure of a protein. This can result in conditions such as sickle cell anaemia (Topic 6.5) or cancer. DNA polymerase corrects most errors as it assembles the new DNA strand.

Leading strand — Lagging strand

DNA polymerase

Direction of DNA polymerase

Direction of DNA polymerase

DNA polymerase synthesises DNA which is complementary to both strands. It works in the 5' – 3' direction of the new strand. As the DNA molecule is antiparallel, the DNA polymerase makes new DNA on the leading strand as a continuous process in the direction of the replication fork. On the lagging strand it makes small sections simultaneously in the opposite direction. These sections are later joined by the enzyme DNA ligase.

Figure 1 *DNA replication fork showing the action of DNA polymerase*

1. *A representative portion of DNA, which is about to undergo replication.*

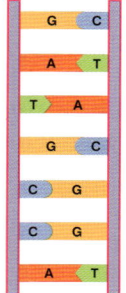

2. *Helicase enzyme causes the two strands of the DNA to separate. Binding proteins then keep the two halves apart.*

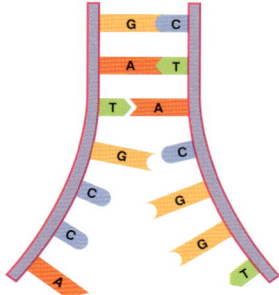

3. *In this portion of DNA, helicase completes the separation of the two strands. Meanwhile, free nucleotides, that have been activated by the addition of two phosphate molecules, are attracted to their complementary bases.*

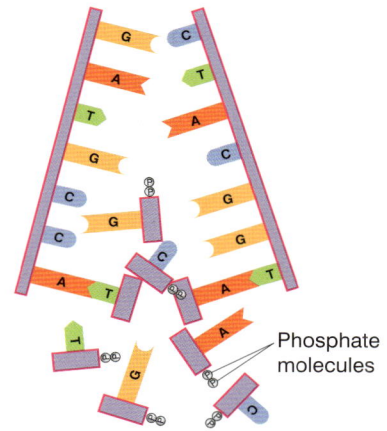

Phosphate molecules

4. *Once the activated nucleotides are lined up, they are joined together by DNA polymerase (bottom three nucleotides) and the phosphate molecules are released. One new strand is continuous. The other new strand is in short segments, which are joined together by DNA ligase.*

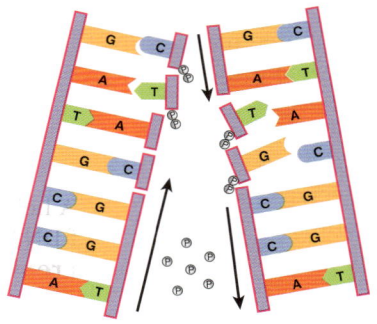

5. *In this way, two identical strands of DNA are formed. As each strand retains half of the original DNA material, this method of replication is called the semi-conservative method.*

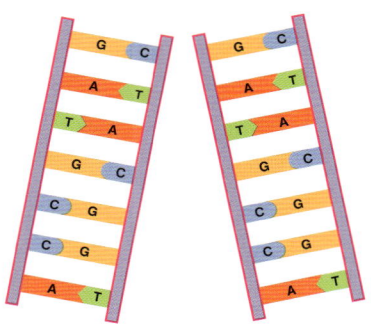

Figure 2 *The semi-conservative replication of DNA*

SUMMARY TEST 6.3

DNA replication involves the separation of the two **(1)** strands that make up the molecule. Firstly the double helix is opened up by an enzyme called **(2)**. The points where the helix splits are called **(3)**. One new strand of DNA is made by the enzyme known as **(4)** in a continuous process. On the other strand the enzyme works in the opposite direction making small sections of **(5)** DNA simultaneously. The sections are then joined together by the enzyme **(6)**. As the new DNA contains one original strand and one newly synthesised strand, the process is known as **(7)** replication.

Evidence for semi-conservative DNA replication

On these pages you will learn to:

• Describe evidence for the semi-conservative replication of DNA during interphase

When James Watson and Francis Crick deduced the structure of DNA in 1953 with the help of Rosalind Franklin's diffraction studies, they remarked in their paper '**It has not escaped our notice that the specific pairing we have postulated immediately suggests a possible copying mechanism for the genetic material**'.

Their idea, namely the semi-conservative method, was, however, only one of three possible mechanisms that needed to be scientifically tested before a definite conclusion could be drawn.

Possible mechanisms of DNA replication

In 1953, the three most feasible explanations for how DNA might replicate were as follows.

• **The conservative model** suggested that the parental DNA remained intact and that a separate daughter DNA copy was built up from new molecules of deoxyribose, phosphate and organic bases. Of the two molecules present, one would be made of entirely new material while the other would be entirely original material (Figure 1).
• **The semi-conservative model** proposed that the DNA molecule split into two separate strands, each of which then replicated its mirror image (i.e. the missing half). Each of the two new molecules would therefore have one strand of new material and one strand of original material (Figure 1).
• **The dispersive model** predicted that parental DNA would be broken down and the **nucleotides** replicated before being dispersed randomly throughout the new molecules. The new molecules would contain both new and original material but this would be randomly distributed and not necessarily with equal amounts of old and new material in each molecule.

The experiments of Meselsohn and Stahl

In 1958, Matthew Meselsohn and Franklin Stahl of the California Institute of Technology evaluated the three proposed mechanisms of DNA replication. If we look at Figure 1, we can see that the distribution of the 'old' DNA after replication is different in each of the three cases. To find out which mechanism was correct was therefore easy, in theory at least – simply label the old DNA in some way and then look at how it was distributed after replication. Meselsohn and Stahl achieved this in a neat and elegant experiment using the **isotope** heavy ^{15}N nitrogen. Their work can be summarised as follows.

• Cells of the gut bacterium *Escherichia coli* were grown on a medium containing a nitrogen source made up of the common form of nitrogen ^{14}N. This sample acted as the first control.
• Some bacterial cells were transferred to a medium where the nitrogen (in the form of ammonium chloride) was of the heavier isotope of nitrogen (^{15}N).
• The bacteria were grown, during which time the nitrogen they needed to make new DNA came from the growth medium and was therefore of the ^{15}N type. The resultant DNA was therefore heavier than that formed by the bacteria grown on the medium with ^{14}N. This was the second control.
• After many generations, the DNA of the cells was almost exclusively of the heavy type.
• Samples of the 'heavy DNA' bacteria were then transferred to a medium in which the nitrogen source (ammonium chloride) was of the light (^{14}N) type.

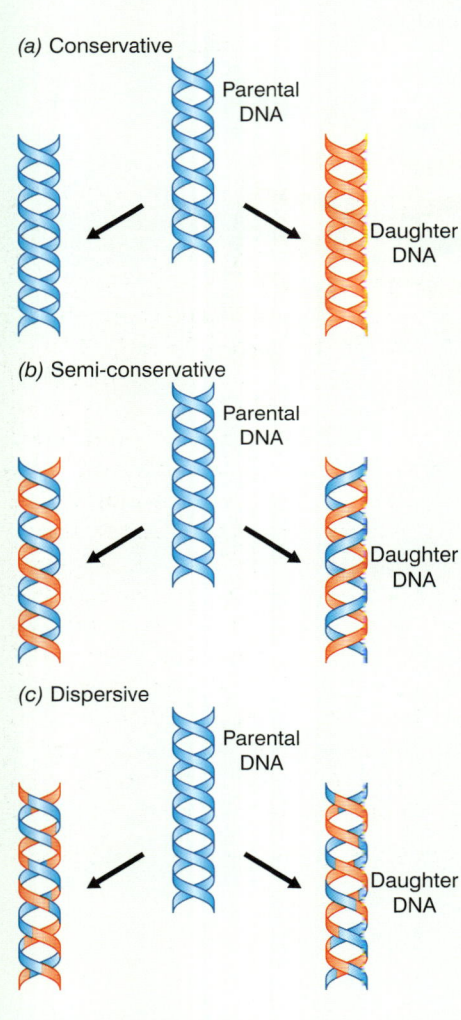

(a) Conservative

Parental DNA

Daughter DNA

(b) Semi-conservative

Parental DNA

Daughter DNA

(c) Dispersive

Parental DNA

Daughter DNA

= original (parental DNA)　　= new DNA

Figure 1 *Possible methods of DNA replication*

- The bacteria were grown just long enough for the cells to divide once (20–50 minutes, depending on temperature).
- A sample was removed and the DNA of the bacterial cells was extracted and placed in a solution of caesium chloride and centrifuged.
- Because the caesium atom is heavy, it sinks towards the bottom of the tube during **centrifugation** and a density gradient is established in the tube.
- DNA molecules sink in the gradient until they reach a level where their density equals that of the caesium chloride. They then 'float' at this level in the tube.
- The DNA containing ^{15}N is denser than that containing ^{14}N and therefore sinks to a lower level in the tube.

- After one generation in the ^{14}N medium, bacteria were found to have produced DNA with a density mid-way between that of the light (^{14}N) DNA and the heavy (^{15}N) DNA. This meant that one strand of each molecule of this DNA contained ^{14}N while the other strand contained ^{15}N.
- An extract of bacteria was taken after two generations and treated in the same way. Two bands were formed – one with the density of light DNA (indicating it contained only nitrogen of the ^{14}N type) and one of intermediate density (half ^{14}N and half ^{15}N).
- After three generations, there was three times as much DNA in the 'light' band than in the 'intermediate' band.

All these results are consistent with the semi-conservative theory of DNA replication. These experiments and the results are illustrated in Figure 2.

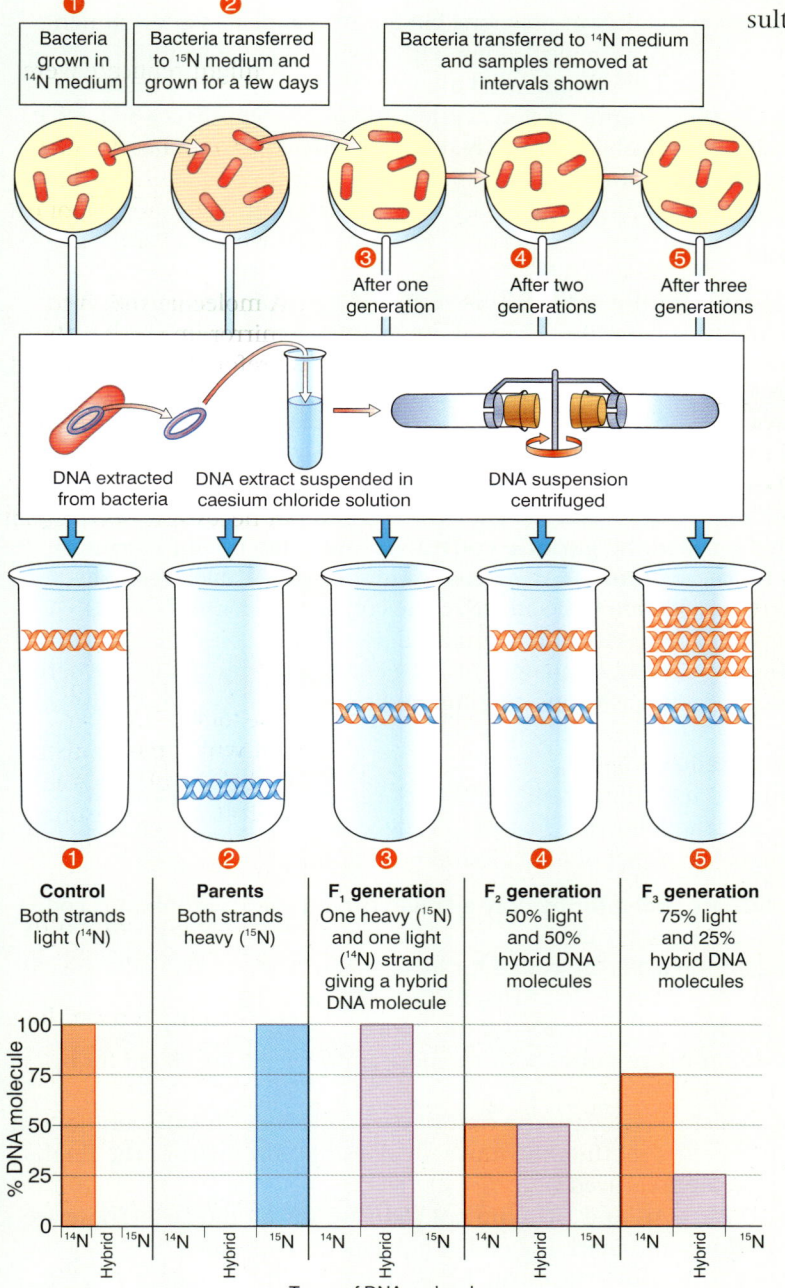

SUMMARY TEST 6.4

In experiments to explain DNA replication, Meselsohn and Stahl used two forms of nitrogen. Light nitrogen written as **(1)** and heavy nitrogen written as **(2)**. Heavy nitrogen is an **(3)** of light nitrogen. Some bacteria were grown only in light nitrogen, others only in heavy nitrogen. These were the **(4)** experiments. Some bacteria grown in heavy nitrogen were transferred to a medium containing light nitrogen and left to divide once. A sample of the bacteria produced was then placed in a solution of caesium chloride and then **(5)**. The DNA of these bacteria had an intermediate density showing that they had hybrid DNA made up of half heavy and half light nitrogen. After a second division, the next generation produced two types of DNA molecule, a **(6)** and a **(7)**. After three generations there were two types of DNA molecule, 75% were of the **(8)** DNA and 25% of the **(9)** DNA. These results support the **(10)** theory of DNA replication.

Figure 2 *Summary of Meselsohn–Stahl experiments on the semi-conservative replication of DNA*

6.5

The genetic code

On these pages you will learn to:

- State that a polypeptide is coded for by a gene and that a gene is a sequence of nucleotides that forms part of a DNA molecule
- State that a gene mutation is a change in the sequence of nucleotides that may result in an altered polypeptide
- Describe the way in which the nucleotide sequence codes for the amino acid sequence in a polypeptide with reference to the nucleotide sequence for normal and sickle cell alleles of the gene for the β-globin polypeptide

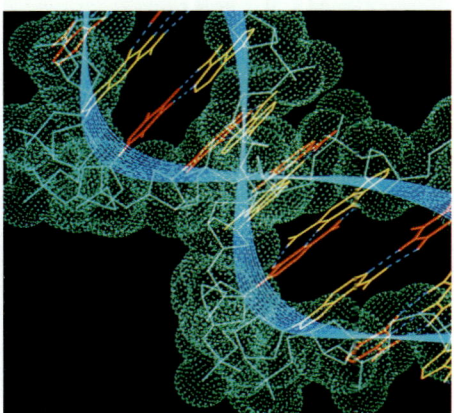

Figure 1 *Computer representation of part of a DNA molecule*

REMEMBER

A codon is a sequence of three adjacent nucleotides in mRNA that codes for an amino acid.

Once the structure of DNA had been discovered, and it had been established beyond doubt that it was the means by which genetic information was passed from generation to generation, scientists puzzled as to exactly how DNA determined the features of organisms.

What is a gene?

Genes are sequences of nucleotides on a DNA molecule that code for a polypeptide or a functional RNA. Polypeptides make up proteins and so genes determine the proteins of an organism. Enzymes are proteins. As enzymes control chemical reactions they are responsible for an organism's development and activities. In other words, genes, along with environmental factors, determine the nature and development of all organisms. A gene is a sequence of DNA nucleotides that determines a polypeptide and a polypeptide is a sequence of amino acids (Topic 2.6). So how exactly does a sequence of DNA bases determine a sequence of amino acids?

The triplet code

Proteins show almost infinite variety. This variety likewise depends upon the sequence of amino acids in each protein. There are just 20 amino acids that regularly occur in proteins, and each must have its own sequence of nucleotide bases on the DNA. As there are only four different bases (adenine, guanine, cytosine and thymine) present in DNA, if each coded for a different amino acid, only four different combinations could be coded for. Using a pair of bases, 16 different codes are possible – which is still not enough. A triplet of bases produces 64 combinations, more than enough to satisfy the requirements of 20 amino acids. This is called the **genetic code** and can be described in terms of triplets of RNA bases, or **codons**. All 64 codes are shown in Table 1. Although there are only four nucleotide bases in DNA, the considerable length of a DNA molecule means that there is an almost unlimited variety of combinations of these bases. In eukaryotes, in the length of DNA corresponding to a gene, there are coding portions known as exons and non-coding portions known as introns.

Features of the genetic code

Further experiments, including frame-shift ones carried out by Watson and Crick, have revealed the following features of the genetic code. In each case, the codon referred to is the triplet of bases found on mRNA.

- Two amino acids are coded for by only a single triplet, e.g. tryptophan is coded only by UGG.
- The remaining amino acids are coded for by between two and six codons each, e.g. leucine has six – UUA, UUG, CUU, CUC, CUA and CUG.
- The codon is always read in the 5′ → 3′ direction.
- The code is a **degenerate code**, because most amino acids are coded for by more than one triplet.
- The start of a sequence is always the codon AUG. This codes for the amino acid methionine. If this first methionine molecule does not form part of the final polypeptide, it is later removed.
- Three codons, UAA, UAG and UGA, do not code for any amino acid. These are called **STOP codons** and mark the end of a polypeptide chain.
- The genetic code is **non-overlapping**, i.e. each base in the sequence is read only once: six bases numbered 123456 are read as triplets 123 and 456, rather

than as triplets 123, 234, 345, 456. Non-overlapping codes need more bases but are less likely to be affected by error. Some viruses, with limited amounts of DNA, use overlapping codes, but this is extremely rare.

- The code is **universal**, i.e. with a few minor exceptions, each triplet codes for the same amino acid in all organisms.

Mutations

A mutation is a change in the sequence of nucleotide bases of DNA. This change may alter a codon causing it to code for a different amino acid. This may result in a different polypeptide and a different primary structure in a protein. In turn this may result in a different tertiary structure of the protein. If this protein is an enzyme it could mean a change in shape of the active site. The substrate might no longer fit the active site preventing the enzyme from functioning. If the alteration to the codon produces a STOP codon, then the polypeptide may be shortened and the final protein may be non-functional.

The human disease sickle cell anaemia, shows how a single base change in DNA can have major effects. Sickle cell anaemia is the result of a gene mutation in the gene producing normal haemoglobin (Hb^A) becoming mutated to form (Hb^S) that produces sickle cells. This causes the following sequence of events:

- In the DNA molecule that produces the β-globin polypeptide chain, a single nucleotide base, thymine, replaces the nucleotide base adenine.
- The normal DNA triplet on the non-template strand is changed to GTG from GAG.
- The triplet code on the template strand of DNA is therefore CAC rather than CTC.
- As a result, the mRNA produced has the codon GUG rather than GAG.
- This mRNA codes for the amino acid valine (GUG) rather than glutamic acid (GAG).
- This minor change produces a molecule of haemoglobin (called haemoglobin-S) that has a 'sticky patch'.
- Haemoglobin-S molecules stick to one another and form long fibres.
- These fibres distort the red blood cells, making them sickle (crescent) shaped.
- These sickle cells are unable to carry oxygen and may block small capillaries because their diameter is greater than that of capillaries.
- People with sickle cell anaemia become tired, listless and have a shortened lifespan.

Table 1 *The genetic code*
The base sequences shown are those on mRNA. A codon is made up of three nucleotide bases read in the sequence shown. For example, UGC codes for the amino acid Cys (cysteine). The first letter (U) is in the 'first position' column, the second letter (G) is in the 'second position' column and the third letter (C) is in the 'third position' column.

First position	Second position				Third position
	U	C	A	G	
U	Phe	Ser	Tyr	Cys	U
	Phe	Ser	Tyr	Cys	C
	Leu	Ser	STOP	STOP	A
	Leu	Ser	STOP	Trp	G
C	Leu	Pro	His	Arg	U
	Leu	Pro	His	Arg	C
	Leu	Pro	Gln	Arg	A
	Leu	Pro	Gln	Arg	G
A	Ile	Thr	Asn	Ser	U
	Ile	Thr	Asn	Ser	C
	Ile	Thr	Lys	Arg	A
	Met	Thr	Lys	Arg	G
G	Val	Ala	Asp	Gly	U
	Val	Ala	Asp	Gly	C
	Val	Ala	Glu	Gly	A
	Val	Ala	Glu	Gly	G

SUMMARY TEST 6.5

The sequence of nucleotides that determines the amino acid sequence of a polypeptide is called a **(1)**. Each amino acid in the sequence is coded for by a total of **(2)** nucleotide bases on a DNA molecule. The complementary sequence of these bases on a messenger RNA molecule is called a **(3)**. An immense variety of proteins can be made in an organism because the length of DNA in each cell is very large. In humans it totals **(4)** base pairs. Most amino acids have more than one codon and the genetic code is therefore described as **(5)**. Each base sequence is read only once and the code is therefore said to be **(6)**. Three codons do not correspond to an amino acid. These are called **(7)** codons and show where one polypeptide ends and the next begins. Using a genetic code table, we can find that the sequence UAU on mRNA codes for the amino acid named **(8)**. The two base sequences on the template strand of DNA that could give rise to this same amino acid are **(9)** and **(10)**. A change to the sequence of nucleotide bases of DNA is called a **(11)**.

Protein synthesis – transcription

Proteins, especially enzymes, are essential to all aspects of life. Every organism needs to make its own, sometimes unique, proteins. Each cell is capable of making any and every protein from just 20 amino acids. Exactly which proteins it manufactures depends on the instructions that are provided, at any given time, by the DNA in the cell's nucleus. DNA cannot leave the nucleus. It is too large to pass through the nuclear pores and by remaining in the nucleus it is protected from the rest of the cell. This means that the genetic information in a gene, carried by the sequence of bases in the DNA, must be copied so that the copy can leave the nucleus. The copy, or transcript, of the gene is mRNA. The process of copying is known as **transcription**. At the ribosome, mRNA is used to produce a polypeptide chain in a process known as **translation**.

Production of messenger RNA (mRNA) – transcription

Transcription (Figure 1) is the process of making mRNA (Topic 6.1) using one strand of the section of DNA as a template. This mRNA then carries the information out of the nucleus to the ribosomes, in the cytoplasm, that are the site of protein synthesis. The process is as follows.

- The enzyme **DNA helicase** acts on a specific region of the DNA molecule that corresponds to a gene to break the **hydrogen bonds** between the bases, causing the two strands to separate and expose the **nucleotide** bases in that region.
- The nucleotide bases on one of the two DNA strands, known as the **template strand**, pair with their complementary RNA nucleotides from the pool that is present in the nucleus. The enzyme RNA polymerase then moves along the strand and joins the nucleotides together.
- In this way an exposed guanine base on the DNA is linked to the cytosine base of a free nucleotide. Similarly, cytosine links to guanine, and thymine links to adenine. The exception is adenine, which links to uracil rather than thymine.
- As the RNA polymerase adds the nucleotides one at a time, to build a strand of pre-mRNA, so the DNA strands rejoin behind it. As a result, only around 12 base pairs on the DNA are exposed at any one time.
- When the RNA polymerase reaches a particular sequence of bases on the DNA which it recognises as a 'stop' signal, it detaches, and the production of pre-mRNA is then complete.
- The pre-mRNA is processed within the nucleus to form the final mRNA molecule.

The mRNA molecules are too large to diffuse out of the nucleus and so, having been processed, they leave via a nuclear pore. Once in the cytoplasm they attach to ribosomes for the next stage of protein synthesis, translation.

Once the introns have been removed, the remaining exon sections can be rejoined in a variety of different combinations (Figure 3). This means that a single section of DNA (gene) can code for up to a dozen different proteins, depending on the order in which the exons are recombined. Mutations can affect the splicing of pre-mRNA. Certain disorders such as Alzheimer's disease, are the result of splicing failures that lead to non-functional polypeptides being made.

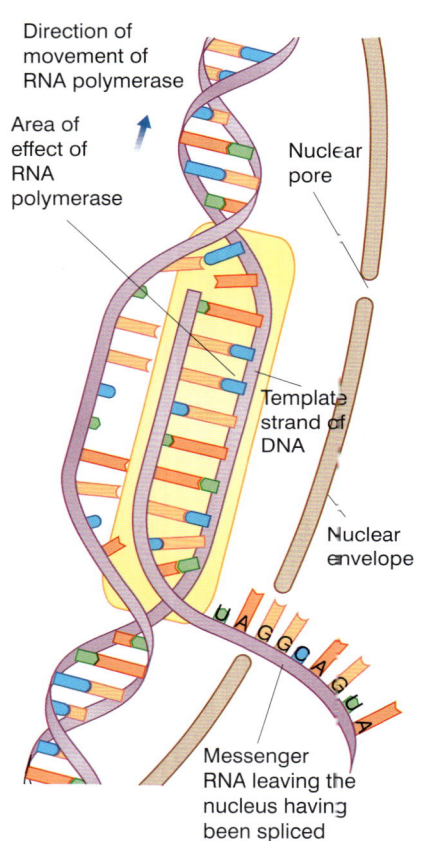

Direction of movement of RNA polymerase

Area of effect of RNA polymerase

Nuclear pore

Template strand of DNA

Nuclear envelope

Messenger RNA leaving the nucleus having been spliced

Figure 1 *Summary of transcription*

EXTENSION

Processing of pre-messenger RNA (pre-mRNA) – post-transcriptional modification

Before leaving the nucleus, the pre-mRNA produced during transcription is modified as follows:

- A guanine nucleotide is added to one end of the pre-mRNA. This 'cap' is used to set off the process of translation when the mRNA reaches a ribosome.

- Around 100 adenine nucleotides are added to the other end of the pre-mRNA. It is thought that this 'tail' may prevent the breakdown of the mRNA in the cytoplasm by nucleases because mRNA without a 'tail' is rapidly destroyed.

- DNA is made up of sections called exons that code for proteins and sections called introns that do not. These intervening introns would interfere with the synthesis of a polypeptide. In the pre-mRNA of **eukaryotic cells**, the base sequences corresponding to the introns are removed and the functional exons are joined together in a process called **splicing**. This process is shown in Figure 2.

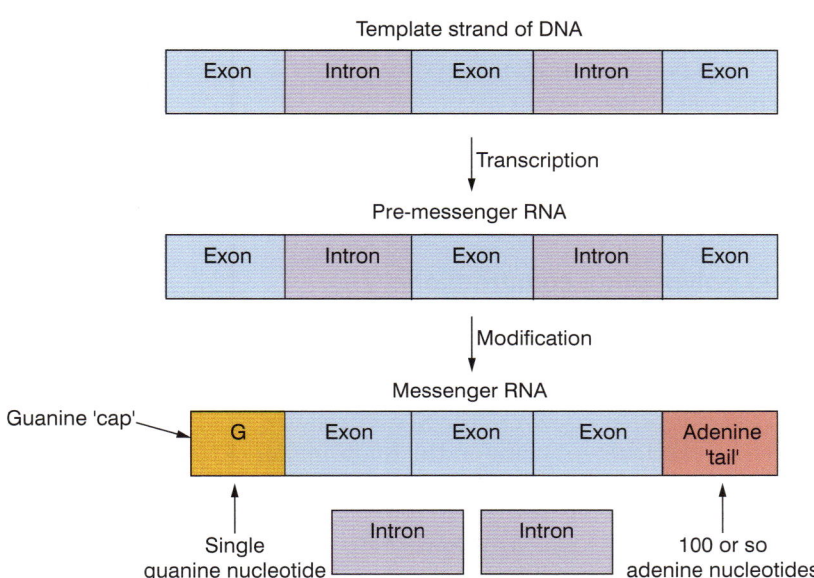

Figure 2 *Processing of pre-mRNA*

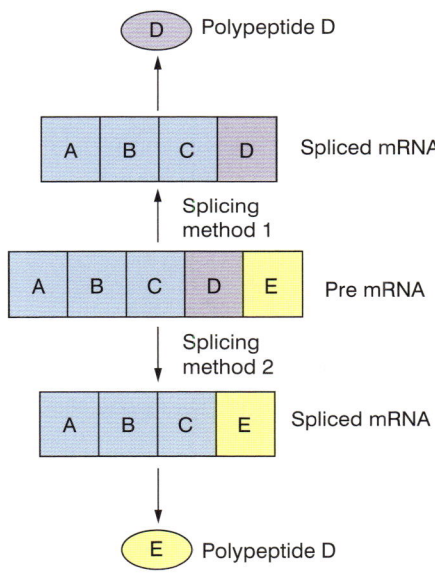

Figure 3 *Splicing pathways. mRNA from a single gene can be spliced differently to provide two different polypeptides. (Introns have been excluded from the figure.)*

SUMMARY TEST 6.6

DNA controls protein synthesis by the formation of a transcript known as **(1)**, which is formed when a specific region of the DNA, called a **(2)**, is opened up by the enzyme **(3)**. Along the strand called the **(4)** strand, the enzyme **(5)** moves, causing the bases on the strand to link with their adjacent nucleotides. Complementary bases pair; in the case of adenine of a DNA nucleotide the complementary base is **(6)**. The template is then modified by the addition of a **(7)** nucleotide at one end and about 100 **(8)** nucleotides at the other. Non-functional portions of the transcript, called **(9)**, are removed before it leaves the **(10)** and enters the cytoplasm.

On these pages you will learn to:

- Describe how the information in DNA is used during translation to construct polypeptides, including the role of messenger RNA (mRNA), transfer RNA (tRNA) and ribosomes

Translation is the process by which the messenger RNA from the nucleus of a cell forms a polypeptide; the sequence of nucleotide bases along the length of mRNA is used to produce the sequence of amino acids in the polypeptide. The process begins with the activation of the amino acids that will make up the polypeptide.

Amino acid activation

The amino acids present in cells must first be **activated** before they can be assembled into a polypeptide. This occurs in two steps:

- The amino acid first forms an intermediate with **ATP**.
- The intermediate then combines with transfer RNA to form an amino acid–tRNA complex called **amino-acyl tRNA** (Figure 3). The reaction is controlled by the enzyme, amino-acyl tRNA synthetase.
- Much of the energy provided by ATP is conserved for peptide bond formation later.

Although the basic structure of tRNA is always the same (Topic 6.1), a sequence of three bases on the anticodon loop varies. There are at least 60 variants, each of which is complementary to a codon of three bases on the messenger RNA. At the other end of the tRNA molecule, there is always the sequence of bases adenine-cytosine-cytosine, and it is to this end that the amino acid attaches (Figure 1). Each amino acid therefore has its own tRNA molecule, with its own unique anticodon.

Starting polypeptide construction

- The small subunit of the ribosome becomes attached to one end of the mRNA molecule (5' end).
- The starting point on the mRNA is normally the triplet of bases **(codon)**, AUG (AUG is often known as the START codon).
- There is an attachment point on the ribosome for tRNA.
- The amino-acyl tRNA molecule with the anticodon sequence of UAC moves to the ribosome and pairs up with the AUG sequence on the mRNA by complementary base pairing (Figure 5).

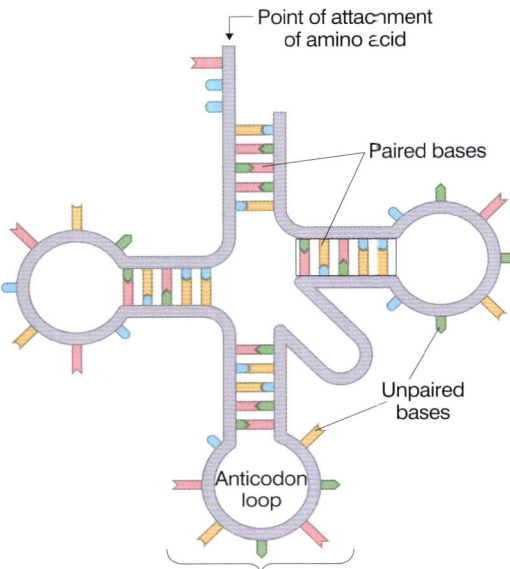

Figure 1 *Clover-leaf structure of tRNA*

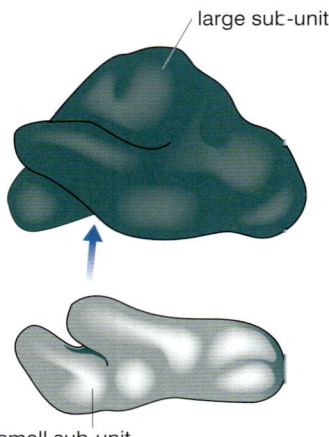

Figure 2 *Structure of a ribosome. The smaller sub-unit fits into a depression on the surface of the larger one*

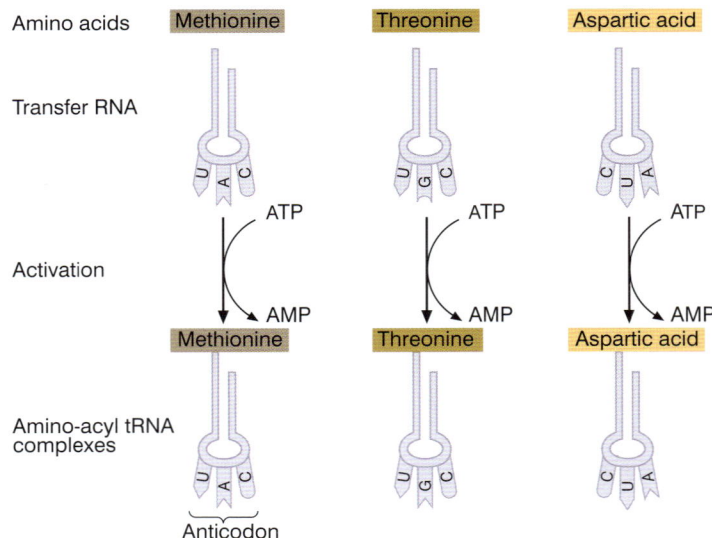

Figure 3 *Amino acid activation*

- As the tRNA which pairs with the AUG sequence on the mRNA always carries the amino acid methionine, polypeptides initially have methionine as the first amino acid.
- However, if this methionine molecule does not make up part of the finished polypeptide, it is removed at the end of the synthesis.

Making the polypeptide

The following explanation of how a polypeptide is made is shown in Figure 5.

- The ribosome moves along the mRNA, bringing together two tRNA molecules at any one time, each pairing up with the complementary two codons on the mRNA.
- By means of an enzyme, the two amino acids on the tRNA are joined by a peptide bond.
- The ribosome moves on to the third codon in the sequence on the mRNA, so that the amino acids on the second and third tRNA molecules can be linked.
- As this happens, the first tRNA is released from its amino acid (methionine) and is free to collect another methionine molecule from the amino acid pool in the cell.
- The process continues in this way, with up to 15 amino acids being linked each second, until a complete polypeptide chain is built up.
- Up to 50 ribosomes can pass immediately behind the first, so that many identical polypeptides can be made at the same time (Figure 4). A group of ribosomes acting in this way is known as a **polysome**.
- The process continues until the ribosome reaches a **STOP codon**. These are UGA, UAG and UAA, and do not attract a tRNA. At this point, therefore, the ribosome, mRNA and the last tRNA molecule all separate and the polypeptide chain is complete.

Assembling the protein

Sometimes a single chain becomes a functional protein. Often, a number of polypeptides are linked to give a functional protein (quarternary structure).

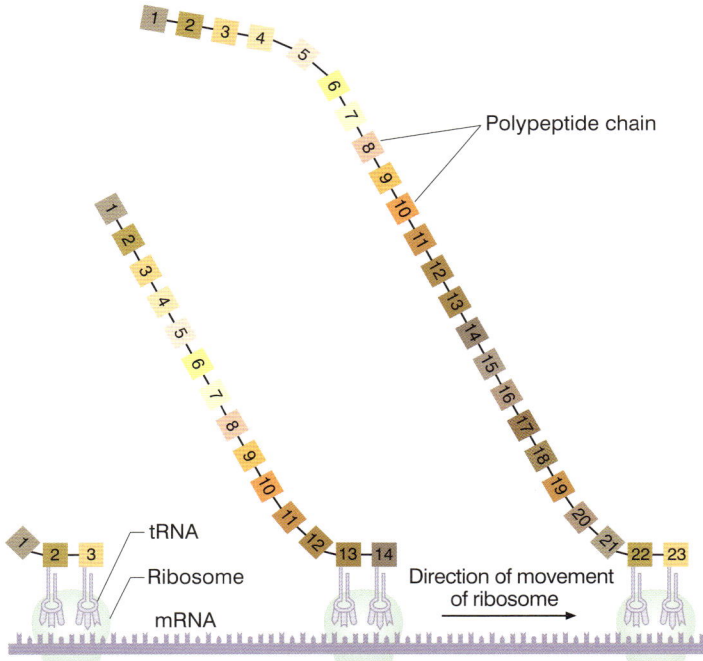

Figure 4 Polypeptide formation

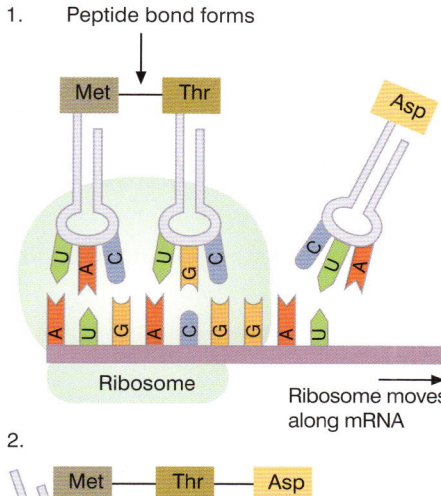

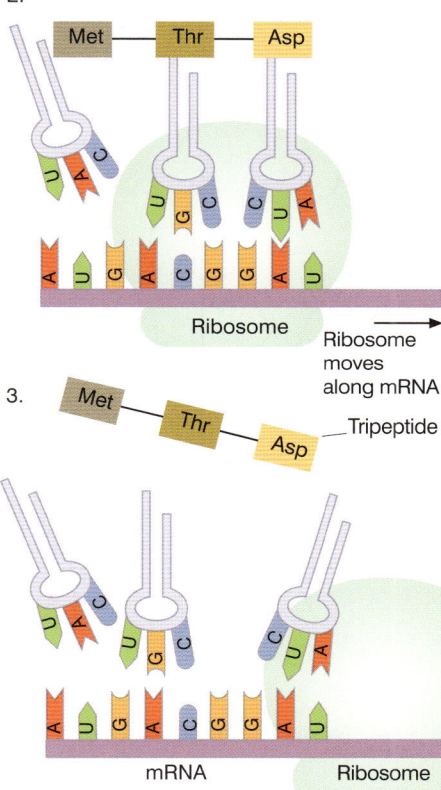

Figure 5 Translation

SUMMARY TEST 6.7

The formation of a polypeptide from the sequence of bases on messenger RNA is known as (1). It begins with the addition of (2) to amino acids, in a process known as activation. These amino acids then combine with transfer RNA to form a complex called (3). To form the polypeptide, a (4) becomes attached to the mRNA molecule and the tRNA attaches to three bases, called a (5) on mRNA.

6 Examination Questions

1 **a** Name the stage during the mitotic cell cycle when replication of DNA occurs. *(1 mark)*

b Figure 1 shows details of DNA replication.

Figure 1

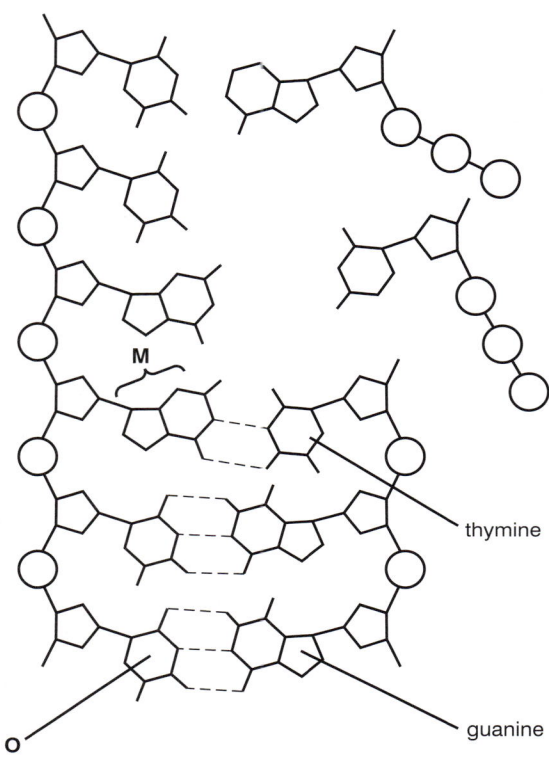

i Name the bonds shown by the dashed lines on Figure 1. *(1 mark)*

ii Name the nitrogenous bases, **M** and **O**. *(1 mark)*

c Explain why DNA replication is described as *semi-conservative*. *(2 marks)*

d The enzyme that catalyses the replication of DNA checks for errors in the process and corrects them. This makes sure that the cells produced in mitosis are genetically identical.

Explain why checking for errors and correcting them is necessary. *(2 marks)*

(Total 7 marks)

Cambridge International AS and A Level Biology 9700 Paper 21 Q5 June 2010

2 DNA and RNA are important biological molecules that are involved in the production of polypeptides.

a Figure 2 shows two nucleotides joined by a covalent bond.

Figure 2

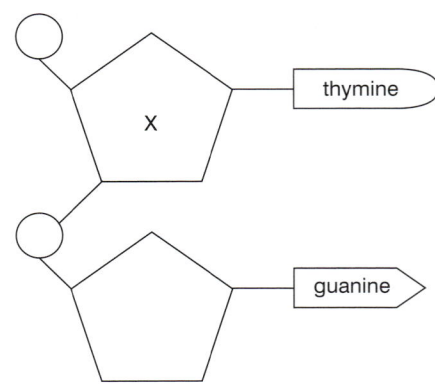

i Figure 2 represents part of a DNA molecule, not part of an RNA molecule. Explain why. *(1 mark)*

ii Name the covalent bond between the two nucleotides. *(1 mark)*

iii Name component X. *(1 mark)*

b Outline the role of transfer RNA (tRNA) in the production of a polypeptide. *(2 marks)*

c Describe how a peptide bond is formed between two amino acids during polypeptide production.

(3 marks)

(Total 8 marks)

Cambridge International AS and A Level Biology 9700 Paper 22 Q4 June 2012

6 Practice Questions

3 **a** What are the three basic components of a nucleotide?

b In terms of the structure of the DNA molecule, explain why the base pairings are not adenine with guanine and thymine with cytosine.

c Suggest a reason why the base pairings of adenine with cytosine and guanine with thymine do not occur.

d The bases on one strand of DNA are TGGAGACT, what is the base sequence on the other strand?

e If 19.9% of the base pairs in human DNA is guanine, what percentage of human DNA is thymine. Show your reasoning.

4 A section of DNA has the following sequence of bases along it: TACGCTCCGCTGTAC.

All of the bases are part of the code for amino acids. The first base in the sequence is the start of the code.

a How many amino acids does the section of DNA code for?

b Two of the amino acids coded for will be the same. Which two?

c Explain how a length of mRNA containing 10 codons that all have a different base sequence can produce a polypeptide chain containing only 6 different amino acids.

d Explain how a change in one base along a DNA molecule might result in an enzyme becoming non-functional.

5 a Describe the role of RNA polymerase in transcription.

b Which other enzyme is involved in transcription and what is its role?

c Why is splicing of pre-mRNA necessary?

d A sequence of bases along the template strand of DNA is ATGCAAGTCCAG.
i What is the sequence of bases on a messenger RNA molecule that has been transcribed from this part of the DNA molecule?
ii How many amino acids does the sequence code for?

e A gene is made up of 756 base pairs. The mRNA that is transcribed from this gene is only 524 nucleotides long. Explain why there is this difference.

6 a Explain why the genetic code is described as:
i universal
ii degenerate
iii non-overlapping.

b State three ways in which the molecular structure of RNA differs from DNA.

c Distinguish between a codon and an anticodon.

d Explain why:
i DNA needs to be chemically very stable.
ii mRNA needs to be easily broken down (chemically unstable).

7 The table below lists some amino acids. Alongside is a codon for each amino acid found on mRNA. The strand of DNA against which mRNA is transcribed is called the template strand.

Amino acid	Codon
Tyrosine	UAC
Serine	AGU
Aspartic acid	GAC
Glutamine	GAG
Histidine	CAU
Leucine	CUA
Alanine	GCA
Lysine	AAA
Proline	CCU
Glycine	GGC

a Using the table, state the:
i tRNA anticodon for histidine
ii triplet on the DNA template strand that codes for serine.

b Name the amino acid coded by the tRNA anticodon GAU.

c A sequence of bases on the template strand of DNA is CTCCGTGGAATGCGT. List the sequence of amino acids that would appear in a polypeptide coded for by this DNA.

d The sequence of amino acids in a section of polypeptide is histidine, proline, aspartic acid, leucine. List the sequence of bases on the DNA template strand that codes for this polypeptide section.

8 a Name the cell organelle involved in translation.

b A codon found on a section of messenger RNA has the sequence of bases AUC. List the sequence of bases found on:
i the transfer RNA anticodon that attaches to this codon
ii the template strand of DNA that formed the mRNA codon.

c Describe the role of tRNA in the process of translation.

d A strand of mRNA has 64 codons but the protein produced from it has only 63 amino acids. Suggest a reason for this difference.

7

7.1

Transport in plants

The need for transport systems in organisms

On these pages you will learn to:

• Identify the reasons for having transport systems in large organisms

Figure 1 *Giant redwoods can transport water to a height of 100 m without the expenditure of metabolic energy*

Figure 2 *The veins of a leaf form a network that allows material to be transported to and from its cells*

Transport over short distances, e.g. between adjacent cells, is adequately achieved by processes such as diffusion (Topic 4.2), osmosis (Topic 4.3) or active transport (Topic 4.5). The situation in larger organisms is, however, altogether different.

Why large organisms need a transport system?

All organisms exchange materials between themselves and their environment. In small organisms, this exchange takes place over the surface of the body to meet the needs of the organism. With increasing size, however, the surface area to volume ratio decreases to a point where the needs of the organism cannot be met (Topic 7.2). A specialist exchange surface therefore evolved to absorb nutrients, to exchange respiratory gases and to remove excretory products. These exchange surfaces are located in specific regions of the organism. As materials enter from the external environment, or as they are produced by particular cells, they can be taken to the cells that need them by a transport system. A transport system can also be used to remove waste materials and transfer them to specialised areas that process them or pass them to the external environment. As organisms have evolved into larger and more complex structures, the tissues and organs of which they are made have become more specialised and dependent upon one another, making a transport system all the more essential. Whether or not there is a specialised transport medium, and whether or not it is circulated by a pump, depends on two factors – the surface area to volume ratio and how active the organism is. The lower the surface area to volume ratio, and the more active the organism, the greater is the need for a specialised transport system with a pump.

Features of transport systems

Whether plant or animal, any large organism encounters the same problems in transporting materials within itself. The transport systems of organisms have many common features:

• A suitable medium, e.g. blood, in which to carry materials. This is normally a liquid based on water because water readily dissolves substances and can be moved around relatively easily.
• A form of mass flow transport in which the transport medium is moved around in bulk over large distances.
• A system of tubular vessels that contains the transport medium and forms a branching network (Figure 2) to distribute it, along specific routes, to all parts of the organism.
• A mechanism for moving the transport medium within vessels. This requires a pressure difference between one part of the system and another. It is achieved in two main ways:
  – Animals use muscular contraction either of the body muscles or of a specialised pumping organ such as the heart (Topic 8.7).
  – Plants do not possess muscles and so rely on passive natural physical processes such as the evaporation of water or differences in solute concentrations.
• A mechanism to maintain the mass flow movement in one direction, e.g. valves.
• A means of controlling the flow of the transport medium to suit the changing needs of different parts of the organism.

Transport systems in plants

The range of materials transported by plants is more limited than that moved by animals. Plants, for example, have no need to transport respiratory gases in bulk because:

- Most gases they use or make are produced or required by leaves, which have a large surface area for the capture of light, and these gases can therefore diffuse directly in and out of them through stomata.
- Plants do not move from place to place and their energy requirements are therefore low, which means a reduced need for **ATP** and hence respiratory gases.
- In light, the oxygen that plants require for respiration can mostly be supplied from photosynthesis, while the carbon dioxide they produce during respiration is used up in photosynthesis.
- No photosynthetic or respiratory tissue is far from the surface, and so diffusion can meet the needs of cells. The furthest from the surface is the central part of the trunk of trees, and this is composed of dead, non-respiring tissue.

Plants need to transport water and mineral salts in the xylem, in one direction from roots to the rest of the plant. Evaporation and the diffusion of water vapour out through open stomata in the leaves creates a water potential gradient which draws water across the leaf from the xylem (Topic 7.6). The xylem forms an uninterrupted column of water and so it is moved up this column in much the same way as water can be moved up through a straw, i.e. as water leaves the xylem a tension is created and results in water being 'pulled' up the xylem (transpiration pull). Water moves across the root due to a **water potential gradient**. Phloem sieve tubes transport assimilates such as sucrose and amino acids from where they are manufactured or stored (sources) to regions where they are used or stored (sinks). These sinks include roots, buds, flowers, fruits and storage organs. The mechanism of movement of assimilates in phloem sap from source to sink relies on differences in hydrostatic pressure to create pressure gradients for mass flow.

Transport systems in animals

Animals have circulatory systems that move the transport medium, the blood, around the body. Smaller animals, such as insects, use an open blood system where blood flows freely over the cells and tissues. Large organisms have a closed blood system where blood remains inside blood vessels. A muscular pump called the heart circulates the blood around the body. A greater diversity of substances is transported in animals compared to plants. These include nutrients like glucose, amino acids, minerals and vitamins, respiratory gases like oxygen and carbon dioxide, as well as hormones and waste metabolites like urea. The blood system of animals also carries white blood cells that are involved in providing immunity and protecting against disease. The ability to clot, and so prevent leakage, which results from blood being under pressure, is another feature. Mammals and birds have a double circulatory system (Figure 3) in which the blood, having had its pressure reduced as it passes through the lung capillary network, is returned to the heart to boost its pressure before being circulated to the rest of the body. This assists the rapid delivery of material which is necessary as birds and mammals have a higher rate of metabolism, due to their constant body temperature.

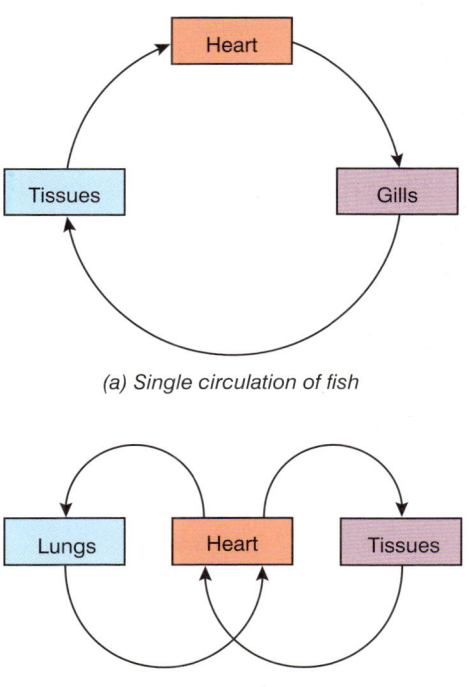

(a) Single circulation of fish

(b) Double circulation of other vertebrates

Figure 3 Single and double circulatory systems

Exchanges between organisms and their environment

On these pages you will learn to:

- Calculate surface areas and volumes of simple shapes (e.g. cubes) to illustrate the principle that surface area to volume ratios decrease with increasing size
- Describe the pathways and explain the mechanisms by which water and mineral ions are taken up by plants

For survival, organisms must transfer materials between themselves and their environment. Examples of things which need to be interchanged include:

- respiratory gases (oxygen and carbon dioxide)
- nutrients (glucose, fatty acids, amino acids, vitamins, minerals)
- excretory products (urea)
- heat.

Except for heat, this exchange can take place in two ways:

- passively (no external energy is required) – by diffusion and **osmosis**
- actively (energy is required) – by **active transport**, **pinocytosis** and **phagocytosis**.

Surface area to volume ratio

Exchange takes place at the surface of an organism, but the materials absorbed are used by the cells that mostly make up its volume. For exchange to be effective, the surface area of the organism must therefore be large compared with its volume.

Small organisms like **protoctists** have a surface area that is large enough, compared with their volume, to allow efficient exchange across their body surface. However, as organisms become larger, their volume increases at a faster rate than their surface area (Table 1), and so simple diffusion of substances across the surface can only meet the needs of the most inactive organisms. Even if the surface could supply enough substances, it would still take too long for it to reach the middle of the organism if diffusion alone was the method of transport. Organisms have evolved one or both of the following features:

- **a flattened shape** so that no cell is ever far from the surface (e.g. flatworm)
- **specialised exchange surfaces** with large areas to increase the surface area to volume ratio (e.g. lungs in mammals, gills in fish).

Features of specialised exchange surfaces

To allow effective transfer of materials across them by diffusion or active transport, exchange surfaces have the following characteristics:

- **large surface area to volume ratio** – to increase the rate of exchange
- **very thin** – so that the diffusion distance is short and therefore materials cross the exchange surface rapidly
- **partially permeable** – to allow selected materials to cross without obstruction
- **movement of the environmental medium, e.g. air** – to maintain a concentration gradient
- **movement of the internal medium, e.g. blood** – to maintain a concentration gradient.

Being thin, specialised exchange surfaces are easily damaged, and therefore are often located inside an organism for protection. Where an exchange surface is located inside the body, the organism needs to have a means of moving the external medium over the surface, e.g. a means of ventilating the lungs in a mammal.

Table 1 How the surface area to volume ratio gets smaller as an object becomes larger

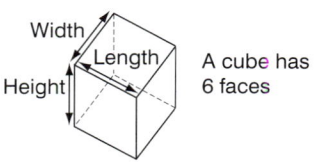

A cube has 6 faces

Length of edge of a cube/ cm	Surface area of whole cube (area of one face × 6)/cm²	Volume of cube (length × width × height)/ cm³	Ratio of surface area to volume (surface area ÷ volume)
1	1 × 6 = 6	1 × 1 × 1 = 1	$\frac{6}{1}$ = 6.0:1
2	4 × 6 = 24	2 × 2 × 2 = 8	$\frac{24}{8}$ = 3.0:1
3	9 × 6 = 54	3 × 3 × 3 = 27	$\frac{54}{27}$ = 2.0:1
4	16 × 6 = 96	4 × 4 × 4 = 64	$\frac{96}{64}$ = 1.5:1
5	25 × 6 = 150	5 × 5 × 5 = 125	$\frac{150}{125}$ = 1.2:1
6	36 × 6 = 216	6 × 6 × 6 = 216	$\frac{216}{216}$ = 1.0

Features of root hairs

Root hairs are an example of a specialised exchange surface in plants. They are responsible for the absorption of water and mineral salts. Plants constantly lose water by the process of transpiration (Topic 7.7). This loss can be up to $700\,dm^3$ a day in a large tree – all of which must be replaced by water that is absorbed through the root hairs.

Each root hair is a tiny extension of a root epidermal cell. These root hairs remain functional for a few weeks before dying back to be replaced by others nearer the growing tip. They are efficient surfaces for the exchange of water and mineral ions because:

- they provide a large surface area because they are very long extensions and occur in their thousands on each of the branches of a root
- they have a thin surface, the cell surface membrane, across which materials can move
- they are permeable – the epidermal cell is not covered by a waxy cuticle and the thin cellulose cell wall is no barrier to the movement of water and ions
- the cell surface membrane has specialised protein channels called **aquaporins** to allow water to pass across it more easily.

Water enters root hair cells by a passive process of osmosis (Topic 4.3). The soil solution is a very low concentration of mineral ions in water and surrounds the particles that make up soil. The soil solution therefore has a high **water potential**. Root hair cells, by comparison, have a relatively high concentration of ions, sugars and organic acids within their vacuoles and cytoplasm. The cells have a lower water potential. Because the root hairs are in direct contact with the soil solution, water moves by osmosis from the higher (less negative) water potential of the soil solution to the lower (more negative) water potential within the root hair cells. Figure 1 shows this process. From the root hair cell, the water moves further into the root due to the water potential gradient that exists between it and the xylem in the centre of the root (Topic 7.5).

The absorption of mineral ions by root hairs is an altogether different situation. The concentration of ions inside the root hair cell is normally greater than that in the soil solution. The uptake of mineral ions is therefore against the concentration gradient and, as a result, requires active transport. This is achieved using special **carrier proteins** that use ATP to provide energy to transport particular ions from the soil solution, where they are in low concentrations, to the root hair cytoplasm and vacuole, where they are in higher concentrations. The details of this process are given in Topic 4.5. On the rare occasions when a particular ion is in a greater concentration in the soil than in the root hair cell, the ion simply moves into the cell by the passive process of facilitated diffusion.

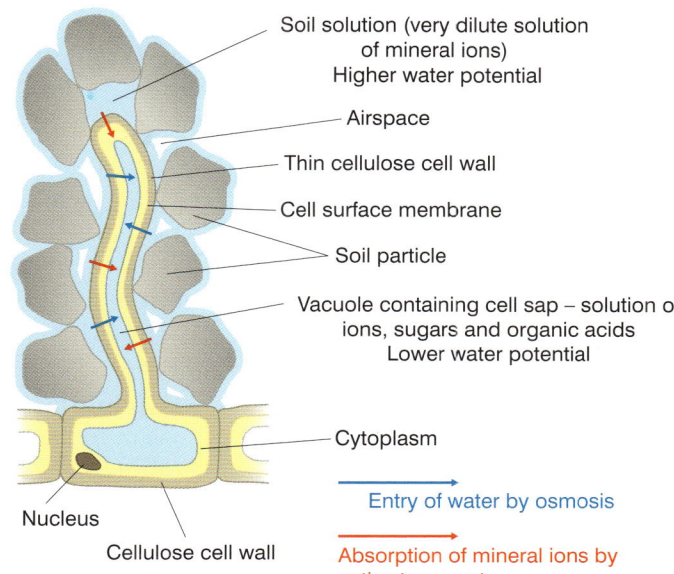

Soil solution (very dilute solution of mineral ions) Higher water potential

Airspace

Thin cellulose cell wall

Cell surface membrane

Soil particle

Vacuole containing cell sap – solution of ions, sugars and organic acids Lower water potential

Cytoplasm

Nucleus

Cellulose cell wall

→ Entry of water by osmosis

→ Absorption of mineral ions by active transport

Figure 1 *Absorption of water and mineral ions by a root hair cell*

SUMMARY TEST 7.2

The surface area to volume ratio of a cube with an edge length of 10 mm is **(1)**. If the edge length is doubled to 20 mm then the surface area to volume ratio is **(2)**. Root hair cells are efficient at absorbing water and mineral salts for a number of reasons including having a thin **(3)** which is not covered by a **(4)**. They also have specialised protein channels called **(5)** to allow water to pass into them more easily. In root hairs, water is absorbed by the process of **(6)** because the water potential of the soil solution is **(7)** than within the root hair. Mineral ions are absorbed by the process of **(8)**.

Distribution of vascular tissues in dicotyledonous plants

On these pages you will learn to:

- Draw and label from prepared slides plan diagrams of transverse sections of stems, roots and leaves of herbaceous dicotyledonous plants
- Label plan diagrams of transverse sections of stems, roots and leaves of herbaceous dicotyledonous plants

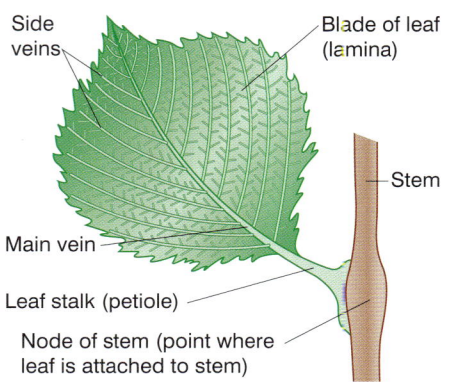

Figure 1 *Leaf of a dicotyledonous plant, e.g. elm, showing arrangement of veins*

Side veins
Blade of leaf (lamina)
Stem
Main vein
Leaf stalk (petiole)
Node of stem (point where leaf is attached to stem)

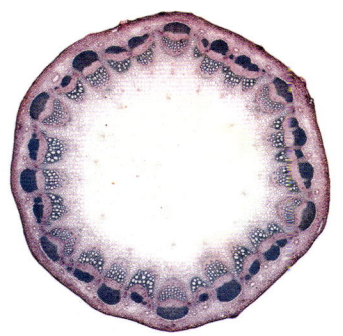

Figure 2 *Transverse section (TS) dicotyledonous stem*

Figure 3 *TS dicotyledonous root*

A flowering plant can be thought of as having two main functional areas: the leaves, which manufacture sugars by photosynthesis at one end, and the roots, which absorb water and minerals at the opposite end. Each relies on the other – the leaves needing water and minerals to photosynthesise, and the roots requiring sugar to respire and keep alive. Equally important, therefore, is the transport system between the two, the specialised **vascular tissue**, of which there are two types:

- **xylem** – for the transport of water and mineral ions from the roots, up the plant to the aerial parts
- **phloem** – for the transport of sugars produced by leaves to other parts of the plant.

The two tissues occur together throughout the plant, sometimes with associated tissues, such as **sclerenchyma** fibres, to form distinct areas, known as **vascular bundles**.

Distribution of vascular tissues in a leaf

The vascular tissues in a **dicotyledonous** leaf form a network of tiny vascular bundles throughout the blade, or **lamina**, of the leaf. These vascular bundles form a series of **side veins** that run parallel with one another. These side veins then merge into a central **main vein**. The main vein runs along the centre of the leaf, increasing in diameter towards the petiole, or leaf stalk. Within each vein, or vascular bundle, there is an area of xylem towards the upper surface of the leaf and an area of phloem towards the lower surface. This arrangement is shown in the section through the leaf shown in Figure 4(a).

Distribution of vascular tissue in a stem

The xylem and phloem in a dicotyledonous stem form vascular bundles that are arranged towards the outside of the stem. The reason for this is that the vascular bundles, together with associated sclerenchyma fibres, provide support in herbaceous stems as well as transport material. The main forces acting on stems are lateral ones caused by the action of wind on them. Such forces are best resisted by an outer cylinder of supporting tissue. Hence the vascular bundles form a non-continuous ring towards the edge of them (Figures 2 and 4(b)). Being non-continuous, this ring of supporting tissues allows the stem to be flexible and to bend in the wind. Within the vascular bundles, the xylem is to the inside of the stem and the phloem towards the outside. Between the two is a thin layer of dividing cells called **cambium**, which gives rise to both xylem and phloem.

Distribution of vascular tissue in a root

The vascular tissue in the root of a dicotyledonous plant (Figure 3) is situated centrally rather than towards the outer edge, as in a stem. This is because roots are subject only to pulling forces. Vertical forces are better resisted by a central column of supporting tissues, such as xylem, rather than an outer cylinder of tissue. The xylem is typically arranged in a single star-shaped block of tissue at the centre of the root, with the phloem situated in separate groups between each of the points of the star-shaped xylem. Around both is the **pericycle** and **endodermis**, more details of which are given in Topic 7.5.

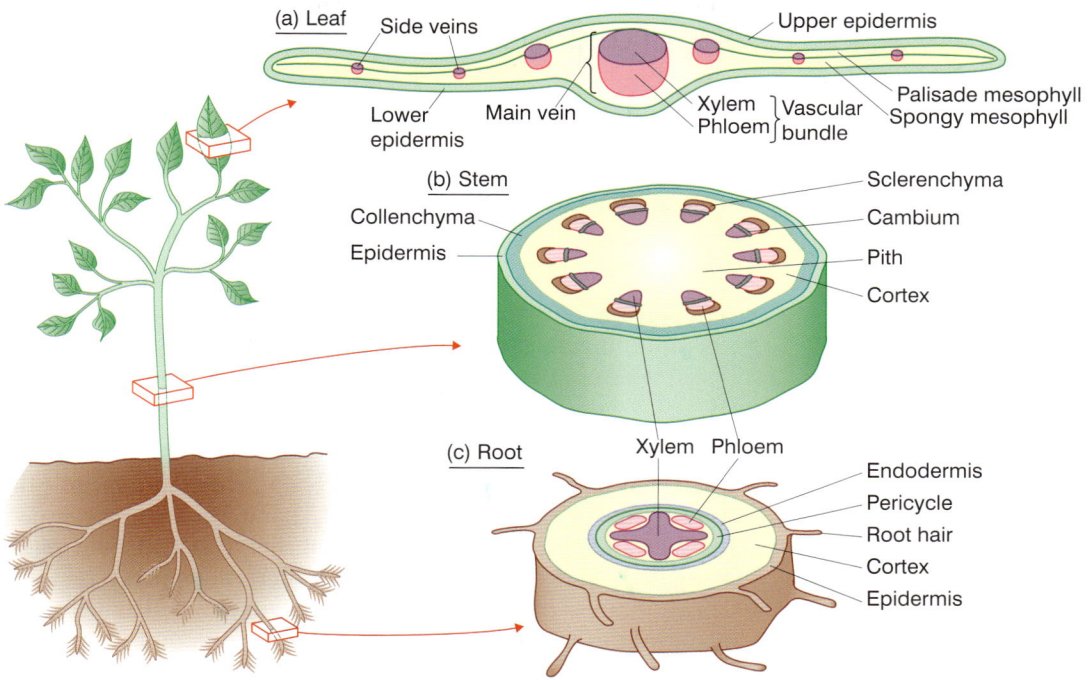

Figure 4 *Distribution of vascular tissues in dicotyledonous plants*

In plants, leaves, stems and roots are organs. A leaf is an organ made up of the following tissues:

- **palisade mesophyll** – which carries out photosynthesis
- **spongy mesophyll** – provides a large surface area for gas exchange and has large air spaces between cells for efficient diffusion of gases
- **epidermis** – to protect the leaf and allow diffusion of gases
- **phloem** – to transport organic materials away from the leaf
- **xylem** – to transport water and ions into the leaf.

You can use an eyepiece graticule to help you to draw each of the tissues in a low power plan in proportion.

To make a high power drawing, draw a few representative cells in sequence. Always draw the cell wall in plant cells and show the nucleus, drawn to shape but without shading it in.

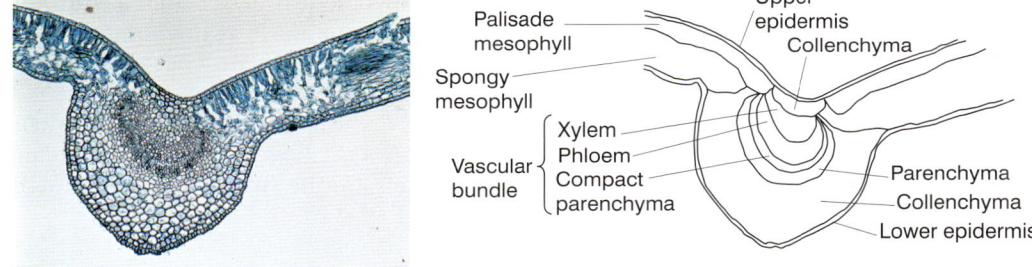

Figure 5 *Low power plan diagram (right) of the tissues shown in the photograph (left) of a TS through the midrib of a dicotyledonous leaf*

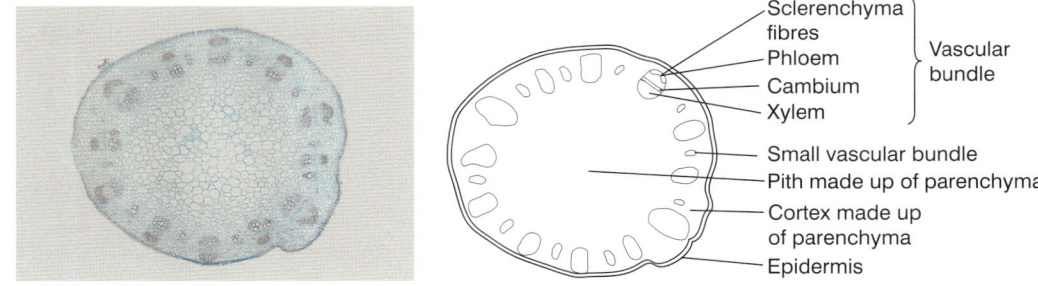

Figure 6 *Low power plan diagram (right) of the tissues shown in photograph (left) of a TS through a dicotyledonous stem*

Figure 7 *Low power plan diagram (right) of the tissues shown in photograph (left) of a TS through a dicotyledonous root*

Structure and function of xylem

Xylem is the main water-conducting tissue in vascular plants. It also provides support for plants.

Structure of xylem

Xylem performs the functions of both supporting the plant and transporting water and minerals. **Sclerenchyma fibres** in the xylem all contribute to support, whereas the **vessels** and **tracheids** have both support and transport roles.

- **Xylem fibres** (Figure 4) are elongated sclerenchyma cells with walls that are thickened with **lignin**; these features suit them to their role of support.
- **Xylem vessel elements** (Figures 1, 2 and 5) vary in structure, depending on the type and amount of thickening of their cell walls, but are all hollow and elongated. As they mature their walls become impregnated with lignin, which causes them to die. The end walls break down, which allows the cells to form a continuous tube. (The word 'element' is sometimes used rather than 'cell' because a cell is a living structure, whereas mature xylem vessels are dead). Sometimes the lignin forms rings (annular thickening) around the vessel; in other cases it forms a spiral or a network (reticulate thickening), see Figure 3. This arrangement is better than a continuous thickening, because it allows elongation of the vessels as the plant grows. There are areas of the lignified wall where lignin is absent. These non-lignified regions are called **pits** (Figure 5). They are not completely open as there is still a cellulose cell wall across them – it is just that the wall is not lignified at these points. Pits allow for lateral (sideways) movement of water. In angiosperms (flowering plants), xylem vessels are the structures through which the vast majority of water is transported.

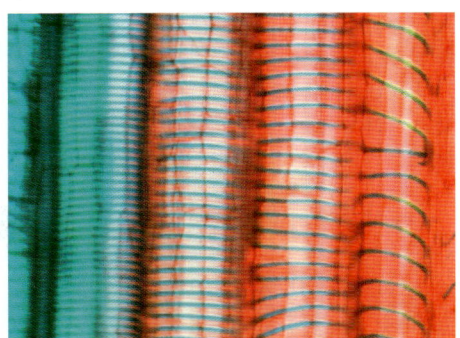

Figure 1 *Xylem in LS as seen using a light microscope*

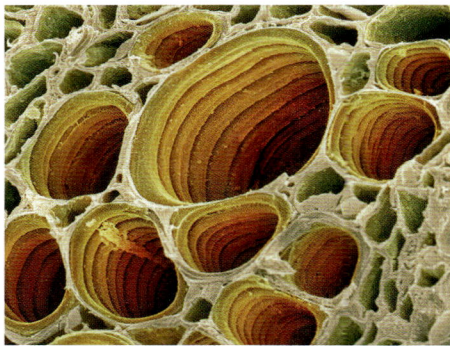

Figure 2 *Leaf of tobacco showing xylem vessels (SEM) (×500 approx.)*

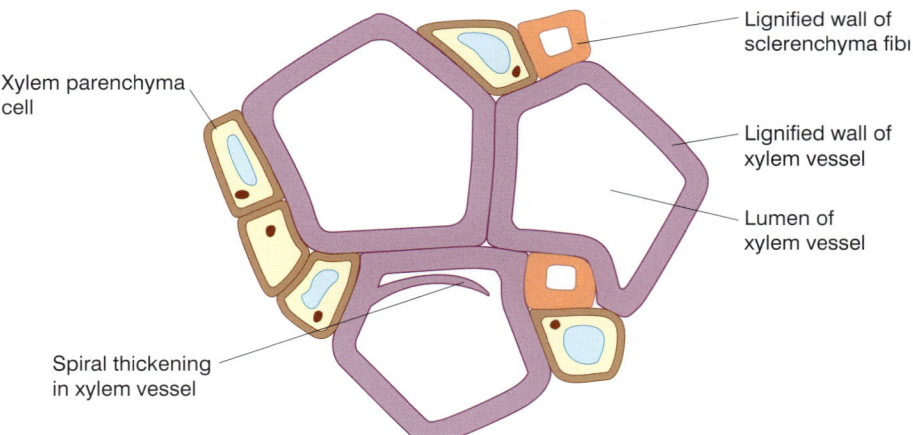

Lignified wall of sclerenchyma fibre

Xylem parenchyma cell

Lignified wall of xylem vessel

Lumen of xylem vessel

Spiral thickening in xylem vessel

Figure 4 *Xylem in TS as seen using a light microscope*

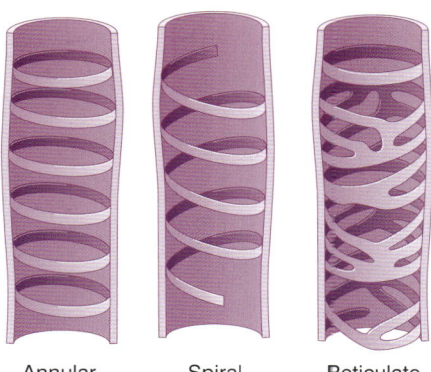

Annular   Spiral   Reticulate

Figure 3 *Types of thickening in xylem vessels*

How xylem's structure is related to its functions

The structure of xylem vessels is well suited to its function in the transport of water and mineral ions:

* The cells are long and arranged end to end to form a continuous tube.
* The cell dies when mature, which means that:
  – there is no cytoplasm or nucleus to hinder water flow
  – the end walls can break down, so that a larger volume of water per unit time can be transported as there is an empty lumen.
* Cell walls are thickened with lignin, which
  – makes them more rigid and therefore less likely to collapse under the tension created by the transpiration pull (Topic 7.6)
  – is waterproof to prevent water escaping.
* Annular, reticulate and spiral thickening allow xylem vessels to elongate during growth, and make them more flexible, so that branches can bend in the wind.
* There are pits throughout the cells, to supply surrounding cells with water.
* Cellulose walls increase the **adhesion** of water molecules, which helps to resist the effects of gravity and keep the column of water moving up (Topic 7.6).

EXTENSION
Other xylem tissues

Vessels and sclerenchyma are not the only xylem tissues found in plants. Other tissues are:

* **Xylem parenchyma** is composed of unspecialised cells that act as packing tissue around the other components of the xylem. They are roughly spherical in shape, but when they are turgid they press upon and flatten each other in places. In this way they provide support.

* **Tracheids** have a similar structure to vessels, except they are longer and thinner, and have tapering ends. They, too, are thickened with lignin and therefore die when mature. As with vessels, the end walls break down, and their side walls possess pits which allow lateral movement of water between adjacent cells. Tracheids are found in all plants and are the main water conducting tissue in ferns and conifers.

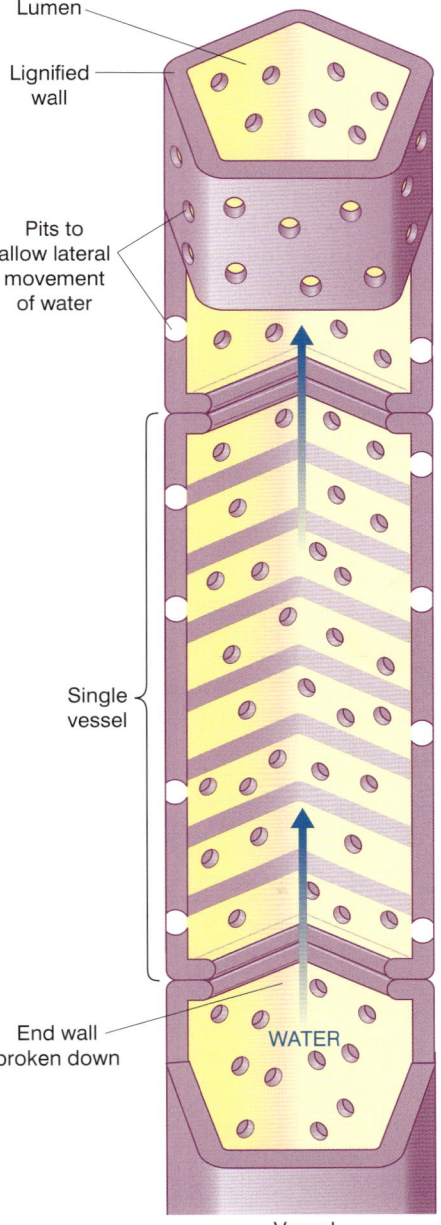

Figure 5 *The structure of a xylem vessel*

SUMMARY TEST 7.4

Xylem transports water within the plant in elongated structures called **(1)**. The walls of both are thickened with a substance called **(2)**, which forms three different patterns of thickening, known as **(3)**, **(4)** and **(5)**. Water moves through the central cavity, called the **(6)**, of the xylem elements but can move sideways between adjacent elements through structures called **(7)**. Also found within the xylem is supporting tissue, called **(8)**.

On these pages you will learn to:

- Explain the movement of water between plant cells, and between them and their environment, in terms of water potential
- Describe the pathways and explain the mechanisms by which water and mineral ions are transported from soil to xylem, including reference to the symplastic pathway and apoplastic pathway and Casparian strip

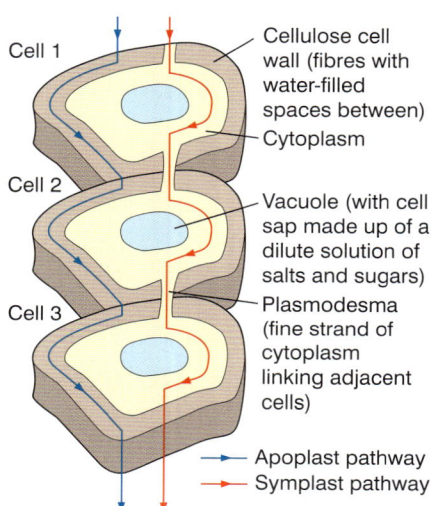

Cell 1

Cellulose cell wall (fibres with water-filled spaces between)

Cytoplasm

Cell 2

Vacuole (with cell sap made up of a dilute solution of salts and sugars)

Cell 3

Plasmodesma (fine strand of cytoplasm linking adjacent cells)

→ Apoplast pathway
→ Symplast pathway

Figure 1 *Apoplast and symplast pathway across the root cortex*

EXTENSION
Aquaporins

As water passes across the cortex of the root, some water molecules will enter the cells from the apoplast pathway and vice versa. Specialised protein channels called aquaporins span the cell surface membranes. These water channels allow water molecules to pass from cell to cell. This can help to maximise the volume of water that enters the xylem. They are also present in the tonoplast of the vacuole. It is known that cells are able to increase and decrease the number of aquaporins in their cell surface membranes so that this might be one way to control the movement of water across the root.

In Topic 7.2 we saw how root hairs are adapted to their role of absorbing water and mineral **ions**. These root hairs take up water, which then passes across the root cortex into the water-conducting tissue of the plant, the xylem. This, in turn, carries it up the stem to the aerial parts of the plant. The general structure of roots and stems is covered in Topic 7.3.

Uptake of water by root hairs

Root hairs arise from **epidermal** cells a little way behind the tips of young roots. These hairs grow into the spaces around soil particles. In damp conditions they are surrounded by a soil solution which contains small quantities of mineral ions, but which is mostly water and therefore has a very high **water potential** – only slightly less than zero. By contrast, the root hairs, and other cells of the root, have sugars, amino acids and mineral ions dissolved within them. These cells therefore have a much lower water potential. As a result, water moves into the epidermal cell by osmosis or moves into the cell wall of the epidermal cell down the water potential gradient.

Water then continues its journey across the root cortex in two ways, as shown in Figures 1 and 2:

- the apoplast pathway
- the symplast pathway

The apoplast pathway

As water is drawn into endodermal cells, it pulls more water along behind it, due to the cohesive properties of the water molecules. This creates a tension that draws water along the cell walls of the cells of the root cortex (cortical cells). The mesh-like structure of the cellulose cell walls of the cortical cells has many water-filled spaces and so there is little or no resistance to this pull of water along the cell walls. The apoplast route can be described as a non-lving pathway as water does not enter the cytoplasm of cells.

The symplast pathway

This takes place across the cytoplasm of the cortical cells as a result of differences in the water potential between cells. Water moves from one cell to the next cell through plasmodesmata, down the water potential gradient. Plasmodesmata are strands of cytoplasm that directly connect one cell to the adjacent cell so that water can take a cytoplasmic route across the cortex. There is therefore, in effect, a continuous column of cytoplasm extending from the root hair cell to the xylem at the centre of the root. Water moves along this column as follows:

- Water entering the root hair cell by osmosis (Topic 4.3) makes its water potential higher.
- The root hair cell now has a higher water potential than the adjacent cortical cell.
- Water therefore moves from the root hair cell to the cortical cells via the plasmodesmata, down the water potential gradient.
- This first cortical cell now has a higher water potential than the cell nearer to the centre of the root.
- Water therefore moves into this neighbouring cortical cell via plasmodesmata along the water potential gradient.

- This second cortical cell now has a higher water potential than the next cell nearer to the centre of the root, and so water moves from the second to the third cell via plasmodesmata down the water potential gradient.
- At the same time, this loss of water from the first cortical cell lowers its water potential, causing more water to enter it from the root hair cell.
- In this way, a water potential gradient is set up across all the cells of the cortex, which carries water along the cytoplasm from the root hair cell to the endodermis, surrounding the central xylem.
- In its passage through a cell, water can also enter and exit the vacuole, but as it has to cross the tonoplast, the vacuolar membrane, this will involve osmosis.

Passage of water into the xylem

Water reaching the endodermis by the apoplast pathway finds its further progress along the cell wall prevented by the waterproof band of **suberin** that makes up the Casparian strip in endodermal cells. At this point, water must enter by osmosis the living protoplast of the cell, where it joins water that has reached there by the symplast pathway. Water leaving the endodermis to enter the xylem now takes the apoplastic, non-living pathway.

Movement of mineral ions across the root

Mineral ions are actively taken up by the root hair cell. They enter via specific carrier membrane proteins and can then travel across the cortex dissolved in the water that is taking a symplast pathway. Dissolved ions in the apoplast pathway, on reaching the endodermis, must also enter the endodermal cell to continue across the pericycle and enter the xylem.

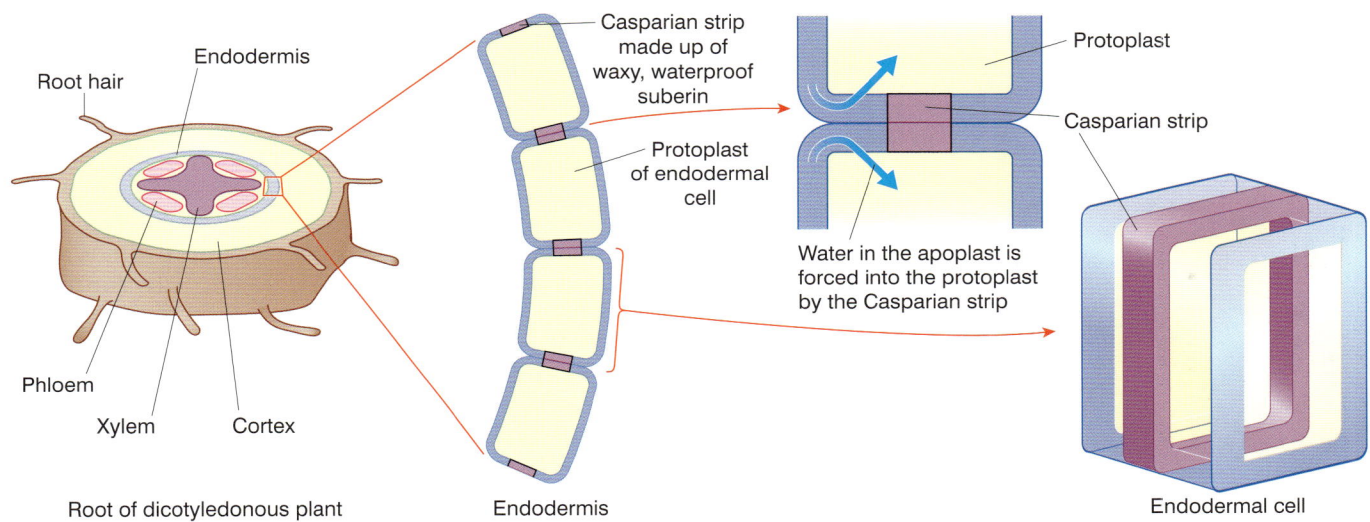

Figure 2 *Movement of water across the endodermis*

SUMMARY TEST 7.5

Water is absorbed from the soil solution by tiny extensions of root epidermal cells called (1). This water is taken up because these cells have a (2) water potential than the soil solution that surrounds them. Water now passes into the cortical cells of the root in two ways: through the water-filled spaces of the cellulose cell wall, a route called the (3) pathway; and through the cytoplasm alone, a route called the (4) pathway, in which water crosses the cell walls through strands of cytoplasm called (5). Around the vascular tissue at the centre of the root is a one-cell-thick ring of cells, called the (6), the cell walls of which are impregnated with a band called the (7), which is made up of the waxy, waterproof substance called (8). Water from all pathways is now forced through the living portion, or (9), of the endodermal cells. The entry of (10) ions into the xylem (11) the water potential. Water enters the xylem down the water potential gradient. This helps to create a force known as (12).

Movement of water up the stem and across the leaf

On these pages you will learn to:

- Explain how hydrogen bonding of water molecules is involved with movement in the xylem by cohesion–tension in transpiration pull and adhesion to cellulose cell walls
- Describe the pathways and explain the mechanisms by which water is transported from roots to leaves

The main force that pulls water up the stem of a plant is the evaporation of water vapour from surfaces of the spongy mesophyll cells of leaves into the air spaces, from where the water vapour diffuses out to the atmosphere through open stomata. The loss of water from the aerial parts of a plant (mostly the leaves) is termed transpiration. It is therefore logical to begin this topic from the point where water molecules evaporate from spongy mesophyll cells and diffuse through the tiny pores, called **stomata**, on the surface of a leaf.

Movement of water across the leaf

The humidity of the atmosphere is usually less than that of the sub-stomatal air space and so, provided that the stomata are open, water vapour diffuses out of the air spaces into the surrounding air, which has a lower water potential. Water vapour lost from the air spaces is replaced by water vapour evaporating from the cell walls of the surrounding **spongy mesophyll** cells. By closing stomatal pores, plants can control water loss. Water vapour evaporating from the cell walls of spongy mesophyll cells is replaced by water reaching them from the xylem by either the apoplast or symplast pathways (Topic 7.5). In the case of the symplast pathway, the water movement occurs because, once the spongy mesophyll cells have lost water to the sub-stomatal air-space, they have a lower (more negative) **water potential**. Water therefore enters from the adjacent cells. The loss of water from these adjacent cells causes them to have a lower (more negative) water potential and so they, in turn, take in water from their neighbours. In this way, a water potential gradient is established that pulls water from the xylem, across the leaf mesophyll, and finally out into the atmosphere. These events are summarised in Figures 1 and 2.

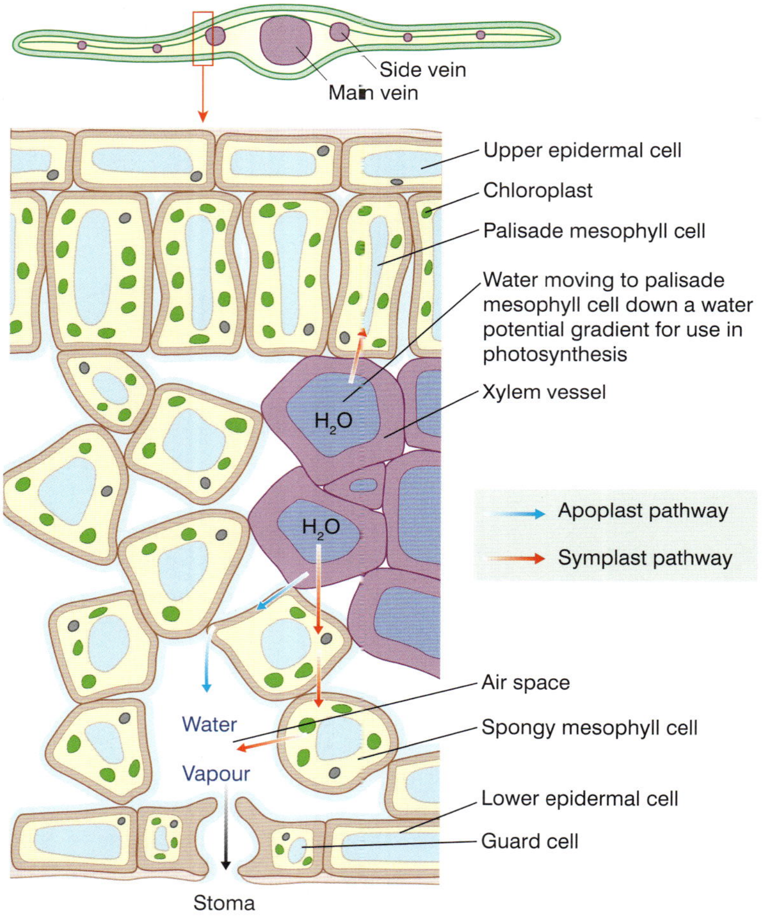

Movement of water up the stem in the xylem

The main mechanism by which water moves up the xylem is known as the **cohesion-tension theory**. It operates as follows:

- Water vapour evaporates from spongy mesophyll cells due to heat from the sun, leading to transpiration.
- Water molecules form **hydrogen bonds** between one another and hence tend to stick together – this is known as **cohesion.**
- Water forms a continuous, unbroken column across the mesophyll cells and down the xylem.
- As water vapour evaporates from mesophyll cells in the leaf into the sub-stomatal air space, more molecules of water are drawn up behind it as a result of this cohesion.
- A column of water is hence pulled up the xylem as a result of transpiration. This is called the **transpiration pull**.
- The transpiration pull puts the xylem under **tension**, i.e. there is a negative pressure within the xylem – hence the name cohesion-tension theory (Figure 3).

Figure 1 *Movement of water across a leaf*

Labels: Side vein; Main vein; Upper epidermal cell; Chloroplast; Palisade mesophyll cell; Water moving to palisade mesophyll cell down a water potential gradient for use in photosynthesis; Xylem vessel; H₂O; H₂O; Apoplast pathway; Symplast pathway; Air space; Water; Spongy mesophyll cell; Vapour; Lower epidermal cell; Guard cell; Stoma

REMEMBER

Cohesion is the attraction of the **same** type of molecules for one another. In other words, it is the ability of molecules (in this case water molecules) to stick to one another.

Evidence for the cohesion-tension theory includes:

- The change which occurs in the diameter of trees according to the rate of transpiration. During the day, when transpiration is at its greatest, there is more tension (more negative pressure) in the xylem. This causes the trunk to shrink in diameter. At night, when transpiration is at its lowest, there is less tension in the xylem and so the diameter of the trunk increases.
- When a xylem vessel is broken, water does not leak out which would be the case if it were under pressure, but rather air is pulled in, which is consistent with it being under tension.

The cellulose lining of the xylem vessels is hydrophilic. Adhesion of water molecules to the cellulose lining and to the hydrophilic groups in lignin in the xylem wall also help to keep the column of water from collapsing and moving up the xylem by the cohesion-tension mechanism. **Root pressure** (Topic 7.5) makes some contribution to the movement of water up the xylem, especially in small **herbaceous** plants.

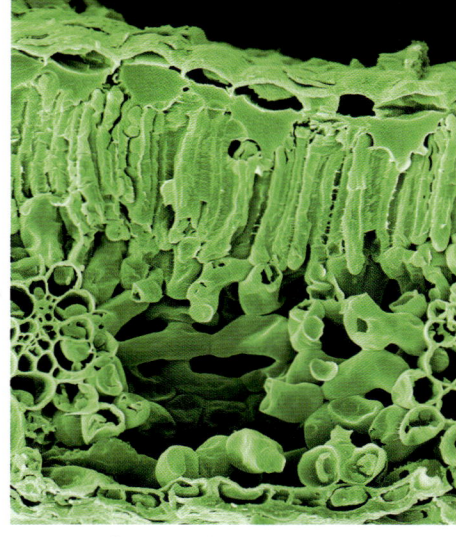

Figure 2 *Colourised scanning electron micrograph of a leaf showing palisade mesophyll above and spongy mesophyll below*

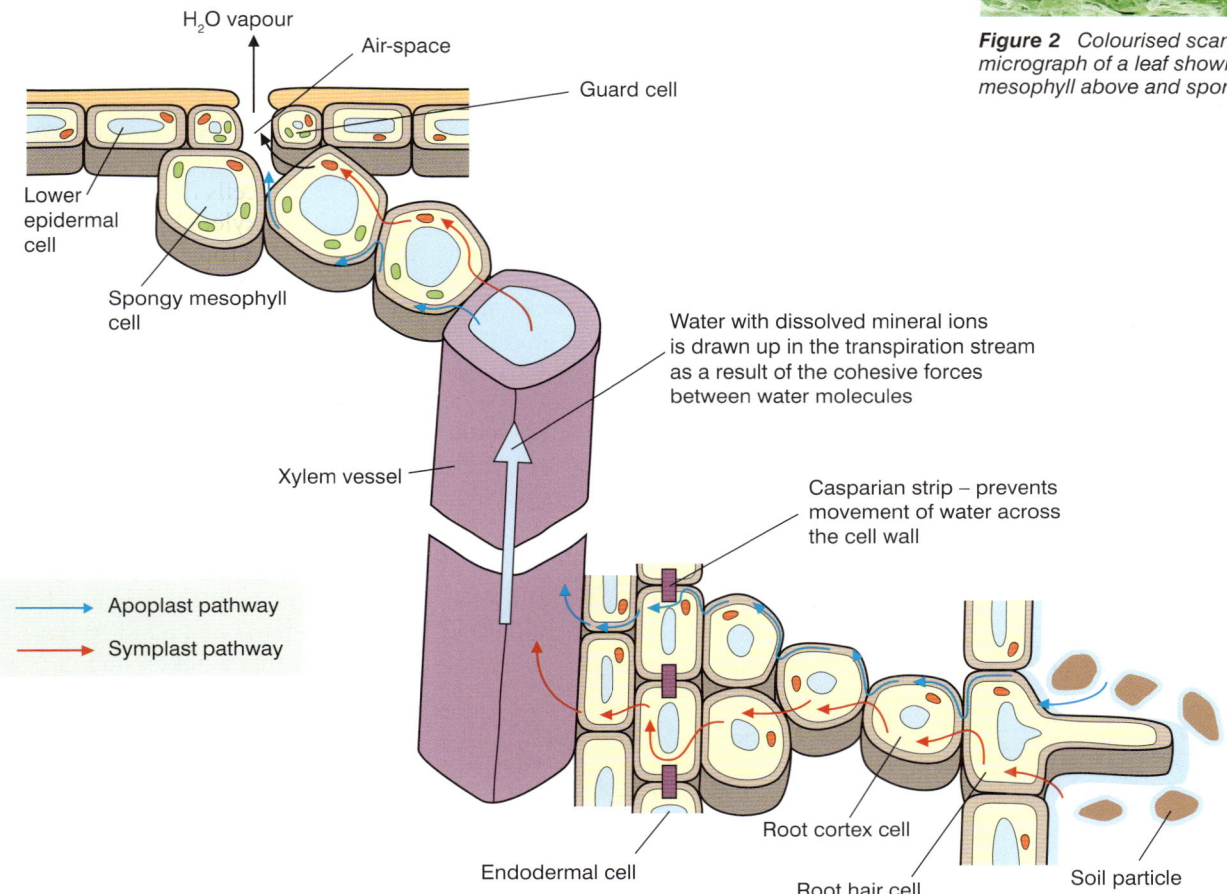

Figure 3 *Summary of water transport through a plant*

SUMMARY TEST 7.6

Water vapour **(1)** from the surfaces of spongy mesophyll in the plant leaf. This water vapour then diffuses through pores called **(2)** in the epidermis of the leaf, each of which is surrounded by a pair of **(3)**. Water vapour evaporates into the air-spaces from spongy mesophyll cells. As a result these cells have a **(4)** water potential and so draw water from adjacent cells. In this way, a **(5)** gradient is set up that draws water from the xylem. Water is pulled up the xylem because water molecules stick together – a phenomenon called **(6)**. The attraction of water molecules to the **(7)** lining of the xylem vessel is known as **(8)**.

7.7 Transpiration

On these pages you will learn to:

- Define the term 'transpiration' and explain that it is an inevitable consequence of gas exchange in plants
- Investigate experimentally and explain the factors that affect transpiration rate using simple potometers

Transpiration is the loss of water vapour from the aerial parts of a plant, mainly the leaves. It is a process that occurs as a result of the evaporation of water vapour into the air spaces from the cell walls of spongy mesophyll cells. This water vapour diffuses out to the atmosphere through the open stomata.

- **Stomata** occur in leaves and **herbaceous** stems, and account for 90% of water loss.
- **Cuticle** is a waxy external layer on plant surfaces that limits water loss from the cell walls of the epidermal cells, although up to 10% of water nevertheless escapes by this route.

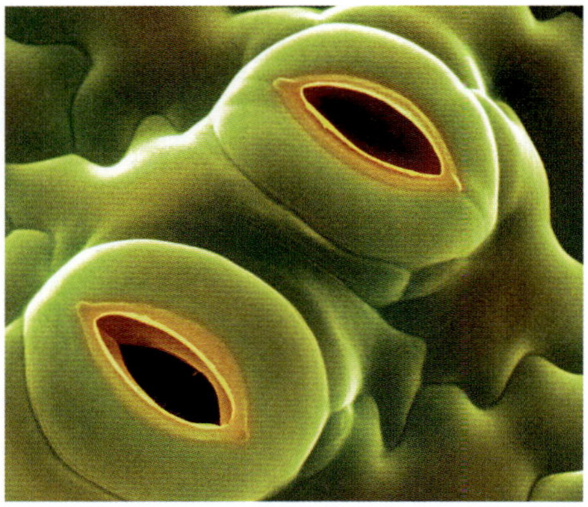

Figure 1 Colourised scanning electron micrograph of open stomata on a tobacco leaf

Transpiration and gas exchange

Although transpiration is universal in flowering plants, it is the unavoidable result of plants having leaves adapted for photosynthesis. Leaves have a large surface area and also have stomata that open during the day to allow the carbon dioxide required for photosynthesis to diffuse into the leaf. Both features, the large surface area and the open stomata, result in an immense loss of water – up to 700 dm^3 a day in a large tree.

Benefits of transpiration

Water is brought up the plant by the transpiration pull. It is needed for hydrolysis reactions and as a reactant in photosynthesis. It is used to maintain the turgor of leaf cells to support the leaf. The cells also receive mineral ions, dissolved in the water. Organic molecules, such as sucrose and amino acids, plant hormones and mineral ions need to be dissolved in water in phloem sap to be transported to sinks. Without transpiration, water would not be so plentiful and the transport of materials would not be as rapid. The evaporation of water from the surfaces of the mesophyll cells has a cooling effect.

Under natural conditions, the transpiration rate of both species increases in the middle of the day because the higher temperature speeds up the outward diffusion of water vapour and with the increased light intensity all stomata will be fully open. The species living in a dry environment is better adapted at preventing water loss.

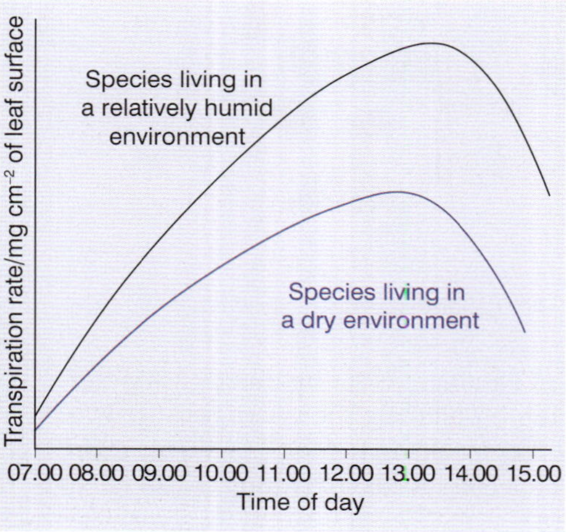

Figure 2 Transpiration rates of two species of plants

Measurement of water uptake under different conditions

It is very difficult to measure transpiration as it is extremely difficult to condense and collect all the water vapour that is lost from all the parts of a plant. What we can easily measure, however, is the volume of water that is taken up in a

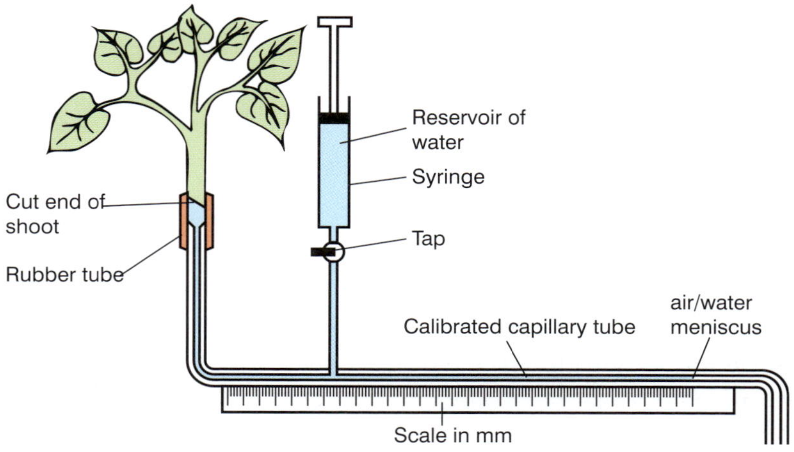

Figure 3 A potometer – an instrument that measures water uptake which is more or less equivalent to water loss

given time by a part of the plant such as a leafy shoot. As around 99% of the water taken up by a plant is lost during transpiration, this measure is almost the same as the rate at which transpiration is occurring. We can then measure water uptake by the same shoot under different conditions, e.g. various humidities, air movements or temperatures. In this way we get a reasonably accurate measure of the effects of these conditions on the rate of transpiration.

The rate of water loss in a plant can be measured using a **potometer** (Figure 3). The experiment is carried out in the following stages:

- A leafy shoot is cut under water to prevent air entering the xylem. Care is taken not to get water on the leaves.
- The potometer is filled completely with water, making sure there are no air bubbles.
- Using a rubber tube the leafy shoot is fitted to the potometer under water.
- The potometer is removed from under the water and all joints are sealed with waterproof jelly.
- The distance moved by the air / water meniscus in a given time is measured a number of times and the mean is calculated.
- The volume of water lost is calculated ($\pi r^2 h$) and plotted against the time in minutes.
- Once the air / water meniscus nears the junction of the reservoir tube and the capillary tube, the tap on the reservoir is opened and the syringe is pushed down until the meniscus is pushed back to the start of the scale on the capillary tube. Measurements then continue as before.

The experiment can be repeated to compare the rates of water loss under different conditions, e.g. at different temperatures, humidities, light intensity, or differences in water loss between different species under the same conditions (Figure 2). To obtain reliable results, we need to take the following precautions when carrying out the experiment.

- The leafy shoot is cut under water to prevent air-bubbles being drawn into the xylem (which is under tension) and blocking the flow of water through the shoot.
- All joints must be watertight to prevent air breaking the continuous column of water and reducing the rate or stopping the movement of water. Tight-fitting rubber tubing and / or waterproof jelly is therefore used at all joints.
- When making comparisons or when repeating experiments, e.g. the rate of transpiration at different temperatures, all other factors, e.g. humidity, air movement, etc. must be kept constant.

Factors affecting transpiration

Both internal and external factors influence the rate of transpiration (Table 1). Plants living in conditions that lead to high transpiration rates but where the supply of water is limited often have adaptations which allow them to reduce water loss. Such plants are called **xerophytes**; their features are considered in Topic 7.8.

Table 1 Summary of factors affecting transpiration

Type	Factor	How factor affects transpiration
External	Light	Stomata open in the light and close in the dark
	Humidity	Decreasing humidity (that is, drier external atmosphere) increases the steepness of the water potential gradient between the air spaces in the leaf and in the atmosphere
	Temperature	Increasing the temperature increases the kinetic energy of the water molecules, increasing the rate of evaporation (from surfaces of mesophyll cells) and diffusion out of water vapour
	Air movement	A greater air movement will remove moist air from around the leaf more quickly and will lead to a steeper water potential gradient
Internal	Leaf area	Some water vapour is lost over the whole surface of the leaf so a smaller leaf area leads to less water vapour being lost than a large leaf area
	Cuticle	Forms a waterproofing layer over the leaf surface. A thick cuticle leads to less water loss than a thin one
	Number of stomata	Most water vapour is lost by diffusion through stomata
	Distribution of stomata	Upper surface is more exposed to environmental factors that increase the rate of transpiration. Having stomata almost entirely on the lower surface of the leaf leads to less water vapour loss than having a greater proportion of stomata on the upper surface

The Practical skills section on the accompanying CD covers the factors that affect transpiration rate using leaf impressions, epidermal peels, and grids for determining surface area.

SUMMARY TEST 7.7

Transpiration is the **(1)** of water vapour from the ariel parts of the plant, especially the leaves. Around 90% of water loss occurs through **(2)**, with the remainder occurring mostly through the waxy **(3)** on plant surfaces. The volume of water taken up by a plant can be measured using a piece of apparatus called a **(4)**. To produce reliable results it is essential that all seals are made **(5)**. Various external and internal conditions affect the rate of transpiration. In each of the following cases state whether the rate of transpiration is increased or decreased: reduced temperature **(6)**, higher humidity **(7)**, thicker cuticle **(8)**, more stomata **(9)**, lower light intensity **(10)**, greater air movement **(11)**.

Xerophytes

Figure 1 *TS through a leaf of* Nerium oleander *showing pits containing hairs on the underside. The stomata are located in these pits. Both pits and hairs help to trap water vapour and reduce transpiration*

Figure 2 Agave *has leaves with a thick waxy cuticle to reduce water loss*

Figure 3 *Conifers such as this have needle-like leaves to reduce the surface area to volume ratio and reduce water loss*

Xerophytes (xero = 'dry', phyte = 'plant') are plants that are adapted to living in areas where their water losses due to **transpiration** may be higher than their water uptake. Without these adaptations these plants would become desiccated (dry out) and die.

Xerophytic plants

Xerophytes are typically thought of as desert plants, showing a whole range of adaptations to cope with hot, dry conditions. However, similar adaptations may also be seen in plants found in sand dunes or other dry, windy places where rainfall is high and temperature relatively low. These adaptations are essential because the rainfall quickly drains away through the sand and out of the reach of the roots, making it difficult for these plants to obtain water. At the same time, coastal areas where sand dunes typically occur have salty soils and this lowers the **water potential** of the soil solution, reducing the water potential gradient between the soil solution and the root hair cells. Water uptake by **osmosis** is therefore very slow. Plants living on salt marshes may have their roots drenched in water but its saltiness means that this water is hard to absorb and so only xerophytic plants survive in these conditions. In addition, coastal regions are exposed to greater wind speeds, which increase transpiration rates. Plants living in cold regions often have difficulty obtaining water because it is frozen in the soil for much of the year. These plants also show xerophytic modifications to enable them to survive. There are therefore many habitats where plants show structural and physiological modifications designed to increase water uptake, store water and reduce transpiration. These modifications are called xerophytic features.

Xerophytic adaptations of leaves that reduce transpiration

One way of surviving in habitats with an unfavourable water balance is to reduce the rate at which water vapour can be lost through transpiration. As the vast majority of transpiration occurs through the leaves, it is these organs that show most modifications. Examples include:

- **Having a thick cuticle.** Although the waxy cuticle on leaves forms a waterproofing barrier, up to 10% of water loss can still occur by this route. The thicker the cuticle is, the less water vapour can escape by this means. Many plants, such as *Agave sp*, have thick cuticles to reduce water vapour loss.
- **Curling up of leaves.** Most leaves have their **stomata** largely, or entirely, located on the lower epidermis. The curling of leaves in a way that protects this lower epidermis from the outside helps to trap a region of still air within the curled leaf. This region becomes saturated with water vapour and so there is only a very slight water potential gradient between the sub-stomatal air space and the outside, and so transpiration is considerably reduced. Esparto grass is an example of a plant with curled leaves. Plants such as *Ammophila* (marram grass) roll their leaves when transpiration rates are high, e.g. in hot or windy conditions.
- **Having trichomes (hairs) on leaves.** A thick layer of hair on leaves, especially on the lower epidermis, traps moist air next to the leaf surface. The water potential gradient between the inside and the outside of the leaves is reduced and therefore less water vapour is lost by transpiration. *Nerium oleander* (oleander) is a plant with this modification.

- **Having stomata in pits or grooves.** These again trap moist air next to the leaf and reduce the water potential gradient. Examples of plants using this mechanism include *Nerium oleander* (oleander).
- **Having sunken stomata.** In some plants, such as the shrub Hakea, the stomata are located just below the surface of the epidermis, so they are not directly exposed to air movements.
- **Having needle-like leaves.** We saw in Topic 7.2 that the smaller the surface area to volume ratio, the slower the rate of diffusion. By having leaves that are small and roughly circular in cross-section, as in conifers e.g. *Pinus sylvestris* (Scot's pine), rather than ones that are broad and flat, the rate of water loss can be considerably reduced. This reduction in surface area must always be balanced against the need for a sufficient area for photosynthesis to meet the needs of the plant.
- **Closing stomata when transpiration rates are very high.** Some xerophytes can close stomata during the hottest parts of the day. Some plants, called C4 plants, use a modified form of photosynthesis that makes more efficient use of carbon dioxide, and so this closure of stomata does not greatly affect rates of photosynthesis. Other plants produce a plant hormone, abscisic acid, in response to the stress of dehydration and this causes the stomata to close. The mechanism of how stomata open and close is given in Topic 14.10.
- **Multi-layered epidermis.** Having many epidermal layers will reduce water loss through the cuticle. *Nerium* has a mutilayered lower and upper epidermis.

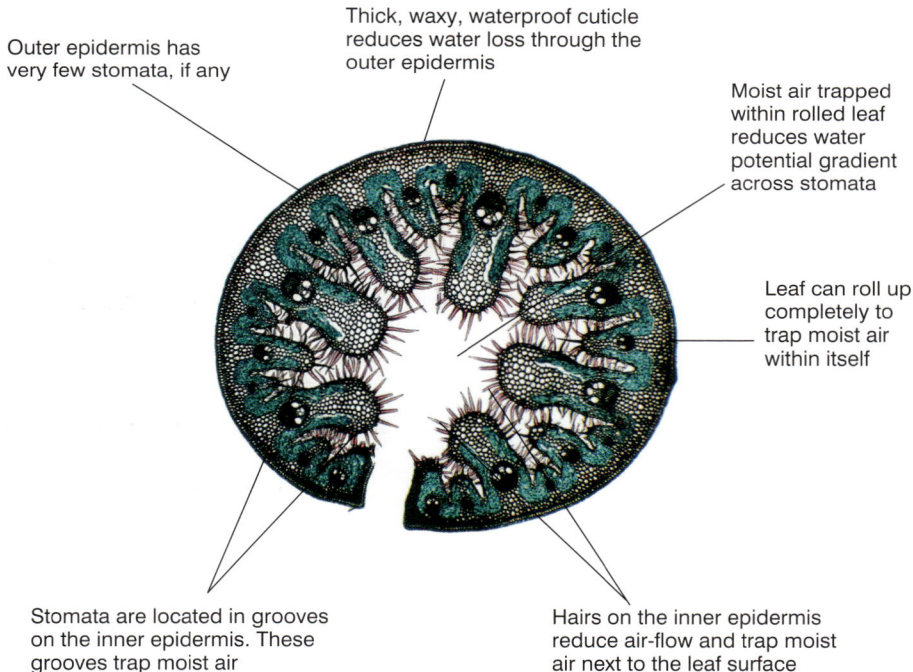

Outer epidermis has very few stomata, if any

Thick, waxy, waterproof cuticle reduces water loss through the outer epidermis

Moist air trapped within rolled leaf reduces water potential gradient across stomata

Leaf can roll up completely to trap moist air within itself

Stomata are located in grooves on the inner epidermis. These grooves trap moist air

Hairs on the inner epidermis reduce air-flow and trap moist air next to the leaf surface

Figure 4 *Xerophytic modifications of the leaf of marram grass*

SUMMARY TEST 7.8

Xerophytes are plants adapted to living in areas where transpiration rates may be higher than water uptake. In addition to hot and cold deserts, xerophytes can be found in other habitats such as **(1)** or **(2)**. The leaves of xerophytic plants show adaptations to reduce water loss. These include having a thick, waxy **(3)** or closing stomata when transpiration rates are high. Increasing humidity around stomata is an effective way of reducing transpiration and three methods by which leaves achieve this are **(4)**, **(5)** and **(6)**.

7.9 Structure and function of phloem

On these pages you will learn to:

- Label phloem sieve tube elements and companion cells and be able to recognise these using the light microscope
- Relate the structure of phloem sieve tube elements and companion cells to their functions

Phloem is the main transport tissue of organic material in vascular plants. It carries organic material, such as sugars and amino acids, from leaves and storage organs to other parts of the plant.

Structure of phloem

Phloem is composed of a number of cell types:

- **Sieve tube elements** are elongated cells that are joined end to end to form long tubes, as shown in Figures 1 and 3a. The cells are living and retain a thin layer of cytoplasm under their cell surface membrane (peripheral cytoplasm), which lies against the cellulose cell wall. In the cytoplasm are mitochondria and a modified form of endoplasmic reticulum. However, unlike most cells, there is no nucleus or Golgi body and there are no ribosomes. These structures are broken down in order to have fewer cell structures within the sieve tubes and so reduce resistance to the flow of liquid within them. The end walls of the sieve tubes are perforated by large pores, 2–6 µm in diameter. These perforated end walls are called **sieve plates** (Figure 2). The central space within the sieve tube is called the **lumen**.
- **Companion cells** are always associated with sieve elements (Figure 3) and both come from the same cell division. As the sieve tube elements lack structures such as a nucleus, ribosomes and Golgi body, they are unable to carry out many of the metabolic processes essential for their survival. The companion cells are the site of these processes. With all the required organelles, dense cytoplasm and thin cellulose cell wall, they perform the metabolic activities for both themselves and the sieve tube elements. In the areas where assimilates are loaded or unloaded there are many plasmodesmata (Figure 4) between the companion cell and the phloem sieve tube element. In other areas where there are fewer plasmodesmata (to maintain pressure gradients; see Topic 7.10) the companion cell is likely to use specific membrane carrier proteins for the lateral transfer of materials. At the tips of veins in the leaf, companion cells have very folded cell walls and cell surface membranes. These special types of companion cells are called **transfer cells** and their large surface area increases the rate of transfer of sucrose into the sieve tube elements.

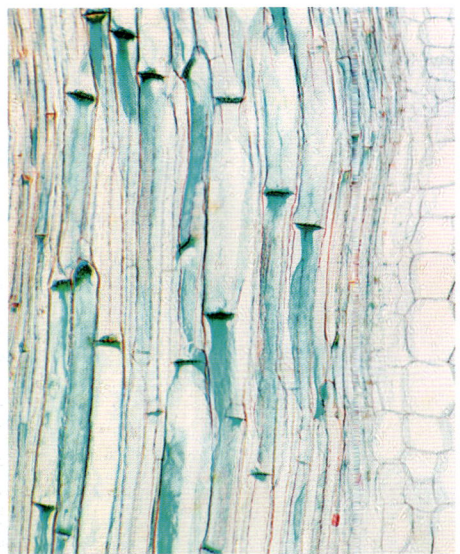

Figure 1 Phloem sieve tube as seen under a light microscope (×400 approx.)

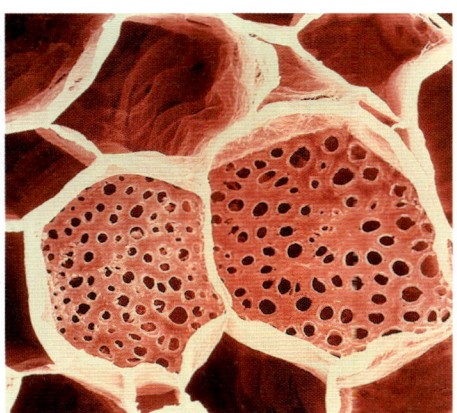

Figure 2 Colourised scanning electron micrograph of sieve plates

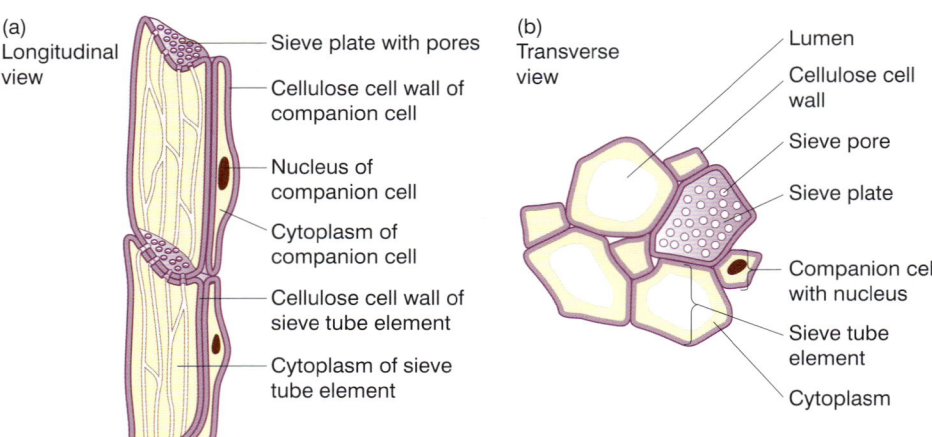

Figure 3 Phloem as seen under a light microscope

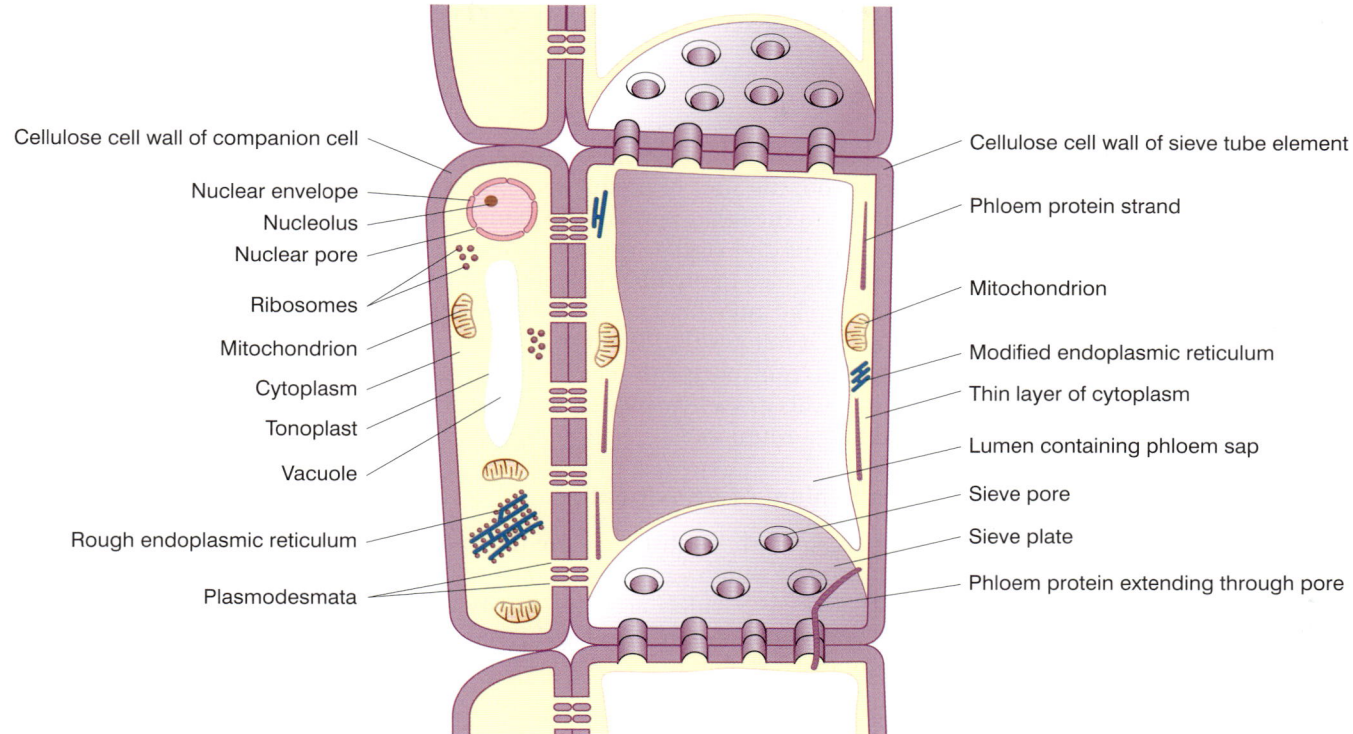

Cellulose cell wall of companion cell

Nuclear envelope

Nucleolus

Nuclear pore

Ribosomes

Mitochondrion

Cytoplasm

Tonoplast

Vacuole

Rough endoplasmic reticulum

Plasmodesmata

Cellulose cell wall of sieve tube element

Phloem protein strand

Mitochondrion

Modified endoplasmic reticulum

Thin layer of cytoplasm

Lumen containing phloem sap

Sieve pore

Sieve plate

Phloem protein extending through pore

Figure 4 *Sieve tube element and companion cell structure as shown by using an electron microscope*

How phloem's structure is related to its function

The structure of sieve tubes has evolved to suit their function of transporting organic materials in solution.

- Sieve tube elements are elongated and arranged end to end to form a continuous tube.
- The nucleus and many of the organelles are located in the companion cells, leaving the lumen of the sieve tube elements more open so reducing resistance to the flow of liquid.
- Sieve plates are perforated with sieve pores, reducing resistance to liquid flow.
- Sieve plates hold the walls of sieve tube elements together and prevent them from bursting.
- The walls contain cellulose microfibrils that run around the cells, giving strength and preventing the tubes bursting under pressure.
- The walls are thin to allow easy entry of water at the source which helps to build up pressure.
- Companion cells have many mitochondria to release the ATP needed for translocation of organic materials (Topic 7.10).
- Plasmodesmata in the areas of loading and unloading allow easy movement of substances to and from companion cells.
- Phloem proteins are a variety of different proteins that are thought to have a role in defence against pathogens and in sealing wounds.

SUMMARY TEST 7.9

Phloem carries organic materials such as **(1)** and **(2)** from the leaves and **(3)** regions to other parts of the plant. It is made up of two main cell types. The sieve tube elements form long, vertical tubes. Each element has perforated end walls called **(4)**. The pores within them have strands of **(5)** running through them. Sieve tube elements lack important structures found in other cells. These include **(6)**, **(7)** and **(8)**. Associated with sieve tube elements are cells that carry out the metabolic activities of sieve tube elements. These cells are called **(9)** and, in the areas of loading and unloading of assimilates, transfer materials to and from sieve tube elements via connecting strands of cytoplasm known as **(10)**.

On these pages you will learn to:

- State that assimilates, such as sucrose and amino acids, move between sources (e.g. leaves and storage organs) and sinks (e.g. buds, flowers, fruits, roots and storage organs) in phloem sieve tubes
- Explain how sucrose is loaded into phloem sieve tubes by companion cells using proton pumping and the co-transporter mechanism in their cell surface membranes
- Explain mass flow in phloem sap down a hydrostatic pressure gradient from source to sink

Having produced sugars during photosynthesis, the plant needs to transport them from the sites of production, known as **sources** (e.g. leaves and storage organs) to the other parts of the plants, known as **sinks** (e.g. buds, flowers, fruits, roots and storage organs) where they will be used directly or stored for future use (see Remember box about storage organs on opposite page). As sinks can be anywhere in a plant – sometimes above and sometimes below the source – it follows that the translocation of molecules in phloem can be in either direction, transported in separate sieve tubes. The phloem transports sucrose and amino acids as well as inorganic **ions** such as potassium, chloride, phosphate and magnesium ions.

Mechanism of translocation

It is accepted that substances are transported in the phloem (Topic 7.9) and that the rate of movement is too fast to be explained by diffusion. The mass flow theory is a widely accepted theory to explain the mechanism of how translocation of assimilates is achieved. It can be divided into three phases, summarised in Figure 1:

Transfer of sucrose into sieve tube elements from photosynthesing tissue.

One way that many plants appear to load sucrose into phloem sieve tubes is by a process known as apoplastic loading:

- Sucrose is manufactured from the products of photosynthesis in cells with chloroplasts.
- Hydrogen ions (protons) are actively pumped from companion cells into the apoplast (cell walls and spaces between cells) using ATP.
- The hydrogen ion concentration builds up and these hydrogen ions (protons) then flow down a concentration gradient through carrier proteins (facilitated diffusion) into the sieve tube elements.
- Sucrose molecules are transported along with the hydrogen ions (protons) in a process known as **co-transport**. The carrier proteins are therefore also known as co-transport proteins. This movement is against the concentration gradient for sucrose and is powered by the flow of protons back into the companion cell.
- Sucrose molecules then move by diffusion through the plasmodesmata from the companion cell into the phloem sieve tube element.

Movement of phloem sap in sieve tube elements from source to sink

'Mass flow' is the bulk movement of a substance through a given channel or area in a specified time. Mass flow of sucrose through sieve tube elements takes place as follows:

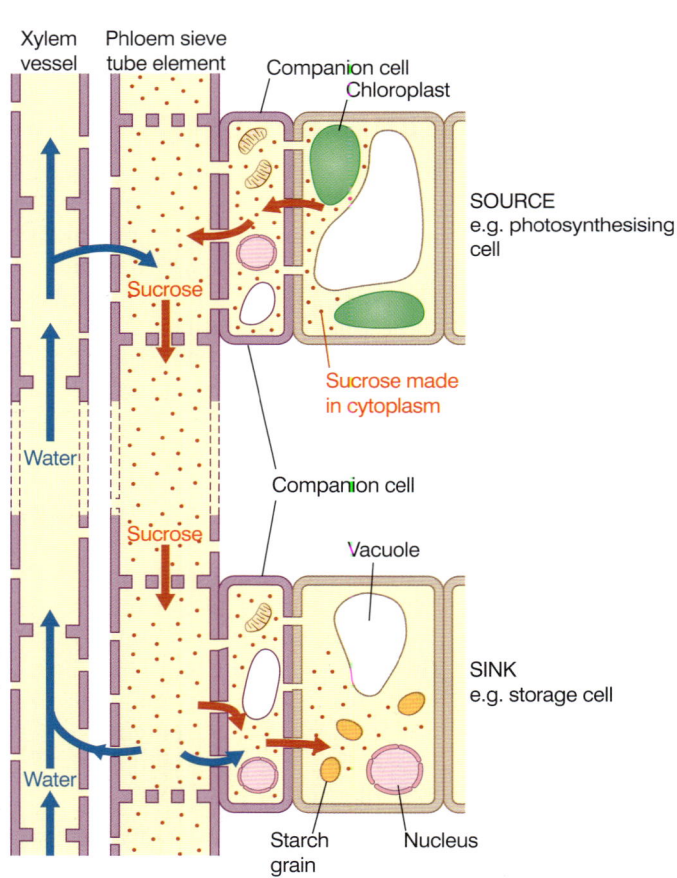

Figure 1 *Movement of sucrose from source to sink through the phloem of a plant*

- The loading of sucrose into the sieve tubes is as described above. It is sometimes termed active loading, as ATP is required to pump out hydrogen ions in order for cotransport of sucrose to occur.
- This causes the sieve tubes to have a lower (more negative) water potential.
- As the xylem has a much higher (less negative) water potential (see Topic 7.8), water moves from the xylem into the sieve tubes by osmosis, creating a high hydrostatic pressure within them.
- At the respiring cells (sink), sucrose is either used up during respiration or converted to starch for storage.
- These cells therefore have a low sucrose content and so sucrose is actively transported into them from the sieve tubes, lowering their water potential.
- Owing to this lowered water potential, water also moves into these respiring cells, from the sieve tubes, by osmosis.
- The hydrostatic pressure of the sieve tubes in this region is therefore low.
- As a result of water entering the sieve tube elements at the source and leaving at the sink, there is a high hydrostatic pressure at the source and a low one at the sink.
- There is therefore a mass flow of sucrose solution down this hydrostatic gradient in the sieve tubes.

Transfer of sucrose from the sieve tube elements into storage or other sink cells. The sucrose is actively transported by companion cells, out of the sieve tubes and into the sink cells.

EXTENSION
Evidence that translocation of organic molecules occurs in phloem

- When phloem is cut, a solution of organic molecules (sap) is exuded.

- Plants provided with radioactive carbon dioxide can be shown to have radioactively labelled carbon in phloem after a short time.

- Aphids which have pierced the phloem with their needle-like mouthparts can be used to extract the contents of the sieve tubes. These contents show diurnal (daily) variations in the sucrose content of leaves that show the same trend a little later by identical changes in the sucrose content of the phloem (Figure 2).

- The removal of a ring of phloem from around the whole circumference of a stem leads to the accumulation of sugars above the ring and their disappearance from below it.

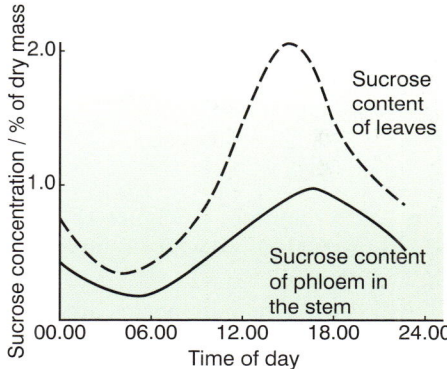

Figure 2 *Diurnal variation in sucrose content of leaves and phloem*

SUMMARY TEST 7.10

Transport of sucrose in plants occurs in the tissue called **(1)**, from places where it is produced, known as **(2)**, to places where it is used up or stored, called **(3)**. One theory of how it is translocated is called the **(4)** theory. Initially hydrogen ions are pumped out of the **(5)** into the **(6)**. As hydrogen ions flow back into the cell, sucrose molecules are also transported in through **(7)** proteins. The sucrose molecules then diffuse into the phloem sieve tube. The contents of the sieve tubes now have a **(8)** water potential due to this sucrose. Water therefore moves into them from the nearby **(9)** tissue that has a **(10)** water potential. The opposite occurs in those cells (sinks) using up sucrose, and water therefore leaves them by the process of **(11)**. Water entering at the sources and leaving at the sinks creates a **(12)** hydrostatic pressure gradient that causes the mass flow of sucrose solution along the phloem.

7 Examination Questions

1 Thale cress, *Arabidopsis thaliana*, is used to study the roles of genes and proteins in plants.

The cell membranes of the root hairs of *A. thaliana* contain proteins called aquaporins that allow the movement of water between the soil and the cytoplasm as shown in Figure 1.

Figure 1

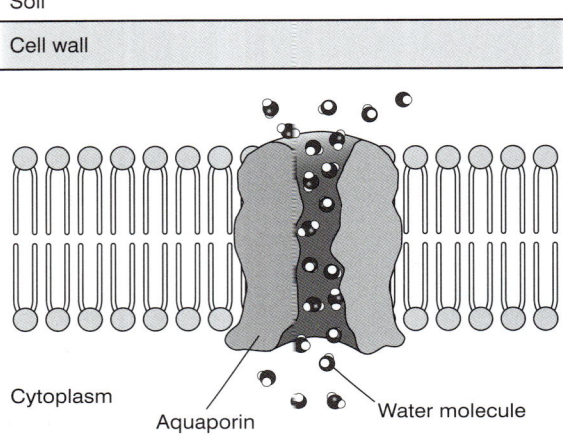

a With reference to Figure 1:
   i explain how water is absorbed by root hairs of *A. thaliana* *(3 marks)*
   ii state why aquaporins are necessary in cell surface membranes. *(1 mark)*

b Describe the pathway taken by water from the cytoplasm of the root hair cell to a xylem vessel in the centre of the root. *(3 marks)*

An investigation was carried out to find the effect of an enzyme in *A. thaliana* on the composition of the cuticle. The enzyme is involved in the production of lipid that accumulates in the cuticle.

Plants were discovered with a mutation of the gene that codes for the enzyme.

Some of these mutant plants (Group **A**) were grown in pots and their rate of transpiration was determined over three days. They were compared with control plants (Group **B**) in which the gene was switched on and the enzyme present. The results are shown in Figure 2.

c With reference to Figure 2, explain:
   i why the rate of transpiration is higher during the day than at night in both groups of plants *(1 mark)*

Figure 2

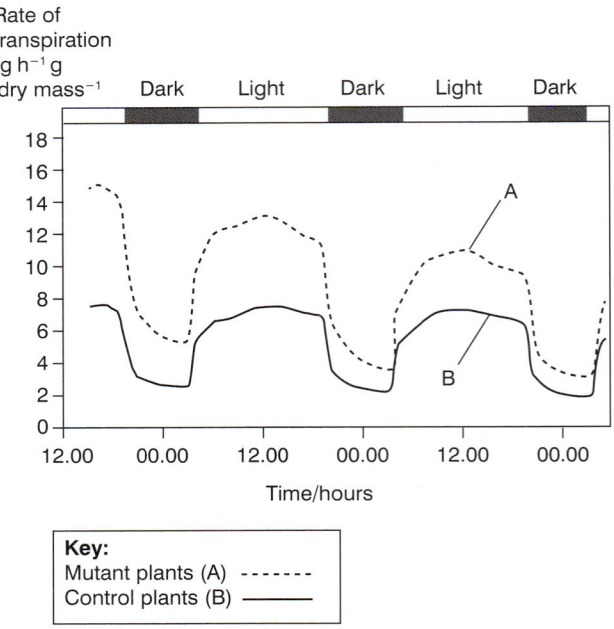

Key:
Mutant plants (A) - - - - - - -
Control plants (B) ————

   ii how the results show that the cuticle is less effective in the mutant plants. *(3 marks)*
   (11 marks)

Cambridge International AS and A Level Biology 9700 Paper 21 Q2 November 2012

2 Phloem transfer cells are specialised companion cells that load sucrose into sieve tube elements.

Figure 3 is an electron micrograph of a transverse section showing phloem tissue from a leaf of *Senecio vulgaris*. The section shows two sieve tube elements and four phloem transfer cells. The sieve tube elements are small in this section because it is taken at the end of a vein in the leaf.

It is thought that the many ingrowths of the cell walls visible in Figure 3 are related to the movement of large quantities of sucrose.

a Describe how companion cells load sucrose into phloem sieve tubes. *(4 marks)*

b Transfer cells move large quantities of sucrose into phloem sieve tubes.

   Suggest why these cells have cell wall ingrowths as shown in Figure 3. *(2 marks)*

c i Explain the advantage of studying cells, such as transfer cells, with the electron microscope rather than the light microscope. *(2 marks)*
   ii Describe the appearance of the phloem sieve tubes when viewed in longitudinal section. *(2 marks)*

Figure 3

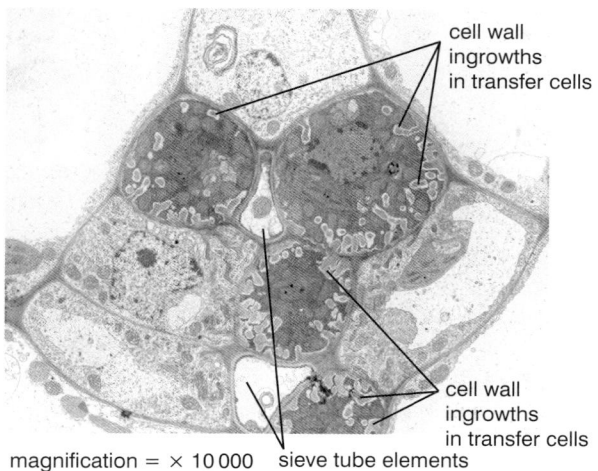

cell wall
ingrowths
in transfer cells

cell wall
ingrowths
in transfer cells

magnification = × 10 000   sieve tube elements

(Total 10 marks)

Cambridge International AS and A Level Biology 9700 Paper 2 Q4 October 2008

7 Practice Questions

3 a Give two reasons why plants growing on sand dunes in temperate regions need to have xerophytic features even though there is plentiful rainfall.

b Explain in terms of water potential why salt marsh plants have difficulty absorbing water despite having plenty around their roots.

c Why do plants in cold regions 'have difficulty obtaining water from the soil for much of the year'?

d Plants living in cold regions often reduce water loss by having needle-like leaves with a small surface area to volume ratio. This reduces the surface area available to capture light for photosynthesis. Photosynthesis is, in part, an enzyme-controlled process. Suggest a reason why a smaller leaf area does not reduce the rate of photosynthesis in the way it would for plants in warmer climates.

4 Figure 4 shows the rate of water flow up a tree and the diameter of the tree trunk over a 24-hour period.

a At what time of day is the transpiration rate greatest? Explain your answer.

b Describe the changes in the rate of flow of water during the 24-hour period.

c Explain in terms of the cohesion–tension theory the changes in the rate of flow of water during the 24-hour period.

d Explain the changes in the diameter of the tree trunk over the 24-hour period.

e If the tree were sprayed with ammonium sulfamate, a herbicide that kills living cells, the rate of water flow would be unchanged for some time. Explain why.

Figure 4

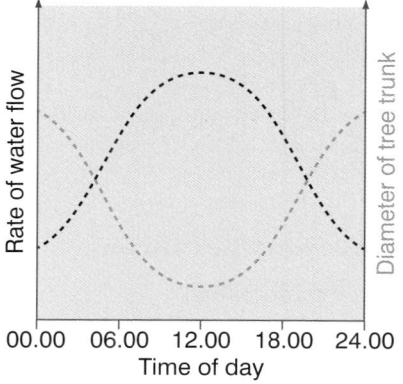

5 Figure 5 shows a potometer, an instrument used to measure the rate of transpiration.

Figure 5

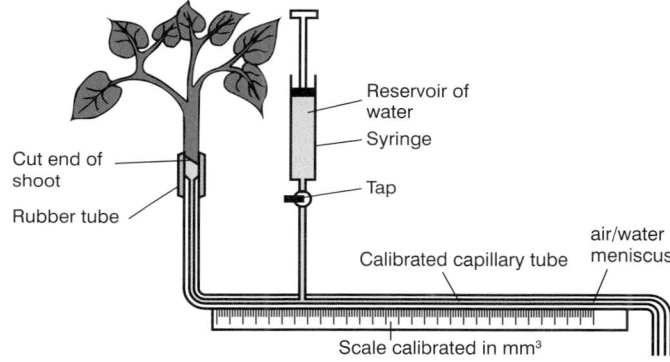

Reservoir of water

Syringe

Tap

Cut end of shoot

Rubber tube

Calibrated capillary tube

air/water meniscus

Scale calibrated in mm³

a From your knowledge of how water moves up the stem, suggest a reason why each of the following procedures is carried out when using the potometer:
  i the leafy shoot is cut under water rather than in the air
  ii all joints are sealed with waterproof jelly.

b What assumption must be made if a potometer is used to measure the rate of transpiration?

c The volume of water taken up in a given time can be calculated using the formula $\pi r^2 l$ (where $\pi = 3.142$, r = radius of the capillary tube, and l = the distance moved by the air bubble). In an experiment the mean distance moved by the air bubble in a capillary tube of radius 0.5 mm during 1 min was 15.28 mm. Calculate the rate of water uptake in $mm^3\,h^{-1}$. Show your working.

d If a potometer is used to compare the transpiration rates of two different species of plant, suggest one feature of both plant shoots that should, as far as possible, be kept the same.

e Suggest why the results obtained from a laboratory potometer experiment may not be representative of the transpiration rate of the same plant in the wild.

109

Transport in mammals

Arteries, veins and capillaries

Active organisms have evolved a **closed circulatory system** in which blood is kept within **vessels**. In mammals, the blood loses pressure as it passes through the capillaries of the lungs. It is therefore returned to the heart to boost its pressure and ensure its rapid circulation to the rest of the body. Blood passes twice through the heart in each complete circuit of the body. This is known as a **double circulatory system**. There are three types of blood vessel in a closed circulatory system:

- **Arteries** carry blood away from the heart. Smaller arteries are called **arterioles**.
- **Veins** carry blood towards the heart. Smaller veins are called **venules**.
- **Capillaries** are smaller vessels that link arteries to veins.

The walls of arteries and veins have the same basic structure. There are three main layers:

- The tunica intima (the lining layer), which is a single layer of endothelial cells attached to a basement membrane.
- The tunica media (the middle layer), which contains smooth muscle and elastic fibres. The proportion of smooth muscle and elastic fibres varies greatly, depending on the type of blood vessel, elastic fibres and the one that is most varied depending on the type of blood vessel.
- The tunica adventitia (the outer layer), which is composed mainly of collagen fibres.

Artery structure related to function

Figure 1 illustrates the structure of an artery. Arteries function in transporting blood rapidly, under high pressure from the heart to the tissues. Their structure is related to this function in the following ways:

- **The tunica media is relatively thick s**o that the artery is well adapted to withstand the high pressure of the blood flowing within it. In the arteries close to the heart, such as the aorta, there is a high proportion of elastic fibres. This is for two main reasons. Firstly, when blood is forced into the arteries following contraction of the ventricles (Topic 8.8), it creates a **pulse** of very high pressure. The elastic wall allows the arteries to expand rather than burst under this pressure. Secondly, it is important that blood pressure in arteries is kept high if blood is to reach the extremities of the body. When the elastic wall is stretched by the pressure within it, it springs back in the same way as a stretched elastic band. This **recoil action** creates another surge of pressure that carries blood forwards in a series of pulses and helps to maintain the blood pressure and the flow of blood forwards when the heart relaxes. In arteries further away from the heart, there are fewer elastic fibres and a higher proportion of smooth muscle. The flow of blood is less pulse-like but does not smooth out completely until it reaches the smallest arteries. The tunica media also contains some collagen fibres. In the smaller arteries, contraction of the smooth muscle causes the vessels to constrict, narrowing the diameter of the lumen and allowing the regulation of blood flow to the tissues.
- **The tunica adventia, with its collagen fibres, provides a tough outer layer.** This outer layer also contains some elastic fibres to allow for stretching as blood flows through.

Figure 1 Comparison of arteries, veins and capillaries

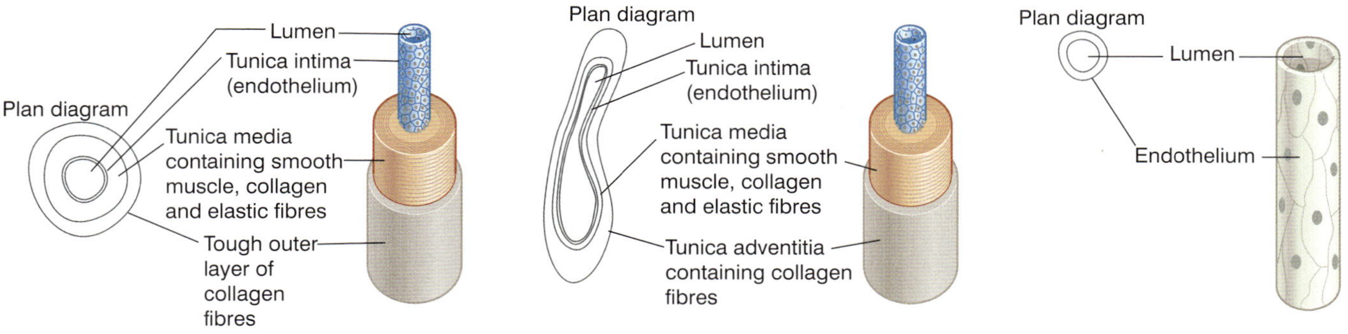

Artery	Vein	Capillary
Thick muscular wall	Thin muscular wall	No muscle
Much elastic tissue	Little elastic tissue	No elastic tissue
Small **lumen** relative to diameter	Large lumen relative to diameter	Large lumen relative to diameter
Capable of constriction	Not capable of constriction	Not capable of constriction
Valves in aorta and pulmonary artery only	Valves throughout all veins	No valves
Transports blood from the heart	Transports blood to heart	Links arteries to veins
Oxygenated blood except in pulmonary artery	Deoxygenated blood except in pulmonary vein	Blood changes from oxygenated to deoxygenated
Blood under high pressure (10–16 kPa)	Blood under low pressure (1 kPa)	Blood pressure reducing (4–1 kPa)
Blood moves in pulses	No pulses	No pulses
Blood flows rapidly	Blood flows slowly	Blood flow slowing

- **The overall thickness of the wall is large**. This again helps prevent arteries bursting under pressure. Arteries have a relatively small lumen in proportion to the thickness of the wall.
- **There are no valves** except in the arteries leaving the heart because blood is under constant high pressure due to the heart pumping blood into the arteries. It therefore tends not to flow backwards.

Vein structure related to function

Figure 1 illustrates the structure of a vein. Veins function in transporting blood, under low pressure, from the tissues to the heart. Their structure is related to this function in the following ways:

- **The tunica media is thin** as the low pressure of the blood will not cause them to burst. There are very few elastic fibres because they do not need to stretch and recoil and there is less smooth muscle because veins carry blood away from tissues and therefore their constriction and dilation cannot control the flow of blood to the tissues.
- **The tunica adventitia, with its collagen fibres, provides a tough outer layer** in order to prevent the veins bursting – more from external physical forces (they are nearer the skin surface than arteries) than from the blood pressure within them. In larger veins there is also a small amount of smooth muscle.
- **The overall thickness of the wall is small** because there is no need for a thick wall as the pressure within the veins is too low to create any risk of bursting. It also allows them to be flattened easily, aiding the flow of blood within them (see below). The lumen is relatively large compared to the thickness of the wall.

111

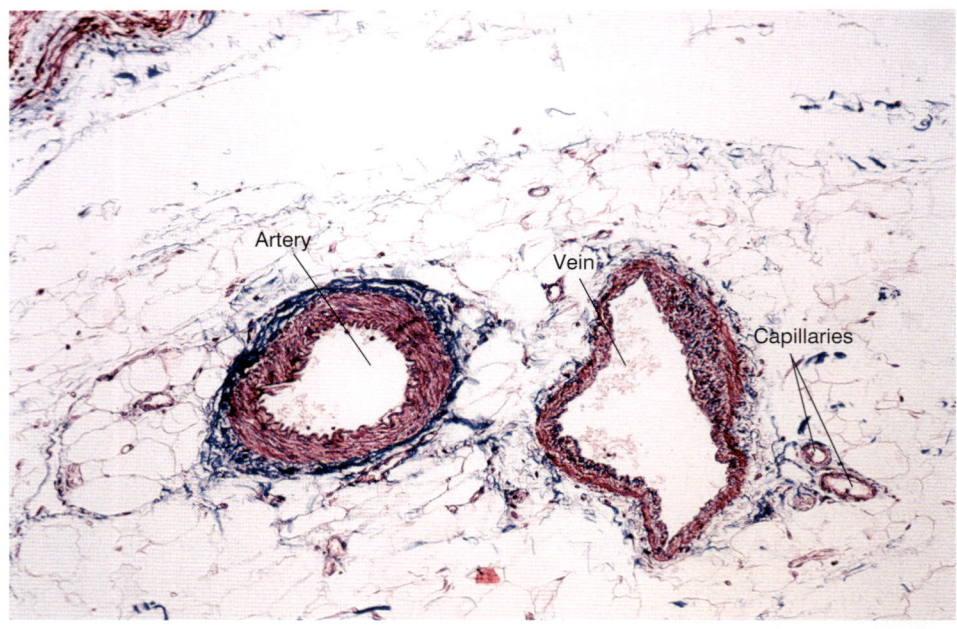

Figure 2 *TS artery (thick-walled vessel on the left) and TS vein (thin-walled vessel on the right) as viewed using a light microscope. The thin capillary wall is a single layer of endothelial cells.*

- **There are semi-lunar valves throughout** (in all but the largest veins) to ensure that blood does not flow backwards, which it might otherwise do because the pressure is so low. When the muscles of the body contract during movement, veins are compressed, pressurising the blood within them. The valves ensure that this pressure directs the blood in one direction only – towards the heart.

Capillary structure related to function

Figure 1 shows the structure of a capillary. The function of capillaries is to exchange materials such as oxygen, carbon dioxide and glucose between the blood and the cells of the body. Their structure is related to their function as follows:

- **Their walls consist only of endothelium** making them extremely thin (one cell thick). This allows for rapid diffusion of materials between the blood and the cells, due to the short distance over which diffusion takes place.
- **They are numerous and highly branched** thus providing a large surface area for diffusion.
- **They have a narrow diameter** and so can reach all body tissues, which means that no cell is far from a capillary.
- **Their lumen is so narrow** – around 7 μm in diameter – that red blood cells are squeezed flat against the side of a capillary. This brings them even closer (as little as 1 μm) to the cells to which they supply oxygen. This again reduces the diffusion distance between the air in the alveoli and the red blood cell.
- **There are spaces between endothelial cells** (known as fenestrations or endothelial pores) which allow white blood cells to escape in order to combat infections within tissues. Also, this means that components of blood or of the tissue fluid surrounding cells do not have to pass through the endothelial cells. This can speed up the delivery or collection of materials, although they still have to cross the basement membrane, which can act as a selective layer. The degree to which material can escape from capillaries varies from tissue to tissue, being greatest in the kidney and least in the brain, where the capillaries have no fenestrations.

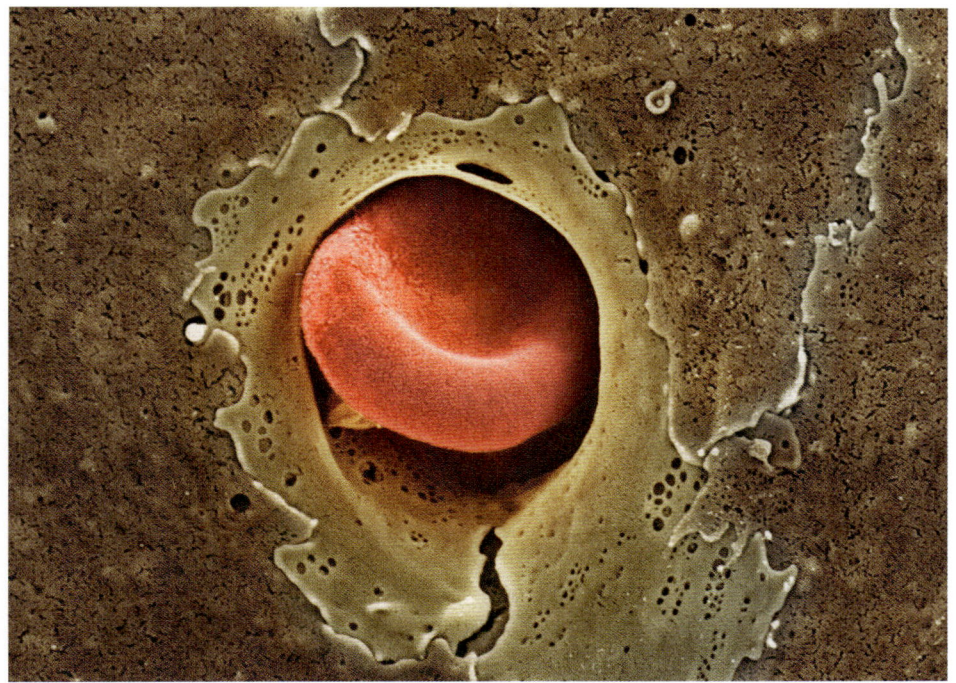

Figure 3 *Colourised scanning electron micrograph of blood capillary with a red blood cell*

SUMMARY TEST 8.1

Where blood is circulated within vessels it is called a **(1)** circulatory system. The vessels are of three types – arteries, veins and capillaries. Arteries carry blood **(2)** the heart and they have more **(3)** and **(4)** in their walls compared with the others. The only vessels that are permeable are the **(5)** while only **(6)** are capable of constricting. The space at the centre of all blood vessels is called the **(7)** and is smallest relative to the vessel's diameter in **(8)**. The inner lining of blood vessels is called the **(9)**. Blood going to the head leaves the **(10)** chamber of the heart, passes along the **(11)** and completes its journey via the **(12)**. Blood from the small intestine passes first to the organ called the **(13)** via the vessel known as **(14)**, and then passes along the **(15)** and the **(16)** before entering the **(17)** chamber of the heart.

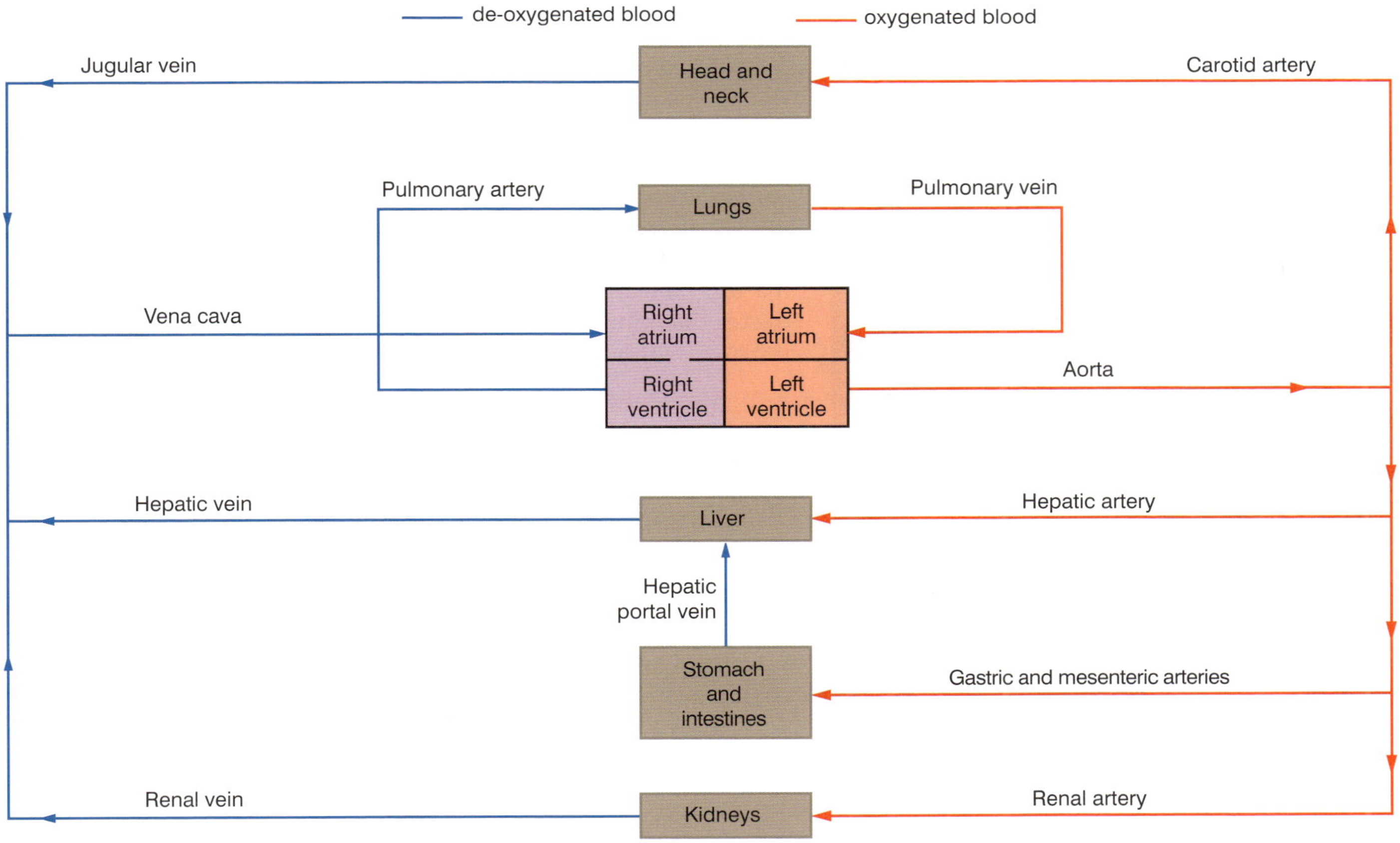

— de-oxygenated blood   — oxygenated blood

Jugular vein   Head and neck   Carotid artery

Pulmonary artery   Lungs   Pulmonary vein

Vena cava

| Right atrium | Left atrium |
| Right ventricle | Left ventricle |

Aorta

Hepatic vein   Liver   Hepatic artery

Hepatic portal vein

Stomach and intestines   Gastric and mesenteric arteries

Renal vein   Kidneys   Renal artery

Figure 4 *Plan of the mammalian circulatory system*

The structure and functions of blood

Blood is the medium by which materials are transported between different parts of the body. Humans have between 4 dm³ and 6 dm³ of blood. It is made up of a liquid – the **plasma** (55%) and three types of cells (45%) – **red cells**, **white cells** and **platelets** (small cells involved in blood clotting; see Figure 2).

The plasma

Blood plasma is 90% water and 10% chemicals, which are either dissolved or suspended in it. The function of the plasma is to transport these chemicals from where they are produced or absorbed to the cells that use or excrete them. These chemicals include:

- **nutrients**, e.g. dissolved glucose, amino acids and vitamins
- **waste products**, e.g. urea
- **mineral ions**, e.g. calcium, iron
- **hormones**, e.g. insulin, adrenaline
- **plasma proteins**, e.g. fibrinogen, prothrombin, albumin
- **respiratory gases**, e.g. oxygen, carbon dioxide.

Between them, these chemicals make the plasma slightly alkaline – around pH 7.4. The removal from the plasma of the proteins that are involved in clotting (prothrombin and fibrinogen) results in a liquid called **serum**, which does not clot.

Red blood cells

Red blood cells are bi-concave discs, i.e. they are like a doughnut with a hole that does not quite go all the way through (Figure 1). Around 7–8 μm in diameter, there are 5 million in each mm³ of blood and each lives for around 120 days. This means that, in adult humans, to maintain their numbers, the bone marrow of certain bones (cranium, sternum, vertebrae and ribs) needs to make over 2 million red blood cells each second. These red blood cells are unusual in having no nucleus, mitochondria, rough endoplasmic reticulum or Golgi body when mature – a feature which, although it gives them a shorter life-span, makes them more efficient in their role of transporting oxygen because, without the nucleus and other organelles:

- They are much thinner in the middle and so form a bi-concave shape which gives them a larger surface area to volume ratio.
- They can more easily change shape, allowing them to be flattened against the capillary walls, thereby reducing the distance across which diffusion takes place, and so speeding up the process (Topic 4.2).
- Without the nucleus and mitochondria, there is more room for the pigment **haemoglobin** which carries oxygen.

It is the pigment **haemoglobin** that gives red blood cells their characteristic colour. The structure of haemoglobin is illustrated in Topic 2.8 and its role in oxygen transport is described in Topic 8.4.

Surface view

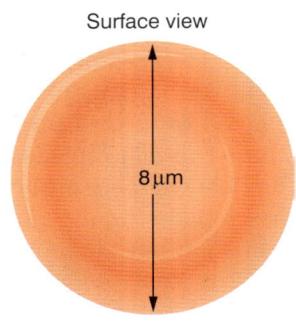

8 μm

Transverse section

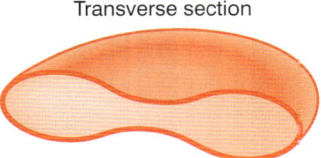

Figure 1 *Red blood cell*

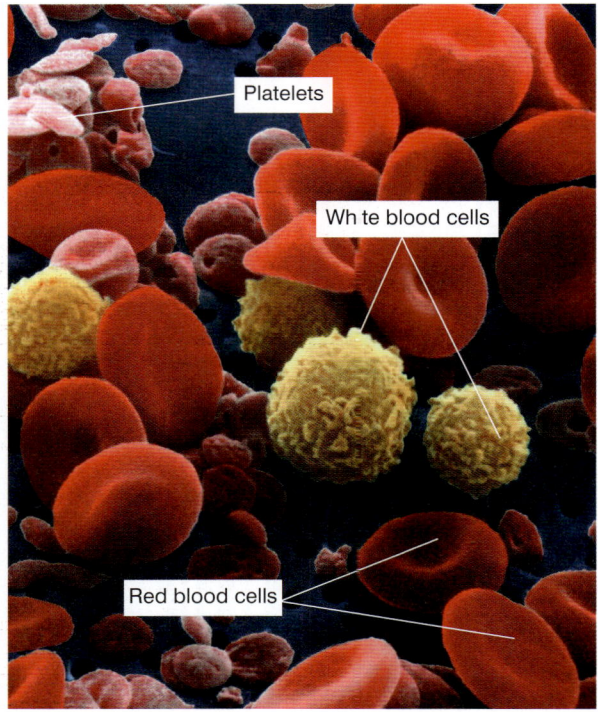

Platelets

White blood cells

Red blood cells

Figure 2 *Colourised scanning electron micrograph of blood showing red cells (red), white cells (yellow) and platelets (pink)*

White blood cells

White blood cells exist in a variety of forms, three of which are shown in Figure 4. They all contain a nucleus. Most are larger than red blood cells. Unlike red blood cells, white blood cells can pass through the gaps in the capillary endothelium into the fluid that surrounds the cells of tissues. Made in the marrow of the limb bones, the white blood cells function to protect the body against infection. When a sample of blood is viewed using a light microscope, some of the white blood cells have a spherical shape and a large, compact spherical nucleus. These are known as lymphocytes. Other white blood cells, known as neutrophils, have a less regular shape and a lobed nucleus. Larger cells known as monocytes, also with an irregular shape, have a large, kidney-shaped nucleus. These cells mature into cells with a granular cytoplasm and are known as macrophages. In terms of their function, the cells can be divided into two groups.

- **Phagocytes**, such as neutrophils and macrophages, remove microorganisms, other foreign material and dead cells by the process of phagocytosis (Topic 11.1). Phagocytosis is non-specific and occurs whatever the infection.
- **Lymphocytes** – act against microorganisms, with some lymphocytes secreting **antibodies** – that immobilise microorganisms and make them ready for phagocytes to engulf. More about this process can be found in Topic 11.2. Each type of lymphocyte acts against one particular pathogen, i.e. they are specific. They can provide long-term immunity to future infections.

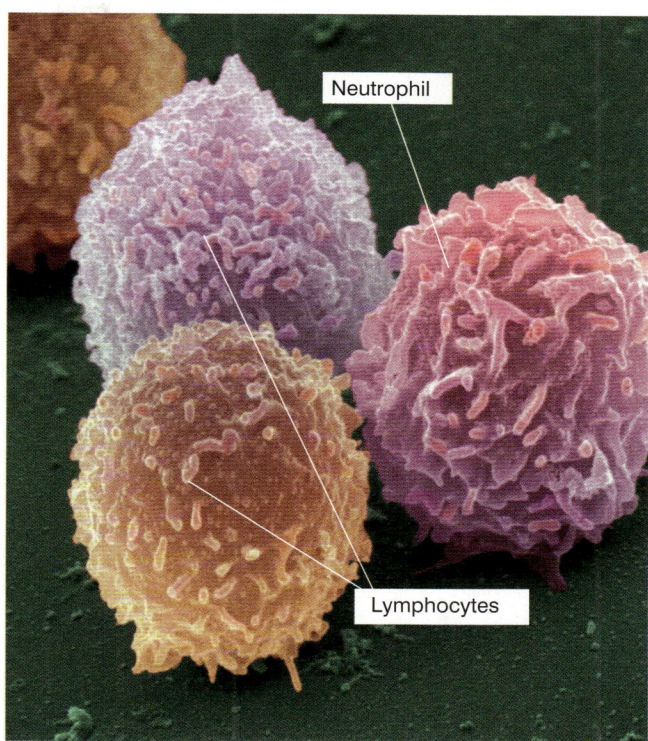

Figure 3 *Colourised scanning electron micrograph of white blood cells*

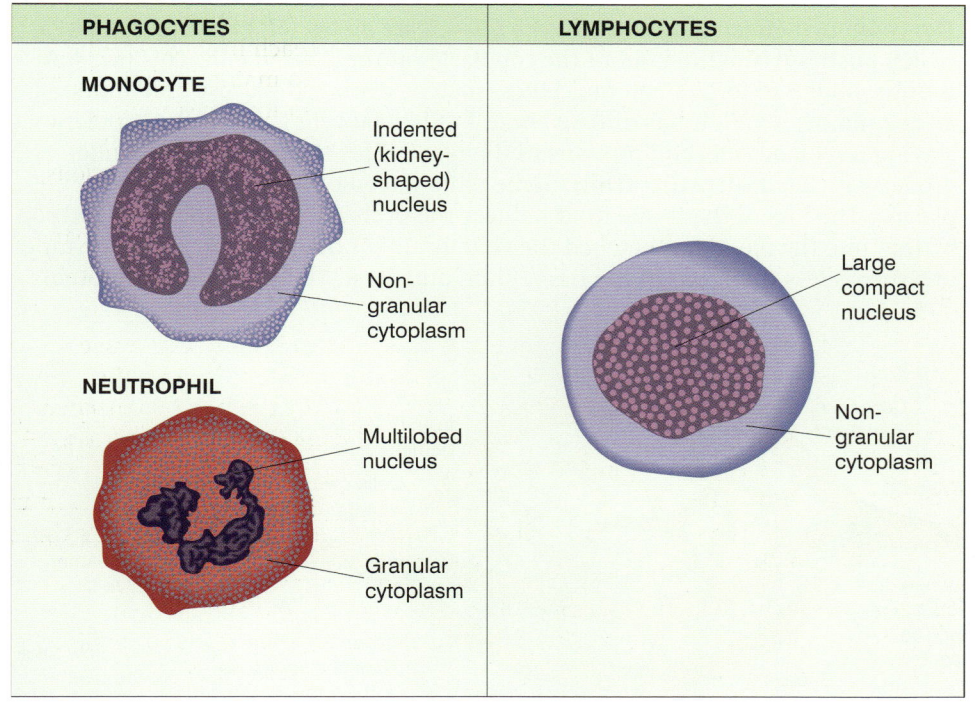

Figure 4 *Three of the many types of white blood cell*

SUMMARY TEST 8.2

Blood is made up of a watery liquid called **(1)**, in which lie a variety of cells. The most numerous are the **(2)**, which are bi-concave discs about **(3)** in diameter. They live for about **(4)** and contain a red pigment called **(5)** that carries **(6)**. A second type of cells are the **(7)**, which have a variety of forms. Those that engulf bacteria are called **(8)**, an example of which is a cell type called **(9)**. Those that secrete chemicals called **(10)** are known as **(11)**.

Tissue fluid and lymph

On these pages you will learn to:

- State and explain the differences between blood, tissue fluid and lymph

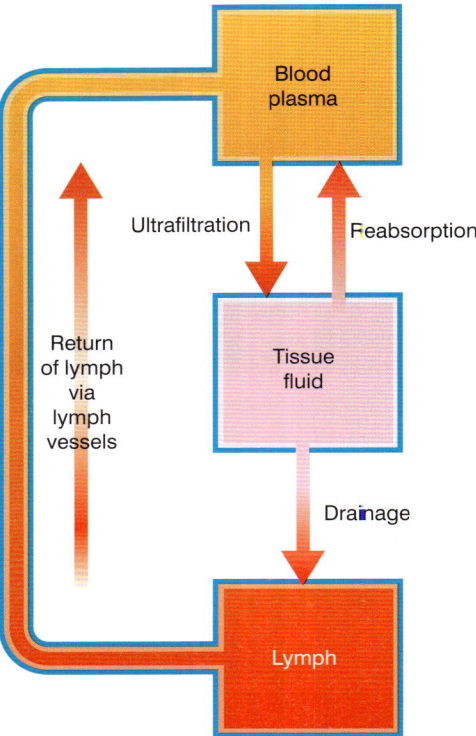

Figure 1 *Relationship between plasma, tissue fluid and lymph*

Blood supplies nutrients to the tissues of the body via tiny vessels called capillaries. Small though they are, these cannot serve every single cell directly, and therefore the final stage of the nutrient's journey is made in solution in a liquid that bathes the tissues – **tissue fluid**.

Tissue fluid

Tissue fluid is formed from the plasma of the blood. It is a watery liquid which contains dissolved glucose, amino acids, fatty acids, mineral ions and oxygen, all of which it supplies to the tissues. In return, it receives carbon dioxide and other waste materials from the tissues. Tissue fluid is the means by which materials are exchanged between blood and cells and it bathes all cells of the body. Compared to plasma, which arrives at the arterial end of the capillary network, tissue fluid contains less oxygen and glucose but has more carbon dioxide.

Formation of tissue fluid

Blood pumped by the heart passes along arteries, then the narrower arterioles and, finally, the even narrower capillaries. By this time the pressure of blood, known as the **hydrostatic pressure**, is around 4.8 kPa at the arterial end of the capillaries, which causes tissue fluid to move out of the blood. The outward pressure is, however, opposed by two other forces:

- hydrostatic pressure of the tissue fluid outside the capillaries, which resists outward movement of liquid
- the lower water potential of the blood (due to plasma proteins) that causes water to move back into the capillaries.

The combined effect of all these forces is to create an overall pressure of 1.7 kPa, which pushes tissue fluid out of the capillaries at the arterial end. This pressure is only enough to force small molecules, such as dissolved glucose and amino acids (and oxygen that has diffused out of red blood cells), out of the capillaries, leaving red blood cells and proteins in the blood. This type of filtration under pressure is called **ultrafiltration**. Some white blood cells may also leave the plasma. The loss of the tissue fluid reduces the pressure in the capillaries and so, by the time the blood has reached the venous end of the network, its hydrostatic pressure is less than that of the tissue fluid outside it. The large plasma proteins

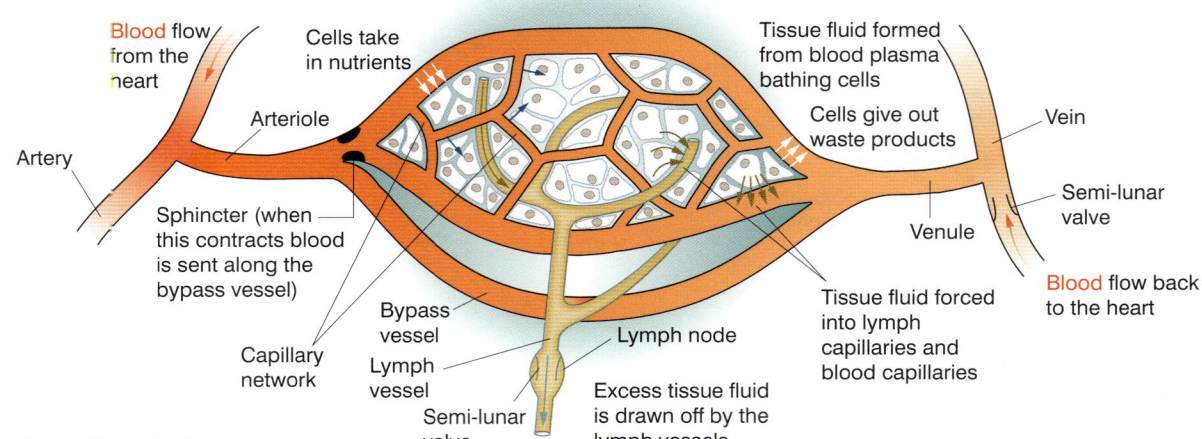

Figure 2 *Formation and return of tissue fluid*

remaining in the capillary help to lower the water potential, so water has a tendency to move back into the plasma by osmosis. There is an overall negative pressure of 1.5 kPa drawing tissue fluid back into the capillaries. This fluid has lost much of its oxygen and nutrients by diffusion to the cells it bathed, but has gained carbon dioxide and excretory products in return. These events are summarised in Figure 2. Not all the tissue fluid can return to the capillaries; the remainder is carried back via the lymphatic system.

Lymph and the lymphatic system

Lymph is a milky liquid made up of material from three sources:

- **tissue fluid** that has not been reabsorbed at the venous end of the capillary network
- **fatty substances** absorbed in the ileum
- **lymphocytes** which have either left the lymph nodes or have left capillaries to fight infection.

Lymph is carried in the **lymphatic system**, which is made up of capillaries resembling blood capillaries except that they are closed at one end (blind-ended). These merge into larger vessels that form a network around the body. The lymph vessels drain their contents back into the blood stream, into the large veins leading to the heart.

At points along the lymph vessels are a series of **lymph nodes**, which produce and store lymphocytes. As lymph passes through these nodes, lymphocytes and proteins are added to it. Lymph nodes filter from the blood any bacteria and other foreign material, which are then engulfed by phagocytes. This causes the nodes to swell with dead cells, and is the reason for the tenderness often felt in the groin, armpits and neck during an infection.

Lymph is moved along lymph vessels in three ways:

- **hydrostatic pressure** of the tissue fluid leaving the capillaries
- **contraction of body muscles** squeezes the lymph vessels. Valves ensure that the fluid inside them moves away from the tissues in the direction of the heart
- **enlargement of the thorax during breathing in** reduces pressure in the thorax, drawing lymph into this region and away from the tissues.

Table 1 Comparison of various body fluids

	Blood	Plasma	Tissue fluid	Lymph
Cells	Red and white blood cells and platelets	None	None	Lymphocytes
Plasma proteins	Present	Present	Absent	Added as it passes through lymph nodes
Location	Within heart, arteries, veins and capillaries	Within heart, arteries, veins and capillaries	Outside vessels, bathing cells	Within lymph vessels
Moved by	Pumping of heart, muscle contraction and breathing action	Pumping of heart, muscle contraction and breathing action	Hydrostatic and osmotic forces	Hydrostatic forces, muscle contraction and breathing action
Function	Transport and defence	Transport over long distances	Transport over short distances	Transport and defence

SUMMARY TEST 8.3

Tissue fluid is formed from the **(1)** due to the **(2)** pressure at the **(3)** end of the capillary network. Although other forces oppose the formation of tissue fluid, there is still sufficient pressure to force water and small molecules out of the blood, but not red blood cells and large molecules such as plasma proteins. This process is called **(4)**. Lymph is made up of tissue fluid, **(5)** and **(6)**.

8.4

Transcript of oxygen

On these pages you will learn to:

- Describe the role of haemoglobin in carrying oxygen

Measuring oxygen concentration

The amount of a gas that is present in a mixture of gases is measured by the pressure it contributes to the total pressure of the gas mixture. This is known as the **partial pressure** of the gas and, in the case of oxygen, is written as pO_2. It is also known as **oxygen tension** and is measured in the usual unit for pressure, namely **kilopascals (kPa)**. Normal atmospheric pressure is 100 kPa. As oxygen makes up 21% of the atmosphere, its partial pressure is normally 21 kPa.

Use of abbreviation – Hb

Hb is commonly used as an abbreviation for haemoglobin. However, in the context of oxygen transport, it is used to represent a single haem group. As a haemoglobin molecule has four haem groups, each of which carries an oxygen molecule (O_2), oxyhaemoglobin is shown as Hb_4O_8.

As organisms evolved, they became more complex and, in many cases, much larger. Metabolic rates increased and with them the demand for oxygen increased. Specialised gaseous exchange surfaces, such as lungs, developed to meet this need. This solved the problem only if there was a mechanism to transport the oxygen from these surfaces to the cells requiring it. Even with blood vessels and a heart to pump the blood around them, the transport of oxygen would be totally inadequate if the gas were simply dissolved in the plasma. Only the evolution of specialised molecules capable of carrying large quantities of oxygen could adequately supply the tissues. These molecules are called **respiratory pigments**, the best known of which is **haemoglobin**.

Haemoglobin

Haemoglobin (Figure 1) is a red pigment that has a relative molecular mass of 65 000. As such, it could be lost from the body during ultrafiltration in the kidneys. It is therefore contained within the red blood cells that carry it around the body, separated from the plasma. Its structure is described in Topic 2.8. One oxygen molecule can combine with each of its four haem groups to form oxyhaemoglobin.

$$4Hb \quad + \quad 4O_2 \quad \rightleftharpoons \quad Hb_4O_8$$

haemoglobin   oxygen   oxyhaemoglobin

To be efficient, haemoglobin must:

- readily associate with oxygen at the gaseous exchange surface (loading)
- readily dissociate from oxygen at those tissues requiring it (unloading).

These two requirements may appear contradictory, but are achieved by the remarkable property of haemoglobin changing its affinity for oxygen in different concentrations of carbon dioxide (Table 1). This is known as the Bohr effect and is explained in Topic 8.5.

Table 1 *Affinity of haemoglobin for oxygen under different conditions*

Region of body	Oxygen tension (concentration)	Carbon dioxide tension (concentration)	Affinity of haemoglobin for oxygen	Result
Gaseous exchange surface	High	Low	High	Oxygen is loaded
Respiring tissues	Low	High	Low	Oxygen is unloaded

Haemoglobin has an affinity for carbon monoxide some 250 times greater than for oxygen – a feature which makes carbon monoxide potentially lethal. Once it is attached to haemoglobin to form carboxyhaemoglobin, the carbon monoxide molecule remains permanently, and so prevents oxygen molecules being loaded. As carbon monoxide is found in fumes from the burning of fuels and in tobacco smoke, both reduce the oxygen-carrying capacity of the blood because haemoglobin is less saturated with oxygen.

Oxygen dissociation curves

The relationship between the loading and unloading of haemoglobin with oxygen and the partial pressure of oxygen (see margin box) is known as the **oxygen dissociation curve**. You will see from this curve, which is shown in Figure 1, that haemoglobin does not load oxygen evenly. The explanation of the shape of the oxygen dissociation curve is as follows:

- The shape of the haemoglobin molecule makes it difficult for the first oxygen molecule to bind to one of the haem groups on its four polypeptide chains because they are closely united. Therefore at low oxygen concentrations, little oxygen binds to haemoglobin. The gradient of the curve is shallow (not very steep) initially.

- However, the binding of this first oxygen molecule changes the quaternary structure of the haemoglobin molecule, causing it to change shape. This change makes it easier for another haem group (on another polypeptide chain) to bind an oxygen molecule. This is known as an allosteric effect.

- It therefore takes a smaller increase in the partial pressure of oxygen to bind the second oxygen molecule than it did to bind the first one. This is known as **positive cooperativity**, because binding of the first molecule makes binding of the second easier, and so on. The gradient of the curve steepens.

- At high oxygen concentrations most, but not all, of the haemoglobin molecules are completely loaded with oxygen molecules and the gradient of the curve reduces and begins to level off. There is never completely 100% saturation as, with the majority of binding sites occupied, it is less likely that a single oxygen molecule will find an empty site to bind to.

- The curve shows that haemoglobin is an ideal molecule to load oxygen in the lungs, where the partial pressure of oxygen is high, and to unload oxygen to the respiring tissues, where the oxygen partial pressure is low. In the range of partial pressures found in the lungs, nearly all the haem groups have bound oxygen, so that there is 98% saturation of haemoglobin with oxygen. Very little oxygen is dissociated from this oxyhaemoglobin as the blood passes from the lungs to the tissues. When the blood enters areas of lower oxygen partial pressures, such as in the respiring tissues, the reverse happens and oxygen is unloaded, with a small decrease in partial pressure leading to a large dissociation of oxygen from oxyhaemoglogin. This means that actively respiring tissues (e.g. partial pressure of 3 kPa) will receive a much greater quantity of oxygen than tissues at rest (e.g. 5 kPa). Although it is relatively easy for the first three oxygen molecules to dissociate as the partial pressure of oxygen decreases, it is much harder for the final molecule to dissociate so that percentage saturation of haemoglobin never reaches 0%.

There are many different oxygen dissociation curves because:

- there are a number of different respiratory pigments
- haemoglobin exists in a number of different forms
- the behaviour of each pigment changes under different conditions.

The many different oxygen dissociation curves, and the ability of haemoglobin to associate with or dissociate from oxygen, are better understood if two facts are always kept in mind:

- the more to the left the curve is, the more readily the pigment associates with oxygen but the less easily it dissociates from it
- the more to the right the curve is, the less readily the pigment associates with oxygen but the more easily it dissociates from it.

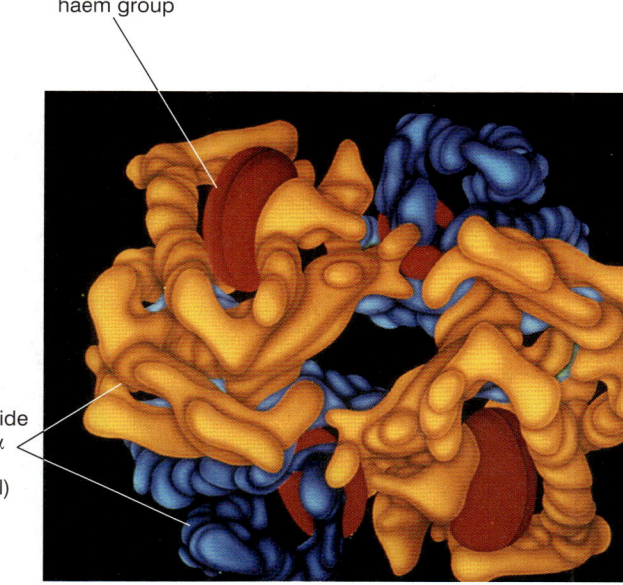

Iron-containing haem group

Two different polypeptide chains (α and β) (4 in total)

Figure 1 *Computer graphic representation of a haemoglobin molecule showing two pairs of polypeptide chains (orange and blue) associated with the haem groups (red)*

SUMMARY TEST 8.4

Haemoglobin in humans is an example of a **(1)**. One molecule has a relative molecular mass of **(2)** and possesses a total of **(3)** haem groups, each of which possesses a single atom of **(4)**. The 'globin' part of the molecule comprises four **(5)**. Each haemoglobin molecule can carry a total of **(6)** molecules of oxygen. The amount of oxygen present in a mixture of gases is known as the oxygen **(7)** and the graph of the relationship between this and the amount of oxygen taken up by haemoglobin is called the **(8)**.

Transport of carbon dioxide

The carbon dioxide produced by respiratory tissues must be carried back to the gaseous exchange surface for removal from the body because its accumulation (build-up) is harmful. This carbon dioxide is, however, essential to the efficient dissociation of oxygen from **oxyhaemoglobin** to the tissues. This is the **Bohr effect**, named after the person who discovered it in 1904, Christian Bohr.

The Bohr effect

Haemoglobin has a reduced affinity for (ability to bind to) oxygen in the presence of carbon dioxide. The greater the concentration of carbon dioxide, the more readily it releases its oxygen. This is the Bohr effect, and explains the differing behaviour of haemoglobin in different regions of the body and in the same region of the body under different conditions.

- At the gaseous exchange surface (e.g. lungs), the concentration of carbon dioxide is low because it diffuses across the exchange surface and is expelled from the organism. Haemoglobin's affinity for oxygen is increased. Together with the high concentration of oxygen in the lungs this means that haemoglobin readily binds oxygen. The reduced carbon dioxide concentration has shifted the oxygen dissociation curve to the left (Figure 1).
- In the respiratory tissues (e.g. muscles), the concentration of carbon dioxide is high in the blood because of its production during respiration. Haemoglobin's affinity for oxygen is reduced. Together with the low concentration of oxygen in the muscles this means that oxygen readily dissociates from oxyhaemoglobin. The increased carbon dioxide concentration has shifted the oxygen dissociation curve to the right (Figure 1). This is especially important during exercise because the more carbon dioxide that is produced, the more readily oxygen is supplied from the haemoglobin to meet extra energy demands due to exercise.

The Bohr effect is a consequence of the acidic nature of dissolved carbon dioxide: it forms hydrogen ions and hydrogencarbonate ions. It is the hydrogen ions that lower the pH and the affinity of haemoglobin for oxygen. Low pH caused by other chemicals, e.g. lactate, therefore also reduces haemoglobin's affinity for oxygen in the same way.

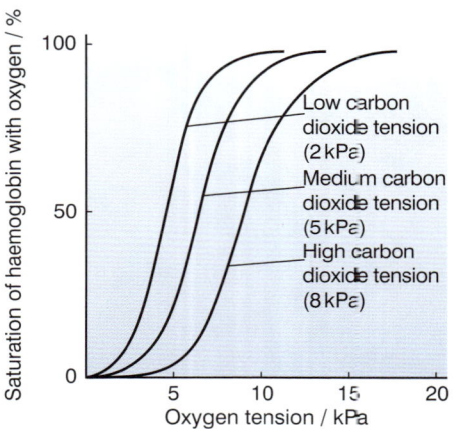

Figure 1 The Bohr effect

Transport of carbon dioxide

Around 5% of carbon dioxide is carried from the tissues to the gaseous exchange surface in solution in blood plasma. The transport of the remaining 95% involves haemoglobin. The carbon dioxide is transported in two ways:

- **In combination with haemoglobin** – carbon dioxide can combine with amino groups in the protein part of the haemoglobin molecule:

$$Hb{-}N\!\!\begin{array}{c}H\\H\end{array} \;+\; CO_2 \;\rightleftharpoons\; Hb{-}N\!\!\begin{array}{c}H\\COO^-\end{array} \;+\; H^+$$

haemoglobin   carbon dioxide   carbaminohaemoglobin   hydrogen ions

About 10% of the total carbon dioxide is carried in this way.

- **As hydrogencarbonate ions** – 85% of the total carbon dioxide is transported in this form. The carbon dioxide combines with water to form carbonic acid, which then dissociates (splits) into hydrogen ions (H^+) and hydrogencarbonate ions (HCO_3^-). The reaction is catalysed by the enzyme carbonic anhydrase and is summarised as:

$$H_2O \; + \; CO_2 \; \xrightarrow{\text{carbonic anhydrase}} \; H_2CO_3 \; \longrightarrow \; H^+ \; + \; HCO_3^-$$

| water | carbon dioxide | | carbonic acid | hydrogen ion | | hydrogencarbonate ion |

This reaction takes place in red blood cells. The hydrogen ions produced combine with haemoglobin to form **haemoglobinic acid** and so cause it to release its oxygen, which diffuses out of the cell into the plasma, to then enter the tissue fluid and diffuse into respiring cells. In this way, haemoglobin acts as a **buffer**, helping to keep the pH of the blood around 7.4. The transport of carbon dioxide is summarised in Figure 2.

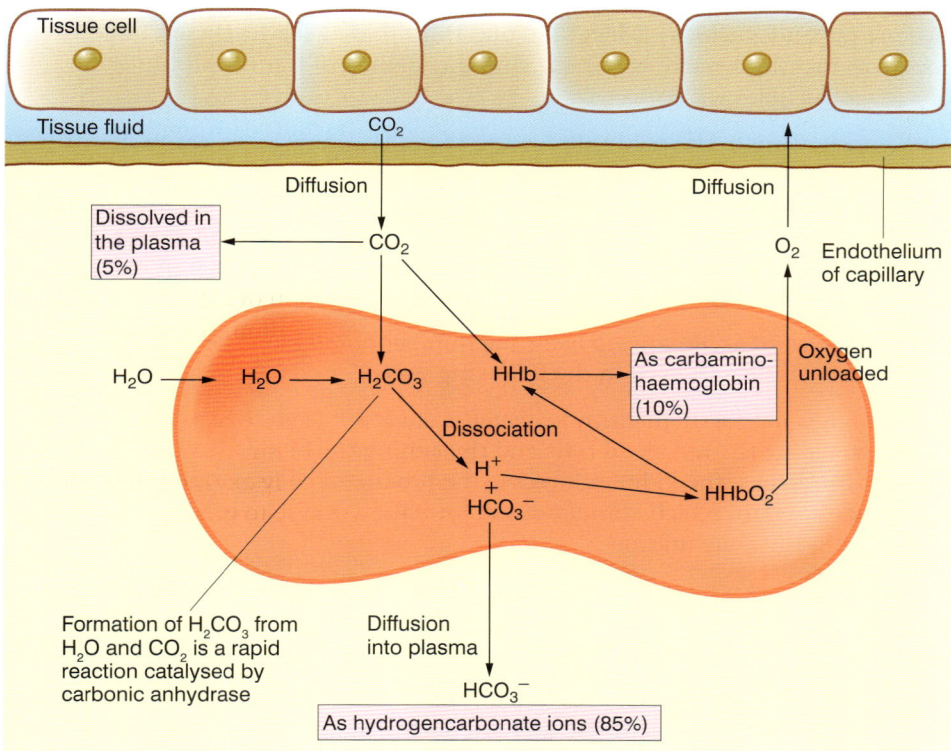

Figure 2 *Transport of carbon dioxide*

SUMMARY TEST 8.5

Carbon dioxide produced by tissues during the process of **(1)** is transported to the gaseous exchange surface in three ways. Firstly, it combines with haemoglobin to form **(2)** and **(3)** ions; this accounts for about **(4)**% of the total carbon dioxide carried. Secondly, around **(5)**% is transported in solution in the **(6)**. The remaining **(7)**% is carried in the form of **(8)**, which are formed from the dissociation of **(9)**, which has been formed from carbon dioxide and water in a reaction catalysed by the enzyme **(10)**. One product of this reaction is hydrogen ions, which then combine with haemoglobin to form **(11)**, which acts as a **(12)** by helping to keep the blood pH around neutral. The affinity of haemoglobin for oxygen is reduced in the presence of carbon dioxide. This change is known as the **(13)**.

Transport of oxygen at high altitude

On these pages you will learn to:

- Describe and explain the significance of the increase in the red blood cell count of humans at high altitude

Figure 1 *Low atmospheric pressure at high altitude makes it difficult to load haemoglobin with oxygen. Climbers, such as this Sherpa near the summit of Mount Everest, frequently breathe oxygen from cylinders to compensate*

Figure 2 *Native highlanders, like these Tibetans, are better acclimatised than those living at low altitude*

In Topic 8.4, we saw how oxygen is transported under normal circumstances. There are, however, some particular circumstances in which adaptations have to be made to ensure efficient transport of oxygen. One example is at high altitude.

Oxygen transport at high altitude

Most people are aware of the problems associated with high altitude and, in particular, the need to use oxygen when climbing in areas such as the Himalayas. What changes at high altitude is the air pressure, which decreases as we ascend from sea-level. Put simply, the higher we go, the less air there is above us weighing down on our bodies. As a result, the air pressure on the top of Mount Everest (altitude 8848 m) is less than one-third that at sea-level (see Table 1). Such low pressures make breathing, and the loading of **haemoglobin** with oxygen, difficult.

Sudden exposure to the reduced pressure at high altitude could result in death within 10 minutes. This is because, if haemoglobin cannot be loaded with oxygen, the level of oxygen in the blood and tissues falls, resulting in a condition called **hypoxia**. However, if ascent is slow, i.e. over many days or weeks, the body can adjust by trying to maximise, in other ways, the amount of oxygen delivered to the tissues. This process is called **acclimatisation**.

One main way of acclimatising is for the number of red blood cells in circulation to increase. This is achieved in the following ways:

- After a few days at high altitude, water is absorbed from the circulation, concentrating the red blood cells and thickening the blood.
- After 1–2 weeks, the kidneys increase the production of a hormone called **erythropoietin**, which stimulates the formation and release of more red blood cells from the bone marrow.

The result of these changes is that, after 2–3 weeks at high altitude, the proportion of the total blood that is made up of red blood cells increases from an average of 45% to an average of 60%.

Due to reduced atmospheric pressure at high altitude, haemoglobin cannot be fully loaded with oxygen. The increase in the number of red blood cells compensates for this as there are now more haemoglobin molecules and this increases the rate of oxygen transport from the lungs to the tissues.

Table 1 *Changes in atmospheric pressure with altitude*

Altitude / m	Atmospheric pressure / kPa	Partial pressure of O_2 / kPa
0	100.0	21.0
3500	65.5	13.8
8500	31.9	6.7

Other changes during acclimatisation

Apart from an increase in red blood cell count other changes that maximise the amount of oxygen delivered to the tissues are:

- **an increase in cardiac output** in the first few days, i.e. the heart beats more rapidly and pumps more blood at each beat

- **hyperventilation** – increasing the rate and depth of breathing, so delivering more air, and therefore oxygen, to the alveoli. Until acclimatisation is complete, this creates a conflict, because hyperventilation also removes larger than normal volumes of carbon dioxide. This results in reduced ventilation and so the overall pattern of breathing can be irregular

- **an increase in the rate of exchange between alveoli and lung capillaries** by increasing the lung volume and the flow of blood through the capillaries

- **an increase in the concentration of haemoglobin in each red blood cell** by as much as 20%

- **an increase in the level of myoglobin in the muscles**.

Native highlander adaptations

There is evidence that some populations of native highlanders, such as Tibetan highlanders, have evolved adaptations to allow them to live successfully at high altitudes.

Acute mountain sickness

More than half the people trekking to 5000 m will experience the altitude-related illness called **acute mountain sickness (AMS)**. Its many symptoms include:

- headaches and dizziness

- nausea and vomiting

- insomnia

- general lethargy

- dry, irritating cough

- breathlessness.

If ignored, these symptoms may worsen, leading to the following:

- **Mental impairment** – at moderate altitude, many people are shown to have slower reactions than at sea-level. As they climb higher, they tend to lack concentration and they will find it difficult to calculate and make judgements. Still higher and they may suffer hallucinations and have a dangerous sense of well-being.

- **Redistribution of body fluids** – severe hypoxia leads to an increased production of ADH (antidiuretic hormone), which reduces urine output and causes more water to remain in the blood. In cold weather at altitude, because the blood flow to the extremities is reduced, extra fluid accumulates outside blood vessels, causing swelling, or **oedema**. This may become apparent in swelling of the feet and legs, as well as the face, especially around the eyes. More seriously, fluid may accumulate in the lungs (**pulmonary oedema**), causing breathlessness and frothing at the mouth. An accumulation of fluid in the brain (**cerebral oedema**) causes it to swell and push against the cranium. Severe headaches may be followed by unconsciousness, and even death, if a return to low altitude does not occur soon enough.

SUMMARY TEST 8.6

The concentration of oxygen at high altitude is **(1)**% of the total atmosphere. Its partial pressure is lower at high altitude, being only about the fraction **(2)** at 8500 m compared with its partial pressure at sea-level. Mountaineers therefore need to **(3)** slowly to the conditions if they are to avoid a drop in blood oxygen concentrations, known as **(4)**. To maintain a high oxygen concentration, after a few weeks the kidneys produce the hormone **(5)**, which results in an increase in **(6)** and hence an increase in haemoglobin.

8.7 Structure of the heart

On these pages you will learn to:

- Describe the external and internal structure of the mammalian heart
- Explain the differences in the thickness of the walls of the different chambers in terms of their functions with reference to resistance to flow

The heart is a muscular organ which, in humans, pumps $13\,000\,dm^3$ of blood each day. It operates continuously and tirelessly throughout the life of an organism. Lying in the thoracic cavity between the two lungs, the heart is made up of a unique type of muscle called cardiac muscle.

Structure of the human heart

The human heart is really two separate pumps lying side by side. The left-hand pump deals with oxygenated blood from the lungs, while the right-hand one deals with deoxygenated blood from the body. Each pump has two chambers.

REMEMBER

Atria are linked to **V**eins, and **A**rteries are linked to **V**entricles. In other words, **A** and **V** always go together.

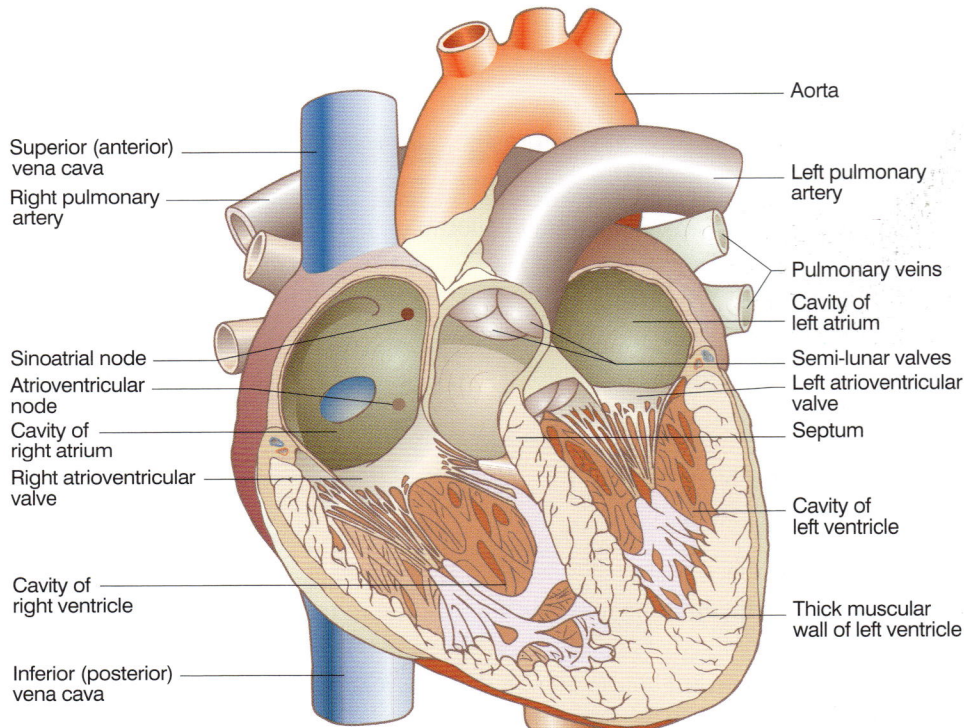

Figure 1 *Section through the human heart (VS)*

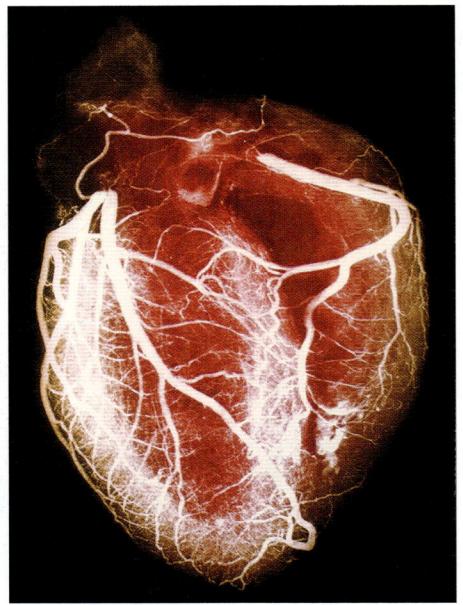

Figure 2 *X-ray of the heart showing coronary arteries which have been injected with an X-ray opaque dye*

EXTENSION

Structure of cardiac muscle

The wall of the heart is almost entirely made up a special type of muscle called cardiac muscle. **Cardiac muscle** is capable of rhythmic contraction and relaxation over a long period without fatigue. It appears striped under a light microscope and is made up of the proteins **actin** and **myosin**. The molecular mechanism of contraction is also the same as that of striated muscle (body muscle) and both have many mitochondria to supply **ATP**. Cardiac muscle is made up of short fibres. Each cardiac muscle fibre is made up of **myofibrils** up to $0.08\,mm$ in length and around $15\,mm$ in diameter.

- **The atrium** is thin-walled and elastic and expands as it collects blood. There is very little resistance to blood flow as it passes the very short distance into the ventricles so less cardiac muscle is required.
- **The ventricle** has a much thicker muscular wall as it has to pump blood some distance, either to the lungs or the rest of the body. The walls of the blood vessels present a resistance to the blood flowing through them. This means more cardiac muscle is required to generate a force to overcome this resistance and allow blood to reach its destination.

As the right ventricle pumps blood at low pressure to the lungs, which are only a short distance away, it has a thinner muscular wall than the left ventricle because there is very little resistance to blood flow in the pulmonary circulation (circulation to the lungs). The left ventricle, in contrast, has a thick muscular wall, enabling it to contract forcefully to create a high pressure to overcome the greater resistance in the systemic circulation (circulation to the rest of the body) and pump blood to the extremities of the body, a distance of about 1.5 m. Although the two sides of the heart are separate pumps and there is no mixing of the blood in each after birth, they still pump in time with each other: both atria contract together, followed by both ventricles contracting together, pumping the same volume of blood.

Between the atrium and ventricle are valves that prevent the backflow of blood into the atria when the ventricles contract. There are two sets of valves.

- **Left atrioventricular (bicuspid) valves** are formed of two cup-shaped flaps on the left side of the heart.
- **Right atrioventricular (tricuspid) valves** are formed of three cup-shaped flaps on the right side of the heart.

To prevent these valves turning inside out under pressure, they are attached to special pillars of muscle (papillary muscles) on the heart wall by fibres called the tendinous cords (chordae tendinae). Each of the four chambers of the heart is served by large blood vessels that carry blood into or away from the heart. Vessels connecting the heart to the lungs are called **pulmonary** vessels. The vessels connected to the four chambers are as follows.

- **The aorta** is connected to the left ventricle and carries oxygenated blood to all parts of the body except the lungs.
- **The vena cava** is connected to the right atrium and brings deoxygenated blood back from the tissues of the body.
- **The pulmonary artery** is connected to the right ventricle and carries deoxygenated blood to the lungs. Unusually for an artery, it carries deoxygenated blood.
- **The pulmonary vein** is connected to the left atrium and brings oxygenated blood back from the lungs. Unusually for a vein, it carries oxygenated blood.

Although oxygenated blood passes through the left side of the heart in vast quantities, the heart does not use this oxygen to meet its own great respiratory needs. Instead, the

heart muscle is supplied by its own blood vessels, called the **coronary arteries**, which branch off the aorta shortly after it leaves the heart.

The structure of the heart and its associated blood vessels is shown in Figures 1 and 3.

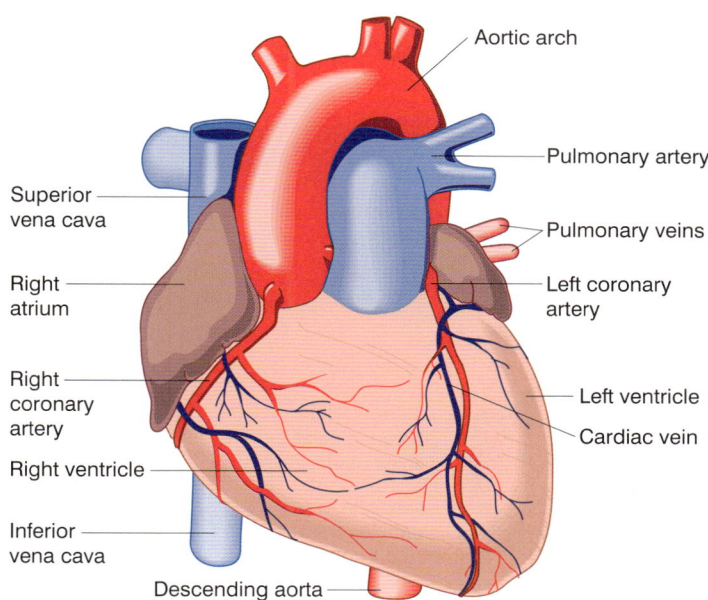

Figure 3 *External appearance of the human heart showing the blood supply to the heart muscle*

EXTENSION
Supplying the cardiac muscle with oxygen
Blockage of the coronary arteries, e.g. by a blood clot, leads to **myocardial infarction**, or **heart attack**, because an area of the heart muscle is deprived of oxygen and so dies, leaving scar tissue.

SUMMARY TEST 8.7

The mammalian heart is made up of four chambers: a pair of thin-walled elastic ones called **(1)** and a pair of thick muscular ones called **(2)**. Between the chambers on the left side of the heart are the **(3)** valves, while those on the right side are called **(4)** valves. Blood from the lungs passes to the heart by the **(5)** and into the chamber called the **(6)**; it leaves the heart via the vessel called the **(7)**. Glucose absorbed from the intestines will first enter the **(8)** chamber of the heart.

8.8

The cardiac cycle

On these pages you will learn to:

- Describe the cardiac cycle (including blood pressure changes during systole and diastole)

1.
Blood enters atria and ventricles from pulmonary veins and venae cavae

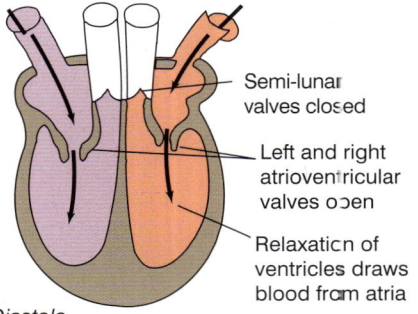

- Semi-lunar valves closed
- Left and right atrioventricular valves open
- Relaxation of ventricles draws blood from atria

Diastole
Atria are relaxed and fill with blood. Ventricles are also relaxed.

2.

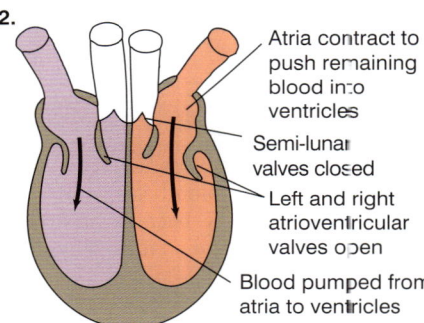

- Atria contract to push remaining blood into ventricles
- Semi-lunar valves closed
- Left and right atrioventricular valves open
- Blood pumped from atria to ventricles

Atrial systole
Atria contract, pushing blood into the ventricles. Ventricles remain relaxed.

3.
Blood pumped into pulmonary arteries and the aorta

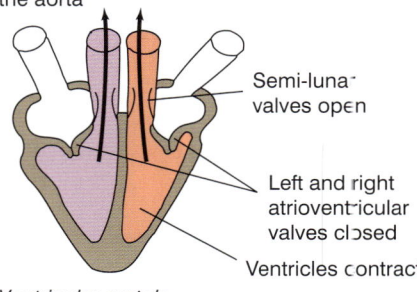

- Semi-lunar valves open
- Left and right atrioventricular valves closed
- Ventricles contract

Ventricular systole
Atria relax. Ventricles contract, pushing blood away from heart through pulmonary arteries and the aorta.

Figure 1 *The cardiac cycle*

The heart undergoes a sequence of events that is repeated in humans around 70 times each minute when at rest. This is known as the **cardiac cycle**. There are two basic components to the beating of the heart – contraction, or **systole**, and relaxation, or **diastole**. Systole occurs separately in the ventricles and the atria and is therefore described in two phases but, for some of the time, diastole takes place at the same time in all chambers of the heart and is therefore treated as a single phase in the account below. The cardiac cycle is illustrated in Figure 1.

Diastole (relaxation of the heart)

Blood returns to the atria of the heart through the pulmonary vein (from the lungs) and the vena cava (from the body) and the atria begin to fill. As the ventricles relax, the pressure within them becomes lower than the pressure in the atria and the atrioventricular valves open, allowing the blood to pass into the ventricles. The **cardiac muscle** of both the atria and ventricles is relaxed at this stage. The relaxation of the ventricle's cardiac muscle causes them to recoil and reduces the pressure within the ventricle. This causes the pressure to be lower than that in the aorta and the pulmonary artery, and so the semi-lunar valves in the aorta and the pulmonary artery close.

Atrial systole (contraction of the atria)

The contraction of the atrial walls forces the remaining blood into the ventricles from the atria. During this stage, the cardiac muscle of the ventricle walls remains relaxed (ventricular diastole).

Ventricular systole (contraction of the ventricles)

After a short delay, the ventricles contract simultaneously. This increases the blood pressure within them, forcing shut the atrioventricular valves and preventing backflow of blood into the atria. With the atrioventricular valves closed, the pressure in the ventricles rises. Once it is higher than that in the aorta and pulmonary artery, the semi-lunar valves open and blood is forced from the ventricles through the semi-lunar valves into these vessels. Topic 8.9 provides more information on the changes in volume and blood pressure in the heart.

Valves in the control of blood flow

It is important to keep blood flowing in the right direction through the heart and around the body. This is achieved mainly by the pressure created by the heart muscle. Blood, as with all liquids and gases, will always move from a region of higher pressure to one of lower pressure. There are, however, situations within the circulatory system when pressure differences would result in blood flowing in the opposite direction from that which is desirable. In these circumstances valves are used to prevent any unwanted backflow of blood. Valves in the cardiovascular system are designed so that they open whenever the difference in blood pressure either side of them favours the movement of blood in the desired direction. When pressure differences are reversed, i.e. when blood would tend to flow in the opposite direction to that which is desirable, the valves are designed to close. Examples of such valves include:

- **The atrioventricular valves** between the left atrium and ventricle (bicuspid valves) and the right atrium and ventricle (tricuspid valves). These prevent backflow of blood when contraction of the ventricles means that ventricular pressure exceeds atrial pressure. Closure of these valves ensures that, when the ventricles contract, blood within them moves to the aorta and pulmonary arteries rather than back to the atria.

- **The semi-lunar valves** in the aorta and pulmonary arteries. These prevent backflow of blood into the ventricles when the recoil action of the elastic walls of these vessels (Topic 8.1) creates a greater pressure in the vessels than in the ventricles.
- **Pocket (semi-lunar) valves** in veins, which occur throughout the venous system. These ensure that when the veins are squeezed, e.g. when body muscles contract, blood flows back to the heart rather than away from it.

The design of all these valves is basically the same. They are made up of a number of flaps of tough, but flexible, fibrous tissue, which are cusp-shaped, i.e. like deep saucers or bowls. When pressure is greater on the convex side of these cusps, rather than on the concave side, they move apart to let blood pass between the cusps. However, when pressure is greater on the concave side than on the convex side, blood collects within the 'bowl' of the cusps, pushing them together to form a tight fit that prevents the passage of blood (Figure 2). So great are the pressures created within the ventricles of the heart that the atrioventricular valves are at risk of becoming inverted. To prevent this, the valves have string-like tendons called the chordae tendinae (heart strings) that are attached to pillars of muscle in the ventricle wall.

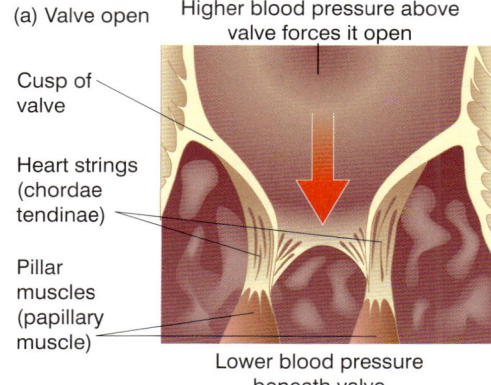

(a) Valve open

Higher blood pressure above valve forces it open

Cusp of valve

Heart strings (chordae tendinae)

Pillar muscles (papillary muscle)

Lower blood pressure beneath valve

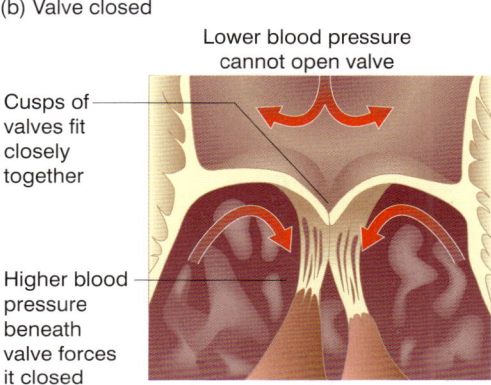

(b) Valve closed

Lower blood pressure cannot open valve

Cusps of valves fit closely together

Higher blood pressure beneath valve forces it closed

Figure 2 *Action of valves*

EXTENSION

Cardiac output

It is often useful to be able to measure the rate at which the heart pumps out blood. This is known as the cardiac output.

Cardiac output is the volume of blood pumped by the heart in a given time. It is usually measured in $dm^3 min^{-1}$ and depends upon two factors:

- the heart rate (the rate at which the heart beats)
- the stroke volume (volume of blood pumped out at each beat).

Cardiac output = heart rate × stroke volume

REMEMBER

Although the left ventricle has a thicker wall than the right ventricle, their internal volumes are the same.

SUMMARY TEST 8.8

In humans the cardiac cycle repeats itself about **(1)** times each minute when the heart is at rest. The phase of the cycle when the atria and ventricles are relaxed is called **(2)**. When the atria contract during the phase called **(3)**, the remaining blood in them is pushed past the **(4)** valves into the chambers called **(5)**. Contraction of these chambers forces open the **(6)** valves and pushes blood into the **(7)**, which then goes to the lungs, and the **(8)**, which supplies blood to the rest of the body.

Control of the cardiac cycle

On these pages you will learn to:

- Explain how heart action is initiated and controlled with reference to the sinoatrial node, the atrioventricular node and the Purkyne tissue

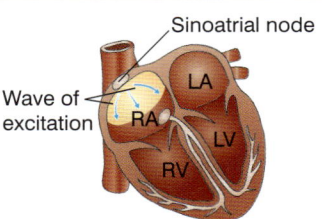

a Wave of excitation spreads out from the sinoatrial node

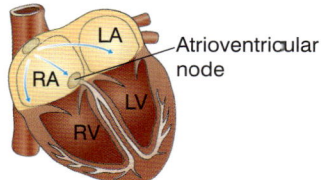

b Wave of excitation spreads across both atria causing them to contract and reaches the atrioventricular node

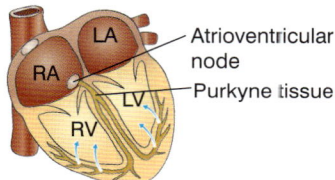

c Atrioventricular node passes a wave of excitation down the septum along the Purkyne fibres to the apex, and up through the ventricle walls, causing the ventricles to contract

Figure 1 Control of the cardiac cycle

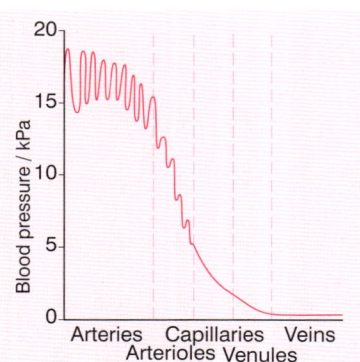

Figure 2 Blood pressure in arteries, capillaries and veins

For the heart to function efficiently, there must be careful control of the sequence of events that takes place during the **cardiac cycle**.

How is the cardiac cycle controlled?

Cardiac muscle is **myogenic**, i.e. its contraction is initiated from within the muscle itself, rather than by nervous impulses from outside (neurogenic), as is the case with other muscle. Within the wall of the right atrium of the heart is a distinct group of cells known as the **sinoatrial node (SAN)**. It is from here that the initial stimulus for contraction originates. The sinoatrial node has a basic rhythm of stimulation that determines the beat of the heart. For this reason it is often referred to as the **pacemaker**. The sequence of events (Figure 1) is as follows.

- A wave of excitation spreads out relatively slowly from the sinoatrial node across both atria, causing them to contract gradually towards the ventricles.
- A fibrous ring of non-conductive tissue (atrioventricular septum) prevents the wave crossing to the ventricles.
- The wave of excitation is allowed to pass through a second group of cells called the **atrioventricular node (AVN)**, which lies between the atria.
- The atrioventricular node, after a short delay of 0.1–0.2 seconds to ensure the ventricles contract after the atria and not at the same time, passes a wave of excitation between the ventricles along a series of specialised muscle fibres called the **Purkyne tissue**.
- The Purkyne tissue conducts the wave very rapidly through the atrioventricular septum to the base of the ventricles to allow near instantaneous stimulation of cardiac muscle.
- The wave of excitation is released from the Purkyne tissue, and passes more slowly across the cardiac muscle causing the ventricles to contract, from the apex of the heart towards the aorta and pulmonary artery.

EXTENSION

Electrocardiogram

During the cardiac cycle, the heart undergoes a series of electrical current changes related to the waves of excitation created by the sinoatrial node and the heart's response to these. If picked up by a cathode ray oscilloscope, these changes can produce a trace known as an **electrocardiogram**. An example, related to the stages of the cardiac cycle, is shown as part of Figure 3. Doctors can use this trace to provide a picture of the heart's electrical activity and hence its health. The electrocardiograms (ECG) below illustrate the difference between a normal ECG, one produced during a heart attack, and one in a person suffering **fibrillation** of the heart. In fibrillation, the cardiac muscle contracts in a disorganised way, causing different sections to contract and relax independently in an irregular manner. This effectively paralyses the heart and causes death if not treated immediately.

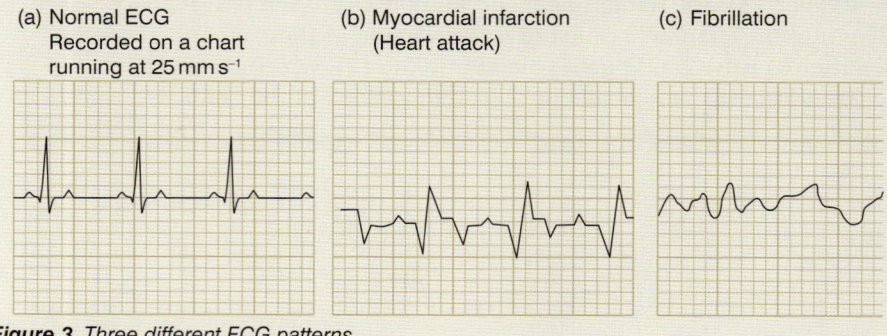

Figure 3 Three different ECG patterns

Pressure and volume changes of the heart

Mammals have a closed, double circulatory sytem, i.e. blood is kept within blood vessels and this allows the pressure within them to be maintained and regulated. Figure 2 shows the pressure within blood vessels. Blood pressure varies between a maximum (systolic blood pressure) when the ventricles contract and a minimum (diastolic blood pressure) when the heart is relaxed. In humans it is normally 16 kPa (120 mmHg) (systolic) and 10.7 kPa (80 mmHg) (diastolic). Figure 4 illustrates the pressure and volume changes that take place in the heart during a typical cardiac cycle.

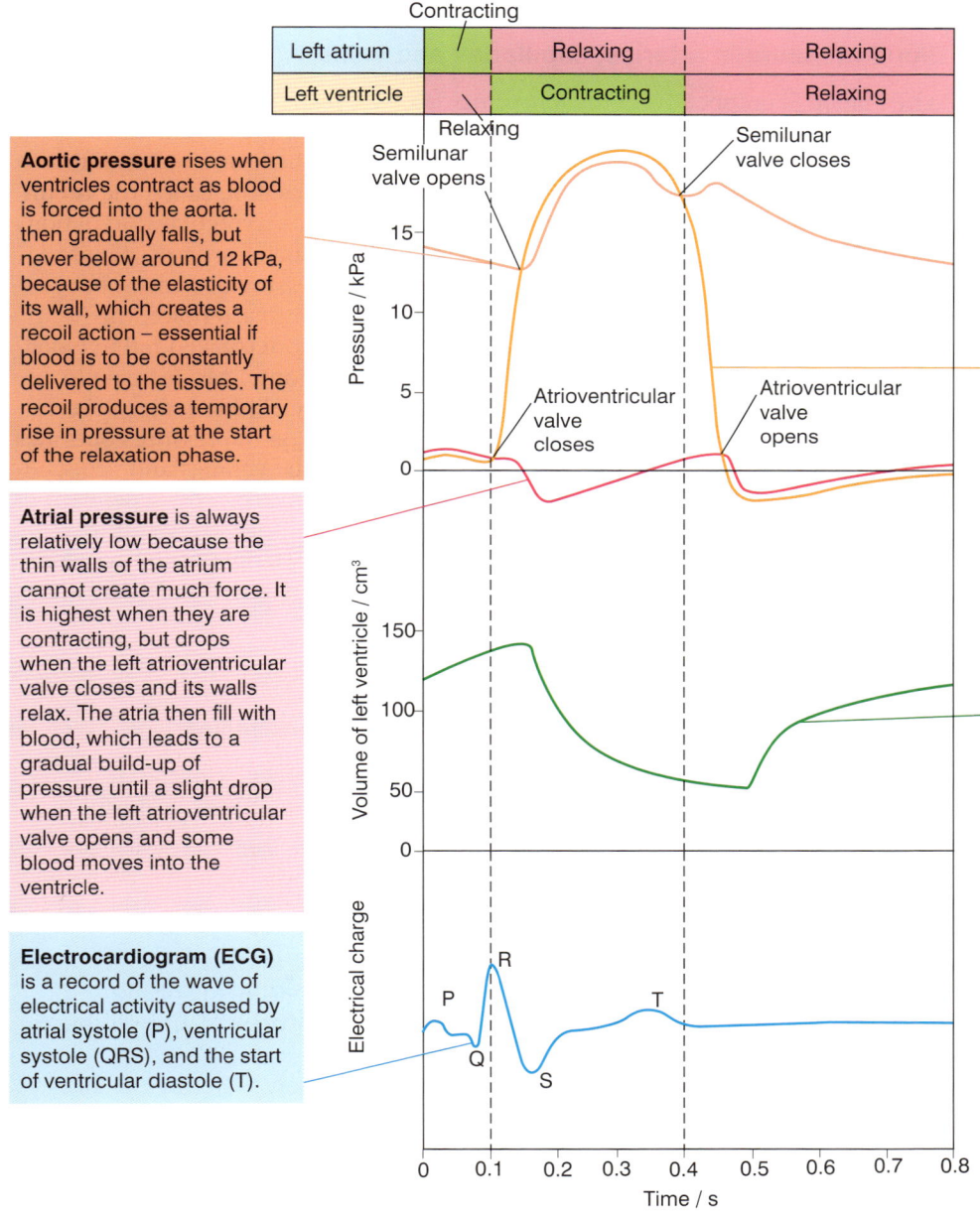

Aortic pressure rises when ventricles contract as blood is forced into the aorta. It then gradually falls, but never below around 12 kPa, because of the elasticity of its wall, which creates a recoil action – essential if blood is to be constantly delivered to the tissues. The recoil produces a temporary rise in pressure at the start of the relaxation phase.

Atrial pressure is always relatively low because the thin walls of the atrium cannot create much force. It is highest when they are contracting, but drops when the left atrioventricular valve closes and its walls relax. The atria then fill with blood, which leads to a gradual build-up of pressure until a slight drop when the left atrioventricular valve opens and some blood moves into the ventricle.

Electrocardiogram (ECG) is a record of the wave of electrical activity caused by atrial systole (P), ventricular systole (QRS), and the start of ventricular diastole (T).

Ventricular pressure is low at first, but gradually increases as the ventricles fill with blood as the atria contract. The left atrioventricular valves close and pressure rises dramatically as the thick muscular walls of the ventricle contract. As pressure rises above that of the aorta, blood is forced into the aorta past the semilunar valves. Pressure falls as the ventricles empty and the walls relax.

Ventricular volume rises as the atria contract and the ventricles fill with blood, and then drops suddenly as blood is forced out into the aorta when the semilunar valve opens. Volume increases again as the ventricles fill with blood.

Figure 4 *Pressure, volume and ECG changes in the left side of the heart during the cardiac cycle*

SUMMARY TEST 8.9

The mammalian heart beat is initiated from within the heart muscle itself, which is therefore termed (1). The pacemaker of the heart is the (2), which lies in the wall of the chamber called the (3). A wave of excitation causes both (4) to contract. The wave is picked up by another group of specialised cells, called the (5), which in turn pass it down to the apex of the ventricles and out into the muscle via small branches of specialist muscle called (6).

Blood pressure and pulse

To function normally, mammals must maintain a minimum pressure of blood within their closed double circulation system (Topic 8.1). This pressure must then be increased as necessary to meet the demands of exertion. Blood pressure is still measured by the medical profession in millimetres of mercury (mmHg) and at rest this is between 120 and 140 mmHg (**systolic**) and 80 and 90 mmHg (**diastolic**), depending on age (blood pressure increases with age). The SI unit which should be used is kilopascals (kPa) and the equivalent values are 16.0–18.5 kPa (systolic) and 10.5–12.0 kPa (diastolic).

Blood pressure in arteries, capillaries and veins

When blood is pumped from the left ventricle of the heart it enters the circulation around the body via the aorta. In the aorta the pressure exerted causes extension of its elastic walls. The pressure falls a little as a result of this extension. The elastic wall of the aorta 'rebounds' in a reaction, called the **recoil action**, that increases the pressure again. This recoil leads to regular fluctuations in blood pressure as it moves along the aorta and other major arteries (Figure 1). As the blood moves further from the heart its overall pressure decreases, but only slightly at first. However, when the arteries divide into arterioles, the overall volume of the vessels increases. This leads to a more marked fall in blood pressure and also to a reduction of the pressure fluctuations. This decrease in pressure is essential to prevent the bursting of the one-cell-thick walls of the capillaries into which the blood passes next. This reduction in pressure continues in the capillaries so that, by the time the blood reaches the venules and veins, it has fallen to almost zero (Figure 1). Its return is therefore mostly dependent on the body muscles squeezing the thin-walled veins and the valves within them, ensuring that the blood moves towards the heart. The pattern of pressure changes is the same for the pulmonary circulation but the actual pressures are only around 25% of those in the circulation to the body.

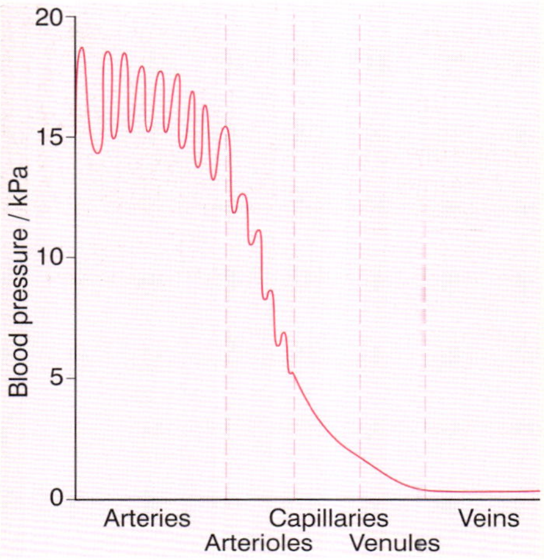

Figure 1 *Blood pressure in arteries, capillaries and veins*

What factors affect blood pressure?

The pressure of blood depends upon a number of factors including:

- how much blood the ventricles pump in a given time (cardiac output)
- the resistance to the flow of blood (e.g. narrowing of blood vessels during vasoconstriction)
- the total volume of the blood.

Blood pressure is increased by:

- increased cardiac output, e.g. during exercise
- increased resistance to blood flow, e.g. caused by the narrowing of the arterioles during **vasoconstriction** or the reduced elasticity of arteries due to atherosclerosis (Topic 9.5)
- increased blood volume, e.g. due to the retention of more water by the kidneys under the influence of antidiuretic hormone (ADH).

Blood pressure is reduced by:

- decreased cardiac output, e.g. during rest or sleep
- decreased resistance to blood flow, e.g. during the widening of arterioles during **vasodilation**
- decreased blood volume, e.g. during loss of blood due to haemorrhage or increased loss of water by the kidneys.

Figure 2 *Taking blood pressure*

Hypertension

Hypertension is the term used to describe sustained (continued) high blood pressure when at rest. It is normal for blood pressure to increase temporarily both during and after exercise. Hypertension is not always easy to define as there is much individual variation in 'normal' blood pressure, depending on age, general health and degree of activity. For example, blood pressure increases as one gets older as a result of the walls of the arteries becoming less elastic. The World Health Organization defines hypertension as a systolic pressure of 20.5 kPa (160 mmHg) and a diastolic pressure of 12.5 kPa (95 mmHg). Hypertension is a major risk factor associated with heart disease, as well as increasing the risk of **atherosclerosis**, stroke and kidney failure. It is very much the 'silent killer' because the disease itself has no symptoms to warn of a forthcoming heart attack or stroke. For reasons that are not fully understood, some individuals are prone to hypertension. In others, it is the consequence of some other medical condition, such as kidney failure or hormone imbalance. There are, in addition, a number of other factors that can contribute to hypertension. These include:

- smoking
- excessive alcohol intake
- too much salt in the diet
- stress
- obesity
- lack of exercise.

There is also a genetic link because there is a tendency for high blood pressure to run in families. Hypertension can be treated by behavioural changes, such as stopping smoking, reducing alcohol or salt intake, increasing exercise (Figure 3), etc. and/or medication.

Figure 3 *Exercise reduces your resting pulse rate and helps prevent high blood pressure (hypertension)*

What is a pulse?

We have seen that the surge of pressure created by the contraction of the walls of the left ventricle causes the elastic walls of the aorta to stretch, creating a 'bulge'. This extended portion of the aorta wall then rebounds due to the elastic fibres in its wall. With the semi-lunar valves preventing blood returning to the heart, this recoil action of the aorta wall causes a new pressure bulge to be created in the aorta a little further away from the heart. This process then continues along the aorta and into the rest of the arterial system. It is these pressure bulges, passing in sequence along arteries, which we recognise as the pulse. The pulse rate is the number of times a person's heart beats in one minute.

SUMMARY TEST 8.10

The normal systolic blood pressure measured in SI units is in the range **(1)**. The equivalent diastolic range is **(2)**. As a person gets older, blood pressure usually **(3)**. Blood pumped from the left ventricle enters the **(4)** blood system via the main artery, called the **(5)**. The recoil action of the walls of this vessel leads to fluctuations in blood pressure called pulses. Blood pressure falls most steeply in the blood vessels called **(6)** and, by the time it reaches the **(7)**, it is almost non-existent. If the blood vessels called arterioles dilate, the pressure of the blood will **(8)**. If the body retains more water than usual, then blood pressure will **(9)**. Sustained high blood pressure when at rest is termed **(10)**.

8 Examination Questions

1   In mammals, haemoglobin is used to transport oxygen and myoglobin is used to store oxygen in muscles.

Figure 1 shows the oxygen dissociation curves for myoglobin, fetal haemoglobin and adult haemoglobin.

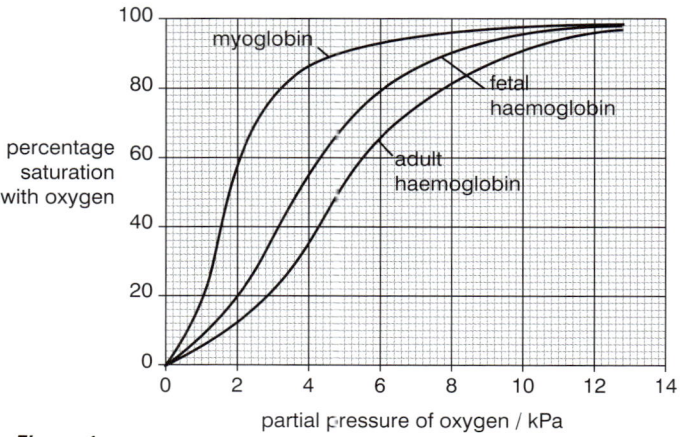

Figure 1

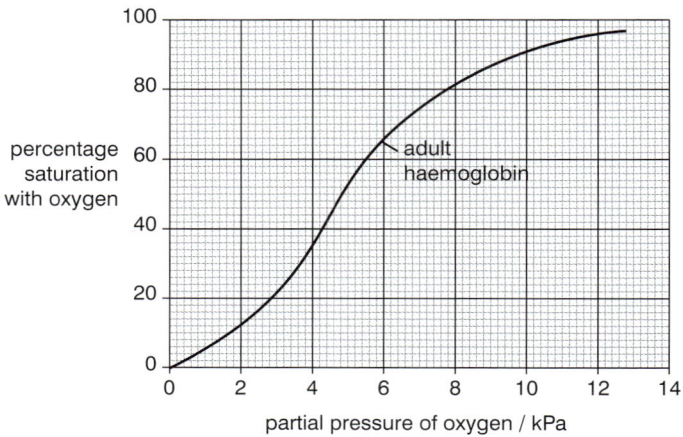

Figure 2

Cambridge International AS and A Level Biology 9700 Paper 2 Q4 June 2011

a i   Name the cells in which haemoglobin is found.
(1 mark)

ii   Use Figure 1 to determine the percentage saturation of myoglobin and adult haemoglobin when the partial pressure of oxygen is 3 kPa.
(1 mark)

iii There is a large difference between the percentage saturation of myoglobin and that of adult haemoglobin at low partial pressures of oxygen. Suggest reasons for this.   (2 marks)

b Fetal haemoglobin has a different oxygen binding affinity to that of adult haemoglobin, as shown in Figure 1. Normally, after birth, the production of the fetal form stops and the adult form is produced. In a rare condition known as Hereditary Persistence of Fetal Haemoglobin (HPFH), fetal haemoglobin continues to be produced well into adulthood in addition to adult haemoglobin. This condition, however, usually lacks any symptoms.
i   Explain, with reference to Figure 1 the significance of the difference in oxygen binding affinity between fetal and adult haemoglobin.   (2 marks)
ii   Suggest why HPFH usually lacks symptoms.   (1 mark)

c Sketch on Figure 2 the dissociation curve you would expect for adult haemoglobin if the concentration of carbon dioxide is increased.
(2 marks)
(Total 9 marks)

2   Mammals have *closed*, *double* circulatory systems.

a Explain what are meant by the terms closed and double as applied to mammalian circulatory systems.
(2 marks)

Figure 3 shows a longitudinal section through a mammalian heart.

Figure 3

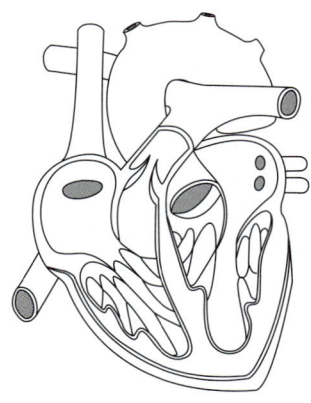

b Use label lines and the letters **P**, **Q**, **R** and **S** to label the following on Figure 3:

P – the right atrium, **Q** – a semilunar valve, **R** – a blood vessel that carries deoxygenated blood, **S** – the position of Purkyne tissue   (4 marks)

Catheters are small tubes that are inserted into blood vessels. A catheter was inserted into an artery in the arm and then moved into the aorta and then into the left ventricle during a diagnostic investigation. The catheter contained a device to measure the blood pressure in the aorta and in the left ventricle. The results are shown in Figure 4.

Figure 4

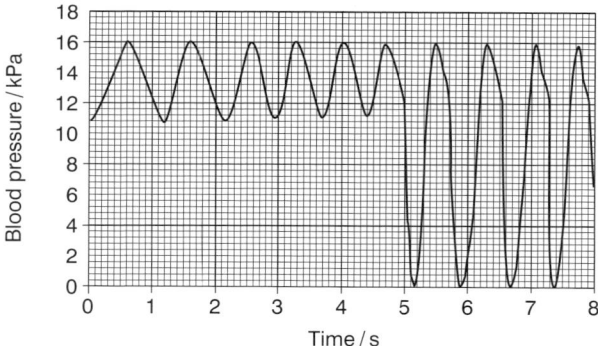

c i Calculate the heart rate during the period of the investigation. Show your working. *(2 marks)*
 ii Describe **and** explain the differences in pressure as the catheter moves from the aorta into the left ventricle. *(4 marks)*

Figure 5 is an X-ray showing narrowing in the blood vessels supplying muscles in the heart. A catheter is used to insert a dye into the blood vessels so that they appear clearly in the X-ray. The arrows indicate where there is narrowing of the blood vessels.

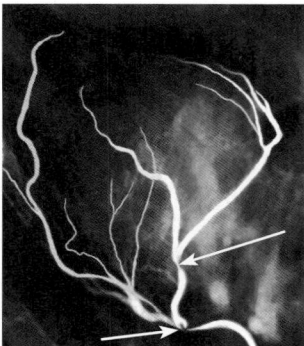

Figure 5

d i Name the blood vessels shown in Figure 5. (*1 mark*)
 ii State the likely effect of narrowing of these blood vessels. (*1 mark*)

e Suggest ways in which the condition shown in Figure 8 may be treated. *(2 marks)*

(Total 16 marks)

Cambridge International AS and A Level Biology 9700 Paper 21 Q5 November 2012

8 Practice Questions

3 The lugworm is an organism that spends almost all its life in a U-shaped burrow made on muddy seashores. Most of the time it is covered by sea water, which it circulates through its burrow. Oxygen diffuses into the lugworm's blood from the water and it uses haemoglobin to transport oxygen to its tissues. Figure 6 shows the oxygen dissociation curve of lugworm haemoglobin compared to that of adult human haemoglobin.

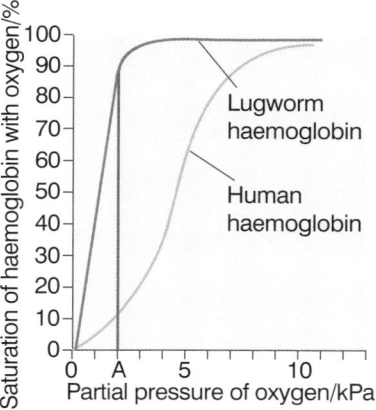

Figure 6

a In Figure 6, line A is drawn at a partial pressure of oxygen of 2 kPa. This is the partial pressure of oxygen found in lugworm burrows after the sea no longer covers them. Using figures from the graph, explain why a lugworm can survive at these concentrations of oxygen while a human could not.

b Using the graphs in Figure 6, explain how the lugworm is able to obtain sufficient oxygen from an environment that contains so little.

c Haemoglobin usually loads oxygen less readily when the concentration of carbon dioxide is high (the Bohr effect). The haemoglobin of lugworms does not exhibit this effect. Explain why to do so could be harmful.

d Suggest a reason why lugworms are not found higher up the seashore.

e Llamas are animals that live at high altitudes. At these altitudes the atmospheric pressure is lower and so the partial pressure of oxygen is also lower. It is therefore difficult to load haemoglobin with oxygen. Suggest where the oxygen dissociation curve of llama haemoglobin is shifted to, relative to human haemoglobin.

Gas exchange and smoking

Human gas exchange system

On these pages you will learn to:

- Describe the gross structure of the human gas exchange system
- Describe the functions of cartilage, cilia, goblet cells, mucous glands, smooth muscle and elastic fibres and recognise these cells and tissues in prepared slides, photomicrographs and electron micrographs of the gas exchange system

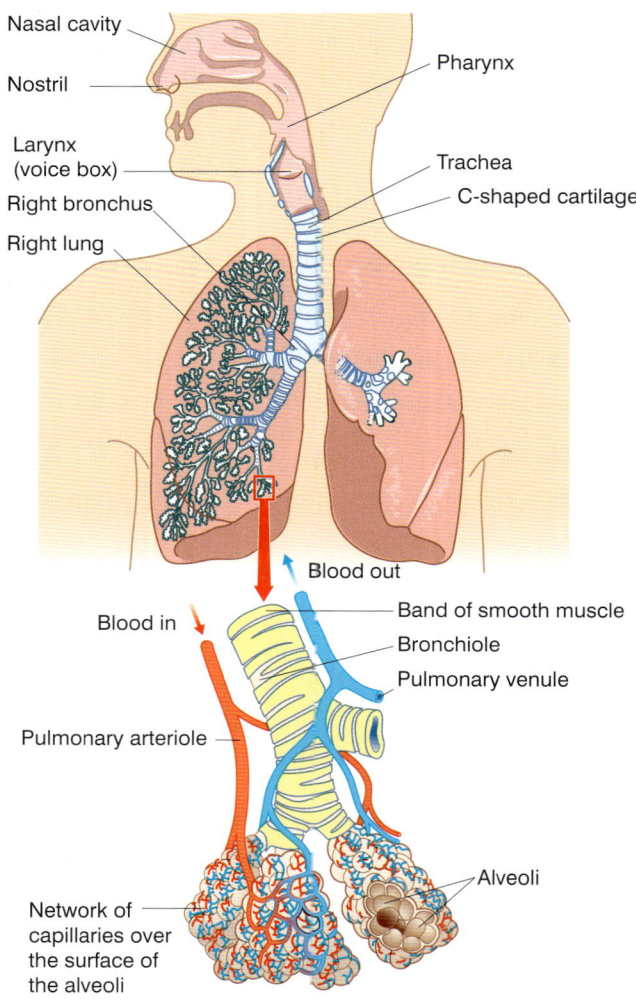

Figure 1 *Human gaseous exchange system showing positions of trachea, bronchi and bronchioles as well as the arrangement of blood vessels and alveoli in the lungs*

All **aerobic** organisms require a constant supply of oxygen in order to produce **ATP** during respiration and supply energy for the needs of the organism. The carbon dioxide produced in the process needs to be removed from the body. The volume of oxygen that has to be absorbed and the volume of carbon dioxide that must be removed are large in mammals because:

- they are relatively large organisms with a large volume of living cells
- they maintain a high body temperature and therefore have high metabolic and respiratory rates.

As a result, mammals have evolved specialised surfaces, called lungs, to ensure efficient gas exchange between the air and their blood. The lungs are located deep within the thorax. Air is warmed and moistened before it reaches the lungs, and so loss of heat and water are minimised. The lungs are supported and protected by a bony box called the rib cage. The ribs can be moved by the muscles between them, called the intercostal muscles. This enables the lungs to be ventilated so that the air within them is constantly changed.

Distribution of tissues within the trachea, bronchi, bronchioles and alveoli

Once air has been drawn in through the mouth or nose, it passes through a region at the back of the throat called the pharynx, before entering the trachea. The **trachea** is a flexible airway that is supported by pieces of **cartilage** (Figure 1). These pieces of cartilage do not form complete rings, but are C-shaped. This allows the adjacent oesophagus to expand without obstruction when food passes along it on its way to the stomach. The wall of the trachea is made up of **smooth muscle** and elastic fibres. It is lined with **ciliated epithelium** which is composed of ciliated epithelial cells and **goblet cells**.

The trachea branches into two bronchi, one **bronchus** entering each of the two lungs (Figure 1). The structure of the bronchi is similar to that of the trachea, only they are smaller in diameter. They too are supported by cartilage (plates of cartilage, not C-shaped rings), have walls of smooth muscle and elastic fibres and are lined with ciliated epithelium, containing ciliated epithelial cells and goblet cells. The bronchi divide repeatedly to give a series of airways, called **bronchioles**. These are similar in appearance to the bronchi, although they lack cartilage. They end in a group of alveoli (Figure 1) which are tiny hollow sacs lined with **squamous epithelium** and elastic fibres and surrounded by a network of **blood capillaries**. The distribution of these tissues is summarised in Table 1.

Functions of tissues within the gas exchange system

- **Cartilage** is a rigid, but flexible supporting material. Its incomplete rings in the trachea and plates of cartilage in the bronchus support the smooth muscle of these airways, keeping them in an open position. In this way, it prevents the trachea and bronchi from collapsing when the air pressure inside them is lowered when breathing in. Being flexible and arranged in rings with spaces between each ring, it allows the trachea to bend and extend – essential when bending or stretching the neck as a whole.

- **Smooth muscle** is tissue which is capable of contraction, but which is not under voluntary control. Although found in the trachea, bronchi and bronchioles, it is in the bronchioles where it has its effect. As the bronchioles are not supported by cartilage, contraction of the rings of smooth muscle around them causes the bronchioles to constrict. In this way, the flow of air to and from the alveoli can be restricted and therefore controlled.

- **Elastic fibres** are flexible fibres that recoil if stretched. The presence of elastic fibres in the walls of the alveoli means that when the alveoli expand during inhalation, the elastic fibres stretch. This is followed by their elastic recoil, which allows the air to be forced out of the alveoli during exhalation.

- **Goblet cells** are so called because they have long, thin stems and rounded tops, resembling a wine goblet in shape (see Figure 2). They are part of the epithelium of trachea and bronchi and produce mucus that forms a thin layer over the whole inner surface of these structures.

- **Mucous glands** are found in the lining of the trachea, bronchi and bronchioles. Like goblet cells, they produce mucus. Mucus acts as a barrier, preventing pathogens entering cells. Its thickness increases during infection to increase the distance between pathogens and cells. Mucus is sticky, and so bacteria, other pathogens, allergens such as pollen and dust particles are trapped in it as the air that contains them moves along these airways. White blood cells, called macrophages, migrate from blood vessels and engulf bacteria and other particles trapped in the mucus. In this way the risk of lung infections is reduced.

- **Ciliated epithelium** is a thin layer of epithelial cells that have hair-like organelles called cilia on one surface (Figures 2 and 3). These cilia move in a synchronised (coordinated) manner and so transport mucus containing the trapped microorganisms, allergens and dust that covers them upwards towards the pharynx. Once at the pharynx, the mucus passes down the oesophagus to the stomach, usually unnoticed. Here the acid conditions kill any bacteria not consumed by the macrophages.

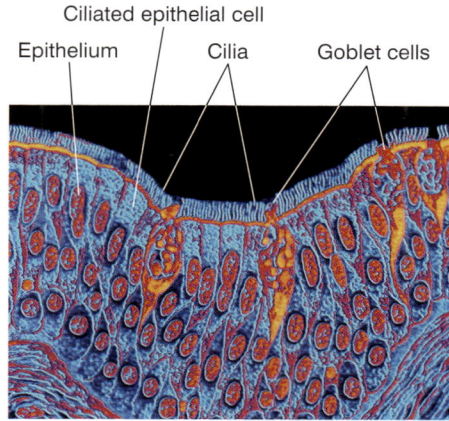

Figure 2 *Ciliated epithelium of human trachea showing goblet cells*

Figure 3 *False-colour SEM of epithelium of the trachea showing ciliated epithelium (× 3570)*

Table 1 *Distribution of tissues within the human gas exchange system*

| Structure | Features | | | | |
	Cartilage	Ciliated epithelium	Elastic fibres	Goblet cells	Smooth muscle
Trachea	present	present	present	present	present
Bronchus	present	present	present	present	present
Bronchiole	absent	present	present	present (but becoming fewer)	present
Alveolus	absent	absent	present	absent	absent

You should be able to observe and draw plan diagrams of the structure of the walls of the trachea, bronchi, bronchioles and alveoli.

Further details about the distribution of tissues within the human gas exchange system, and the drawing of plan diagrams, can be found in the Practical skills section on the accompanying CD.

SUMMARY TEST 9.1

Air enters the body through the mouth or nose and enters the trachea, which is supported by **(1)** shaped pieces of cartilage. The trachea branches into two bronchi. The trachea and bronchi are lined with epithelium, which produces mucus from **(2)** cells. This mucus is sticky and traps **(3)** and **(4)** present in the air. The mucus is moved towards the mouth by **(5)**, present on the epithelium. The bronchi, in turn, divide to form **(6)**, which can constrict when the **(7)** they contain contracts. These structures are dilated by the action of **(8)**.

Gas exchange in the alveoli

On these pages you will learn to:

- Describe the process of gas exchange between air in the alveoli and the blood

The alveoli are the site of gas exchange in mammals. They are tiny hollow sacs, each with a diameter of between 100 μm and 300 μm. Each lung of an adult human contains around 300 million of them, giving them a surface area of around 70 m². This is about half the area of a tennis court. Each alveolus is lined with squamous epithelium (Topic 1.9) and together with the thin basement membrane and small amount of connective tissue containing the elastic fibres, they form the wall of the alveolus, or alveolar wall. As each cell is only between 0.1 μm and 0.5 μm thick, the alveolar wall is very thin. The external appearance of a group of alveoli is shown in Figure 1.

REMEMBER

The diffusion pathway is short because the alveoli have only a single layer of epithelial cells and the pulmonary blood capillaries have only a single layer of endothelial cells.

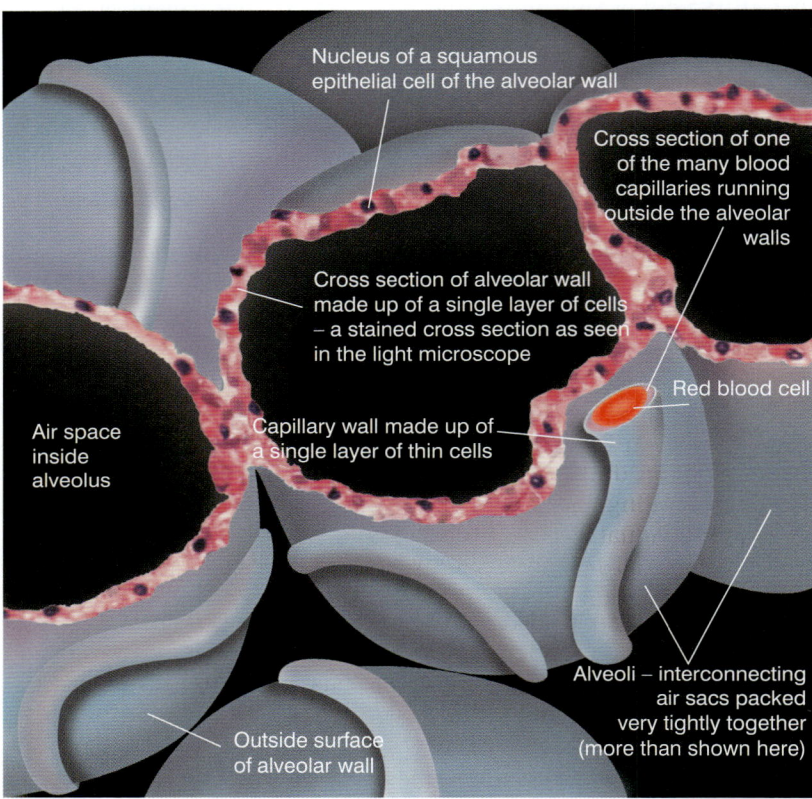

Nucleus of a squamous epithelial cell of the alveolar wall

Cross section of one of the many blood capillaries running outside the alveolar walls

Cross section of alveolar wall made up of a single layer of cells – a stained cross section as seen in the light microscope

Red blood cell

Air space inside alveolus

Capillary wall made up of a single layer of thin cells

Alveoli – interconnecting air sacs packed very tightly together (more than shown here)

Outside surface of alveolar wall

Figure 1 *External appearance of a group of alveoli*

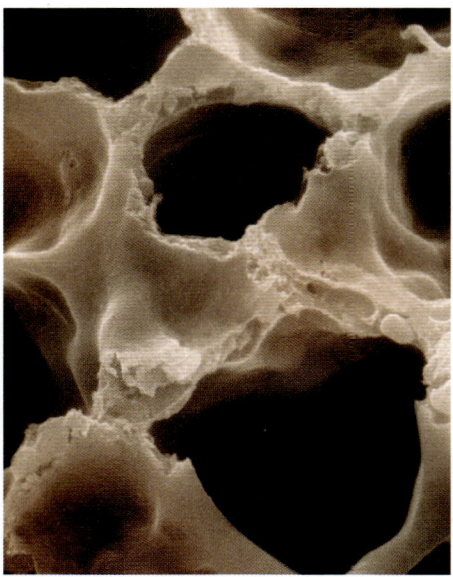

Figure 2 *Colourised scanning electron micrograph of human lung showing thin-walled alveoli*

Diffusion of gases between the alveoli and the blood is very rapid because:

- the walls of both alveoli and capillaries are very thin and therefore the distance over which diffusion takes place is very short
- alveoli and pulmonary capillaries have a very large total surface area
- red blood cells are slowed as they pass through pulmonary capillaries, allowing more time for diffusion
- the distance between alveolar air and red blood cells is reduced as the red blood cells are flattened against the capillary walls
- breathing movements constantly ventilate the lungs, and the action of the heart constantly circulates blood around alveoli. Together, these ensure that a steep concentration gradient of the gases to be exchanged is maintained.

Figure 3 shows the diffusion of gases in an alveolus.

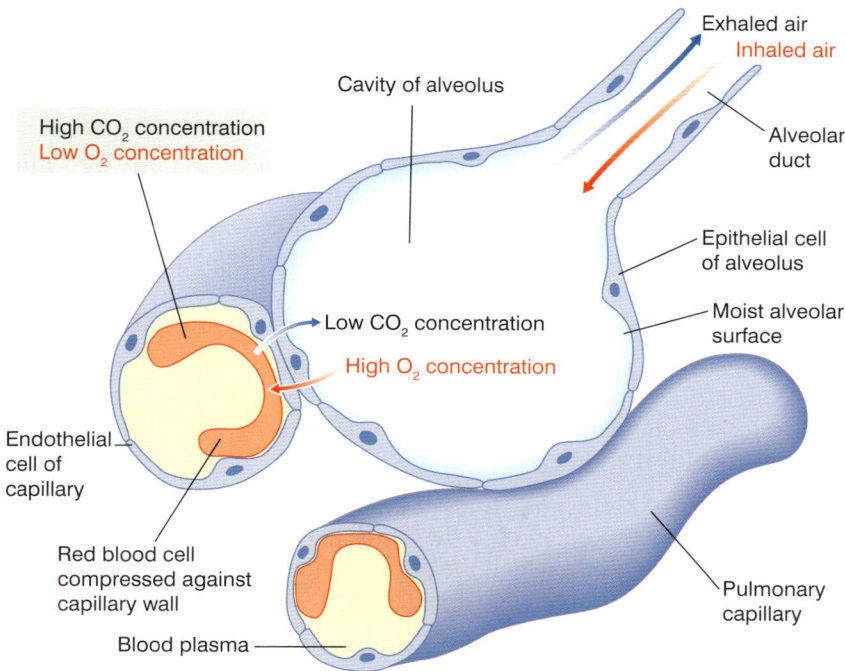

High CO_2 concentration
Low O_2 concentration

Cavity of alveolus

Exhaled air
Inhaled air

Alveolar duct

Epithelial cell of alveolus

Moist alveolar surface

Low CO_2 concentration

High O_2 concentration

Endothelial cell of capillary

Red blood cell compressed against capillary wall

Blood plasma

Pulmonary capillary

Figure 3 *Diffusion of gases in an alveolus*

SUMMARY TEST 9.2

The site of gaseous exchange in mammals is the alveoli, which have a diameter of **(1)** and have walls made of **(2)** which is very thin, being only **(3)** in thickness. Diffusion across the alveolar epithelium is rapid because it is **(4)** and it has a large **(5)**. The rate of diffusion is also increased because red blood cells passing through **(6)** capillaries are both **(7)** and **(8)**.

EXTENSION

Lung capacities

A range of terms is used to describe the changes in the volume of the lungs during breathing. The actual volumes vary a little depending on the size, sex, age, health and fitness of an individual. During normal breathing at rest, you will exchange 0.5 dm³ (500 cm³) at each breath. This is called **tidal volume**. If you breathe in normally and then, rather than breathing out, continue to breathe in until your lungs are as full of air as you can make them, you will have taken in a further 1.5 dm³ (1500 cm³) of air (inspiratory reserve volume). If you carry out a similar process on breathing out, you will find that you can expel around an additional 1.5 dm³ (1500 cm³) of air if you force out of the lungs as much as you can (expiratory reserve volume). You may think you have completely emptied the lungs of air at this point. However, you would be wrong because around 1.5 dm³ (1500 cm³), called the **residual volume**, remains in the lungs. Without this residual volume, the moist walls of the alveoli would stick together and it would be impossible to re-inflate the lungs. If you take a maximum breath out, followed by a maximum breath in, you will inhale some 3.5 dm³ (3500 cm³) of air. This is called the **vital capacity**. These volumes can be measured using an instrument called a **spirometer** and are illustrated in the graph in Figure 4. Although 0.5 dm³ of air (tidal volume) is exchanged during normal breathing, only 0.35 dm³ of this air actually reaches the alveoli and is hence available for gas exchange.

The remaining 0.15 dm³ is left in the trachea, bronchi and bronchioles, where no gas exchange occurs.

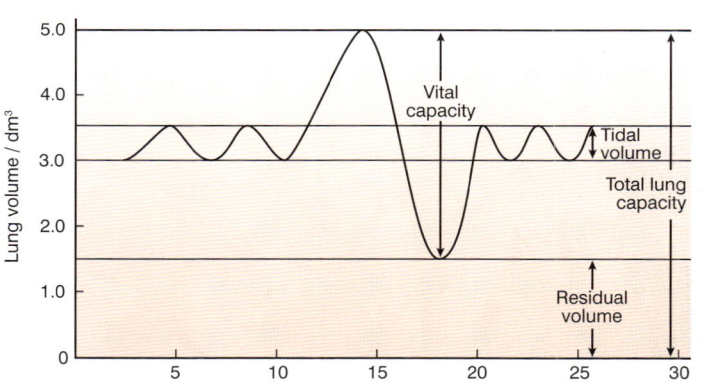

Figure 4 *Graph showing lung capacities as measured using a spirometer*

To summarise:

- Tidal volume (0.5 dm³) – the volume of air normally exchanged at each breath at rest.
- Residual volume (1.5 dm³) – the volume of air that is always left in the lungs.
- Vital capacity (3.5 dm³) – the volume of air that can be exchanged between maximum inspiration and maximum expiration.

Tobacco smoke and its effects on gas exchange

On these pages you will learn to:

- Describe the effects of tar and carcinogens in tobacco smoke on the gas exchange system

On average, smoking a single cigarette lowers one's life expectancy by 7 minutes – longer than it takes to smoke the cigarette! While this is a statistical deduction rather than a scientific one, there is now clear scientific evidence to support the view that smoking cigarettes damages your health and reduces life expectancy.

Who smokes and how much?

Most people who smoke begin smoking before they are adults. Nearly a quarter of young smokers had their first cigarette before they were ten years old. The greatest proportion of young smokers is in parts of India, some Western Pacific islands and in Central and Eastern Europe.

Around 250 million women and almost a billion men smoke worldwide. Smoking rates have, however, peaked and are slowly declining. In general, educated people give up smoking first which means that the habit is more common amongst poorer, less educated people.

Despite this, the number of cigarettes smoked is increasing as a result of the expanding world population and smokers smoking more cigarettes. Each year, manufacturers produce nearly 1000 cigarettes for every man, woman and child in the world. Asia, Australia and the Far East are the largest consumers with America, Eastern Europe, Former Soviet Countries and Western Europe close behind.

What is in tobacco smoke?

Tobacco smoke is a mixture of up to 4000 different chemicals and is released into the atmosphere in two ways:

- **mainstream smoke** from the filter or mouth end of the cigarette/cigar/pipe
- **sidestream smoke** from the burning tip of the cigarette/cigar/pipe tobacco.

Around 85% of tobacco smoke in a room will be sidestream smoke – the form which contains higher concentrations of many toxins than mainstream smoke. Breathing in this smoke is called **passive smoking** and presents a health hazard to people nearby who inhale it. Of the thousands of chemicals in tobacco smoke, three important ones are:

- **Carbon monoxide** – a poisonous gas also found in car exhaust fumes and in the smoke from burning many fuels.
- **Nicotine** – a poisonous alkaloid drug that is addictive. 60 mg of nicotine placed on the tongue would kill an individual within minutes. It is absorbed by the body very rapidly, reaching the brain in less than 30 seconds.

Figure 1 *It has been calculated that smoking each one of these cigarettes would, on average, reduce your life-expectancy by seven minutes.*

- **Tar** – a sticky, brown substance responsible for the staining of the fingers and teeth of smokers. It appears in tobacco smoke as minute droplets.

Effects of carcinogens in tobacco smoke on the gas exchange system

Over 80% of all lung cancer deaths are caused by smoking and a quarter of all smokers die from this cause. The additional risk of getting lung cancer is related to the number of cigarettes smoked. Tobacco smoke contains a number of carcinogens (factors that cause cancer). These cause considerable damage to the **genes** of the epithelial cells which line the bronchi and bronchioles and are therefore exposed to these carcinogens and may cause mutations of some genes. Among these genes where mutations have occurred are ones that control normal cell division. It is these mutations that give rise to lung cancer. One identified carcinogen is benzopyrene (BP) found in the tar of tobacco smoke. A derivative of BP binds directly to the tumour-suppressor gene p53, which normally stops cell division and destroys cells that have mutations. The binding inactivates the gene and epithelial cells divide by **mitosis** in an uncontrolled manner leading to the formation of a **tumour**. Growth of this tumour may be slow and it may take over 20 years for any symptoms to develop.

Mutations in another type of gene, protooncogenes, may cause them to become oncogenes and lead to uncontrolled division. Protoconcogenes are active in early development of an individual and have a role in cell growth and division. Normally these genes are switched off, but once they become oncogenes following a mutation, they are active and can cause uncontrolled cell division.

Figure 2 *Smoking causes a range of lung diseases*

Effects of tar in tobacco smoke on the gas exchange system

Tar in tobacco smoke is a mixture of chemicals that enter the respiratory tract as an aerosol of minute droplets. About 70% of this tar is deposited on the airways and alveoli. It is an irritant that causes inflammation of the mucous membranes (the epithelium and connective tissue containing the mucous glands) lining the trachea, bronchi and bronchioles, resulting in them producing more mucus. At the same time, the tar paralyses the cilia on its surface (Topic 9.1). As a consequence these cilia cannot remove the mucus secreted by the epithelial lining. The mucus, laden with dust and microorganisms, therefore accumulates in the lungs, leading to infections and damage. The cough, typical of many smokers, is the result of trying to remove this build-up of mucus from the lungs. These various responses, either directly or indirectly, reduce the rate of gaseous exchange in a number of ways.

- Build-up of mucus on the walls of the airways reduces their diameter and so limits the rate at which air can reach the alveoli.
- Coughing can cause damage to the airways and alveoli; scar tissue builds up which again reduces air movements and rates of diffusion.
- Infections arise because the cilia no longer remove mucus and pathogens remain in the airways to give more time to gain entry into epithelial cells or the circulation. Bacteria will also replicate and increase their population size, increasing the chance of infection.
- **Allergens** such as pollen also accumulate, leading to further inflammation of the airways, reduced air-flow in and out of the lungs, and possible **asthma** attacks.

In addition to its effects on gaseous exchange, tars also contribute to chronic obstructive pulmonary disease (COPD) and lung cancer (Topic 9.4) as well as cardiovascular disease (Topic 9.5).

SUMMARY TEST 9.3

Tobacco smoke contains thousands of chemicals but three are of particular importance to health: carbon monoxide, tar and a poisonous and addictive alkaloid called **(1)**. 70% of the tar inhaled is deposited on airways such as the trachea, **(2)** and **(3)**. This leads to the paralysis of the **(4)** on the epithelium of these airways, causing **(5)** containing bacteria, dirt and tar to accumulate in the lungs. The bacteria may be **(6)** that cause infectious disease. Substances which cause cancer are called **(7)** and are found in tobacco smoke. They cause cells to divide by **(8)** in an uncontrolled way.

Pulmonary diseases and smoking

On these pages you will learn to:

- Describe the effects of tar and carcinogens in tobacco smoke on the gas exchange system with reference to lung cancer and chronic obstructive pulmonary disease (COPD)

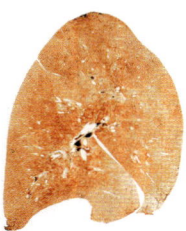

Figure 1 *Section through a healthy human lung*

Figure 2 *Section through the cancerous lung of a smoker*

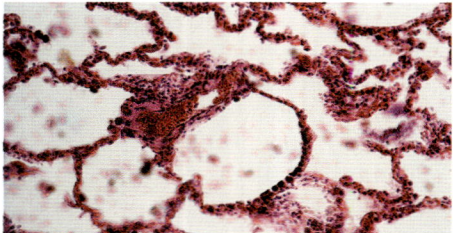

Figure 3 *Normal lung tissue as seen under a light microscope*

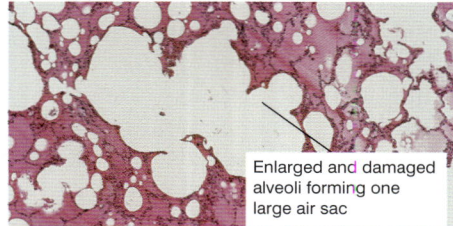

Enlarged and damaged alveoli forming one large air sac

Figure 4 *Lung tissue damaged by emphysema as seen under a light microscope. Note the enlarged, damaged alveoli reducing the surface area for gas exchange*

It is estimated that 5.4 million premature deaths worldwide occur each year as a result of smoking. Almost one million of these deaths are in India. In China smoking kills 2000 people every day. About 20% of these deaths are due to two diseases, emphysema and chronic bronchitis, and 25% are a result of lung cancer (Figure 2). Chronic bronchitis and emphysema frequently occur together, in which case the term **chronic obstructive pulmonary disease (COPD)** is used to describe this combined condition.

Chronic bronchitis

Chronic bronchitis is a long-term, life-threatening disease. It results from irritation of the lining of the bronchi and bronchioles by dust, fumes, atmospheric pollution and, most significantly, tars in tobacco smoke. Acute bronchitis differs from chronic bronchitis in that it is caused by a viral or bacterial infection and is usually a short-term illness, when the person makes a full recovery. A person with chronic bronchitis is more at risk from infectious diseases such as acute bronchitis. Symptoms of bronchitis include:

- inflammation of the mucous membranes of airways, restricting the passage of air and causing breathlessness
- excessive production of mucus from the goblet cells (Topic 9.1) of the epithelium
- damage to the cilia of the epithelium, causing a slowing or complete halt to their action of moving mucus containing pathogens and other trapped particles out of the lungs
- accumulation of this mucus in the lungs, reducing air movement and gas exchange
- coughing, as reflex actions cause the forced expulsion of mucus
- build-up of scar tissue on the bronchi and bronchioles, as a result of the coughing, leading to even thicker walls and greater restriction of air-flow
- infections, such as pneumonia, due to bacteria accumulating rather than being 'swept away' along with mucus by the cilia.

Chronic bronchitis develops slowly and the main symptoms of coughing, wheezing and breathlessness only are noticed when the disease is persistent (does not go away). It is difficult to reverse the damage caused once the disease is established.

Emphysema

One in every five smokers will develop serious lung disease called **emphysema**, which in its later stages severely affects the mobility of the individual and the ability to carry out normal activities. In its early stages the only symptom is a slight breathlessness but, later a chronic cough and blue skin colouration develop. Other later symptoms include difficulty in breathing (shortness of breath and difficulty breathing out), wheezing and fatigue (tiredness). Death due to emphysema is usually a result of respiratory failure, often accompanied by infection. A small number die of heart failure as the heart becomes enlarged and over-worked pumping more blood through arteries to compensate for the reduced concentration of oxygen it contains.

Healthy lungs (Figures 1 and 3) contain large quantities of elastic fibres comprised predominantly of the protein elastin. This tissue stretches when we breathe in and recoils, returning to its former size when we breathe out. In emphysematous lungs the elastic fibres are damaged and when air enters the alveoli they stretch as the alveolus increases in size but are unable to recoil. This means that the lungs are no longer able to force out all the air from the alveoli. Some of the alveoli may burst, creating one larger air sac and reducing the surface area available for gas exchange (Figure 4). As a result, little if any exchange of gases can take place. The bronchioles become inflamed and scarred, leaving a narrower lumen.

The damage to the elastic fibres is brought about by abnormally high levels of elastase, an enzyme formed in some of the white blood cells, which breaks down elastin. Elastase also degrades other proteins so that, in the later stages of the disease, breakdown of lung tissue results in large, non-functional holes in the lung. In healthy lungs elastin is not broken down because a protein inhibitor (PI) inhibits the action of the enzyme elastase. However, in smokers, it has been suggested that the oxidants in cigarette smoke inactivate PI, resulting in greater elastase activity and hence a breakdown of elastin.

Elastase is produced by **phagocytic** cells which need it so they can migrate through tissue to reach sites of infection. This is part of the body's normal inflammatory response. In smokers, where a large number of phagocytic cells are attracted to the lungs by the particles in smoke, a combination of the release of elastase and a low level of its natural inhibitor lead to a lot of tissue degradation. The only way to minimise the chances of getting emphysema is not to smoke at all, or to give up – the function cannot be restored to smoke-damaged lungs but giving up can significantly slow down the progress of the disease.

Lung cancer

We saw in Topic 9.3 that tobacco smoke contains a number of **carcinogens**. These cause **mutations** that lead to uncontrolled **mitosis** and the formation of tumours. The symptoms of lung cancer include:

- persistent cough, which is a reflex action to the obstruction to the airways caused by the tumour
- blood in the sputum, resulting from damage to the lung tissues caused by the tumour and coughing
- shortness of breath because the tumour is obstructing the airways and replacing the alveoli
- hoarseness or other changes to the voice due to pressure of the tumour on the airways, larynx or nerves serving the larynx
- wheezing noises as air is forced along airways obstructed by the tumour.

Cancer cells may break away from the original tumour and spread the disease to other parts of the body (metastasis), leading to a wide variety of symptoms as a result of these secondary tumours.

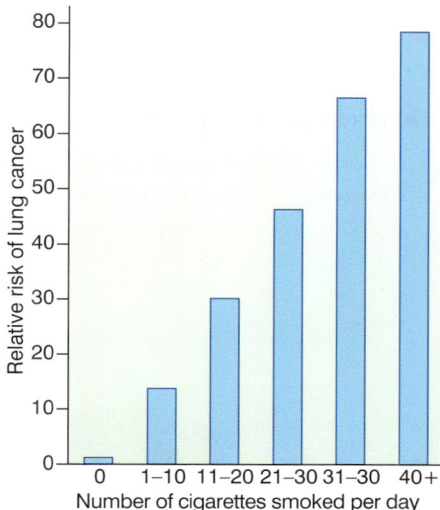

Figure 5 *Relative risk of lung cancer according to daily cigarette consumption (Source: Souhami & Tobias, Cancer and its management, 1986)*

SUMMARY TEST 9.4

Of the 5.4 million premature deaths worldwide caused by smoking, 25% result from lung cancer and 20% from emphysema and chronic bronchitis, which together are called **(1)**. Emphysema's early symptom is **(2)**, although this may take up to 20 years to develop. Emphysema results from white blood cells producing the enzyme **(3)** so that they can migrate through the lung tissues to fight infections and engulf particles from tobacco smoke. The enzyme breaks down the protein **(4)** in the alveoli, resulting in the production of non-functional holes. Chronic bronchitis is due to swelling of the **(5)** membranes of the airways and the production of excess mucus from the **(6)** cells within them. Damage is also caused to the **(7)** on the membranes and, as a result, mucus containing pathogens and other particles accumulates in the lungs. Tar in tobacco smoke contains cancer-causing chemicals called **(8)** which can cause gene mutations in the cells lining the airways. Cells divide to form a **(9)** from which cells may become separated and spread the disease to other parts of the body – a process called **(10)**. Symptoms of lung cancer include **(11)**, **(12)** and **(13)** as well as chest pain, hoarseness and a wheezing noise when breathing.

Cardiovascular disease and smoking

On these pages you will learn to:

- Describe the short-term effects of nicotine and carbon monoxide on the cardiovascular system

Nicotine and carbon monoxide in tobacco smoke have short-term effects in the body. Some of these effects contribute to longer-term effects associated with smoking, such as cardiovascular disease.

Effects of carbon monoxide on the cardiovascular system

Carbon monoxide in tobacco smoke combines easily, and permanently. As haemoglobin has a higher affinity for carbon monoxide than oxygen, carbon monoxide reduces the oxygen-carrying capacity of the blood. To supply the equivalent volume of oxygen to the tissues, the heart must therefore work harder. This can lead to hypertension (high blood pressure), described in Topic 8.10, which in the long term increases the risk of coronary heart disease and strokes. The reduction in the oxygen-carrying capacity of the blood means that it may be insufficient to supply the heart muscle during exercise in people who already have cardiovascular disease. This could lead to angina or, in severe cases, a myocardial infarction. Carbon monoxide can also cause damage to the endothelial lining of blood vessels and can speed up the deposition of cholesterol and LDLs in atheroma formation.

Effects of nicotine on the cardiovascular system

Nicotine in tobacco smoke stimulates the production of the hormone adrenaline by the adrenal glands, leading to an increase in heart rate and raised blood pressure (hypertension). It can cause constriction of arteries (vasoconstriction), which further increases blood pressure. As a consequence there is a greater risk of smokers suffering coronary heart disease (2.5 times greater than in a non-smoker) or a stroke (1.5 times greater than a long-term non-smoker or a light smoker, but rising to 3 times for those smoking 20 or more cigarettes a day). Nicotine also makes the platelets in the blood more 'sticky', and this leads to a higher risk of thrombosis and hence of strokes or myocardial infarction (heart attack). Platelets are blood cells that have a role in the clotting of blood. Like carbon monoxide, nicotine can also damage the endothelial lining, which if not repaired, can lead to the start of atheroma formation (see Extension).

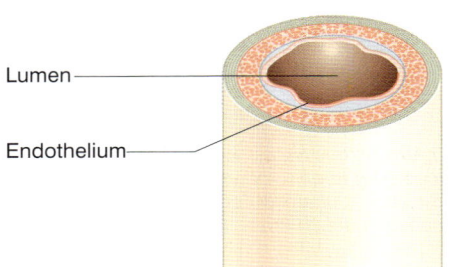

Lumen
Endothelium

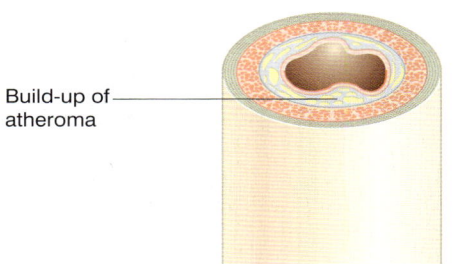

Build-up of atheroma

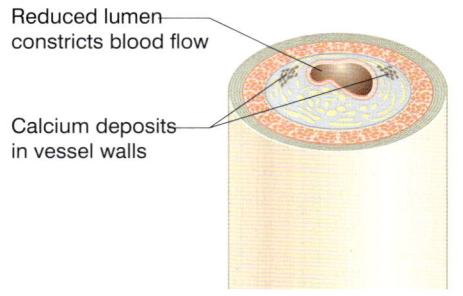

Reduced lumen constricts blood flow

Calcium deposits in vessel walls

Figure 1 *Development of arteriosclerosis*

REMEMBER

To use the term 'atheroma' to describe fatty deposits in an artery. Writing about 'furred up arteries' is too imprecise.

EXTENSION

Cardiovascular disease

Cardiovascular disease consists of diseases of the heart and circulatory system. There are many forms, of which the three most significant are:

- atherosclerosis
- coronary heart disease
- stroke.

It is the leading cause of death in many countries of the world. Cardiovascular disease is multifactorial (has many causes), many of which are self-inflicted, including smoking. In particular it is the carbon monoxide and nicotine components of tobacco smoke that increase the risk of cardiovascular disease amongst smokers.

Atherosclerosis

Atherosclerosis is a condition in which the walls of blood vessels thicken as a result of **cholesterol** and other fatty substances being deposited (jointly termed 'atheroma'). These walls, over time, harden ('sclerosis'). When atherosclerosis occurs in arteries, it is termed arteriosclerosis. Arteriosclerosis begins with damage to the endothelial lining and the formation of streaks made up of white blood cells that have taken up low density lipoproteins, which contain cholesterol. These then increase in size to form patches, known as **plaques**, that eventually thicken the wall and narrow the lumen (opening) of the artery (Figure 1). These thickenings, called **atheromas**, most commonly occur in the larger arteries and are the result of fibres, dead smooth muscle cells and cholesterol being deposited. These may eventually rupture the artery lining, leaving it rough and uneven, which causes platelets in the blood to begin the clotting process. The blood clot that forms further prevents blood flow and starves of oxygen the tissues that the artery serves. Atheromas are uneven and disturb the flow of blood, which can then gradually form a clot, known as a **thrombus**. This thrombus can become so large that it blocks the blood vessel where it forms. Small pieces, called **emboli**, may break off and block smaller vessels elsewhere. These blockages cause the blood supply to particular tissues to be reduced (ischaemia). Over many years, calcium may also become deposited in the atheroma, causing the artery wall to harden. This condition is known as **arteriosclerosis** and is particularly associated with aging (Figure 1).

Coronary heart disease

Coronary heart disease (CHD) is a disease affecting the coronary arteries which supply blood to the cardiac muscle of the heart (Topic 8.7). It is also called ischaemic heart disease because, as a result of an atheroma or thrombus, the supply of blood to the heart muscle is restricted. This causes a condition called **angina** in which symptoms such as chest pain and difficulty in breathing occur, especially when exercising. If the blockages of the coronary arteries are very large, part of the heart may be completely deprived of oxygen and die. This is called a **myocardial infarction**, or heart attack (Figures 2 and 3).

Strokes

A **stroke** occurs when the supply of arterial blood to part of the brain is interrupted in some way and, as a result, cells in that region die. It is also called a **cerebrovascular accident** and is of two types.

- **A thrombosis** (blood clot) develops either on an atheroma in an artery supplying the brain with blood, or an atheroma develops elsewhere and a piece of it breaks away to block an artery supplying the brain. In either case the blockage prevents blood, and therefore oxygen, reaching a particular region of the brain.
- **An aneurysm** is a weakness in the wall of an artery, often the result of atherosclerosis. These aneurysms frequently burst, leading to a **haemorrhage** and therefore the loss of blood supply to the region of the brain served by that artery.

The effects of a stroke depend on the region of the brain that is deprived of blood. Most commonly a stroke causes paralysis on one side of the body.

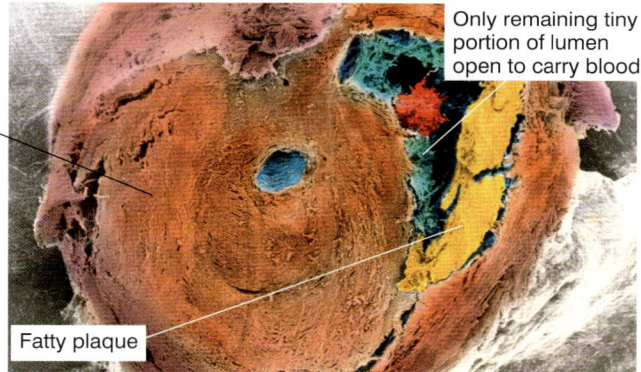

Arterial wall thickened with fibres, dead smooth muscle cells and fatty deposits

Only remaining tiny portion of lumen open to carry blood

Fatty plaque

Figure 2 *Human coronary artery with an atheroma almost completely obstructing its lumen*

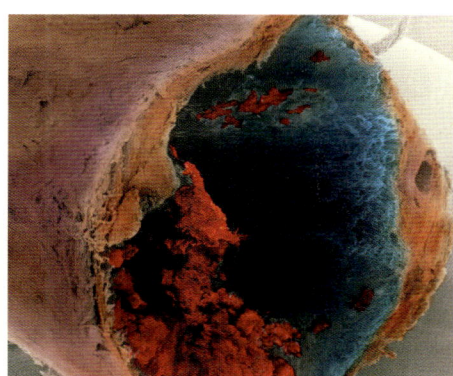

Figure 3 *Human coronary artery with a thrombus (bottom left) partly obstructing the lumen*

SUMMARY TEST 9.5

Cardiovascular diseases are degenerative diseases of the heart and circulatory system which are the main cause of death in many countries. In arteriosclerosis the walls of the arteries thicken due to deposits of **(1)** and other fatty substances. Initially patches called **(2)** develop, which eventually increase in size to form **(3)**, which may block the **(4)** of the artery. As a result the flow of blood is disturbed and so a **(5)** may develop. If this happens in a coronary artery the blood supply to a region of the heart may be cut off, depriving it of **(6)**. This region of heart muscle may die, something called a **(7)**, or heart attack. If, however, the blockage is in an artery serving the brain, then a stroke, also called a **(8)**, is the result. One type of stroke occurs when a weakness in the artery wall, called an **(9)**, bursts, leading to a **(10)**. Some cardiovascular disease is the result of the gas **(11)** in tobacco smoke, which combines permanently with **(12)**. Another constituent of tobacco smoke, nicotine, increases the production of the hormone **(13)**, which in turn leads to a rise in **(14)**.

Drug (substance) abuse

What is a drug? There are many definitions but they usually have in common the idea that a drug in some way modifies the normal mental or physical functions of the body. The term is usually restricted to those substances which are taken into the body rather than ones produced naturally by it.

Legal and illegal drugs

Legal drugs are ones that are lawful according to the legislation of a particular country, illegal ones are unlawful. It follows that which drugs are legal and which are not, differs from country to country. Marijuana and alcohol are acceptable in some countries, but not in others. Some drugs, such as amphetamines and barbiturates, have a perfectly legal and acceptable use as prescribed treatments, but become illegal when taken without the doctor's advice by the person or by people who have not been prescribed the drug. The acceptability of drugs also changes with time. Tobacco smoking, once popular and encouraged, is now considered antisocial in many countries while alcohol was banned in the USA from 1920 to 1933.

Drug dependency

Drug dependence arises when an individual has a particular need to take a drug and removal of it leads to a certain degree of discomfort. Two types of dependency are recognised:

- **Psychological dependency** – occurs when there is an emotional need for a drug and any discomfort is restricted to an altered state of mind such as depression.
- **Physical dependency** – occurs when the body has become adapted to the drug in such a way that it, or its products, are necessary for its systems to function normally. Withdrawal of the drug leads to physical symptoms which can cause severe discomfort or pain.

Alcohol abuse

The alcohol in most alcoholic drinks is ethanol. Once consumed, ethanol is rapidly absorbed by the stomach and then metabolised by the liver into ethanal. Short-term effects of this ethanal are:

- Hepatitis (inflammation of the liver) due to ethanal altering the shape of proteins to the extent that the body no longer recognises them as its own. Antibodies are then produced against these proteins leading to hepatitis.
- A fatty liver as a consequence of the excess alcohol being metabolised instead of fat which therefore builds up in the liver.

Longer term effects of the ethanal produced from alcohol consumption are:

- Cirrhosis of the liver (Figure 1) which is the result of collagen production causing fibrous scars that prevent the normal functioning of liver cells.
- An increased risk of liver cancer.
- Demyelination of nerve cells – damage to the myelin sheath around nerve cells.
- Impaired (damaged) nervous transmission due to inhibition of **neurotransmitter** production as well as demyelination. This causes brain damage resulting in memory loss, poor judgement and difficulty in learning.
- Dehydration of brain cells which similarly affects brain function.

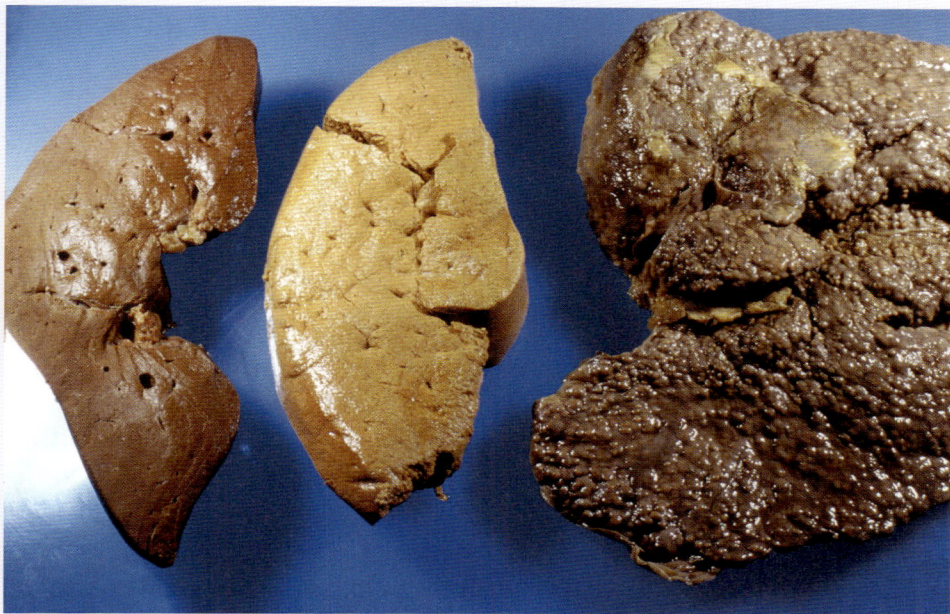

Figure 1 *A healthy liver (left), a fatty liver (centre), and a liver with cirrhosis (right)*

The effects of excessive alcohol use are not restricted to physical damage. The social consequences can be at least as severe. These include:

- Aggressive and violent behaviour because judgement is impaired.
- Family violence between partners and other relatives with women and children being especially at risk.
- Family breakdown.
- Drink driving that may cause injury and death.
- Petty crime such as threatening behaviour, minor theft and vandalism.
- Serious crime including serious assault, rape and even murder.
- Economic consequences of additional policing and medical care.

Units of alcohol and safe limits

A unit of alcohol is usually measured as 10 cm³ (8 g) of pure alcohol although there are slight differences in different countries. This is equivalent to:

- half a pint (280 cm³) of ordinary strength beer, lager or cider (3–4% alcohol by volume), or
- a small measure (25 cm³) of spirits (40% alcohol by volume), or
- a measure (50 cm³) of fortified wine such as sherry or port (20% alcohol by volume), or
- a small glass (80 cm³) of ordinary strength wine (12% alcohol by volume).

Governments, acting on medical advice, set out safe limits known as Daily Alcohol Limits (DAL). These are generally:

- **Men** should drink no more than four units of alcohol in any one day and no more than 21 units per week.
- **Women** should drink no more than three units of alcohol in any one day and no more than 14 units per week
- **Pregnant women**. It is known that alcohol can damage a developing baby. Therefore, advice is that pregnant women and women trying to become pregnant should not drink at all. If women do choose to drink when pregnant they are strongly advised to limit it to one or two units, once or twice a week.

Obesity

Table 1 Body mass index (BMI)

BMI	Category
Under 20	Underweight
20–24.9	Acceptable
25–29.9	Overweight
30–39.9	Obese
Over 40	Very obese

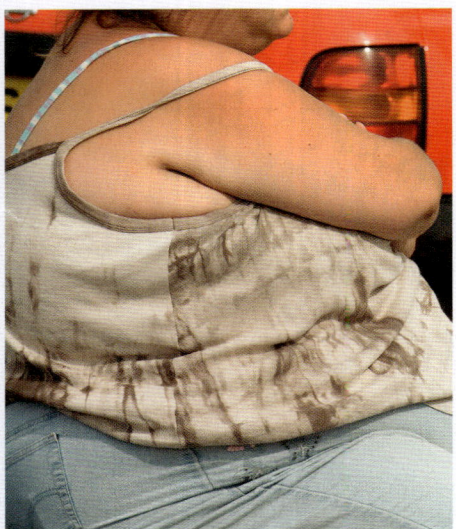

Figure 1 Obesity is on the increase in both developed and developing countries

Over-nutrition is as much a form of malnutrition as under-nutrition. If people regularly consume more energy than they use, they will put on mass and if they continue to do so, they will become obese.

Body mass index (BMI)

Body mass index (BMI) is a mathematical calculation used to find out whether an individual is underweight or overweight. It is calculated by dividing a person's body mass in kilograms by his/her height in metres squared.

$$\text{Body mass index (BMI)} = \frac{\text{body mass in kilograms}}{(\text{height in metres})^2}$$

Table 1 shows how this index can be used to determine whether a person is underweight, overweight or obese. The BMI should only be used for adults because younger people store fat as part of their normal growth and this distorts the figure produced. Some consider the ratio of waist to hip circumference to be a better measure. A ratio above 0.9 in men and above 0.85 in women indicates obesity.

How common is obesity?

Obesity is very much more common in the developed countries, especially those of North America, Europe and Australasia. Even among the poorer households in these countries there is a tendency to spend scarce income on high-calorie fat and carbohydrate foods, rather than on fruit and vegetables. There is worrying evidence that obesity is becoming increasingly common in the developing nations, especially those with emerging economies, e.g. China. In the Caribbean, over 50% of females are in the overweight category with over 20% being obese. The equivalent values for men are 40% overweight and 10% obese.

- Globally, women have higher rates of obesity than men, although men have a higher incidence (number of new cases) of being overweight.
- Obesity in children and adolescents is on the rise in both developed and developing countries.
- Obesity is uncommon in African and Asian developing countries; where it does occur, it is more common in urban than rural areas.
- In the majority of countries obesity has increased by between 10% and 40% in the past 10 years.
- As a country's economic development increases, so does the BMI of its people.
- In some developing countries, obesity occurs along with starvation.

Why is obesity on the increase?

Some of the possible reasons for the increasing number of cases of obesity in developed countries are:

- **lack of physical exercise** – entertainment is often sedentary (inactive), e.g. watching TV, playing computer games
- **sedentary occupations** – machines now carry out much physical labour that was previously performed manually
- **well-heated homes and work places** – mean that little energy is lost as heat and therefore less food needs to be consumed
- **better transportation** – people drive or take public transport rather than walk, and take lifts (elevators) or escalators rather than climb stairs

SUMMARY TEST 9.7

Body mass index (BMI) is calculated by dividing a person's **(1)** by his/her **(2)** squared. A person is considered to be obese if he/she has a BMI above **(3)**, and underweight if their BMI is below **(4)**. People who are obese are more likely to suffer from coronary heart disease due to increased **(5)** and to high levels of **(6)** in the blood that are associated with obesity. They are also prone to a form of diabetes called **(7)**. Two degenerative diseases that are associated with obesity are **(8)** and **(9)**. High blood cholesterol levels can increase the risk of CHD, as can a high intake of **(10)** and fatty acids – especially **(11)** ones.

- **increased wealth** – food is readily available and easily afforded
- **convenience / fast foods** – beefburgers and pizzas, for example, are relatively cheap, readily available, heavily marketed and very high in calories
- **increased stress** – leads to the use of convenience foods (they are quick to prepare in a busy world) and 'comfort eating'
- **increased alcohol consumption** – alcohol is nutritionally poor, but high in calories
- **urbanisation** – means food stores / burger bars, etc. are close by and often open 24 hours a day. Fear of crime in cities can prevent people, especially women and the elderly, from walking outside, especially at night.

Health problems associated with obesity

People who are obese are at greater risk of developing one or more serious medical conditions, and of dying prematurely. There is scientific evidence that obesity is associated with more than 30 medical conditions. These conditions include:

- **Coronary heart disease** – may be caused by increased blood pressure and blood **cholesterol** levels that commonly occur with obesity.
- **Type II (mature onset) diabetes** – food intake, especially carbohydrate, in obese individuals can be greater than normal insulin production can cope with. As a result they are unable to control their blood sugar levels properly. As many as 90% of type II diabetes sufferers are overweight.
- **Cancer** – breast cancer, especially in post-menopausal women, is more common if they are overweight. Cancers of the oesophagus, colon, rectum and the kidney are also more common in the obese, especially men.
- **Osteoarthritis** – a degenerative disease that affects especially the hips, spine and knees of obese individuals because the extra body weight puts additional strain on the bones in these regions.
- **Rheumatoid arthritis** – another degenerative disease which is more likely to arise when people are obese because extra weight puts stress on joints.
- **Hypertension** – around three-quarters of all cases of high blood pressure occur in people who are obese.

Diet and coronary heart disease

Tables 2 and 3 show that there is an increased risk of **coronary heart disease** (CHD) in both men and women who are overweight, with those who are obese being about twice as likely to suffer from CHD than those with an acceptable BMI. CHD arises when a fatty substance called atheroma builds up inside the coronary arteries that supply blood to heart muscle. This narrows these arteries, reducing the flow of blood and causing angina (pain or pressure in the chest) and heart attacks. There are a number of links between diet and CHD in addition to obesity:

- **High levels of salt** produce hypertension (increased blood pressure) that increases the risk of CHD.
- **High blood cholesterol** contributes to the formation of atheroma in the coronary arteries, especially the form known as low density lipoprotein (LDL). LDLs are used to transport cholesterol from the liver to various tissues, including artery walls where they leave their cholesterol – especially at sites of damage. High density lipoproteins (HDLs) transport cholesterol in the opposite direction, i.e. from the tissues to the liver, ready for excretion. HDLs therefore help guard against CHD.
- **High fatty acid intake**, especially saturated fatty acids, increases the risk of CHD. For example, an increase in trans fatty acids from the recommended 2% of food intake to 4% increases the incidence of CHD by a quarter.

On the other hand, there are a number of aspects of the diet which seem to reduce the risk of coronary heart disease. These include:

- **Eating dietary fibre (non-starch polysaccharides)** protects against obesity and reduces insulin levels in the blood, so reducing CHD.
- **Moderate consumption of alcohol**, especially wine and beer, has been shown by some studies to reduce the risk of CHD.
- **Eating oily fish**, such as mackerel and herring.

Table 2 Prevalence (number of cases) of medical conditions by body mass index (BMI) for men (source: NHANES III, 1988–1994)

Condition	% Prevalence of medical conditions at different BMIs			
BMI	18.5–24.9	25–29.9	30–34.9	≥40
Type II diabetes	2.03	4.93	10.10	10.65
Coronary heart disease	8.84	9.60	16.01	13.97
High blood pressure	23.47	34.16	48.95	64.53
Osteoarthritis	2.59	4.55	4.66	10.04

Table 3 Prevalence of medical conditions by body mass index (BMI) for women (source: NHANES III, 1988–1994)

Condition	% Prevalence of medical conditions at different BMIs			
BMI	18.5–24.9	25–29.9	30–34.9	≥40
Type II diabetes	2.38	7.12	7.24	19.89
Coronary heart disease	6.87	11.13	12.56	19.22
High blood pressure	23.26	38.77	47.95	63.16
Osteoarthritis	5.22	8.51	9.94	17.19

Exercise – its effects and consequences

In Topic 10.1 we will see that health is more than simply the absence of disease; it is a state of mental, physical and social well-being. Regular exercise contributes to these states of well-being by helping to keep people physically fit and therefore their bodily systems working efficiently and coordinating with one another.

Aerobic fitness

Aerobic fitness is the ability of an individual to take in and use oxygen. An individual's aerobic fitness depends on three factors:

- effective ventilation of the lungs
- effective transport of oxygen from the lungs to the respiring cells, e.g. muscle cells
- effective use of oxygen within the cells.

Aerobic fitness is therefore affected by the efficiency of the respiratory and cardiovascular systems. Although it is largely determined by an individual's **genes** (e.g. the extent and type of different muscle fibres present), aerobic fitness can be improved by training. It is normally measured by finding out the maximum volume of oxygen that an individual's body can take in and use. This is known as VO_2 max and is usually expressed as $dm^3 min^{-1} kg^{-1}$ or $cm^3 min^{-1} kg^{-1}$, where kg refers to body mass.

As individuals cannot work at their maximum aerobic capacity for long periods, a more useful measure is the percentage of their VO_2 max that an individual can operate at for prolonged periods. A well-trained endurance athlete can work at around 85% of their VO_2 max for long periods, while a non-athlete would struggle to maintain 65% of VO_2 max.

Short-term effects of exercise

During exercise, the demand for oxygen by muscle cells goes up. To meet the demand, the ventilation rate must be increased by:

- increasing the number of breaths taken each minute, from about 15 at rest to 45 during exercise
- increasing the volume of air exchanged at each breath, from around $0.5 \, dm^3$ at rest to $3.5 \, dm^3$ during exercise.

During moderate exercise like jogging, the amount of oxygen absorbed by the lungs can increase 13 times, from $0.3 \, dm^3 min^{-1}$ to $4.0 \, dm^3 min^{-1}$.

The short-term effects of exercise on the cardiovascular system are as follows:

- **Cardiac output** increases from around $5 \, dm^3 min^{-1}$ at rest to a maximum of $30 \, dm^3 min^{-1}$.
- **Vasodilation** or **vasoconstriction** of different blood arterioles redistributes the blood towards muscles and away from organs such as the kidney and intestines, whose need for blood is less immediate. The heart needs more blood to maintain the higher cardiac output and the brain must continue to function normally in order to coordinate activities – its supply is therefore largely unchanged during exercise. As exercise generates heat, the blood supply to the skin is increased during exercise to help dissipate this heat to the environment. Table 1 describes these changes in the distribution (transfer) of blood.
- There is a rise in **systolic** blood pressure although **diastolic** pressure is largely unchanged.

Table 1 *Rate of blood flow to different parts of the body at rest and during strenuous exercise*

Part of the body	Rate of blood flow/ $cm^3 min^{-1}$	
	At rest	During exercise
Liver and intestines	2500	90
Body muscle	1000	16000
Kidneys	1000	300
Brain	750	750
Skin	500	1000
Heart muscle	250	1200

Long-term effects of exercise

Regular exercise over a period of time produces changes to the heart which are designed to improve the rate at which blood transports oxygen from the lungs to the tissues, especially muscles. These changes include:

- **Hypertrophy of the heart** – this increase in size of the heart is often called athlete's heart. As the amount of **cardiac muscle** increases so does the force it can generate, allowing it to pump out more blood at each beat.
- **Increase in stroke volume** – the heart not only contracts more forcefully, but also holds more blood, allowing more to be pumped at each beat. The stroke volume for a trained athlete at rest is around 90 cm^3, compared to 65 cm^3 for an untrained individual.
- **Decrease in resting heart rate** – this is the result of cardiac output remaining the same but the stroke volume increasing in a trained athlete. The heart can therefore beat more slowly but still pump the same volume of blood. As a result of aerobic training, the resting heart rate may fall from 73 to 52 beats min^{-1}.
- **Increased maximum cardiac output** – while the cardiac output at rest is largely unchanged, training can increase the maximum output by around 50%.

In the same way as it affects the heart, regular exercise can produce adaptations of the blood and vascular system that improve the circulation of the blood. These include:

- **Increased volume of blood** – mostly as a result of an increase in the volume of blood plasma. However, there is also a small increase in the number of red blood cells. This provides more **haemoglobin** and hence a greater quantity of oxygen carried by the blood.
- **Decrease in blood pressure when at rest** – as a result of increased elasticity of the arterial walls.

- **Increase in the number of blood capillaries** – this occurs both in the lungs and in muscle. This improves the rate of gaseous exchange in both tissues.

General health benefits of regular and appropriate exercise include:

- **Reduced risk of coronary heart disease** both directly as a consequence of strengthening the heart and increasing the blood flow through it, and indirectly through lowering blood pressure and blood **cholesterol**, and reducing obesity. Exercising three times a week can halve the risk of a heart attack.
- **Reduced risk of hypertension** – as a result of improved elasticity of the artery walls and greater dilation of them, which reduces the resistance to blood flow within them.
- **Less obesity** – exercise uses up energy, e.g. one hour of jogging can use 4000 kJ – the equivalent of 140 g of fat. Losing weight also reduces the strain put on joints.
- **Lowering of blood cholesterol** and therefore the build-up of fatty **plaques** on the artery linings, which cause **atherosclerosis**.
- **Slowing the process of atherosclerosis** – arteries thicken less rapidly due to improved elasticity and lower cholesterol levels.
- **Maintaining sensitivity of cells to insulin** – this helps to keep blood sugar levels within the normal range, and so reduces the risk of type II **diabetes**.
- **Reduced risk of strokes** due to increased blood flow through the brain, lower blood pressure and less atherosclerosis.
- **Lower risk of osteoporosis**, especially in females, because exercise builds up the calcium content of bones, making them harder and less liable to break.
- **Stronger ligaments and tendons** and greater body flexibility with less risk of falls and strains.
- **Better mental health** as a consequence of being in a more relaxed psychological state and feeling good about yourself and your performance. Exercise can also improve sleep and reduce stress.

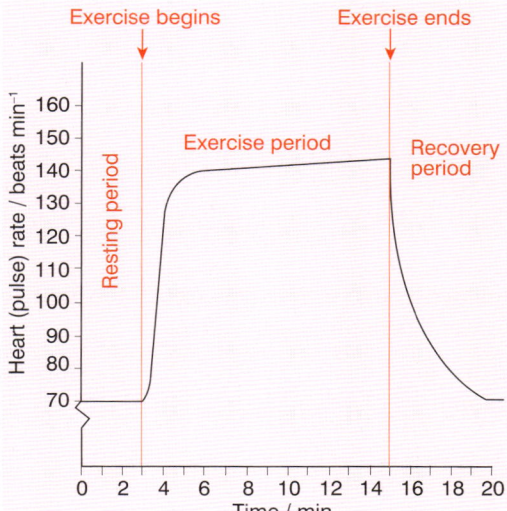

Figure 1 Heart (pulse) rate over a 20-minute period including 12 minutes of exercise

Figure 2 Taking regular exercise has important short- and long-term benefits to health

9 Examination Questions

1 Figure 1 is a diagram of part of the human gas exchange system.

Figure 1

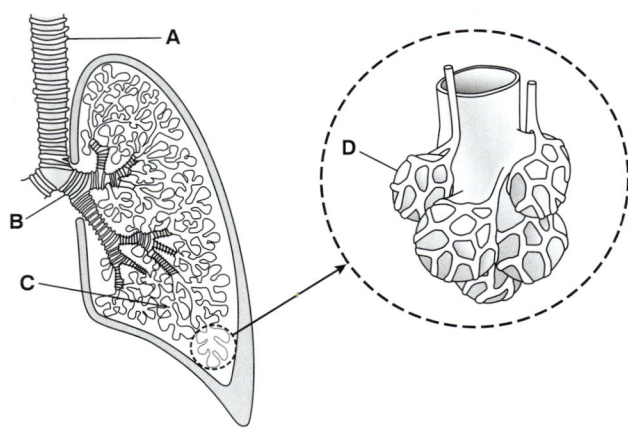

2 Figure 2 is a section of an alveolus and surrounding tissue.

Figure 2

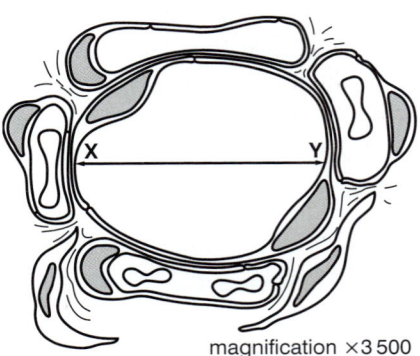

magnification ×3 500

a Calculate the actual diameter of the alveolus along the line **X–Y**. **Show your working and give your answer to the nearest micrometre**. *(2 marks)*

b i Describe the role of elastic fibres in the wall of the alveolus. *(2 marks)*

ii With reference to Figure 2, explain how alveoli are adapted for gas exchange. *(4 marks)*

c Chronic obstructive pulmonary disease (COPD) is a progressive disease that develops in many smokers. COPD refers to two conditions:
* chronic bronchitis
* emphysema.

i State two ways in which the lung tissue of someone with emphysema differs from the lung tissue of someone with healthy lungs. *(2 marks)*

ii State two symptoms of emphysema. *(2 marks)*

(Total 12 marks)

Cambridge International AS and A Level Biology 9700 Paper 21 Q2 June 2010

a Complete the table to show the distribution of the structural features within the parts of the gas exchange system, **A** to **D**, shown in Figure 1.

Use a tick (✓) if the feature is present and a cross (✗) if the feature is absent. Some of the boxes have been completed for you.

structure	features				
	cartilage	ciliated epithelium	elastic fibres	goblet cells	smooth muscle
A		✓		✓	
B			✓		
C				✓	✓
D	✗				✗

(4 marks)

b Explain the role of goblet cells and cilia in the maintenance of a healthy gas exchange system.

(4 marks)

(Total 8 marks)

Cambridge International AS and A Level Biology 9700 Paper 2 Q5 June 2010

3 Figure 3 is an electron micrograph of a cancer cell in the process of dividing by mitosis.

a The stage of mitosis visible in Figure 3 is metaphase.

State which features of the cell shown in Figure 3 indicate that it is at metaphase and not at anaphase. *(2 marks)*

b People who have smoked cigarettes for many years are at risk of developing lung cancer.

Describe how cigarette smoke is responsible for the development of lung cancer. *(4 marks)*

Figure 3

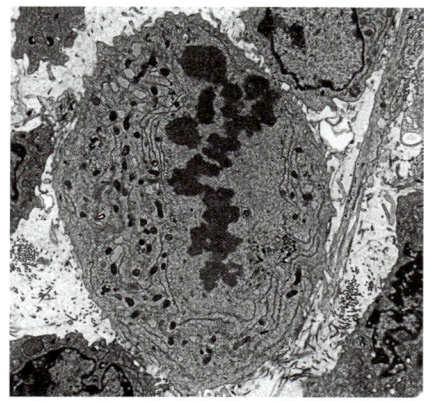

c Figure 4 shows the change in the percentage of smokers in the male population of the UK between 1950 and 2005.

Figure 5 shows the change in mortality rate in the UK in men aged 75 to 84 between 1950 and 2005.

Figure 4

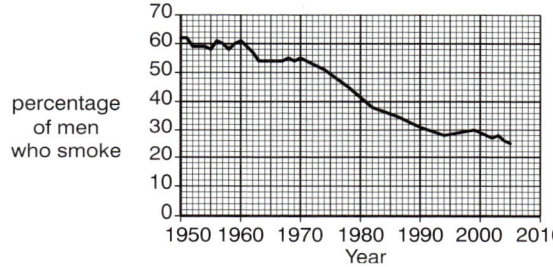

Figure 5

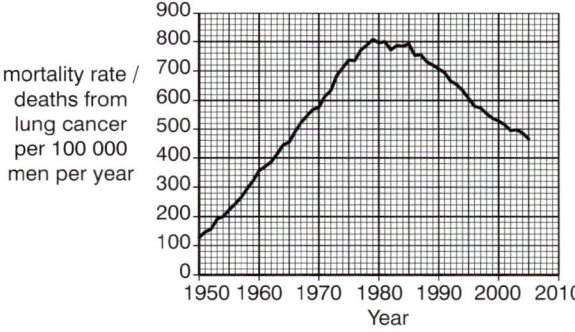

Figure 4 and Figure 5 appear to show that there is no link between the percentage of the population that smoke and the death rate from lung cancer.

Explain why the mortality rate from lung cancer among men increased and then decreased over the period shown in Figure 5, even though the percentage of smokers decreased over the same period of time.

(3 marks)

(Total 9 marks)

Cambridge International AS and A Level Biology 9700 Paper 2 Q6 June 2009

9 Practice Questions

4 The following graph shows the volume and pressure changes that occurred in lungs of a person during breathing while at rest.

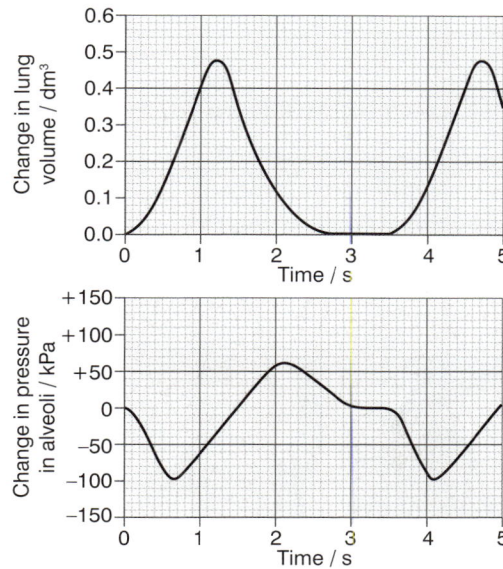

a From the graph, calculate the rate of breathing of this person. Give your answer in breaths per minute. Show how you arrived at your answer.

b If the volume of air in the lungs when the person had inhaled was 3000 cm³ what would the volume of air in the lungs be after the person had exhaled? Show your working.

5 a Name four risk factors associated with lung disease.

Life expectancy related to the number of cigarettes smoked

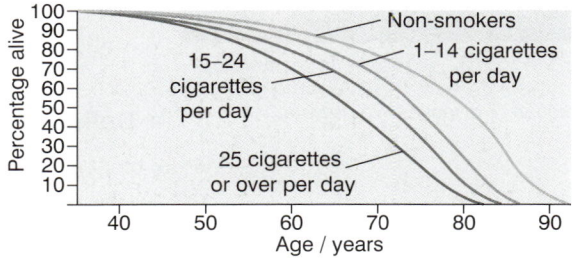

b Using the figure above, state what percentage of non-smokers are likely to survive to age 80.

c How many times greater is the likelihood of a non-smoker living to age 70 than someone who smokes over 25 cigarettes a day living to the same age?

d 10 to 15 years after giving up smoking the risk of death approaches that of non-smokers. Using this information and that in the figure above explain to a 40-year old who smokes 30 cigarettes a day the likely impact on her life expectancy of giving up smoking immediately.

Health and disease

On these pages you will learn to:

- Define the term 'disease' and explain the difference between an infectious disease and a non-infectious disease

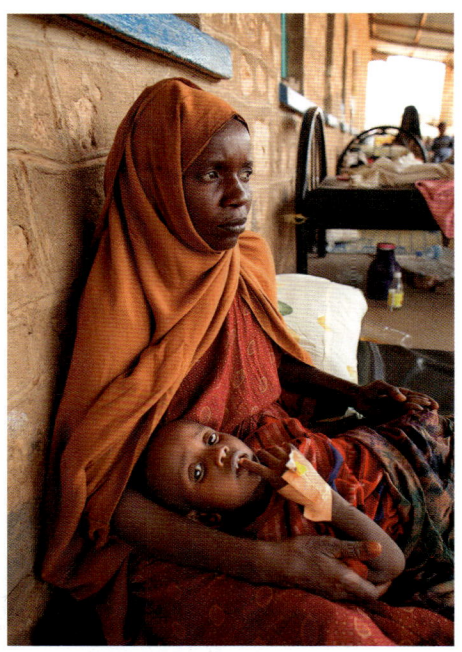

Figure 1 *A mother looks after her baby who is suffering from cholera. Health is more than just the absence of diseases like cholera.*

Before we define disease, it is helpful to look firstly at what we mean by health.

Definition of health

It is not easy to define health. Some might say that it is the absence of disease. However, would it be correct to describe a person whose bodily systems are functioning normally but who feels depressed as healthy? What if they are just 'unhappy'? Can an alcoholic or drug addict whose habits have not yet caused any physical harm be called healthy? The World Health Organization (WHO) defines health as: '**a state of complete physical, mental and social well-being and not merely the absence of disease or infirmity**'. In other words, to be healthy, a person should not only have all body organs working efficiently, but should also feel well. By the WHO definition, health includes:

- **physical well-being** – the body is able to function to its maximum potential and individuals are physically fit through taking exercise, having a balanced diet and taking rest and sleep they need
- **mental well-being** – individuals are free of mental illness and enjoy a high level of personal contentment (are satisfied and happy with their lives) as well as feeling good about themselves
- **social well-being** – individuals have their basic social needs met, e.g. proper housing and sanitation, and are properly integrated into the society to which they belong
- **absence of disease** – there are no symptoms of any disease that may prevent the efficient functioning of the body systems.

Often the 'feel good' factor that is important to good health comes from having control over one's life. This usually involves having the means to exercise this control and often includes having employment or some other means of income. Overall, there are three main factors which determine health – genetics, lifestyle and environment.

Definition of disease

If it is hard to say what is meant by 'health', to define 'disease' is even more difficult. Disease is not so much one thing, but rather a description of certain symptoms, either physical or mental, or both. Disease suggests a malfunction of body or mind which leads away from good health. Like health, it has mental, physical and social aspects. Some diseases, like malaria, have a single cause, but others, like heart disease, have a number of causes and are said to be **multifactorial**. Diseases may be divided into two forms depending upon their duration:

- **acute** – a sudden onset but short-duration condition from which the person recovers quickly
- **chronic** – an on-going (continuing) condition which may recur (keep occurring) over a number of years.

These terms do not describe the severity of a disease. A chronic case of indigestion may be hardly noticeable, while an acute attack of hepatitis may be fatal. There are many ways of classifying diseases, but whichever is used, the groups tend to overlap, so that any one disease may be put in a number of different categories. One way of classifying diseases is into two groups, infectious and non-infectious.

Infectious diseases

The human body makes an ideal habitat for many microorganisms. It provides a warm environment of constant temperature, a near-neutral pH, a ready supply of food and water and mechanisms for removing wastes. Not surprisingly therefore, our bodies are naturally colonised by a wide variety of microorganisms. Many live more or less permanently in or on our bodies, causing us no harm; others, however, cause disease. Any organism that lives on or in a host organism and gains an advantage while causing harm to the host is called a **parasite**. Parasites include worms such as *Wucheria spp.*, which causes elephantiasis, and insects like fleas. Parasitic disease-causing microorganisms, such as viruses, bacteria, fungi and protoctists, are called **pathogens**.

Diseases that can be spread from person to person, or from animals to people, are called infectious or communicable diseases. The parasites may be transmitted in a variety of ways, e.g. through water, food, sexual contact or other social interactions. Examples of infectious diseases include smallpox, measles (Figure 2), cholera (Figure 1), tuberculosis, malaria and HIV/AIDS.

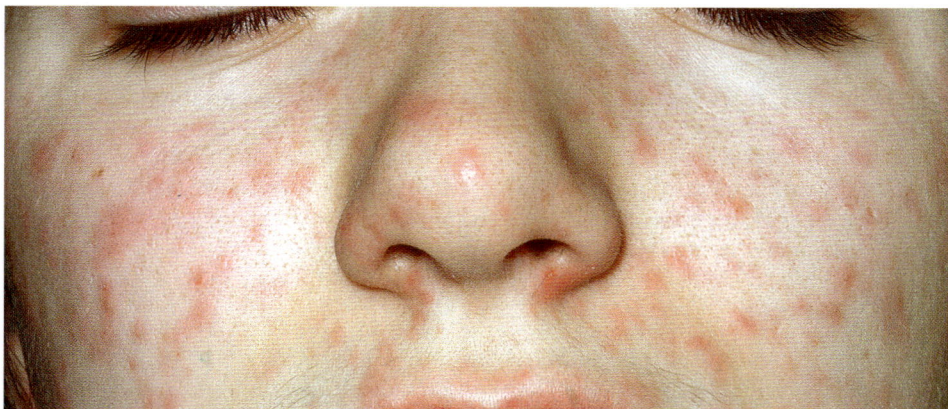

Figure 2 Measles is an example of an infectious disease

Non-infectious diseases

This is a category which includes all those diseases that are not caused by pathogens. Also called non-communicable diseases, they include sickle cell anaemia and lung cancer. Sickle cell anaemia is a genetic disease with a single cause – a change of a single nucleotide base in DNA (Topic 6.5). Lung cancer, by contrast, can be caused by a number of different factors.

EXTENSION
Genetic or inherited diseases

The terms 'genetic' and 'inherited' are often considered to have the same meaning when referring to disease. Strictly speaking, however, not all genetic diseases are inherited. The form of Down's syndrome where a child has three copies of chromosome 21, for example, is a genetic disorder but it is not inherited. All **cancers** could be considered as genetic diseases because they involve uncontrolled cell division (Topic 5.3) and yet very few forms of cancer are inherited. Many, possibly all, diseases could be said to have a genetic component because our genes partly determine such factors as our susceptibility to infectious disease and our personality, which in turn affect our lifestyle.

SUMMARY TEST 10.1

Disease is not merely the absence of health, but also the complete **(1)** well-being, **(2)** well-being and **(3)** well-being of an individual. Some diseases, such as heart disease, have many causes and so are said to be **(4)**. Regardless of its severity, a disease may either be short-lived, in which case it is said to be **(5)**, or be on-going over many years, in which case it is called **(6)**. Any organism that lives in or on a host organism, causing harm while gaining an advantage itself, is called a **(7)**. Disease-causing microorganisms are known as **(8)**. Where these diseases are spread from individual to individual, the disease is said to be **(9)**.

Types of disease

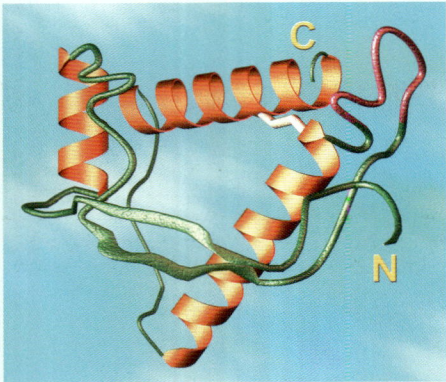

Figure 1 *Computer-generated model of a prion molecule, the abnormal form of a normal cell protein that causes Creutzfeldt–Jacob Disease (CJD). CJD shows the problem of classifying disease because it has mental and physical symptoms, is infectious and degenerative and, in some cases, can be inherited.*

We have seen that diseases can be broadly divided into infectious and non-infectious ones. There are however many other categories of disease as we shall see in this topic.

Physical diseases

Almost all diseases have some physical component. In other words, they involve permanent or temporary damage to some part of the body. An example is arthritis, which damages the joints. The only diseases that have no physical component are those psychological disorders that do not involve any physical damage to the brain.

Mental diseases

Mental diseases cover a broad range of disorders that cause psychological, personality or behavioural symptoms. What constitutes a mental illness may depend upon the 'normal' behaviour of a particular society. What is considered abnormal behaviour in one group might be perfectly acceptable in another. There is no clear distinction between physical and mental disease because each type often displays symptoms of the other. One example of a mental disease is schizophrenia. People with schizophrenia display severe distortion and disorder of thought, leading to delusions, hallucinations (unreal thoughts), strange behaviour and social withdrawal. The causes of schizophrenia are not fully understood but may be linked to the brain neurotransmitter, dopamine. A mental disease with a known cause is new variant Creutzfeldt–Jacob disease (CJD), the human form of BSE (bovine spongiform encephalopathy) in cattle. CJD shows the overlapping nature of disease classification because it has mental and physical symptoms, is infectious as well as degenerative and, in 15% of cases, is inherited.

Social diseases

Some diseases result from the social environment in which individuals live or the behaviour which they exhibit. They can include almost all infectious diseases because poor sanitation and over-crowding increase the risk of infections spreading in a population. Pulmonary tuberculosis can be spread in this way. Sexually transmitted diseases, such as acquired immune deficiency syndrome (HIV/AIDS), which are the result of sexual behaviour, are often referred to as social diseases. Social conditions may add to the risk of contracting (becoming ill from) certain diseases. Cardiovascular disease is more common amongst the poorer groups of the developed world, while deficiency diseases, such as kwashiorkor, are the result of a poor diet amongst the poor of the developing world.

Degenerative diseases

Degenerative diseases are the result of the gradual breakdown in the functioning of tissues or organs as the result of deterioration (gradually getting worse) (Figure 2). This deterioration may be the result of ageing as in the case of senile dementia. One form of senile dementia is Alzheimer's disease, where there is a series of degenerative changes in the brain leading to memory loss and confusion. Another example is rheumatoid arthritis, a common disease of joints. This begins with inflammation of the synovial membrane of a joint. This thickens and becomes filled with white blood cells that attack and erode away the cartilage at the ends of the bones in the joint. Examples like this, where the body's own immune system is mistakenly directed against its own tissues rather than foreign ones, are known as autoimmune diseases.

Table 1 *Main groups of disease and examples of each*

Category of disease	Scurvy	Measles	Lung cancer	Coronary heart disease	Alzheimer's disease	CJD	Cystic fibrosis
Physical	✓	✓	✓	✓	✓	✓	✓
Mental					✓	✓	
Social			✓	✓			
Infectious		✓				✓	
Non-infectious	✓		✓	✓	✓		✓
Degenerative				✓	✓	✓	
Inherited						✓	✓
Self-inflicted			✓	✓			
Deficiency	✓						

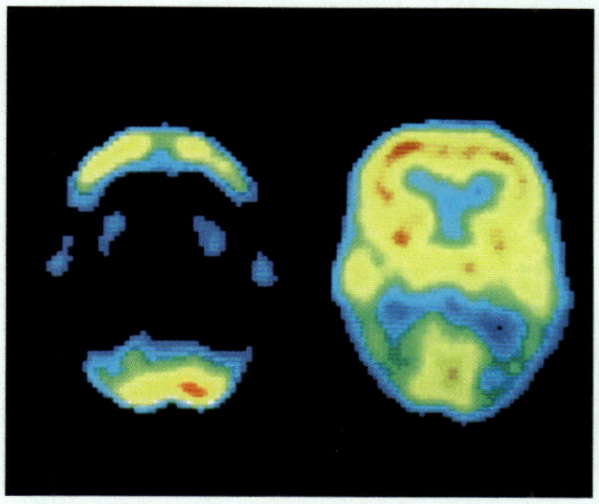

Figure 2 *PET scan of the brain, showing the degenerative changes in the brain of a person with Alzheimer's disease (left) compared to a normal brain (right)*

Inherited diseases

Inherited diseases are those that are caused by genes and can be passed from parents to their children. An example of a genetically inherited disease is sickle cell anaemia (Topic 6.5). It results from a single nucleotide base, adenine, being substituted by the nucleotide base thymine. The normal DNA triplet on the template strand is hence changed from CTC to CAC. As a result, the mRNA produced has a different codon.

The changed mRNA codon codes for the amino acid valine rather than glutamic acid. This minor difference produces a molecule of haemoglobin (called haemoglobin-S) that has a 'sticky patch'. When the haemoglobin molecules are not carrying oxygen (i.e. at low oxygen concentrations) they tend to adhere to one another by their sticky patches and become insoluble, forming long fibres within the red blood cells. These fibres distort the red blood cells, making them inflexible and sickle (crescent) shaped. These sickle cells are unable to carry oxygen and may block small capillaries because their diameter is greater than that of capillaries. Those with two mutant alleles often die, while those with one mutant allele easily become tired.

Self-inflicted diseases

There are a number of disorders that result from an individual's own actions and behaviour. In some cases the harmful consequences are known at the outset – few people who begin smoking can be unaware of the increased risk of lung cancer and emphysema that can result. Self-inflicted diseases cover a variety of conditions, ranging from attempted suicide, linked to poor mental health, to the misuse of drugs such as alcohol, nicotine or heroin. Obesity and heart disease may result from a high intake of fatty food, and skin cancer may be the consequence of excessive sunbathing.

Deficiency diseases

Deficiency diseases are caused by the shortage of some essential nutrient in the diet. Perhaps the most dramatic is kwashiorkor, which results from a deficiency of protein in children. Another example is vitamin D deficiency. Vitamin D is also manufactured naturally through the action of sunlight on the skin. Problems may arise for dark-skinned people in temperate countries, or for those who, for whatever reason, do not expose their skin to the sun. In these cases, calcium absorption is poor, leading to softer bones and a bowing of the legs in the young – a condition known as rickets, or osteomalacia (soft bones that are easily fractured) in adults.

SUMMARY TEST 10.2

Physical diseases cause temporary or permanent damage to some part of the body. For example, the disease called (1) can damage joints. (2) diseases cause psychological, personality or behavioural symptoms. Social diseases result from the environment in which people live. Examples include the sexually transmitted disease called (3) and the deficiency disease known as (4) that results in a swollen abdomen. Some diseases are the result of deterioration due to old age and these are termed (5) diseases. One example is a form of senile dementia, called (6). Some genetic disorders are examples of (7) diseases. An example of a genetic disorder is a condition called (8) that is the result of a change in a single nucleotide base of a DNA molecule. Diseases that result from one's own abuse of the body are known as (9) diseases. The lack of some essential nutrient in the diet may lead to a (10) disease, such as (11), which results from a lack of vitamin D in children.

Health statistics

The incidence of diseases alters in time and space. In our efforts to fight disease, it is therefore helpful to collect statistics on the patterns and distribution of disease.

Epidemiology

Epidemiology is the study of the spread of disease and the factors that affect it. It involves an analysis of the incidence and patterns of disease with a view to finding various means of prevention and control. Originally it was applied to infectious diseases but modern epidemiology is equally concerned with environmental and lifestyle factors in the incidence of diseases such as cancer and cardiovascular diseases.

Epidemiologists try to identify the cause of any new disease – for example HIV/AIDS, which first appeared in the 1980s, and severe acute respiratory syndrome (SARS), which appeared in 2003. The collection of statistics over many years may help to establish a link between a certain factor and disease, e.g. smoking and lung cancer, mesothelioma and asbestos. Such links are then followed up with scientific experiments to confirm whether there is a biological basis for such a link and that it is more than a statistical coincidence.

Terminology

In order to refer clearly and accurately to health statistical data, a number of terms need to be understood.

- **Incidence** – this is the number of new cases of a disease in a population in a given time, for example in one month or one year. It is expressed as follows:

$$\text{Incidence} = \frac{\text{number of new cases of a given disease in a certain period}}{\text{number of individuals in a population}}$$

- **Prevalence** – this is the number of people in a population with a particular disease at a particular time, regardless of when the disease was first recognised.
- **Mortality rate** – this is the number of people who have died of a particular disease in a given time, e.g. in one month or one year. This is usually expressed as:

$$\text{Mortality rate} = \frac{\text{number of deaths as a result of a given disease in a certain period}}{\text{number of individuals in a population}}$$

It can, however, be calculated using only those individuals who are known to have the disease, in which case it is expressed as:

$$\text{Mortality rate} = \frac{\text{number of deaths as a result of a given disease in a certain period}}{\text{number of individuals in the population suffering from the same disease}}$$

- **Morbidity rate** – this is the number of people who have a particular disease in a given time. It is normally expressed as the number of people with the disease divided by the number of individuals in the population.
- **Endemic** – infectious diseases are referred to as endemic if the disease is always present in the population. Tuberculosis is endemic in many countries, and most people carry the bacteria that cause it, even though they do not show any symptoms of the disease.

Table 1 *Millions of deaths worldwide due to various infectious diseases in 2011*

Infectious disease	Deaths / millions
Acute respiratory infections (e.g. pneumonia and influenza)	3.5
Diarrhoeal diseases (e.g. cholera, typhoid and dysentery)	2.5
HIV/AIDS	1.6
Tuberculosis	1.3
Malaria	0.7
Measles	0.13

Figure 1 *Health statistics helped to establish the link between smoking and lung cancer*

- **Epidemic** – refers to any disease that spreads suddenly to a large number of people over a widespread area. Influenza epidemics arise periodically, affecting large numbers of people, often across different parts of the globe. This, however, is distinct from the regular, expected rise in influenza cases during the winter in many countries.
- **Pandemic** – this term is used to describe worldwide epidemics, such as acquired immune deficiency syndrome (HIV/AIDS). Severe occurrences of influenza may affect much of the globe and so are referred to as a pandemic.

Using health statistics

Statistics on the health of populations, rather than of individuals, are collected and expressed in a form that allows fair comparisons to be made, e.g. death rate per thousand of the population per year. The information gives a valuable insight into the health of a population or a country. Comparisons can be made between different geographical areas, ethnic groups, ages or occupations. This enables governments to set priorities for spending, targeting particular diseases or areas, or setting up screening programmes. The World Health Organization (WHO) operates similar procedures on a global scale, identifying areas with specific problems and coordinating international action. The collection of accurate data not only enables suitable action to be taken, but also enables the effectiveness of various programmes to be monitored. Such data inform us that no more than six deadly infectious diseases (Table 1) caused half of all premature deaths in 2011.

Analysing health statistics

Mild food poisoning can be caused by various bacteria, including the bacterium *Campylobacter*. Humans usually become infected with this bacterium by eating contaminated meat, especially chicken, or by drinking contaminated milk. The incidence of food poisoning caused by *Campylobacter* in England and Wales over a typical 3-year period is given in Figure 2. What does this graph tell us and what are the explanations for the fluctuations in the cases of food poisoning?

Firstly the graph shows an annual variation, with the highest incidence in the summer months and the lowest in the winter months. Clearly, it is warmer in the summer and therefore:

- bacteria can multiply more rapidly and build up in chicken and milk that is not kept refrigerated
- more barbecues take place in the summer months. Meat may not be as thoroughly cooked on barbecues as it would be in a thermostatically controlled oven, and so bacteria such as *Campylobacter* may not be killed during cooking.

The collection of health statistics for diseases such as food poisoning can therefore provide patterns of disease incidence that cause questions to be asked, such as those above. The answers to these questions allow health professionals to provide guidance, e.g. keep milk refrigerated and cook meat thoroughly on barbecues, which can lead to a reduction in the incidence of disease.

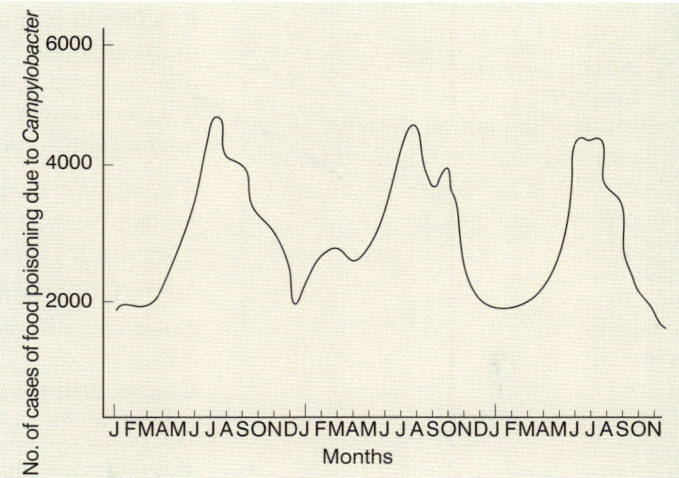

Figure 2 *Health statistics for food poisoning caused by* Campylobacter

Presenting and comparing health statistics

When making comparisons between the numbers of deaths due to a disease in one **population** and another, it can be misleading simply to state the number of people affected or who die. This is because it takes no account of the total numbers in each population. Population A may have twice as many people affected by a disease as population B, but if population A is 10 times larger than population B, then the proportion of individuals affected is actually much lower. For this reason, health statistics are usually expressed as the number of deaths, or people affected, per 100 000 of the population.

SUMMARY TEST 10.3

The study of the spread and distribution of disease is called **(1)**. The number of people in a population with the disease at any one time is called its **(2)**, while the number dying from it is called its **(3)**. Where a disease is always present in a population, it is said to be **(4)**, and an example of such a disease is **(5)**. Diseases such as **(6)**, which occur all over the world, are said to be **(7)**. To make health statistics easy to compare, the number of people dying or affected by a disease is usually expressed per **(8)** of the population.

Cholera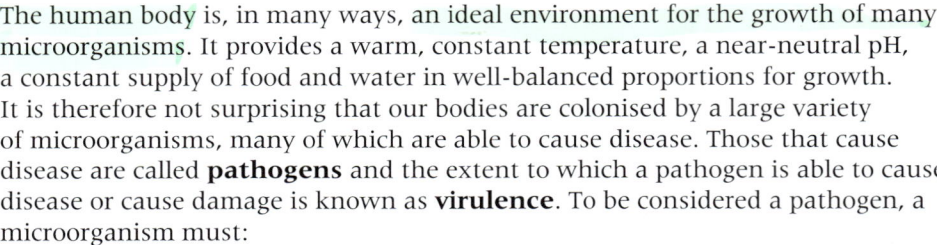

On these pages you will learn to:

- State the name and type of causative organism (pathogen) of cholera
- Explain how cholera is transmitted
- Discuss the biological, social and economic factors that need to be considered in the prevention and control of cholera

The human body is, in many ways, an ideal environment for the growth of many microorganisms. It provides a warm, constant temperature, a near-neutral pH, a constant supply of food and water in well-balanced proportions for growth. It is therefore not surprising that our bodies are colonised by a large variety of microorganisms, many of which are able to cause disease. Those that cause disease are called **pathogens** and the extent to which a pathogen is able to cause disease or cause damage is known as **virulence**. To be considered a pathogen, a microorganism must:

- gain entry to the host
- colonise the tissues of the host
- resist the defences of the host
- cause damage to the host's tissues.

One such pathogen causes the disease known as cholera.

Cause and means of transmission

The pathogen causing cholera is a curved, rod-shaped bacterium named *Vibrio cholerae*, which has a flagellum at one end (Figure 1). Cholera is transmitted by ingestion of water and, more rarely, food, that has been contaminated with faecal material containing the pathogen. Such contamination can be caused when:

- drinking water is not properly purified (Figure 2)
- untreated sewage leaks into water-courses
- food to be eaten is contaminated by those preparing and serving it
- organisms, especially shellfish, feed on untreated sewage released into rivers or the sea.

The main symptoms of cholera are diarrhoea, which also leads to dehydration. Up to 75% of those with cholera have few if any symptoms. Consequently they act as carriers, not knowing that they are spreading the disease to uninfected people. Less than 10% of infected people develop the moderate or severe dehydration associated with cholera.

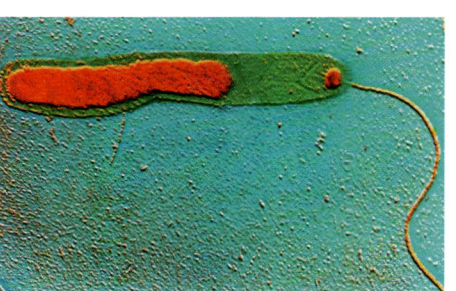

Figure 1 *Colourised scanning electron micrograph of* Vibrio cholerae – *the bacterium causing cholera*

Figure 2 *Cholera is most easily transmitted where there is a lack of treated water and where it is difficult to practise good hygiene*

EXTENSION
Worldwide importance

There have been several **pandemics** of cholera. The last one began in 1961 in Indonesia and spread rapidly to Bangladesh (1963), India (1964), the USSR, Iran and Iraq (1965–66). In 1970 it reached West Africa – an area which had not experienced a cholera outbreak for more than a century. It has now become **endemic** to much of the continent of Africa. Cholera spread rapidly around the world, reaching Latin America in 1991 (from where it had also been absent for over 100 years). It has now spread throughout South America.

The strain of *Vibrio* initially responsible for the present pandemic is serogroup 01 and is of the variety called 'El Tor'. In 1992, however, a new strain, serogroup 0139 'Bengal', emerged in Bangladesh and has been responsible for more recent outbreaks of the disease. There is evidence that 'Bengal' is more **virulent** than 'El Tor'.

Globally, cholera is of great importance, affecting 3–5 million people and killing an estimated 120 000 people each year. In some areas fatality rates may exceed 20% of those contracting (becoming ill from) the disease; with proper treatment, this should not go above 1%. In 2011, there were nearly 600 000 cases reported to the World Health Organisation (WHO) from 56 countries. The worst affected continent is Africa.

Prevention and control

To prevent the transmission of cholera, it is clearly necessary to prevent faecal matter containing *V. cholerae* from contaminating food and water. This can be achieved by:

- ensuring water supplies are clean, uncontaminated and treated, e.g. treatment with chlorine
- proper sanitation and sewage treatment
- personal hygiene, e.g. washing hands after using the toilet
- food hygiene, e.g. wearing gloves when handling food.

In developing countries, as the population of cities grows rapidly, there is often not enough money or resources to provide the necessary infrastructure. As a result, there may be no treated water supply or sewerage system. Housing may be of a poor standard, with no running water or toilet facilities. Without the means to cook, food such as shellfish may be eaten raw, increasing the risk of contracting (becoming ill from) cholera. Cholera outbreaks frequently occur or worsen following natural disasters, such as earthquakes or war, because:

- services such as water supply and sewage disposal are disrupted
- health services are overstretched
- residents are made homeless
- fleeing refugees spread the disease to neighbouring areas.

Outbreaks of cholera in Basra and Baghdad followed the 2003 war in Iraq and outbreaks in Haiti followed the 2010 earthquake.

It is clearly not possible to build water treatment plants and sewerage systems and to re-house millions of people overnight. The main prevention method, in the short term, therefore has to be **vaccination**. A vaccine based on inactivated *V. cholerae* 01 has been available for over 40 years but has had limited effect for two main biological reasons:

- it does not prevent transmission of the bacterium
- it only gives **immunity** for a period of about six months.

Two oral vaccines, which are easier to administer (deliver), are now available:

- a genetically modified form of *V. cholerae* that produces few, if any, symptoms but still causes the body to produce **antibodies** against it
- a mixture of dead *V. cholerae* bacteria and a harmless form of the toxin produced by the bacteria.

The continuing evolution of new strains of *V. cholerae* makes the development of effective vaccines difficult.

V. cholerae colonises the epithelium of the gut, causing diarrhoea. It is therefore beyond the reach of the body's immune system. **Antibiotics** are of little use in controlling the disease because they are passed out of the gut before they can be absorbed. Antibiotics may be given intravenously in severe cases, but overuse of these has led to new strains of the bacterium that are resistant to one or more antibiotics. Cholera is an easily treatable disease. Up to 80% of sufferers can be successfully treated by replacing the lost fluids through oral rehydration therapy (see extension box). Apart from the social advantages of controlling cholera by providing better sanitation and housing with piped, treated water and flush toilets, there are economic ones too. In countries where the disease is endemic, the poor health of workers means that economic output is reduced. Tourism suffers because people are afraid of catching cholera and even food exports are hit because countries do not want to risk importing contaminated produce. Eradication of cholera would reverse these economic disadvantages.

SUMMARY TEST 10.4

Organisms that enter, colonise and damage the tissues of a host organism are called **(1)**, and the extent of the damage they cause is known as **(2)**. One such organism is the bacterium called **(3)**, which causes cholera. This disease is transmitted mostly through **(4)** that is contaminated with faecal material containing the bacterium, although, to a lesser extent, contaminated **(5)** may be responsible. Prevention in the long term depends on better housing, clean water and proper **(6)**, as well as better hygiene. In the short term, **(7)** is the best preventative measure. Treatment may involve injecting **(8)** and replacing fluids using **(9)**.

- *Plasmodium*, which has developed resistance to drugs and adapts genetically to new situations.

The effective prevention and control of malaria needs to be aimed at all three of these biological factors:

- **Control of mosquitoes** – through reducing their populations by:
  - draining marshes and other areas of water where the mosquitoes lay their eggs and the larvae develop
  - introducing fish, which consume mosquito larvae, into marshes, ponds, etc. (biological control). However, these fish may also eat beneficial aquatic life
  - spraying fresh-water areas with a parasite that kills mosquito larvae but is harmless to other wildlife (biological control)
  - spraying fresh-water areas with insecticides to kill mosquito larvae (chemical control). These insecticides often kill beneficial aquatic organisms as well and may accumulate in food chains, harming other wildlife and even humans
  - spraying oil over fresh-water areas to prevent the mosquito larvae breathing the air they need to survive (chemical control).
- **Control by humans** – through avoiding being bitten by mosquitoes by:
  - keeping doors and windows closed as far as possible
  - using insect repellents on the skin
  - wearing clothing that covers most, if not all, the skin
  - sleeping under mosquito nets.
- **Control of *Plasmodium*** – for *P. falciparum* malaria, especially where there are problems of resistance, WHO recommends using ACT (artemisinin-based combination therapy). This involves using a combination of drugs that have different effects.
  - protect those not yet affected, by taking preventa-tive (prophylactic) drugs, such as chloroquine and atovaquone/proguanil, for at least a week before expo-sure to the parasite and for four weeks after exposure.

The social and economic consequences of malaria are huge. For example, it is estimated that a single bout of malaria costs a sum equivalent to over 10 days wages in Africa. International efforts to control malaria were successful in the late 1950s and early 1960s but the disease has since returned and is a major concern. Prevention and control of malaria is still proving difficult. Some of the reasons for this include the following:

- poor education and a failure to follow preventative measures properly
- wars and political unrest preventing governments and health authorities carrying out anti-malarial measures
- migrants, both political and economic, spreading the disease to previously unaffected areas
- poverty in affected countries leaving few resources for medical treatments or education about how to prevent the spread of the disease
- ease of movement, e.g. air travel, spreading the disease globally

- mosquitoes breeding in untreated water, e.g. puddles
- insecticide resistance in mosquitoes
- banning of certain insecticides, e.g. DDT
- drug resistance in *Plasmodium*
- **global warming** increasing the breeding range and life-span of *Anopheles* mosquitoes
- no vaccine being available.

Difficulties with developing a malarial vaccine

For years scientists have sought an effective vaccine to control malaria. Recently, in 2014, there was an announcement that a vaccine had been developed that, during trials, appeared to be effective in providing immunity to children from the disease. Why then is it so difficult to produce an effective vaccine when vaccines exist for many other diseases? *Plasmodium* has the ability to change the proteins that make up its surface antigens– it has **antigenic variability**. This variability is a result of a number of factors:

- It is a eukaryotic organism and therefore has many genes which can code for a wide variety of surface **antigens**.
- It has many stages in its life cycle within humans, each one having different surface antigens.
- It divides by meiosis which increases variety.
- It has haploid stages that allow **recessive alleles** to be expressed.

This constant changing of antigens is called **antigen shifting** and is the main reason why a vaccine has proved so difficult to develop. No sooner has a vaccine been developed against one set of antigens, than they change to another form and a new vaccine is required.

Plasmodium also lives inside red blood cells. Surrounded by the membrane of the red blood cell, it does not cause an antigenic response from the lymphocytes of the body's immune system (**antigenic concealment**) (Topic 11.7). Equally, any antibodies present in the blood cannot work against the stages of *Plasmodium* that occur within cells.

SUMMARY TEST 10.5

Malaria is caused by four species of the genus *Plasmodium*. *Plasmodium* belongs to the group **(1)** and it has two hosts: humans, and mosquitoes belonging to the genus **(2)**. As mosquitoes spread the disease to the main host, humans, they are called a **(3)**. Around the world some **(4)** million people suffer from malaria and up to 650 000 die each year, mostly in **(5)**. Control of mosquitoes can help prevent malaria. This is carried out by draining marshes, where the **(6)** stage of the mosquito lives, or by spraying chemical agents such as **(7)** or **(8)** on the water to kill them or using biological methods like **(9)** or **(10)** to have the same effect. Humans can help to avoid being bitten by mosquitoes by using **(11)** to cover them when sleeping or putting **(12)** on the skin to discourage them. Drugs such as **(13)** and **(14)** may be used to prevent contracting malaria when visiting a country where it is endemic.

Tuberculosis (TB)

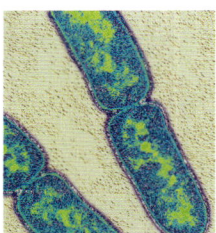

Figure
1 *Transmission electron micrograph of two pairs of recently divided* Mycobacterium tuberculosis, *the bacteria that causes tuberculosis*

SUMMARY TEST 10.6

Tuberculosis (TB) is caused by two types of bacterium. The one found only in humans is **(1)** while the other, called **(2)**, is found in humans and cattle. The disease is mainly transmitted through the **(3)** by droplet infection. Around **(4)**% of the world's population is thought to be infected by tuberculosis bacteria but, of these, only around **(5)**% have symptoms and pass the disease on to others. Certain groups of people are more at risk, including those living in **(6)** conditions, those who are malnourished, and those whose immune systems are weakened by medicine and diseases like silicosis or who are **(7)**. It can be prevented by the **(8)** vaccine and can be treated by drugs. One problem with the drugs is that they need to be taken for **(9)** months and many people fail to complete the treatment, leading to **(10)** strains of the bacteria that cause TB.

Tuberculosis (TB) is an infectious disease that can affect any part of the body but is usually found in the lungs because these are the first site of infection, causing persistent (continued) coughing, shortness of breath, tiredness, loss of weight, loss of appetite, fever and sweating. It is second only to HIV/AIDS as the infectious disease that kills most people worldwide.

Causes and transmission

Tuberculosis is caused by one of two rod-shaped bacteria: *Mycobacterium tuberculosis* or *Mycobacterium bovis* (Figure 1). It is estimated that up to 30% of the world's population have one or other form of the bacterium within their bodies. These people form two groups which differ **epidemiologically**.

- Most do not suffer any symptoms and the infection is controlled by the body's immune system. These individuals cannot pass on the disease. The inactive bacteria may, however, be activated, often after many years, especially when the person's immune system is weakened by other infections, or by HIV/AIDS.
- Some develop the disease because the bacteria overcome the body's defences, causing the symptoms listed above. These individuals can transmit the disease to others.

Despite its **prevalence** (number of cases) around the world, it is not all that easy to contract TB compared to many other infectious diseases. It is spread through the air by **droplet infection** when infected individuals cough, sneeze, laugh or even just talk and uninfected individuals breathe in the droplets containing the bacterium. However, it normally takes close contact with an infected person over a period of time, rather than a casual meeting in the street, to transmit the bacteria. TB is therefore usually spread between family members, close friends or work colleagues, especially in crowded and poorly ventilated conditions. TB can also be spread from cows to humans because *M. bovis* also infects cattle. Meat and especially milk may contain the bacterium. Some groups are at greater risk of contracting TB. These include people who:

- are in close contact over long periods with infected individuals, e.g. living, and especially sleeping, in over-crowded conditions
- are infected with HIV and have a lowered immunity as a result of the infection
- have other medical conditions that make the body less able to resist the disease, e.g. diabetes, lung disease such as silicosis
- are undergoing treatment with immuno-suppressant drugs (e.g. following transplant surgery)
- are malnourished (do not have a balanced diet)
- are working or living in long-term care facilities where relatively large numbers of people live close together, e.g. old people's homes, care homes, hospitals and prisons
- are alcoholic, injecting drug-users and/or homeless.

Distribution and worldwide importance

TB is a global disease which, at one time, appeared to be under control, at least in developed countries (Figure 2). However, a major return of the disease led the World Health Organization (WHO) to declare a global emergency in 1993. There has been a recent small decline in the number of cases but it is still estimated that:

- 8.6 million people developed TB and 1.3 million died of it in 2012
- 30% of the world's population is currently infected with *Mycobacterium*
- 5–10% of people infected with *Mycobacterium* will become ill and infectious with TB at some time during their lives

- nearly 1% of the world's population is newly infected with *Mycobacterium* each year
- someone in the world is newly infected with *Mycobacterium* every second.

In Eastern Europe and Africa the disease is now on the increase after 40 years of decline. Over 60% of cases occur in South East Asia and sub-Saharan Africa, largely as a result of the HIV/AIDS **epidemic** in these regions.

Reasons why TB is on the increase include the following.

- The spread of HIV (Topic 10.7) throughout the world means that there are many more individuals with a weakened (compromised) immune system and so TB develops when normally it would be controlled.
- Increased movement of people as a result of global trade and tourism has spread the disease worldwide.
- War and political unrest has led to mass movements of refugees, who are often housed in densely populated camps and shelters. Even where treatment programmes are carried out, refugees often move on before the 6-month treatment is complete.
- Poorly managed TB treatment and prevention programmes lead to incomplete treatment of TB or the wrong dose or drug being given.
- Drug-resistant forms of *Mycobacterium* have developed throughout the world, largely as the result of incomplete treatment.
- A greater number of people are homeless. They are therefore often in poor health and may live temporarily in crowded sheltered accommodation.
- There is a larger proportion of elderly people in the population and older people often have less effective immune systems.

Prevention and control

The main biological preventative measure for tuberculosis (TB) is **vaccination**. Children can be tested for their **immunity** to TB. Vaccination of those individuals who are already immune is unnecessary and dangerous. Those without immunity are given the **Bacille-Calmette-Guerin (BCG) vaccine**. This is an attenuated (weakened) strain of *M. bovis*, the organism that causes bovine TB. While this form of mass **immunisation** has been the major reason for the fall in cases of TB in many countries, there is no doubt that improved social conditions have also played a part (Figure 2). Vaccination has not eradicated TB because in some people it does not appear to be effective, it does not always give lasting protection, the bacteria remain in the body for a long time and many people have the disease without knowing (they are carriers).

Other biological means of controlling TB include treatment with drugs, such as isoniazid. The drug must be taken for six to nine months to be effective. One problem with all drug treatment for TB is the long period over which the drugs must be taken. When individuals are ill, they willingly take the drugs as they are keen to recover. The drugs initially destroy

the least-resistant strains of *Mycobacterium*. After a number of months, the patients feel better because the vast majority of *Mycobacterium* have been killed. Some people consider themselves cured and stop taking the drugs. This is almost the worst course of action because the few bacteria that remain are those that are most resistant to the drug. These resistant strains survive, multiply and spread to others. There is therefore a selection pressure that leads to the development of strains of *Mycobacterium* that do not respond to the drug. There is also a reservoir of bacteria that remain in the body, increasing the chance that mutations may occur that lead to antibiotic resistance (Topic 10.9). To overcome the problem of multi-drug resistance, a combination of three or four drugs is used to ensure that at least one will be effective. In addition, in 1991, WHO introduced a scheme for the detection and cure of TB called DOTS (Direct Observation of Therapy). This combines:

- a political commitment from the government of each country involved
- services of laboratories that can test for the presence of *Mycobacterium* in sputum samples taken from the population
- supplies of a range of anti-TB drugs
- monitoring and recording of each patient's treatment
- direct observation of the treatment by health and community workers and volunteers to ensure that drugs are taken properly and the course of treatment is completed.

Another biological control method is the pasteurisation (heat treatment) of milk, which kills any *M. bovis* present. Cattle herds are also regularly checked by veterinary surgeons for any sign of TB or the bacterium causing it.

In addition to the biological means of control discussed above, there are social and economic measures that can be introduced to reduce the number of TB cases. These include:

- better education about how TB is spread and how antibiotic resistance develops, particularly the need to complete all courses of drugs
- more and better housing, leading to less over-crowding
- improved health facilities and treatments, including more effective drugs, vaccination programmes and contact tracing
- better nutrition to ensure that immune systems are not weakened by poor diet.

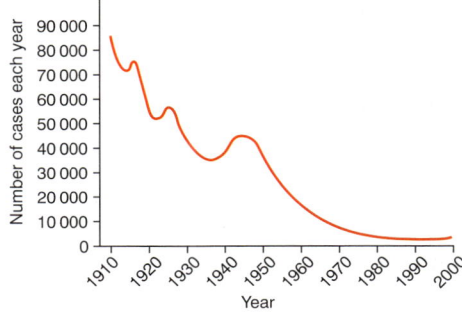

Figure 2 *Graph showing the number of TB cases in the UK during the period 1910–2000*

On these pages you will learn to:

- State the name and type of causative organism (pathogen) of HIV/AIDS
- Explain how HIV/AIDS is transmitted
- Discuss the biological, social and economic factors that need to be considered in the prevention and control of HIV/AIDS
- Discuss the factors that influence the global patterns of distribution of HIV/AIDS and assess the importance of this disease worldwide

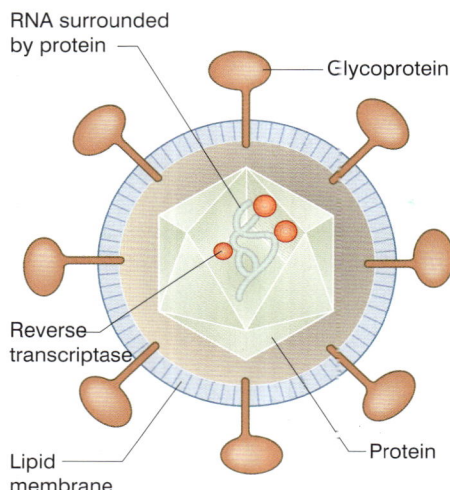

Figure 1 *Structure of human immunodeficiency virus*

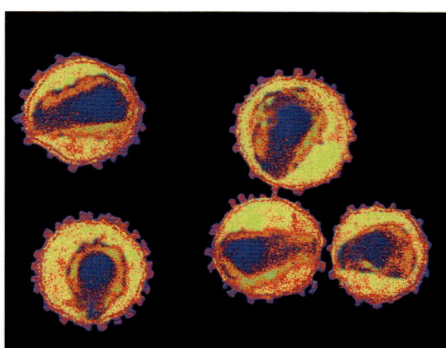

Figure 2 *Colourised transmission electron micrograph of the human immunodeficiency virus (HIV)*

Unlike many other infectious diseases, HIV/AIDS is a relative newcomer, having first been diagnosed in 1981. It is already one of the most serious diseases in human history and kills more people worldwide than any other single infectious disease. Without treatment, HIV/AIDS is a fatal condition. Treatment only prolongs life; it does not provide a cure.

Causes and means of transmission

HIV/AIDS is caused by the **human immunodeficiency virus (HIV)**, a spherical retrovirus whose structure is shown in Figures 1 and 2. Once infected with HIV, an individual is said to be **HIV positive**, a condition which lasts throughout life. As the virus remains dormant for about eight years on average, an HIV-positive person does not show any symptoms during this period but can act as a carrier, often not knowing that they could be spreading the disease. The virus can be detected in virtually all body fluids of an HIV-positive individual. However, since it is only in blood, semen or vaginal fluid that its concentration is high, the virus is usually spread through sexual intercourse or transfer of contaminated blood from one person to another – as when drug-abusers share a hypodermic needle – or from mother to baby during childbirth. There is evidence that HIV can be transmitted from mother to baby across the placenta as well as in breast milk. In the past, the transfusion of HIV-contaminated blood has spread HIV/AIDS, although blood is now routinely screened to avoid any risk to people, such as haemophiliacs, who depend on transfused blood. An infected person has such a low presence of HIV in the faeces, urine, sweat, saliva and tears that contact with these does not present a risk. In any case, the virus quickly dies outside the human body, and therefore even blood, semen and vaginal secretions must be transferred directly. Having entered the blood, HIV infects white blood cells, such as T helper cells (Topic 11.5), as well as macrophages and brain cells, to which the virus readily binds. HIV is known as an enveloped virus. This is because it has an outer envelope containing sections of host cell surface membrane that it takes with it when it is released from a host cell. These sections of membrane readily fuse with the cell surface membrane of certain white blood cells, allowing it to easily gain entry to a new host cell. Because it has host cell surface membrane on its surface, it is not recognized as 'foreign' and so does not stimulate an immune response from the **lymphocytes** in the blood. Replication of the virus is controlled and it frequently becomes dormant. It is months or years later the dormant virus replicates, so that it is on average eight years before HIV/AIDS develops. When the virus becomes active, the number of T-cells decreases, increasing the chance of infection by other pathogens (opportunist pathogens). This can lead to a person having a number of different diseases, collectively known as HIV/AIDS.

When people with HIV start to become ill and show the symptoms of HIV/AIDS, many may die within two years, usually as a result of opportunistic infections caused by the weakened immune system.

Distribution and worldwide importance

HIV/AIDS is truly a global disease. Unlike malaria, which only occurs in regions where the *Anopheles* mosquito is found, HIV/AIDS has affected almost every country in the world and certainly all the major regions (Table 1). The current **pandemic** has spread rapidly since its origins and in 2012, there was estimated to be 35 million people infected with HIV worldwide and 1.6 million died of an HIV/AIDS-related illness.

Table 1 *Regional statistics for HIV/AIDS, end of 2012 (Source: UNAIDS)*

Region	Adults & children living with HIV/AIDS (millions)	HIV/AIDS-related deaths in adults & children
Sub-Saharan Africa	25.00	1 200 000
South and South-East Asia	3.90	220 000
East Asia	0.88	41 000
Latin America	1.50	52 000
Caribbean	0.25	11 000
Eastern Europe and Central Asia	1.30	91 000
North America	1.30	20 000
Western and Central Europe	0.86	7900
Worldwide	35.00	1 600 000

HIV/AIDS is far more common in sub-Saharan Africa than elsewhere. In some countries of this region 25% of the population are infected with HIV.

Because people with HIV/AIDS are more prone to other infections, the number of cases (**prevalence**) of the disease in Africa is especially significant because it also has the highest number of new cases (**incidence**) of other diseases, such as cholera and malaria. Also the rise in TB cases worldwide has been linked to an increase in the number of HIV/AIDS cases. This has a very harmful effect on the economy of African countries because:

- HIV/AIDS affects mostly people in the 20–40 years age range and this is usually the most economically productive group of people of any population
- scarce financial resources have to be spent on expensive drugs, leaving little for economic development.

Prevention and control

There is, as yet, no cure for HIV/AIDS. A lot of effort has gone into developing a vaccine but progress is being hindered by the rapid rate at which HIV mutates, the fact that HIV conceals itself within the lymphocyte cell membrane and the risk that a vaccine from attenuated HIV could cause cancers. Furthermore, as HIV/AIDS is a disease only of humans, there are no suitable animals on which to test new drugs. Current approaches to finding a suitable treatment for HIV/AIDS involve development of:

- drugs which inhibit part of the replication cycle of HIV
- a vaccine to stimulate an immune response and provide immunity to HIV or to kill the virus in an HIV-positive person
- medicines which boost the immune system of people with HIV/AIDS
- treatments for the other infections which develop in people with HIV/AIDS.

Drugs are expensive and many people in Africa and other developing countries cannot afford them. Many of these drugs are now increasingly effective so that, in the

UK, deaths dropped from 1723 in 1995 to 490 in 2012. These drugs, however, often have unpleasant side effects, including headaches, diarrhoea and even permanent nerve damage. They are not a cure, but rather a way of delaying the onset of AIDS. Without a cure or a vaccine available at present, preventive measures remain the best means of trying to control the disease. Such measures include:

- **Advising HIV-positive mothers not to breast-feed** as the virus can be transferred from mother to child in breast milk. This must however be balanced against the benefits of breast-feeding which may outweigh the risks of HIV.
- **Contact tracing** to avoid the spread of HIV: anyone who is found to be HIV positive is asked to contact others who he/she might have infected, e.g. sexual partners, so that they may be tested and treated as necessary.
- **Education** is very important and involves informing the population of the risks and how to minimise them.
- **Needle-exchange schemes** for drug-abusers who can exchange used needles for new ones so that needle-sharing becomes unnecessary.
- **Screening of blood from donors** now occurs routinely in many countries and it is heat treated to kill HIV so that the risk to haemophiliacs and others using blood products is almost zero in developed countries.
- **HIV testing for individuals** at particular risk, e.g. injecting drug-abusers, sex-workers and male homosexuals.
- **Using condoms** or other barriers, such as femidoms and dental dams, for all forms of sexual contact. These act as a physical barrier and prevent the mixing of body fluids.

The biggest problem with trying to control the HIV/AIDS pandemic is that it takes a long time for symptoms of disease to appear. During this period, HIV-positive individuals can spread the disease widely, often without knowing. Another problem is that HIV-positive individuals may feel rejected by, or isolated from, society and are unwilling to ask for treatment. HIV-positive individuals may feel ostracised and isolated and therefore are deterred from seeking treatment.

On these pages you will learn to:

- State the name and type of causative organism (pathogen) of smallpox (*Variola*) and measles (*Morbillivirus*)
- Explain how measles is transmitted
- Discuss the biological, social and economic factors that need to be considered in the prevention and control of measles

Measles – cause and means of transmission

Measles is caused by the *Morbillivirus*, which infects the respiratory system. It is therefore usually transmitted through an infected person coughing and sneezing and air droplets containing the virus being breathed in by an uninfected person (droplet infection). It is highly infectious with around 90% of susceptible close contacts of an infected person likely to get the disease. The first symptoms include a fever, sore eyes, a cough and runny nose. These are followed by the characteristic red rash that begins on the head and moves down the body.

Prevention and control

Measles is a global disease that kills around 330 children each day, mostly in sub-Saharan Africa and South East Asia. The mortality (death) rate has fallen from 2400 a day in 1999 thanks largely to a successful vaccination programme. The World Health Organization has stated that they aim to work for the complete eradication of measles but has not set a target date. It aims to achieve at least 80% vaccination coverage worldwide, and to reduce measles mortality by 95% compared to 2000 levels by 2015.

By far the most important method of preventing measles is immunisation using the MMR vaccine (Figure 1). It is a live, attenuated (weakened), vaccine that protects against the measles, **mumps**, and **rubella** viruses. It is given to children at around 13–18 months of age. It is not given earlier because up to this age babies keep the anti-measles **antibodies** given to them by their mother during pregnancy. Giving vaccines to these babies could cause a dangerous reaction as the antibodies respond to the vaccine. A second dose at age four to five years can be given to raise immunity levels. Where immunisation programmes are not carried out, epidemic cycles arise every 2 to 3 years with some 90% of the population contracting the disease by the time they are 15 years old. Where successful vaccination programmes are in place, the number of cases has dropped by 90%. The vaccination is more effective when it is given to children that are well nourished.

Not everyone takes up the offer of immunisation because of religious and political objections. In the UK and other countries, immunisation rates fell following the publication of a paper in 1998 linking the MMR vaccine to autism. It was later found that the author did not follow ethical codes in his research and his work is not considered to be scientifically valid. Further studies have failed to find any link.

The isolation of those with measles from those who have not been immunised can help to prevent transmission of the disease. However, this is not always practical, and given the highly infectious nature of the virus, not always successful.

Smallpox

Smallpox is caused by one of two types of the *Variola* virus. The virus infects small blood vessels of the skin, throat and mouth causing a rash of raised blisters. The disease is highly infectious with around a third of those who contract it dying. Smallpox is remarkable for being the first disease to be eradicated from the world through the process of vaccination. The last naturally acquired case of smallpox was in Somalia in 1977 (although a year later a worker was accidentally infected by a laboratory sample).

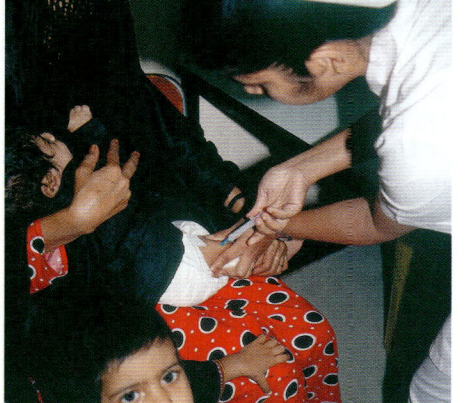

Figure 1 *Immunisation with the MMR vaccine is the most effective way of preventing measles and controlling its spread*

EXTENSION

EXTENSION
DENGUE FEVER

Dengue fever is an infectious tropical disease caused by the dengue fever virus (DENV). The virus is transmitted by the mosquito, *Aedes aegypti,* which lives in tropical and sub-tropical regions of the world and bites during the day.

Symptoms

Some of those infected with dengue fever show few if any symptoms, for others the disease is life-threatening. Typical symptoms include fever, headache, muscle and joint pains, the latter giving rise to its alternative name, breakbone fever.

Transmission

If a mosquito takes blood from someone with the dengue fever virus, it is taken into the mosquito's gut where the virus infects the lining cells. Eight to ten days later the virus migrates to other tissues including the salivary glands. From here it can be transmitted to other humans when the mosquito injects saliva while taking a further meal of blood. Other, less common, means of transmission include the transfusion of infected blood or transplantation of organs and from mother to child during pregnancy.

Figure 2 *Colourised SEM of the* Aedes aegypti, *the mosquito vector of dengue fever. Note the long thin proboscis for penetrating the skin.*

Impact

Dengue fever is considered to be the next most important tropical disease after malaria. Most people with dengue recover without any ongoing problems, but for some the effect is severe and life-threatening. The mortality is 1–5% without treatment, but less than 1% with treatment. It is thought that between 50 and 100 million people worldwide are infected with it each year, causing 12 500–25 000 deaths. The incidence of dengue increased 30 times between 1960 and 2010 as a result of population growth, more international travel and urbanisation. Global warming has increased the range of *Aedes aegypti*, the vector of the virus, leading to more cases of the fever. In those regions where dengue fever is endemic, the high levels of illness have an economic and social impact. People are unable to work to full capacity and this leads to a strain on families and medical resources.

Control

The control of mosquitoes and control by humans are the same as those for malaria, details of which are given in Topic 10.5. There are currently no approved vaccines for dengue fever, and there are no effective medicines or antibiotics to cure the disease. Oral rehydration medicines, pain killers and anti-inflammatory drugs are given to relieve the symptoms.

SUMMARY TEST 10.8

Measles is an infectious disease that infects the **(1)** system and is caused by a **(2)**. It kills around 330 children every day, mostly in the regions of **(3)** and **(4)**. The main method of controlling the disease is by immunisation using the **(5)** vaccine. This is not given before the age of **(6)** because babies younger than this still have **(7)** against measles that were provided by their mothers during pregnancy. Smallpox is caused by the **(8)** virus, but the disease has now been completely eradicated.

10.9

Antibiotics

On these pages you will learn to:

- Outline how penicillin acts on bacteria and why antibiotics do not affect viruses
- Explain in outline how bacteria become resistant to antibiotics with reference to mutation and selection
- Discuss the consequences of antibiotic resistance and the steps that can be taken to reduce its impact

REMEMBER

Preventing the formation of cross-linkages in bacterial cell walls is rather like removing the nails/screws from a wooden box. Without the linkages that hold it together any force applied to the box causes it to fall apart.

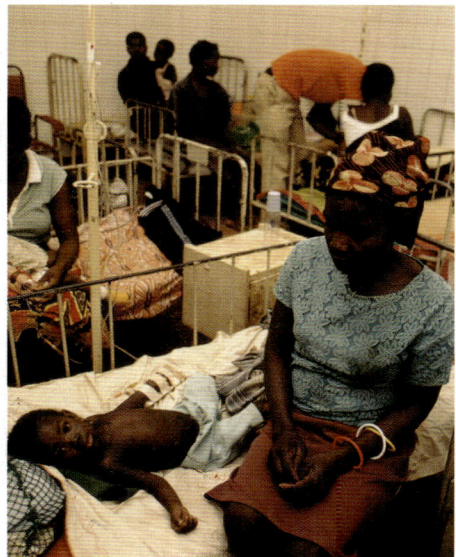

Figure 1 *The use of antibiotics is common in hospitals and increases the chance of resistance developing in bacteria*

Antibiotics are substances produced by microorganisms that destroy other microorganisms or inhibit their growth. Let us consider how antibiotics work and look at the causes and effects of antibiotic resistance (Figure 1).

How the antibiotic penicillin works

Penicillin works by preventing bacteria from making normal cell walls. In bacterial cells, as in plant cells, water constantly enters by osmosis (Topic 4.3). This entry of water would normally cause the cell to burst – a process called **osmotic lysis**. It does not burst because of the wall that surrounds all bacterial cells. This wall is made of **peptidoglycan (murein)**, a tough material that is not easily stretched. As water enters the cell by osmosis, it expands and pushes against the cell wall. Being relatively inelastic, the cell wall resists expansion and so prevents further entry of water. It therefore prevents osmotic lysis.

The peptidoglycans of bacterial cell walls are long molecules made of a mixture of amino acids and sugars. These long molecules are held together by short peptide molecules that form cross-linkages between them. As a young bacterium develops its cell wall, it secretes enzymes known as **autolysins**. These break down areas of the cell wall forming tiny holes. The holes allow the cell wall to stretch as it grows and they are normally filled in as new peptidoglycan chains form across them. Penicillin inhibits certain enzymes required for the synthesis and assembly of the peptide cross-linkages in the new bacterial cell walls. This weakens the walls, making them unable to withstand pressure. As water enters naturally by osmosis, the cell bursts and the bacterium dies. Penicillin is therefore only effective when bacteria are growing.

Why antibiotics do not affect viruses

Viruses, unlike bacteria, are not affected by antibiotics. There are a number of reasons for this, including:

- Viruses do not have their own metabolism but rather use that of their host cells. Antibiotics that work by disrupting metabolism are therefore not effective on a virus.
- Viruses are extremely simple with very few structures of their own, as they use those of the host cell to carry out their replication. Antibiotics that target cellular organelles are again not effective.
- Viruses have a protein coat rather than a peptidoglycan (murein) cell wall. There are therefore no sites for antibiotics like penicillin to work on.
- Viruses live inside the cells of their host, out of reach of antibiotics.

Antibiotic resistance

Shortly after the discovery of antibiotics it became clear that the effectiveness of some of them was reduced. It was found that these populations of bacteria had developed resistance to antibiotics, such as penicillin. The resistance was not due to a build-up of tolerance to the antibiotic, but rather a chance mutation within the bacteria. You will see in Topic 16.12 that a mutation is a change in DNA that results in different characteristics usually due to a change to some protein. In the case of resistance to penicillin, the mutation resulted in certain bacteria being able to make a new protein. The new protein was an enzyme, which broke down the antibiotic penicillin before it was able to kill bacteria. The enzyme was given the name penicillinase.

Mutations occur randomly and are very rare. However as there are so many bacteria around, the total number of mutations is large. Many of these mutations will be of no advantage to a bacterium. Indeed most will be harmful, in which case the bacterium will probably die. Very occasionally a mutation will be advantageous. Even then it depends upon the situation. For example, a mutation that leads to the production of penicillinase is only an advantage when the bacterium is in the presence of penicillin. If this is the case, then the penicillin will kill all the normal bacteria without penicillinase, but not the mutant type with penicillinase. Only the mutant individual will survive and divide. This means that all bacteria produced from this survivor will be of the mutant type and therefore be resistant to penicillin. The gene coding for penicillinase, and hence antibiotic resistance, is passed from one generation to the next – **vertical gene transmission**. As a result, the resistant form is selected for rather than the non-resistant form when exposed to penicillin. These penicillin-resistant bacteria therefore gradually form the largest part of the population. The frequency of the allele for penicillin resistance increases in the population.

The allele for antibiotic resistance can be carried on the small circular loops of DNA called **plasmids**. These plasmids can be transferred from cell to cell by a process called **conjugation**. Resistance can therefore find its way into bacteria of the same species that do not have the plasmid or into other bacterial species – **horizontal gene transmission**. Horizontal gene transmission can lead to certain bacteria accumulating DNA that gives them resistance to a range of antibiotics (Figure 2). These are the so-called multi-drug resistant bacteria ('superbugs').

Consequences of antibiotic resistance

New mutations that give bacteria resistance to antibiotics occur randomly all the time. However, the more we use antibiotics the greater the chance that the mutant bacterium will gain an advantage over the normal variety. In time and with continued use of the antibiotic, the chance that the mutant will out-compete and replace the normal variety becomes greater.

Certain steps can be taken to reduce the impact of antibiotic resistance.

- Limit antibiotic treatments to cases where they are essential rather than to treat minor illnesses with symptoms that are not serious or are short-lived.
- Encourage patients to always complete the course of antibiotics as prescribed.
- Try to prevent patients building up a supply of unused antibiotics from previous prescriptions and then using them later in smaller doses than they should.
- Avoid the use of antibiotics in the treatment of minor illnesses in domesticated animals.
- Limit antibiotic use in the prevention of disease among intensively reared animals such as chickens.
- Limit use of antibiotics by farmers and companies to reduce disease and hence increase productivity of animals.
- Hold back certain antibiotics for future use when others have been made ineffective by resistance.

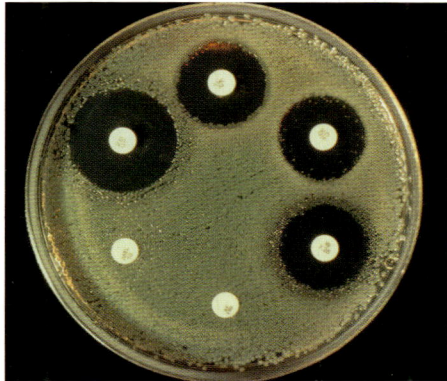

Figure 2 Antibiotic resistance in Escherichia coli. *The six white discs, each possessing a different antibiotic, were placed on a Petri dish with a growing culture of E. coli. Around four of the discs there is a lack of bacterial growth (inhibition zone), indicating that the bacterium is sensitive to these antibiotics. The growth of the bacterium around the two other discs is unaffected, however, indicating that it is resistant to these antibiotics.*

SUMMARY TEST 10.9

Young bacterial cells produce enzymes called **(1)** which make holes in their cell walls to allow the walls to **(2)** as they grow. These holes are normally covered over by chains of **(3)** molecules but penicillin can prevent this by inhibiting **(4)** that help to synthesise the **(5)** cross-linkages between these molecules. As a result, the wall is weakened and water entering the cell causes it to burst – a process called **(6)**. Resistance to penicillin is due to certain bacteria developing the ability to produce an enzyme called **(7)**. This happened suddenly as a result of a **(8)**. The gene for this enzyme is found on circular portions of DNA called **(9)** which can be inherited by each succeeding generation in a process known as **(10)**. These portions of DNA can also be transferred to other bacterial species by the process of conjugation. This transfer is known as **(11)**.

10 Examination Questions

1 a i Describe how TB is spread from infected to uninfected people.

   ii Suggest reasons why TB is sometimes referred to as a 'disease of poverty'.

   iii Suggest why heating milk to 72°C for 15 seconds (pasteurisation) might help to control the spread of TB.

b The main biological preventative measure for tuberculosis (TB) is vaccination. All children in the UK are routinely tested for their immunity to TB. Vaccination of those individuals who are already immune is unnecessary and dangerous. Those without immunity are given the vaccine. This is an attenuated (weakened) strain of *Mycobacterium bovis*, the organism that causes TB in cattle.

Suggest a reason why the bacteria in the vaccine are weakened before they are injected

c In addition to biological means of prevention, there are social and economic measures that can be introduced to reduce the number of TB cases.

Explain how 'more and better housing' can help prevent TB.

d Other means of controlling TB include treatment with drugs.

   i State two factors that reduce the effectiveness of treating TB with drugs.

   ii To treat TB, a combination of four drugs is often given. Explain why this is necessary.

e Despite all these measures, there has been a recent increase in TB in many developed countries.

Suggest a reason why an increasingly elderly population might have led to an increase in TB infections.

f Suggest a possible reason why the widespread use of condoms might help reduce the incidence of TB in a population.

2 a State the name of the organism that causes cholera.

(1 mark)

b NOR is an important respiratory enzyme located in the cell surface membrane of the bacterium that causes cholera.

A student suggested that an inhibitor of the enzyme NOR could be used as a drug in the prevention and control of cholera.

Suggest and explain how this inhibitor would function.

(3 marks)

c Table 1 shows the statistics for cholera reported to the World Health Organization (WHO) in four regions of the world in 2008.

Table 1

region	number of cases	number of deaths	fatality rate/%
Africa	179 323	5074	2.83
Asia	10 778	69	0.64
Europe	22	0	0.00
North America	7	0	0.00
Total	190 130	5143	

   i Calculate the total cholera fatality rate for 2008. Show your working.

     answer %   *(2 marks)*

   ii Apart from differences in total population size in each of the regions, suggest explanations for the differences shown in Table 1.   *(4 marks)*

(Total 10 marks)

Cambridge International AS and A Level Biology 9700 Paper 22 Q5 June 2012

3 An estimated 300 to 500 million cases of malaria occur worldwide each year resulting in 1 to 3 million deaths. 80% of these cases are in children under the age of five.

There are four species of malarial parasite, of which *Plasmodium falciparum* is responsible for most of the deaths from this disease.

a Describe how the malarial parasite is transmitted.

(3 marks)

b Several potential vaccines against malaria have been developed. Some of these make use of proteins from the surface membrane of *P. falciparum*.
  i Explain how using such a vaccine may give long-term immunity to malaria. *(4 marks)*
  ii Researchers have been trying to develop a successful vaccine against malaria for about 20 years. Explain why it has proved so difficult to develop such a vaccine. *(2 marks)*

c Proteins on the surface of the parasite are responsible for binding to surface receptors on the red blood cells. These are removed when the parasites enter the red blood cells.

An enzyme has recently been discovered in *P. falciparum* that is responsible for the removal of these proteins. If the enzyme does not function then the parasites cannot enter red blood cells.

It has been suggested that a drug could be developed to inhibit this enzyme.

Describe **one** possible way in which such a drug might act on the enzyme to prevent it from functioning. *(3 marks)*
(Total 12 marks)

Cambridge International AS and A Level Biology 9700 Paper 2 Q5 June 2008

10 Practice Questions

4 a Cholera is transmitted by food and water that is contaminated with faecal matter. Suggest three measures that might be used to limit the spread of the disease.

b Suggest a reason why, in countries where cholera is common, babies who are breast-fed are affected by cholera far less often than babies who are bottle-fed.

c Suggest how inhibiting the development of a flagellum in the bacterium that causes cholera might prevent the disease.

d Suggest a reason why injecting antibiotics into the blood can be effective in killing the cholera bacterium while the same antibiotic taken orally (by mouth) is not.

5 a What is an antibiotic?

b Some antibiotics prevent the synthesis of cross-linkages in bacterial cell walls. Explain how this may lead to the death of a bacterium.

c Explain why antibiotic resistance is more likely to develop the more antibiotics are used.

d Give a reason why a patient might stop taking antibiotics prematurely.

e Why would patients with tuberculosis be more likely to stop taking antibiotics prematurely than patients with other diseases?

f Why are strains of bacteria that are resistant to many antibiotics more likely to arise in hospitals?

6 Below is a list of four infectious diseases:
  P Acquired Immune Deficiency Syndrome (HIV/AIDS)
  Q Malaria
  R Tuberculosis
  S Cholera

For each of the following statements give the letter corresponding to those diseases to which the statement refers. There may be more than one letter for each answer, and each letter may be used once, more than once or not at all.

a Antibiotics are used in its control.
b A vector is involved in its transmission.
c It is spread through the air.
d It can be controlled by vaccination.
e Treatment involves oral rehydration therapy.
f The organism causing the disease is a prokaryote.
g Infected persons are prone to opportunistic infections.
h It occurs particularly where there is poor sanitation.
i Symptoms include a persistent cough.
j It is caused by a pathogen.
k It affects almost every country in the world.

Immunity

Defence against disease and phagocytosis

On these pages you will learn to:

- State that phagocytes (macrophages and neutrophils) have their origin in bone marrow and describe their mode of action

We have seen in Chapter 10 some examples of infectious diseases and the damage they can do. Tens of millions of humans die each year from such infections. Many more survive and others appear never to be affected in the first place. Why then are there these differences? Any disease is, in effect, a battle between the **pathogen** and the body's various defence mechanisms. Sometimes the pathogen overcomes the host and death is the result. Sometimes the body's defence mechanisms overcome the pathogen and the individual recovers from the disease. Having overcome the pathogen, however, the body's defences seem to be better prepared for any second invasion from the same pathogen and so prevent it before it can cause any harm. This is known as **immunity** and is the main reason why certain people are unaffected by certain pathogens. There is a complete range of intermediates between the stages described above. Much depends on the overall state of health of an individual. A fit, healthy adult will rarely die from an infection. Those in ill health, the young and the elderly are usually more at risk.

Defence mechanisms

The human body has a range of defences to protect itself from infection. There is a first line of defence to prevent the entry of pathogens, and if this is not successful, a second line of defence can be used. The second line of defence involves white blood cells: phagocytes that are non-specific and will respond to any foreign invader(Figure 1); and lymphocytes that are highly specific. The aim of the second line of defence is to:

- neutralise any toxins produced by the pathogen
- prevent the pathogen multiplying
- kill the pathogen
- remove any remains of the pathogen.

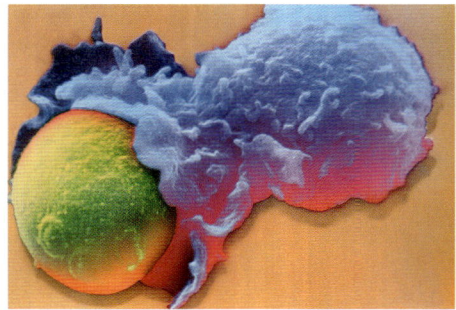

Figure 1 *False-colour scanning electron micrograph of a white blood cell engulfing a yeast cell by phagocytosis*

EXTENSION

Physical barriers

The first line of defence, the physical barrier, takes a number of different forms:

- **A protective covering** – the skin covers the body surface, providing a physical barrier that most pathogens find hard to penetrate. Some pathogens, such as the malarial parasite *Plasmodium*, use a vector, the mosquito, to penetrate this covering and so gain entry to the body (Topic 10.5).

- **Epithelia covered in mucus** – as the body needs to obtain certain substances by diffusion, e.g. oxygen and digested food, there are parts of the body which cannot be covered by a thick layer like the skin. Exchange regions, such as the alveoli in the lungs and the lining of the gut, are covered only by a thin epithelial layer. While this still acts as a barrier to pathogens, it is more easily crossed than the skin. Many epithelial layers therefore produce mucus, which acts as a further defence against invasion. In the lungs, pathogens stick to this mucus, which is then transported away by cilia (Topic 9.1), up the trachea, to be swallowed into the stomach where the acidic conditions kill the pathogens.

- **Hydrochloric acid in the stomach** – this provides such a low pH that the enzymes of most pathogens are **denatured** and therefore the organisms are killed.

Phagocytosis

Phagocytosis is the process by which large particles are taken up by cells, in the form of vacuoles (vesicles) formed from the cell surface membrane. In the blood and other body tissues, two types of white blood cell that carry out phagocytosis are **macrophages** and **neutrophils**. These cells are known as **phagocytes** and are produced in the marrow of the long bones. The process of phagocytosis by a neutrophil is shown in Figure 2 and is summarised as follows:

- **Antibodies** attach themselves to **antigens** on the surface of the bacterium.
- Proteins, found in the plasma, attach themselves to the antibodies.
- As a result of a series of reactions, the surface of the bacterium becomes coated with proteins called **opsonins**. This process is called **opsonisation**.
- Complement proteins and any chemical products of the bacterium act as attractants, causing neutrophils to move towards the bacterium.
- Neutrophils attach themselves to the opsonins on the surface of the bacterium.
- Neutrophils engulf the bacterium to form a vesicle, known as a **phagosome**.
- Lysosomes (Topic 1.7) move towards the phagosome and fuse with it.
- The enzymes within the lysosomes break down the bacterium into smaller, soluble material.
- The soluble products from the breakdown of the bacterium are absorbed into the cytoplasm of the neutrophils and waste products are released from the cell.

EXTENSION
Inflammation

Inflammation occurs at the site of infection. This is caused by the release of **histamine** from another type of white blood cell. Histamine causes the dilation of blood vessels for increased blood flow to the area and makes the capillaries more leaky. This means that more phagocytes move out of the bloodstream into the infected area. Histamine makes nerve endings more sensitive. Together, these responses explain the swelling, redness and pain associated with inflammation. Pus, which is found in inflamed areas, contains dead bacteria and phagocytes.

SUMMARY TEST 11.1

The first line of defence against disease is to prevent the entry of pathogens. The skin provides the main physical barrier, but the thinner coverings, such as the **(1)** layer in the lungs, also secrete **(2)** to provide a further barrier. Other means of preventing entry by pathogens include the secretion of **(3)** by the stomach. Pathogens that do invade the body may be engulfed by cells which carry out **(4)**. These cells are of two types, known as **(5)** and **(6)**, both of which are produced in the **(7)** of the **(8)** bones. In the process the pathogens are coated with proteins called **(9)**. Once engulfed the pathogen is broken down by enzymes released from organelles called **(10)**.

1. The neutrophil is attracted to the bacterium by chemoattractants. It moves towards the bacterium along a concentration gradient

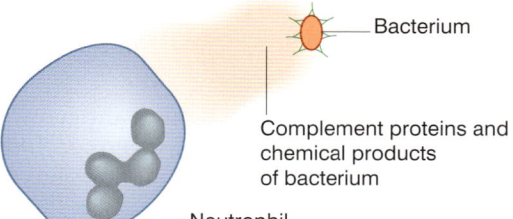

2. The neutrophil binds to the bacterium

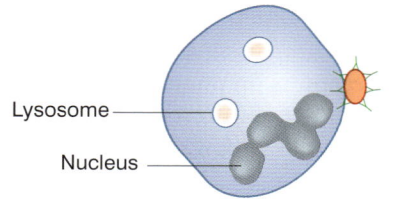

3. Lysosomes within the neutrophil migrate towards the phagosome formed by pseudopodia engulfing the bacterium

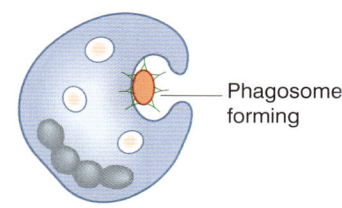

4. The lysosomes release their lytic enzymes into the phagosome, where they break down the bacterium

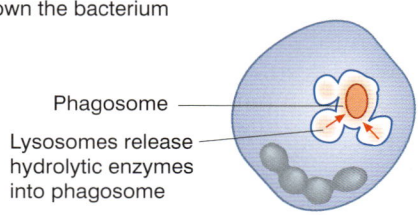

5. The breakdown products of the bacterium are absorbed by the neutrophil or are released from the cell

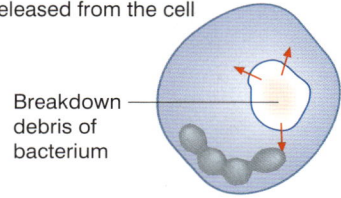

Figure 2 *Summary of phagocytosis of a bacterium by a neutrophil*

Principles of immunity – antigens and antibodies

- Describe and explain the significance of the increase in white blood cell count in humans with infectious diseases and leukaemias
- Explain the meaning of the term 'immune response', making reference to the terms 'antigen', 'self' and 'non-self'
- Explain, with reference to myasthenia gravis, that the immune system sometimes fails to distinguish between self and non-self
- Relate the molecular structure of antibodies to their functions

The variable region differs with each antibody. It has a shape which exactly fits an antigen. Each antibody therefore can bind to two antigens.

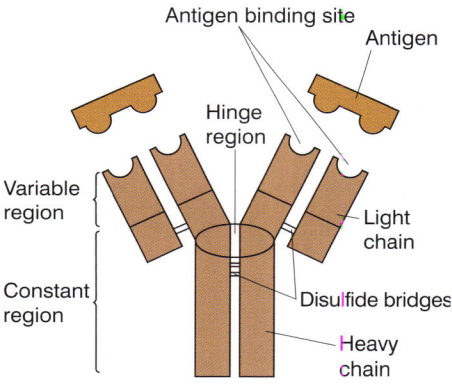

Figure 1 *Structure of an antibody*

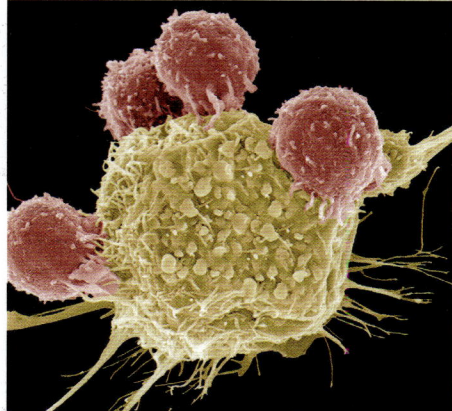

Figure 2 *Lymphocytes (pink) attached to a cancer cell*

Immunity is the ability of organisms to resist infection by protecting against disease-causing microorganisms that invade their bodies. It involves the recognition of foreign material and the production of chemicals that help destroy it.

Antigens

An **antigen** is any substance that is recognised as non-self (foreign) by the immune system and provokes an **immune response**. Antigens are usually proteins that make up the cell surface membranes of invading cells or the protein coat of viruses, microorganisms, or diseased cells, such as **cancer** cells. The presence of an antigen triggers the production of an antibody as part of the body's defence system. Glycoproteins and polysaccharides can also act as antigens.

Antibodies

Antibodies are also known as **immunoglobulins (Ig)**. They are proteins synthesised by cells of the immune system known as plasma cells, a type of **B lymphocyte**. (Topic 11.4). When the body is infected by non-self (foreign) material, a B lymphocyte produces antibodies, which act against antigens on the surface of the foreign material by binding to them precisely, in the same way as a key fits a lock. They are therefore very specific, each antigen having its own separate antibody. A massive variety of antibodies is possible because they are proteins – molecules that occur in an almost infinite number of forms. Antibodies are made up of four polypeptide chains. The chains of one pair are long and are called **heavy chains**; the chains of the other pair are shorter and are known as **light chains**. The chains are held together by disulfide bridges (Topic 2.7). The hinge region is a flexible area that allows slight movement of the molecule when binding antigens. Antibodies have two sites, called **binding sites**, which fit very precisely onto the antigen (Figure 1). The binding sites are different on different antibodies and are therefore called the **variable region**. They consist of a sequence of amino acids that form a specific three-dimensional shape which binds directly to a single type of antigen. The rest of the antibody is the same in all antibodies and is known as the **constant region**. This binds to receptors on phagocytes, making **phagocytosis** of **pathogens** easier. There are a number of different groups of antibodies. Each group functions in one or more of the following ways.

An antibody molecule has two antigen binding sites and can bind to antigens on different pathogens. Also, a pathogen can have a number of antigens each of which is bound by an antibody. These lead to clumping together of pathogens for phagocytosis.

The concept of self and non-self

To be able to defend the body from invasion by foreign material, B lymphocytes must be able to distinguish the body's own cells and chemicals (self) from those that are foreign (non-self). If they could not do this, B lymphocytes would produce antibodies that would destroy the organism's own tissues. How then do B lymphocytes recognise their own cells?

- Each specific antibody is produced by a specific lymphocyte. Held on the surface of B lymphocytes, these act as receptors.
- There are more than 10 million different lymphocytes, each capable of recognising a different chemical shape.
- In the fetus, these lymphocytes are constantly colliding with other cells.

- Infection in the fetus is rare because it is protected from the outside world by the mother and, in particular, the placenta.
- Lymphocytes will therefore collide almost exclusively with the body's own material (self).
- Some of the lymphocytes will have receptors that exactly fit those of the body's own cells.
- These lymphocytes either die or are suppressed.
- The only remaining lymphocytes are those that have receptors that fit foreign material (non-self), and therefore the only antibodies produced are those that respond to foreign material.

Occasionally B lymphocytes fail to distinguish between self and non-self. One example is a condition called **myasthenia gravis**, in which B lymphocytes produce antibodies against proteins naturally occurring in the body (self). In this case, antibodies act on **acetylcholine** receptors on the surface of muscle fibres (muscle cells). To stimulate muscle fibre contraction the transmitter substance, acetylcholine, released from the end of a motor neurone (nerve cell), must bind to receptors on the muscle fibre. If the antibodies are bound to the receptors, muscles are not stimulated to contract normally, leaving the person weak, especially after exercise. The muscles of the eye are especially prone, making it difficult to control eye and eyelid movement. A disease that results from the body's immune system attacking its own cells and tissues is known as an **autoimmune disease**.

Significance of an increase in white blood cell count

As white blood cells play an important role in the immune response to invasion by viruses and bacteria, it follows that a high white blood cell count can indicate an infection. The significance of the larger number of white blood cells is that phagocytosis and antibody production are increased to kill the viruses or bacteria and prevent them from causing further harm. There will also be an increase in T lymphocytes, cells of the immune system that have a number of roles in specific immunity (Topic 11.5).

One of the most serious causes of a high white blood cell count is a form of cancer of the bone marrow called leukaemia. In this case, many of the white cells are abnormal and therefore non-functional. The significance is that these abnormal cells are more common because of uncontrolled mitosis, grow faster and live longer than normal white blood cells. As a result they become very numerous in the blood, and the population of other blood cells becomes smaller because more resources in the bone marrow are used in production of the abnormal cells. A person with leukaemia is therefore more prone to infection than usual, as the abnormal cells do not produce antibodies or carry out phagocytosis. The reduction in red blood cell count leaves them anaemic and lacking in energy, and the reduction in platelets reduces the ability of blood to clot, which means that they are easily bruised.

SUMMARY TEST 11.2

An antigen is any substance that is recognised as **(1)** by the immune system. These antigens trigger the production of antibodies, which are **(2)** and are also known as **(3)**. Antibodies are synthesised by **(4)** and are made up of a pair of long chains called **(5)** chains and a pair of short ones called **(6)** chains. The two pairs of chains are held together by **(7)**. Antibodies have two sites called **(8)** that fit a specific antigen very precisely. There are a number of groups of different antigens which function in one or more different ways. Some antibodies precipitate out soluble antigens, while others neutralise the **(9)** produced by pathogens, and others attract compounds that break down foreign cells in a process called **(10)**.

Monoclonal antibodies

On these pages you will learn to:

- Outline the hybridoma method for the production of monoclonal antibodies
- Outline the use of monoclonal antibodies in the diagnosis of disease and in the treatment of disease

A pathogen invading the body has many hundreds of different antigens on its surface. Each different antigen will induce a different B lymphocyte to multiply and clone itself and so a mixture of different antibodies is produced (polyclonal antibodies). It is of great medical benefit to produce a single type of antibody from just one clone rather than a mixture of them. These are known as **monoclonal antibodies**.

Producing monoclonal antibodies

The production of monoclonal antibodies has long been recognised as useful. The problem had always been that B lymphocytes are short-lived in culture and only divide inside a living body. There was much competition amongst scientific research teams to overcome the problem of getting B lymphocytes to grow indefinitely outside of the body. Many varied methods were investigated. Cesar Milstein and Georges Kohler evaluated these methods and observed the behaviour of cancer cells. This led them to develop the following solution to the problem in 1975.

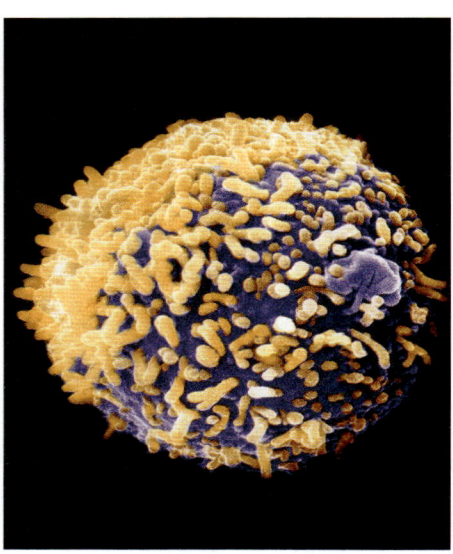

Figure 1 *False-colour SEM of a hybridoma cell used to produce monoclonal antibodies*

- A mouse is exposed to non-self material that carries the antigen against which an antibody is required.
- The different B lymphocytes within the mouse produce a mixture of antibodies (polyclonal antibodies) and the cells are extracted from the spleen of the mouse.
- To enable these B lymphocytes to divide outside the body, they are mixed with cells that divide readily outside the body e.g. cells from a cancer tumour.
- A fusogen (a detergent such as polyethylene glycol) is added to the mixture to allow the cell surface membranes of both types of cell to fuse together. These fused cells are called **hybridoma** cells.
- The hybridoma cells are separated out and each single cell is grown into a group of genetically identical cells from the single ancestor hybridoma cell (clone). Each clone is tested to see if it is producing the required antibody.
- Any group producing the required antibody is grown on a large scale and the antibodies extracted from the growing medium.
- As these antibodies come from cells cloned from a hybridoma cell, they are monoclonal antibodies (collectively termed monoclonal antibody).
- As a mouse was used as the host organism to produce the B lymphocytes, the monoclonal antibodies produced have to be modified to make them like human cells before they can be used – a process called **humanisation**.

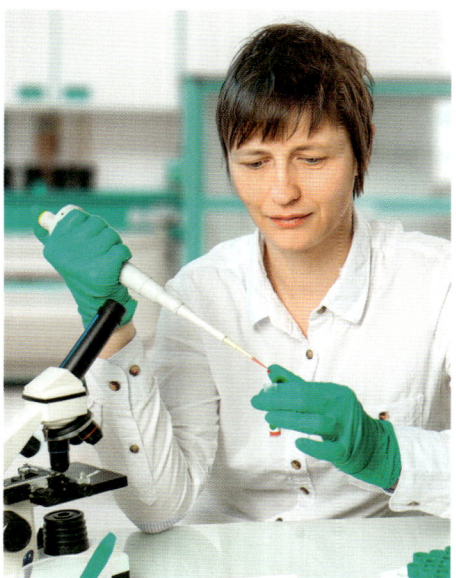

Figure 2 *Technician testing for cancer by adding monoclonal antibodies to human tissue samples*

Use of monoclonal antibodies in diagnosing disease

Monoclonal antibodies are a very valuable tool in diagnosing disease with over a hundred different diagnostic products based on them. They are used for the diagnosis of influenza, hepatitis and *Chlamydia* infections where they produce a much more rapid result than conventional methods of diagnosis. They are important in diagnosing certain cancers. For example, men with prostate cancer often produce more of a protein called prostate specific antigen (PSA) leading to unusually high levels of it in the blood. By using a monoclonal antibody which interacts with this antigen, it is possible to obtain a measure of the level of PSA in a sample of blood. While a higher than normal level of PSA is not itself diagnostic of the disease, it gives an early warning of its possibility and the need for further tests. Regular PSA tests to monitor levels of the antigen lead to earlier diagnosis and therefore a better chance of cure than waiting for the symptoms to develop.

Monoclonal antibodies can be used to directly bind to an antigen, confirming the presence of a specific pathogen. In other cases, it is easier to test an individual for the presence of an antibody against a particular pathogen, indirectly confirming the presence of the pathogen. For these tests, monoclonal antibody specific to the antibody can be used.

Use of monoclonal antibodies in treatment of disease

As antibodies are very specific to particular proteins (antigens), monoclonal antibodies can be used to target specific substances and specific cells. One type of cell they can target is cancer cells. Monoclonal antibodies can be used to treat cancer in a number of ways. By far the most successful is direct monoclonal antibody therapy.

- Monoclonal antibodies are produced that are specific to an antigen on cancer cells.
- The person with cancer is treated with these monoclonal antibodies.
- The antibodies attach to the receptors on the cancer cells.
- They then block the chemical signals that stimulate their uncontrolled growth.

An example is herceptin, a monoclonal antibody used to treat breast cancer. The advantage of direct monoclonal antibody therapy is that since the antibodies are not toxic, they lead to fewer side effects than other forms of therapy.

Another method, called indirect monoclonal antibody therapy, involves attaching a radioactive or cytotoxic drug (a drug that kills cells) to the monoclonal antibody. When the antibody attaches to the cancer cells the cells are killed. In a variation on this treatment, drugs can be given in two stages:

- In stage 1, a monoclonal antibody is chemically linked to an enzyme and the antibody-enzyme complex is given to the patient and attaches itself to the surface of the cancer cells.
- In stage 2, the patient is given an inactive form of a cytotoxic drug. This drug is activated by the enzyme. As the enzyme is attached to the monoclonal antibody, and this is only found on the cancer cell, only these cells, and not the rest of the body cells, are killed by the drug.

For obvious reasons, monoclonal antibodies used in this way are referred to as 'magic bullets' and can be used in smaller doses, as they are targeted on specific sites. Using them in smaller doses is not only cheaper but also reduces any side effects the drug might have.

Use of monoclonal antibodies for pregnancy testing

It is important that a mother knows as early as possible that she is pregnant, not least because there are certain actions she can take to ensure the unborn baby's welfare. The use of pregnancy testing kits that can easily be used at home has made possible the early detection of pregnancy. These kits rely on the fact that the placenta produces a hormone called human chorionic gonadatrophin (hCG) and that this is excreted in the mother's urine. One type of kit works as follows:

- The test strip of a home pregnancy test is dipped into a urine sample which moves across the strip by capillarity. Stretched across the test strip is a membrane made of nitrocellulose, a material used because it assists the movement of proteins.

- As the urine comes into contact with the membrane, it first mixes with antibodies linked to coloured particles that are present in the nitrocellulose.

- If hCG is present in the urine it combines with the antibodies that are linked to coloured particles.

- The hGC-antibody-colour complex moves along the strip until it reaches a narrow region where another type of monoclonal antibody is immobilised.

- The hCG-antibody-colour complex is trapped by these second antibodies creating a coloured line.

- As the liquid moves towards the end of the strip, it meets a region that has another immobilised monoclonal antibody. This one traps the coloured particles whether or not the solution contains hCG.

This second line will appear for both positive and negative results and acts as a control. If both lines appear, the test is positive.

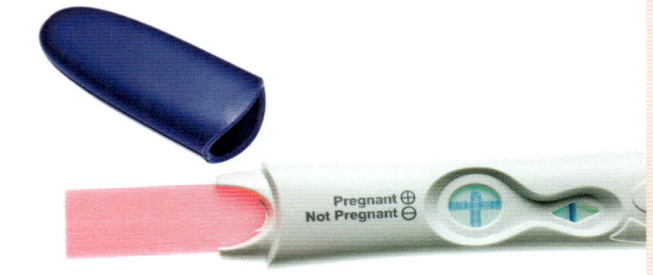

Figure 3 *Home pregnancy testing kit showing a positive result*

SUMMARY TEST 11.3

Monoclonal antibodies are produced by fusing a **(1)** lymphocyte with a cancer cell to form a **(2)** cell. These cells are allowed to divide to form a **(3)** of identical cells, which produce one type of antibody. Monoclonal antibodies are valuable in **(4)** and treating diseases. In the treatment of cancer they can be used to directly stop cancer cells dividing as in the case of **(5)** used to treat breast cancer. Alternatively, they can be used to deliver either a radioactive or a **(6)** drug to cancer cells. Monoclonal cells are also used in pregnancy testing kits. These detect the hormone **(7)** produced by the **(8)** during pregnancy.

On these pages you will learn to:

- Describe the modes of action of B-lymphocytes
- Explain the role of memory cells in long-term immunity

Immune responses such as phagocytosis (Topic 11.1) are non-specific and occur whatever the infection. The body also has specific responses that fight individual forms of infection. These are slower in action at first, but they can provide long-term immunity. This specific immune response depends on a type of white blood cell called a **lymphocyte**. There are two types of lymphocyte, each with its own immune response:

- B lymphocytes (B cells) – humoral immunity (involves antibodies which are present in body fluids or 'humours')
- T lymphocytes (T cells) – cell-mediated immunity (involves cells).

Both types of lymphocyte are formed from stem cells found in the bone marrow. Their names, however, indicate where they develop and mature:

- B lymphocytes mature in the **B**one marrow
- T lymphocytes mature in the **T**hymus gland.

The maturation process takes place in the fetus and, in the case of B lymphocytes, results in more than 10 million different types, each capable of responding to a different antigen.

Humoral immunity

We saw in Topic 11.2 that, when the body is invaded by foreign (non-self) material, this material possesses antigens that stimulate B lymphocytes to produce antibodies. These antibodies are soluble in the blood and tissue fluid of the body. Another word for body fluids is 'humour' and hence the

production of antibodies in this way is known as **humoral immunity**. There are many different types of B lymphocytes, possibly as many as 10 million, and each type produces a different antibody which responds to one specific antigen. When an antigen, e.g. a protein on the surface of a **pathogen** cell, enters the blood or tissue fluids, one type of B lymphocyte will have antibodies on its surface that exactly fit it and therefore it attaches to the antigen. The recognition of, and binding to, the antigen activates the B lymphocyte. This type of B lymphocyte grows and divides by mitosis (Topic 5.2) to form a clone of identical B lymphocytes, all of which produce antibody specific to the foreign antigen. In practice, a typical pathogen, e.g. *Mycobacterium tuberculosis*, has many different proteins on its surface, all of which act as antigens. Pathogens, such as the bacterium that causes tetanus, also produce toxins, each of which will act as an antigen. Therefore many different B lymphocytes make **clones**, each of which produces its own type of antibody. This is known as **polyclonal activation**. For each clone, the cells produced develop into one of two types of cell:

- **Plasma cells**, which secrete antibodies. The plasma cells survive only a few days, but each can make around 2000 antibodies every second during its brief life-span. These antibodies destroy the pathogen, and any toxins it produces, in the ways described in Topic 11.2. The specific B lymphocytes and plasma cells, and the antibodies produced, together with the production of memory cells (below), are all features of the **primary immune response**. The primary immune response is necessary if the person is to recover from the disease caused by the pathogen.
- **Memory cells** live considerably longer than plasma cells – often for decades. These memory cells do not produce antibodies directly, but rather remain in the circulation (blood, tissue fluid, lymph and lymph nodes) until they come across the same antigen at some future date. With the greater number of specific cells present in the body, the chances of coming across the antigen more quickly than the first time are increased. When they do so, they rapidly divide and develop into plasma cells and more memory cells. The plasma cells produce the antibodies needed to destroy the pathogens, while the new memory cells circulate and are ready for a further infection at some time in the future. In this way, memory cells provide long-term immunity against the original infection. This is known as the **secondary immune response**. It is both more rapid and of greater intensity than the primary immune response and the pathogens are destroyed before they have a chance to multiply and cause any harm to the body. This means the person may not even know that they have been infected again. Figure 1 shows the relative quantities of antibody produced in the primary and secondary immune responses. Here, the other ways that the specific B lymphocyte can

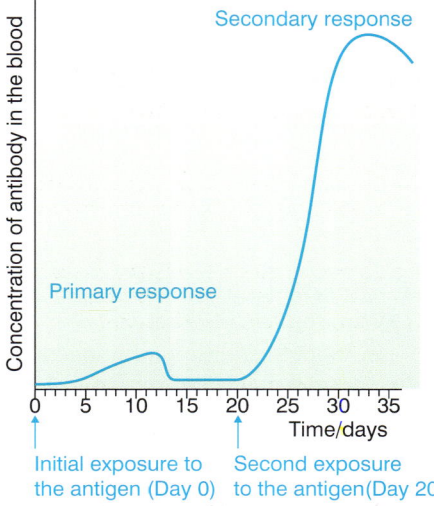

Figure 1 *Primary and secondary responses to an antigen*

become activated are shown. Macrophages are often the first cells to come across pathogens and destroy them following phagocytosis. The macrophages and some B lymphocytes can process antigens to form a complex with specific cell surface proteins known as MHC proteins. This complex can be attached to the cell surface and the cell can act as an antigen presenting cell (APC). APCs can stimulate the humoral response by activating B lymphocytes. Recognition and binding to an antigen on an APC can activate T-helper cells to bind to a B-lymphocyte and further stimulate the response. Other ways that the T-helper cell responds are discussed in Topic 11.5. Figure 2 summarises the role of B lymphocytes in immunity.

The way that memory cells function explains why most of us only develop diseases such as chickenpox once during our life. The pathogens causing each of these diseases are of a single type and so are quickly identified by the memory cells when they invade the body on subsequent occasions. Cold viruses, by contrast, have over 100 different strains, which are constantly changing. New infections are therefore highly unlikely to be the same as a previous one. With no specific memory cells to stimulate antibody production, we have to wait for our slower, less intense, primary response to overcome the infection – during which time we suffer a sore throat and runny nose.

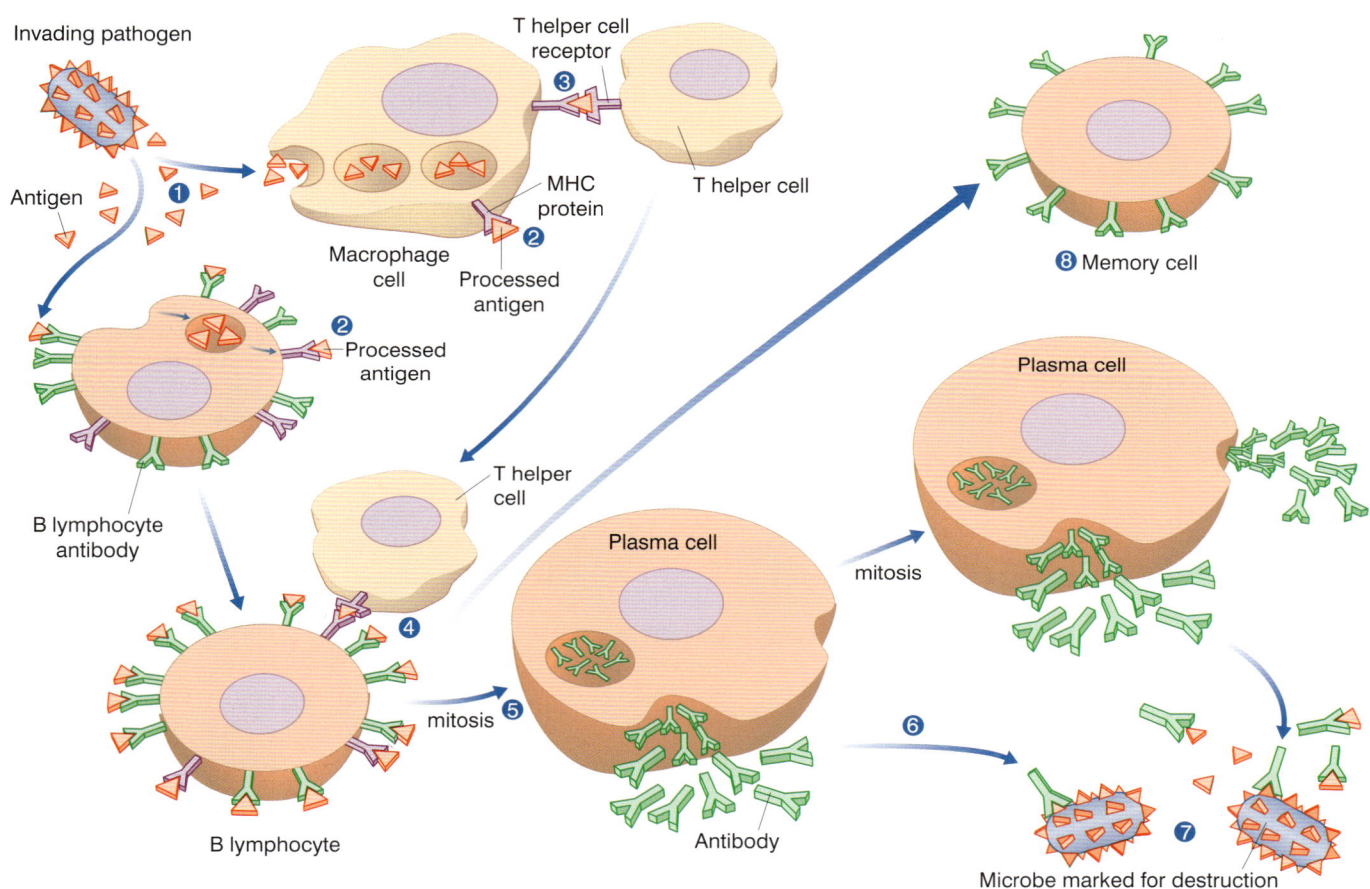

1. Invading pathogen produces antigens that are taken up by macrophages by phagocytosis. Antigen bound to receptors on B lymphocytes can be taken into the cell for processing.

2. Both the macrophage cells and the B lymphocytes process the antigens and bind them to a MHC protein (MHC = major histocompatibility complex). The MHC protein presents the processed antigen on the surface of the cells, which are therefore referred to as **antigen-presenting cells**.

3. A T helper cell attaches to the processed antigen on the macrophage cell and becomes activated, making it capable of interacting with B lymphocytes.

4. T helper cells attach to the MHC proteins with antigens on the surface of the B lymphocyte.

5. The B lymphocyte is activated to divide by mitosis (clonal expansion) to give a clone of plasma cells. Specific memory B lymphocytes are also formed as a result of clonal expansion.

6. The cloned plasma cells produce antibodies that exactly fit antigens on the pathogen's surface.

7. The antibodies attach to antigens on the pathogen, causing agglutination and lysis of the pathogen, thereby destroying it (= **primary response**).

8. Some B lymphocytes develop into memory cells that survive for long periods. Future invasions by the same pathogen lead to rapid division of memory cells, some of which develop into plasma cells that produce antibodies. The process is then repeated from stage 6 (= **secondary response**).

Figure 2 *Summary of the role of B lymphocytes in the immune response (humoral immunity)*

T lymphocytes and cell-mediated immunity

Cell-mediated immunity

While B lymphocytes respond to non-self (foreign) cells and the foreign products, e.g. toxins, that they produce, T lymphocytes respond to an organism's own cells that have been invaded by non-self material, e.g. a virus, a bacterium or a **cancer** cell. They also respond to transplanted material, which is genetically different. How then can T lymphocytes distinguish these invader cells from normal ones? It is made possible because:

• macrophage cells that have engulfed a pathogen and broken it down, present some of the proteins produced on their own outer surface
• body cells invaded by a virus also manage to present some of the viral proteins on their own cell surface membrane, as a sign of distress
• cancer cells likewise display non-self proteins on their cell surface membranes.

The non-self materials on the surface of all these cells act as **antigens** and therefore the term **antigen-presenting cells** is used to describe them (Topic 11.4). There are many different versions of the two main types of T lymphocytes in the body,

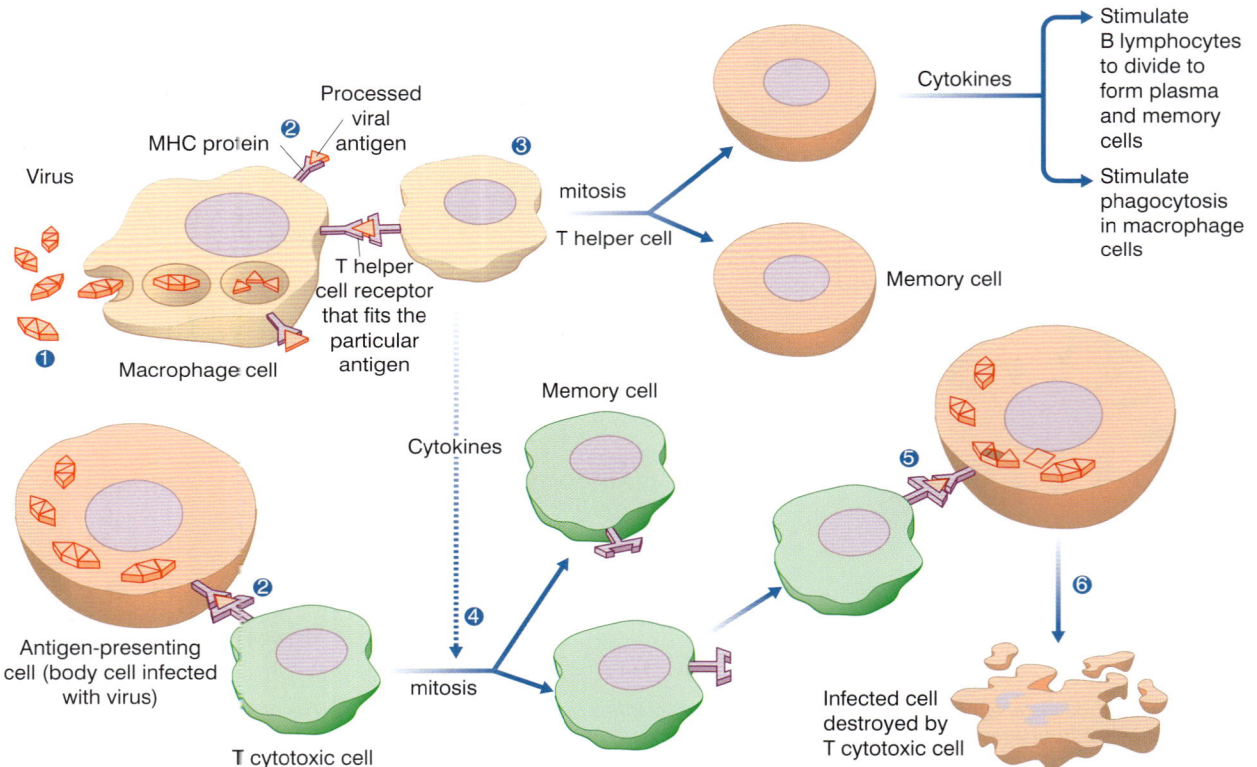

1. Viruses both invade body cells and are taken in during phagocytosis by macrophage cells.

2. Both the body cells and the macrophage cells process the viruses and bind antigens from them to a MHC protein. The MHC protein presents the antigen on the surface of the cells for recognition by and binding to specific T lymphocytes..

3. A T helper cell attaches to the antigen on the surface of the macrophage cells and

is thereby stimulated to divide by mitosis (clonal expansion). Some of the new T helper cells develop into memory cells that survive for long periods and respond immediately to any new infection by the same virus. Other T helper cells produce cytokines that stimulate B lymphocytes and macrophage cells.

4. The cytokines also cause T cytotoxic cells to divide by mitosis. Some of these T cytotoxic

cells form memory cells that survive and respond immediately to any new infections by the same virus.

5. Other T cytotoxic cells attach to any body cell presenting the viral antigen (i.e. those that are infected by the virus).

6. The attached T cytotoxic cells produce perforins to make holes in the cell membrane and so destroy the cell, along with the viruses it contains.

Figure 1 *Summary of the role of T lymphocytes in the immune response (cell-mediated immunity)*

each of which has a different receptor protein on its surface. Although these receptors function in a similar way, they are not **antibodies** because they remain attached to the cell rather than being released into the blood plasma.

As T lymphocytes will only respond to antigens that are attached to a body cell (rather than ones that are within body fluids), this type of response is called **cell-mediated immunity**. Figure 1 summarises how T lymphocytes respond to a viral infection.

Types of T lymphocytes

There are two main types of T lymphocyte:

- **T helper cells**, which play a key role in immunity. When they attach to an antigen-presenting cell, T helper cells secrete chemicals called **cytokines**. These cytokines:
  - stimulate macrophage cells to engulf pathogens by phagocytosis
  - stimulate B lymphocytes to divide and develop into antibody-producing plasma cells
  - activate T cytotoxic cells (T killer cells).
- **T cytotoxic cells (T killer cells)**, which kill body cells that are infected by non-self (foreign) material. They kill not by phagocytosis but by making holes in the cell surface membrane using proteins called **perforins**. These holes allow water to rush into the cell, causing it to burst. As viruses need living cells in which to reproduce, this sacrifice of body cells prevents viruses multiplying.

Both T helper cells and T cytotoxic cells produce their own type of **memory cells**, which circulate in the blood in readiness to respond to future invasions by the same pathogen. Another type of T lymphocyte is the **T suppressor cell**. This, as its name suggests, turns off the actions of the various other lymphocytes once the pathogens have been eliminated from the body. Figure 2 summarises the origin and roles of lymphocytes in immunity.

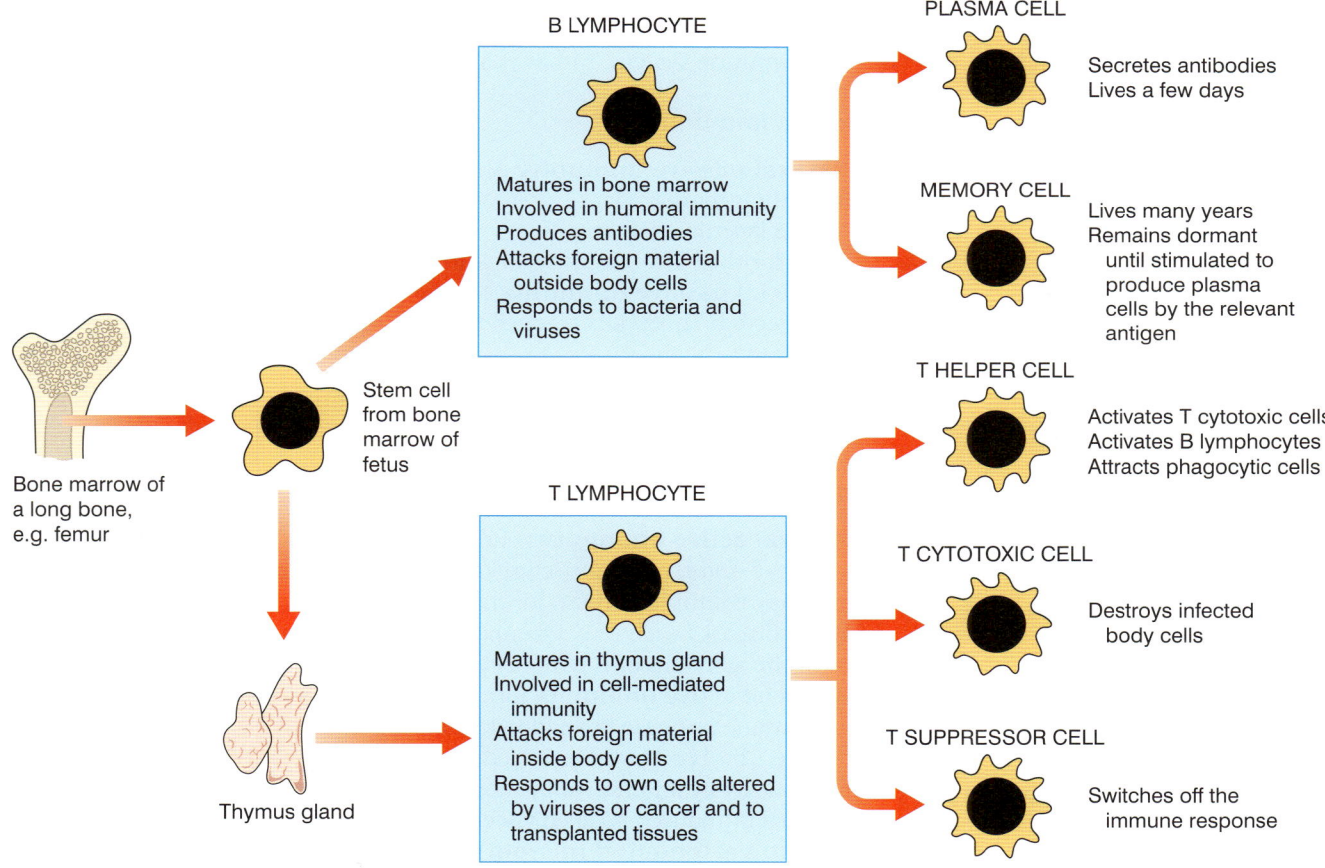

Figure 2 *The origin and roles of lymphocytes in immunity*

SUMMARY TEST 11.5

T lymphocytes respond to their own cells which have been infected by pathogens such as **(1)**. This type of response is called **(2)** immunity. There are a number of different types of T lymphocytes. One type, called **(3)** cells, secretes chemicals called **(4)**, which stimulate **(5)** cells to engulf pathogens by **(6)** and also stimulate **(7)** cells to divide to form antibody-producing cells called **(8)** cells. They also stimulate another type of T lymphocyte called **(9)** cells, which destroy pathogens.

On these pages you will learn to:

- Distinguish between active and passive, natural and artificial immunity

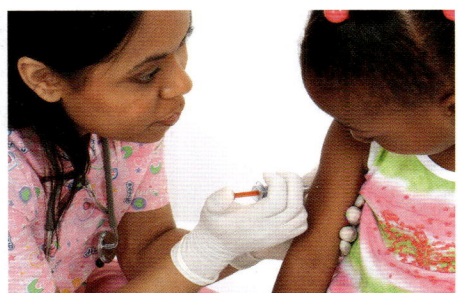

Figure 1 *Vaccination programmes for children have considerably reduced deaths from infectious diseases*

Immunity is the ability of an organism to resist infection. This immunity may be naturally acquired (obtained) or artificially induced (caused). The process of artificially inducing immunity is known as **immunisation**.

Types of immunity

- **Natural immunity** is immunity which is either inherited, or acquired as part of normal life processes, e.g. as a result of having had a disease.
- **Artificial immunity** is immunity acquired as a result of the deliberate exposure of the body to **antibodies** or **antigens** in non-natural circumstances (situations), e.g. **vaccination** (Figure 1).

Both natural and artificial immunity may be passively or actively acquired.

- **Passive immunity** is immunity acquired from the introduction of antibodies from another individual, rather than one's own immune system. It is generally short-lived.
- **Active immunity** is immunity resulting from the activities of an individual's own immune system, rather than an outside source. It is generally long lasting.

Passive immunity

- **Natural passive immunity** occurs when an individual receives antibodies from their mother via:
  - the placenta as a fetus
  - the mother's milk during suckling (breast feeding).
- **Artificial passive immunity** occurs when antibodies from another individual are injected. This takes place in the treatment of diseases such as tetanus and diphtheria.

Figure 2 shows the changes in antibody concentration over time with passive and active immunity.

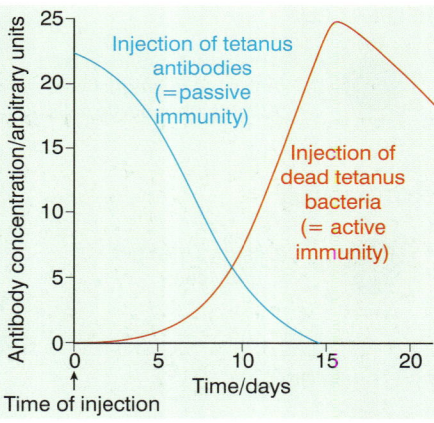

Figure 2 *Graph showing the concentration of tetanus antibodies, over time, with active and passive immunity*

Active immunity

- **Natural active immunity** results from an individual becoming infected with a disease under normal circumstances. The body produces its own antibodies, and may continue to do so for many years. It is for this reason that many people suffer diseases such as chickenpox only once in a lifetime. The immunity results from the activities of B lymphocyte memory cells (Topic 11.4).
- **Artificial active immunity** forms the basis of immunisation. It involves inducing an immune response in an individual, without them having the symptoms of the disease. This is achieved by introducing the appropriate disease antigens into the body, either by injection or by mouth. The process is called **vaccination**, and the material introduced is called **vaccine**. There are different forms of vaccine:
  - **Living attenuated microorganisms** are living microorganisms which have been treated, e.g. by heat, so that they do not cause symptoms, but still multiply. Although harmless, they stimulate the body's immune system. Measles, TB and poliomyelitis can be vaccinated against in this way.
  - **Dead microorganisms** have been killed by some means. Again, they are harmless, but induce immunity. Typhoid, cholera and whooping cough can be controlled by this means.
  - **Genetically engineered microorganisms** can be produced in which the genes for antigen production are transferred from a harmful organism to a harmless one. These are then grown in fermenters and the extracted antigen is separated and purified before injection. Hepatitis B vaccine is of this type.

IMMUNITY IN CHILDREN

As most immunity has to be acquired, either naturally or artificially, it follows that children, and especially newborn babies, are particularly at risk from infection. It takes many years for an individual to build up immunity to a wide variety of diseases. Why then is it that deaths from infections in young humans are not as common as might be expected? There are a number of reasons:

- Even before birth, a fetus has immunity to certain diseases. This is because the placenta allows some antibodies, e.g. anti-measles ones from the mother to pass into the fetus (natural passive immunity). This immunity only lasts for a short period after birth – in the case of measles for around four months.

- The milk formed by the breasts of the mother during the first few days after birth contains antibodies. This early milk, called **colostrum**, has a high concentration of antibodies belonging to the class Immunoglobulin A (IgA). These antibodies both remain in the intestines and are absorbed into the blood providing temporary immunity to a variety of pathogens both inside the intestines and within the body (natural passive immunity).

- In many countries there are programmes of vaccination designed to artificially induce immunity at the most appropriate stage of a child's development.

Table 1 *Summary of different types of immunity*

	Natural	Artificial
	Inherited or acquired naturally, not deliberately	Acquired deliberately by exposure to causative agent
Passive Results from the introduction of antibodies from another organism's immune system, rather than one's own Short-lived but immediate protection	Antibodies pass from mother • to fetus via placenta • to baby during suckling	Antibodies from a different individual or organism are injected
Active Results from the activities of an individual's own immune system Long-lasting	Antibodies acquired as a result of a previous infection producing B lymphocyte memory cells, which are reactivated on the second infection	Antigens are injected or given by mouth as a vaccine. They induce the body to produce its own antibodies to the disease. Memory cells are also formed. Vaccine may contain • dead pathogen • attenuated pathogen • genetically engineered antigens

SUMMARY TEST 11.6

Where immunity is inherited or acquired naturally, it is referred to as **(1)** immunity. It may be acquired from the introduction of antibodies from someone else rather than an individual's own immune system, in which case it is said to be **(2)** immunity. Where immunity is given by deliberately exposing the body to antibodies, it is known as artificial immunity. Artificial active immunity forms the basis of immunisation, where antigens are introduced into the body as part of the process known as **(3)**. There are three main forms of microorganism that can be used to produce immunity. These are **(4)**, **(5)** and **(6)**. Of the various forms of immunity, the full name of each of the following is: via the placenta as a fetus, called **(7)** immunity; injecting microorganisms weakened by heat treatment, called **(8)** immunity; and injecting antibodies, such as those from the tetanus bacterium, called **(9)** immunity.

Control of disease by vaccination

On these pages you will learn to:

- Explain how vaccination can control disease
- Discuss the reasons why vaccination programmes have eradicated smallpox, but not measles, tuberculosis (TB), malaria or cholera

Figure 1 *The production of new vaccines is a highly technological process requiring an extremely high standard of hygiene*

When smallpox was eradicated in 1977, HIV/AIDS was unknown – the first case was diagnosed in 1981. Had HIV/AIDS been around, it is unlikely that the vaccine used against smallpox would have been completely effective. One major advantage of the smallpox vaccine was that it was a live attenuated (weakened) vaccine. As such it multiplied within the body until the immune system overcame it. This took longer than would have been the case with a dead vaccine and so the immunity produced was more lasting and effective. People with HIV/AIDS have weakened immune systems and it is unlikely that they would have been able to overcome the vaccine. Far from giving protection, the vaccine would probably have caused the disease in HIV/AIDS sufferers. On the other hand, had HIV/AIDS sufferers not been vaccinated they would be at risk of smallpox and could have provided a reservoir of infection. Either way eradication would have been much more difficult, if not impossible.

Vaccination can only be used to fight diseases caused by pathogens. It is therefore not effective against genetic diseases such as sickle cell anaemia. The programme of vaccination against various diseases has had considerable success in controlling them. In cases such as smallpox, programmes have eradicated the disease altogether. Yet in other instances similar measures have had less success. To explore why there are these differences we need first to look at what is necessary for effective control of a disease through vaccination and then examine why such measures give different results with different diseases.

Features of a successful vaccination programme

To be effective, a programme of vaccination depends upon:

- **A suitable vaccine** being economically available and enough of the vaccine being available to immunise all of the vulnerable (at risk) population.
- **Few, if any, side effects** from vaccination. Unpleasant side effects may discourage individuals in the population from being vaccinated.
- **The mechanisms to produce, store and transport the vaccine**. This normally involves technologically advanced equipment (Figure 1), hygienic conditions and refrigerated transport.
- **The means of administering (giving) the vaccine properly at the appropriate time**. This involves training staff, at different centres, to be able to give the vaccine in the correct way and at the right time.
- **The ability to vaccinate the vast majority (all, if possible) of the vulnerable population**. This is best done at one time so that the transmission of the **pathogen** is interrupted because for a certain period there are no individuals in the population with the disease. This is known as **herd immunity**.

Why vaccination does not eliminate a disease

Even where these criteria for successful vaccination are met, it can still prove extremely difficult to eradicate a disease. The reasons for this include:

- Vaccination fails to induce immunity in certain individuals, e.g. ones with defective immune systems that do not produce the necessary **clones** of B and T lymphocytes (Topics 11.4 and 11.5) or ones whose diet lacks protein.
- Individuals may develop the disease immediately after vaccination, but before their immunity levels are high enough to fight it. These individuals may harbour the pathogen and reinfect others.
- The disease-causing agent (pathogen) may mutate frequently, so that its **antigens** change suddenly (antigenic shift) as opposed to gradually (antigenic drift). This means that vaccines suddenly become ineffective as the **antibodies** they induce the immune system to produce no longer recognise the new antigens on the pathogen. This happens with the influenza virus, which changes its antigens frequently.
- There may be so many varieties of a particular pathogen that it is all but impossible to develop a vaccine that is effective against them all. The common cold has over 100 types, for example.
- There are, as yet, no established vaccines against pathogens that are **eukaryotes**. Diseases such as malaria and sleeping sickness cannot therefore be tackled in this way.
- Certain pathogens 'hide away' from the body's immune system, either by concealing themselves inside cells, or by living in places out of reach (antigenic concealment), such as within the intestines, e.g. the cholera pathogen (Topic 10.4).

- Some pathogens suppress the body's immune system, so stimulating it through vaccination is ineffective, e.g. human immunodeficiency virus (Topic 10.7).
- There may be objections to vaccination for religious, ethical or medical reasons. For example, concerns over the Measles, Mumps and Rubella (MMR) triple vaccine led a number of parents to opt for separate vaccinations for their children, or to avoid vaccination altogether.

Figure 2 *Posters promoting child vaccination programmes are used to encourage parents to have their children immunised*

Smallpox – the last Jenneration!

The first ever vaccinations were carried out by Edward Jenner in 1794, against smallpox, and this very same disease was the first to be completely eradicated from the world, by the same process, in 1977. Why have we been able to remove smallpox when many other diseases such as measles, tuberculosis, malaria and cholera are still around? There are many reasons, which include:

- There was a simple, safe, easily stored vaccine against smallpox.
- The smallpox vaccine was easily and economically produced and simply administered.
- As a live vaccine, it was especially effective in producing immunity because the pathogen that was introduced into the body could reproduce and therefore remained in the body longer, enabling lasting immunity to build up.
- The smallpox virus was genetically stable, and so it did not mutate or change its antigens and so the vaccine was always effective. This also meant that the same vaccine could be used anywhere across the globe.
- The lethal nature of the disease (up to 30% of the victims died) encouraged people to be involved in the vaccination programme.
- The symptoms of the disease were easily recognised, and so infected patients could be isolated and treated before they spread the disease further.
- The virus did not remain in the body after an infection and so did not form a potential source of future infections.
- No other host organism was involved, making it easier to break the transmission cycle.

- There was a concerted worldwide vaccination programme, coordinated by the World Health Organization.

The problems of controlling measles by vaccination

In contrast to smallpox, the control of measles by vaccination has proved much more difficult because:

- To eradicate measles effectively it is necessary to ensure almost all of the population (95%) is immune at the same time. This herd immunity interrupts transmission of the pathogen, so preventing susceptible individuals from acquiring it. As the vaccine for measles is only 95% effective, this means virtually everyone in the population must be immune at any one time to eradicate the disease. As this is almost impossible, measles is unlikely to be eradicated.
- The measles antigens are complex and it is therefore difficult to make a wholly effective vaccine.
- Some children do not respond effectively to the measles vaccine, especially those who are malnourished, and therefore need a series of boosters. It is difficult, especially where the population is shifting, to ensure these boosters are administered.
- Parental concern over the side effects of the MMR vaccine has led to a reduced level of vaccination in some countries.

The problems of controlling cholera and tuberculosis by vaccination

Control of cholera by means of vaccination is difficult because:

- Cholera is an intestinal disease and therefore not easily reached by the immune system.
- The antigens of the cholera pathogen change rapidly (antigen shifting), making it difficult to develop an effective and lasting vaccine.
- Mobile populations as a result of global trade, tourism and refugees fleeing wars spread cholera and make it difficult to ensure that individuals are vaccinated and herd immunity is achieved.
- Many people require more than one dose of the vaccine to gain immunity and there may be problems following up with booster doses.

Control of TB by vaccination is difficult because:

- The increase in HIV infections means that many people will already have a weakened immune system, making a TB vaccination less effective.
- Mobile populations as a result of tourism, global trade and refugees have spread the disease worldwide and make it difficult to ensure that individuals are vaccinated and that herd immunity is achieved.
- The proportion of elderly people in the population is increasing and they often have less effective immune systems and so vaccination is less effective at stimulating immunity.

11 Examination Questions

1 a Outline the hybridoma method for the production of a monoclonal antibody. *(4 marks)*

b Herceptin is a monoclonal antibody used in the treatment of some breast cancers. It binds strongly to molecules of a receptor protein, HER2, that is produced in abnormally large quantities in the plasma (cell surface) membranes of about 30% of human breast cancers.

Investigations have been made into the most effective way to use Herceptin to treat breast cancer.

One experiment investigated the ability of different treatments to induce cell death in breast cancer cells.

Herceptin and X-ray treatment were used both separately and together. The results are shown in Figure 1.

Figure 1

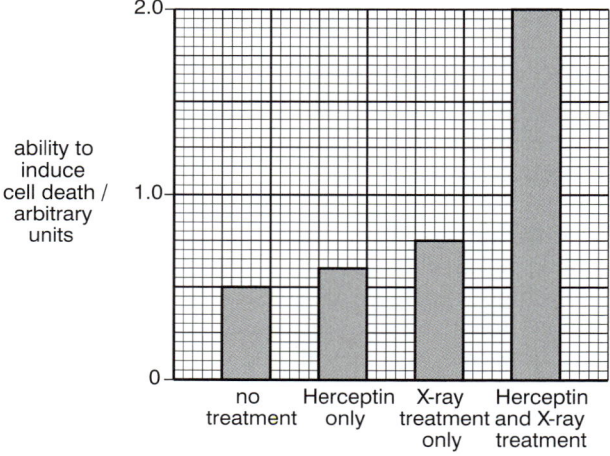

With reference to Figure 1,

i compare the effects on breast cancer cells of the different treatments. *(3 marks)*

ii calculate the percentage increase in the ability to induce cell death of using Herceptin **and** X-ray treatment compared with using Herceptin only. Show your working. *(2 marks)*

c A second experiment investigated the effect of increasing the doses of X-rays on the survival of breast cancer cells in the presence and absence of Herceptin. The results are shown in Figure 2.

Figure 2

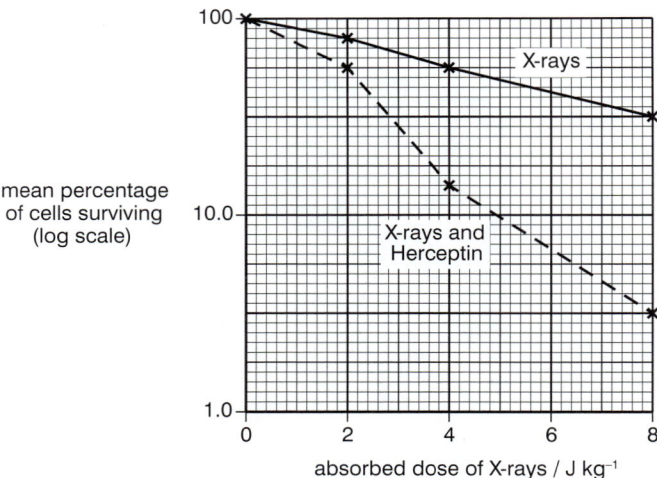

With reference to Figure 2,

i compare the effects of increasing doses of X-rays on cells in the presence and absence of Herceptin. *(3 marks)*

ii suggest an explanation for the effect of Herceptin. *(2 marks)*

(Total 14 marks)

Cambridge International AS and A Level Biology 9700 Paper 4 Q4 November 2008

2 Receptor proteins are part of the fluid mosaic structure of cell surface (plasma) membranes of T lymphocytes. Each type of receptor protein is specific to a particular antigen.

Figure 3 shows a receptor protein and the surrounding phospholipids of a cell surface membrane of a T lymphocyte.

a i Copy Figure 3 and draw a bracket (}) to indicate the width of the phospholipid bilayer. *(1 mark)*

ii Explain the term *fluid mosaic*. *(2 marks)*

iii Describe how the **structure** of the receptor molecule shown in Figure 3 is similar to the structure of an antibody molecule. *(2 marks)*

b Describe the roles of T-lymphocytes in a primary immune response. *(4 marks)*

c Describe three functions of cell surface membranes, **other than** the recognition of antigens. *(3 marks)*

(Total 12 marks)

Figure 3

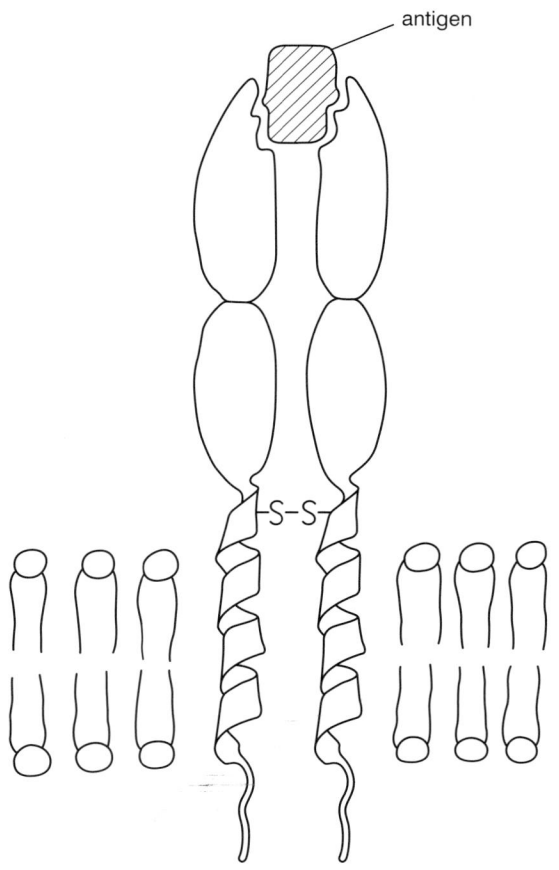

antigen

-S-S-

Cambridge International AS and A Level Biology 9700 Paper 2 Q1 November 2008

11 Practice Questions

3 a State two differences between a specific and a non-specific defence mechanism.

b Distinguish between self and non-self.

c After a pathogen gains entry to the body it is often a number of days before the body's immune system begins to control it. Why is this so?

d In the above case, why would it be inaccurate to say that the body takes days to 'respond' to the pathogen.

4 a Explain why the secondary immune response is much more rapid than the primary one.

b What are the differences between cell-mediated and humoral responses to a pathogen?

c Plasma cells can produce around 2000 protein antibodies each second. Suggest three cell organelles that you might expect to find in large quantities in a plasma cell, and explain why.

d Suggest why proteins, rather than carbohydrates or fats, have evolved as the molecules of which antibodies are made.

5 a What is an antigen?

b State two similarities of T cells and B cells.

c State two differences between T cells and B cells.

6 Avian (bird) flu is caused by one of many strains of the influenza virus. Although it is adapted primarily to infect birds, the H5N1 strain of the virus can infect other species including humans. Avian (bird) flu affects the lungs and can cause the immune system to go into overdrive. This results in a massive overproduction of T cells.

a From your knowledge of cell-mediated immunity and lung structure suggest why humans infected with the H5N1 virus may sometimes die from suffocation.

b Suggest a reason why any spread of bird flu across the world is likely to be very rapid.

7 The figure below shows part of the process of cell-mediated immunity.

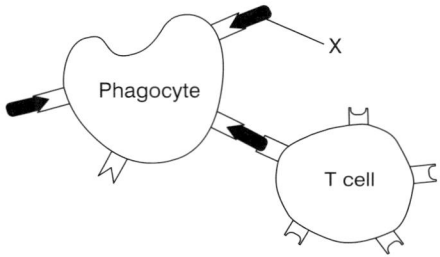

X

Phagocyte

T cell

a What is the name of the structure labelled X?

b Where did structure X originally come from?

c Give an alternative name for the cell labelled phagocyte.

d What type of T cell is shown in the figure?

e This type of T cell divides by mitosis to form long-lasting cells. What is this type of cell?

f The T cell secretes chemicals called cytokines. State two effects of cytokines.

8 a Immunity is the ability of organisms to resist infection. It can be divided into two types – active and passive. Distinguish between active and passive immunity.

b One way to provide immunity is to vaccinate people with a living attenuated microorganism. What is meant by 'a living attenuated microorganism'?

c Many diseases can be controlled through vaccination and some, like smallpox, have been eliminated by this means. In the case of influenza, however, the vaccine against it is not always effective. Suggest a reason why this is so.

d Give three reasons why it has proved difficult to control tuberculosis by vaccination.

Energy and respiration

Energy

On these pages you will learn to:

- Outline the need for energy in living organisms, as illustrated by anabolic reactions, such as DNA replication and protein synthesis, active transport, movement and the maintenance of body temperature

All living organisms require energy in order to remain alive. This energy comes initially from the Sun (or in a few instances from chemicals). Plants use solar energy to combine water and oxygen into complex organic molecules by the process of **photosynthesis**. Both plants and animals then break down these organic molecules to make **adenosine triphosphate (ATP)** that is used as the energy source to carry out processes that are essential to life.

What is energy?

Energy is defined as 'the ability to do work'. It can be considered to exist in two states:

- **Kinetic energy** is the energy of motion. Moving objects perform work by making other objects move.
- **Potential energy** is stored energy. An object that is not moving may still have the capacity to do so and therefore possesses potential energy. A stone on a hillside has potential energy. If it is set in motion, gravity will cause it to roll downhill and some of its potential energy will be converted into kinetic energy.

Other facts about energy include:

- it takes a variety of different forms, e.g. light, heat, sound, electrical, magnetic, mechanical, chemical and atomic
- it can be changed from one form to another
- it cannot be created or destroyed
- it is measured in joules (J).

Why do organisms need energy?

Without some input of energy, natural processes tend to break down in randomness and disorder. Living organisms are highly ordered systems that require a constant input of energy to prevent them becoming disordered – a condition that would lead to their death. More particularly energy is needed for:

- **Anabolism**, in which smaller, more simple substances are built up into larger, more complex ones, e.g. during DNA replication, in which nucleotides are joined by condensation reactions to form polynucleotides, and protein synthesis, in which amino acids are joined together to form polypeptides.
- **Movement** both within an organism (e.g. circulation of blood) and of the organism itself (e.g. locomotion due to muscular contraction or movement of cilia and flagella).
- **Active transport** of ions and molecules against a concentration gradient across membranes, such as the cell surface membrane and the tonoplast, e.g. the **sodium–potassium pump**.
- **Maintenance, repair and division** of cells and the organelles within them.
- **Maintenance of body temperature** in birds and mammals. These organisms are **endothermic** and need energy to replace that lost as heat to the surrounding environment.

EXTENSION

Laws of thermodynamics

The first law of thermodynamics states that energy cannot be created or destroyed, but only converted from one form to another. The amount of energy in the universe is always the same, although that on Earth may fluctuate slightly. The Earth obtains the majority of its energy in the form of light from our nearest star – the Sun.

Most of the energy used by mankind comes initially from the sunlight via a series of energy conversions.

The second law of thermodynamics states that disorder (more technically called **entropy**) in the universe is continuously increasing. In other words, disorder is more likely than order. For example, a heap of bricks is more likely to fall down and become scattered than it is to arrange itself into a neat column because there is less energy in a disordered system than in an ordered one. When the universe was formed, it possessed the maximum potential energy it has ever had. Since then, it has become increasingly disordered because at each energy conversion the amount of entropy increased.

Energy and metabolism

The flow of energy through living systems occurs in three stages:

- The Sun's light energy is converted by plants to chemical energy during photosynthesis.
- The chemical energy from photosynthesis, in the form of organic molecules, is converted into ATP.
- ATP is used by cells to perform useful work.

All the reactions that take place within organisms are collectively known as **metabolism**. These reactions are of two types:

- **Anabolism** is the build up of larger, more complex molecules from smaller, simpler ones – a process requiring energy.
- **Catabolism** is the breakdown of complex molecules into simpler ones, with the release of energy.

Free energy

When one form of energy is converted to another, e.g. when electrical energy is converted to light energy in a light bulb, not all the energy is converted into its intended form; some is lost as heat. By 'lost' we mean the energy is no longer available to do useful work because it is distributed evenly. Energy that is available to do work under conditions of constant temperature and pressure is called **free energy**. If the products of a reaction contain more energy than the reactants and free energy must be supplied to make the reaction happen, it is known as an **endergonic (endothermic) reaction**. Reactions in which the products have less energy than the reactants and therefore energy is released, are known as **exergonic (exothermic) reactions**.

Activation energy

A typical chemical reaction may be represented as:

$$A \rightarrow B + C$$

In this case, A represents the **substrate** and B and C are the **products**. Before any chemical reaction can proceed it must be initially activated, i.e. its energy must be increased. The energy required is called the **activation energy**. Catalysts, such as enzymes, lower this activation energy and enable reactions to take place more rapidly and/or at lower temperatures. These events are summarised in Figures 1 and 2.

Once provided, the activation energy allows the products to be formed with a release of free energy. Chemical reactions can be reversible and therefore C and B can be synthesised into A. Such a reaction is not spontaneous, however, and requires an external source of energy if it is to proceed. Most biological processes are a cycle of reversible reactions. As there is inevitably some loss of free energy in the form of heat each time the reaction is reversed, the process cannot continue without a substantial input of energy from outside the organisms. The ultimate source of this energy is the light radiation of the Sun.

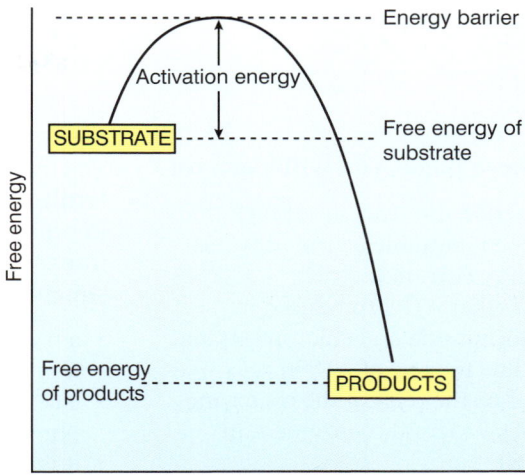

Figure 1 Concept of activation energy

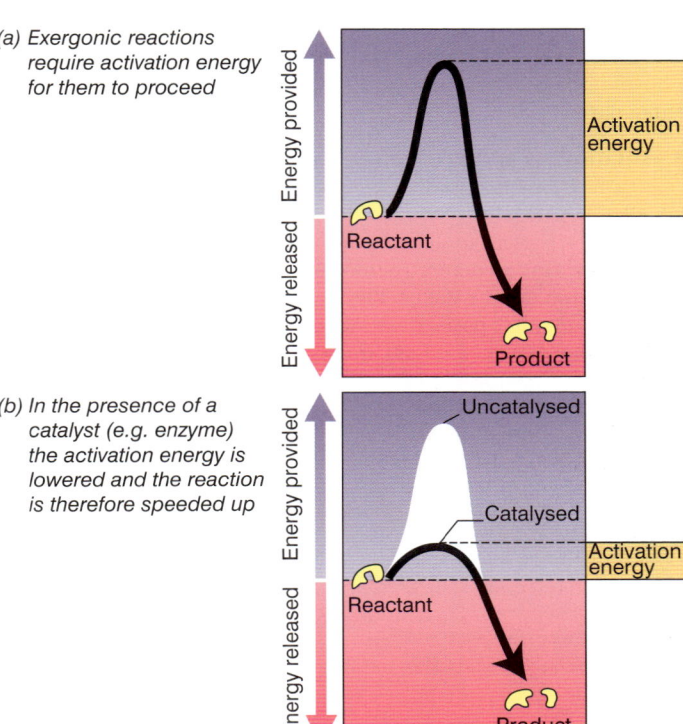

Figure 2 Effects of catalysts on activation energy

SUMMARY TEST 12.1

Energy is defined as the ability to do **(1)**. The energy of motion is known as **(2)** energy, whereas **(3)** energy is stored energy. Living organisms need energy for many reasons, including **(4)** reactions in which simple molecules are built up into complex ones and the movement of material by **(5)** against a concentration gradient. Living organisms also use energy for movement and the maintenance of **(6)** in birds and mammals. Before a chemical reaction can take place energy must be provided; this is known as **(7)** energy.

189

On these pages you will learn to:

- Describe the features of ATP that make it suitable as the universal energy currency
- State that ATP is produced in mitochondria and chloroplasts and outline the role of ATP in cells
- Outline the roles of the coenzymes NAD, FAD and coenzyme A in respiration

Structure of adenosine triphosphate (ATP)

The ATP molecule (Figure 1) is a phosphorylated nucleotide and it has three parts:

- **Adenine** – a nitrogen-containing organic base belonging to the group called purines.
- **Ribose** – a sugar molecule with a 5-carbon ring structure (pentose sugar) that acts as the backbone to which the other parts are attached.
- **Phosphates** – a chain of three phosphate groups.

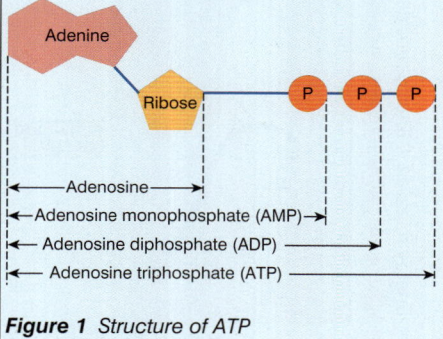

Figure 1 *Structure of ATP*

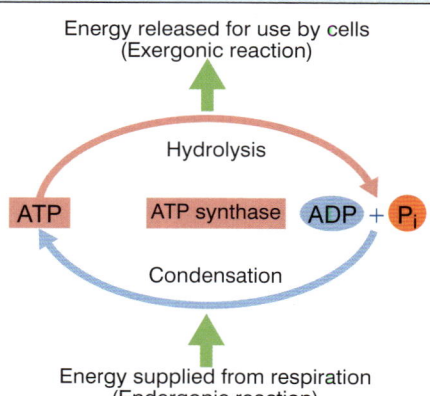

Figure 2 *ATP cycle*

The universal energy currency of all cells is a molecule called **adenosine triphosphate (ATP)** (see Figure 1). Almost every energy-requiring process in cells uses ATP. It is a small water-soluble molecule and therefore easily transported around the cell. Some of the features that help to explain why ATP is suitable as the universal energy currency include:

- a one-step reaction provides an immediate source of energy (see below)
- it is easily hydrolysed to release energy
- a constant supply of ATP is possible as it is recycled from ADP, which is easily phosphorylated (see Figure 2)
- it is a relatively small molecule that can move around the cell with ease
- it is a water-soluble molecule so it can take part in metabolic reactions
- the quantity of energy released and the efficiency of recycling ATP means that the needs of the cell can be satisfied.

It is the three phosphate groups that are the key to how ATP is the energy currency of the cell. Each one is very negatively charged and so they repel one another. This makes the **covalent bonds** that link them rather unstable. These unstable covalent bonds have a low activation energy, which means they are easily broken. When they do break they release a considerable amount of energy – $30.5 \, \text{kJ mol}^{-1}$ for each of the first two phosphates removed and $14.2 \, \text{kJ mol}^{-1}$ for the removal of the final phosphate. The terminal phosphate is removed according to the enzyme-catalysed reversible equation:

$$\underset{\substack{\text{adenosine} \\ \text{triphosphate}}}{\text{ATP}} + \underset{\text{water}}{\text{H}_2\text{O}} \rightleftharpoons \underset{\substack{\text{adenosine} \\ \text{diphosphate}}}{\text{ADP}} + \underset{\substack{\text{inorganic} \\ \text{phosphate}}}{\text{P}_i} + \underset{\text{energy}}{30.5 \, \text{kJ}}$$

Synthesis of ATP

The conversion of ATP to ADP is a reversible reaction (Figure 2) and therefore energy can be used to add an inorganic phosphate to ADP to re-form ATP. The interconversion rate of ATP and ADP is phenomenal. Although there are only around 50 g of ATP in the human body at any point in time, it is thought that, even at rest, a single human uses 65 kg of ATP in a 24-hour period. This means that, on average, a single ATP molecule undergoes around 1300 cycles of synthesis and **hydrolysis** each day. As the synthesis of ATP from ADP involves the addition of a phosphate molecule, it is a **phosphorylation** reaction. This phosphorylation is catalysed by the enzyme ATP synthase (sometimes called ATP synthetase) and it occurs in three ways:

- **Photophosphorylation** that takes place in grana of the chloroplasts during photosynthesis (Topic 13.4).
- **Oxidative phosphorylation** that takes place on the inner mitochondrial membranes of plant and animal cells, and the cell surface membrane of bacteria, during the process of electron transport (Topic 12.5).
- **Substrate-level phosphorylation** that takes place in plant and animal cells when phosphate groups are transferred from donor molecules to ADP to make ATP. For example, in the formation of pyruvate at the end of glycolysis (Topic 12.3).

In the first two, ATP is synthesised using energy released during the transfer of electrons along a chain of electron-carrier molecules in either the chloroplasts or the mitochondria. There is a difference in hydrogen ion concentration either side of certain **phospholipid** membranes in chloroplasts and mitochondria and it is essentially the flow of these ions across these membranes that generates ATP.

The process is referred to as the **chemiosmotic theory of ATP synthesis**. Although it takes place in a similar way in both chloroplasts and mitochondria, the summary account that follows and which is shown in Figure 3 describes the process in mitochondria.

- Hydrogen atoms produced during respiration are carried to the electron transport chain where they are split into **protons** (hydrogen ions – H^+) and electrons.
- As electrons pass along the electron carriers of the electron transport chain, each one being at a lower energy level than the one before, the energy released is used to pump the protons (H^+) into the space between the inner and outer mitochondrial membranes.
- Protons accumulate (build up) in the inter-membranal space, leading to a concentration gradient of protons (H^+) between the space and the matrix. This also means that there is an electrochemical gradient between the inter-membranal space and the matrix.
- As the inner mitochondrial membrane is almost impermeable to protons, they can only diffuse back through the **chemiosmotic** channels in the ATP synthase complexes.
- As protons flow through these channels their electrical potential energy is used to combine ADP with inorganic phosphate (P_i) to produce ATP.
- The phosphorylation reaction is catalysed by **ATP synthase** found in the head piece (Figure 3) of the ATP synthase complexes.
- Once in the matrix the protons recombine with the electrons on carriers on the inner membrane to form hydrogen atoms, which in turn combine with oxygen to form water.

Role of ATP

ATP is **not** a good long-term energy store. Fats, and carbohydrates such as glycogen, serve this purpose far better. ATP is therefore the **immediate energy source** of a cell. As a result, cells do not store large quantities of ATP, but rather just maintain a few seconds' supply. This is not a problem, as ATP is rapidly re-formed from ADP and inorganic phosphate (P_i) and so a little goes a long way. ATP is the source of energy for:

- **Anabolic processes** – It provides the energy needed to build up macromolecules from their basic units, e.g.
  - polysaccharide synthesis from monosaccharides
  - polypeptide synthesis from amino acids
  - DNA/RNA synthesis from nucleotides.
- **Movement** – ATP provides the energy for muscle contraction, ciliary and flagellar action and movement of vesicles along microtubules within the cell. In muscle contraction, ATP provides the energy for the filaments of striated muscle to slide past one another and therefore shorten the overall length of a muscle fibre.
- **Active transport** – ATP provides the energy necessary to move molecules or ions against a concentration gradient. This is an essential role, as every cell must maintain a precise ionic content.
- **Secretion** – ATP is needed to form the vesicles necessary for the secretion of cell products.
- **Activation of chemicals** – ATP makes chemicals react more readily, e.g. the phosphorylation of glucose at the start of glycolysis (Topic 12.3), e.g. activated nucleotides in DNA and mRNA synthesis.

Stereogram of the mitochondrion
- Outer membrane ⎫ Mitochondrial
- Inner membrane ⎬ envelope
- Cristae
- ATP synthase complex

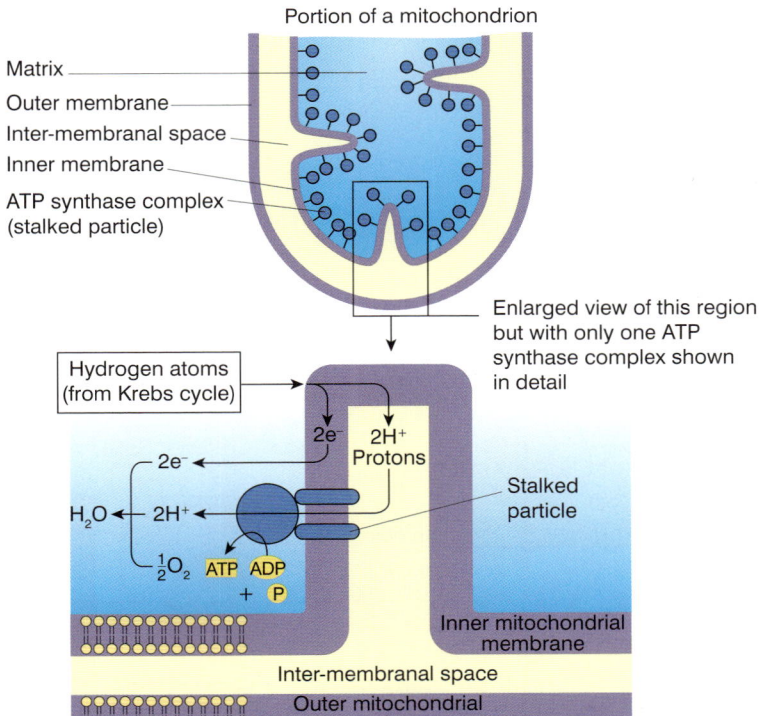

Portion of a mitochondrion
- Matrix
- Outer membrane
- Inter-membranal space
- Inner membrane
- ATP synthase complex (stalked particle)

Enlarged view of this region but with only one ATP synthase complex shown in detail

Hydrogen atoms (from Krebs cycle)

$2e^-$ $2H^+$ Protons

$2e^-$

$H_2O \leftarrow 2H^+$

Stalked particle

$\frac{1}{2}O_2$ ATP   ADP + P

Inner mitochondrial membrane

Inter-membranal space

Outer mitochondrial membrane

ATP synthase complex (stalked particle)

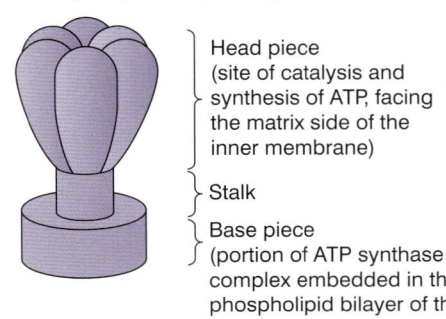

- Head piece (site of catalysis and synthesis of ATP, facing the matrix side of the inner membrane)
- Stalk
- Base piece (portion of ATP synthase complex embedded in the phospholipid bilayer of the inner membrane)

Figure 3 *Role of mitochondria in the synthesis of ATP according to the chemiosmotic theory*

On these pages you will learn to:

- List the four stages in aerobic respiration (glycolysis, link reaction, Krebs cycle and oxidative phosphorylation)
- State where glycolysis occurs in eukaryotic cells
- Explain that ATP is synthesised in substrate-linked reactions in glycolysis
- Outline glycolysis as phosphorylation of glucose and the subsequent splitting of fructose 1,6-bisphosphate (6C) into two triose phosphate molecules, which are then further oxidised to pyruvate with a small yield of ATP and reduced NAD

REMEMBER

In eukaryotic cells glycolysis takes place in the cytoplasm while the link reaction, Krebs cycle and oxidative phosphorylation all take place in the mitochondria.

Cellular respiration (also just called 'respiration') is the process by which the energy in food is converted into the energy for an organism to do biological work. Glucose is the main respiratory substrate and the overall equation for the process in aerobic conditions is:

$$C_6H_{12}O_6 \quad + \quad 6O_2 \quad \rightarrow \quad 6CO_2 \quad + \quad 6H_2O \quad + \quad energy$$

glucose   oxygen   carbon dioxide   water

Overview of respiration

Respiration in aerobic conditions can be divided into four stages:

- **Glycolysis** – an enzyme-controlled pathway in which one molecule of 6-carbon glucose is converted into two 3-carbon pyruvate molecules.
- **Link reaction (pyruvate oxidation)** – the 3-carbon pyruvate molecule is converted into carbon dioxide and a 2-carbon molecule called acetyl coenzyme A.
- **Krebs cycle** – the introduction of acetyl coenzyme A into a cycle of eight enzyme-catalysed reactions that yield reduced coenzymes NAD and FAD and some ATP.
- **Oxidative phosphorylation (electron transport system)** – oxidation of reduced NAD and FAD as part of an electron transport chain and ATP synthesis by chemiosmosis. Oxygen is required as a final electron acceptor and water is produced.

The main respiratory pathways are summarised in Figure 1. Respiration in anaerobic conditions involves glycolysis but not the other three stages described above. The pyruvate produced is further metabolised to lactate in mammalian tissue and to ethanol (and carbon dioxide) in yeast.

Glycolysis

Glycolysis occurs in the cytoplasm of all cells of multicellular organisms and in many unicellular organisms and is the process by which a hexose (6-carbon) sugar, usually glucose, is converted into two molecules of the 3-carbon molecule,

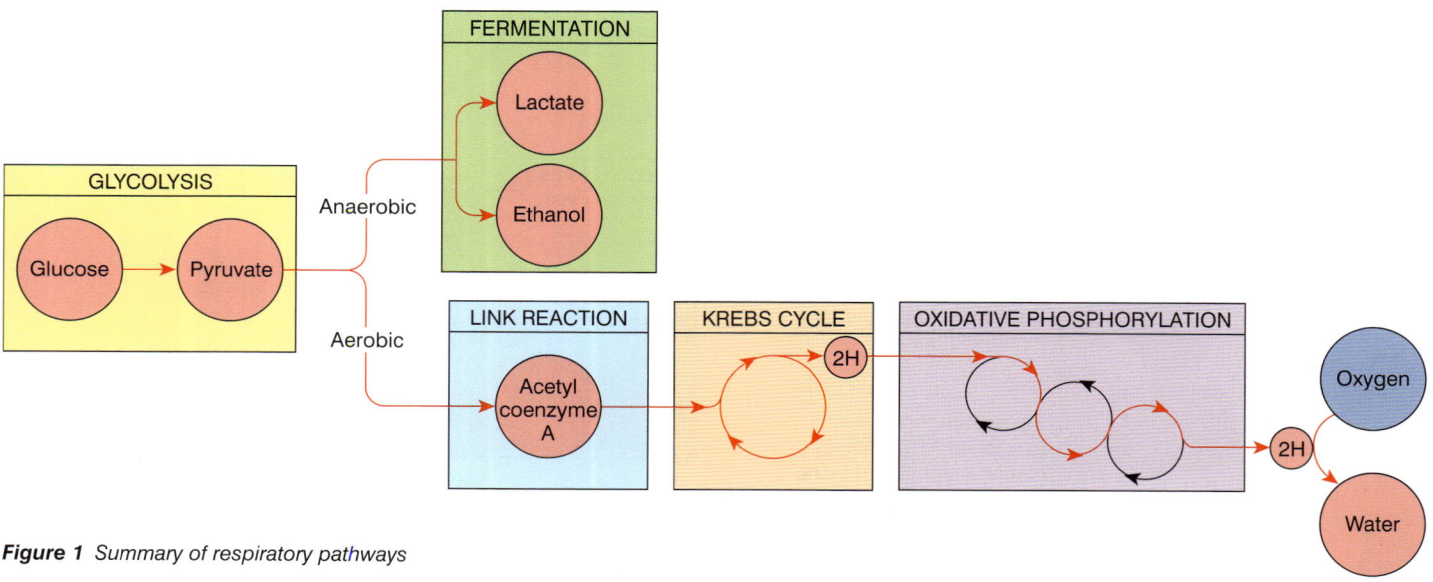

Figure 1 *Summary of respiratory pathways*

pyruvate. Although there are 10 smaller enzyme-controlled reactions in glycolysis, these can be conveniently grouped into four stages:

- **Activation of glucose by phosphorylation**. Before glycolysis can proceed, glucose must first be made more reactive by the addition of two phosphate molecules = **phosphorylation**. The phosphate molecules come from the **hydrolysis** of two ATP molecules to ADP. This provides the energy to activate glucose (**activation energy**) and also prevents glucose from being transported across the cell surface membrane and out of the cell. In the process glucose is converted to fructose 1,6-bisphosphate.
- **Splitting of the phosphorylated hexose sugar**. Fructose 1,6-bisphosphate is then split into two 3-carbon molecules known as triose phosphate.
- **Oxidation of triose phosphate**. Hydrogen is removed from each of the two triose phosphate molecules and transferred to a hydrogen carrier molecule called nicotinamide adenine dinucleotide (NAD$^+$) to form reduced NAD.
- **The production of ATP**. Four enzyme reactions convert each triose phosphate into another 3-carbon molecule called pyruvate. In the process, two molecules of ATP are formed from ADP and two more hydrogens are produced, which become attached to a molecule of NAD$^+$ to give reduced NAD. It must be remembered, however, that for each molecule of glucose at the start of the process, there are two molecules of triose phosphate produced. Therefore these yields must be doubled, i.e. 4 × ATP and 2 × reduced NAD.

The events of glycolysis are summarised in Figure 2.

The overall yield from one glucose molecule undergoing glycolysis is therefore:

- two molecules of ATP (four molecules of ATP are produced, but two were used up in the initial phosphorylation of glucose and so the net increase is two molecules)
- two molecules of reduced NAD (these have the potential to produce more ATP)
- two molecules of pyruvate.

Glycolysis was one of the earliest biochemical processes to evolve. It occurs in the cytoplasm of cells and does not require any organelle or membrane for it to take place. As it does not require oxygen it can proceed in both **aerobic** and **anaerobic** conditions. In the absence of oxygen the pyruvate produced by glycolysis can be converted into either lactate or alcohol by a process called fermentation. This is necessary in order to re-oxidise NAD so that glycolysis can continue. This is explained, along with details of the reactions, in Topic 12.6. These reactions, however, yield only a small fraction of the potential energy stored in the pyruvate molecule. In order to release this energy, most organisms use oxygen to break down pyruvate further in a process called the Krebs cycle.

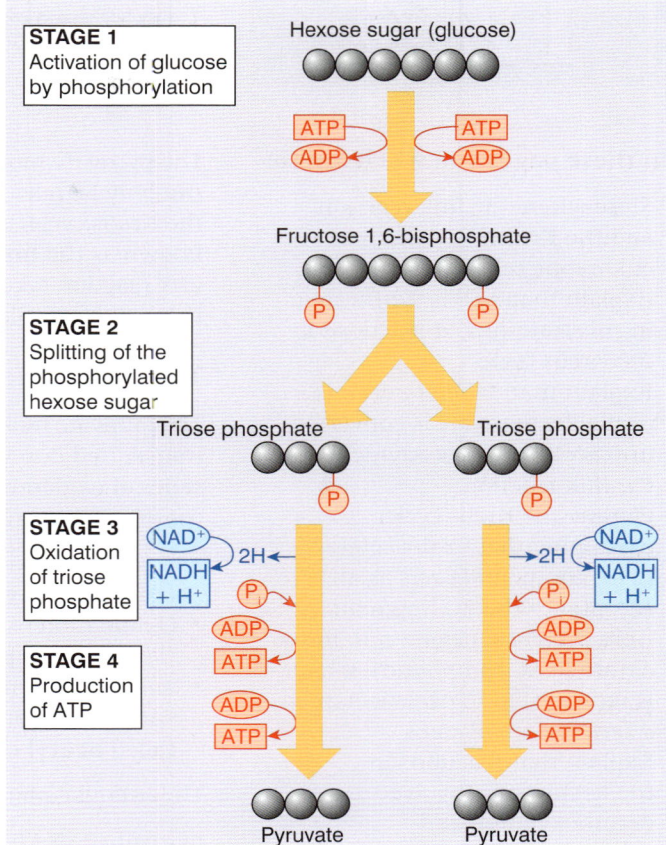

Figure 2 *Summary of glycolysis*

SUMMARY TEST 12.3

Glycolysis takes place in the **(1)** of cells and begins with the activation of the main respiratory substrate, namely the hexose sugar called **(2)**. This activation involves the addition of two **(3)** molecules provided by two molecules of **(4)**. The resultant molecule is known as **(5)** and in the next stage of glycolysis it is split into two molecules called **(6)**. The third stage entails the oxidation of these molecules by the removal of **(7)**, which is transferred to a carrier called **(8)**. The final stage is the production of the 3-carbon molecule **(9)**, which also results in the formation of two molecules of **(10)**.

EXTENSION
Enzyme complex – pyruvate dehydrogenase

The oxidation of pyruvate to acetyl coenzyme A in the link reaction involves a complex series of steps with three intermediate stages. The reaction is catalysed by a multienzyme complex called pyruvate dehydrogenase. This is one of the largest enzyme complexes in organisms, consisting of 60 sub-units. Enzyme complexes like this occur frequently in organisms and are made up of individual enzymes organised in sequence so that the product of one enzyme acts as the substrate for the next enzyme in the chain. This arrangement is more efficient because, rather than depending upon a chance meeting of enzyme and substrate as would be the case if the enzymes floated freely around, the substrate is not released but rather 'handed on' to the next enzyme in the biochemical sequence.

The pyruvate molecules produced during **glycolysis** possess potential energy that can only be released in a process called the **Krebs cycle**. Before they can enter the Krebs cycle, these pyruvate molecules must first be oxidised in a procedure known as the **link reaction**. In **eukaryotic cells** both the Krebs cycle and the link reaction take place exclusively inside mitochondria and these will only occur if oxygen is available.

The link reaction

The pyruvate molecules produced in the cytoplasm during glycolysis are actively transported into the matrix of mitochondria. Here pyruvate undergoes a complex series of oxidation-reduction reactions that are catalysed by a multienzyme complex (see extension). During these reactions the following changes take place:

- A carbon dioxide molecule is removed from each pyruvate (= **decarboxylation**) by means of the enzyme pyruvate decarboxylase.
- Oxididation of pyruvate results in the reduction of NAD to form reduced NAD (later used to produce ATP).
- The 2-carbon molecule that is formed is called an acetyl group and combines with a cofactor called coenzyme A (CoA) to produce a 2-carbon compound called **acetyl coenzyme A**.

The overall equation can be summarised as:

$$\text{pyruvate} + \text{NAD}^+ + \text{CoA} \rightarrow \text{acetyl CoA} + \text{reduced NAD} + CO_2$$

EXTENSION
The importance of acetyl coenzyme A

Coenzyme A is made up of vitamin B_5, the organic base adenine and the sugar ribose. It carries the acetyl group made from pyruvate into the Krebs cycle in the form of acetyl coenzyme A. The acetyl coenzyme A molecule is important because most molecules that are used by living organisms for energy are made into acetyl coenzyme A before entering the Krebs cycle. Most carbohydrates and fatty acids can be metabolised into acetyl coenzyme A to release energy. In the case of fats, these are first hydrolysed into glycerol and fatty acids. The glycerol can then be converted into triose phosphate that can be broken down during glycolysis, while the fatty acids are progressively broken down in the matrix of the mitochondria into 2-carbon fragments that are converted into acetyl coenzyme A. The reverse is also true, namely that excess carbohydrate can be made into fats via acetyl coenzyme A, making it a pivotal molecule in the interconversion of major substances in eukaryotic cells.

Krebs cycle

The Krebs cycle was named after the biochemist, Hans Krebs, who worked out its sequence in 1937. A central element of aerobic respiration, the Krebs cycle involves a series of eight small enzyme-catalysed steps that take place in the matrix of mitochondria. Its events are shown in Figure 1 and can be summarised as:

- The 2-carbon acetyl coenzyme A from the link reaction combines with a 4-carbon acceptor molecule (oxaloacetate) to produce a 6-carbon molecule (citrate).
- This 6-carbon molecule (citrate) is decarboxylated (loses CO_2) and dehydrogenated (loses two hydrogens) to give a 5-carbon compound, carbon dioxide and reduced NAD.

- Further decarboxylation and dehydrogenation produces a 4-carbon molecule (oxaloacetate), carbon dioxide, reduced NAD and reduced FAD and a single molecule of ATP produced as a result of substrate-level phosphorylation (Topic 12.2).
- The 4-carbon molecule (oxaloacetate) can now combine with a new molecule of acetyl coenzyme A to begin the cycle again.

For each molecule of pyruvate, the link reaction and Krebs cycle therefore produces:

- four molecules of reduced NAD. These have the potential to produce a total of 10 ATP molecules (Topic 12.5)
- one molecule of reduced FAD which has the potential to produce 1.5 ATP molecules (Topic 12.5)
- one molecule of ATP
- three molecules of carbon dioxide.

As two pyruvate molecules are produced for each original glucose molecule, these quantities must be doubled when the yields from a single glucose molecule are being calculated.

Roles of coenzymes

There are a number of coenzymes found in cells whose role it is to carry hydrogen atoms, and hence also electrons, from one compound to another. The two important hydrogen-carrying coenzymes in respiration are:

- **nicotinamide adenine dinucleotide (NAD)**, which is important in respiration
- **flavine adenine dinucleotide (FAD)**, important in aerobic respiration.

In respiration, NAD is the most important carrier. It works with dehydrogenase enzymes that catalyse the removal of hydrogen atoms from substrates like citrate and transfer

them to other molecules such as the hydrogen carriers involved in oxidative phosphorylation (Topic 12.5). The process works as follows:

- The two hydrogen atoms removed by the dehydrogenase enzyme dissociate into hydrogen ions (**protons**) and **electrons**:

$$2H \rightleftharpoons 2H^+ + 2e^-$$
hydrogen atoms   hydrogen ions (protons)   electrons

- Each NAD molecule in a cell exists in a form in which it has lost an electron, i.e it is oxidised and therefore exists as NAD^+.
- NAD^+ combines with the hydrogen ions (protons) and electrons to form NADH and a hydrogen ion (proton):

$$NAD^+ + 2H^+ + 2e^- \rightarrow \text{ reduced NAD (NADH + H}^+)$$

- When the hydrogen atom is transferred to a new molecule, the reduced NAD is re-oxidised, by the reversal of the above process, to re-form NAD^+.

In respiration, coenzyme A has a role as a carrier of an acetyl group in the formation of acetyl CoA from pyruvate. The 2-carbon acetyl CoA can then enter the Kreb's cycle.

The significance of the Krebs cycle

The Krebs cycle performs an important role in biochemistry for four reasons:

- It breaks down macromolecules into simpler ones; pyruvate is broken down into carbon dioxide.
- It produces hydrogen atoms that are carried by NAD and FAD to the electron transport chain for oxidative phosphorylation and the production of ATP by chemiosmosis, which provides metabolic energy for the cell.
- It regenerates the starter material (oxaloacetate), which would otherwise be completely used up.
- It is a source of intermediate compounds used by cells in the manufacture of other important substances such as fatty acids, amino acids and chlorophyll.

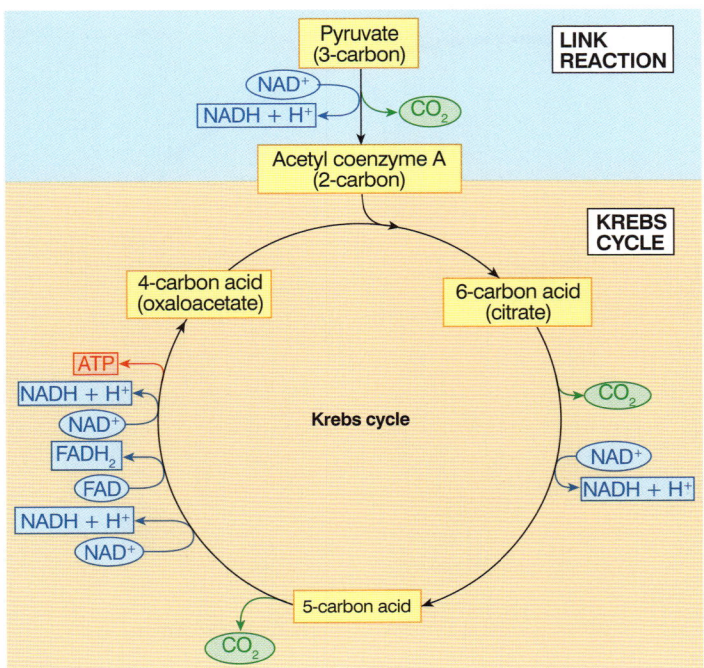

Figure 1 *Summary of link reaction and Krebs cycle*

SUMMARY TEST 12.4

Pyruvate molecules produced during (**1**) are moved into the (**2**) of the mitochondria by the process of (**3**). Before pyruvate can enter the Krebs cycle, it first has a carbon dioxide molecule removed – a process known as (**4**), and also has a pair of hydrogen atoms removed – a process called (**5**). The resultant molecule that enters the Krebs cycle is called (**6**). This molecule is important because it can also be made from alternative respiratory substrates such as other carbohydrates and (**7**). This 2-carbon molecule enters the Krebs cycle and combines with a 4-carbon molecule called (**8**) to produce a 6-carbon molecule called (**9**). The progressive loss of two carbon dioxide molecules and eight hydrogen atoms produces the original 4-carbon molecule. Most of the hydrogen atoms are transferred to a hydrogen carrier called (**10**).

12.5 Oxidative phosphorylation

On these pages you will learn to:

- State where oxidative phosphorylation occurs in eukaryotic cells
- Outline the process of oxidative phosphorylation including the role of oxygen as the final electron acceptor
- Explain that during oxidative phosphorylation:
  - energetic electrons release energy as they pass through the electron transport system
  - the released energy is used to transfer protons across the inner mitochondrial membrane
  - protons return to the mitochondrial matrix by facilitated diffusion through ATP synthase providing energy for ATP synthesis
- Describe the relationship between structure and function of the mitochondrion using diagrams

So far in the process of respiration, we have seen how hexose sugars such as glucose are converted to pyruvate (**glycolysis**) and how the 3-carbon pyruvate is fed into the **Krebs cycle** to yield carbon dioxide and hydrogen atoms. The carbon dioxide is a waste product and is removed during the process of gaseous exchange (Topic 9.2). The hydrogen atoms, or more precisely the **electrons** they contain, are however valuable as a potential source of energy. **Oxidative phosphorylation** (Figure 2) is the final stage in respiration and it is during this stage that most ATP molecules are synthesised.

Krebs cycle, oxidative phosphorylation and mitochondria

Mitochondria are rod-shaped organelles, between 1 μm and 10 μm in diameter, that are found in all but a few eukaryotic cells. Details of their structure are given in Topic 1.6, and in summary here. Each mitochondrion is bounded by a smooth outer membrane and an inner one that is folded into extensions called **cristae** (Figure 1). On the large surface area provided by the cristae are ATP synthase complexes (Figure 4). The inner space, or **matrix**, of the mitochondrion contains 70S ribosomes and is made up of a semi-rigid material of protein, lipids and traces of circular DNA.

Mitochondria are the sites of the link reaction, Krebs cycle and **oxidative phosphorylation**. More specifically:

- **The inner folded membrane (cristae)** has attached to it the proteins involved in the electron transport chain and therefore enables oxidative phosphorylation to take place.
- **The ATP synthase** complexes located in the membrane of the cristae contain ATP synthase for the synthesis of ATP by the chemiosmosis method (Topic 12.2).

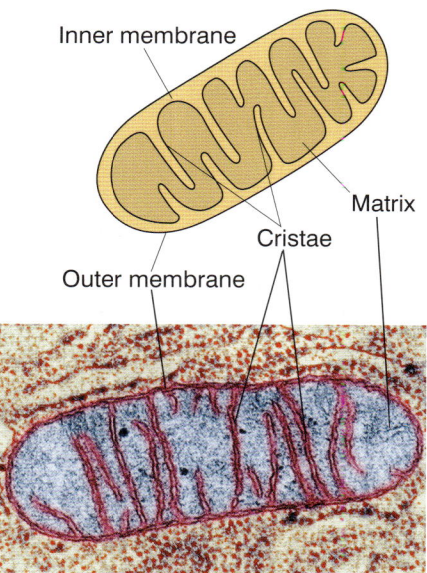

Figure 1 *The basic structure of a mitochondrion (top); false-colour transmission electron micrograph (TEM) of a mitochondrion (bottom)*

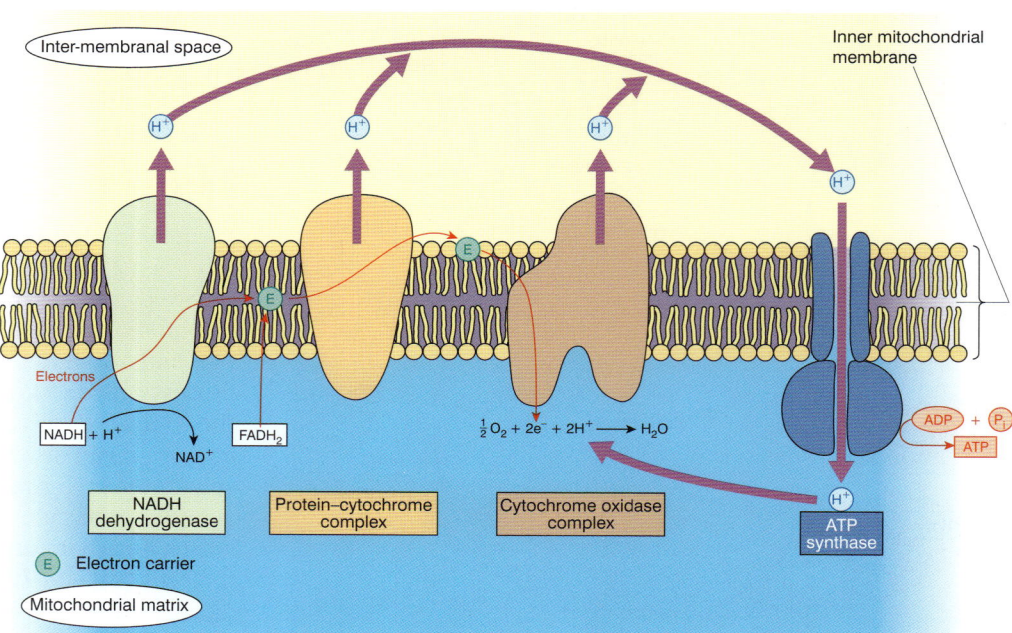

Figure 2 *Summary of oxidative phosphorylation*

- **The matrix** has the enzymes needed for the link reaction and Krebs cycle (Topic 12.4) and is where these processes occur. In addition, the energy required to synthesise ATP results from the hydrogen ion (pH) gradient that exists between the matrix and inter-membranal space.

As mitochondria play such a vital role in respiration and the release of energy, it is hardly surprising that they occur in greater numbers in metabolically active cells such as those of the muscles, liver and epithelial cells that carry out active transport. The mitochondria in these cells also have more densely packed cristae that provide a greater surface area for the attachment of proteins and enzymes involved in oxidative phosphorylation.

The process of oxidative phosphorylation

- The hydrogen atoms produced during glycolysis, the link reaction and Krebs cycle are combined with special molecules called **carriers** that are attached to the mitochondrial membranes. Most hydrogen atoms are combined with **nicotinamide adenine dinucleotide (NAD)**, although one pair from the Krebs cycle combines with **flavine adenine dinucleotide (FAD)**. The hydrogen atoms split into their protons (H^+) and electrons (e^-).
- The first carrier to accept electrons is a complex called **NADH dehydrogenase**.
- The electrons then pass via a carrier to a protein–cytochrome complex and finally, via another carrier to the cytochrome oxidase complex.
- The sequence of transfer of electrons from one carrier to the next is called the **electron transport chain**. The enzymes that catalyse these reactions are called **oxidoreductases** (see extension).
- Each of the three complexes in the chain acts as a proton pump using energy released from the energetic electrons to drive the protons (H^+) from the mitochondrial matrix into the inter-membranal space.
- The protons (H^+) build up in the inter-membranal space before they return by facilitated diffusion into the mitochondrial matrix through ATP synthase.

- As the protons pass through ATP synthase, ADP is combined with inorganic phosphate (P_i) to produce ATP. This ATP is formed using a diffusion force similar to osmosis: the process is called **chemiosmosis** (Topic 12.2).
- At the end of the chain the protons and electrons recombine and the hydrogen atoms so formed link with oxygen to form water. Oxygen is the final electron acceptor.

These events are summarised in Figure 2. There were thought to be 3 ATP molecules produced for each reduced NAD and 2 ATP produced for each reduced FAD molecule (fewer because reduced FAD enters further along the electron transport chain). Recent research suggests that these figures are more accurately 2.5 and 1.5 ATP molecules respectively. Not all the potential energy in the 32 ATP molecules produced for each glucose molecule (Topic 12.6) is a net yield. Around 25% of the energy produced is needed to transport ADP into the matrix of the mitochondrion so that it can combine with inorganic phosphate to form ATP.

The importance of oxygen in respiration is to act as the final acceptor of the hydrogen atoms produced in glycolysis and Krebs cycle. Without its role in removing hydrogen atoms at the end of the chain, the hydrogen ions and electrons would 'back up' along the chain and the process of respiration would come to a halt. This point is shown by the effect of cyanide on respiration. Most people are aware that cyanide is a very potent poison that causes death rapidly. It is lethal because it is an inhibitor of the final dehydrogenase enzyme in the electron transport chain, cytochrome oxidase. This enzyme catalyses the addition of the hydrogens to oxygen to form water. Its inhibition causes hydrogen ions and electrons to accumulate on the carriers, bringing the Krebs cycle to a halt and leaving pyruvate from glycolysis to accumulate. This pyruvate is converted to lactate as we shall see in the next topic.

EXTENSION
Oxidation, reduction and oxidoreductases

When a substance combines with oxygen the process is called **oxidation**. The substance to which oxygen has been added is said to be **oxidised**. When one substance gains oxygen from another, the one losing the oxygen is said to be **reduced**, the process being known as **reduction**. In practice, when a substance is oxidised it loses electrons and when it is reduced it gains electrons. This is the more usual way to define oxidation and reduction. Oxidation results in energy being given out, whereas reduction results in it being taken in. Oxidation and reduction always take place together and we call such reactions **redox** reactions (**red**uction and **ox**idation). The enzymes that catalyse these reactions are called **oxidoreductases**.

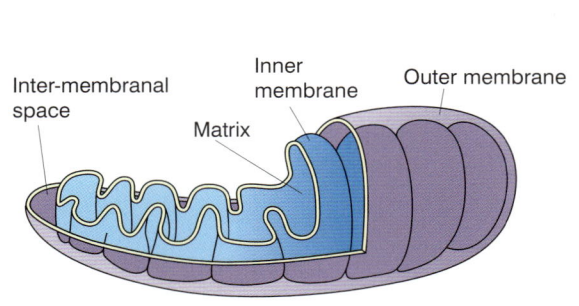

Inter-membranal space

Inner membrane

Outer membrane

Matrix

Figure 3 *Structure of a mitochondrion*

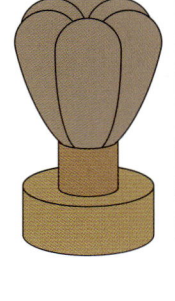

Head piece (site of catalysis and synthesis of ATP, facing the matrix side of the inner membrane)

Stalk

Base piece (portion of ATP synthase complex embedded in the phospholipid bilayer of the inner membrane)

Figure 4 *ATP synthase complex*

On these pages you will learn to:

- Explain the relative energy values of carbohydrate, lipid and protein as respiratory substrates and explain why lipids are particularly energy-rich
- Distinguish between respiration in aerobic and anaerobic conditions in mammalian tissue and in yeast cells, contrasting the relative energy released by each
- Explain the production of a small yield of ATP from respiration in anaerobic conditions in yeast and in mammalian muscle tissue, including the concept of oxygen debt
- Explain how rice is adapted to grow with its roots submerged in water in terms of tolerance to ethanol from respiration in anaerobic conditions and the presence of aerenchyma

We saw in Topic 12.5 that oxygen is needed if the hydrogen atoms produced in **glycolysis** and **Krebs cycle** are to be converted to water and thereby drive the production of ATP. What happens if oxygen is temporarily or permanently unavailable to a tissue or a whole organism?

In the absence of oxygen, neither the Krebs cycle nor **oxidative phosphorylation** can take place, leaving only the **anaerobic** process of glycolysis as a potential source of ATP. For glycolysis to continue, its products of pyruvate and hydrogen must be constantly removed. In particular, the hydrogen must be released from the reduced **NAD** in order to regenerate NAD^+. Without this the already tiny supply of NAD^+ in cells will be entirely converted to reduced NAD, leaving no NAD^+ to take up newly produced hydrogen from glycolysis. Glycolysis will then stop. The regeneration of NAD^+ is achieved by the pyruvate molecule from glycolysis accepting the hydrogen from reduced NAD in a process called fermentation.

In eukaryotic cells there are two main types of fermentation: alcoholic fermentation and lactate fermentation.

Alcoholic fermentation

Alcoholic fermentation occurs in certain bacteria and fungi (e.g. yeast) as well as in some cells of higher plants, e.g. root cells under waterlogged conditions.

The pyruvate molecule formed at the end of glycolysis first loses a molecule of carbon dioxide (decarboxylation) to form ethanal.

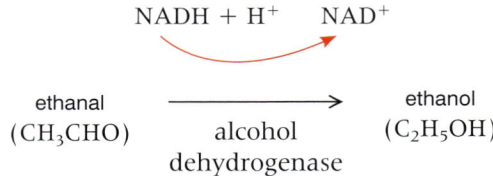

The ethanal accepts hydrogen from reduced NAD to produce ethanol. The summary equation for this is:

$$NADH + H^+ \qquad NAD^+$$

$$\underset{(CH_3CHO)}{\text{ethanal}} \xrightarrow[\text{alcohol dehydrogenase}]{} \underset{(C_2H_5OH)}{\text{ethanol}}$$

The overall equation using glucose as the starting point is:

$$\underset{\text{glucose}}{C_6H_{12}O_6} \longrightarrow \underset{\text{ethanol}}{2C_2H_5OH} + \underset{\text{carbon dioxide}}{2CO_2}$$

Alcoholic fermentation in yeast has been exploited by humans for thousands of years, in both the brewing and baking industries. In brewing, ethanol is the important product. Yeast is grown in anaerobic conditions in which it ferments natural carbohydrates in plant products such as grapes (wine production) or barley seeds (beer production) into ethanol. The ethanol produced kills the yeast cells that make it when its concentration reaches around 15%.

Adaptation of rice to anaerobic conditions

When soils are flooded or are waterlogged, the concentration of oxygen in the soil soon decreases and creates conditions where there is very little, or no, oxygen. If plant tissues are to survive submerged in water, there needs to be a way to receive oxygen from other areas of the plant that are not submerged. Alternatively, the submerged tissues, which now have to respire in anaerobic conditions, must either remove or show tolerance to the ethanol that is produced. This ethanol is normally toxic when it builds

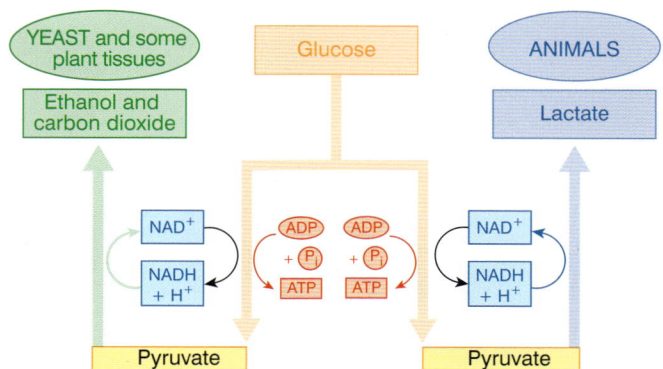

Figure 1 *How the NAD^+ needed for glycolysis is regenerated during fermentation in yeast and animals*

up. Rice has a number of adaptive features that allows it to grow with its roots submerged in water.

- The cells of the embryo are tolerant to the high concentrations of ethanol that build up as a result of anaerobic respiration.
- The stems and roots have many air spaces between the cells, which allow oxygen to diffuse from aerial parts not submerged and allow the cells in the root tissue to respire aerobically. This type of tissue is called **aerenchyma**.
- If the roots remain short of oxygen, they respire anaerobically and tolerate the build-up of ethanol. They also produce an abundance of the enzyme alcohol dehydrogenase, which breaks the ethanol down.
- Some varieties of rice have a higher rate of anaerobic respiration to increase the rate of ATP production.

Lactate fermentation

Lactate fermentation occurs in animals as a means of overcoming a temporary shortage of oxygen. Clearly, such a mechanism has considerable survival value, for example for a baby mammal in the period immediately after birth or an animal living in water where the concentration of oxygen fluctuates. However, lactate fermentation occurs most commonly in muscles as a result of strenuous exercise. In these conditions oxygen may be used up more rapidly than it can be supplied and therefore an **oxygen deficit** occurs. It may be essential, however, that the muscles continue to work despite the lack of oxygen – for example if the organism is fleeing from a predator. In the absence of oxygen, each pyruvate molecule produced takes up the two hydrogen atoms from glycolysis to form lactate as shown below:

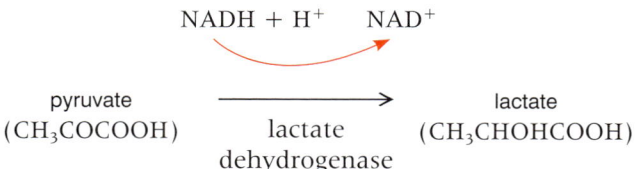

The lactate produced will cause cramp and muscle fatigue if it is allowed to accumulate in the muscle tissue. Although muscle has a certain tolerance to lactate, it is nevertheless important that it is removed by the blood and taken to the liver. Here it is converted to glycogen in a process called the Cori cycle. Some lactate may also be oxidised to pyruvate, in the reverse of the above equation, and then enter the Krebs cycle. The individual incurs an **oxygen debt**, which is later repaid when oxygen is available again. Figure 1 shows how the NAD^+ needed for glycolysis to continue is regenerated in both common forms of fermentation.

Energy yields from anaerobic and aerobic respiration

Energy from respiration in aerobic conditions is obtained by substrate level phosphorylation (in glycolyis and the Krebs cycle) and by oxidative phosphorylation. The number of ATP molecules produced for each glucose molecule is 32. However, the equivalent of one ATP is used in transporting some of the chemicals involved, giving a net yield of 31 ATP. Each ATP releases $30.5\,kJ\,mol^{-1}$ on hydrolysis, which gives a total yield of $945.5\,kJ\,mol^+$.

In anaerobic conditions, for each glucose molecule, only 2 ATP molecules are produced by substrate level phosphorylation in glycolysis. This is a total yield of $61.0\,kJ\,mol^{-1}$.

The theoretical maximum energy yield for the complete breakdown of a glucose molecule is $2870\,kJ\,mol^{-1}$. This means that respiration in aerobic conditions is 33% efficient, whereas respiration in anaerobic conditions is only approximately 2% efficient.

Energy yields from other respiratory substrates

Although we normally think of glucose as the main respiratory substrate, other carbohydrates, as well as lipids and protein, may also be used in certain circumstances, without first being converted to glucose.

As lipids contain relatively more C–H bonds than an equivalent mass of carbohydrate, their breakdown produces more hydrogen atoms for the electron transport chain and hence more energy. As a result, lipids release more than twice as much energy ($39.4\,kJ\,g^{-1}$) than the same mass of carbohydrate ($15.8\,kJ\,g^{-1}$). Once the immediate stores of carbohydrate, such as glycogen in the liver, have run out, organisms start to metabolise lipids for energy. Protein is normally only metabolised for energy in extreme situations such as starvation. When all carbohydrate and lipid reserves have been exhausted, organisms will, as a last resort, break down their protein into amino acids. These then have the amino groups (NH_2) removed before entering the respiratory pathway at a number of different points depending on their carbon content. The four-carbon and five-carbon amino acids are converted to Krebs cycle intermediates, whereas three-carbon amino acids are converted to pyruvate. Amino acids with large numbers of carbons are first converted to three-, four- and five-carbon amino acids. Although the energy yield depends on the exact composition of each protein, they normally yield around $17.0\,kJ\,g^{-1}$ – slightly more energy than carbohydrates.

On these pages you will learn to:

- Carry out investigations, using simple respirometers, to determine the RQ of germinating seeds or small invertebrates
- Carry out investigations, using simple respirometers, to measure the effect of temperature on the respiration rate of germinating seeds or small invertebrates
- Define the term 'respiratory quotient' (RQ) and determine RQs from equations for respiration

The rate of respiration in an organism can be determined either by measuring the volume of oxygen taken in or the volume of carbon dioxide produced. Measurements can be taken using a **respirometer**.

A simple respirometer

One type of simple respirometer is shown in Figure 1. It consists of two identical chambers – an experimental one containing the respiring organisms and a control one containing an equal volume of non-respiring material such as glass beads. The two chambers are connected by a U-shaped manometer tube that contains a coloured fluid. This type of respirometer is sometimes referred to as a **differential respirometer** because it has a built-in control chamber that makes sure that any fluctuation in temperature or pressure affects both sides of the manometer equally and so they cancel each other out. An equal volume of some carbon dioxide-absorbing material such as soda-lime is added to each chamber. In Figure 1 the apparatus has been set up to measure the rate of respiration of small invertebrates such as woodlice at different temperatures.

A simple respirometer is used as follows:

- The apparatus is left in the water bath for about 10 minutes to allow it to reach the desired temperature (equilibrate).
- Screw clips A and B are kept open during this time to allow air to escape as it expands.
- Both screw clips are closed at the end of the 10 minutes.
- The woodlice respire, absorbing oxygen and giving out the same volume of carbon dioxide.
- The carbon dioxide given off by the woodlice in chamber B is absorbed by the soda-lime. There is hence a reduction in the volume of air in chamber B, due to the oxygen being absorbed by the woodlice.
- As chamber B is airtight, this leads to a reduction in pressure within it.

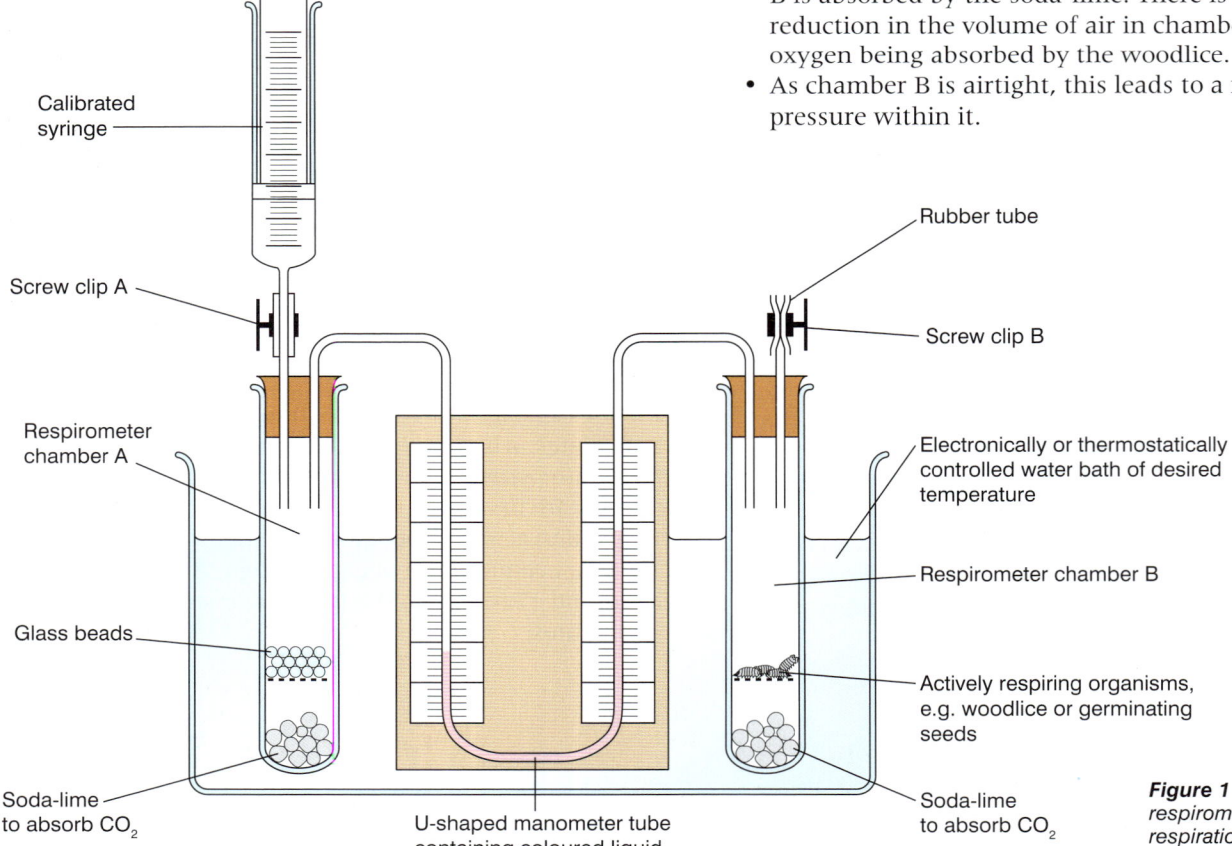

Calibrated syringe

Screw clip A

Rubber tube

Screw clip B

Respirometer chamber A

Electronically or thermostatically controlled water bath of desired temperature

Respirometer chamber B

Glass beads

Actively respiring organisms, e.g. woodlice or germinating seeds

Soda-lime to absorb CO_2

U-shaped manometer tube containing coloured liquid

Soda-lime to absorb CO_2

Figure 1 *A simple respirometer used to measure respiration rate at different temperatures*

- The pressure in chamber A (control) is now greater than that in chamber B (experimental) and so air moves towards chamber B, pushing the liquid in the manometer towards chamber B as it does so.
- The distance moved by the liquid in a given time is measured.
- At the end of this period, the screw clip A is opened and the syringe attached to chamber A is drawn upwards to reduce the pressure in it until it is again equal to that in chamber B, i.e. the liquid in the manometer is at an equal height in both tubes.
- The complete process is repeated two or three times and a mean of the measurements is calculated.
- Both screw clips are re-opened to allow oxygen to diffuse into the chambers and to allow the air to expand/contract when the temperature of the water bath is raised/lowered to a new value.
- The experiment is repeated at the new temperature.

The actual volume of oxygen taken in by the woodlice can be found in two ways:

- We can calculate the volume using the equation:

    volume $= \pi r^2 h$

    (where h is the distance moved by the liquid in the manometer and r is the internal radius of the manometer tube).
- We can simply measure on the calibrated syringe the volume of air needed to equalise the levels in the manometer tubes.

Respiratory quotients

The **respiratory quotient (RQ)** is a measure of the ratio of carbon dioxide given out by an organism to the oxygen consumed over a given period:

$$RQ = \frac{\text{volume of } CO_2 \text{ given out}}{\text{volume of } O_2 \text{ taken in}}$$

Although the term respiratory quotient is used, the ratio should more accurately be called the **respiratory exchange ratio (RER)**. This is because some of the gases being expired may come from non-respiratory sources. The RQ should not include these gases. In most cases the RQ and the RER are the same.

Different RQs give an indication of the type of substrate being respired. For example, a typical sugar such as glucose is oxidised according to the following equation:

$$C_6H_{12}O_6 + 6O_2 \longrightarrow 6CO_2 + 6H_2O$$

The RQ is therefore: $\quad \dfrac{6CO_2}{6O_2} = 1.0$

Lipids however, have less oxygen relative to carbon and hydrogen. A greater volume of oxygen is therefore required to oxidise a lipid completely and their RQs are therefore lower than those of carbohydrates:

$$C_{18}H_{36}O_2 + 26O_2 \longrightarrow 18CO_2 + 18H_2O$$

The RQ is therefore: $\quad \dfrac{18CO_2}{26O_2} = 0.7$

Proteins have a very varied structure depending on the number and types of each amino acid in the protein molecule. Their RQs are therefore equally varied but most have values around 0.9.

The usefulness of RQs in determining the substrate being respired is limited because:

- substances are rarely completely oxidised and partial oxidation gives a different value
- organisms rarely, if ever, respire a single food substance and the RQ therefore reflects the proportions of the different substrates being respired.

Most resting animals have an RQ of between 0.8 and 0.9 (Figure 2). Although these values would suggest that protein was being respired, we know that protein is used only in extreme situations such as starvation. We must assume, therefore, that these values are due to a mixture of carbohydrate (1.0) and lipid (0.7) being respired.

In anaerobic respiration, carbon dioxide is produced but no oxygen is taken in. If only anaerobic respiration takes place, then the RQ will be infinity. Where there is a mixture of aerobic and anaerobic respiration, RQ values are greater than 1.0. Some of these various values and their explanations are shown in Table 1, which shows the RQ of germinating seeds.

Figure 2 *Most resting animals such as this Mediterranean tree frog and lioness have an RQ between 0.8 and 0.9 by respiring a mixture of fat and carbohydrate*

Table 1 *Showing the RQ values of germinating seeds*

Time	RQ	Possible explanations
Seeds soaked in water	7.2	With little dissolved oxygen in water, respiration is a mixture of aerobic($\leqslant$1.0) and anaerobic (infinity)
After 14 hours in soil	1.5	As oxygen becomes available, the amount of aerobic respiration ($\leqslant$1.0) increases, whereas anaerobic respiration (infinity) decreases
After 48 hours in soil	0.7	A mixture of lipids (0.7) and carbohydrate (1.0) from the stores in the seed is being respired. The conversion of stored lipid to carbohydrate (0.35) is also taking place
After 14 days	1.0	The leaves have emerged and photosynthesis is producing carbohydrate (1.0), which is being respired

12 Examination Questions

1 Figure 1 shows the structure of ATP.

Figure 1

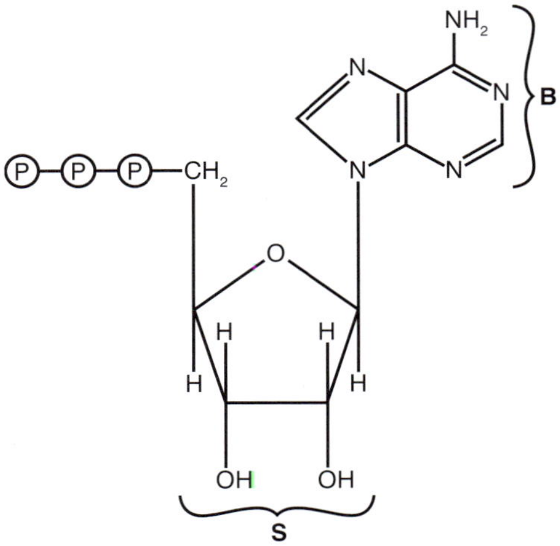

a i Name the nitrogenous base labelled **B**.   *(1 mark)*

   ii Name the sugar labelled **S**.   *(1 mark)*

b ATP is described as having a universal role as the energy currency in all living organisms. Explain why it is described in this way.   *(4 marks)*

c State **precisely** two places where ATP is synthesised in cells.   *(2 marks)*

                            (Total 8 marks)

Cambridge International AS and A Level Biology 9700 Paper 4 Q6 November 2008

2 a The respiratory quotient (RQ) is used to show what substrates are being metabolised in respiration.

   The RQ of a substrate may be calculated using the formula below:

$$RQ = \frac{\text{molecules of } CO_2 \text{ given out}}{\text{molecules of } O_2 \text{ taken in}}$$

   When the unsaturated fatty acid linoleic acid is respired aerobically the equation is:

$$C_{18}H_{32}O_2 + 25O_2 \longrightarrow \ldots\ldots CO_2 + 16H_2O + \text{energy}$$

   i Calculate how many molecules of carbon dioxide are produced when one molecule of linoleic acid is respired aerobically.   *(1 mark)*

   ii Calculate the RQ for linoleic acid.   *(1 mark)*

b Hummingbirds feed on nectar from flowers only during daylight hours. Nectar is rich in sugars.

Figure 2 shows a hummingbird.

Figure 2

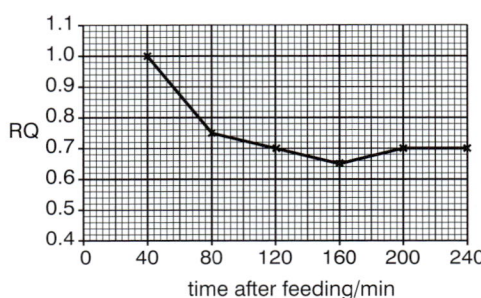

A study of aerobic respiration in captive hummingbirds was carried out. The hummingbirds were allowed to feed freely and then made to fast for four hours in constant conditions. During this time their RQ values were calculated every 40 minutes.

Figure 3 shows the results from this study.

Figure 3

Describe **and** explain the results shown in Figure 3.   *(4 marks)*

c Hummingbirds regulate their body temperature whereas butterflies do not regulate their body temperature.

Explain briefly the effect of an increase in temperature on the rate of respiration of a butterfly.   *(2 marks)*

                            (Total 8 marks)

Cambridge International AS and A Level Biology 9700 Paper 4 Q1 June 2009

12 Practice Questions

3 Cyanide is a non-competitive inhibitor of an enzyme in the electron transport chain. It therefore prevents the transfer of electrons along this chain. To determine where in the cell some of the respiratory pathways take place, scientists carried out the following experiment involving cyanide.

Mammalian liver cells were broken up (homogenised) and the resulting homogenate was centrifuged.

Incubated with	Complete homogenate		Nuclei only		Ribosomes only		Mitochondria only		Remaining cytoplasm only	
	Carbon dioxide	Lactate	Carbon dioxide	Lactate	Carbon dioxide	Lactate	Carbon dioxide	Lactate	Carbon dioxide	Lactate
Glucose	✓	✓	✗	✗	✗	✗	✗	✗	✗	✓
Pyruvate	✓	✓	✗	✗	✗	✗	✓	✗	✗	✓
Glucose + cyanide	✗	✓	✗	✗	✗	✗	✗	✗	✗	✓
Pyruvate + cyanide	✗	✓	✗	✗	✗	✗	✗	✗	✗	✓

Portions containing only nuclei, ribosomes, mitochondria and the remaining cytoplasm were separated out.

Samples of each portion, and of the complete homogenate, were incubated as follows:

- with glucose
- with pyruvate
- with glucose and cyanide
- with pyruvate and cyanide.

After incubation the presence or absence of carbon dioxide and lactate in each sample was recorded.

The results are shown in the table above in which ✓ = present and ✗ = absent.

a Briefly describe how the different portions of the homogenate might have been separated out by centrifuging.

b From the results of this experiment, name **two** organelles that appear not to be involved in respiration. Explain your answer.

c i In which cell organelle would you expect to find the enzymes of the Krebs cycle?
ii Explain how the results in the table support your answer.

d Which portion of the homogenate contains the enzymes that convert pyruvate into lactate?

e Explain why lactate production occurs in the presence of cyanide but carbon dioxide production does not.

f Explain why carbon dioxide can be produced by the complete homogenate when none of the separate portions can do so.

g If glucose were incubated with cytoplasm from yeast cells, which **two** products would be formed?

h Which **three** of the following might you expect to be rich in mitochondria: xylem vessel, liver cell, mature red blood cell, epithelial cell, muscle cell?

4 Coenzymes such as NAD are important in respiration. They help enzymes to function by carrying hydrogen atoms from one molecule to another. Scientists can model the way coenzymes work in cells using a blue dye called DCPIP. It can accept hydrogen atoms and so become reduced. Reduced DCPIP is colourless.

$$\text{DCPIP} + \text{hydrogen} \longrightarrow \text{reduced DCPIP}$$
$$\text{(blue colour)} \qquad\qquad \text{(colourless)}$$

In an investigation into respiration in yeast, three test tubes were set up as follows:

Tube A	Tube B	Tube C
$2\,cm^3$ yeast suspension	$2\,cm^3$ distilled water	$2\,cm^3$ yeast suspension
$2\,cm^3$ glucose solution	$2\,cm^3$ glucose solution	$2\,cm^3$ distilled water
$1\,cm^3$ DCPIP	$1\,cm^3$ DCPIP	$1\,cm^3$ DCPIP

All three tubes were incubated at a temperature of 30 °C. The colour of each tube was recorded at the start of the experiment and after 5 and 15 minutes. The results are shown in the table below:

Time/mins	Colour of tube contents		
	Tube A	Tube B	Tube C
0	blue	blue	blue
5	colourless	blue	blue
15	colourless	blue	pale blue

a Tube B acts as a control. Explain why this control was necessary in this investigation.

b Using your knowledge of respiration, suggest an explanation for the colour change after 15 minutes in:
i tube A
ii tube C.

c How might the results in tube A after 15 minutes have been different if the experiment had been carried out at 60 °C? Explain your answer.

d After 20 minutes the contents of tube A were mixed with air by shaking it vigorously turning the DCPIP back to a blue colour. Suggest a reason for this.

e Suggest why conclusions made only on the basis of the results of this experiment may not be reliable.

Photosynthesis

An overview of photosynthesis

On these pages you will learn to:

- Carry out investigations on the effects of light intensity, carbon dioxide and temperature on the rate of photosynthesis using whole plants, e.g. aquatic plants such as *Elodea* and *Cabomba*

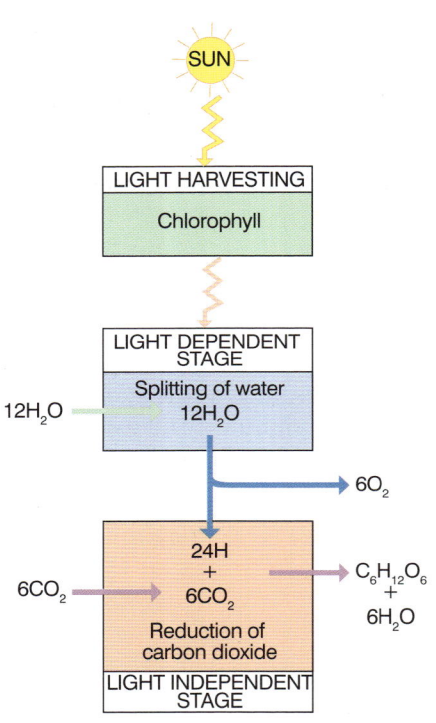

Figure 1 *Overview of photosynthesis*

Figure 2 *Green plants use sunlight to produce complex organic molecules from simple ones which then provide food for heterotrophs such as this bird and insect*

Humans, along with almost every other living organism, owe their very existence to photosynthesis. The energy we use, whether from the food we respire or from the wood, coal, oil or gas that we burn in our homes, has been captured from sunlight by photosynthesis. Photosynthesis also produces the oxygen we breathe by releasing it from water molecules.

Autotrophic nutrition

The nutrition of organisms can be divided into two broad categories:

- **Autotrophic nutrition** involves the build-up of simple inorganic molecules such as carbon dioxide and water into complex organic ones like lipids, carbohydrates and proteins using energy from light or from chemical reactions. Plants, algae and some bacteria are autotrophs.
- **Heterotrophic nutrition** involves the breakdown of complex organic molecules into simple soluble ones. Animals, fungi and some bacteria are heterotrophs (Figure 2).

The word autotroph means '*self-feeding*' and refers to those organisms such as plants that have no obvious means of obtaining or digesting food – no mouth, teeth, alimentary canal, etc. Instead of obtaining their food by consuming complex organic molecules, they manufacture their own from simple inorganic substances using energy from two possible sources:

- **Photoautotrophs** use light as their source of energy to drive the process of photosynthesis. Examples of photoautotrophs include green plants, algae and photosynthetic bacteria (e.g. cyanobacteria).
- **Chemoautotrophs** use energy from certain chemical reactions. The process is far less common than photosynthesis, but takes place in the **nitrifying** and **denitrifying bacteria** that are important in the nitrogen cycle (Topic 18.7).

An outline of photosynthesis

The overall equation for photosynthesis is:

$$6CO_2 + 6H_2O + \text{energy} \rightarrow C_6H_{12}O_6 + 6O_2$$

$$\text{carbon dioxide} \quad \text{water} \quad \text{light} \quad \text{glucose} \quad \text{oxygen}$$

Experiments using radioactive **isotopes** show that all the oxygen produced ($6O_2$) comes from water molecules and not the carbon dioxide molecules. However, the $6H_2O$ in the equation only provides six oxygen atoms, rather than the 12 produced. What happens in practice is that 12 water molecules are used to produce the oxygen, and the hydrogens from the water are used to reduce the carbon dioxide and produce the six water molecules. The equation for photosynthesis is therefore more accurately represented by the equation:

$$6CO_2 + 12H_2O + \text{energy} \rightarrow C_6H_{12}O_6 + 6O_2 + 6H_2O$$

Photosynthesis is a process in which the light energy, by a series of steps, is converted into chemical energy. There are two main stages to photosynthesis (Figure 1):

- **Capturing of light energy (light harvesting)** by chloroplast pigments such as chlorophyll, carotene and xanthophyll.

- **The light dependent stage**, is a process in which light energy is captured by chloroplast pigments such as chlorophyll, carotene and xanthophyll. An electron flow is created by the effect of light on chlorophyll, known as photoactivation. Photoactivation also causes water to split (**photolysis**) into hydrogen ions and oxygen. The useful products of the light dependent stage are ATP from chemiosmosis, and reduced NADP (Topic 12.2).
- **The light independent stage**, in which carbon dioxide is reduced to produce sugars and other organic molecules using the reduced NADP and ATP from the light dependent stage.

Measuring photosynthesis

The rate of photosynthesis is usually found by measuring the volume of oxygen produced by an aquatic plant such as Canadian pondweed (*Elodea*). This does not give an altogether accurate measure because:

- dissolved oxygen, nitrogen and other gases are often released from the leaf and surrounding water and become included in the gas volume measured
- some oxygen produced in photosynthesis will be used up in respiration.

The following account outlines how the rate of photosynthesis at different light intensities can be measured.

- The apparatus, known as an **Audus photosynthometer** is set up as in Figure 3, taking care not to introduce any air bubbles into it and checking that the apparatus is completely air-tight.
- The water bath is used to maintain a constant temperature throughout the experiment and should be adjusted as necessary. Better still, an electronically or thermostatically controlled water bath should be used.
- Potassium or sodium hydrogencarbonate solution can be used around the plant to provide a source of carbon dioxide – especially important if the experiment is to extend over a long period.
- A source of light that can have its voltage adjusted to change its intensity, is arranged close to the apparatus, which is kept in a dark room to prevent other light (which may vary in intensity) falling on the plant.
- The apparatus is kept in the dark for two hours to prevent photosynthesis and allow oxygen already produced by the plant to disperse.
- The light source is switched on and a stop clock started.
- Oxygen produced by the plant during photosynthesis collects in the funnel end of the capillary tube above the plant.

- After 30 minutes this oxygen is drawn up the capillary tube by gently withdrawing the syringe until its volume can be measured on the scale. This can be done directly (if the scale is calibrated in mm^3) or, if the scale is not calibrated, calculated using the formula $\pi r^2 h$ (where r is the internal radius of the tube and h is the length of the column of oxygen collected).
- The gas is drawn up into the syringe, which is then pushed in again before the process is repeated at the same light intensity four or five times and the mean volume of oxygen produced per hour is calculated.
- The apparatus is left in the dark for two hours before the procedure is repeated with the light source set at a different light intensity. The actual light intensity can be measured by a light meter placed in the same position relative to the light source as the plant was during the experiment.
- An alternative method of varying the light intensity is to change the distance of the light source relative to the plant. The light intensity is inversely proportional to the square of the distance from the plant to the light source, i.e. doubling the distance apart reduces the light intensity by a quarter.

The experiment can be modified to measure the effect of other factors on the rate of photosynthesis as follows:

- **Wavelength of light** – the experiment is repeated using different light sources that emit specific wavelengths or using filters of different colours between the light source and the photosynthometer.
- **Temperature** – the temperature of the water bath can be varied and the rates of photosynthesis compared.
- **Carbon dioxide concentration** – different concentrations of potassium or sodium hydrogencarbonate can be used to compare the rate of photosynthesis at different carbon dioxide concentrations.

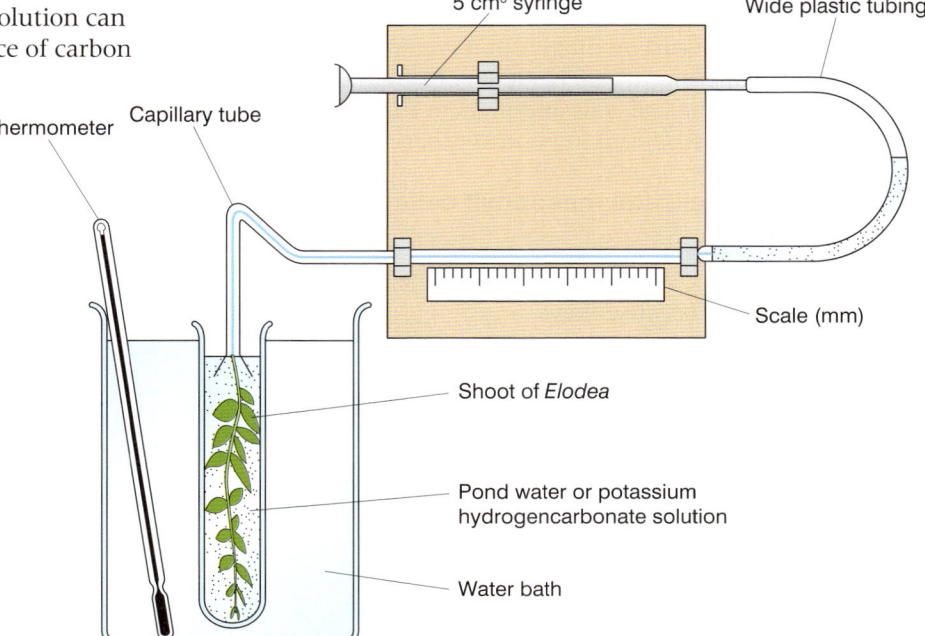

Figure 3 *Apparatus used to measure the rate of photosynthesis under various conditions*

On these pages you will learn to:

- State the sites of the light dependent and the light independent stages in the chloroplast
- Describe the relationship between structure and function in the chloroplast using diagrams and electron micrographs

The leaf is the main photosynthetic organ of the plant. Within the leaf, photosynthesis occurs in the cells of the palisade and spongy mesophyll tissues. In these cells, the organelles where photosynthesis takes place are the chloroplasts.

Structure of the leaf

Photosynthesis takes place largely in the leaf, the structure of which is shown in Figure 1(a) and (b). Leaves are adapted to bring together the three raw materials of photosynthesis (water, carbon dioxide and light) and remove its products (oxygen and glucose). These adaptations include:

- a large surface area that collects as much sunlight as possible
- a thin lamina (leaf blade), to keep the diffusion distance short
- a transparent cuticle and epidermis that let light through to the photosynthetic palisade cells beneath
- numerous stomata for gaseous exchange that open and close in response to changes in light intensity
- many air spaces, especially in the spongy mesophyll, to allow diffusion of carbon dioxide and water vapour
- a network of vascular tissue made up of xylem that brings water to the leaf cells and phloem that carries away the sugars produced in photosynthesis.

Adaptations of palisade mesophyll cells for photosynthesis

Palisade mesophyll cells (Figure 1(c)) are adapted to carry out photosynthesis because they:

- are closely packed and thin-walled to absorb maximum light
- are arranged vertically so there are fewer cross walls that could filter out the light
- are packed with numerous chloroplasts that move within the cells and are arranged in the best positions to collect the maximum quantity of light
- have a large vacuole that pushes the cytoplasm and chloroplasts to the edge of the cell allowing them to absorb maximum light and leave a short diffusion pathway for carbon dioxide

- have a large surface area and moist, thin walls for rapid diffusion of gases.

Structure and role of chloroplasts in photosynthesis

Photosynthesis takes place within cell organelles called **chloroplasts**, the structure of which is shown in Figure 1d. These vary in shape and size but are typically disc-shaped, 3–10 µm long and 1 µm in diameter. They are surrounded by a double membrane called the **chloroplast envelope**. The inner membrane is highly selective in what it allows to enter and leave the chloroplast. Inside the chloroplast envelope are two distinct regions:

- **The stroma** is a fluid-filled matrix where the light independent stage of photosynthesis takes place. Within the stroma are a number of other structures such as starch grains and lipid droplets. A small, circular piece of DNA and 70S ribosomes are also present.
- **The grana** are stacks of up to 100 disc-like structures called **thylakoids** (Figure 1(e)), where the light dependent stage of photosynthesis takes place. Within the thylakoids are the chloroplast pigments, which are arranged in a structured way and form complexes called photosystems I and II. Some thylakoids have tubular extensions that join up with thylakoids in adjacent grana. These are called **inter-granal lamellae**. There are far fewer photosytem I complexes in the thylakoids than photosystem II. In the inter-granal lamellae most of the complexes are photosystem I.

Chloroplasts (Figure 2) are adapted to their function of harvesting sunlight and carrying out the light dependent and light independent stages of photosynthesis in the following ways:

- The granal membranes provide a large surface area for the attachment of the photosynthetic pigments, electron carriers and enzymes that carry out the light dependent reaction.
- A network of proteins in the grana hold the photosynthetic pigments in a very precise manner that forms special units called **photosystems** (Topic 13.3), allowing maximum absorption of light.
- The granal membranes have many ATP synthase complexes attached to them, which manufacture ATP by chemiosmosis.
- The fluid of the stroma has all the enzymes needed to carry out the light independent stage (Calvin cycle).
- The stroma fluid surrounds the grana and so the products of the light dependent stage in the grana can readily pass into the stroma.
- Chloroplasts contain both circular DNA and 70S ribosomes so they can quickly and easily manufacture some of the proteins needed for photosynthesis.

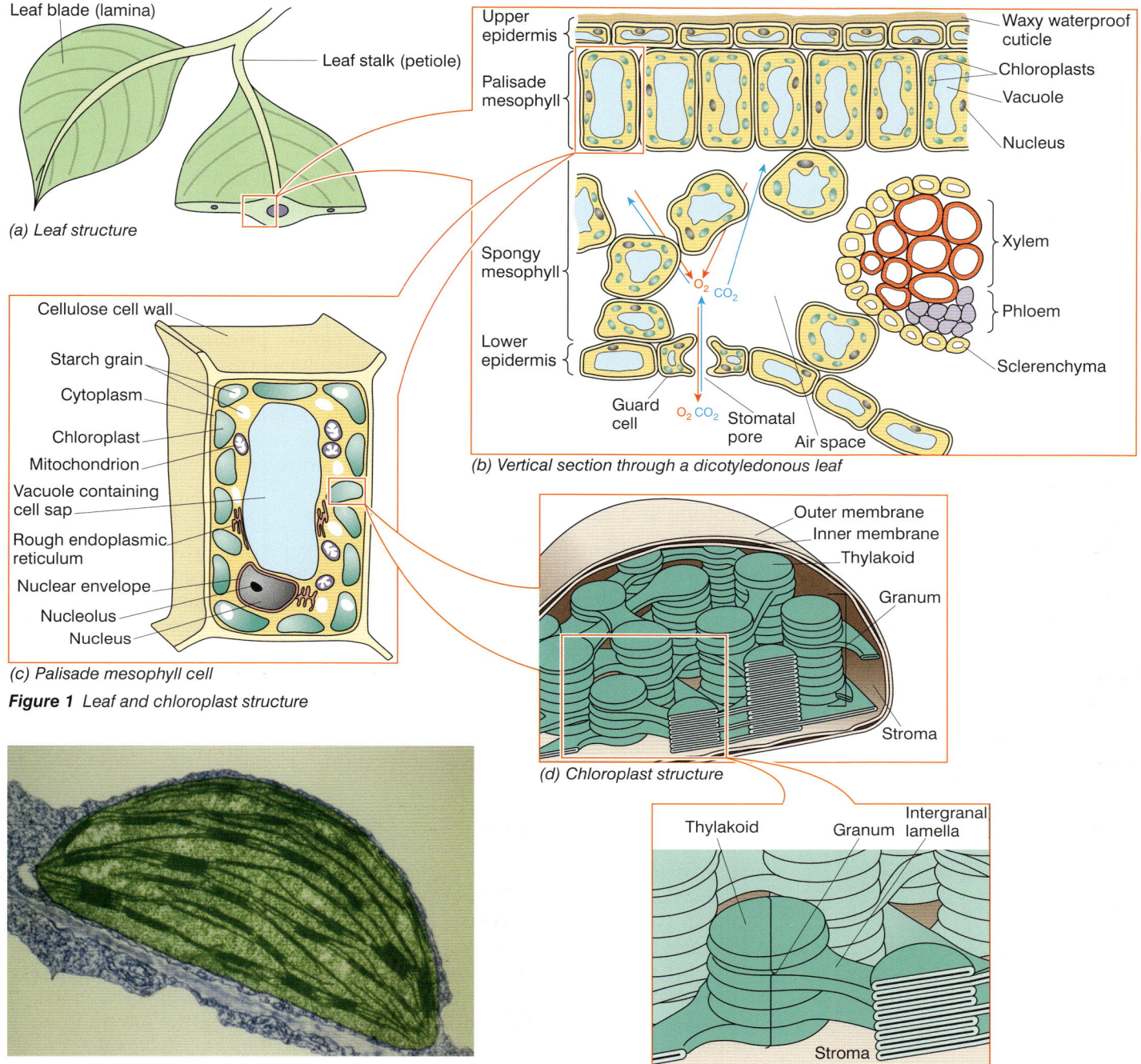

(a) Leaf structure

Leaf blade (lamina)

Leaf stalk (petiole)

Upper epidermis

Palisade mesophyll

Spongy mesophyll

Lower epidermis

Guard cell

O₂ CO₂

Stomatal pore

Air space

Waxy waterproof cuticle

Chloroplasts

Vacuole

Nucleus

Xylem

Phloem

Sclerenchyma

(b) Vertical section through a dicotyledonous leaf

Cellulose cell wall

Starch grain

Cytoplasm

Chloroplast

Mitochondrion

Vacuole containing cell sap

Rough endoplasmic reticulum

Nuclear envelope

Nucleolus

Nucleus

(c) Palisade mesophyll cell

Figure 1 Leaf and chloroplast structure

Outer membrane

Inner membrane

Thylakoid

Granum

Stroma

(d) Chloroplast structure

Thylakoid

Granum

Intergranal lamella

Stroma

(e) Grana and thylakoids

Figure 2 Colourised TEM of a chloroplast

SUMMARY TEST 13.2

Leaves are the main site of photosynthesis, which occurs mainly in their **(1)** cells. They have numerous **(2)** that allow exchange of gases between themselves and the air around them and a network of veins made up of the tissue called **(3)** that brings water into the leaf and the tissue called **(4)** that carries away **(5)** produced in photosynthesis. Chloroplasts are the organelles that carry out photosynthesis. They are surrounded by a double membrane or **(6)** and possess both **(7)** and **(8)** that enable them to make their own proteins. Inside, there is a fluid-filled matrix called the **(9)** that carries out the **(10)** reaction and also contains other structures like **(11)** grains and oil droplets. Within this matrix are disc-like structures called **(12)** that are stacked in groups of up to 100 to form structures called **(13)** where the **(14)** reaction of photosynthesis takes place.

On these pages you will learn to:

- Describe the role of chloroplast pigments (chlorophyll a, chlorophyll b, carotene and xanthophyll) in light absorption in the grana
- Interpret absorption and action spectra of chloroplast pigments
- Use chromatography to separate and identify chloroplast pigments and carry out an investigation to compare the chloroplast pigments in different plants

EXTENSION

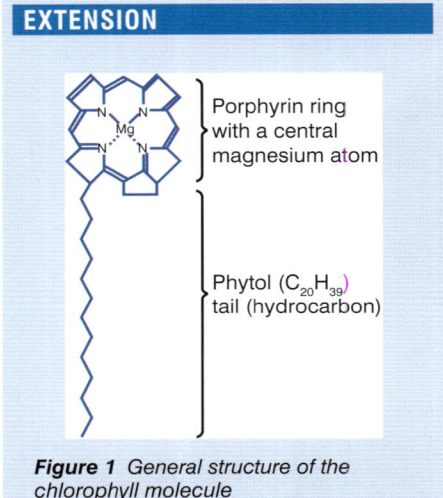

Figure 1 *General structure of the chloroplast molecule*

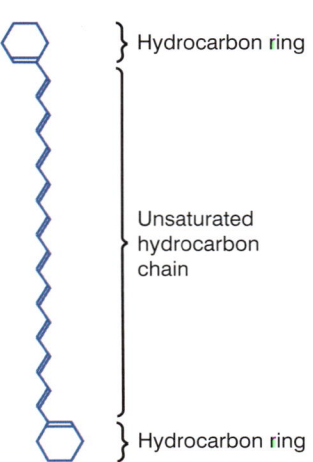

Figure 2 *General structure of a carotene and xanthophyll molecule*

There are a number of pigments found in chloroplasts that act together to capture the light necessary for photosynthesis. The three most important groups of these pigments are the **chlorophylls**, **carotene** and **xanthophyll**. Apart from carbon, hydrogen and oxygen, they also contain the elements magnesium and nitrogen, which are obtained from minerals taken up from the soil by the roots.

Chlorophyll

Chlorophyll is not a single substance, but rather a group of similar green pigments of which chlorophyll *a*, the most important photosynthetic pigment, and chlorophyll *b* are the most common. These pigments strongly absorb light in the blue and red wavelengths of the spectrum (Figure 3). Photoactivation of chlorophyll involves chlorophyll *a*. Chlorophylls are made up of a complex ring called a **porphyrin** ring, which has the same basic structure as the 'haem' group of the blood pigment **haemoglobin** but at its centre there is a magnesium atom.

Carotene and xanthophyll

Carotene and xanthophyll have a basic structure comprising two small rings linked by a long hydrocarbon chain (Figure 2). They range in colour from pale yellow, through orange to red. The greater the number of double bonds in the hydrocarbon chain, the deeper the colour. Carotene and xanthophyll are known as **accessory pigments** because they are not directly involved in the **light-dependent reaction** of photosynthesis. Instead they absorb light wavelengths that are not efficiently absorbed by chlorophyll *a* and pass the energy they capture to chlorophyll *a* for use in the light dependent stage.

Chromatography

The various photosynthetic pigments can be separated from one another by means of chromatography. It involves moving the mixture of molecules over a stationary phase. The separation of the molecules depends on their solubility and molecular mass. To separate photosynthetic pigments, the mixture of pigments is concentrated in a spot at one end of a paper strip and then dipped in a solvent which moves up the paper by capillarity, carrying the molecules with it. The different pigments separate out at different distances from the original spot. Each pigment can be identified by its Rf value, calculated by dividing the distance travelled by the pigment by the distance travelled by the solvent front. For any particular solvent used, each pigment has a characteristic Rf value.

See also the Practical skills section on the accompanying CD.

Absorption and action spectra

Radiant energy comes in discrete packages called quanta. A single quantum of light is called a **photon**. Light also has a wave nature and so forms part of the electromagnetic spectrum. Visible light is made up of different wavelengths. The shorter the wavelength the greater the quantity of energy it possesses. A pigment, such as one of the chlorophylls, will absorb some wavelengths of light more than others. If the amount of light it absorbs at each wavelength is plotted on a graph, we obtain what is called the **absorption spectrum**.

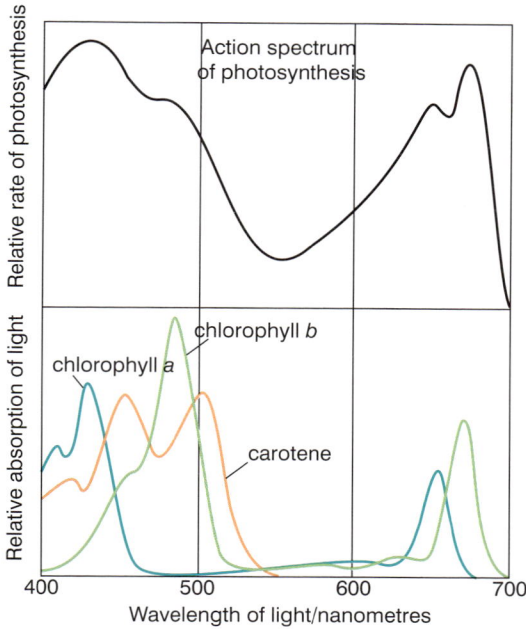

Figure 3 *Action spectrum for photosynthesis and absorption spectrum of common plant pigments*

We can also plot the effectiveness of different wavelengths of light in bringing about photosynthesis. This is called the **action spectrum** and shows that blue (450 nm) and red (650 nm) light are most effective in bringing about photosynthesis.

These two spectra are shown in Figure 3 and can be seen to follow a similar pattern, suggesting that the pigments shown are responsible for absorbing the light used in photosynthesis.

Light harvesting

In 1932, plant physiologists Emerson and Arnold discovered that it took hundreds of chlorophyll molecules to produce a single molecule of oxygen. This led them to conclude that light for photosynthesis, rather than being absorbed by independent pigment molecules, is captured by groups of chlorophyll molecules along with their accessory pigments. These groups are now known as **photosystems** and are located in the photosynthetic membranes (thylakoids and intergranal lamellae). They operate as follows:

- Each photosystem is a collection of chlorophyll *a* molecules, accessory pigments and associated proteins all fixed within a protein matrix.
- One particular pair (primary pigments) of chlorophyll a molecules, often termed the special pair, acts as a **reaction centre** for each photosystem.

- The remaining pigment molecules (**accessory pigments**) of the photosystem absorb light energy (**photons**). These molecules are called the **antenna complex**. They are held tightly together by proteins that act as a framework holding the pigment molecules in the best positions to allow energy to be transferred between them.
- The photon absorbed by an accessory pigment creates an excitation energy that is passed along a chain of pigment molecules to the reaction centre.
- Energy from many pigment molecules in the antenna complex is funnelled in this way to the reaction centre.
- Energy from one photon excites an electron in each of the primary pigment molecules (special pair) of the reaction centre. These electrons play an important part in the light dependent stage (Topic 13.4).

Figure 4 illustrates the role of photosystems in light harvesting.

There are two different photosystems involved in photosynthesis:

- **Photosystem I (PSI)** has a reaction centre with a light absorption peak of 700 nm and is therefore known as P700. Photosystem I occurs mostly on inter-granal lamellae of the chloroplast.
- **Photosystem II (PSII)** has a reaction centre with a light absorption peak of 680 nm and is therefore known as P680. Photosystem II occurs mostly on the granal lamellae of the chloroplast.

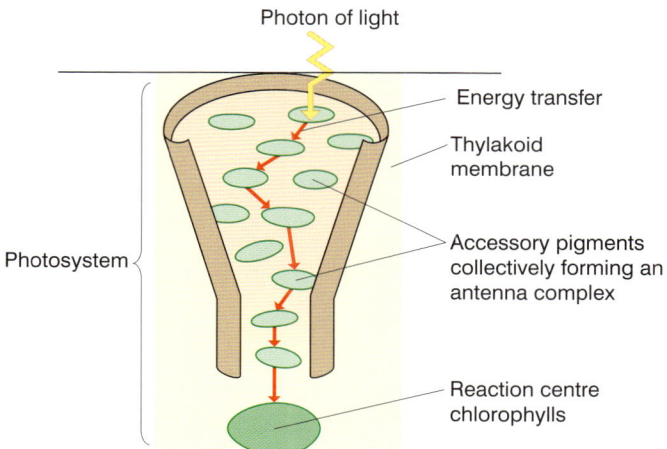

Figure 4 *Light harvesting system*

Photosynthesis – the light dependent stage

On these pages you will learn to:

- Describe the light dependent stage as the photoactivation of chlorophyll resulting in the photolysis of water and the transfer of energy to ATP and reduced NADP

The two photosystems in plants use light of different wavelengths. Photosystem I uses light with a peak absorption of 700 nm, whereas photosystem II uses light with a peak absorption of 680 nm. When these two wavelengths of light are provided together, the rate of photosynthesis is greater than the sum of the rates when each wavelength is provided separately (Figure 1). This is known as the **enhancement effect** and shows that the two systems act together rather than independently.

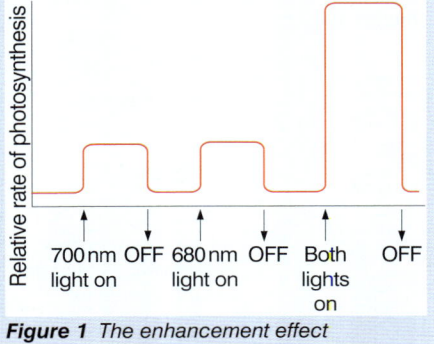

Figure 1 *The enhancement effect*

Table 1 *Comparison of cyclic and non-cyclic photophosphorylation*

	Cyclic	Non-cyclic
Electrons returned to chlorophyll molecule directly	Yes	No
Photosystems involved	I	I and II
Photolysis of water involved	No	Yes
Products	ATP	Reduced NADP + oxygen

The light dependent stage of photosynthesis takes place in the thylakoids of the chloroplasts. It involves the capture of light whose energy is used for two purposes:

- To add an inorganic phosphate molecule to ADP, to make **ATP**. As this process of phosphorylation is brought about by light it is known as **photophosphorylation**.
- To split water into protons, electrons and oxygen. As the splitting is caused by light, it is known as **photolysis**.

Photoactivation of chlorophyll

When light is passed to the reaction centre chlorophyll molecules, a pair of electrons is raised to a higher energy level. This is known as the photoactivation of chlorophyll. These electrons are said to be in an **excited state** and are taken up by a molecule called an **electron carrier** or **electron acceptor**. Each chlorophyll molecule has been oxidised while the carrier, which has gained electrons, has been reduced. The electrons are now passed along a number of electron carriers in a series of redox reactions (Topic 12.5). Each new carrier is at a slightly lower energy level than the previous one, and so the electrons lose energy at each stage. This energy is used to transfer hydrogen ions (protons) across the thylakoid membrane into the thylakoid space (lumen). ATP is produced as a result of chemiosmosis, as described in Topic 12.2. Protons that build up in the thylakoid space flow through the ATP synthase complex of the thylakoid membrane, thereby providing the energy to combine inorganic phosphate with ADP to form ATP. This process is called photophosphorylation, as it requires light (not oxygen as in oxidative phosphorylation). The events are shown in Figure 2. The question now is, what happens to the electrons? There are two alternative processes they can enter: either cyclic photophosphorylation or non-cyclic photophosphorylation. These are compared in Table 1.

Cyclic photophosphorylation

Cyclic photophosphorylation uses only photosystem I. When light raises an electron in a reaction centre chlorophyll molecule to an excited state, the electron is taken up by an electron acceptor and simply passed back to the same chlorophyll molecule via a sequence of electron carriers, i.e. it is recycled. While this does not produce any reduced NADP, it does generate sufficient energy to combine inorganic phosphate with ADP. The ATP so produced is then used in the light independent stage (Topic 13.5) or is used directly, as in guard cells where it is used to pump potassium ions into the guard cells, thereby reducing water potential and leading to water entering them by osmosis and increasing their turgidity, with the result that the stoma opens.

Non-cyclic photophosphorylation

Non-cyclic photophosphorylation uses both photosystem I and photosystem II. Electrons raised to an excited state in photosystem II are taken up by an electron acceptor and passed along a sequence of electron carriers to then replace the electrons lost in photosystem I. The electrons raised to an excited state from photosystem I are taken up by an electron acceptor and are then taken up by NADP$^+$ (nicotinamide adenine dinucleotide phosphate) and passed into the light independent stage of photosynthesis. This leaves the reaction centre chlorophyll molecules of photosystem II short of electrons and therefore positivey charged. Before the photosystem can operate again these electrons must be replaced. The replacement electrons are provided from water molecules that are split using light energy. This photolysis of water also yields hydrogen ions (protons) into the thylakoid lumen, where they contribute to the build-up of the proton gradient. Hydrogen ions in the stroma can be used for the reduction of NADP.

Photolysis

Photolysis is the splitting of water as a direct consequence of the photoactivation of chlorophyll. It occurs only in photosystem II, which is associated with an enzyme known as the **oxygen evolving complex**. Having lost an electron, the chlorophyll molecule needs to replace it. In the case of non-cyclic photophosphorylation, it does this using electrons from water molecules that are split by the oxygen evolving complex into protons, electrons and oxygen according to the following equation:

$$2H_2O \rightarrow 4H^+ + 4e^- + O_2$$
$$\text{water} \quad \text{protons} \quad \text{electrons} \quad \text{oxygen}$$

- The electrons replace those lost by the chlorophyll molecules. The protons reduce NADP to NADPH + H$^+$, which then enters the light independent stage where it reduces carbon dioxide.
- The oxygen by-product is either used in respiration or diffuses out of the leaf as a waste product of photosynthesis.

The Z-scheme

All the processes of the light independent stage are closely linked. These events are summarised in Figure 3, which illustrates the zig-zag energy levels of the electrons. As the diagram resembles a Z on its side, the complete process is called the Z-scheme.

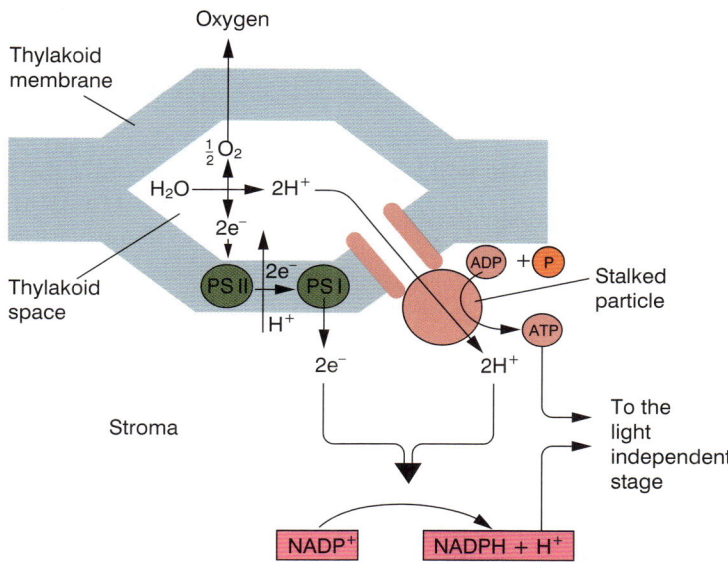

Figure 2 ATP production during the light dependent stage of photosynthesis

REMEMBER

There are three ways in which oxidation and reduction can be described.

Oxidation – loss of electrons or loss of hydrogen or gain of oxygen.

Reduction – gain of electrons or gain of hydrogen or loss of oxygen.

1. Light energy is trapped in photosystem II and boosts electrons to a higher energy level.
2. The electrons are received by an electron acceptor.
3. The electrons are passed from the electron acceptor along a series of electron carriers to photosystem I. The energy lost by the electrons is captured by converting ADP to ATP. Light energy has thereby been converted to chemical energy.
4. Light energy absorbed by photosystem I boosts the electrons to an even higher energy level.
5. The electrons are received by another electron acceptor.
6. The electrons which have been removed from the chlorophyll are replaced by pulling in other electrons from a water molecule.
7. The loss of electrons from the water molecule causes it to dissociate into protons and oxygen gas.
8. The protons from the water molecule combine with the electrons from the second electron acceptor and these reduce **nicotinamide adenine dinucleotide phosphate. (NADP)**
9. Some electrons from the second acceptor may pass back to the chlorophyll molecule by the electron carrier system, yielding ATP as they do so. This process is called **cyclic photophosphorylation.**

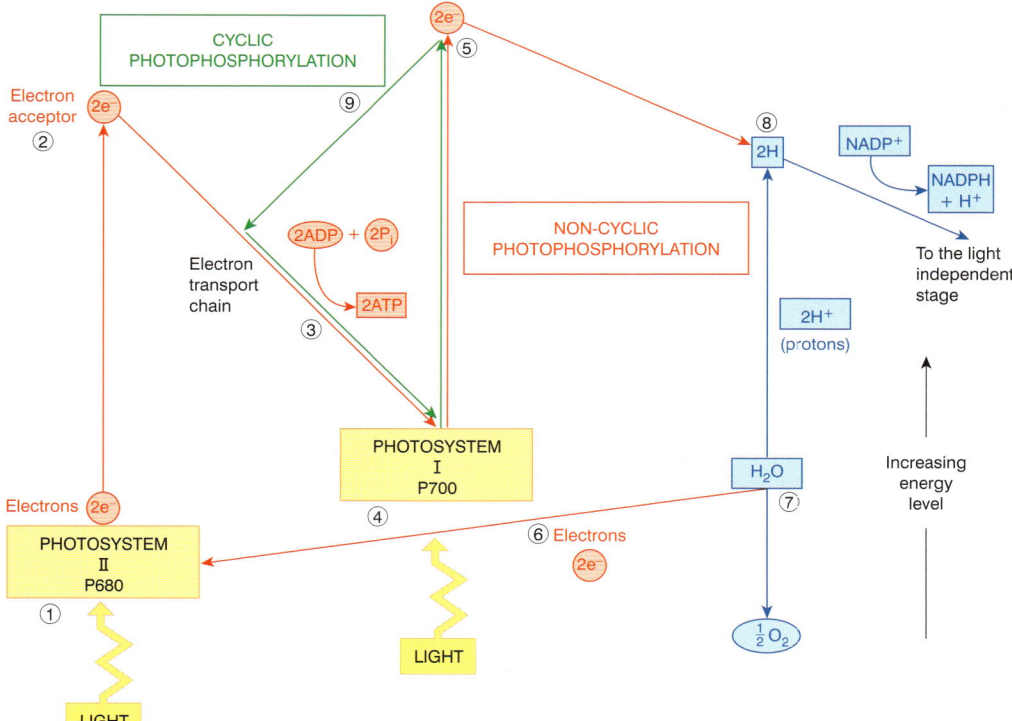

Figure 3 Summary of light dependent stage of photosynthesis

13.5

Photosynthesis – the light independent stage

- Explain that energy transferred as ATP and reduced NADP from the light dependent stage is used during the light independent stage (Calvin cycle) of photosynthesis to produce complex organic molecules
- Outline the three main stages of the Calvin cycle:
  - fixation of carbon dioxide by combination with ribulose bisphosphate (RuBP), a 5C compound, to yield two molecules of GP (PGA), a 3C compound
  - the reduction of GP to triose phosphate (TP) involving ATP and reduced NADP
  - the regeneration of ribulose bisphosphate (RuBP) using ATP
- Describe, in outline, the conversion of Calvin cycle intermediates to carbohydrates, lipids and amino acids and their uses in the plant cell

EXTENSION

Rubisco – the most common protein on Earth?

The enzyme that combines carbon dioxide with ribulose bisphosphate (RuBP) is called ribulose bisphosphate carboxylase (rubisco). By normal enzyme standards rubisco is a slow operator, combining only about three molecules of CO_2 with RuBP each second, compared with 1000 substrate molecules per second for a typical enzyme. To compensate for this, and to ensure a fast conversion rate of CO_2 into sugar, the enzyme is present in huge quantities. Typically, over half the protein of a leaf is rubisco. Given the considerable **biomass** of the world that is leaves, it is not surprising that rubisco is thought to be the most abundant protein on the planet.

The products of the light dependent stage of photosynthesis, namely ATP and reduced NADP (NADPH H$^+$) are used to reduce carbon dioxide in the second part of photosynthesis. Unlike the first stage, this stage does not require light directly and is therefore called the **light independent stage**. In practice, it requires the products of the light dependent stage and so rapidly stops when light is absent. The light independent reaction stage takes place in the **stroma** of the chloroplasts. The details of this stage were worked out by Melvin Calvin and his co-workers. The process is therefore often known as the **Calvin cycle**.

The Calvin cycle

In the following account of the Calvin cycle, the numbered stages are shown in Figure 1. Each step in the Calvin cycle is enzyme controlled.

1. Carbon dioxide from the atmosphere diffuses into the leaf through **stomata** and dissolves in water around the walls of the **palisade** cells. It then diffuses through the cell surface membrane, cytoplasm and chloroplast envelope into the stroma of the chloroplast.
2. In the stroma, the carbon dioxide combines with the five-carbon compound **ribulose bisphosphate (RuBP)** using the enzyme **ribulose bisphosphate carboxylase (rubisco)**, to form an unstable six-carbon compound.
3. The unstable six-carbon compound immediately breaks down into two molecules of the three-carbon **glycerate 3-phosphate (GP)**.

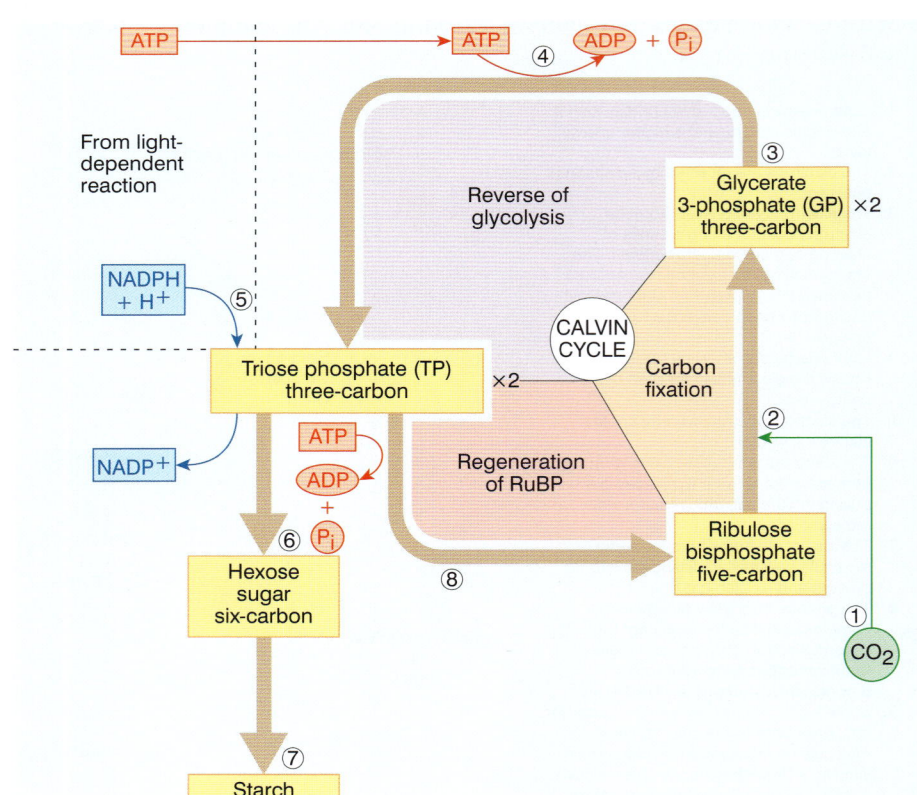

Figure 1 Summary of light independent stage of photosynthesis

4. Using one of the ATP molecules from the light dependent reaction, the GP (glycerate 3-phosphate) is converted into a 3 carbon sugar **triose phosphate (TP)**.
5. Reduced NADP from the light dependent stage provides hydrogen for the reduction of GP to TP (glycerate-3-phosphate to triose phosphate).
6. Triose phosphate molecules combine in pairs to form six-carbon (hexose) sugars.
7. The six-carbon sugars can be **polymerised** into starch.
8. Five out of every six triose phosphate molecules produced are used to regenerate ribulose bisphosphate using the remainder of the ATP from the light dependent stage as the source of energy.

A summary of both stages of photosynthesis is given in Figure 2.

Formation of other substances for use by the plant

Plants, like other organisms, are made up of a range of complex organic molecules. The bulk of these are carbohydrates, lipids and proteins. Unlike animals and other **heterotrophic** organisms, plants cannot obtain these substances by taking them in from the outside. They must synthesise them from the various compounds of the Calvin cycle.

- Carbohydrates, e.g. sucrose (the carbohydrate which is transported in the phloem) are made by combining the two hexose sugars, glucose and fructose. Glucose is used as a respiratory substrate.
- Starch (the storage carbohydrate) and cellulose (the essential component of cell walls) are made by polymerising glucose in different ways.
- Lipids are made up of glycerol and fatty acids. Plants make glycerol from triose phosphate and fatty acids from glycerate 3-phosphate (GP). Lipids are used in plant cells for storage and to form phospholipids in their cell membranes.
- Proteins are made up of amino acids that in turn can be produced from glycerate 3-phosphate (GP) via acetyl coenzyme A and the intermediates of **Krebs cycle**. Proteins are important components of cell membranes and all enzymes are proteins.

REMEMBER

Any substance whose name ends in 'ose' is a sugar. The ending 'ate' usually means that the substance is an acid (in solution).

SUMMARY TEST 13.5

The light independent stage is also known as the **(1)** cycle after the person who determined its biochemical sequence. In the process, carbon dioxide combines with a five-carbon compound called **(2)** to form a six-carbon intermediate that immediately splits into two molecules of **(3)**. By the addition of **(4)** and **(5)** formed in the **(6)** stage, each of the molecules is then converted into a **(7)** molecule. These combine in pairs to form a **(8)** sugar that is then made into starch by a process called **(9)**.

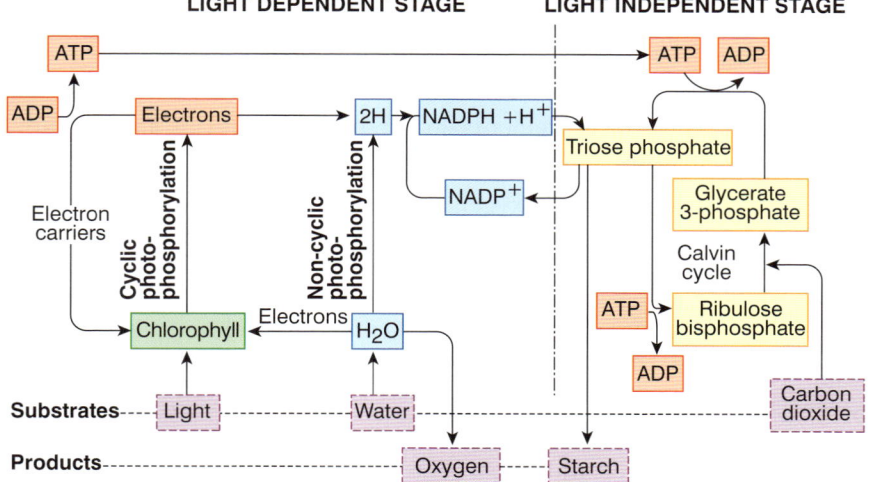

Figure 2 *Summary of photosynthesis*

Details of how to carry out an investigation to determine the effect of light intensity or wavelength on the rate of photosynthesis using a redox indicator, for example, DCPIP, and a suspension of chloroplasts (the Hill reaction) can be found in the Practical skills section on the accompanying CD.

13.6   Limiting factors affecting photosynthesis

On these pages you will learn to:

- Explain the term 'limiting factor' in relation to photosynthesis
- Explain the effects of changes in light intensity, carbon dioxide concentration and temperature on the rate of photosynthesis
- Explain how an understanding of limiting factors is used to increase crop yields in protected environments, such as glasshouses

Before we consider some of the factors affecting photosynthesis, it is necessary to understand the **concept of limiting factors**.

Limiting factors

In any complex process such as photosynthesis, the factors that affect its rate all operate simultaneously. However, the rate of the process at any given moment is **not** affected by a combination of all the factors, but rather by just one – the one whose level is at the least favourable value. This factor is called the **limiting factor** because it alone limits the rate at which the process can take place. However much the levels of the other factors change, they do not alter the rate of the process.

To take the example of light intensity limiting the rate of photosynthesis:

- In complete darkness, it is the absence of light alone that prevents photosynthesis occurring.
- No matter how much we raise or lower the temperature or change the concentration of carbon dioxide, there will be no photosynthesis. Light, or rather the absence of it, is the factor determining the rate of photosynthesis at that moment.
- If we provide light, however, the rate of photosynthesis will increase.
- As we add more light, the more the rate increases. This does not continue indefinitely, however, because there comes a point at which further increases in light intensity have no effect on the rate of photosynthesis.
- At this point some other factor, such as the concentration of carbon dioxide, is in short supply and so limits the process.
- Carbon dioxide is now the limiting factor and only an increase in its level will increase the rate of photosynthesis.
- In the same way as happened with light, providing more carbon dioxide will lead to more photosynthesis.
- Further increases in carbon dioxide levels will fail to have any effect.
- At this point a different factor, e.g. temperature, is the limiting factor and only an alteration in its level will affect the rate of photosynthesis.

These events are shown in Figure 1.

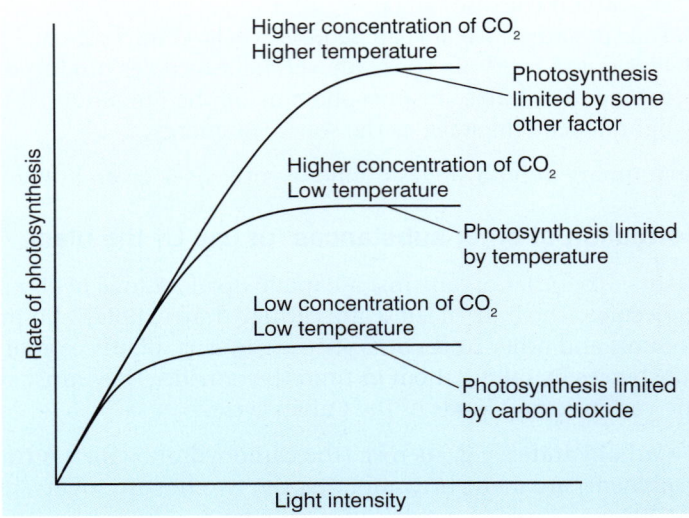

Figure 1 *Concept of limiting factors as illustrated by the effects of levels of different conditions on the rate of photosynthesis. Before the curves begin to level off, light intensity is a limiting factor*

The law of limiting factors can therefore be expressed as: **At any given moment, the rate of a physiological process is limited by the one factor which is at its least favourable value, and by that factor alone.**

Effect of light intensity on the rate of photosynthesis

When light is the limiting factor, the rate of photosynthesis is directly proportional to light intensity. The rate of photosynthesis is usually measured in two ways:

- the volume of oxygen produced by a plant
- the volume of carbon dioxide taken up by a plant.

These measurements do not, however, provide an absolute measure of photosynthesis because:

- some of the oxygen produced in photosynthesis is used in cellular respiration and so never leaves the plant and therefore cannot be measured
- some carbon dioxide from cellular respiration is used up in photosynthesis and therefore the volume taken up from the atmosphere is less than that actually used in photosynthesis.

As light intensity is increased, the volume of oxygen produced and carbon dioxide absorbed due to photosynthesis will increase to a point at which it is exactly balanced by the oxygen absorbed and carbon dioxide produced by respiration. At this point there will be no net exchange of gases into or out of the plant. This is known as the **light compensation point**. Further increases in light intensity will cause a proportional increase in the rate of photosynthesis and increasing volumes of oxygen will

214

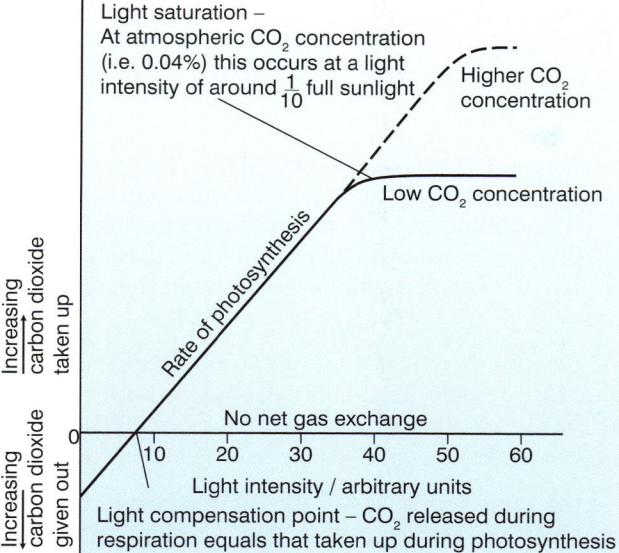

Figure 2 *Graph showing the effect of light intensity on the rate of photosynthesis as measured by the amount of CO_2 exchange*

be given off and carbon dioxide taken up. A point will be reached at which further increases in light intensity will have no effect on photosynthesis. At this point some other factor such as carbon dioxide concentration or temperature is limiting the reaction. These events are shown in Figure 2.

Effect of carbon dioxide concentration on the rate of photosynthesis

Carbon dioxide is present in the atmosphere at a concentration of around 0.04%. This level continues to increase as the result of human activities such as burning fossil fuels and the clearing of rain forests. It is still one of the rarest gases present and is often the factor that limits the rate of photosynthesis under normal conditions. The optimum concentration of carbon dioxide for a consistently high rate of photosynthesis is 0.1% and growers of some glasshouse crops like tomatoes enrich the air in the glasshouses with more carbon dioxide to provide higher yields. Figures 1 and 2 illustrate the effect of enhanced carbon dioxide levels on photosynthesis.

Effect of temperature on the rate of photosynthesis

Provided that other factors are not limiting, the rate of photosynthesis increases in direct proportion to the temperature. Between the temperatures of 0 °C and 25 °C the rate of photosynthesis is approximately doubled for each 10 °C rise in temperature. Above the optimum temperature of 25 °C the rate levels off and then declines – largely as a result of enzyme **denaturation**. Purely photochemical reactions are not usually affected by temperature, and so the fact that photosynthesis is temperature sensitive suggested to early researchers that there was also a totally chemical process involved as well as a photochemical one. We now know that this chemical process is the light independent stage (Topic 13.5) and that there are enzymes involved in the light dependent stage.

Increasing crop yields in glasshouses

Food production depends on photosynthesis. As the rate of photosynthesis is determined by the factor that is in shortest supply (limiting factor) it follows that there is commercial value in determining which factor is limiting photosynthesis at any one time. By supplying more of this factor, photosynthesis, and hence food production, can be increased.

It is not practical to control the environment of crops in natural conditions. Plants grown in glasshouses are a different matter. In the enclosed environment of a glasshouse it is possible to regulate temperature, humidity, light intensity and carbon dioxide concentration. Scientists are able to predict the effects of changing these factors on the rate of photosynthesis. They can then advise commercial growers on the optimum conditions that should be created in order to increase the rate of photosynthesis and hence the growth of their crops.

It may seem logical to simply increase the level of all factors, with a view to increasing photosynthesis to a high value and ensure maximum yield. In practice different plants have different optimum conditions and too high a level of a particular factor may reduce yield or kill the plant altogether. For example, high temperatures may increase the yield of one species but denature the enzymes of another, and kill the plants of that species. It is also uneconomic and wasteful to use energy to raise temperature or to increase carbon dioxide concentrations or light intensity beyond what is necessary. Precise control of the environment is therefore essential. This can be brought about in ways ranging from totally manual control to the use of advanced computerised systems.

To take the example of carbon dioxide concentration, the average concentration in the atmosphere is around 400 parts per million (ppm). It has been shown that by raising this level to 1000 ppm the yields from tomato plants can be increased by 20% or more.

SUMMARY TEST 13.6

At any given moment the rate of a physiological process is restricted by the one factor that is at its least favourable value. This is known as the **(1)**. The rate of photosynthesis can be measured by calculating the volume of **(2)** taken up or the volume of **(3)** produced by a plant. The light intensity at which there is no net exchange of gases into or out of the plant is known as the **(4)**. Carbon dioxide concentration affects the rate of photosynthesis. Normally present in the atmosphere at a proportion of **(5)**%, the optimum for a consistently high rate of photosynthesis is **(6)**%.

Investigations on the effects of light intensity, carbon dioxide and temperature on the rate of photosynthesis, using aquatic plants such as the pondweeds Elodea and Cabomba, can be found in the Practical skills section on the accompanying CD.

Adaptations to photosynthesis – C4 plants

On these pages you will learn to:

- Explain how the anatomy and physiology of the leaves of C4 plants, such as maize or sorghum, are adapted for high rates of carbon fixation at high temperatures in terms of:
  - the spatial separation of initial carbon fixation from the light dependent stage
  - the high optimum temperatures of the enzymes involved

REMEMBER

High temperatures can break hydrogen and other bonds within enzyme molecules. This can alter the shape of the active site and as a result the substrate fits less easily into the active site and the rate of reaction slows or stops altogether = denaturation.

In Topic 13.5 we learnt about the action of ribulose bisphosphate carboxylase (rubisco) during photosynthesis. During the Calvin cycle, this important enzyme combines carbon dioxide with a five-carbon ribulose bisphosphate molecule to form a six-carbon compound. This compound is unstable and immediately splits into two three-carbon compounds. Plants which photosynthesise in this way are therefore called C3 plants.

Rubisco also catalyses a second reaction in which ribulose bisphosphate combines with oxygen rather than carbon dioxide. This process is called **photorespiration** and it releases carbon dioxide. This works against the Calvin cycle in which carbon dioxide is incorporated into molecules rather than released from them.

The two reactions take place at the same active site on the rubisco enzyme and therefore compete with one another. At a temperature of 25 °C, around 20% of carbon dioxide fixed by photosynthesis is lost to photorespiration. The higher the temperature, the greater this loss, with up to 50% of photosynthetically fixed carbon being lost in this way.

To overcome this wastage, some plants have adapted to warmer environments by evolving a different photosynthetic pathway. Among these plants are maize and sorghum (Figure 2) and the pathway they use is called the C4 pathway.

The C4 pathway

Plants which use C4 photosynthesis, add a carbon dioxide molecule to a three-carbon molecule called phosphoenol pyruvate (PEP) instead of the five-carbon ribulose bisphosphate. A molecule of oxaloacetate is formed that has four carbon atoms, hence the name, C4 pathway. The enzyme that catalyses this reaction is called PEP carboxylase and, importantly, it does not carry out oxidation and so there is no photorespiration. In addition, PEP carboxylase has a greater affinity for carbon dioxide than rubisco does. It also operates at a higher optimum temperature without being denatured.

The oxaloacetate formed is converted to malate, which passes from the mesophyll cells to special cells called **bundle sheath cells**. Carbon dioxide is lost from malate and enters the Calvin cycle in the usual way. The three-carbon pyruvate, which forms as a result, passes back into the mesophyll cells and is converted to phosphoenol pyruvate ready to accept another carbon dioxide molecule. A summary of C4 photosynthesis is shown in Figure 1.

Adaptation of the leaf in C4 plants

If you look at the arrangement of the cells in the leaf of a C4 plant as shown in Figure 1, you will see they are different from those of a C3 plant (Topic 13.2). You will notice that around the vascular bundle is a tight ring of bundle sheath cells which itself is surrounded by a ring of tightly fitting mesophyll cells. This arrangement ensures that the bundle sheath cells are isolated from the air inside the leaf. This prevents photorespiration by preventing oxygen reaching the bundle sheath cells. It also prevents carbon dioxide being lost, which therefore accumulates within them. This store of carbon dioxide can be used when supplies from outside the leaf are in short supply. An adaptation of this kind is very useful for plants that grow in hot climates and where stomata may close at midday to prevent excessive water loss.

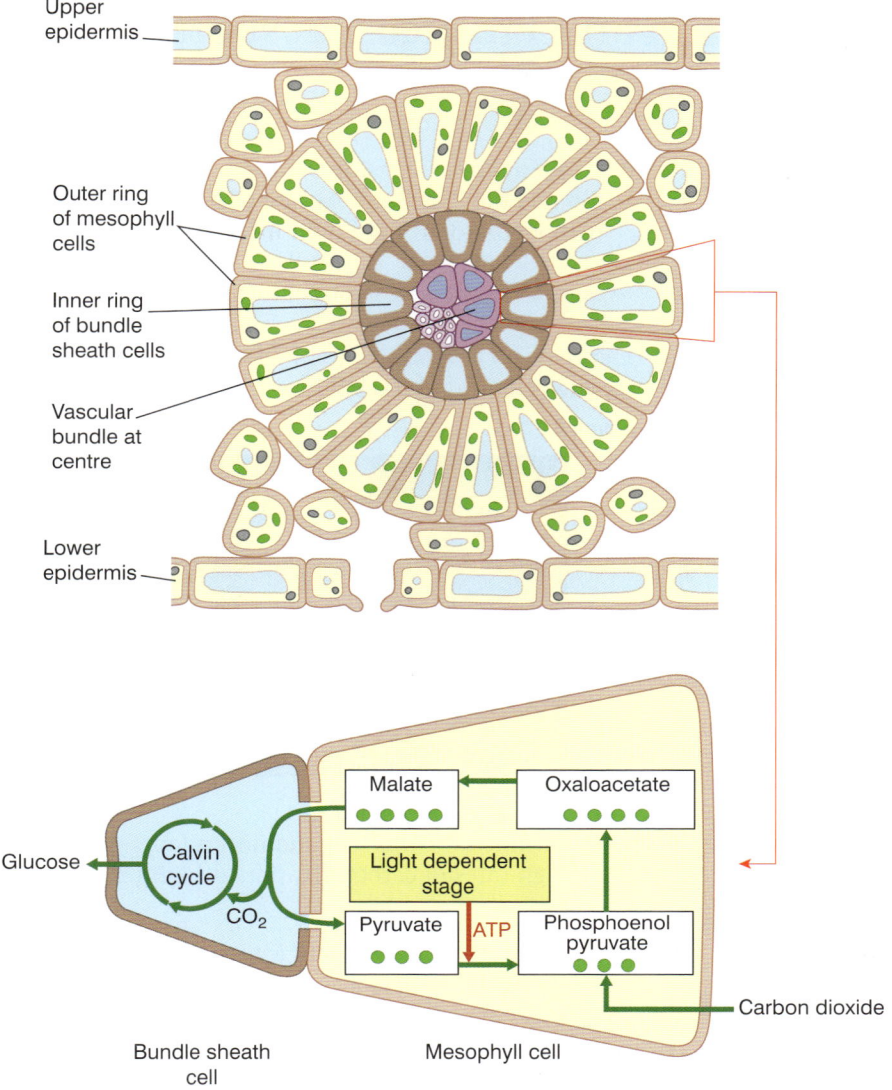

Upper epidermis

Outer ring of mesophyll cells

Inner ring of bundle sheath cells

Vascular bundle at centre

Lower epidermis

Malate
Oxaloacetate

Calvin cycle

Glucose

CO₂

Light dependent stage

Pyruvate

ATP

Phosphoenol pyruvate

Carbon dioxide

Bundle sheath cell

Mesophyll cell

Figure 1 *Arrangement of cells in the leaf of a C4 plant and how the light independent stage of C4 photosynthesis takes place*

The light dependent stage, where oxygen is produced, takes place in mesophyll cells and not bundle sheath cells. This spatial separation of light dependent and the initial carbon fixation of the light independent stages prevents photorespiration taking place by keeping rubisco separated from oxygen. A further adaptation is the presence of numerous plasmodesmata between bundle sheath cells and mesophyll cells. These allow more rapid movement of malate and pyruvate between the two.

Figure 2 *Sorghum is a plant that has adapted to the high temperatures of the regions in which it grows by evolving a different type of photosynthesis called C4 photosynthesis*

SUMMARY TEST 13.7

In C3 plants, the enzyme ribulose bisphosphate carboxylase (rubisco) catalyses the reaction between ribulose bisphosphate and **(1)** to give a six-carbon compound which immediately splits to give two three-carbon molecules. Rubisco also catalyses the oxidation of ribulose bisphosphate – a process called **(2)** which reduces the efficiency of photosynthesis. To overcome this C4 plants such as **(3)** and **(4)** have developed a different pathway in which carbon dioxide is combined with **(5)** to give oxaloacetate. This is then converted to malate, which passes into **(6)** cells. The malate loses a molecule of carbon dioxide which is used in the **(7)** and a molecule of **(8)** is formed.

13 Examination Questions

1 a An **absorption** spectrum is a graph of the absorption of different wavelengths of light by a photosynthetic pigment.

An **action** spectrum is a graph of the rate of photosynthesis at different wavelengths of light.

Figure 1 shows the absorption spectra of chlorophyll a and chlorophyll b as well as an action spectrum.

Figure 1

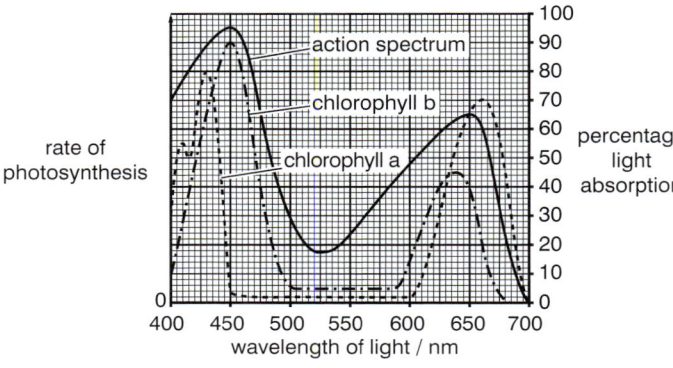

With reference to Figure 1,
i   compare the **absorption** spectra of chlorophyll a and chlorophyll b,   *(3 marks)*
ii   explain the shape of the **action** spectrum, *(3 marks)*
iii   explain why plants appear green.   *(2 marks)*

b Figure 2 is an electron micrograph showing a section through part of a chloroplast.

Figure 2

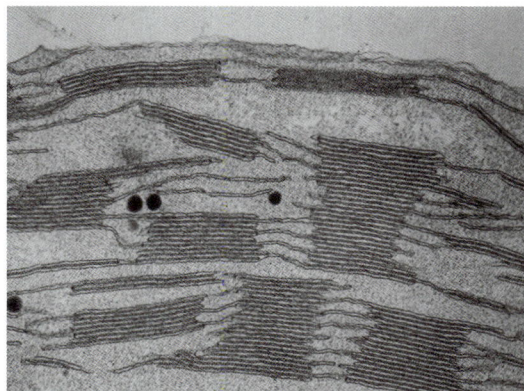

On a copy of Figure 2, draw label lines and use the letters **W** and **Y** to identify the following structures:
•   **W** where the light independent reactions occur
•   **Y** where chlorophyll is found   *(2 marks)*

c Explain why increasing the concentration of carbon dioxide may increase the rate of production of carbohydrates at high light intensities.   *(5 marks)*
                             (Total 15 marks)
Cambridge International AS and A Level Biology 9700 Paper 4 Q7 June 2008

2 Figure 3 is a photomicrograph of a transverse section through the leaf of a C4 plant.

Figure 3

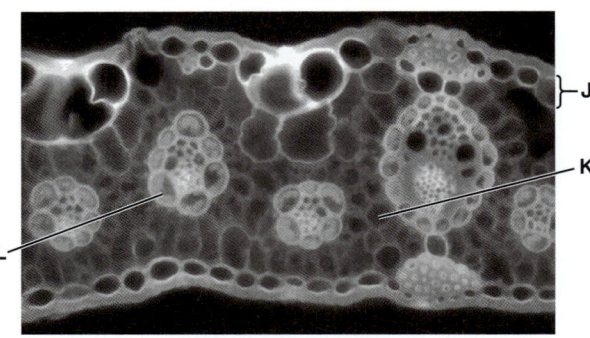

a i   Identify structures **J** to **L**.   *(3 marks)*
   ii   Outline how this leaf anatomy adapts the plant for high rates of carbon fixation at high temperatures.   *(4 marks)*

b Sorghum is a C4 plant and *Sorghum bicolor* is a major food crop in dry tropical regions. The leaves of *S. bicolor* are covered with a layer of wax made up of a mixture of esters and free fatty acids, with a melting point of 77–85 °C. Waxes from the leaves of non-tropical plants tend to have melting points lower than this. For example, wax from the bayberry, *Myrica* sp., has a melting point of 45 °C.

Suggest how the wax on sorghum leaves helps the plant to survive in dry, tropical regions.   *(2 marks)*

c An investigation was carried out into the response of sorghum to being kept at a low temperature for a short period of time. Soybean plants, which are better adapted than sorghum for growth in subtropical and temperate climates, were used for comparison.

Plants of sorghum and soybean were kept at 25 °C for several weeks and then at 10 °C for three days. The temperature was then increased to 25 °C again for seven days. Day length, light intensity and carbon dioxide concentration were kept constant throughout.

The uptake of carbon dioxide, as mg CO_2 absorbed per gram of leaf dry mass, was measured
•   at 25 °C before cooling
•   on each of the three days at 10 °C
•   for seven days at 25 °C.

The results are shown in Table 1.

Table 1

plant	carbon dioxide uptake / mg $CO_2 g^{-1}$				
	at 25 °C, before cooling	at 10 °C			at 25 °C (mean over days 4 to 10)
		day 1	day 2	day 3	
sorghum	48.2	5.5	2.9	1.2	1.5
soybean	23.2	5.2	3.1	1.6	6.4

  i Compare the **changes** in carbon dioxide uptake in sorghum and soybean during the three days at 10 °C. *(2 marks)*

  ii During the cooling period, the ultrastructure of the sorghum chloroplasts changed. The membranes of the thylakoids moved closer together, eliminating the spaces between them. The size and number of grana became reduced.

    Explain how these changes could be responsible for the low rate of carbon dioxide uptake by sorghum even when returned to a temperature of 25 °C.
(4 marks)
(Total 15 marks)

Cambridge International AS and A Level Biology 9700 Paper 41 Q4 June 2010

3 **a** A student investigated the effects of temperature and light intensity on the rate of photosynthesis of an aquatic plant.

Figure 4 shows the results of the investigation.

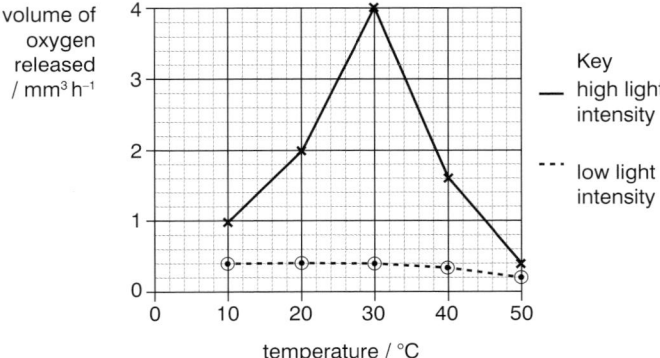

Figure 4

With reference to Figure 4:
  i describe the results of the investigation *(3 marks)*
  ii suggest explanations for the results for high light intensity above 30 °C. *(2 marks)*

b **i** Name the process in the light dependent stage of photosynthesis that produces oxygen. *(1 mark)*
  ii Name the photosystem involved in the production of oxygen in the light dependent stage. *(1 mark)*

  iii Explain why the volume of oxygen released from the plant does not give a true rate of photosynthesis. *(1 mark)*
(Total 8 marks)

4 **a** Describe the structure of a chloroplast. *(9 marks)*

  b Explain how the palisade mesophyll cells of a leaf are adapted for photosynthesis. *(6 marks)*
(Total 15 marks)

Cambridge International AS and A Level Biology 9700 Paper 4 Q10 June 2007

13 Practice Questions

5

  Figure 5

  1 0.1% carbon dioxide at 25 °C

  2 0.04% carbon dioxide at 35 °C

  3 0.04% carbon dioxide at 25 °C

  4 0.04% carbon dioxide at 15 °C

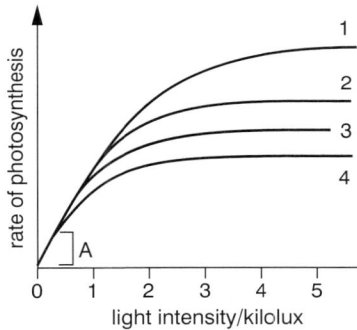

a State one measurement that could be taken to determine the rate of photosynthesis in this experiment.

b Name the factor that is limiting the rate of photosynthesis over the region marked A on the graph. Explain your answer.

c In the spring, a commercial grower of tomatoes keeps her glasshouses at 25 °C and at a carbon dioxide concentration of 0.04%. The light intensity is 4 kilolux at this time of year. Using the graph, predict whether the tomato plants would grow more if the carbon dioxide level was raised to 0.1% or if the temperature were increased to 35 °C. Explain your answer.

d Why is there no point in the grower heating her glasshouses on dull day?

e Using your knowledge of the light independent stage, explain why, at 25 °C, raising the level of carbon dioxide from 0.04% to 0.1% increases the concentration of glucose produced.

Homeostasis in mammals

On these pages you will learn to:

- Discuss the importance of homeostasis in mammals and explain the principles of homeostasis in terms of internal and external stimuli, receptors, central control, coordination systems, effectors (muscles and glands)
- Define the term 'negative feedback' and explain how it is involved in homeostatic mechanisms
- Outline the roles of the endocrine system in coordinating homeostatic mechanisms

Figure 1 *The skills involved in this cheetah catching its gazelle prey requires highly efficient coordination of many of its body systems*

Figure 2 *Homeostasis allows animals such as these penguins in the Antarctic (above) and these camels in the desert (below) to survive in extreme environments*

As species of organisms evolved from simple, single-celled organisms into complex, multicellular ones, these organisms evolved to perform a specialist function. With specialisation in one function came the loss of the ability to perform other functions. Different groups of cells each carried out their own function, and this made the cells dependent on each other. Cells specialising in reproduction, for example, depend on different groups of cells for their survival: one group to obtain oxygen for their respiration, another to provide glucose and another to remove their waste products. This means that the more complex, multicellular organisms have a division of labour that is seen in tissues, organs and organ systems. These different functional systems must be coordinated if they are to perform efficiently.

There are two forms of coordination in most multicellular animals: nervous and endocrine. The nervous system allows rapid communication between specific parts of an organism. The endocrine system usually provides a slower, less specific form of communication. Both systems need to work together. The increased complexity of multicellular organisms meant the development of an internal environment at the same time. This internal environment is made up of extracellular fluids that bathe each cell, supplying nutrients and removing wastes. By maintaining this fluid at levels which suit the cells, the cells are protected from the changes that affect the external environment and so give the organism a degree of independence.

What is homeostasis?

Homeostasis is the maintenance of a constant state. More specifically, it refers to the internal environment of organisms and involves maintaining the chemical make-up, volume and other features of blood and tissue fluid within narrow limits, sometimes called **normal ranges**. Homeostasis ensures that the cells of the body are in an environment that meets their needs and allows them to function normally despite external changes. This does not mean that there are no changes – on the contrary, there are continuous fluctuations brought about by variations in internal and external conditions. These changes, however, occur around a **set point**. Homeostasis is the ability to return to that set point and so maintain organisms in a balanced equilibrium (Figure 2).

The importance of homeostasis

Homeostasis is essential for the proper functioning of organisms because:

- The enzymes that control the biochemical reactions within cells, and other proteins such as membrane channel proteins, are sensitive to changes in pH and temperature (Topic 3.2). Any change to these factors reduces the efficiency of enzymes or may even prevent them working altogether, e.g. may **denature** them. Changes to membrane proteins may mean that substances cannot be transported into or out of cells.
- Changes to the water potential of the blood and tissue fluids may cause cells to shrink and expand (even to bursting point) owing to water leaving or entering by **osmosis**. In both situations the cells cannot operate normally.

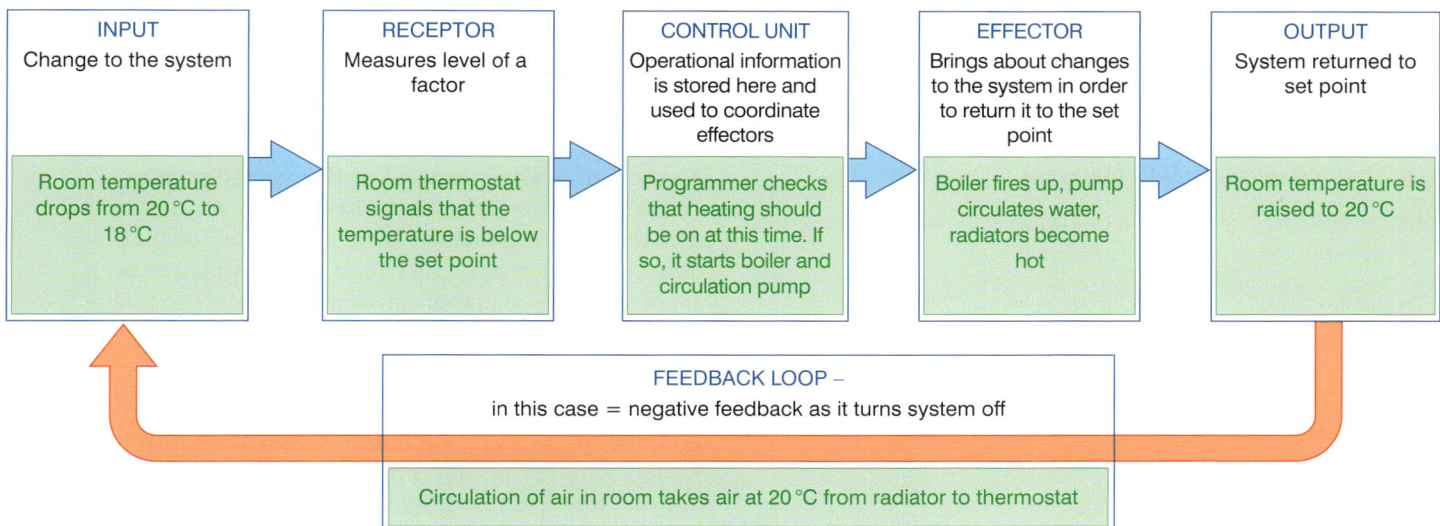

Figure 3 *Components of a typical control system*

- Biochemical reactions in organisms are in a state of dynamic equilibrium between the forward and reverse reactions. Changes to the environment of cells can upset this equilibrium to the harm of the organism.
- Organisms with the ability to maintain a constant internal environment are more independent of the external environment. They have a wider geographical range and therefore have a greater chance of finding food, shelter, etc. Mammals, for example, with their ability to maintain a constant temperature, are found in most habitats from hot arid deserts to cold, frozen poles.
- Maintaining a constant concentration of glucose in the blood means that an organism can release energy needed for various activities at a constant rate. The regulation of blood glucose is covered in Topic 14.8.

Control mechanisms and feedback

The control of any self-regulating system involves a series of stages:

- **set point** – the desired level at which the system operates. This is monitored by a
- **receptor** that detects internal and external stimuli which indicate any deviation from the set point and informs the
- **central control** that coordinates information from various sources and sends instructions to a suitable
- **effector**, often a muscle or gland, that brings about the necessary change needed to return the system to the set point. This return to the desired level creates a
- **feedback loop** that informs the receptor of the changes to the system brought about by the effector.

Figure 3 shows the relationship between these stages using the everyday example of controlling a central heating system.

Most systems, including biological ones, use **negative feedback**, i.e. the information fed back turns the system off. We shall see examples of negative feedback in the following topics.

Positive feedback occurs when a deviation from the set point causes changes that result in an even greater deviation from the normal. Examples are rare, but include:

- In **neurones**, a stimulus causes a small influx (movement into the cell) of sodium ions (Topic 15.4). This influx increases the permeability of the neurone to sodium, more ions enter, causing a further increase in permeability and even more rapid entry of ions. In this way, a small stimulus can bring about a large and rapid response.
- Oxytocin causes contractions of the uterus at childbirth. The contractions stimulate the release of more oxytocin, causing even more contractions. The increasing frequency of contractions leads to the birth of the baby.

Coordination of control mechanisms

Systems normally have many receptors and effectors. It is important to ensure that the information provided by receptors is analysed by the central control before action is taken. Receiving information from a number of sources allows a better degree of control. For example, temperature receptors in the skin may signal that the skin is cold and that body temperature should be raised. However, information from the hypothalamus in the brain may indicate that blood temperature is already above normal. This situation could occur during strenuous exercise when blood temperature rises but sweating cools the skin. By analysing the information from all the detectors, the brain can decide the best course of action – in this case not to raise the body temperature further. In the same way, the central control must coordinate the action of the effectors so that they operate together. For example, sweating would be less effective in cooling the body if vasodilation did not occur at the same time.

On these pages you will learn to:

- Describe the deamination of amino acids and outline the formation of urea in the urea cycle
- Describe the gross structure of the kidney

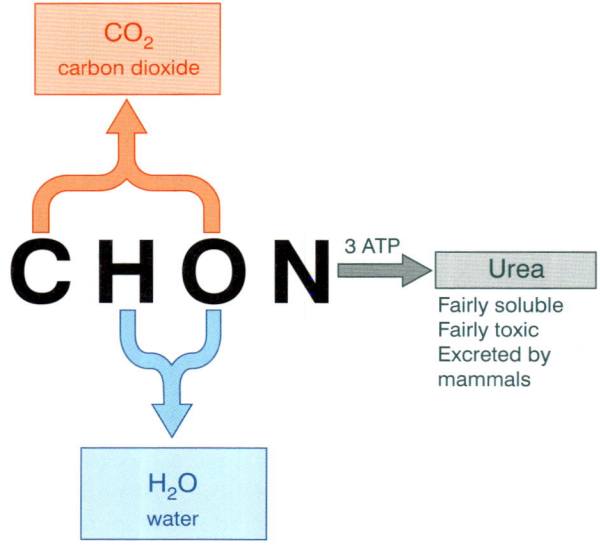

Figure 1 *Waste products formed in mammals from the four main elements in organic compounds*

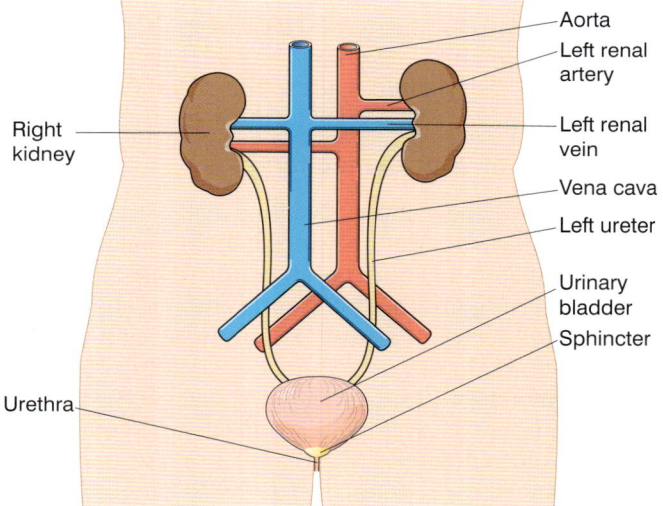

Figure 2 *Position of the kidneys in humans*

Excretion is the removal of the waste products of metabolism from the body. This is distinct from **elimination (egestion)**, which is the removal of substances such as dietary fibre that have never been involved in the metabolic activities of cells.

Excretory substances

An adult human produces about $500 \, dm^3$ of carbon dioxide and $400 \, cm^3$ of water each day as a result of respiration. Other excretory products include bile pigments and mineral salts as well as the nitrogenous excretory product **urea – $CO(NH_2)_2$**.

Urea is used as the nitrogenous excretory product of organisms that have some access to water, but not in large volumes, e.g. animals living on land, such as mammals. Urea is produced in the liver from excess amino acids in three stages:

- Amino groups (NH_2) are removed from the amino acids in a process called **deamination** and made into ammonia.
- The remainder of the amino acid can be respired to give ATP.
- The ammonia is converted to urea by the addition of carbon dioxide in a pathway called the **ornithine cycle**. ATP is required for this process.

Figure 1 summarises how this waste product is formed from the main elements in organic compounds.

EXTENSION

Other excretory substances

Ammonia (NH_3) is the easiest product to form from the amino groups (NH_2) produced when amino acids are oxidised. Its production requires no **ATP** and it is very soluble in water and so is easily dissolved and washed out of the body. It is, however, extremely poisonous – 800 times more so than carbon dioxide. Only organisms such as freshwater fish with access to large volumes of water are therefore able to use ammonia as their nitrogenous excretory product.

Uric acid is almost insoluble in water and cannot diffuse into cells, making it hardly poisonous at all. However, it takes seven ATP molecules to produce it. As almost no water is needed for its removal, it is used by organisms living in very dry conditions. As it is low in mass when stored it is also used by flying organisms. Animals such as birds and reptiles that lay eggs have an additional reason for using it – to remove their nitrogenous waste. As the young develop within the egg, their wastes cannot be removed and so anything more toxic than uric acid would kill them.

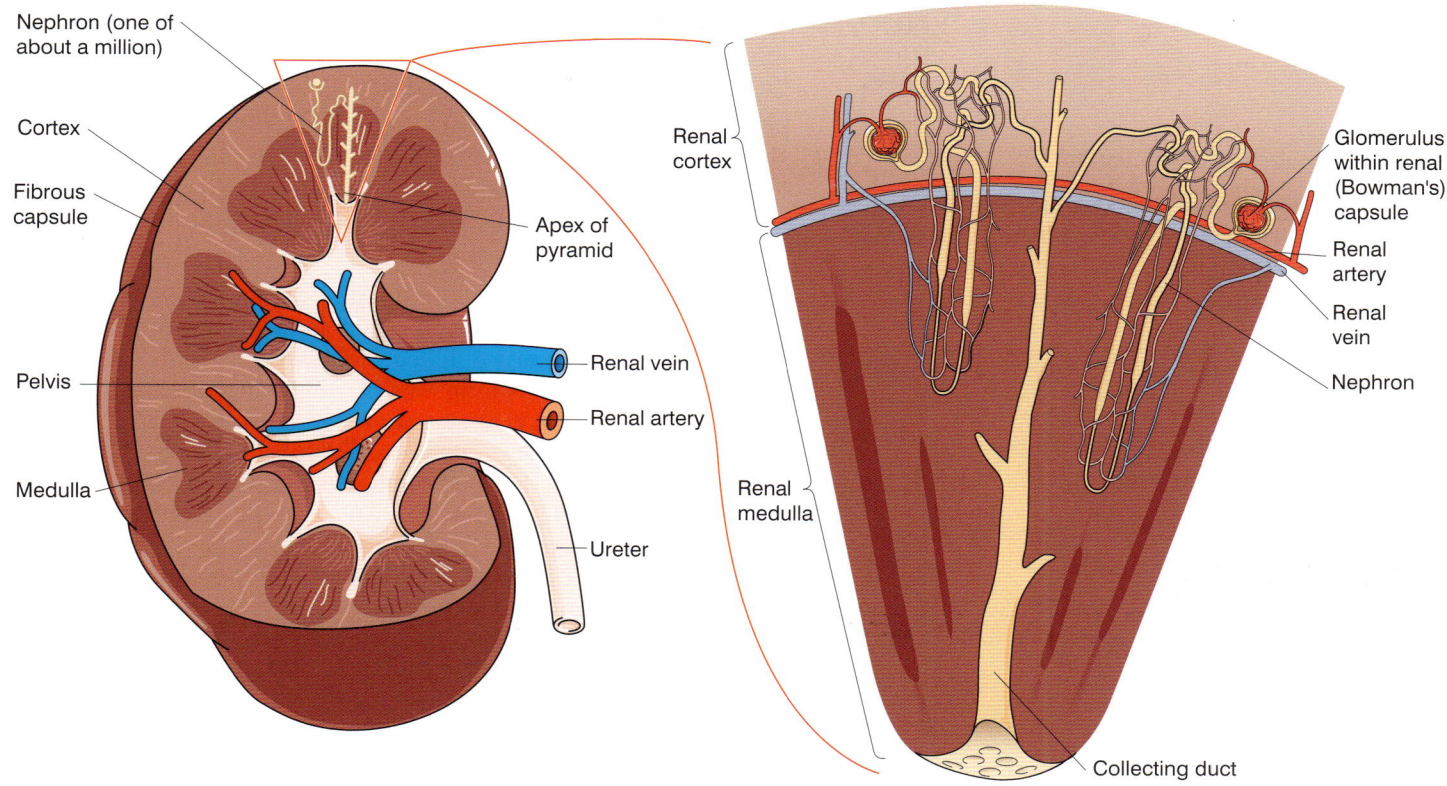

Figure 3 *Detailed structure of mammalian kidney (LS) showing the position of two of the million or more nephrons in each kidney*

Structure of the mammalian kidney

In mammals there are two kidneys found at the back of the abdominal cavity, one on each side of the spinal cord (Figure 2). In humans each kidney is usually surrounded by fat that gives it some physical protection. Weighing only 150 g each, they filter your blood plasma every 22 minutes of your life. A section through the kidney (Figure 3) shows it is made up of the:

- **fibrous capsule** – an outer membrane that protects the kidney
- **cortex** – a lighter coloured outer region made up of **renal (Bowman's) capsules**, convoluted tubules and blood vessels
- **medulla** – a darker coloured inner region made up of **loops of Henlé**, collecting ducts and blood vessels
- **renal pelvis** – a funnel-shaped cavity that collects urine into the ureter
- **ureter** – a tube that carries urine to the bladder
- **renal artery** – supplies the kidney with blood from the heart via the aorta
- **renal vein** – returns blood to the heart via the vena cava.

A microscopic examination of the cortex and medulla reveals around one million tiny tubular structures in each kidney. These are the basic structural and functional units of the kidney – the **nephrons**.

SUMMARY TEST 14.2

Excretion is the removal of metabolic waste products from the body, whereas the removal of non-metabolic material such as roughage is known as **(1)**. Respiration in humans produces around 400 cm³ of **(2)** and 500 dm³ of **(3)** that need to be removed from the body. Other excretory products include mineral salts, bile pigments and nitrogenous wastes. Urea is made by removing amino groups from amino acids and converting them to ammonia – a process called **(4)**. The ammonia is then converted to urea by the addition of **(5)** in a pathway called the **(6)** cycle. The mammalian kidney is surrounded by a protective **(7)** and in cross section is seen to be made up of a lighter coloured outer region called the **(8)** and a darker inner region called the **(9)**. These regions are made up of around a million tubular structures called **(10)**. Blood is brought to the kidney by the vessel called the **(11)**, and urine leaves it via a tube called the **(12)**.

The structure of the nephron

On these pages you will learn to:

- Describe the detailed structure of the nephron with its associated blood vessels using photomicrographs and electron micrographs
- Describe the gross structure of the kidney

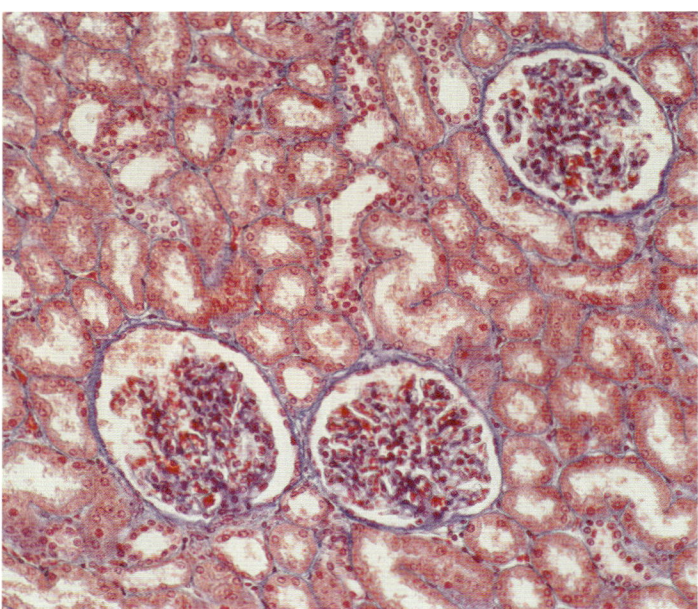

Figure 1 *Photomicrograph of cortex of human kidney (TS). There are three glomeruli: each glomerulus appears as a mass of blood capillaries surrounded by a clear space – the lumen of the renal capsule. The background shows sections through convoluted tubules.*

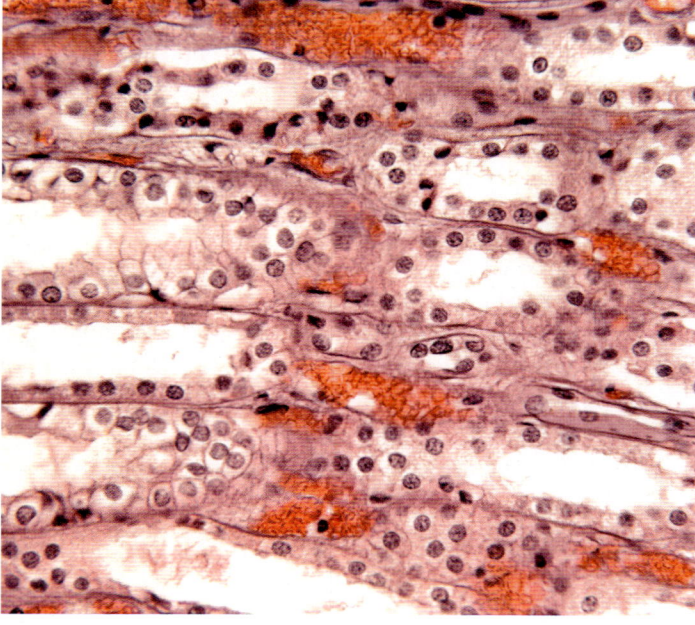

Figure 2 *Photomicrograph of medulla of human kidney showing loops of Henlé. Around them are blood capillaries containing red blood cells (red).*

The nephron is the functional unit of the kidney. It is a narrow tube, closed at one end, with two twisted regions separated by a long hairpin loop. Each nephron is made up of a:

- **Renal (Bowman's) capsule** – the closed end at the start of the nephron. It is cup-shaped and contains within it a mass of blood capillaries known as the glomerulus. Its inner layer is made up of specialised cells called **podocytes** (Figure 1).
- **Proximal (first) convoluted tubule** – a series of loops surrounded by blood capillaries. Its walls are made of cuboidal epithelial cells with microvilli (Figure 1).
- **Loop of Henlé** – a long, hairpin loop that extends from the cortex into the medulla of the kidney and back again. It is surrounded by blood capillaries (Figure 2).
- **Distal (second) convoluted tubule** – a series of loops surrounded by blood capillaries. Its walls are made of cuboidal epithelial cells, but it is surrounded by fewer capillaries than the proximal tubule.
- **Collecting duct** – a tube into which a number of distal convoluted tubules empty. It is lined by cuboidal epithelial cells and becomes increasingly wide as it empties into the pelvis of the kidney.

Associated with each nephron are a number of blood vessels (Figure 3):

- **afferent arteriole** – a tiny vessel that is a branch of the renal artery and supplies the nephron with blood. The afferent arteriole enters the renal capsule of the nephron where it forms the
- **glomerulus** – a many-branched knot of capillaries from which fluid is forced out of the blood (Figure 4). The glomerular capillaries recombine to form the
- **efferent arteriole** – a tiny vessel that leaves the renal capsule. It has a smaller diameter than the afferent arteriole, which causes an increase in blood pressure within the glomerulus. The efferent arteriole carries blood away from the renal capsule and later branches to form the
- **peritubular capillaries** – a concentrated network of capillaries that surrounds the proximal convoluted tubule, the loop of Henlé and the distal convoluted tubule and from where they reabsorb mineral salts, glucose and water. The peritubular capillaries merge together into venules (tiny veins) that in turn merge together to form the renal vein.

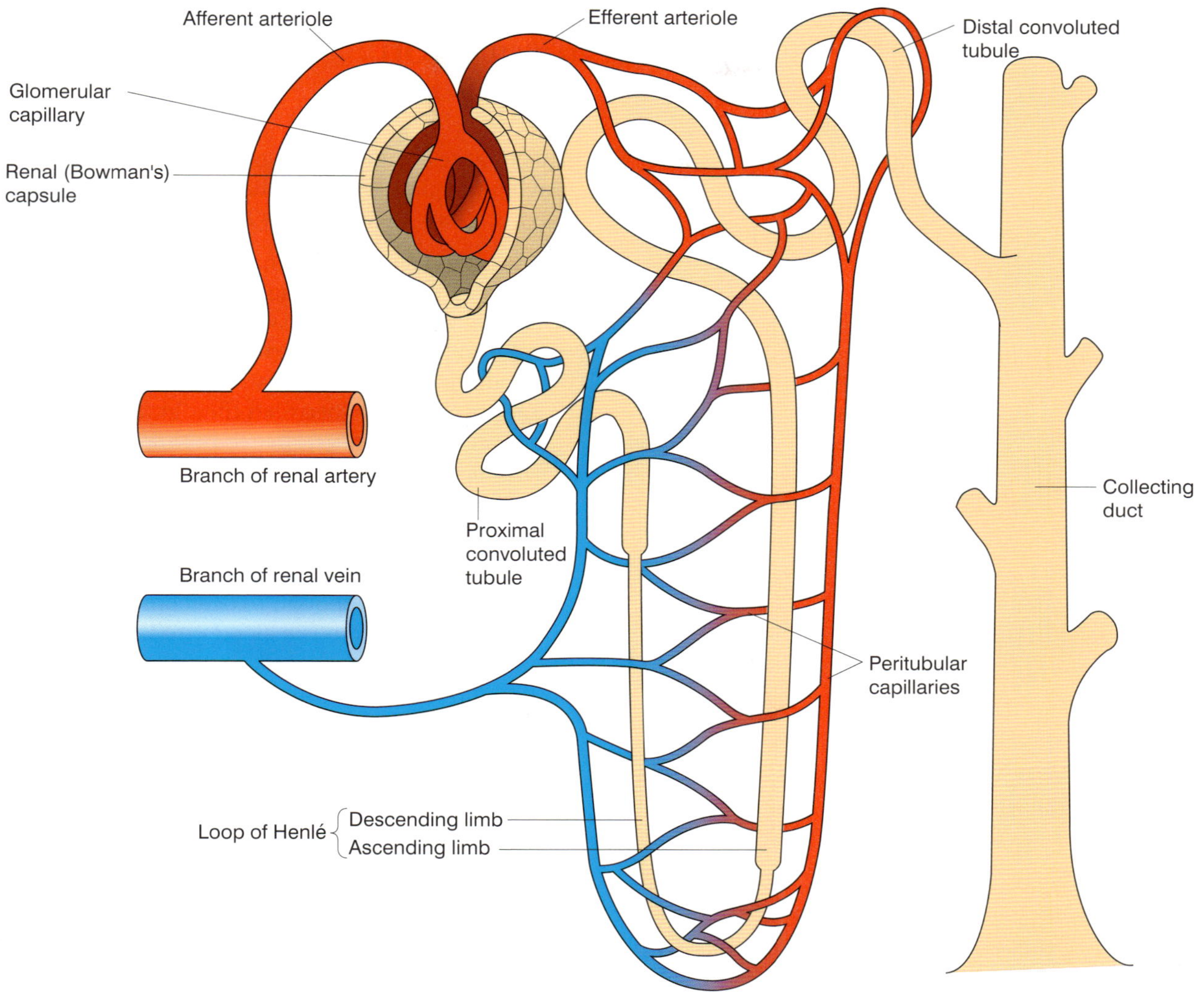

Afferent arteriole

Efferent arteriole

Distal convoluted tubule

Glomerular capillary

Renal (Bowman's) capsule

Branch of renal artery

Branch of renal vein

Proximal convoluted tubule

Collecting duct

Peritubular capillaries

Loop of Henlé { Descending limb
Ascending limb

Figure 3 *Regions of the nephron and associated blood vessels*

SUMMARY TEST 14.3

The nephron is the structural unit of the kidney. It comprises a cup-shaped structure called the **(1)** that contains a knot of blood vessels called the **(2)** which receives its blood from a vessel called the **(3)** arteriole. The inner wall of this cup-shaped structure is lined with specialised cells called **(4)** and from it extends the first, or **(5)**, convoluted tubule whose walls are lined with **(6)** epithelial cells that have **(7)** to increase their surface area. The next region of the nephron is a hairpin loop called the **(8)** which then leads onto the second, or **(9)**, convoluted tubule. This in turn leads onto the **(10)** which empties into the renal pelvis. Around much of the nephron is a dense network of blood vessels called the **(11)** capillaries.

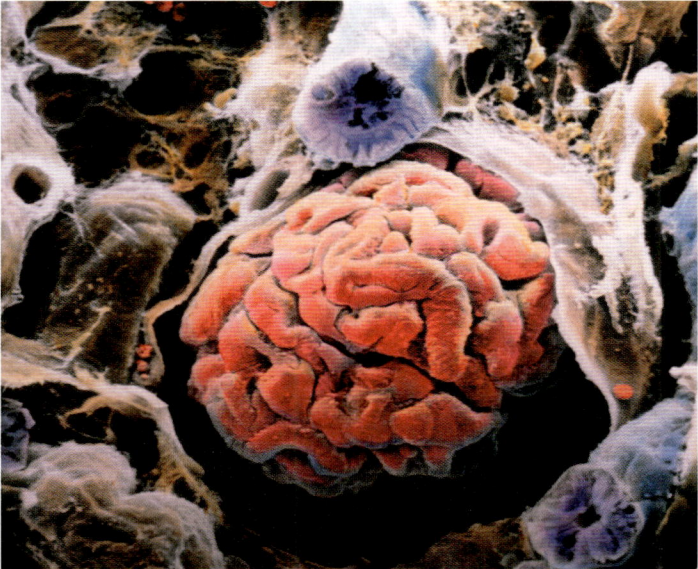

Figure 4 *Colourised scanning electron micrograph of a glomerulus (centre) surrounded by the renal capsule, seen as a white-brown membrane at centre right. Part of the proximal convoluted tubule is seen, coloured blue.*

On these pages you will learn to:

- Describe how the processes of ultrafiltration and selective reabsorption are involved with the formation of urine in the nephron

The functions of the kidney are listed in Table 1. The main one, that of regulating the composition of blood, is carried out by the nephrons in a series of stages – ultrafiltration, selective reabsorption and the reabsorption of water and minerals.

Table 1 *The functions of the kidneys*

- Regulating the composition of the blood and maintaining a constant water potential by:
  - Maintaining a constant volume of water
  - Removing wastes such as urea
  - Maintaining the concentration of mineral ions and other substances constant
- Regulating blood pressure
- Maintaining the body's calcium level
- Stimulating the production of red blood cells

Ultrafiltration

Blood enters the kidney through the renal artery, which branches frequently to give around one million tiny arterioles, each of which enters a **renal (Bowman's) capsule** of a nephron. This is called the **afferent arteriole** and it divides to give a complex of capillaries known as the **glomerulus**. The glomerular capillaries later merge to form the **efferent arteriole**, which then sub-divides again into capillaries (the peritubular capillaries), which wind their way around the various tubules of the nephron before combining to form the renal vein (see extension). The walls of the glomerular capillaries are made up of endothelial cells with pores between them. As the diameter of the afferent arteriole is greater than that of the efferent arteriole, there is a build up of hydrostatic pressure within the glomerulus. As a result, water, glucose, mineral ions and other substances up to a relative molecular mass of up to 68 000 are squeezed out of the capillary to form the **glomerular filtrate**. The movement of this filtrate out of the glomerulus is resisted by the:

- capillary endothelium
- basement membrane of the epithelial layer of the renal (Bowman's) capsule
- epithelial cells of the renal (Bowman's) capsule
- the hydrostatic pressure of the fluid in the renal capsule space – the **intracapsular pressure**
- the low water potential of the blood in the glomerulus.

This total resistance would be sufficient to prevent filtrate leaving the glomerular capillaries, but there are some modifications to reduce this barrier to the flow of filtrate:

- The inner layer of the renal (Bowman's) capsule is made up of highly specialised cells called **podocytes** (Figure 2). These cells, which are illustrated in Figure 1, are lifted off the surface membrane on little 'feet' ('podo' = feet). This allows filtrate to pass beneath them and through gaps between their branches. Filtrate passes between these cells rather than through them.
- The endothelium of the glomerular capillaries has spaces up to 100 nm wide between its cells (Figure 1). Again, fluid can therefore pass between, rather than through, these cells.

As a result, the hydrostatic pressure of the blood in the glomerulus is high enough to overcome the resistance and so filtrate passes from the blood into the renal capsule. This filtration under pressure is known as ultrafiltration. The filtrate has much the same composition as blood plasma, with the exception of the plasma proteins which are too large to pass across the basement membrane. Many of the substances in the 125 cm³ of filtrate passing out of blood each minute are extremely useful to the body and need to be reabsorbed.

EXTENSION
The glomerulus – a unique capillary bed

In mammals, the glomerulus is the only capillary bed in which an arteriole (the afferent arteriole) supplies it with blood and an arteriole (the efferent arteriole) also drains blood away. In all other mammalian capillary beds it is a venule that drains away the blood. Why then do we not make life simpler and call the efferent arteriole a venule? The reason is that the efferent arteriole later divides up into a second capillary bed – the peritubular capillaries – and these are drained by a venule. In any case, the structure of the wall is that of an arteriole and not a venule. The glomerular capillaries need to merge into an efferent arteriole because this increases the hydrostatic pressure within the glomerulus and allows ultrafiltration to occur.

Selective reabsorption

In the proximal convoluted tubule nearly 85% of the filtrate is reabsorbed back into the blood. Why then, you may ask, allow it to leave the blood in the first place? Ultrafiltration operates on the basis of size of molecule: all below 68 000 relative molecular mass are removed. Some are wastes, but most are useful.

About 180 dm³ of water enters the nephrons each day. Of this volume, only about 1 dm³ leaves the body as urine. Eighty-five per cent of the reabsorption of water occurs in the proximal convoluted tubule.

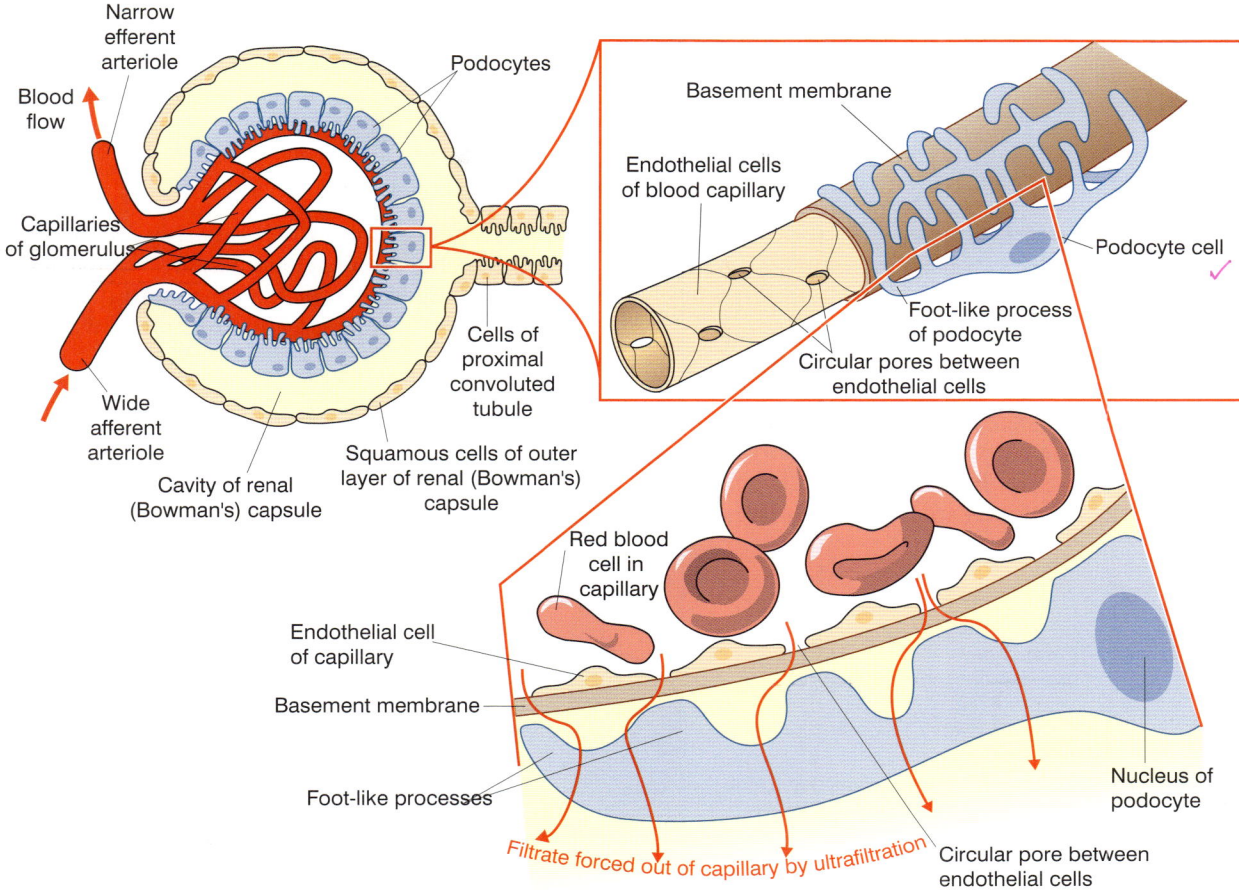

Figure 1 *Podocyte and ultrafiltration*

The cells of the proximal convoluted tubules are adapted to reabsorb substances into the blood by having microvilli which give them a large surface area and many mitochondria (Figure 3) to provide ATP for active transport of sodium ions. The process is as follows:

- Sodium ions are actively transported out of the cells lining the proximal convoluted tubule into blood capillaries which carry them away. This is by the action of a membrane carrier protein, the sodium-potassium pump. The sodium ion concentration of these cells is therefore lowered.
- Sodium ions now diffuse along a concentration gradient from the lumen of the proximal convoluted tubule into the lining cells but only through special carrier proteins.
- These carrier proteins are of different types, each of which carries another molecule (glucose or amino acids or chloride ions, etc.) along with the sodium ions. This is known as co-transport (Topic 4.5). Water follows osmotically down the water potential gradient that is created.
- The molecules which have been co-transported into the cells of the proximal convoluted tubule then diffuse into the blood. As a result, all the glucose, amino acids, chloride ions and most other valuable molecules are reabsorbed as well as water.

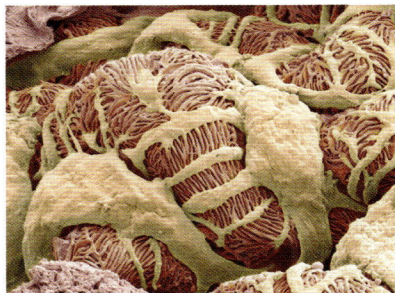

Figure 2 *Colourised scanning electron micrograph of podocyte cells around a glomerulus in a human kidney*

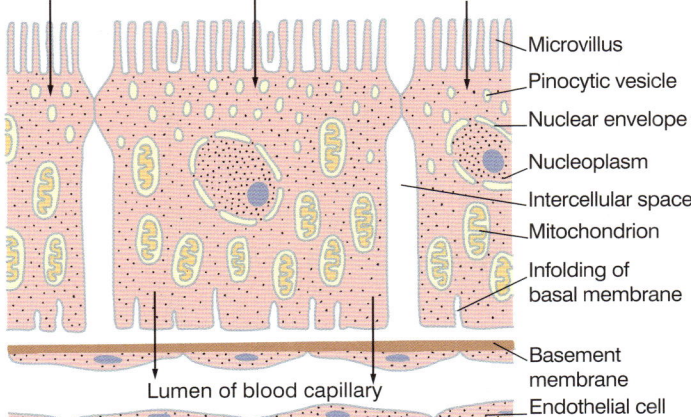

Figure 3 *Details of cells from the wall of the proximal convoluted tubule*

14.5 Kidney function – loop of Henlé and reabsorption of water

The loop of Henlé

The loop of Henlé is a hairpin-shaped tubule that extends into the medulla of the kidney. It is responsible for creating the conditions in the surrounding interstitial fluid that lead to the reabsorption of water from the distal convoluted tubule and the collecting duct. This results in concentrating the urine so that it has a lower **water potential** than the blood. The concentration of the urine produced is directly related to the length of the loop of Henlé. It is short in mammals whose **habitats** are in or by water (e.g. beavers) and long in those whose habitats are dry regions (e.g. kangaroo rat).

The loop of Henlé has two regions:

- The descending limb, which is narrow, with thin walls that are highly permeable to water.
- The ascending limb, which after a short distance is wider, with thick walls that are impermeable to water.

The loop of Henlé acts as a counter-current multiplier. To understand how this works it is necessary to consider the following sequence of events using Figure 1, to which the numbers refer.

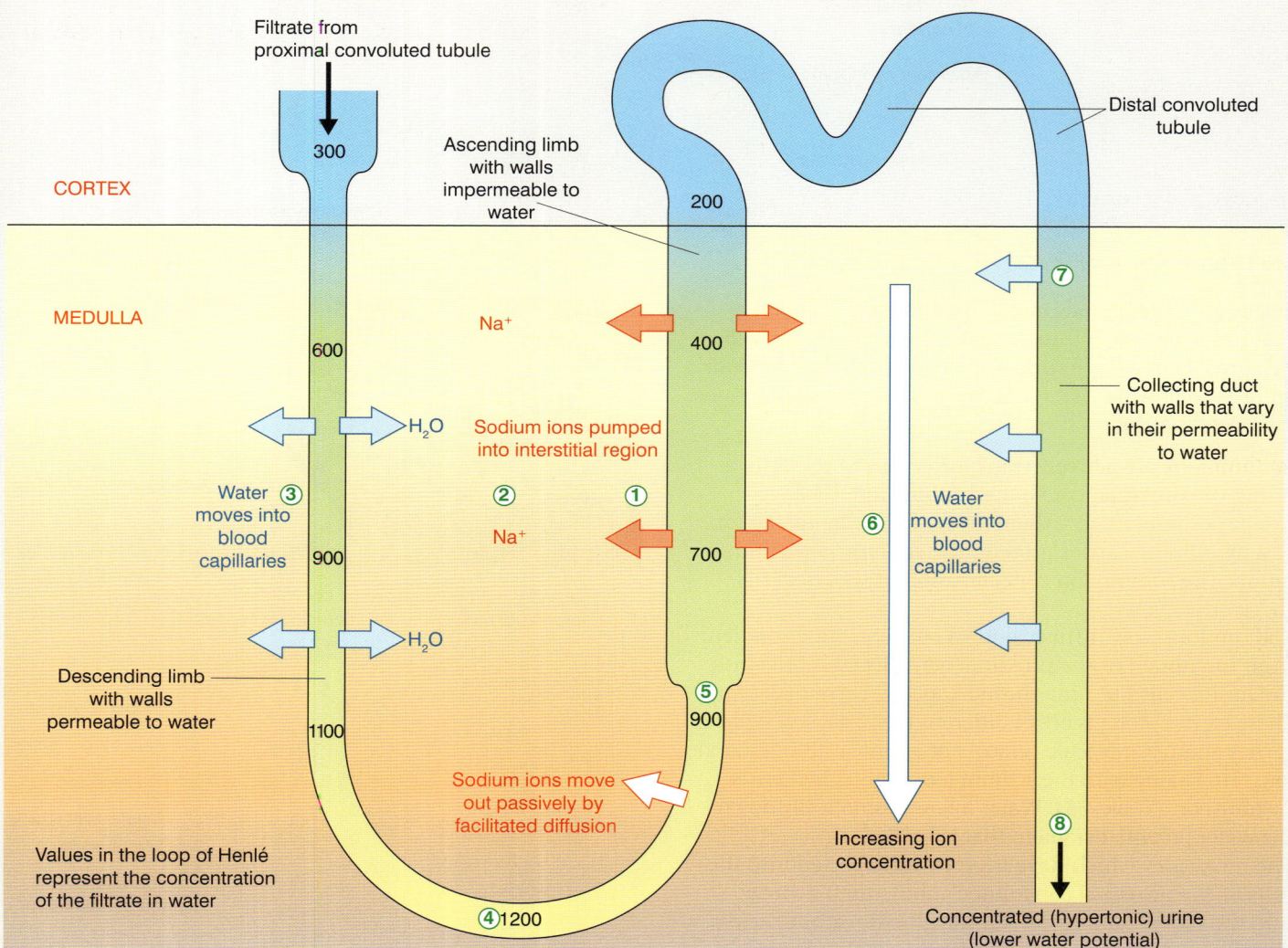

Figure 1 Counter-current multiplier of the loop of Henlé

1. Sodium and chloride ions are actively pumped out of the ascending limb of the loop of Henlé using ATP provided by the many mitochondria in the cells of its wall.
2. This creates a low water potential (high ion concentration) in the region of the medulla between the two limbs (called the interstitial region). In normal circumstances water would pass out of the ascending limb by osmosis. However, the thick walls are almost impermeable to water and so very little, if any, escapes.
3. The walls of the descending limb are, however, very permeable to water and as the cells have many membrane protein channels known as aquaporins it passes out of the filtrate, by osmosis, into the interstitial space. This water enters the blood capillaries in this region by osmosis and is carried away.
4. The filtrate progressively loses water in this way as it moves down the descending limb lowering its water potential. It reaches its minimum water potential at the tip of the hairpin.
5. At the base of the ascending limb, sodium and chloride ions diffuse out of the filtrate and as it moves up the ascending limb these ions are also actively pumped out (see point **1**) and therefore the filtrate develops a progressively higher water potential.
6. In the interstitial space between the ascending limb and the collecting duct there is a gradient of water potential with the highest water potential in the cortex and an increasingly lower water potential the further into the medulla one goes (see Topic 14.6).
7. The collecting duct is permeable to water and so, as the filtrate moves down it, water passes out of it by osmosis. This water passes by osmosis into the blood vessels that occupy this space, and is carried away (see Topic 14.6).
8. As water passes out of the filtrate its water potential is lowered. However, the water potential is also lowered in the interstitial space and so water continues to move out by osmosis down the whole length of the collecting duct. The counter-current multiplier ensures that there is always a water potential gradient drawing water out of the tubule.

The water that passes out of the collecting duct by osmosis does so through aquaporins (water channels). The hormone ADH (Topic 14.6) can alter the number of these channels and so control water loss. By the time the filtrate, now called urine, leaves the collecting duct on its way to the bladder, it has lost most of its water and so it has a lower water potential than the blood.

The distal (second) convoluted tubule

The cells that make up the walls of the distal (second) convoluted tubule have microvilli and many mitochondria that allow them to reabsorb material rapidly from the filtrate, by either diffusion or **active transport**. The main role of the distal tubule is to make final adjustments to the water and salts that are reabsorbed and to control the pH of the blood by selecting which ions to reabsorb. To achieve this, the permeability of its walls becomes altered under the influence of various hormones (Topic 14.6). A summary of the processes taking place in the **nephron** is given in Figure 2.

Counter-current multiplier

When two liquids flow in opposite directions past one another, the exchange of substances (or heat) between them is greater than if they flowed in the same direction next to each other. In the case of the loop of Henlé, the counter-current flow means that the filtrate in the collecting duct with a lower water potential meets interstitial fluid that has an even lower water potential. This means that, although the water potential gradient between the collecting duct and interstitial fluid is small, it exists for the whole length of the collecting duct. There is therefore a steady flow of water into the interstitial fluid, so that around 80% of the water enters the interstitial fluid and hence the blood. If the two flows were in the same direction (parallel) less of the water would enter the blood.

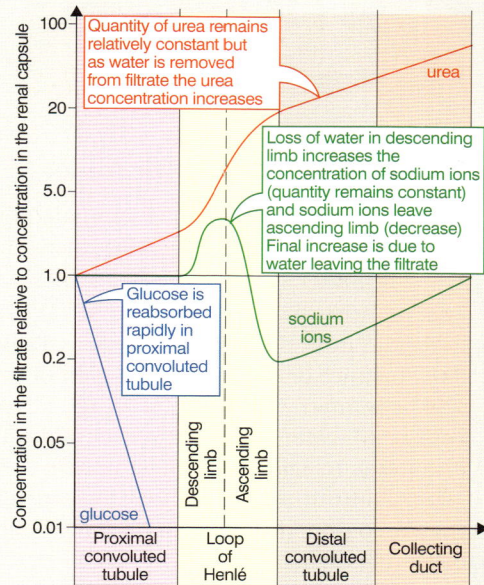

Figure 2 *Relative concentrations of three substances in the filtrate as it passes along a nephron. NB Scale is not linear.*

On these pages you will learn to:

- Describe the roles of the hypothalamus, posterior pituitary, ADH and collecting ducts in osmoregulation

The quantity of water and salts we take in varies from day to day, as does the quantity we lose. Table 1 shows the daily balance between loss and gain of salts and water for a typical human. The blood, however, needs to have a constant volume of water and concentration of salts to avoid osmotic disruption to cells. The **homeostatic** control of water and solute concentrations in the blood is achieved by hormones that act on the distal (second) convoluted tubule and the collecting duct.

Table 1 *Daily water and salt balance in a typical human*

WATER

Volume of water / cm^3 day^{-1}			
Water gain		**Water loss**	
Diet	2300	Urine	1500
Metabolism,	200	Expired air	400
e.g. respiration		Evaporation from skin	350
		Faeces	150
		Sweat	100
TOTAL	2500	TOTAL	2500

SALT

Mass of salt / g day^{-1}			
Salt gain		**Salt loss**	
Diet	10.50	Urine	10.00
		Faeces	0.25
		Sweat	0.25
TOTAL	10.50	TOTAL	10.50

As a result of facilitated diffusion and active transport by cells in the ascending limb of the loop of Henle, sodium and chloride ions are concentrated in the interstitial fluid surrounding the distal convoluted tubule and the collecting duct. There is a gradient of water potential within this interstial region, with the highest water potential in the cortex and the lowest in the medulla region closest to the renal pelvis (see Topic 14.5, Extension, for further details). Some urea passes out of the filtrate in the collecting duct into the interstitial region, so further increasing the concentration of solutes and decreasing the water potential. At all points, the interstitial region has a lower water potential than the filtrate passing down the collecting duct.

The naming of antidiuretic hormone

The name antidiuretic hormone (ADH) may, at first, seem unusual. However, it describes its function precisely. **Diuresis** is the production of large volumes of dilute urine. It is a symptom of a disease called **diabetes insipidus** (so called because the urine from sufferers did not taste sweet!). The disease was successfully treated with pituitary extract. Therefore it was suggested that a hormone existed that was given the name 'antidiuretic hormone'. As the effect of ADH is to increase the permeability of collecting ducts so that more water is reabsorbed into the blood, it causes the production of small volumes of concentrated urine.

This is the opposite of diuresis – hence the name **anti**diuretic hormone.

Regulation of the water potential of the blood

The **water potential** of the blood is determined by the balance of water and salts within it. A rise in solute concentration lowers its water potential. This may be caused by:

- too little water being consumed
- much sweating occurring
- large amounts of salt being taken in (ingested).

The body responds to this decrease in water potential as follows:

- Sensory cells called **osmoreceptors** in the **hypothalamus** of the brain detect the decrease in water potential.
- It is thought that, when the water potential of the blood is low, water is lost from these osmoreceptor cells by osmosis.
- Owing to this water loss the osmoreceptor cells shrink, a change that stimulates the neurosecretory cells in the hypothalamus to produce a hormone called **antidiuretic hormone (ADH)**.
- ADH passes along the neurones (nerve cells) to the posterior **pituitary gland**, from where it is secreted into the capillaries.
- ADH passes in the blood to the kidney, where it increases the permeability to water of the cell surface membrane of the cells that make up the walls of the distal (second) convoluted tubule and the collecting duct.
- Receptors on the cell surface membrane of these cells bind to ADH molecules, activating a second messenger system within the cell (cyclic AMP – Topic 14.7). This results in the activation of a protein kinase, an enzyme that adds phosphate groups to other proteins to activate them.
- The action of protein kinase causes vesicles within the cell to move to, and fuse with, its cell surface membrane. (ADH binding also leads to an increase in transcription of the gene coding for the aquaporin protein, increasing the number of available aquaporins.)

- When the vesicles fuse with the cell surface membrane, the number of aquaporins in the membrane increases greatly, making the cell surface membrane much more permeable to water.
- As it is a small molecule, some urea can cross the phospholipid bilayer of the membrane. ADH binding also leads to an increase in membrane transport proteins for urea, so that the collecting duct becomes more permeable to urea and some will leave the filtrate to further decreases the water potential of the interstitial region.
- The combined effect is that more water leaves the collecting duct by osmosis down a water potential gradient and re-enters the blood.
- As the reabsorbed water came from the blood in the first place, this will not, in itself, increase the water potential of the blood, but merely prevent it from decreasing any further. Therefore the osmoreceptors also stimulate the thirst centre of the brain, to encourage the individual to seek out and drink more water.
- The osmoreceptors in the hypothalamus detect the increase in water potential and ADH secretion from the posterior pituitary is reduced.
- The decrease in ADH concentration in the blood will lead to a decreased permeability of the collecting duct to water and urea so that the permeability returns to its former state. This is an example of homeostasis and the principle of negative feedback (Topic 14.1).

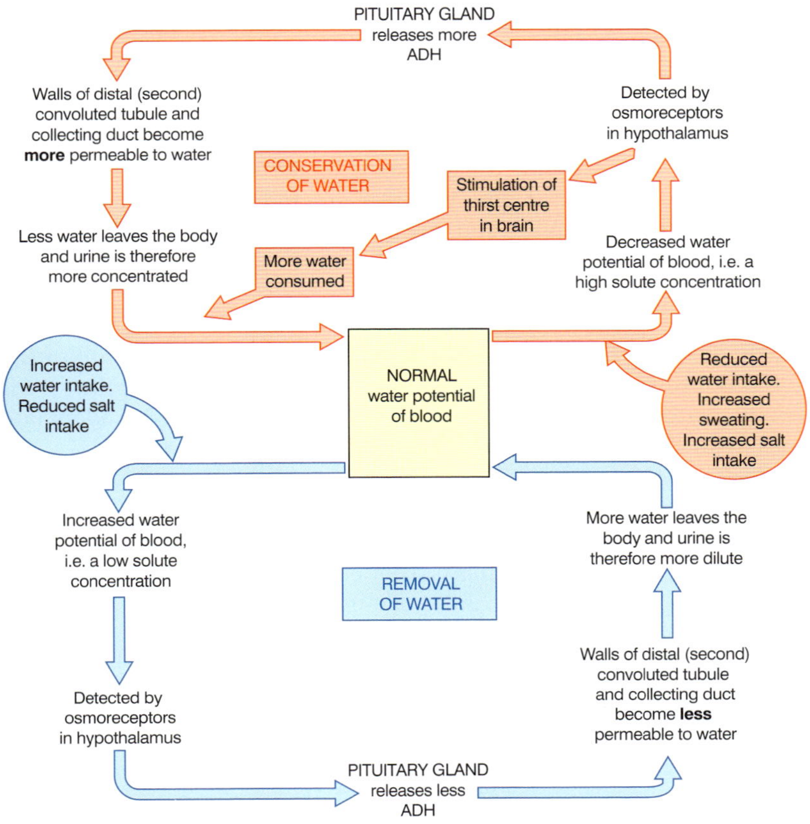

A decrease in the solute concentration of the blood increases its water potential. This may be caused:

- by large volumes of water being consumed
- by salts used in metabolism or excreted not being replaced in the diet.

The body responds to this increase in water potential as follows:

- The osmoreceptors in the hypothalamus detect the increase in water potential and stimulate the pituitary gland to reduce its release of ADH.
- ADH, via the blood, decreases the permeability of the collecting ducts to water and urea.
- Less water is reabsorbed back into the blood from the collecting duct.
- More dilute urine is produced and the water potential of the blood decreases.
- When the water potential of the blood has returned to normal, the osmoreceptors in the hypothalamus cause the pituitary to raise its release of ADH back to normal (= negative feedback).

These events are summarised in Figure 1.

Figure 1 *Regulation of water potential of the blood by antidiuretic hormone (ADH)*

SUMMARY TEST 14.6

A human gains around (1) cm³ of water each day, of which (2) cm³ comes from the diet, with the remainder being produced in metabolic processes such as (3). More than half this water is lost from the body as (4). The same typical human needs around 10.5 g of salt in the diet, of which 10 g is lost in the urine and 0.25 g in faeces. The remainder is lost in (5). Despite daily fluctuations in water and salt intake, the water potential of the blood remains relatively constant as a result of (6) control achieved by hormones that act on the (7) and collecting duct. If too little water or too much salt is consumed, or if (8) is excessive, the water potential of the blood will (9). In response to this, osmoreceptors in the (10) of the brain detect the change and produce antidiuretic hormone (ADH) that passes to the (11) gland from where it is secreted. ADH passes via the blood to the kidney where it increases the (12) of the distal convoluted tubule and collecting duct to water and (13). As a result more water is reabsorbed and enters the blood. The osmoreceptors also stimulate a thirst response and so more water is drunk and the water potential of the blood therefore (14). When the water potential returns to normal, the osmoreceptors detect this and ADH production is reduced to normal – an example of the principle of (15).

On these pages you will learn to:

- Outline the role of cyclic AMP as a second messenger with reference to the stimulation of liver cells by adrenaline and glucagon
- Describe the three main stages of cell signalling in the control of blood glucose by adrenaline as follows:
  - hormone–receptor interaction at the cell surface
  - formation of cyclic AMP, which binds to kinase proteins
  - an enzyme cascade involving activation of enzymes by phosphorylation to amplify the signal

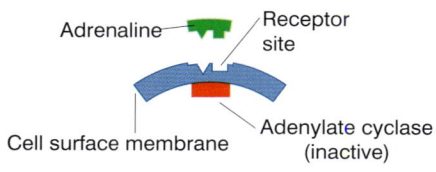

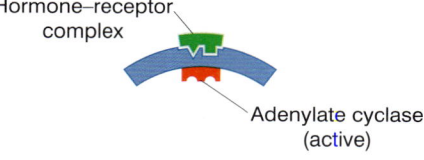

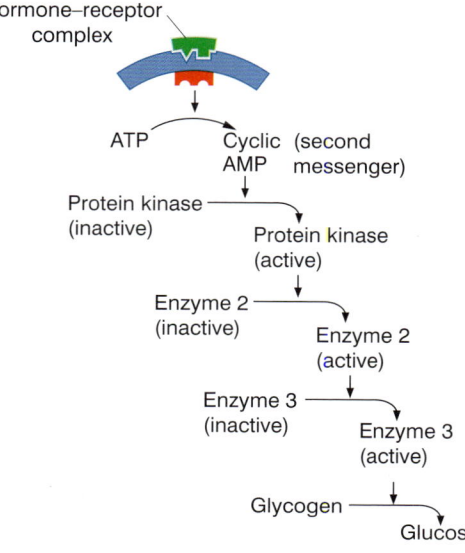

Figure 1 *Second messenger mechanism of hormone action*

Mammals possess two main coordinating systems – the nervous system that communicates rapidly, and the endocrine system that usually does so more slowly. Both systems interact in order to maintain a constant internal environment, at the same time being responsive to a varying external environment. Both systems also use chemical messengers – the endocrine system exclusively so, and the nervous system through the use of **neurotransmitters** in chemical **synapses**. The chemical messenger **adrenaline** may act both as a hormone and a neurotransmitter.

Hormones

A hormone is a regulating chemical produced and secreted by an endocrine gland and is carried in the blood to the cells, tissues or organ on which it acts – known as the **target cell (target tissue, target organ)** – that have complementary receptors on their cell surface membranes or their internal membranes. Hormones may differ chemically from one another, but they share many common characteristics. Some hormones have their action at the cell surface and other hormones are able to enter the cell to have an effect.

Hormones are:

- effective in very small quantities, but often have widespread and permanent effects
- normally relatively small molecules
- often proteins or polypeptides, although some are **steroids**
- transported by the blood system
- produced by endocrine glands.

Mechanisms of hormone action

One mechanism of hormone action is called the **cyclic AMP second messenger system**. An example of this system occurs in the stimulation of liver cells by the hormones adrenaline and glucagon, leading to the conversion of glycogen to glucose.

The process involving adrenaline, shown in Figure 1, has three stages.

1. Adrenaline binds to its complementary receptor on the cell surface membrane of a liver cell. Binding activates a membrane protein, termed a G protein.
2. The G protein activates another membrane protein, an enzyme called adenylate cyclase. The activated enzyme converts ATP to cyclic AMP (the second messenger). Cyclic AMP binds to and activates a kinase protein.
3. There follows an enzyme cascade in which enzymes are activated by phosphorylation. This amplifies the first signal as one enzyme molecule can catalyse the phosphorylation of many other enzyme molecules. The last enzyme in the chain catalyses the breakdown of glycogen to glucose.

The glucose diffuses out of the liver cell and into the blood, through protein carriers called transporter proteins.

Endocrine glands

A gland is a group of cells that produce a particular substance or substances by a mechanism known as **secretion**. Glands can be divided into two groups:

- **Exocrine** glands transport their secretions to the site of action by special ducts. Many digestive secretions e.g. saliva and pancreatic juice, are produced by exocrine glands.
- **Endocrine** glands are ductless glands that secrete hormones directly into the blood.

Endocrine glands may be discrete organs such as the thyroid gland, or groups of cells within other organs, such as the islets of Langerhans in the pancreas.

SUMMARY TEST 14.7

Hormones are produced by endocrine glands and are carried by **(1)** to the cell, tissue or organ on which they act, called the **(2)** cell/organ. Hormones often have widespread and **(3)** effects and are usually proteins although some are **(4)**. The pancreas produces digestive secretions that it passes along its pancreatic duct and, as such, it acts as an **(5)** gland. It is also an endocrine gland, producing the hormone insulin from **(6)** cells found in the groups of endocrine cells called the **(7)**. Another type of cell, called **(8)** cells produce the hormone **(9)**. Both hormones act together to control **(10)**.

EXTENSION

The pancreas

The pancreas is a large, pale coloured gland that is situated in the upper abdomen, behind the stomach. It is both an exocrine and an endocrine gland.

- As an **exocrine gland** it produces the digestive secretion called pancreatic juice that passes along the pancreatic duct and into the small intestine. Among other substances, pancreatic juice contains the digestive enzymes protease, amylase and lipase.
- As an **endocrine gland** the pancreas produces the hormones insulin and glucagon, which pass directly into the blood capillaries that pass through it.

When examined microscopically, the pancreas is made up largely of exocrine cells arranged radially around tiny ducts. These cells produce pancreatic juice, which is secreted into these ducts. Scattered throughout the pancreas, like small islands in the sea of exocrine cells, are the groups of endocrine cells known as **islets of Langerhans**. These cells appear different in structure from the exocrine cells and, upon close examination, are seen to be of two types:

- the larger **α cells** that produce the hormone glucagon
- the smaller **β cells** that produce the hormone insulin.

Figure 2 shows the cellular structure of part of the pancreas. Both types of cells are rich in secretory vesicles and their role in controlling blood glucose concentrations is the subject of our next topic.

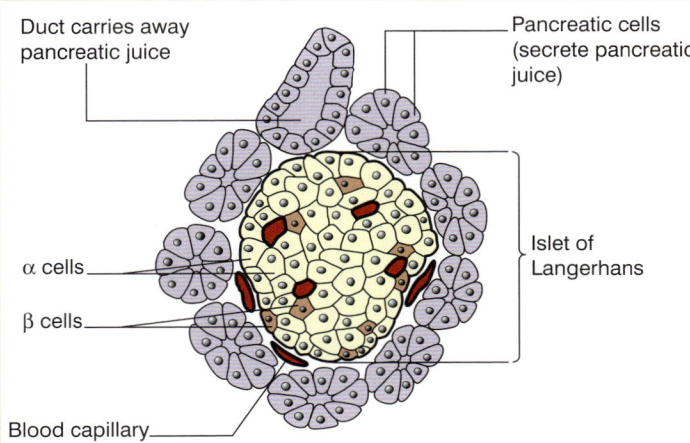

Figure 2 *Section through the pancreas showing an islet of Langerhans*

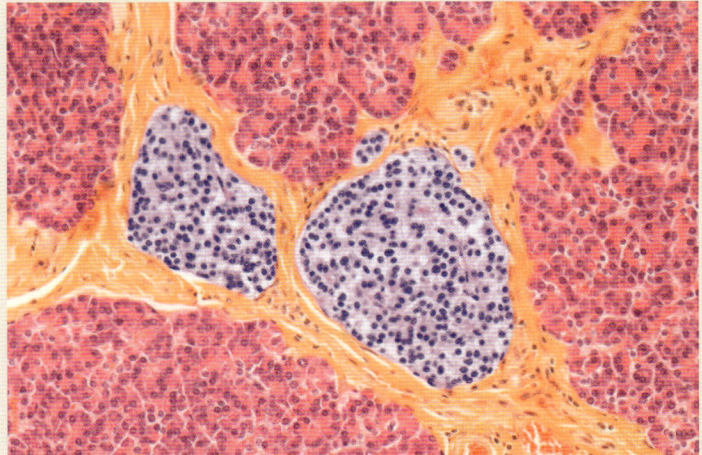

Figure 3 *Photomicrograph of the pancreas showing two islets of Langerhans (centre) containing α cells and β cells. Around the islets are the exocrine pancreatic cells.*

On these pages you will learn to:

- Explain how the blood glucose concentration is regulated by negative feedback control mechanisms, with reference to insulin and glucagon

Glucose is the main substrate for respiration, providing the source of energy for almost all organisms. It is therefore essential that the blood of mammals contains a relatively constant concentration of glucose for respiration. If it falls too low, the energy supply in cells will be too low and the cells will die – brain cells are especially sensitive in this respect because they can only respire glucose. If the concentration rises too high, it decreases the water potential of the blood and creates osmotic problems that can cause dehydration and be equally dangerous. Homeostatic control (Topic 14.1) of blood glucose concentration is therefore essential.

Blood glucose and variations in its concentration

The normal concentration of blood glucose is 90 mg in each 100 cm³ of blood. There are three sources of blood glucose:

- **Directly from the diet** as glucose from the breakdown of other carbohydrates such as starch, maltose, lactose and sucrose.
- **Breakdown of glycogen (glycogenolysis)** from the stores in the liver and muscle cells. A normal liver contains 75–100 g of glycogen, made by converting excess glucose from the diet in a process called glycogenesis.
- **Gluconeogenesis** is the production of new glucose, i.e. from sources other than carbohydrate. The liver, for example, can make glucose from glycerol and amino acids.

As animals may not eat continuously as their diet varies, their intake of glucose fluctuates. Likewise, glucose is used up at different rates depending on the level of mental and physical activity. With changes in the supply and demand of glucose the concentration of glucose in the blood fluctuates. Three main hormones, insulin, glucagon and **adrenaline** operate to maintain a constant blood glucose concentration.

Insulin and the β cells of the pancreas

We saw in Topic 14.7 that in the pancreas there are groups of special cells known as the islets of Langerhans. These cells are of two types: larger alpha (α) cells and smaller beta (β) cells. The β cells detect and respond to a rise in blood glucose concentration by secreting the hormone **insulin** directly into the blood. Insulin is a globular protein made up of 51 amino acids (Figure 1).

Almost all body cells (but not red blood cells) have **glycoprotein** receptors on their membranes that bind with insulin molecules. Binding of insulin leads to an increase in membrane permeability and enzyme action so that the blood glucose concentration is lowered in one or more of the following ways:

- Cellular respiratory rate is increased, using up more glucose and increasing its uptake by cells.
- The rate of conversion of glucose into glycogen (glycogenesis) is increased in the cells of the liver and muscles.
- The rate of conversion of glucose to fat in **adipose tissue** is increased.
- The rate of absorption of glucose into cells increases, especially in muscle cells.

REMEMBER

To help understand some terms in this topic it is worth remembering that:

'gluco' / 'glyco'	=	glucose
'glycogen'	=	glycogen
'neo'	=	new
'lysis'	=	splitting
'genesis'	=	birth / origin

Therefore:

glycogen – o – lysis	=	splitting of glycogen
gluco – neo – genesis	=	formation of new glucose

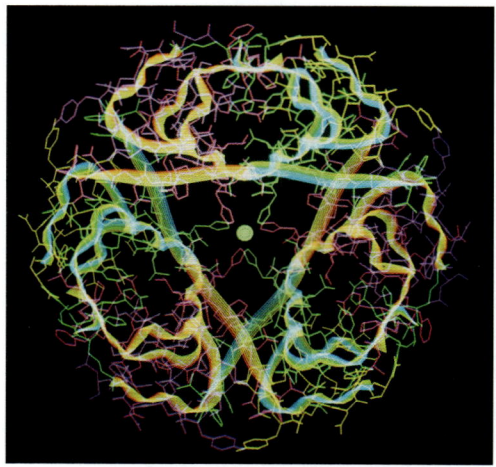

Figure 1 *Molecular graphic of an insulin molecule. Insulin is made up of 51 amino acids arranged in two chains (shown here as yellow and green ribbons) held together by disulfide bridges*

The effect of these processes is to remove glucose from the blood and return its concentration to normal. This lowering of the blood glucose concentration causes the β cells to reduce their secretion of insulin (= **negative feedback**).

Glucagon and the α cells of the pancreas

The α cells of the islets of Langerhans respond to a fall in blood glucose by secreting the hormone glucagon directly into the blood. Only the cells of the liver have receptors that bind to glucagon, so only liver cells respond, by activating the enzyme phosphorylase, which converts glycogen to glucose, and by increasing the conversion of amino acids and glycerol into glucose (= **gluconeogenesis**).

The overall effect is therefore to increase the quantity of glucose in the blood and return it to its normal concentration. This increase in the blood glucose concentration causes the α cells to reduce the secretion of glucagon (= negative feedback).

Adrenaline and other hormones regulating the blood glucose level

There are at least four other hormones besides glucagon that can increase blood glucose level. The best known of these is **adrenaline**. At times of excitement or stress, adrenaline is produced by the adrenal glands that lie above the kidneys. It causes the breakdown of glycogen in the liver, raising the blood glucose concentration. If the glycogen supplies in the liver are used up, the adrenal glands produce the hormone **cortisol**, which causes the liver to convert amino acids and glycerol into glucose.

Hormone interaction in regulating blood sugar

The two hormones, insulin and glucagon, act in opposite directions. Insulin lowers blood glucose concentration, whereas glucagon increases it. The two hormones are said to act **antagonistically**. The system is self-regulating because the concentration of glucose in the blood determines the quantity of insulin and glucagon produced. In this way the interaction of these two hormones allows highly sensitive control of the blood glucose concentration. The concentration of glucose is, however, not constant, but fluctuates around a set point. This is because of the way negative feedback mechanisms work. Only when the blood glucose concentration falls below the set point is insulin secretion reduced (negative feedback), leading to a rise in blood glucose. In the same way, only when the concentration exceeds the set point is glucagon secretion reduced (negative feedback), causing a fall in the blood glucose concentration. The control of blood glucose level is summarised in Figure 2.

SUMMARY TEST 14.8

Glucose is the main **(1)**. It is important that blood glucose concentration is maintained at **(2)** mg in each 100 cm³ of blood by **(3)**, because if it falls too low the energy supply of cells will be too low. The cells of the **(4)** are especially sensitive to low blood glucose concentrations. If it rises too high **(5)** problems occur that may cause dehydration. Blood glucose is formed directly from **(6)** in the diet or from the breakdown of **(7)**, which is stored in the cells of the liver and **(8)**. The liver can also increase blood glucose levels by making glucose from other sources such as glycerol and **(9)** in a process known as **(10)**. Blood glucose is used up when it is absorbed into cells, converted into fat or **(11)** for storage or is used up during **(12)** by cells. To maintain a constant concentration of blood glucose the pancreas acts as an **(13)** gland in producing two hormones from clusters of cells within it called **(14)**. The β cells are **(15)** in size and produce the hormone **(16)**, which causes the blood glucose concentration to **(17)**. The α cells produce the hormone **(18)**, which has the opposite effect. The two hormones are therefore said to act **(19)**. Another hormone called **(20)** can also raise blood glucose concentrations.

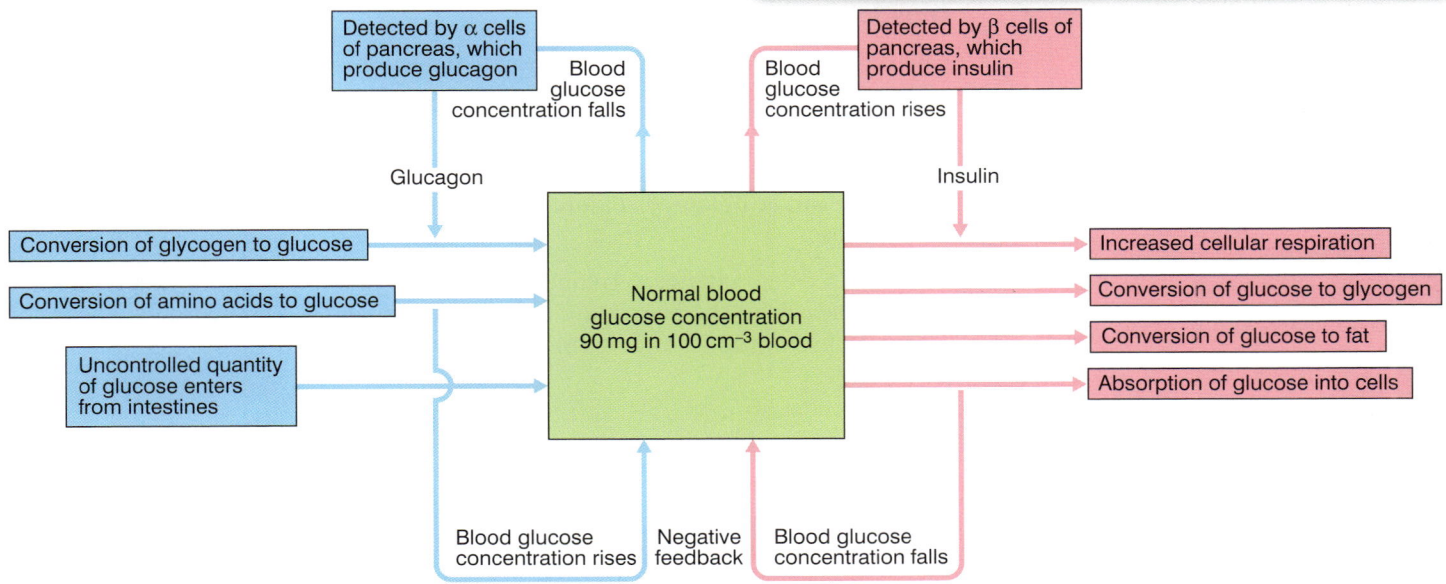

Figure 2 *Summary of regulation of blood glucose*

On these pages you will learn to:

- Explain the principles of operation of dip sticks containing glucose oxidase and peroxidase enzymes, and biosensors that can be used for quantitative measurements of glucose in blood and urine
- Explain how urine analysis is used in diagnosis with reference to glucose, protein and ketones

Biosensors are devices that use immobilised biological molecules, such as enzymes or antibodies, or biological systems, such as whole cells, to detect a specific chemical and in most cases, measure the concentration of the chemical. In its simplest form, a biosensor can be a simple dipstick, but commonly biosensors are now taken to mean those devices that are coupled with microelectronics so that results are rapid, the measurements are extremely accurate, and the chemical can be detected in very small quantities.

Detecting glucose in blood and urine

As explained in Topic 14.8, it is important for healthy functioning that blood glucose concentrations remain at or around a normal concentration of 90 mg 100 cm^{-3} blood. If concentrations rise too high, glucose is excreted in the urine.

Simple dipsticks or more complex biosensors are frequently used by people with diabetes to measure the concentration of glucose in the blood or urine. For people with type I diabetes (a lack of insulin) a check on glucose concentration will help to inform them if they need to adjust their insulin doses. Health care professionals routinely use dipsticks as a quick and reliable method to detect the abnormal presence of glucose in urine.

The method of detection of glucose makes use of immobilised enzymes (see Topic 3.4). For both dipsticks and biosensors the same enzymes can be used:

- glucose oxidase catalyses the conversion of glucose to gluconic acid and hydrogen peroxide

$$\text{Glucose} + \text{O}_2 \xrightarrow{\text{Glucose oxidase}} \text{gluconic acid} + \text{H}_2\text{O}_2$$

- peroxidase catalyses the breakdown of hydrogen peroxide to water and oxygen

$$2\text{H}_2\text{O}_2 \xrightarrow{\text{Peroxidase}} 2\text{H}_2\text{O} + \text{O}_2$$

Dipsticks to detect glucose

Glucose dipsticks are a convenient way to detect glucose in a sample of urine.

They involve a colour change, which could be a change in the intensity of colour, to give a semi-quantitative measure of glucose concentration. The colour change can be compared to a coloured standards chart and an estimate of concentration of glucose can be obtained (Figure 1). The dipsticks are highly specific and detect only glucose.

To obtain a colour change, a colourless hydrogen donor is used to react with the oxygen released in the peroxidase reaction shown above. The hydrogen donor acts as a chromogen, changing colour by being oxidised by the oxygen that is given off.

$$2\text{H}_2\text{O}_2 + \text{DH}_2 \xrightarrow{\text{Peroxidase}} 2\text{H}_2\text{O} + \text{D}$$

hydrogen donor (colourless) → coloured compound

The dipstick is a thin strip of absorbent paper (some dipsticks are a thin strip of plastic with a paper pad stuck on at one end). At the test end of the strip the chromogen and the enzymes glucose oxidase and peroxidase are added. A thin cellulose membrane covers the area so that only small molecules such as glucose can enter the test area. The end of the dipstick is dipped into the urine sample and after a set time any colour change can be compared to the chart. If glucose is present in the urine, the action of the two enzymes will result in a colour change.

Biosensors to detect glucose

Biosensors are extremely sensitive and accurate. As glucose oxidase is a highly specific enzyme, biosensors used to detect glucose in blood samples are insensitive to other chemicals present. They can also be re-used, so are cost effective and also have the advantage of being small and portable so that they carried by a person and used at any time.

In a glucose biosensor, the enzymes glucose oxidase is immobilised onto an inert supporting material to form a biological recognition layer. The layer is separated from the blood sample by a partially permeable membrane that only allows small molecules such as glucose to diffuse through. Glucose molecules present in the blood will bind to the active sites of the glucose oxidase enzymes and the reaction results in the production of gluconic acid and hydrogen peroxide, as shown above.

The next part of the biosensor detects that a reaction has occurred and converts this into an electric current. This conversion is carried out by a transducer. In some biosensors the decrease in oxygen can be detected by a platinum oxygen electrode. In others, the production of hydrogen ions (from the gluconic acid produced) can be detected.

The final part of the biosensor is the amplification of the electrical signal and the production of a digital reading. The reading is proportional to the reaction that has occurred (for example, proportional to the decrease in oxygen, or the increase in hydrogen ions) so it is proportional to the concentration of glucose in the sample.

Urine analysis

Dipsticks can be used to estimate the quantity of substances in urine. The results provide valuable information about the body's metabolism and enable doctors to make medical diagnoses.

Among the substances that can be identified are:

- proteins, which can be detected using the albustix test. As proteins are large molecules, they do not normally leave the **glomerulus** during **ultrafiltration** in the kidneys. Their presence in urine (proteinuria) indicates that they are being forced out of the glomerulus. This might be due to high blood pressure (hypertension). Damage to kidneys due to high blood pressure, diabetes or infection could also result in protein being present in urine. Urinary tract infections can also lead to the presence of protein in urine.
- glucose, which can be detected using a dipstick test such as the Diastix test. Glucose is normally reabsorbed in the proximal convoluted tubules of the kidney. Its presence in urine (glucosuria) could indicate diabetes.
- ketones, which are produced when fatty acids, rather than glucose, are being used as a respiratory substrate. Their presence suggests that the supply of glucose is exhausted. This might be the result of starvation or, where a person is diabetic, inadequate control of the condition.

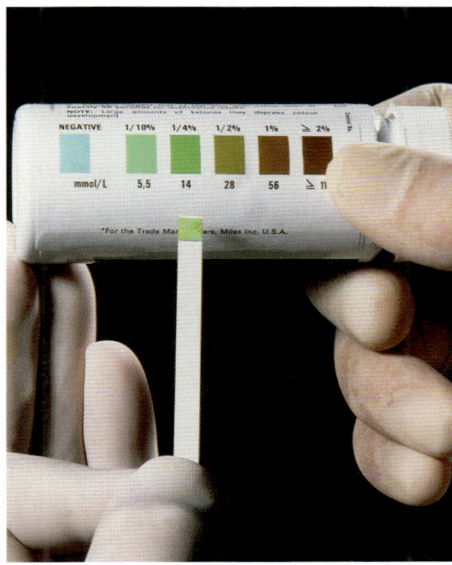

Figure 1 *The Diastix test for glucose carried out on the urine of a person with diabetes. A positive result for glucose is seen. People with diabetes lack the hormone insulin, which is the hormone that leads to a lowering of blood glucose concentration. The results here indicate that the person has a raised blood glucose concentration, as glucose has been excreted in the urine.*

Homeostasis in plants

Plants, as well as animals, have homeostatic mechanisms to ensure that they maintain a constant level of essential materials for their needs. Carbon dioxide uptake and water loss are two examples of processes that must be regulated. This regulation is achieved by controlling the opening and closure of structures called **stomata**.

To take in carbon dioxide from the external atmosphere, the stomata must open. However, open stomata mean water loss. Plants overcome this problem by having daily rhythms of opening and closing their stomata. To obtain carbon dioxide, they open in the light, when photosynthesis is occurring. This means that water loss is an inevitable consequence of photosynthesis. They then close their stomata at night, when there is no need for carbon dioxide, and so minimise water loss. Stomata can also respond to changes in the environment to regulate diffusion of carbon dioxide and water vapour loss by transpiration.

Stomata

Stomata are minute pores that occur mainly on the leaves, especially the underside (lower epidermis). Each stoma (singular) is surrounded by a pair of special, kidney-shaped cells called **guard cells**. When a stoma is open, these cells surround a small opening a few micrometres wide called the **stomatal pore** (Figure 3). Unlike other epidermal cells, guard cells have chloroplasts and dense cytoplasm. The inner cell walls of the guard cells are thicker and less elastic than the outer ones. This means that any increase in the volume of the guard cells, for example due to the osmotic intake of water, causes the outer wall to bend more than the inner wall and so widen the stomatal pore. To 'close' the stoma completely, the reverse occurs and the stomatal pore decreases in size until it is no longer present. In this way they can control the rate of gaseous exchange.

The mechanism of stomatal opening

It is known that stomata open and close in response to certain stimuli. For example, they usually open in the light and close in the dark. One suggested mechanism for the opening and closing of stomata is as follows.

- a particular stimulus such as light activates ATP synthase, an enzyme that increases the production of ATP by the chloroplasts in the guard cells
- these chloroplasts only have photosystem I (Topic 13.3) and no Calvin cycle enzymes, so ATP is produced in cyclic photophosphorylation but not used up in the Calvin cycle
- this ATP is therefore available to provide more energy for the active transport of protons (H^+) from the guard cells
- the reduced H^+ concentration and increased negative charge inside the guard cell causes potassium (K^+) channels in the membrane of these cells to open
- potassium ions (K^+) now diffuse into the guard cells down an electrochemical gradient. To maintain a balance, chloride ions also enter the guard cells
- these K^+ lower the water potential of the guard cells and so water enters them by osmosis down a water potential gradient. There are numerous aquaporins in the cell surface membranes of the guard cell for water movement
- the extra water causes the guard cells to become more turgid and to swell
- the thinner outer and thicker inner walls of the guard cells means that when they swell they bend outwards, and so widen the stomatal aperture.

These events are shown in Figure 2.

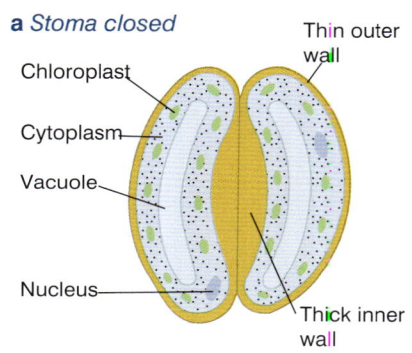

a *Stoma closed*

Chloroplast

Cytoplasm

Vacuole

Nucleus

Thin outer wall

Thick inner wall

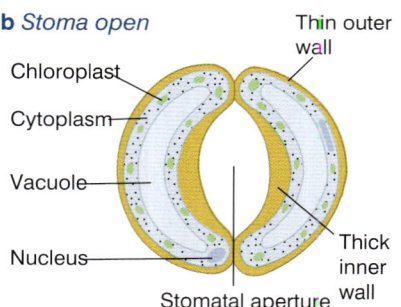

b *Stoma open*

Chloroplast

Cytoplasm

Vacuole

Nucleus

Thin outer wall

Thick inner wall

Stomatal aperture

Figure 1 *Surface view of a stoma closed and open*

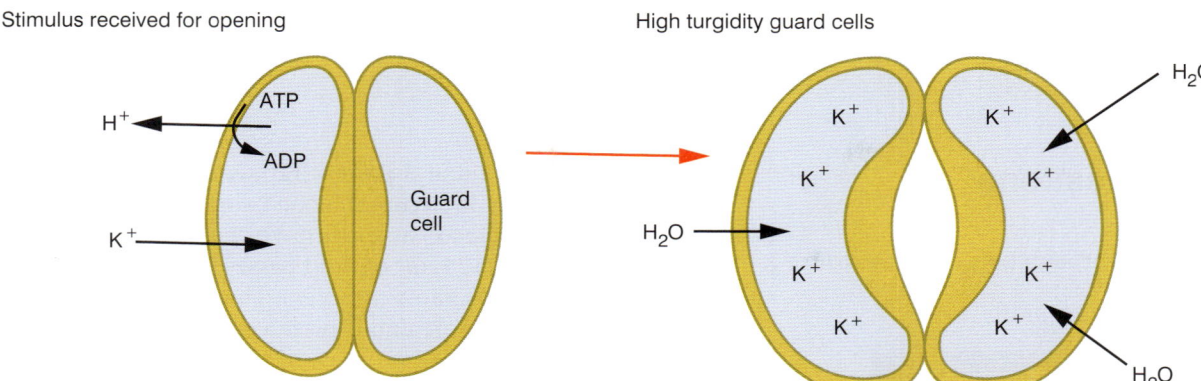

Stimulus received for opening

High turgidity guard cells

Figure 2 *Mechanism of stomatal opening*

Role of abscisic acid in the closing of stomata

Abscisic acid is sometimes referred to as the 'stress hormone' and has been shown to increase in concentration up to 40 times during drought conditions. It is this increase in abscisic acid that causes the closure of **stomata** when water loss needs to be reduced. As stomata are the main means by which plants lose water vapour during **transpiration** (Topic 7.7), it follows that abscisic acid helps to reduce water loss. The response is rapid, as it needs to be if the plant is to survive, with stomata closing within a minute or two of abscisic acid being applied. The speed of this reaction rules out a mechanism involving regulating the expression of a **gene**. Although the exact mechanism of how abscisic acid operates is not yet fully understood, one suggested process is as follows.

- Abscisic acid combines with specific receptors on the cell surface membrane of the **guard cells** that surround stomata.
- This binding of abscisic acid increases the concentration of calcium ions in the guard cells (more enters from outside and some is released from intracellular stores).
- The calcium ions act as a **second messenger** by altering potassium channels in the cell surface membrane in a way that causes potassium ions to diffuse from the guard cells to the epidermal cells.
- Abscisic acid may also inhibit the action of the proton pump that moves hydrogen ions out of the guard cells.
- The solute concentration in the guard cells is therefore reduced and their water potential becomes higher than in the epidermal cells.
- Water therefore leaves by osmosis, so the volume of the guard cells decreases, they become less turgid and therefore the stomatal pore closes.

SUMMARY TEST 14.10

Two examples of substances that need to be homeostatically controlled by stomata are (**1**) and (**2**). Light can activate an enzyme called (**3**) in guard cells to increase production of ATP by structures called (**4**). This ATP is used to provide more energy for the (**5**) of (**6**) ions out of the guard cells, opening potassium ion (gated) channels and causing an influx of potassium ions. As a result, the (**7**) concentration of the guard cells increases, and the water potential (**8**), causing water to enter them by osmosis. The cells become more (**9**) and the stomatal pore opens. Abscisic acid is sometimes referred to as the (**10**). It can cause stomata to close in a process that uses (**11**) ions as a second messenger.

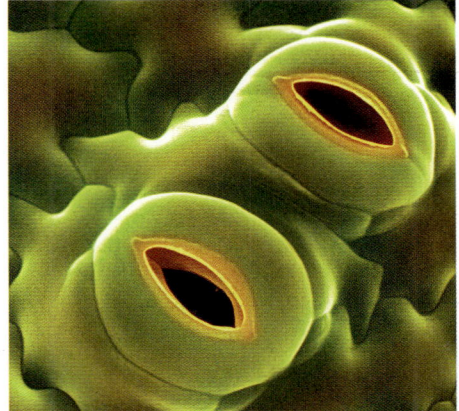

Figure 3 *Colourised scanning electron micrograph of open stomata on a tobacco leaf*

239

14 Examination Questions

1 Figure 1 shows a section through part of the cortex of a kidney.

Figure 1

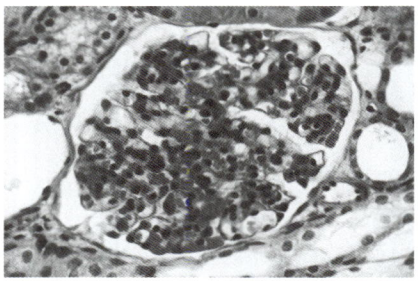

a On a copy of Figure 1, draw label lines and use the letters **G** and **R** to identify:
  • A glomerulus with the letter **G**.
  • A renal capsule with the letter **R**. *(2 marks)*

b State the name of the hormone that is involved in the control of the water potential of the blood. *(1 mark)*

c Table 1 shows the concentration of some compounds in the fluids of a glomerulus, a renal capsule and a collecting duct of the kidney.

Table 1

compound	concentration / g 100 cm⁻³		
	blood plasma entering glomerulus	filtrate in renal capsule	urine in collecting duct
water	90.0	90.0	96.0
proteins	8.0	0.0	0.0
glucose	0.1	0.1	0.0
urea	0.03	0.03	2.0

With reference to Table 1,
 i explain why proteins occur in the blood entering the glomerulus but not in the filtrate in the renal capsule *(2 marks)*
 ii explain why there is glucose present in the filtrate but not in the urine *(2 marks)*
 iii explain the difference in the concentration of urea between the filtrate and urine. *(2 marks)*
  (Total 9 marks)

Cambridge International AS and A Level Biology 9700 Paper 4 Q7 November 2008

2 a The human kidneys process 1200 cm³ of blood every minute. This 1200 cm³ of blood contains 700 cm³ of plasma. As blood passes through the glomeruli of the kidneys, 125 cm³ of fluid passes into the renal capsules (Bowman's capsules). This fluid is called the glomerular filtrate and is produced by a process called ultrafiltration.
   i Calculate the percentage of plasma that passes into the renal capsules.
     Show your working and give your answer to one decimal place. *(2 marks)*
   ii Explain how the structures of the glomerular capillaries and the podocytes are adapted for ultrafiltration. *(4 marks)*

  b The glomerular filtrate then passes through the proximal convoluted tubule.

  Figure 2 a transverse section through part of the proximal convoluted tubule.

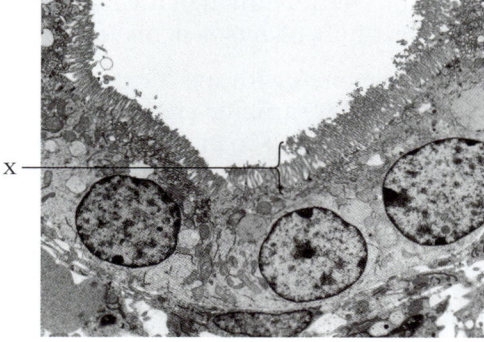

Figure 2
   i Name the structures labelled X. *(1 mark)*
   ii Explain why the epithelial cells of the proximal convoluted tubule have many mitochondria in them. *(2 marks)*
   iii Of the 125 cm³ of glomerular filtrate that enters the renal capsules each minute, only 45 cm³ reaches the loops of Henlé.
     Name **two** substances that are reabsorbed into the blood from the proximal convoluted tubule, **apart from water**. *(2 marks)*
  (Total 11 marks)

Cambridge International AS and A Level Biology 9700 Paper 4 Q6 June 2013

14 Practice Questions

3 The figure below shows some of the homeostatic changes that occur as a result of water being lost from the blood due to sweating.

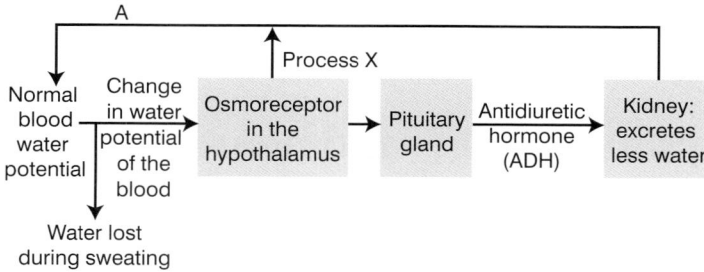

a Describe the change in water potential that occurs in the blood as a result of sweating.

b Which of the structures shown on the figure acts as:
i   a receptor
ii an effector?

c Describe how ADH gets from the pituitary gland to the kidney.

d The kidney conserves the water that is already in the blood. Given that the water potential of the blood returns to its normal level prior to sweating, suggest what is happening in process X.

e State as precisely as possible what mechanism is shown by the line labelled A.

4 An experiment was carried out with two groups of people. Group X had type I diabetes while group Y did not (control group). Every 15 minutes blood samples were taken from all members of both groups and the mean levels of insulin, glucagon and glucose were calculated. After an hour, every person was given a glucose drink. The results are shown in the figure below.

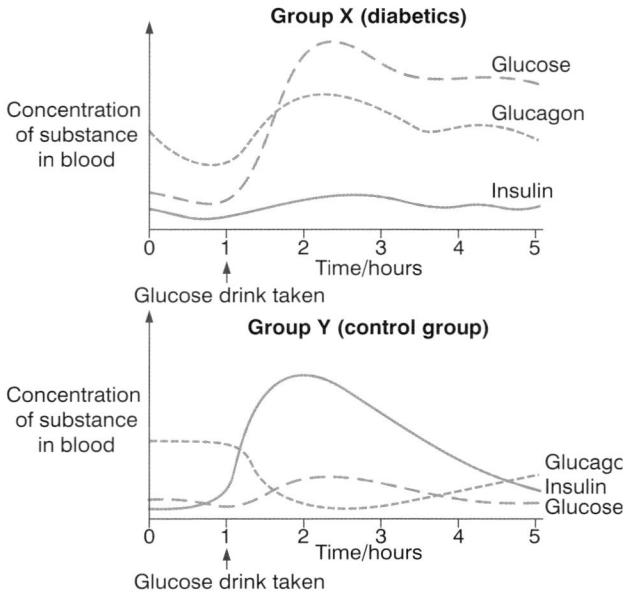

a Name a hormone other than insulin and glucagon that is involved in regulating blood glucose levels.

b State two differences between groups X and Y in the way insulin secretion responds to the drinking of glucose.

c Suggest a reason why the glucose concentration falls in both groups during the first hour.

d Using information from the graphs, explain the changes in the blood glucose concentration in group Y after the glucose is drunk.

e Explain the difference in blood glucose concentration in group X compared to group Y.

f Suggest what might happen to the blood glucose concentration of group X if they had no food intake over the next 24 hours.

5 Figure 5 is a diagram of the surface of a leaf showing a stoma with its guard cells and the surrounding epidermal cells.

Figure 5

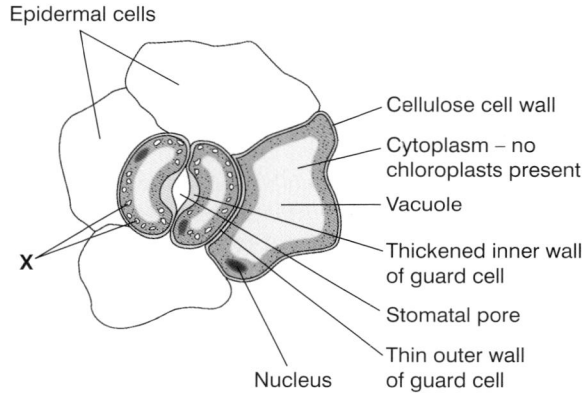

a Name the structures labelled X.

b Explain the importance of guard cells having inner and outer walls of different thicknesses.

c Stomata usually open in the light. During the process of opening:
i   What is the name of the enzyme that is activated by light?
ii What is the name of the ion that is moved into the guard cells?
iii By what process does this ion enter guard cells?
iv Describe the relative water potential of the guard cells and the epidermal cells when a stoma is open.
v   State how the volume of the guard cells changes when a stoma is opening.

d Abscisic acid (ABA) causes stomata to close.
i   Where **exactly** does abscisic acid bind?
ii Name the ion whose concentration increases in guard cells in response to this binding.
iii This ion acts as a secondary messenger, which in turn causes another ion to move out of the guard cells. What is the name of this second ion and by what process does it move out?

On these pages you will learn to:

- Outline the roles of the nervous system in coordinating homeostatic mechanisms
- Compare the nervous and endocrine systems as communication systems that co-ordinate responses to changes in the internal and external environment
- Describe the structure of a sensory neurone and a motor neurone

Stimulus and response

The ability to react to **stimuli** is a basic characteristic of all living organisms. The stimuli may occur internally or externally and they lead to a **response** from the organism. The ability to respond to a stimulus increases the chances of survival for an organism. For example, to be able to detect and move away from harmful stimuli such as predators, extremes of temperature and pH, or to detect and move towards a source of food clearly aid survival.

Table 1 *Comparison of endocrine and nervous systems*

Endocrine system	Nervous system
Communication is by chemicals called hormones	Communication is by nerve impulses (and neurotransmitters)
Transmission is by the blood system	Transmission is by nerve fibres
Transmission is usually relatively slow	Transmission is very rapid
Hormones travel to all parts of the body, but only target organs respond	Nerve impulses travel to specific parts of the body
Effects are widespread	Effects are localised
Response is slow	Response is rapid
Response is often long lasting	Response is short lived
Effect may be permanent and irreversible	Effect is temporary and reversible

Those organisms that survive have a greater chance of raising offspring and of passing their **alleles** to the next generation. There is always, therefore, a **selection** pressure favouring organisms with better responses.

Stimuli are received by **receptors** and the response is carried out by **effectors**. Receptors and effectors are often some distance apart and a form of communication is therefore needed between them if the organism is to respond effectively. This communication may be relatively slow via the endocrine system, which uses hormones (Topic 14.7), or rapid via the nervous system, which uses nerve impulses. Further differences between the endocrine and nervous system are given in Table 1.

As animal species became more complex and the number of receptors and effectors increased, it became more efficient to link each receptor and effector to a central control centre. This is the **central nervous system**, consisting of the brain and spinal cord. The inter-relationships of all these various components are shown in Figure 1.

The structure of neurones

Neurones (nerve cells) are specialised cells adapted to rapidly carry electrochemical changes called nerve impulses from one part of the body to another. Mammalian neurones are made up of:

- **A cell body** that contains a nucleus, mitochondria and large amounts of rough endoplasmic reticulum grouped to form **Nissl's granules**. These are associated with the production of proteins and **neurotransmitters**.
- **Dendrons** – small extensions of the cell body that sub-divide into smaller branched fibres called dendrites that carry nerve impulses towards the cell body.

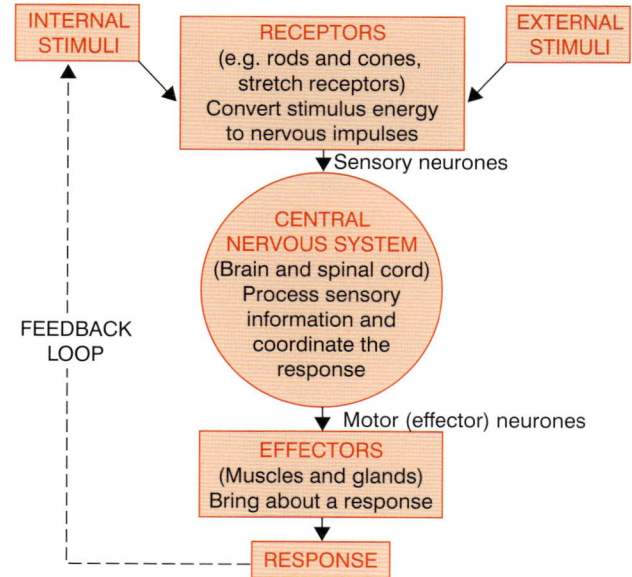

Figure 1 *Inter-relationships of components of the nervous system*

- **Axon** – a single long fibre that carries nerve impulses away from the cell body (the dendron of the sensory neurone is sometimes also termed a peripheral or afferent axon).

Many axons are surrounded by **Schwann cells**, which protect and provide insulation, act as phagocytes to remove cell debris and play a part in peripheral nerve regeneration. These Schwann cells wrap themselves around the axon many times, so that layers of their membranes build up around the axon. These membranes are rich in a lipid known as **myelin** and so form a covering to the axon called the **myelin sheath**. The space between adjacent Schwann cells lacks myelin, forming gaps 2–3 µm long, called **nodes of Ranvier**, which occur every 1–3 mm in humans. Neurones with a myelin sheath are called **myelinated neurones** and transmit nerve impulses faster than neurones without the myelin sheath (unmyelinated neurones) (Topic 15.6). The structure of a typical neurone is illustrated in Figure 3.

Neurones can be classified according to their function:

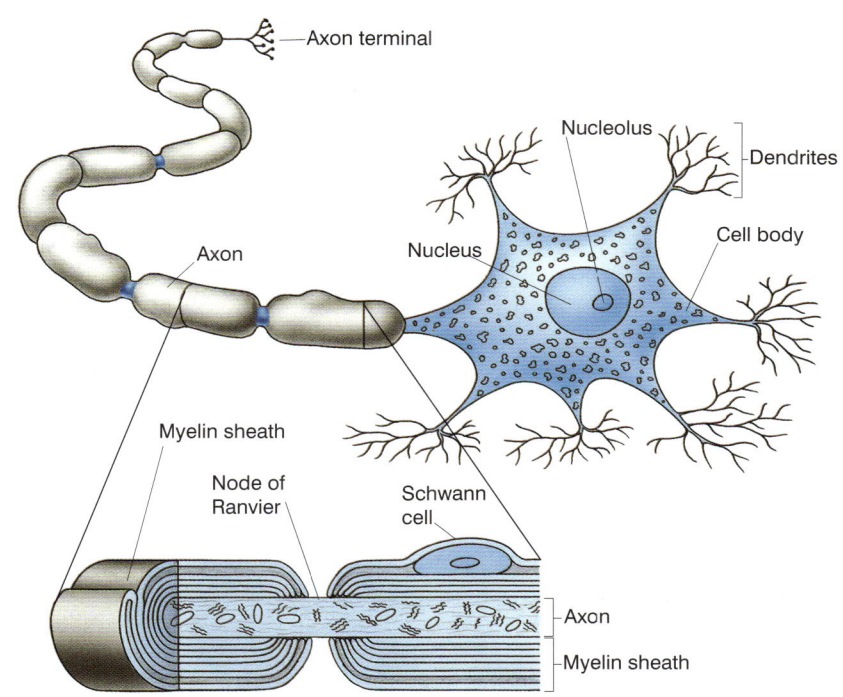

Figure 3 *Motor (effector) neurone*

- **Sensory neurones** transmit nerve impulses from a receptor to a relay or motor neurone. They have one afferent dendron that brings the impulse towards the cell body and one axon that carries it away from the cell body.
- **Relay neurones (intermediate neurones)** transmit impulses between neurones, e.g. from sensory to motor neurones. They have numerous short processes (extensions).
- **Motor neurones (effector neurones)** transmit nerve impulses from a relay or sensory neurone to an effector such as a gland or a muscle. They have a long axon and many short dendrites.

The three different types of neurone are illustrated in Figure 4.

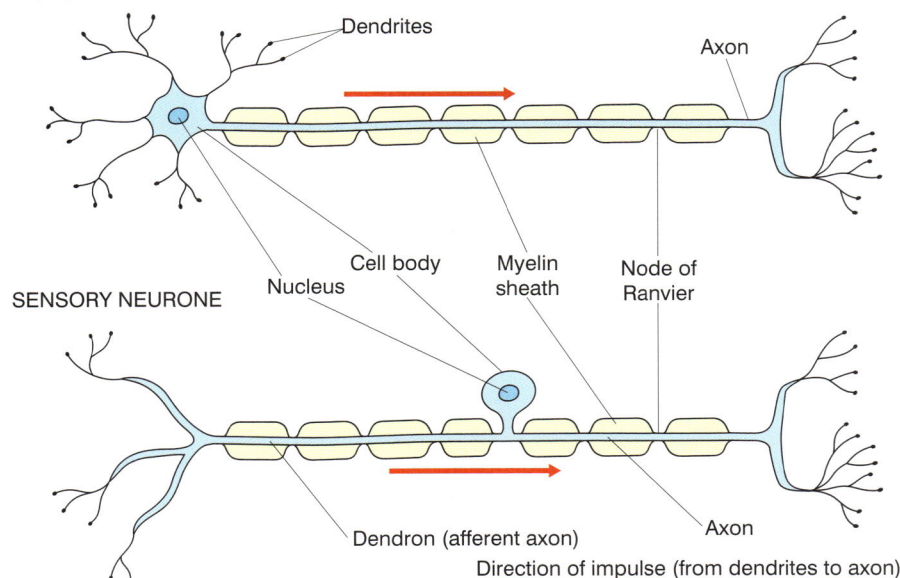

Figure 4 *Types of neurone*

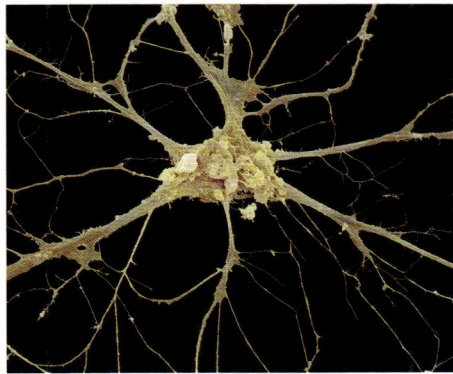

Figure 2 *Scanning electron micrograph of neurone with cell body at its centre and dendrites radiating from it*

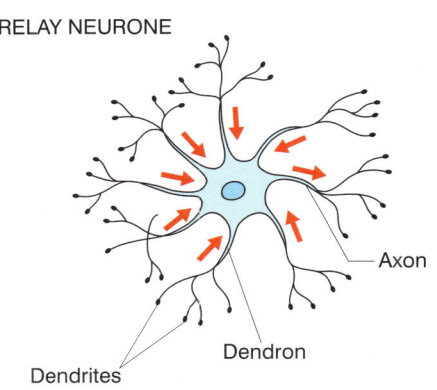

Sensory receptors

The central nervous system receives sensory information from its internal and external environment through a variety of sensory receptors that detect different types of stimuli. These receptors include sense organs and sensory receptor cells, often found within sense organs. Each type of sensory receptor cell detects a specific stimulus. **Sensory reception** is the function of these sense organs, whereas **sensory perception** involves making sense of the information from the receptors. This is largely a function of the brain. In this topic we shall look at how chemoreceptors in taste buds act as sensory receptor cells.

The role of chemoreceptor cells of human taste buds as sensory receptor cells

Taste buds are onion-shaped structures located in the epithelium of the tongue. Within each taste bud there are 50–100 chemoreceptor cells that detect the presence of chemicals associated with taste. Each one has microvilli that project up through an opening at the top of the taste bud, called the taste pore. The microvilli provide a large surface area to allow chemicals dissolved in saliva to contact the chemoreceptor cell. The structure of a taste bud and a chemoreceptor cell are shown in Figure 1. Chemoreceptor cells are thought to detect the chemicals associated with just four tastes – salt, sour, bitter and sweet – although savoury has been suggested as a fifth. As with all sensory receptors, taste chemoreceptor cells:

- **are specific to a single type of stimulus** – in this case to dissolved chemicals only
- **act as transducers** – they convert the energy of the stimulus into a receptor potential, which is a change in the potential difference that exists across the membrane of the chemoreceptor cell (Topic 15.4). Different types of chemoreceptor may have a slightly different sequence of events occurring when a stimulus is present. For some, the chemical binds to a specific membrane receptor, whereas for others the chemical enters the cell through specific membrane transport proteins. Whichever mechanism, the events lead to a change in the cell surface membrane to create the receptor potential. A receptor potential also leads to the release of a chemical transmitter from the end of the chemoreceptor that forms a synapse with a sensory neurone (Figure 1(b)). The stronger the stimulus, the greater the receptor potential and the more chemical transmitter is released (Topic 15.7).
- **produce a generator potential** – as a result of the release of the chemical transmitter (neurotransmitter), the receptor potential of the chemoreceptor cell may be enough to create a generator potential in the sensory neurone with which it synapses (in very close contact).
- **give an all-or-nothing response** – the greater the intensity of the stimulus, the greater the size of the generator potential. If the generator potential reaches or exceeds the set **threshold level**, an action potential is generated in the sensory neurone. Anything less than the threshold level, and no action potential is generated. Anything more than the threshold level, and the same action potential is generated, regardless of by how much the level is exceeded (Topic 15.4).
- **become adapted** – if exposed to a steady stimulus over a period of time, there is a slow decline in the frequency of generator potentials produced and so action potentials in the sensory neurone become less frequent and eventually stop. This is adaptation and prevents the nervous system becoming overloaded with unimportant information.

a Taste bud

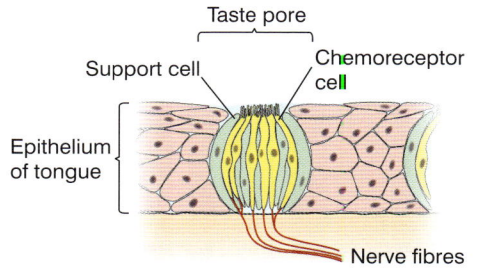

b Single chemoreceptor cell

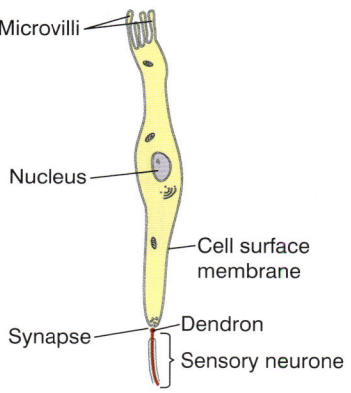

Figure 1 Structure of a taste bud and chemoreceptor cell

Rods and cones as light receptors

Rods and cones are **photoreceptor** cells found in the retina of the mammalian eye. The structure of a single rod cell is illustrated in Figure 2. Both rods and cones are secondary receptors. There are around six million cones, often with their own separate sensory neurone, in each human eye. The rods are more numerous, with 120 million in each eye, but up to 150 of them may share a single sensory neurone. As they share sensory neurones they cannot resolve very well, i.e. they have low **visual acuity**. Rods cannot distinguish different wavelengths of light and therefore produce images only in black and white. Cones, by contrast, need much higher light intensities to respond, but have high visual acuity and detect colour.

Transduction in rod cells

Although a rod cell is used in this account, the basic mechanism of **transduction** is the same in rods and cones. Each rod cell possesses up to a thousand vesicles in its outer segment. These contain the photosensitive pigment called **rhodopsin**, which is made up of the protein **opsin** and a derivative of vitamin A, called **retinal**.

The process of transduction in the rod cell is as follows:

- Light reaching a rod cell changes one isomer of retinal into another.
- This causes the rhodopsin to split into opsin and retinal – a process called **bleaching**.
- The splitting causes a chain of reactions that make the cell surface membrane of the rod cell less permeable to sodium ions.
- As sodium ions cannot now easily diffuse back into the rod cell, but continue to be actively pumped out of it, they accumulate outside, making this positive relative to the inside.
- This redistribution of sodium creates **hyperpolarisation**, which acts as the generator potential.
- If the threshold level is reached or exceeded by this change, then an action potential will be generated in the sensory neurone, which is connected to the brain via the optic nerve.

- Mitochondria found in the inner segment of the rod cell, generate ATP, which provides the energy necessary to recombine retinal and opsin into rhodopsin.

The process in a cone cell is very similar except that the pigment here is **iodopsin**. This is less sensitive to light and so a greater light intensity is required for it to breakdown and so create an action potential in the sensory neurone.

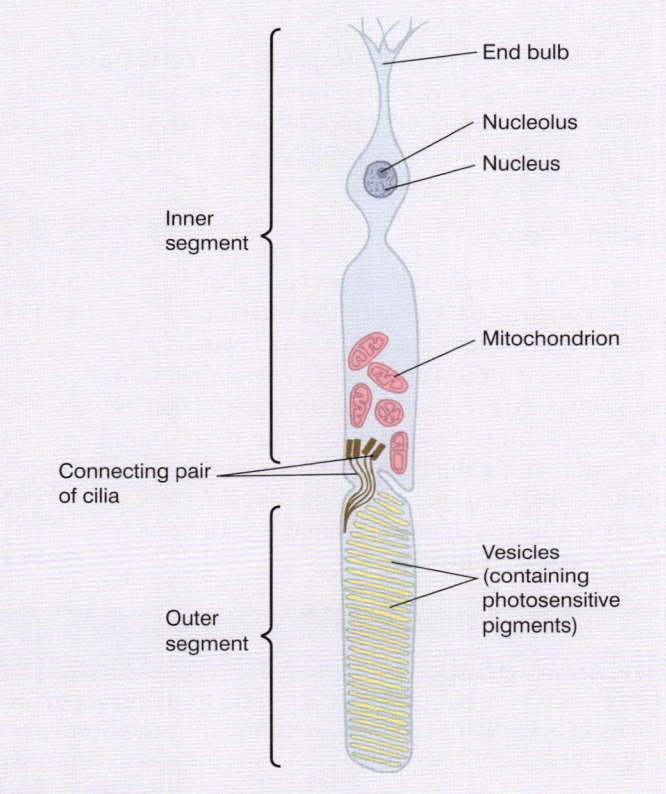

Figure 2 Structure of a single rod cell

SUMMARY TEST 15.2

Receptors are cells and organs that respond to different stimuli. Making sense of the information provided by sensory receptors is called sensory **(1)**. Chemoreceptor cells of taste buds respond to the stimulus of **(2)**, converting the energy of the stimulus into a **(3)** potential – as such they act as **(4)**. In turn a **(5)** potential is set up in the attached associated sensory neurone. If this potential equals or exceeds a threshold level, then an **(6)** is produced, which is the same regardless of how much the threshold level is exceeded. This is known as an **(7)** response.

On these pages you will learn to:

- Describe the functions of sensory, relay and motor neurones in a reflex arc

The simplest type of nervous response is a **reflex arc**. Before considering how a spinal reflex works, it is helpful to understand how the millions of **neurones** in a mammalian body are organised, and to know the structure of the spinal cord.

The spinal cord

The spinal cord is a column of nervous tissue running along the back within the vertebral column for protection. There is a small canal, the spinal canal, at its centre. The central region, called **grey matter**, comprises neurone cell bodies, synapses and unmyelinated relay neurones. Around the grey matter are many myelinated neurones running along the spinal cord. The **myelin** gives this region a lighter appearance and it is therefore known as **white matter**.

Coming out at intervals along the spinal cord are 31 pairs of nerves (in humans). Each one divides into two soon after leaving the spinal cord. The upper division (nearest the back) is called the **dorsal root**, and contains **sensory neurones**, while the lower division is called the **ventral root**, and contains **motor neurones**. The cell bodies of sensory neurones occur within the dorsal root, forming a swelling called the **dorsal root ganglion**.

The structure of the spinal cord is shown in Figure 1.

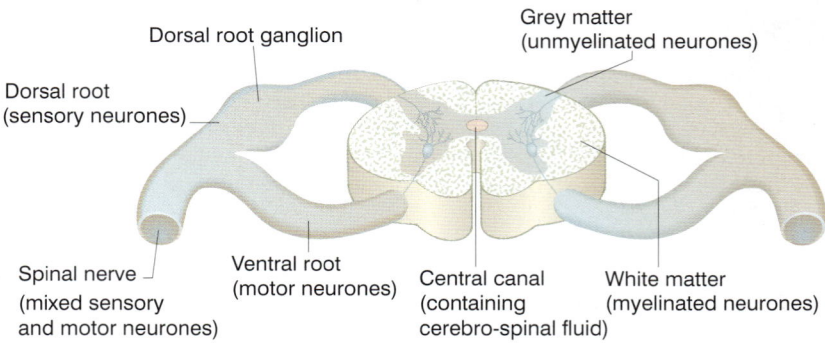

Figure 1 *TS through the spinal cord*

EXTENSION

Nervous organisation

The nervous system has two major divisions: the **central nervous system** (CNS), which is made up of the brain and spinal cord, and the **peripheral nervous system** (PNS), which is made up of pairs of nerves that originate from either the brain or the spinal cord.

The peripheral nervous system is divided into:

- **The sensory (afferent) nervous system**, which carries nerve impulses towards the central nervous system.

- **The motor (efferent) nervous system**, which carries nerve impulses away from the central nervous system.

The motor nervous system can be further sub-divided into:

- **The somatic nervous system**, which carries nerve impulses to skeletal muscles and is under voluntary control.

- **The autonomic nervous system**, which carries nerve impulses to glands, **smooth muscle** and **cardiac muscle** and is not under voluntary control, i.e. it is involuntary.

A summary of nervous organisation is given in Figure 2 and the way its components interact is shown in Figure 3.

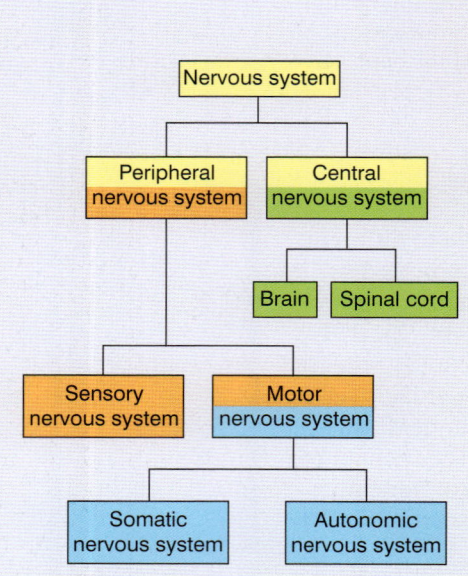

Figure 2 *Nervous organisation*

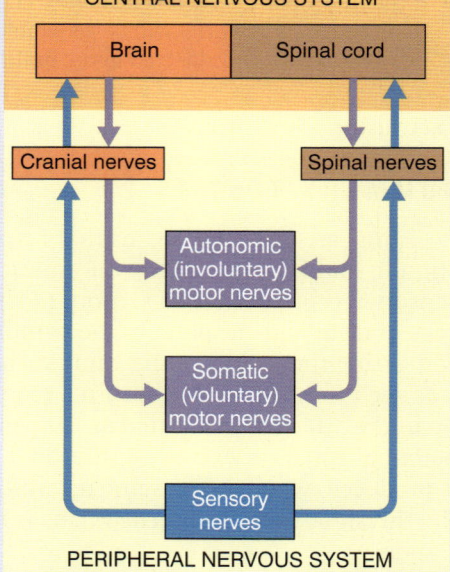

Figure 3 *Interaction between components of the nervous system*

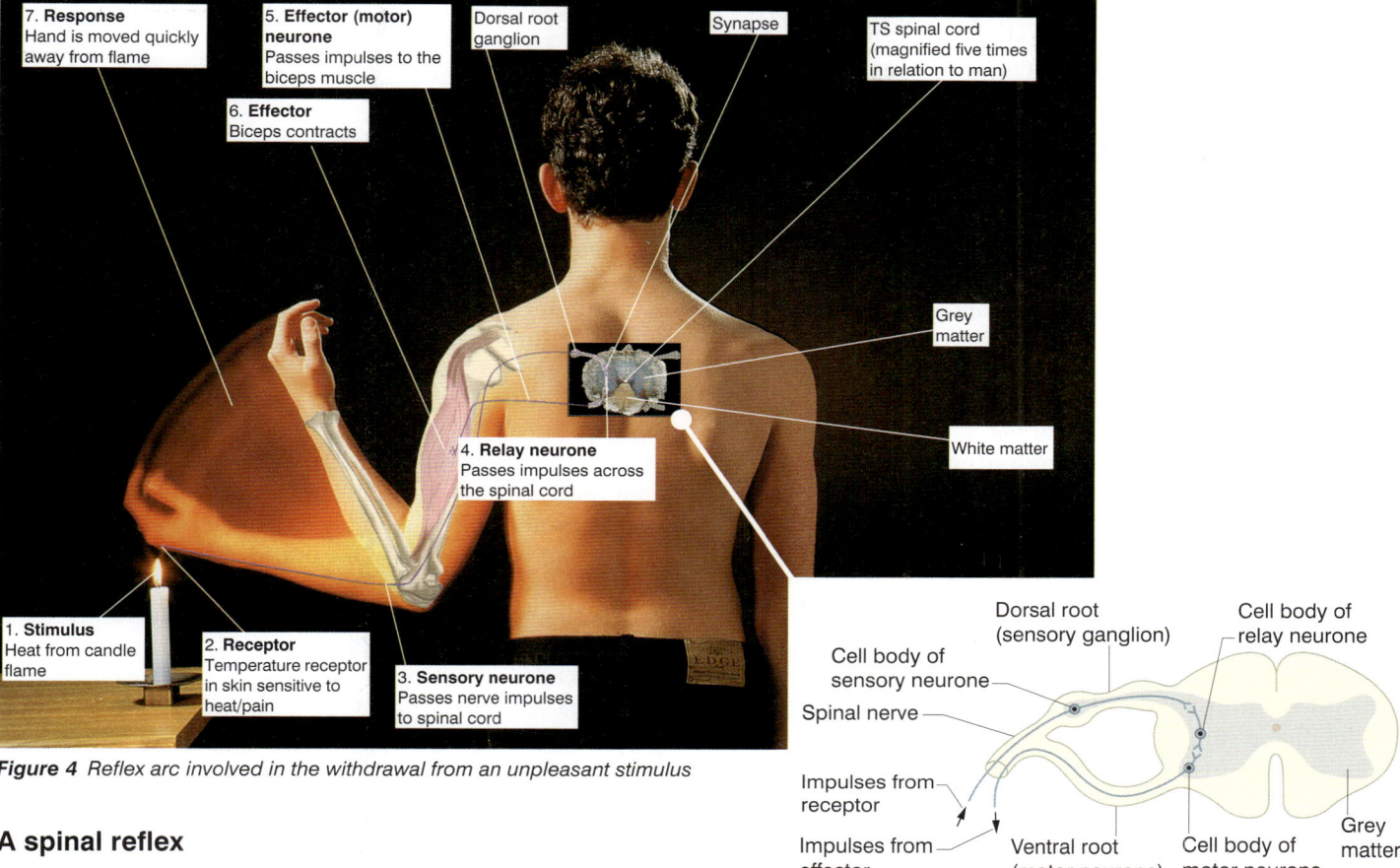

Figure 4 Reflex arc involved in the withdrawal from an unpleasant stimulus

Figure 5 Detail of reflex arc in the spinal cord

A spinal reflex

An involuntary response that follows a sensory stimulus is called a **reflex**. The pathway of neurones involved in a reflex is known as a **reflex arc**. The simplest forms of reflex arc, such as the knee jerk reflex, involve only a sensory and a motor neurone. More complex ones, like the withdrawal reflex, also involve a relay neurone. If a reflex involves the spinal cord but not the brain it is known as a **spinal reflex**. The main stages of a spinal reflex arc such as withdrawing the hand from a hot object are:

- **stimulus** – heat from the hot object
- **receptor** – temperature receptors in the skin of the back of the hand. If the threshold value of the temperature receptor is exceeded, a **generator potential** is established
- **sensory neurone** – the generator potential leads to an **action potential** passing along the sensory neurone to the spinal cord
- **relay (intermediate) neurone** – links the sensory neurone via **synapses** to the motor neurone within the grey matter of the spinal cord
- **motor (effector) neurone** – carries an action potential away from the spinal cord to the biceps muscle in the forearm
- **effector** – the biceps muscle of the forearm is stimulated to contract
- **response** – the hand is raised away from the hot object.

These events are shown in Figure 4.

Adaptive value of reflex arcs

Any action that aids survival is said to have an **adaptive value**. Reflexes are involuntary – the actions they control do not need to be 'considered', because there is only one obvious course of action, e.g. remove the hand from the hot object. The adaptive value of reflex actions include:

- Being involuntary, they do not need the decision-making powers of the brain, leaving it free to carry out more complex responses. In this way the brain is not overloaded with situations in which the response is always the same. Some impulses are sent at the same time to the brain, so that it is informed of what is happening and can over-ride (prevent) the reflex if necessary.
- They protect the body from dangerous stimuli. They are effective from birth as they do not have to be learned.
- They are fast, because the neurone pathway is short with very few, typically one or two, synapses (which are the slowest link in a neurone pathway). This is important in withdrawal reflexes.

On these pages you will learn to:

- Describe and explain the transmission of an action potential and its initiation from a resting potential

A nerve impulse may be defined as a self-propagating wave of electrical disturbance that travels along the surface of the axon membrane. It is not, however, an electrical current, but a temporary reversal of the electrical potential difference across the axon membrane. This reversal is between a state called the **resting potential** and another called the **action potential**.

Resting potential

The movement of sodium ions (Na^+) and potassium ions (K^+) across the axon membrane is controlled in a number of ways:

- The **phospholipid** bilayer of the axon cell surface membrane is impermeable to sodium ions (Na^+) and potassium ions (K^+).
- The channels are specific for either sodium or potassium ions. One type of channel, termed a voltage-gated channel, can be opened to allow the specific ion across or closed to prevent movement of that ion. A different type of channel remains open and allows diffusion of the ion across. There are more of these channels, sometimes termed 'leak' channels, for potassium ions than for sodium ions.
- Some intrinsic proteins **actively transport** potassium ions into the axon and sodium ions out of it. This is called the **sodium–potassium pump (cation pump)**.

As a result of these various controls, the inside of an axon is negatively charged relative to the outside. This is known as the **resting potential** and is in the range 50–90 millivolts (mV), but is usually 65 mV. In this condition the axon is said to be **polarised.** To achieve this potential difference the following events occur:

- Sodium ions are actively transported **out of** the axon by sodium–potassium pumps (specialised carrier proteins).
- Potassium ions are actively transported **into** the axon by sodium–potassium pumps.
- The active transport of sodium ions is faster than that of potassium ions, so that three sodium ions move out for every two potassium ions that move in.
- Although both sodium and potassium ions are positively charged, the outward movement of sodium ions is greater than the inward movement of potassium ions. As a result, there are more sodium ions in the tissue fluid surrounding the axon than in the cytoplasm, and more potassium ions in the cytoplasm than in the tissue fluid. For each ion, this creates a chemical gradient.

- The sodium ions begin to diffuse back naturally into the axon while the potassium ions begin to diffuse back out of the axon.
- As there are more of the leak channel proteins for potassium ions, the result is that the axon membrane is 100 times more permeable to potassium ions, which therefore diffuse back out of the axon faster than the sodium ones diffuse back in. This further increases the potential difference between the negative inside and the positive outside of the axon.
- Apart from the chemical gradient that causes the movement of the potassium and sodium ions, there is also an electrical gradient. As more and more potassium ions diffuse out of the axon, so the outside becomes more and more positive. Further outward movement of potassium ions is therefore made difficult because, being positively charged, they are attracted back into the axon by its overall negative state and repelled from moving outwards by the overall positive state of the surrounding tissue fluid.
- The presence of large, negatively charged proteins within the cytoplasm of the axon contributes to this overall negative state.
- An equilibrium is established whereby there is no net movement of ions and which is a balance between the chemical and electrical gradients.

These events are summarised in Figure 1.

The action potential

The energy conversion that occurs when a stimulus is received by a receptor leads to a temporary reversal of the charges on the axon membrane (Topic 15.2). As a result, the negative charge of −65 mV inside the membrane becomes a positive charge of around +40 mV. This is known as the

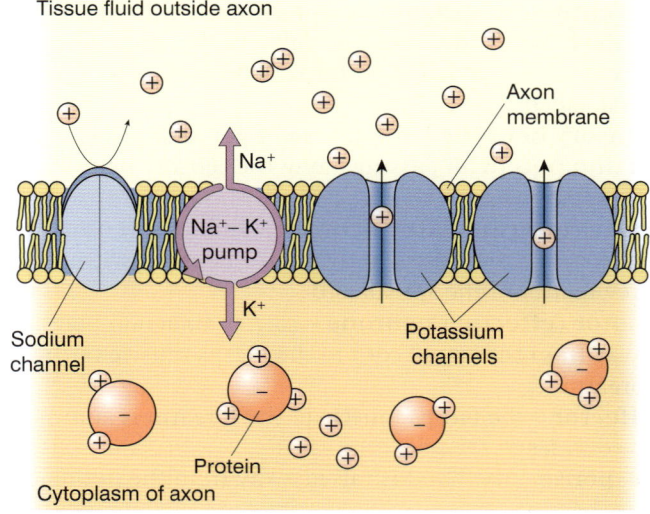

Figure 1 Distribution of ions at resting potential

action potential, and in this condition the membrane is said to be **depolarised**. This depolarisation involves the **voltage-gated** channels. The sequence of events is described below and the numbers relate to the stages shown in Figure 2.

1. At resting potential some potassium ion channels (leak channels) are open but the potassium voltage-gated and sodium voltage-gated channels are closed.
2. As a result of the stimulus some voltage-gated sodium channels in the axon membrane open and therefore sodium ions diffuse in through the channels along their electrochemical gradient. Being positively charged, they begin the reversal in the potential difference across the membrane and the membrane depolarises. Voltage-gated potassium channels remain closed.
3. As sodium ions enter, so more voltage-gated sodium channels open, causing an even greater influx of sodium ions. This is an example of positive feedback (Topic 14.1). An action potential will only occur if the membrane depolarises enough to allow the remaining voltage-gated sodium channels to open. This is known as the threshold potential (approximately −5 to −15 mV less negative than the resting potential).
4. Once the action potential of around +40 mV has been established (depolarisation has occurred), the voltage gates on sodium channels close (so further influx of sodium is prevented) and the voltage gates on the potassium channels begin to open.
5. With some voltage-gated potassium channels now open this causes the other voltage-gated channels to open and more potassium ions diffuse out, causing repolarisation of the axon membrane.
6. The outward movement of these potassium ions and the slight delay in the closing of the gates causes the temporary overshoot of the electrical gradient, with the inside of the axon being more negative (relative to the outside) than usual. This is called **hyperpolarisation**. The gates on the potassium channels now close and the activities of the sodium–potassium (cation) pumps cause sodium ions to be pumped out and potassium ions in, once again. The axon membrane returns to a resting potential and the axon is said to be **repolarised**.

The terms **action** potential and **resting** potential can be misleading, because the movement of sodium ions inwards during the action potential is purely due to diffusion – a passive process – and the resting potential is maintained by active transport – an active process. The term action potential simply means that the axon membrane is transmitting a nerve impulse, whereas resting potential means that it is not.

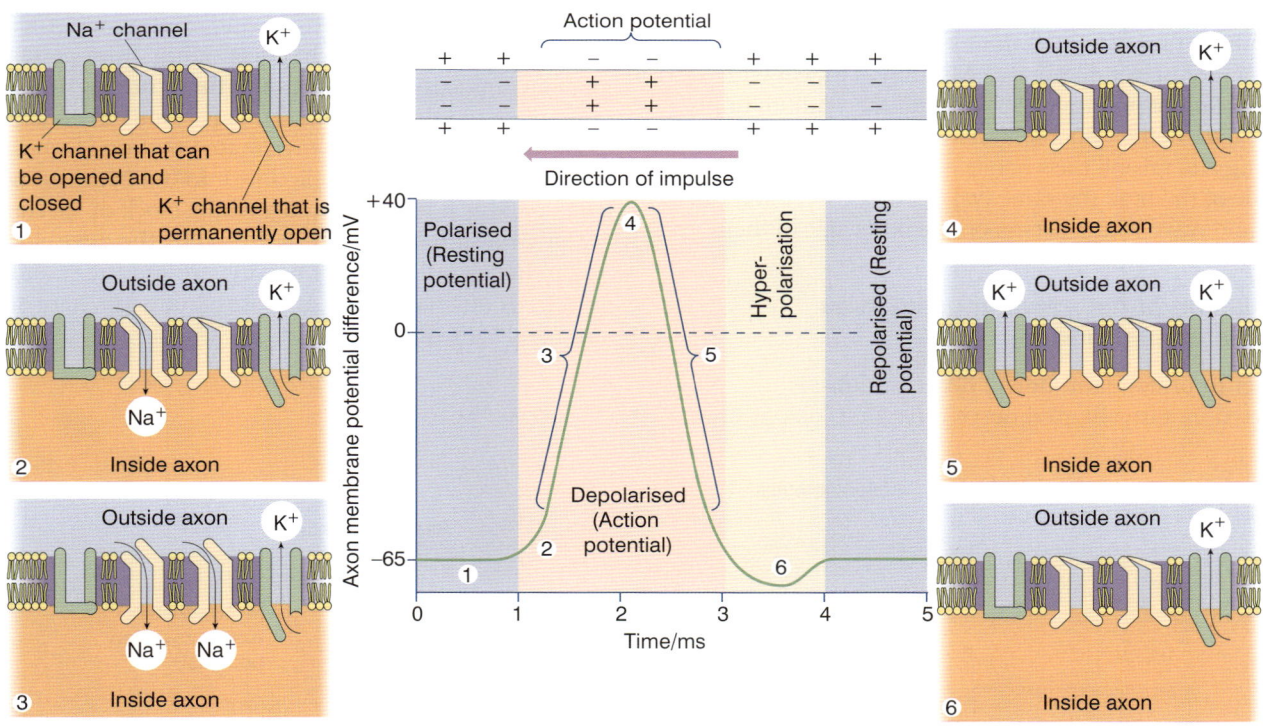

Figure 2 *The action potential*

SUMMARY TEST 15.4

A nerve impulse is the result of a temporary reversal of **(1)** potential difference across the axon membrane. At resting potential the potential difference is in the range **(2)** millivolts but is usually around **(3)** millivolts. In this condition the axon is said to be **(4)** with the outside being **(5)** charged relative to the inside. During an action potential, the charges are reversed with a potential difference of **(6)** millivolts and the membrane is said to be **(7)**.

On these pages you will learn to:

- Describe and explain the transmission of an action potential in a myelinated neurone

As one region of the axon produces an action potential and becomes depolarised, it acts as a stimulus for the **depolarisation** of the next region of the axon. This reversal of electrical charge is reproduced and action potentials are generated along each small region of the axon membrane. As one action potential triggers the next, the previous region of the membrane returns to its **resting potential**, i.e. it is repolarised. The size of the action potential remains the same from one end of the axon to the other. Strictly speaking, nothing physically 'moves' from place to place along the **axon** of the neurone, but rather the reversal of electrical charge is reproduced at different points along the axon membrane.

The process can be likened to the 'Mexican wave' that frequently takes place in a crowded stadium during a sporting event. Although the wave of people standing up and raising their hands (action potential) moves around the stadium, the people themselves do not move from seat to seat with the wave. They do not physically pass around the stadium until they return to their original seat. Rather, their individual action of standing and raising their hands is reproduced by the person to one side of them, in the same way that they were stimulated to stand and wave by the person on the other side of them.

Transmission of the nerve impulse in an unmyelinated neurone

It is easier to understand how a nerve impulse is transmitted in a myelinated nerve if we first look at how

it is transmitted in an unmyelinated one. The process is described and illustrated in Figure 2. This shows how local circuit currents are set up. The movement in of sodium ions during an action potential will lead to some passive movement of ions (current) within the axon. This is enough to begin depolarisation of the adjacent section of membrane. The positively charged ions have a tendency to repel each other and move towards the less positively charged region. A similar situation happens when the ions have moved out of the axon into the surrounding fluid.

Transmission of a nerve impulse in a myelinated neurone

In myelinated neurones, the fatty sheath of myelin around the axon acts as an electrical insulator, preventing action potentials from forming. At intervals of 1–3 mm there are breaks in this insulatory myelin, called **nodes of Ranvier**, where there is a high concentration of voltage-gated ion channels and sodium-potassium pumps (Topic 15.1). Action potentials can only occur at these points. The localised circuits therefore arise between adjacent nodes of Ranvier and the action potentials in effect 'jump' from node to node in a process known as **saltatory conduction** (Latin 'saltare' = to jump) (Figure 1). This results in an action potential passing along a myelinated neurone faster than an unmyelinated one (Topic 15.6). In our Mexican wave analogy, this is equivalent to a whole block of spectators leaping up simultaneously, followed by the next block and so on. Instead of the wave passing around the stadium in hundreds of small stages, it passes around in 20 or so large ones and so is more rapid.

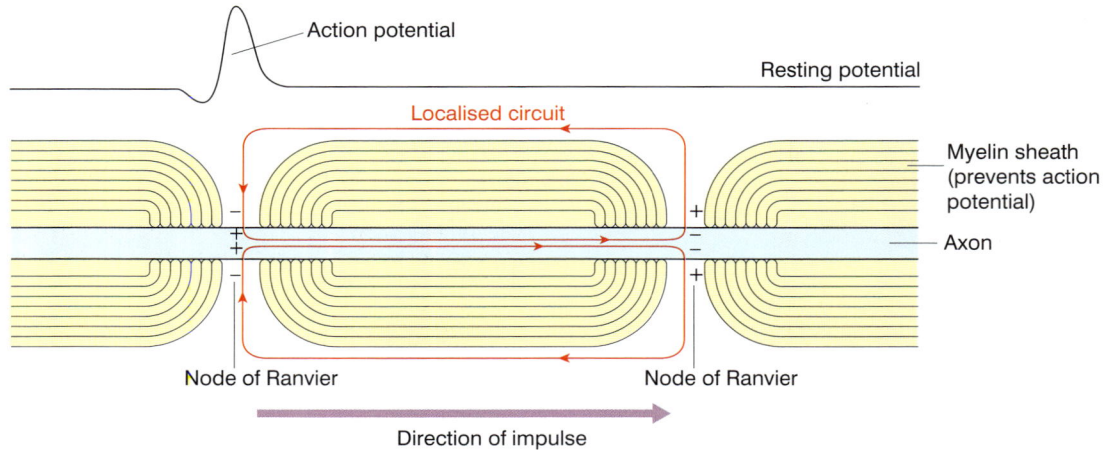

Figure 1 *Propagation of an impulse along a myelinated neurone. Action potentials are produced only at nodes of Ranvier. Depolarisation therefore skips from node to node – saltatory conduction*

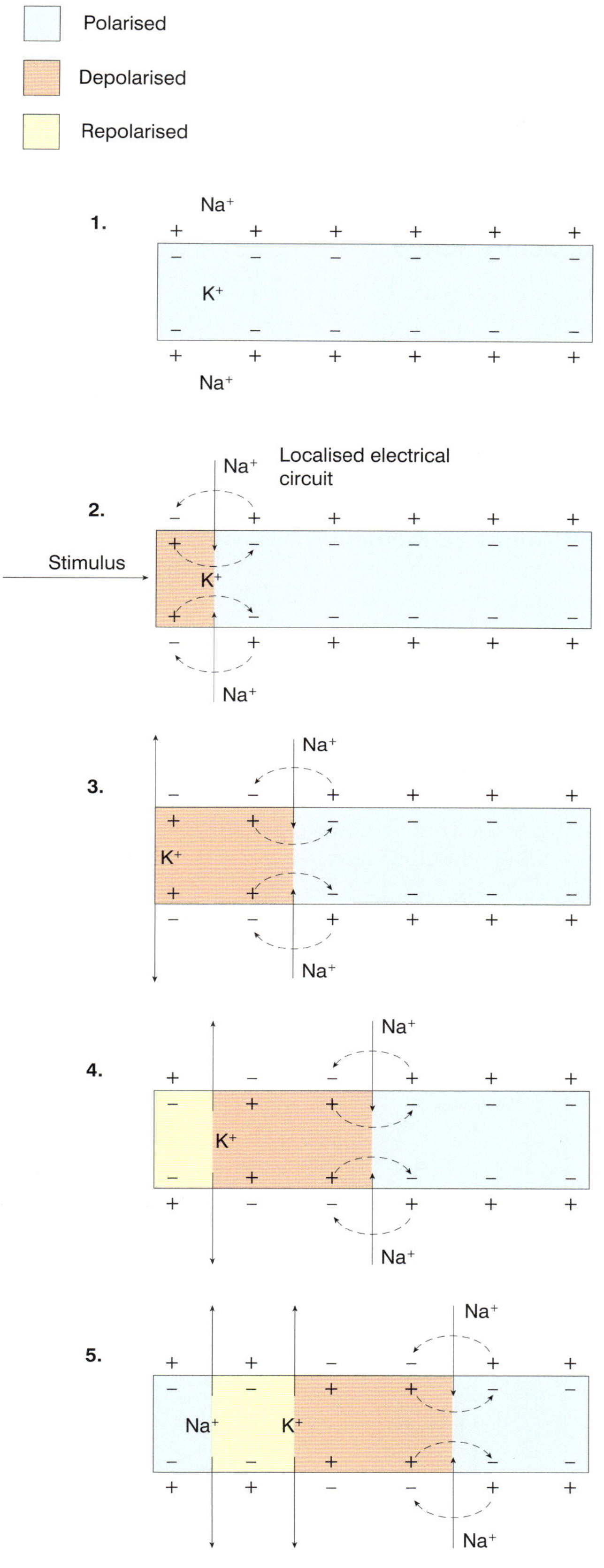

Polarised

Depolarised

Repolarised

1.

Na⁺

$$+ \quad + \quad + \quad + \quad + \quad +$$
$$- \quad - \quad - \quad - \quad - \quad -$$

K⁺

$$- \quad - \quad - \quad - \quad - \quad -$$
$$+ \quad + \quad + \quad + \quad + \quad +$$

Na⁺

1. At resting potential the concentration of sodium ions outside the axon membrane is high relative to the inside, whereas that of the potassium ions is high inside the membrane relative to the outside. The overall concentration of positive ions is, however, greater on the outside, making this positive compared with the inside. The axon membrane is polarised. In our Mexican wave analogy, this is equivalent to all the people in the whole stadium being seated, i.e. at rest.

2.

Na⁺   Localised electrical circuit

Stimulus →

K⁺

Na⁺

2. A stimulus causes the sodium voltage-gated channels to open leading to a sudden influx of sodium ions and hence a reversal of charge on the axon membrane making the inside positive. This is the action potential and the membrane is depolarised. In our analogy, a vertical line of people are stimulated into action and stand up and wave their arms.

3.

Na⁺

K⁺

Na⁺

3. The localised electrical circuits set up by the influx of sodium ions cause the opening of sodium voltage-gated channels a little further along the axon. The resulting influx of sodium ions in this region causes depolarisation. Behind this new region of depolarisation, the sodium voltage-gated channels close and the potassium ones open. Potassium ions begin to leave the axon along their electrochemical gradient. The sight of the person next to them standing and waving stimulates the person in the adjacent seat to stand and wave. A new vertical line of people stands and waves, while the original line of people begin to sit down again.

4.

Na⁺

K⁺

Na⁺

4. The action potential (depolarisation) is propagated in the same way further along the axon. The outward movement of the potassium ions has continued to the extent that the axon membrane behind the action potential has returned to its original charged state (positive outside, negative inside), i.e. it has been repolarised. The second line of people standing and waving stimulates the third line of people to do the same. Meanwhile, the first line have now returned to their original positions as they are now sitting down.

5.

Na⁺

Na⁺   K⁺

Na⁺

5. Repolarisation of the axon allows sodium ions to be actively transported out and potassium ions to be actively transported in, once again returning the axon to its resting potential in readiness for a new stimulus if it comes. The people who have just sat down settle back in their seats and readjust themselves ready to repeat the process should they be stimulated to do so again.

Figure 2 *Propagation of an impulse along an unmyelinated neurone*

On these pages you will learn to:

- Explain the importance of the myelin sheath (saltatory conduction) in determining the speed of nerve impulses and the refractory period in determining their frequency

Once an **action potential** has been set up, it is propagated from one end of the **neurone** to the other without any decrease in amplitude (size). In other words, the final action potential at the end of the **axon** is as large as the first action potential. A number of factors, however, affect the speed at which the action potential passes along the axon. Depending upon these factors, an action potential may travel as little as 0.5 m in a second or as much as 100 m in the same time. Table 1 gives some examples of transmission speeds in different axons.

Table 1 *Transmission speeds of different axons*

Axon	Myelin	Axon diameter / μm	Transmission speed / m s⁻¹
Human motor axon to leg muscle	Yes	20	100
Human sensory axon from skin pressure receptor	Yes	10	50
Squid giant axon	No	500	25
Human motor axon to internal organ	No	1	2

Factors affecting the transmission of action potentials

- **The myelin sheath** – we saw in Topic 15.5 that the myelin sheath acts as an electrical insulator, preventing an action potential forming in the part of the axon covered in myelin. It does, however, jump from **node of Ranvier** to node of Ranvier (**saltatory conduction**), speeding transmission from $0.5\,\mathrm{m\,s^{-1}}$ in a unmyelinated neurone to $100\,\mathrm{m\,s^{-1}}$ in a similar myelinated one.
- **The diameter of the axon** – the greater the diameter of an axon, the faster the speed of transmission. This is due to a combination of less leakage of ions from a large axon (leakage makes membrane potentials harder to maintain) and an increase in current flow within the axon.
- **Temperature** – affects the rate of diffusion of ions and therefore the higher the temperature the faster the nerve impulse. Above a certain temperature, the cell surface membrane proteins are **denatured** and impulses fail to be conducted at all. Temperature is clearly an important factor in response times in **ectothermic** animals, in which body temperature varies with the environment.

- **The refractory period** – we shall look at this in more detail next.

The refractory period

Once an action potential has been created in any region of an axon, there is a period afterwards when inward movement of sodium ions is prevented because the sodium **voltage-gated channels** are closed and temporarily inactivated. During this time it is not possible for a further action potential to be generated. This is known as the **refractory period**. The refractory period is made up of two portions:

- **The absolute refractory** period lasts for about 1 ms, during which no new impulses can be passed, however intense the stimulus. On Figure 1 this can be seen as a neurone excitability of zero.
- **The relative refractory** period lasts around 5 ms, during which a new impulse may be propagated provided the stimulus exceeds the normal threshold value. The degree to which it needs to exceed the threshold value becomes less over the period. On Figure 1 this is shown by an increase in neurone excitability. At normal resting excitability the voltage-gated channels are returned to their resting potential state.

The refractory period serves two purposes:

- The action potential cannot be propagated in the region that is refractory, i.e. it can only move in a forward direction. This prevents the action potential from spreading out in both directions, which it would otherwise do.

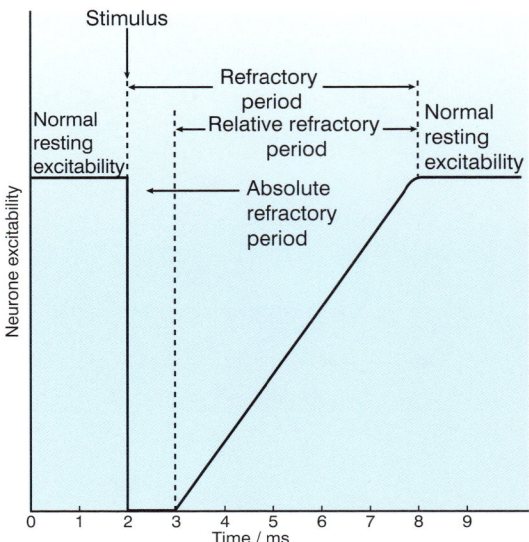

Figure 1 *Graph illustrating neurone excitability before and after a nerve impulse*

- Because a new action potential cannot be formed immediately behind the first one, it ensures that action potentials are separated from one another and therefore limits the number of action potentials that can pass along an axon in a given time, i.e. it determines their frequency.

All or nothing response

Nerve impulses are described as **all or nothing** responses. There is a certain level of stimulus, called the **threshold value**, which triggers an impulse. Below the threshold value no impulse is generated; above the threshold value an impulse is generated. The action potential, however, is the same regardless of how much the stimulus is above the threshold value. How then can an organism determine the size of a stimulus? This is achieved in two ways:

- by the number of impulses passing in a given time. This is known as **frequency coding**. The larger the stimulus, the more action potentials that are generated in a given time (Figure 3).
- by having different neurones with different threshold values. The brain interprets the number and type of neurones that pass impulses as a result of a given stimulus and thereby determines its size.

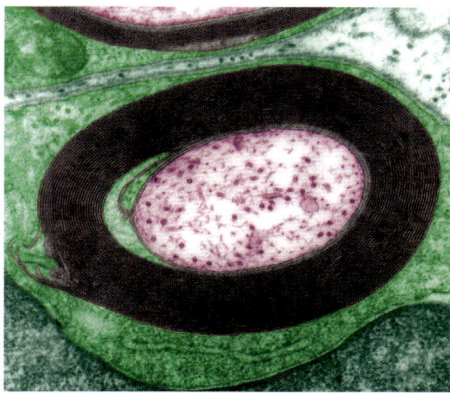

Figure 2 *Colourised transmission electron micrograph of a section through a myelinated neuron. A myelin sheath (black) surrounds the axon (purple). The myelin sheath is surrounded by the Schwann cell (green)*

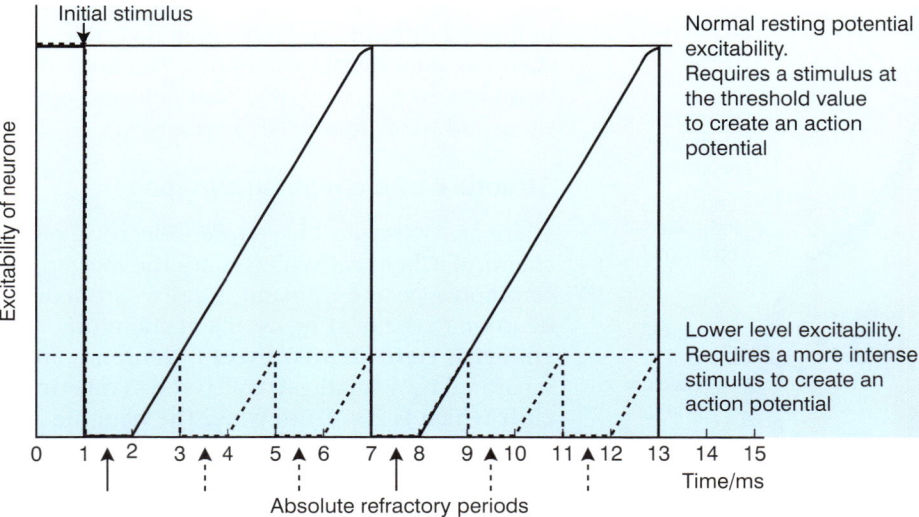

Where the stimulus is at the threshold value the excitability of the neurone must return to normal before a new action potential can be formed. In the time interval shown, this allows just two action potentials to pass, i.e. a low frequency of impulses. Where the stimulus exceeds the threshold value, a new action potential can be created before neurone excitability returns to normal. In the time interval shown, this allows six action potentials to pass, i.e. a high frequency of impulses.

Figure 3 *Determination of impulse frequency*

Structure and function of synapses

On these pages you will learn to:

- Describe the structure of a cholinergic synapse and explain how it functions, including the role of calcium ions
- Outline the roles of synapses in the nervous system in allowing transmission in one direction and in allowing connections between one neurone and many others

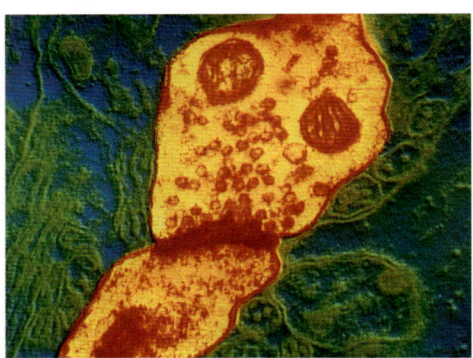

Figure 1 *Transmission electron micrograph of synapse. The synaptic cleft between the two neurones (centre) appears deep red. The cell above the cleft has many small vesicles (red-yellow spheres) containing neurotransmitter, whereas the two larger spheres above the vesicles are mitochondria*

A synapse is the point where the **axon** of one **neurone** connects with the **dendrite** of another or with an effector. They are important in linking different neurones together and therefore coordinating activities. The human brain has 10^{11} ($100\,000\,000\,000$) neurones and a thousand times more (10^{14}) synapses.

Structure of a chemical synapse

There are two types of synapse – electrical and chemical. Chemical synapses are the most common type and transmit impulses from one neurone to the next by means of chemicals known as **neurotransmitters**. Neurones are separated by a small gap called the **synaptic cleft** which is 20–30 nm wide. The neurone that releases the neurotransmitter is called the **presynaptic neurone** and it has a swollen portion of axon, the **synaptic knob**, at its end. This possesses many mitochondria, large quantities of endoplasmic reticulum and **synaptic vesicles** containing the neurotransmitter which, once released from the vesicles by **exocytosis**, diffuses across to the postsynaptic neurone, which possesses **receptor molecules** on its membrane to receive it. The structure of a chemical synapse is shown in Figure 2.

Functions of synapses

Synapses perform a number of important functions:

- **Transmit information between neurones** – synapses convey impulses from one neurone to the next and it is from this basic function that all the others are able to occur.

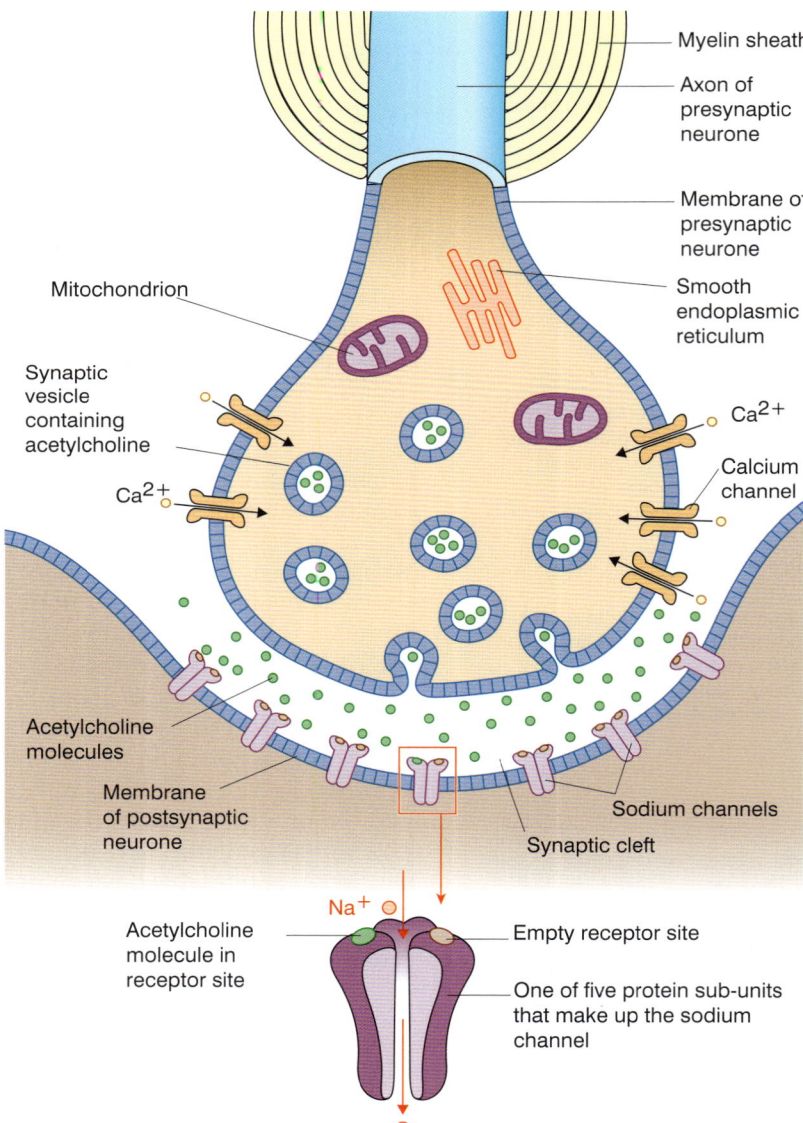

Myelin sheath

Axon of presynaptic neurone

Membrane of presynaptic neurone

Smooth endoplasmic reticulum

Mitochondrion

Synaptic vesicle containing acetylcholine

Ca^{2+}

Calcium channel

Ca^{2+}

Acetylcholine molecules

Membrane of postsynaptic neurone

Sodium channels

Synaptic cleft

Na^+

Acetylcholine molecule in receptor site

Empty receptor site

One of five protein sub-units that make up the sodium channel

Figure 2 *Structure of a chemical synapse*

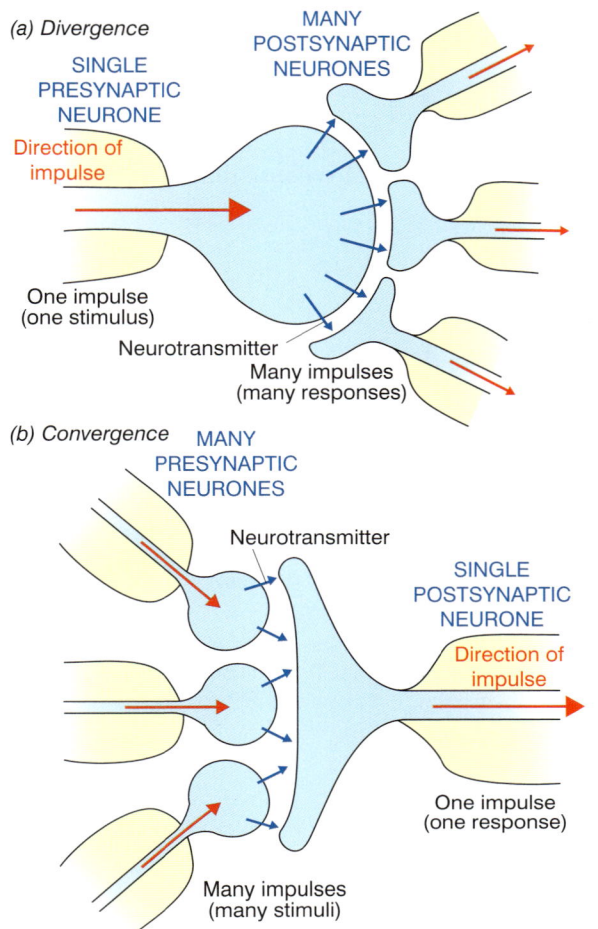

(a) Divergence

SINGLE PRESYNAPTIC NEURONE

MANY POSTSYNAPTIC NEURONES

Direction of impulse

One impulse (one stimulus)

Neurotransmitter

Many impulses (many responses)

(b) Convergence

MANY PRESYNAPTIC NEURONES

Neurotransmitter

SINGLE POSTSYNAPTIC NEURONE

Direction of impulse

One impulse (one response)

Many impulses (many stimuli)

Figure 3 *Divergence and convergence at synapses*

- **Are unidirectional** – synapses can only pass impulses in one direction. This is because only the presynaptic neurone has vesicles containing the neurotransmitter and only the postsynaptic membrane has receptors for the neurotransmitter. Any action potential arriving at the postsynaptic neurone would simply stop at this point.
- **Act as junctions** – allowing nerve impulses to diverge and converge. In divergence a single impulse along one neurone can be conveyed to a number of different neurones at a synapse (Figure 3a). This allows a single stimulus to create a number of simultaneous (all at the same time) responses. In convergence, a number of impulses can be combined into a single impulse (Figure 3b). This occurs in the retina of the eye, for example.

SUMMARY TEST 15.7

Synapses connect the axon of one neurone with the **(1)** of another. The two parts are separated by a gap called the synaptic **(2)** that is around **(3)** nm wide. One neurone, called the **(4)** neurone, releases a chemical messenger known generally as a **(5)**, of which acetylcholine is an example. These messengers are stored within small sacs called **(6)** and, once released, diffuse across to receptor molecules on the **(7)** neurone. Synapses perform a number of functions, all of which derive from their ability to transmit impulses between one neurone and the next. They are **(8)**, as they only allow impulses to pass in one direction across them. Synapses also act as **(9)**, because they allow nerve impulses to diverge and converge.

EXTENSION

Other functions of synapses

- **Filter out low level stimuli** such as the background noise of traffic or machinery. The stimulus produces low frequency impulses that cause the release of only small quantities of neurotransmitter at the synapse. This is insufficient to create a new impulse in the postsynaptic neurone and so there is no response. The absence of a response is rarely, if ever, harmful.

- **Summation** – low frequency impulses that do not produce enough neurotransmitter to trigger a new action potential in the postsynaptic neurone, can be made to do so by a process called **summation**. This needs a build-up of neurotransmitter in the synapse by one of two methods:

  spatial – where a number of different presynaptic neurones together release enough neurotransmitter to trigger a new action potential

  temporal – where a single presynaptic neurone releases neurotransmitter many times over a short period. If the total amount of neurotransmitter exceeds the **threshold value** of the postsynaptic neurone, then a new action potential is triggered.

- **Prevent overstimulation and fatigue** – where a stimulus is powerful and prolonged, the high frequency of impulses in the presynaptic neurone leads to the release of considerable amounts of neurotransmitter. In these circumstances the release of neurotransmitter stops, together with any response to the stimulus. The synapse is said to be **fatigued**. The purpose of such a response is to prevent overstimulation, which might otherwise damage an effector.

- **Involved in memory and learning** – it is thought that synapses have a role in the brain in allowing organisms to recall events and learn to recognise individuals.

- **Inhibition** – on the postsynaptic membrane of some synapses, the protein channels carrying chloride ions (Cl^-) can be made to open. This leads to an inward diffusion of chloride ions making the inside of the postsynaptic membrane even more negative than when it is at resting potential. This is called hyperpolarisation and makes it less likely that a new action potential will be created. For this reason these synapses are called inhibitory synapses..

On these pages you will learn to:

- Describe a cholinergic synapse and explain how it functions, including the role of calcium ions

Figure 1 shows transmission across a chemical synapse that involves the neurotransmitter acetylcholine.

1. The arrival of an action potential at the end of the presynaptic neurone causes voltage-gated calcium channels to open and calcium ions (Ca^{2+}) enter the synaptic knob.

2. The influx of calcium ions into the presynaptic neurone causes synaptic vesicles to fuse with the presynaptic membrane, so releasing acetylcholine by exocytosis into the synaptic cleft. Acetylcholine diffuses across the synaptic cleft to the postsynaptic neurone.

3. Acetylcholine molecules bind with receptor sites on two of the five protein sub-units that make up each sodium channel on the postsynaptic membrane. This causes the sodium channels to open, allowing sodium ions (Na^+) to diffuse in rapidly down a concentration gradient.

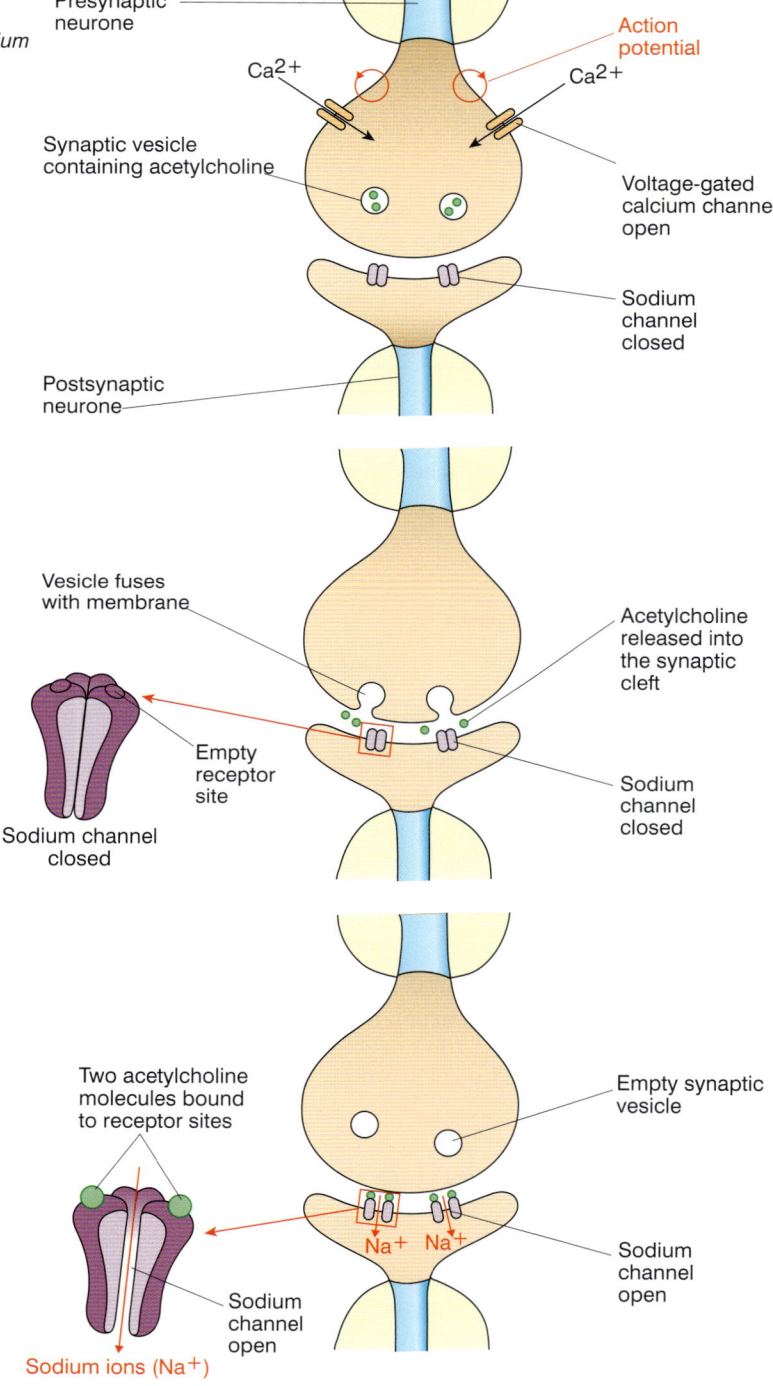

Figure 1 *Mechanism of transmission across a cholinergic synapse*

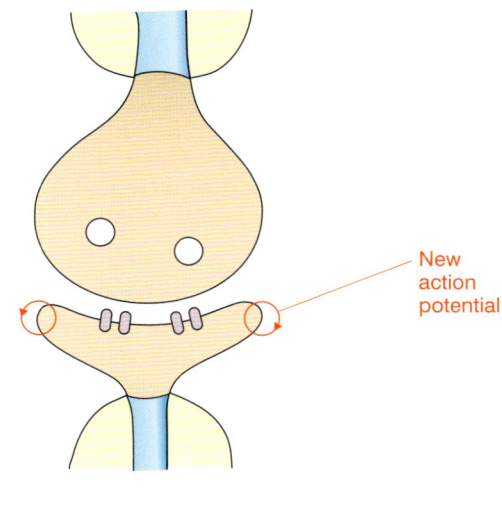

4. *The influx of sodium ions generates a new action potential in the postsynaptic neurone. This is called the* **excitatory postsynaptic potential (EPSP).**

New
action
potential

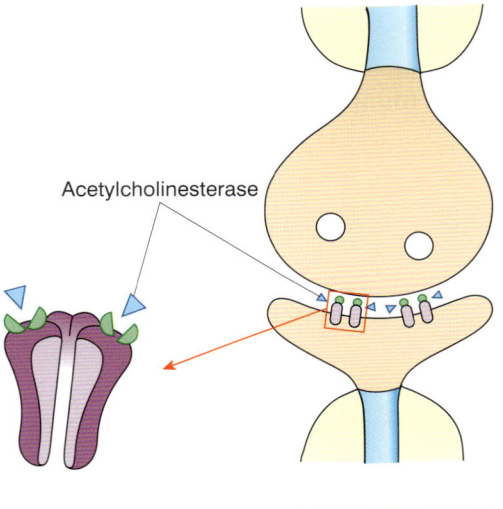

Acetylcholinesterase

5. *The enzyme acetylcholinesterase hydrolyses acetylcholine into choline and acetate, which diffuse back across the synaptic cleft into the presynaptic neurone (=* **recycling***).*

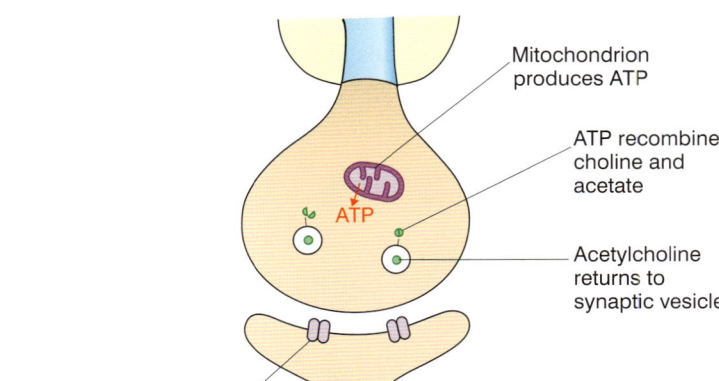

Mitochondrion
produces ATP

ATP recombines
choline and
acetate

Acetylcholine
returns to
synaptic vesicle

Sodium channel closed

6. *ATP released by mitochondria is used to recombine choline and acetate into acetylcholine. This is stored in synaptic vesicles for future use. Sodium channels close in the absence of acetylcholine in the receptor sites.*

SUMMARY TEST 15.8

When an action potential arrives at the end of the presynaptic neurone, **(1)** ions enter the synaptic knob. These ions cause **(2)** to fuse with the presynaptic membrane so releasing acetylcholine, by the process of **(3)**, into the **(4)**. Channels on the postsynaptic neurone have receptor sites to which acetylcholine binds, causing them to open and allowing **(5)** ions to move in by the process of **(6)**. The influx of these ions generates a new action potential known as the **(7)** potential.

Striated muscle

On these pages you will learn to:

- Describe the ultrastructure of striated muscle with particular reference to sarcomere structure

Muscles are effector organs that respond to nervous stimulation by contracting and so bring about movement. **Striated muscle** makes up the bulk of body muscle in vertebrates. It is attached to bone and acts under voluntary, conscious control.

A rope is made up of millions of separate threads. Each thread has very little individual strength and can easily be snapped. Yet grouped together in a rope, these threads can support a mass running into hundreds of tonnes. Individual muscles are made up of thousands of muscle fibres, bundled in groups, and running parallel to the length of the muscle. Each muscle fibre is a mulitnucleate (many nuclei) cell containing many parallel myofibrils (see Figure 1). Early in development, separate muscle cells fuse to give a single muscle fibre. This fusion allows the muscle to contract effeciently and gives the muscle overall strength. Surrounding the myofibrils, is the muscle fibre cytoplasm, known as **sarcoplasm**, containing a large concentration of mitochondria and endoplasmis reticulum.

Microscopic structure of striated muscle

We can see from Figure 1 that each muscle fibre is made up of myofibrils. Myofibrils are made up of two types of protein filament:

- **actin**, which is thinner and consists of two strands twisted around one another
- **myosin**, which is thicker and consists of long rod-shaped fibres with bulbous heads that project to the side.

The arrangement of these filaments is shown in Figure 2.

Myofibrils appear striped due to their alternating light-coloured and dark-coloured bands. The light bands are called **isotropic bands (I-bands).** They appear lighter because the actin and myosin filaments do not overlap in this region. The dark bands are called **anisotropic bands (A-bands).** They appear darker because the actin and myosin filaments overlap in this region.

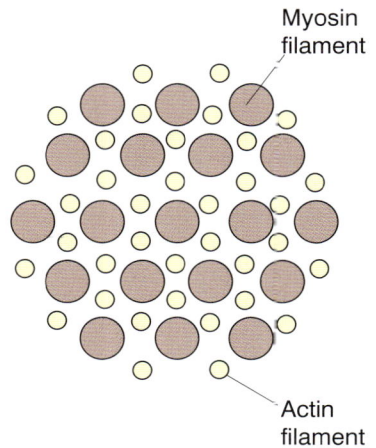

Figure 2 *Transverse section through part of a myofibril, showing the arrangement of actin and myosin filaments*

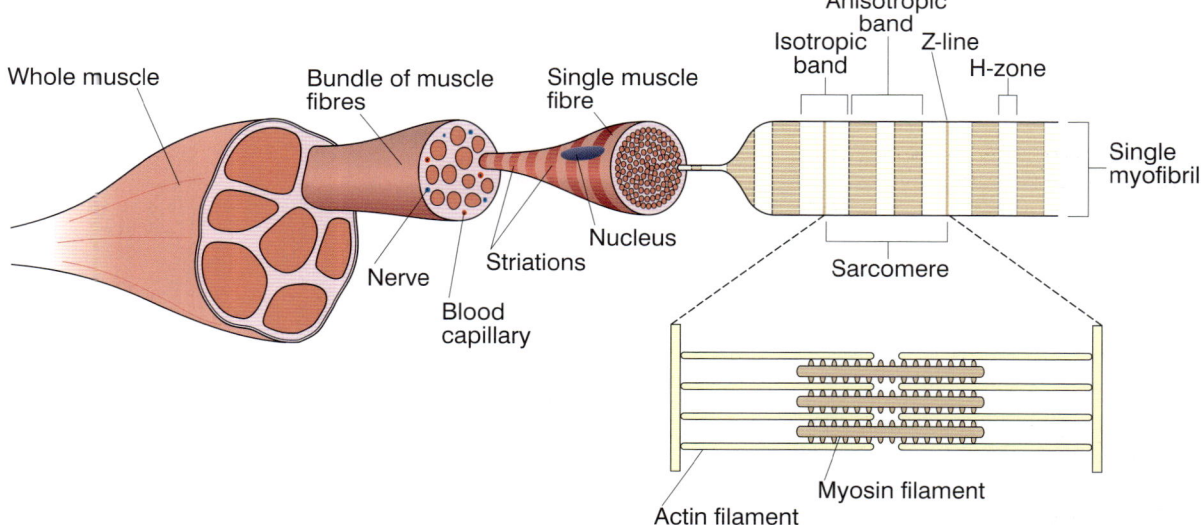

Figure 1 *The gross and microscopic structure of striated muscle*

At the centre of each anisotropic band is a lighter coloured region called the **H-zone**. At the centre of each isotropic band is a line called the **Z-line**. The distance between adjacent Z-lines is called a **sarcomere** (Figure 1). When a muscle contracts, these sarcomeres shorten and the pattern of light and dark bands change.

Two other important proteins are found in muscle:

- **tropomyosin**, which forms a fibrous strand around the actin filament
- **troponin**, a globular protein involved in muscle contraction.

The neuromuscular junction

The neuromuscular junction is the point at which a motor neurone meets a striated muscle fibre. There are many such junctions along the muscle. If there were only one junction of this type it would take time for a wave of contraction to travel across the muscle – in which case, not all the fibres would contract simultaneously and the movement would be slow. As rapid muscle contraction is frequently essential for survival, there are many neuromuscular junctions spread throughout the muscle. This is to ensure that contraction of a muscle is rapid and powerful when it is simultaneously (at the same time) stimulated by action potentials. All muscle fibres supplied by a single motor neurone act together as a single functional unit and are known as a **motor unit**. This arrangement gives control over the force that the muscle exerts. If only slight force is needed, only a few units are stimulated. If a greater force is required, a larger number of units are stimulated.

When a nerve impulse is received at the neuromuscular junction, the synaptic vesicles fuse with the presynaptic membrane and release acetylcholine. The acetylcholine diffuses to the postsynaptic membrane, altering its permeability to sodium ions, which enter rapidly, depolarising the membrane. The description of how this leads to the contraction of the muscle is given in Topic 15.10.

The acetylcholine is broken down by acetylcholinesterase to ensure that the muscle is not over-stimulated. The resulting choline and acetate diffuse back into the neurone, where they are recombined to form acetylcholine using energy provided by the mitochondria found there.

The structure of a neuromuscular junction is shown in Figure 4.

Figure 3 Transmission electron micrograph of striated muscle

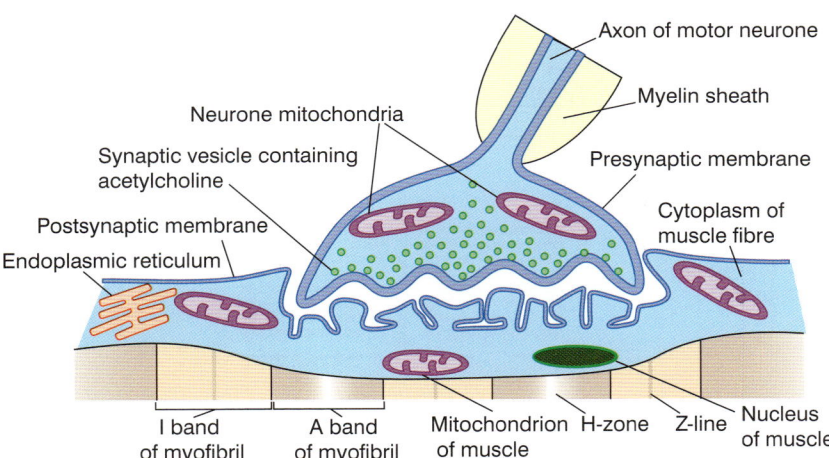

Figure 4 The neuromuscular junction

Figure 5 Photomicrograph of a neuromuscular junction

259

15.10

Muscle contraction

On these pages you will learn to:

- Describe the roles of neuromuscular junctions, transverse system tubules and sarcoplasmic reticulum in stimulating contraction in striated muscle
- Explain the sliding filament model of muscular contraction including the roles of troponin, tropomyosin, calcium ions and ATP

Having looked at the structure of striated muscle in Topic 15.9, let us turn our attention to how exactly the arrangement of the various proteins brings about contraction of the muscle fibre. The process involves the actin and myosin filaments sliding past one another and is therefore called the **sliding filament mechanism**.

Evidence for the sliding filament mechanism

In Topic 15.9 we saw that myofibrils appear darker in colour where actin and myosin filaments overlap and lighter where they do not. If the sliding filament mechanism is correct, then there will be more overlap of actin and myosin in a contracted muscle than in a relaxed one. Look at Figure 1. You will see that when a muscle contracts, the following changes occur to a **sarcomere**:

- the I-band becomes narrower
- the Z-lines move closer together (in other words, the sarcomere shortens)
- the H-zone becomes narrower.

The A-band remains the same width. As the width of this band is determined by the length of the myosin filaments, it follows that the myosin filaments have not become shorter.

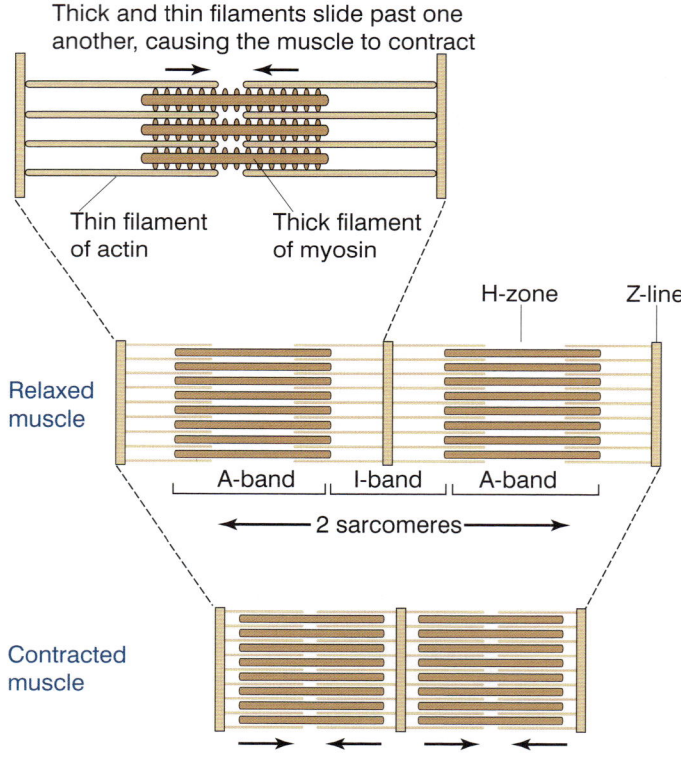

Figure 1 *Comparison of two sarcomeres in a relaxed and a contracted muscle*

Before we look at how the sliding filament mechanism works, let us take a closer look at the four main proteins involved in the process.

• **Myosin** (see Figure 2) each myosin filament is made up of a several hundred myosin molecules. Each myosin molecule is composed of several polypeptide chains and is made up of a globular 'head' section and a long 'tail' region.

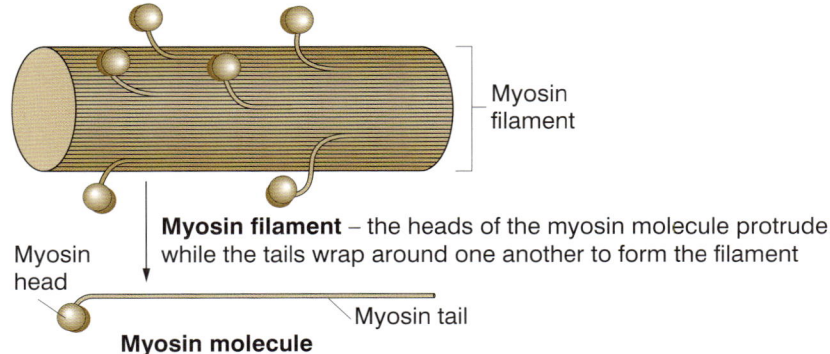

Myosin filament – the heads of the myosin molecule protrude, while the tails wrap around one another to form the filament

Figure 2 *Structure of the myosin filament*

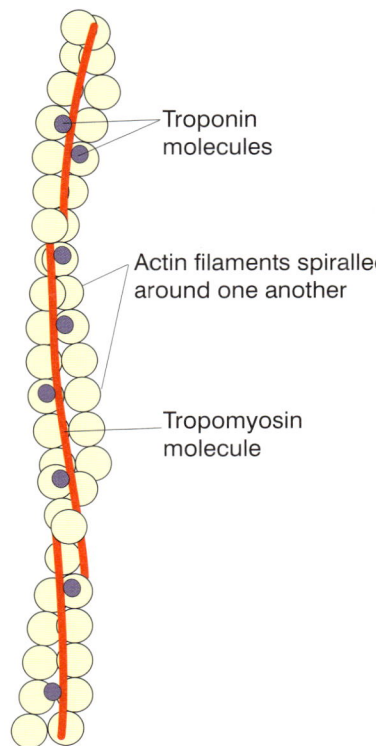

Figure 3 *The relationship of tropomyosin and troponin to an actin filament*

• **Actin** is a globular protein whose molecules are arranged into long chains that are twisted around one another to form a helical strand.
• **Tropomyosin forms** long thin threads that are wound around actin filaments.
• **Troponin** is a globular protein attached to tropomyosin to form a troponin–tropomyosin complex.

The arrangement of the molecules of actin and tropomyosin are shown in Figure 3.

The sliding filament mechanism of muscle contraction

The hypothesis that actin and myosin filaments slide past one another during muscle contraction is supported by the changes in the band pattern on myofibrils. The next question that scientists had to answer was: by what mechanism do the filaments slide past one another? Clues to the answer lie in the shape of the various proteins involved.

To summarise: the bulbous heads of the myosin filaments form cross bridges with actin filaments. They do this by attaching themselves to binding sites on the actin filaments, and then flexing (bending) in unison, pulling the actin filaments along the myosin filaments. They then become detached and, using ATP as a source of energy, return to their original angle and reattach themselves further along the actin filament. This process is repeated up to 100 times a second. The procedure is similar to the way a ratchet operates.

The following account describes the sliding filament of muscle contraction in detail. The process is continuous but for ease of understanding has been divided into muscle stimulation, contraction and relaxation.

Remember

The action of the myosin heads is similar to the rowing action of rowers in a boat. The oars (myosin heads) are dipped into the water, flexed as the rowers pull on them, removed from the water and then dipped back into the water further along. The rowers work in unison and the boat and water move relative to one another.

Summary test 15.10

According to the sliding filament theory of muscle contraction, when a muscle contracts the (**1**) move closer together and the (**2**) and (**3**) become narrower. For contraction to occur an (**4**) must reach the (**5**), where it travels deep into a network of tubules called (**6**) tubules. These tubules are connected to the endoplasmic reticulum of the muscle, which is called the (**7**). Some of the calcium ions from the endoplasmic reticulum attach to a protein molecule called (**8**), which changes shape and causes the displacement of a molecule called (**9**) from the actin filament. This displacement allows the head of a (**10**) molecule to attach to the actin filament and to change the angle of its head and so pull the actin filament along. The molecule detaches from the actin filament when a molecule of (**11**) attaches to it.

Muscle stimulation

- An **action potential** reaches many **neuromuscular junctions** simultaneously, causing calcium channels to open and calcium **ions** to move into the synaptic knob.
- The calcium ions cause synaptic vesicles to fuse with the presynaptic membrane and release their acetylcholine into the synaptic cleft.
- Acetylcholine diffuses across the synaptic cleft and binds with receptors on the postsynaptic membrane, causing it to depolarise.

Muscle contraction

This part of the process is illustrated in Figure 4.
- The action potential travels deep into the fibre through a network of tubules called **transverse system tubules** (T-tubules) that branch throughout the cytoplasm of the muscle (sarcoplasm).
- The tubules are in contact with the endoplasmic reticulum of the muscle, known as the **sarcoplasmic reticulum**, which have actively absorbed calcium ions from the sarcoplasm.
- The action potential opens the calcium channels on the endoplasmic reticulum and calcium ions flood into the muscle cytoplasm down a diffusion gradient.
- Some calcium ions attach to troponin and cause the troponin–tropomyosin complex to change shape.
- The change in shape of the troponin–tropomyosin complex causes the tropomyosin molecules that were blocking the binding sites on the actin filaments to be displaced (Figure 4, stages 1 and 2).
- The ADP molecule attached to the myosin heads means they are now in a state to bind to the actin filament and form a cross-bridge (Figure 4, stage 3). This means that the ATP that was previously attached has been hydrolysed and the energy released can be used for the next stage.
- Once attached to the actin filament, the myosin heads change their angle, pulling the actin filament along as they do so (power stroke) and releasing a molecule of ADP (Figure 4, stage 4).
- An ATP molecule attaches to each myosin head, causing it to become detached from the actin filament (Figure 4, stage 5).
- The calcium ions then activate the myosin molecule, which is also an ATPase that catalyses the hydolysis ATP to ADP (Figure 4, stage 6).
- The myosin head, once more with an attached ADP molecule, then reattaches itself further along the actin filament and the cycle is repeated as long as nervous stimulation of the muscle continues (Figure 4, stage 7).

Muscle relaxation

- When nervous stimulation ceases, calcium ions are actively transported back into the sarcoplasmic reticulum using energy from the hydrolysis of ATP.
- This reabsorption of the calcium ions allows tropomyosin to block the actin filament again.
- Myosin heads are now unable to bind to actin filaments and contraction stops, i.e. the muscle relaxes.

1 Tropomyosin molecule prevents myosin head from attaching to the binding site on the actin molecule.

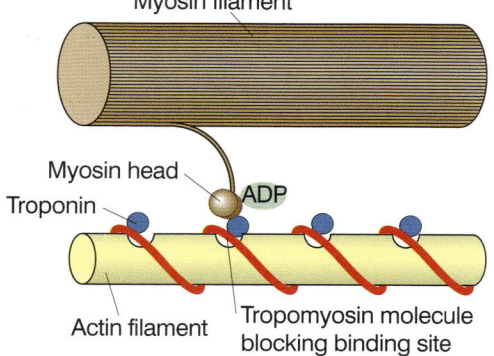

Myosin filament

Myosin head

Troponin

ADP

Actin filament

Tropomyosin molecule blocking binding site

2 Calcium ions released from the endoplasmic reticulum bind with troponin and cause the tropomyosin molecule to be displaced from the binding sites on the actin molecule.

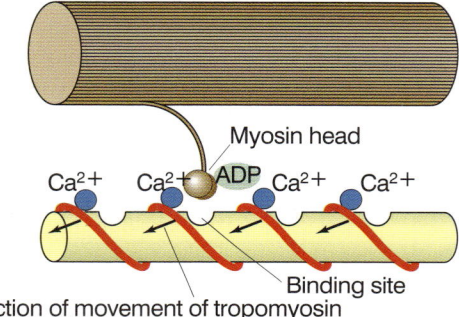

Myosin head

Ca^{2+} Ca^{2+} ADP Ca^{2+} Ca^{2+}

Binding site

Direction of movement of tropomyosin

3 Myosin head now attaches to the binding site on the actin filament. Energy from the hydrolysis of ATP is required.

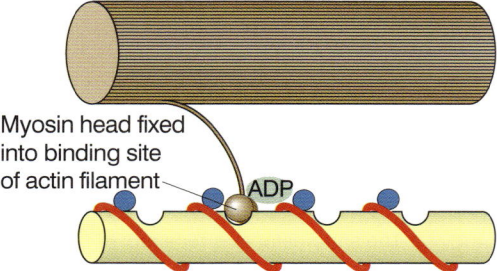

Myosin head fixed into binding site of actin filament

ADP

4 Head of myosin changes angle, moving the actin filament along as it does so. The ADP molecule is released.

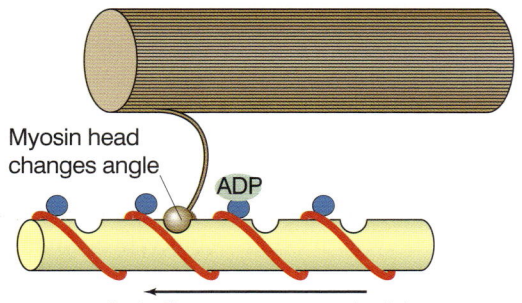

Myosin head changes angle

ADP

Actin filament moves to the left

5 ATP molecule fixes to myosin head, causing it to detach from the actin filament.

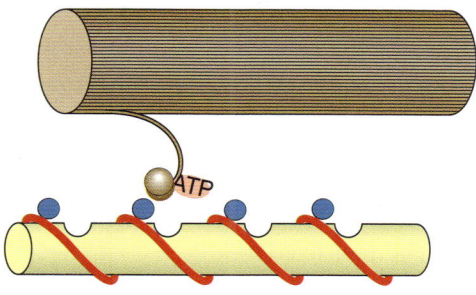

ATP

6 Hydrolysis of ATP to ADP by ATPase provides the energy for the myosin head to resume its normal position.

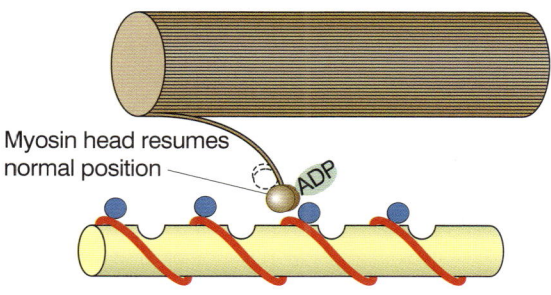

Myosin head resumes normal position

ADP

7 Head of myosin reattaches to a binding site further along the actin filament and the cycle is repeated.

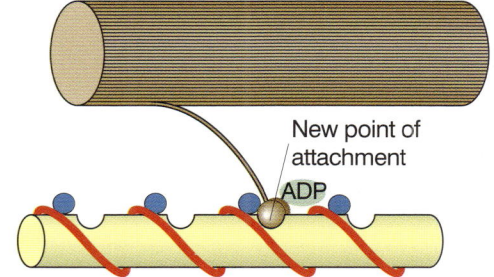

New point of attachment

ADP

Figure 4 *Sliding filament mechanism of muscle contraction (showing only one myosin head throughout)*

The menstrual cycle

On these pages you will learn to:

- Explain the roles of the hormones FSH, LH, oestrogen and progesterone in controlling changes in the ovary and uterus during the human menstrual cycle

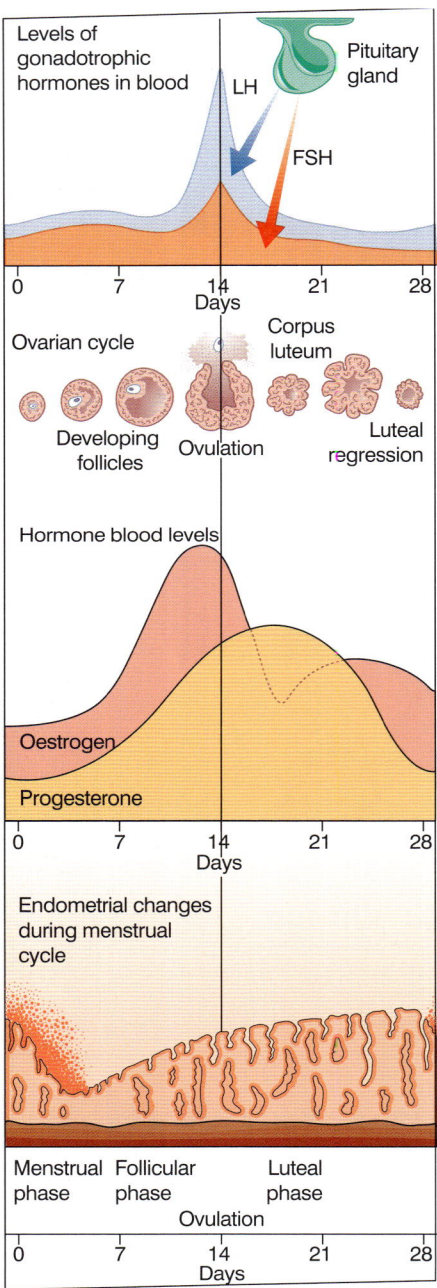

Figure 1 *Events of the menstrual cycle*

Changes during the menstrual cycle

The length of each menstrual cycle varies from individual to individual, although it is usually around 28 days. It is divided into four stages (Figure 2):

- **The menstrual phase** (days 1–5) occurs when the **endometrium** (uterus lining) is shed along with some blood (**menstruation**).
- **The follicular phase** (days 6–13) is when a **Graafian follicle** develops within the ovary and matures, ready for its secondary oocyte to be released. At the same time, the endometrium, which was lost in the menstrual phase, is repaired and thickened.
- **The ovulatory phase** (day 14) is the release of the secondary oocyte from the Graafian follicle in the ovary (**ovulation**).
- **The luteal phase** (days 15–28) is so called because the now empty Graafian follicle develops into a **corpus luteum** (yellow body). If fertilisation does not take place and/or the newly formed ball of cells (blastocyst) does not implant in the endometrium, then the corpus luteum degenerates and the endometrium breaks down, marking the start of the next menstrual cycle. The unfertilised secondary oocyte passes out of the body via the vagina.

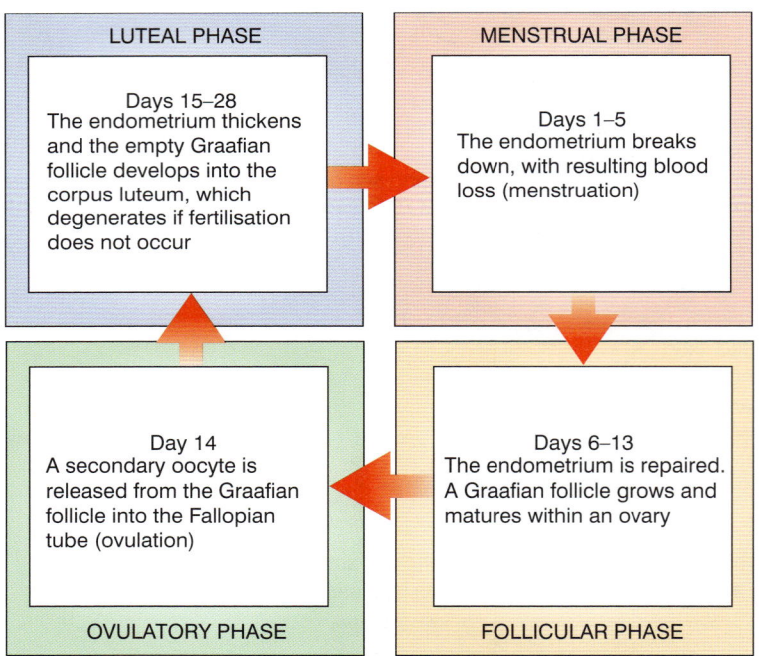

Figure 2 *Summary of changes during the menstrual cycle*

Hormonal control of the menstrual cycle

There are a number of hormones that control the menstrual cycle, each interacting with the other in a way that ensures a regular cycle of events. The cycle begins with production, by the hypothalamus, of a hormone (GnRH) that stimulates the anterior lobe of the pituitary gland to produce its hormonal secretions. The two hormones (glycoproteins) produced by the anterior lobe of the pituitary gland (which lies at the base of the brain) are known as **gonadotrophic stimulating hormones** and function as follows:

- **Follicle stimulating hormone (FSH)** causes development of the follicles in the ovary, and stimulates the ovaries to produce oestrogen.
- **Luteinising hormone(LH)**, together with FSH, stimulates follicle growth and causes ovulation to occur. LH also stimulates the development of the corpus luteum from the empty Graafian follicle and stimulates the corpus luteum to produce progesterone.

The remaining two hormones are produced by the ovaries. These steroid hormones are known as **ovarian hormones** and function as follows:

- **Oestrogen** causes the rebuilding of the endometrium of the uterus after menstruation, and stimulates the pituitary gland to produce LH. It also inhibits the production of FSH.
- **Progesterone** maintains the endometrium of the uterus in readiness to receive the blastocyst (young embryo), and inhibits the production of FSH from the pituitary gland.

In a simplified form the sequence of operation is:

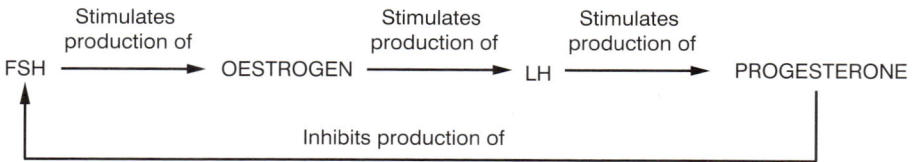

- Progesterone, at the end of the sequence, inhibits the FSH at the beginning.
- In the absence of FSH, production of oestrogen, LH and progesterone also stops.
- In the absence of progesterone, the inhibition of FSH ceases.
- FSH production resumes and the cycle repeats itself.

This alternate switching on and off of these hormones is responsible for the regular sequence of events in the menstrual cycle (Figure 3).

SUMMARY TEST 15.11

The menstrual cycle in humans lasts around **(1)** days. The cycle begins with the initial discharge of blood called **(2)** that lasts around **(3)** days. This is followed by the **(4)** phase, when a Graafian follicle develops in an ovary as a result of the influence of the hormone called **(5)**. This hormone is produced by the **(6)** gland when it is stimulated by **(7)** from the hypothalamus. The ovulatory phase usually occurs on day **(8)** of the cycle and involves the release of a **(9)** from the Graafian follicle, an event stimulated by the hormone known as **(10)**. The final phase is called the **(11)** phase during which the Graafian follicle develops into a **(12)** which degenerates if **(13)** does not occur, leading to the breakdown of the **(14)** – the lining of the uterus – and the start of the next cycle.

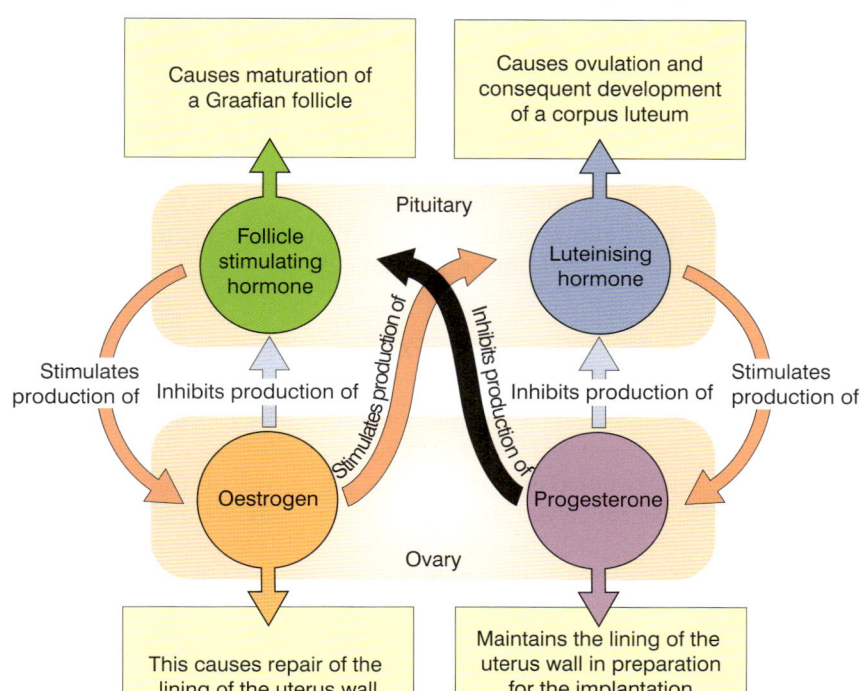

Figure 3 Hormone interaction in the menstrual cycle

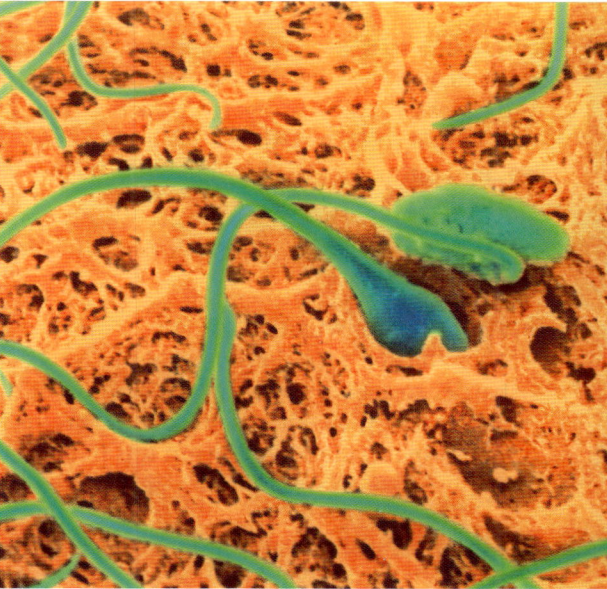

Figure 4 Scanning electron micrograph of sperm on the surface of a human secondary oocyte during fertilisation

15.12   Contraception and birth control

On these pages you will learn to:

- Outline the biological basis of contraceptive pills containing oestrogen and/or progesterone

Contraception

Contraception is the intentional prevention of conception by some means – natural or artificial. Conception is the formation of a viable zygote by the fertilisation of a secondary oocyte by a spermatozoan. It follows that any means by which this fertilisation is intentionally prevented is contraception. It takes a number of forms, including the contraceptive pill.

Birth control

Birth control is slightly different from contraception. Birth control is the voluntary limitation of the number of children conceived. In other words choosing if, and when, to have children. Clearly contraception plays a major role in birth control.

The contraceptive pill

The contraceptive pill has been around since the 1960s. There are a number of different types and they don't always come in pill form. As a reliable and tested male contraceptive pill is a few years away, we shall limit our discussion to those used by females. Contraceptive pills contain synthetic forms of one or both of the hormones progesterone and oestrogen. These hormones may be given as a pill taken by mouth (oral contraceptive), an injection, an implant under the skin or a vaginal ring.

By whatever method the contraceptive is delivered, they work by having each of the following effects to a greater or lesser degree:

- They prevent ovulation.
- They thicken secretions around the cervix and so form a barrier to sperm reaching the uterus.
- They make the endometrium of the uterus thinner, so that implantation of an ovum less likely.

There are two main types of oral contraceptive pill:

- **The mini pill,** which contains only one hormone – a synthetic form of progesterone (taken daily, with no breaks).
- **The combined pill,** which contains two hormones – synthetic forms of both progesterone and oestrogen.

The synthetic progesterone in both types of pill affects the hypothalamus by inhibiting the release of gonadotrophin releasing hormone (GnRH). This decreases the secretion of

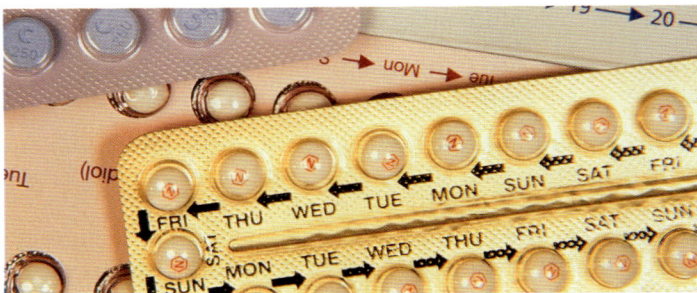

Figure 1 *A variety of contraceptive pills used by women. These pills contain synthetic female hormones that mimic the effects of oestrogen and progesterone in the body.*

both follicle stimulating hormone (FSH) and luteinising hormone (LH) from the anterior pituitary and so prevents the development of follicles in the ovary. Details of the way these hormones work can be found in Topic 15.11.

The inclusion of synthetic oestrogen in the combined pill also helps to prevent ovulation because, as we saw in Topic 15.11, oestrogen has a negative feedback effect by inhibiting FSH production. As FSH causes a Graafian follicle to mature, its inhibition means there are no mature secondary oocytes ready for release during ovulation. We also saw that oestrogen builds up the endometrial lining of the uterus. Its inclusion in the pill therefore helps to prevent menstrual bleeding during those days when the pill is taken (it is usually taken for 21 days followed by seven pill-free days).

The many different types of combined pill vary slightly in the concentration and proportions of each of the two hormones. The pill works in conjunction with the oestrogen and progesterone produced naturally by women. The levels of these naturally produced hormones vary from individual to individual. Having different types of pill allows women to select the one most appropriate to their natural hormone balance.

Combined emergency contraceptive pills (the morning after pill)

The morning after pill, which contains progesterone and oestrogen, can prevent pregnancy where contraception either failed or was not used and a woman thinks she may have conceived. The name is not really accurate as the pill can be taken any time up to 72 hours after sex, although its effectiveness decreases with time.

The morning after pill works by preventing ovulation (sperm can live on average 4–5 days and, exceptionally, up to 7 days, and so might still be able to fertilise an oocyte produced sometime after sex). It also alters the lining of the uterus making implantation of a fertilised oocyte (ovum) very unlikely.

EXTENSION

Implications of birth control

Humans are unique amongst other species in being able to control their environment. This enables them to breed for longer and allows their children to survive long enough to produce their own offspring. The result has been a massive world population increase, especially over the past century. Although the world population is still increasing, the populations of many developed countries have stabilised as birth rates have fallen. This change has been made possible by birth control. The widespread availability of methods of birth control has not only biological implications but also social and ethical ones.

There are undoubted benefits in the use of birth control. These include:

- The world population can be limited so that there is sufficient food and other resources for mankind and therefore less poverty. Critics argue that this can be achieved in other ways such as a fairer distribution of wealth and greater protection of the environment.

- Unwanted pregnancies and births can be prevented along with the distress this causes.

- People can choose if and when to have children to suit their social and economic conditions. The children born are therefore more likely to be better cared for.

- Women have greater control over their own bodies and their own fertility. It therefore promotes gender equality and women's autonomy. They have greater freedom to go to work.

- There are health benefits because birth control reduces the need for abortions, which can be hazardous to the mother. Condoms reduce the incidence of sexually transmitted diseases. Some women would risk their health if they became pregnant.

Some people and groups oppose birth control. Their concerns include:

- It is unnatural because the natural purpose of sex is to produce children and contraception interferes with this.

- Contraception is anti-life, especially those types which prevent implantation of the fertilised oocyte and are, they argue, a form of abortion.

- It carries health risks such as an increased risk of contracting a sexually transmitted disease and cervical cancer as a result of greater sexual freedom and promiscuity.

- It can be misused to engineer the type of population or race that a government desires. In other words as a **eugenic** tool.

- It separates sex from reproduction and can lead to immoral behaviour through promiscuity because the fear of pregnancy and hence being 'found out' is removed. This may lead to an increase in the breakdown of marriages.

Figure 2 *Doctor discussing the oral contraceptive pill with a patient*

SUMMARY TEST 15.12

Contraception is any means, artificial or natural, which prevents a viable **(1)** forming. It takes many forms including the contraceptive pills of which there are two main types. One type, the mini pill, contains only a synthetic form of the hormone called **(2)**. The other type, called the **(3)** pill includes a synthetic form of a second hormone **(4)**. Both types are effective contraceptives because they prevent **(5)** occurring, they also thicken the secretions around the **(6)** and therefore form a barrier to sperm and they make the **(7)** layer of the uterus thinner making **(8)** of an ovum less likely.

Control and coordination in plants

Plants can respond to a stimulus very rapidly: for example the sudden closure of leaves on a Venus flytrap when capturing an insect. For the most part, plant responses are slower than those of animals but they are still complex processes. Different concentrations of a plant hormone can have very different effects.

Role of auxins in elongation growth

Auxins are a group of chemical substances of which indoleacetic acid (IAA) is the most common. The transport of auxin is in one direction, away from the tip of shoots and roots where it is produced. Auxin has a number of effects on plant cells, including altering the state of the cell wall so that permanent elongation of the cells can occur. This is only effective on young cell walls before they have developed greater rigidity through secondary thickening. The proposed explanation of how auxin has a role in this elongation growth is called the **acid growth hypothesis** and is outlined below.

- Auxin binds to receptors, which causes transport proteins (ATPases) in the cell surface membrane to actively transport protons (hydrogen ions) from the cytoplasm into spaces in the cell wall.
- The protons cause the fluid-filled spaces in the cell wall to become more acidic.
- The decrease in pH creates the optimum conditions for proteins known as expansins to weaken the cell wall, temporarily disrupting the hydrogen bonding between cellulose microfibrils and between these and other cell wall polysaccharides.
- This causes a loosening of the cell wall.
- Water entering the cell by osmosis will cause an increase in turgor pressure and the loosening of the cell wall means that during active cell growth its protoplast can expand, causing elongation of the cell.

The hypothesis is supported by experiments in which **buffer** solutions that neutralise the acidity of cell walls are shown to prevent cell elongation. Conversely, agents that acidify cell walls cause cell elongation. It has also been observed that protons are released from cells in response to auxin.

The elongation of cells on one side only of a stem or root can lead to the stems or roots bending. This is the means by which plants respond relatively quickly to environmental stimuli such as light, water and gravity. These responses can be explained in terms of the stimuli causing uneven distribution of auxin as it moves away from the tip of the stem or root.

Response of a Venus flytrap

The Venus flytrap lives in a boggy habitat with few nutrients. In order to obtain nutrients such as nitrogen, needed for growth, it has to capture and digest small insects. When an insect lands on the specialised leaf of the Venus flytrap, the two lobes of the leaf that are usually convex suddenly become concave, snapping shut and so trapping the insect in the cavity that is formed. It is not clear how this rapid response is brought about, but one explanation is as follows.

- The mechanical stimulus of the insect touching hairs on the lobes of the leaves generates an action potential that passes to the lower cells of the midrib, known as hinge cells. It requires at least two hairs, which act as receptors, to be triggered in quick succession cause a response.
- A proton (H^+) pump moves protons out of the hinge cells and into the cell wall spaces between the cells in response to action potentials from the trigger hairs.

Figure 1 Venus flytrap

This may be in response to auxin, the concentration of which has been observed to increase in the hinge cells.

- By the acid growth hypothesis described earlier, the increased acidity dissolves the calcium pectate that holds the cell walls together.
- The loss of protons from the hinge cells makes them more negative, and so positively charged ions such as calcium ions are attracted into the cells, decreasing their water potential.
- There is now a water potential gradient that causes water to enter the cells by osmosis. As the cell walls are much less rigid and more flexible, the cells rapidly expand, causing the lobes of the leaf to become concave, closing together and trapping the insect.

Alternative theories involve loss of turgor in other cells, causing the response. As there is some experimental evidence for each, it may be that both mechanisms are involved.

Role of gibberellins in stem elongation

Gibberellins are a group of over 90 plant growth regulators found not only in flowering plants but also in fungi, algae and some bacteria. They are thought to be made in developing seeds and apical portions of stems and roots. One gibberellin increases the length of stems and so increases the height of plants. If the gibberellin is added to certain genetic dwarf varieties of plant, the plants grow to normal size. This is probably because height is controlled by a gene with two alleles:

- a **dominant** allele (*Le*) that controls the production of an enzyme needed in the synthesis pathway of gibberellin; plants with at least one dominant allele therefore grow to normal height
- a **recessive** allele (*le*) that codes for a non-functioning enzyme involved in the synthesis pathway of gibberellin; plants with both alleles of the recessive form therefore cannot synthesise gibberellin and so develop into dwarf varieties.

Role of gibberellins in the germination of barley seeds

Once plant seeds are formed they often remain dormant for some time before they germinate. This allows them to overcome adverse conditions like the cold temperatures of winter and allows time for them to be dispersed to new regions by wind or animals. This dormancy is, in part, due to the very low water content – between 5 and 10% – of most seeds. What then breaks dormancy and starts the process of germination? To answer this question, we need first to look at the structure of a typical endospermous seed such as wheat or barley. These seeds are made up of:

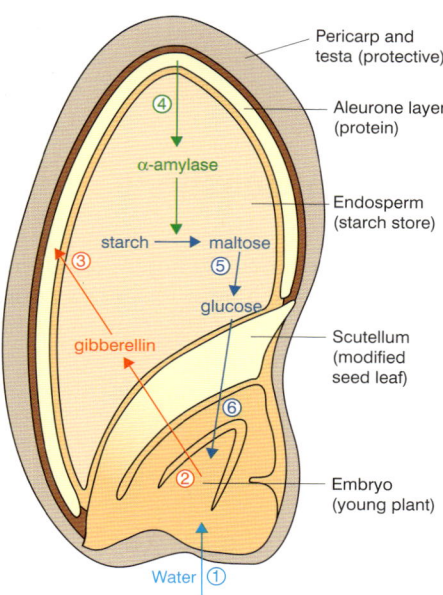

Figure 2 *Structure of a barley seed and the role of gibberellin in its germination*

- **pericarp and testa** – an outer tough, protective layer made up of the testa (seed coat) and pericarp (fruit coat) fused together
- **aleurone layer** – a protein-rich layer just beneath the testa
- **endosperm** – a large region of stored starch that provides an energy source for the growing embryo
- **scutellum** – a modified form of the single cotyledon (seed leaf) of the seed
- **embryo** – the result of **mitotic** division of the zygote, this will develop into the new plant.

The process of germination typically requires the presence of water, oxygen and a favourable temperature. Proteins called DELLA proteins normally inhibit germination (Topic 16.15), as follows (the numbers refer to those on the summary of the process in Figure 1):

1. Water softens the pericarp and testa covering and enters the rest of the seed.
2. The water stimulates the embryo to produce gibberellin.
3. The gibberellin diffuses into the cells of the aleurone layer, where it causes the breakdown of DELLA proteins and so induces **transcription** of the genes producing α-amylase and other enzymes that are manufactured from its protein store in the aleurone layer.
4. The α-amylase and other enzymes diffuse into the endosperm.
5. α-amylase **hydrolyses** the starch in the endosperm into maltose, which in turn is hydrolysed by maltase to glucose.
6. The glucose diffuses, via the scutellum, into the embryo where it is used to provide the ATP (from respiratory breakdown) and raw material (e.g. cellulose) needed for germination and growth.

15 Examination Questions

1 Figure 1 shows the changes in potential difference (p.d.) across the membrane of a neurone over a period of time. The membrane was stimulated at time **A** and time **B** with stimuli of different intensities.

Figure 1

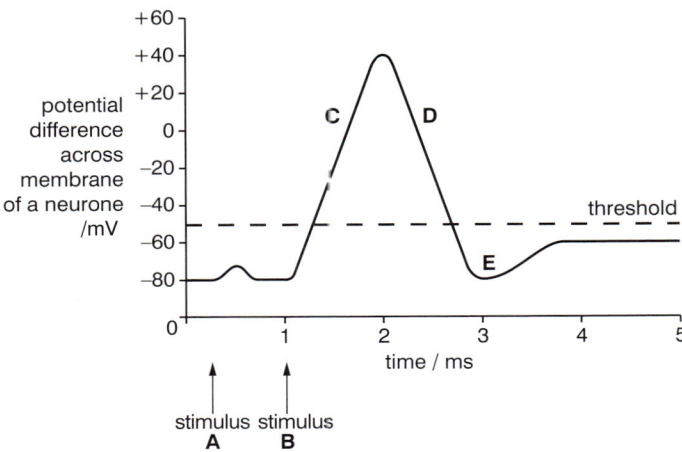

a Stimulus **B** resulted in an action potential. Describe what is occurring at **C**, **D** and **E**.　　*(6 marks)*

b Suggest why stimulus **A** did not result in an action potential being produced whereas stimulus **B** did.
　　　　　　　　　　　　　　　　　　　(2 marks)
　　　　　　　　　　　　　　　　　(Total 8 marks)

Cambridge International AS and A Level Biology 9700 Paper 4 Q8 November 2007

2 a The steroid hormones oestrogen and progesterone are secreted by the ovary.

State precisely the sites of secretion of each. *(2 marks)*

b The most effective oral contraceptives for general use are the so-called combined oral contraceptives (COCs), which contain both oestrogen and progesterone.

Explain how COCs produce their effects.　*(4 marks)*

c Describe two **social** implications of the use of contraceptives.
　　　　　　　　　　　　　　　　　　　(2 marks)
　　　　　　　　　　　　　　　　　(Total 8 marks)

Cambridge International AS and A Level Biology 9700 Paper 4 Q2 June 2009

3 a Copy and complete Table 1 to show, for each of the two hormones, follicle stimulating hormone (FSH) and progesterone,
   • the site of secretion
   • the target tissues(s)
   • the action of the hormone during the human menstrual cycle.　　　　　　　　　*(6 marks)*

Table 1

hormone	site of secretion	target tissue(s)	Action during human menstrual cycle
FSH			
progesterone			

b Explain the biological basis of the oestrogen/progesterone contraceptive pill.　*(3 marks)*
　　　　　　　　　　　　　　　　　(Total 9 marks)

Cambridge International AS and A Level Biology 9700 Paper 4 Q5 November 2008

4 a Describe a reflex arc **and** explain why such reflex arcs are important.　　　　　*(7 marks)*

b Describe the structure of a myelin sheath **and** explain its role in the speed of transmission of a nerve impulse.　　　　　　　　　　*(8 marks)*
　　　　　　　　　　　　　　　　(Total 15 marks)

Cambridge International AS and A Level Biology 9700 Paper 41 Q10 November 2009

5 a Describe how a nerve impulse crosses a cholinergic synapse.　　　　　　　　　　*(9 marks)*

b Explain the roles of synapses in the nervous system.
　　　　　　　　　　　　　　　　　　　(6 marks)
　　　　　　　　　　　　　　　　(Total 15 marks)

Cambridge International AS and A Level Biology 9700 Paper 4 Q9 June 2007

15 Practice Questions

6 The table below shows the speeds by which different axons conduct action potentials.

Axon	Myelin	Axon diameter /μm	Transmission speed / m s⁻¹
Human motor axon to leg muscle	Yes	20	120
Human sensory axon from skin pressure receptor	Yes	10	50
Squid giant axon	No	500	25
Human motor axon to internal organ	No	1	2

a Using data from the table, describe the effect of axon diameter on the speed of conductance of an action potential.

b The data show that a myelinated axon conducts an action potential faster than an unmyelinated one. Explain why this is so.

c What is the name of the cells whose membranes make up the myelin sheath around some types of axon?

d State whether the presence of myelin or the diameter of the axon has the greater influence on the speed of conductance of an action potential. Use information from the table to explain your answer

e The squid is an ectothermic animal. This means that its body temperature fluctuates with the temperature of the waters in which it lives. Suggest how this might affect the speed with which a squid conducts action potentials along its axon.

7 a Suggest a reason why there are numerous mitochondria in the sarcoplasm of muscle.

b If we cut across a myofibril at certain points, we see only thick myosin filaments. If we cut at a different point we see only thin actin filaments. At yet other points we see both types of filament. Explain why.

c How is the shape of the myosin filament adapted to its role in muscle contraction?

d During the contraction of a muscle sarcomere, a single actin filament moved 0.8 μm. If the hydrolysis of a single ATP molecule provides enough energy to move an actin filament 40 nm, how many ATP molecules were needed to move the actin filament 0.8 μm? Show your working.

e Dead cells can no longer produce ATP. Soon after death muscles contract, making the body stiff – a state known as rigor mortis. From your knowledge of muscle contraction, explain why rigor mortis occurs after death.

8 Indoleacetic acid (IAA) is a chemical that belongs to a group of substances called auxins. IAA moves in one direction, away from the tips of shoots and roots where it is produced. An experiment was performed to investigate how IAA is distributed after it leaves the tip of bean shoots. Tips were cut from the shots of bean seedlings and placed on small agar blocks. Groups of these tips on agar blocks were treated in four different ways:

1 left in complete darkness

2 illuminated from one side only and an impermeable glass plate placed in the agar block

3 left in complete darkness and an impermeable glass plate placed through each tip

4 illuminated from one side only and an impermeable glass plate placed through each tip.

The average relative quantity of IAA (in arbitrary units) was measured in the agar blocks of each experimental group. Figure 2 shows how the experiment was set up and the relative quantity of IAA in each of the blocks at the end of the experiment.

Figure 2

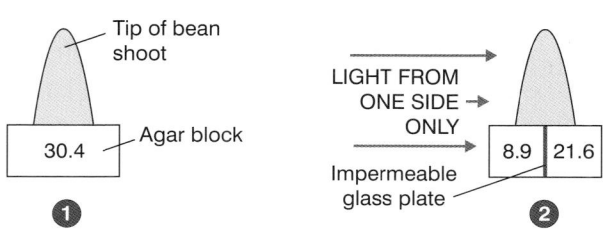

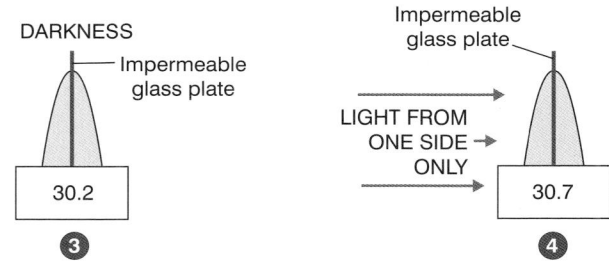

a What was the reason for using the impermeable glass plates?

b Using the information above, describe the effects that light from one side has on IAA.

c Give one piece of evidence that shows that light causes a redistribution of IAA rather than destroys it.

9 The Venus flytrap is a plant that captures and digests small insects. To do so it uses touch-sensitive hairs on its leaves. Describe how an insect touching the hairs can result in the rapid closure of the leaves around it.

Inherited change

The role of meiosis

On these pages you will learn to:

- Explain what is meant by homologous pairs of chromosomes
- Explain the meanings of the terms 'haploid' and 'diploid' and the need for a reduction division (meiosis) prior to fertilisation in sexual reproduction
- Outline the role of meiosis in gametogenesis in humans and in the formation of pollen grains and embryo sacs in flowering plants

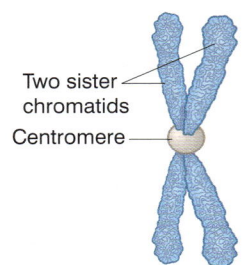

Figure 1 *Structure of a chromosome after DNA replication*

Table 1 *Differences between mitosis and meiosis*

Mitosis	Meiosis
A single division of the nucleus	A double division of the nucleus
The number of chromosomes remains the same	The number of chromosomes is halved
Homologous chromosomes do not associate	Homologous chromosomes do associate
Chiasmata are never formed	Chiasmata may be formed
Crossing over never occurs	Crossing over may occur
Daughter cells are genetically identical to parent cells (if no mutations)	Daughter cells are genetically different from parental ones
Two daughter cells	Four daughter cells
Daughter cells have the same number of chromosome sets as the parents	Daughter cells have half the number of chromosome sets as the parents

Division of the nucleus of cells takes place in two ways:

- **Mitosis** produces two daughter nuclei with the same number of chromosomes as the parent and as each other (Topic 5.2).
- **Meiosis** produces four daughter nuclei each with half the number of chromosomes of the parent. If the parent cell had two complete chomosome sets (diploid) each new daughter cell now has one complete set of chromosomes.

The structure of a chromosome is shown in Figure 1 and Table 1 compares the processes of mitosis and meiosis.

The principles behind meiosis

In sexual reproduction two gametes fuse to form a zygote. It follows that, to maintain the number of chromosomes in the adults of a species constant, the number must be halved at some stage in the life cycle. This halving occurs as a result of meiosis and is not random as each cell formed will have a complete set of chromosomes.

Each of the cells of an adult of a species has a fixed number of chromosomes, usually made up of pairs that carry the same genes. One of each pair is from the chromosomes provided in the egg by the mother (maternal chromosomes) and the other of each pair is from the chromosomes provided in the sperm by the father (paternal chromosomes). In plants, the equivalent to the sperm is the pollen. These are known as **homologous pairs** and the total number is referred to as the **diploid** number; in humans this is 46. During meiosis each member of a homologous pair is separated from its homologue (the other member of the pair), so that only one enters each daughter cell. This is known as the **haploid** number and in humans is 23. When two haploid gametes fuse, the diploid number of chromosomes is restored.

Homologous chromosomes

All diploid cells of organisms have two sets of chromosomes, one set provided by each parent. There are therefore always two sets of genetic information for each feature of an individual. Any two chromosomes that have the same genes are termed a homologous pair. Having the same genes is not the same as being identical. For instance, a homologous pair of chromosomes may each possess information on seed shape and colour in plants, but one chromosome may carry the genes that code for round, yellow seeds, while the other carries the genes that code for wrinkled, green seeds. During meiosis, the halving of the number of chromosomes is not done randomly. Instead, each daughter cell receives one of each homologous pair of chromosomes. In this way each cell has one set of information for each gene possessed by the organism. When these haploid cells (usually gametes) combine, the diploid state, with its paired homologous chromosomes, is restored. The events of meiosis are illustrated and explained in Topic 16.2.

Role of meiosis in gametogenesis

Gametogenesis is the formation of the haploid reproductive cells, known as gametes. In humans the gametes originate from diploid germ cells, which initially multiply by mitosis to build up large numbers of parent cells that will undergo meiosis to form gametes. These later divide by meiosis to give rise to the mature haploid gamete, which, when fused with a gamete from the opposite sex, restores the diploid state. In males the process takes place in the testes, produces sperm and is called **spermatogenesis**. In females it takes place in the ovaries, produces ova and is called **oogenesis**.

The situation in plants is more complex, although the basic idea of building up numbers by mitosis and halving the number of chromosomes by meiosis is the same. However, in plants the diploid mother cell divides initially by meiosis to give haploid cells, and these then divide by mitosis to increase numbers. Nuclear division is not always followed by cell division, and so the female part of a plant produces embryo sacs containing eight haploid nuclei, whereas the male parts produce pollen grains each with two haploid nuclei. These events are shown in Figure 2.

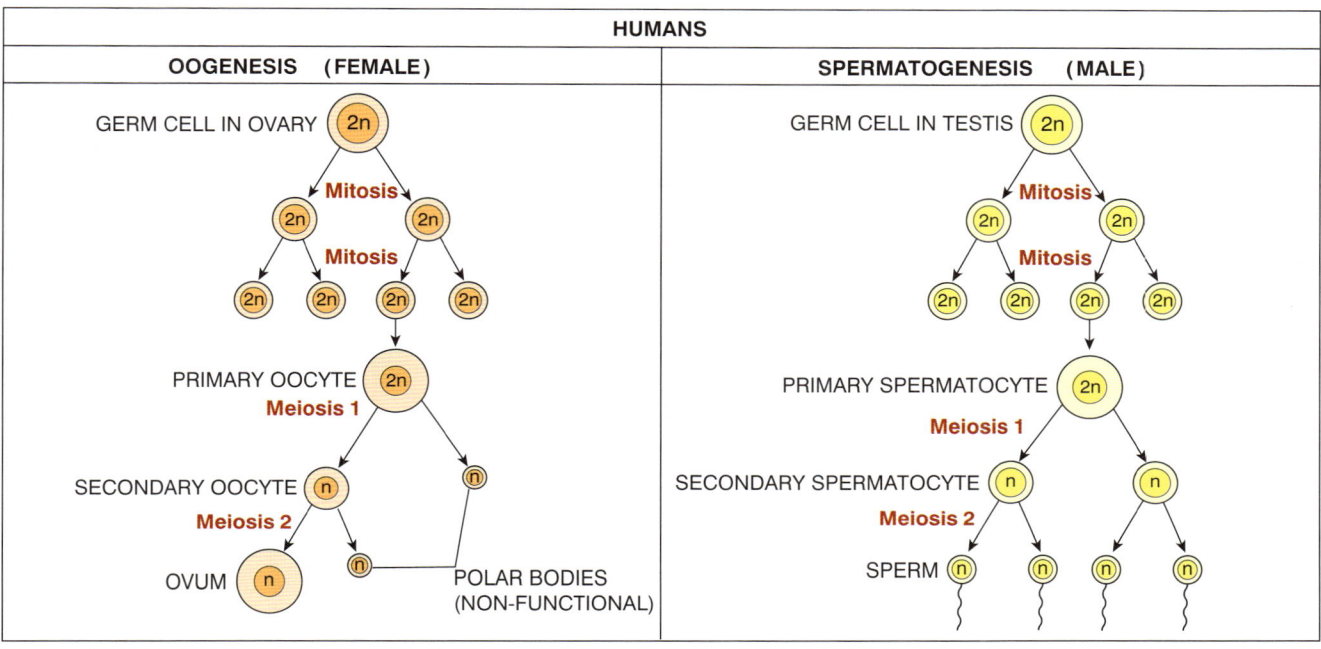

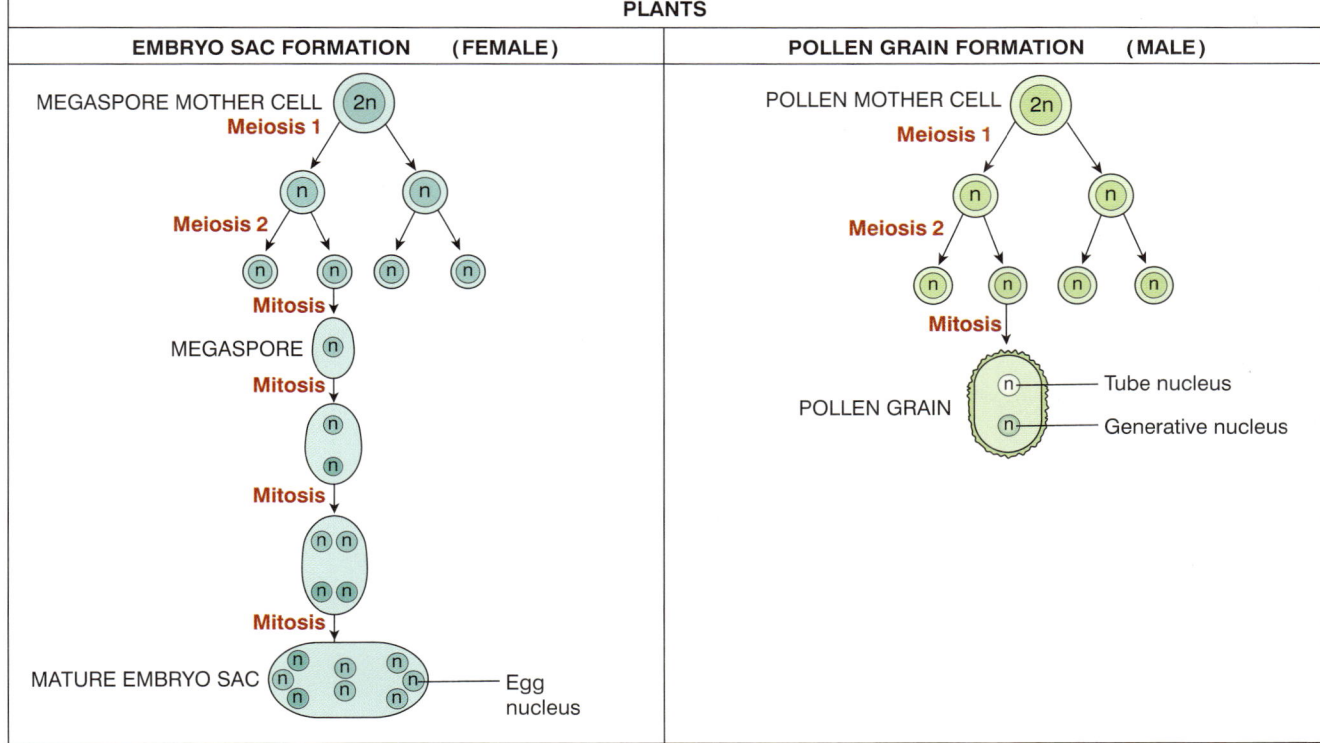

Figure 2 *Formation of gametes in humans and plants*

On these pages you will learn to:

- Describe, with the aid of photomicrographs and diagrams, the behaviour of chromosomes in plant and animal cells during meiosis, and the associated behaviour of the nuclear envelope, cell surface membrane and the spindle

The stages of meiosis (reduction division) are described in Figure 1. The parent cell has two sets of chromosomes (diploid). For convenience, only four chromosomes are shown, two in each set.

Interphase
In late interphase (S phase), before meiosis, DNA replicates so that the cell now contains four, rather than the original two, copies of each chromosome. In animal cells the pair of centrioles replicates.

Prophase I
The chromosomes shorten and fatten (condensation) and come together in their homologous pairs to form a bivalent. The chromatids wrap around one another and non-sister chromatids of homologous chromosomes attach at points called chiasmata. The chromatids may break at these points and swap similar sections of chromatids with one another in a process called crossing over. Finally the nucleolus disappears and the nuclear envelope disassembles. The centromeres are attached to spindle fibres.

Metaphase I
With centromeres attached to the spindle, the bivalents arrange themselves randomly on the equator of the cell with each of a pair of homologous chromosomes facing opposite poles.

Anaphase I
One of each pair of homologous chromosomes is pulled by spindle fibres to opposite poles.

Telophase I and cytokinesis
Microtubules pull two sides of the cell surface membrane together so that the cell becomes narrower towards its centre until the opposite parts of the membrane fuse to give two separate cells. In most animal cells a nuclear envelope re-forms around the chromosomes at each pole, but in most plant cells there is no telophase I and the cell goes directly into metaphase II.

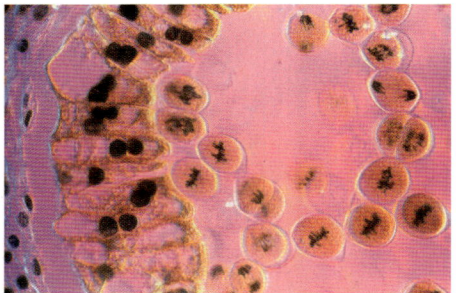

Figure 2 *Photomicrograph of meiosis in bluebell anther cells. In the right half of the image, cells mostly in prophase I, metaphase I and anaphase I of meiosis can be seen*

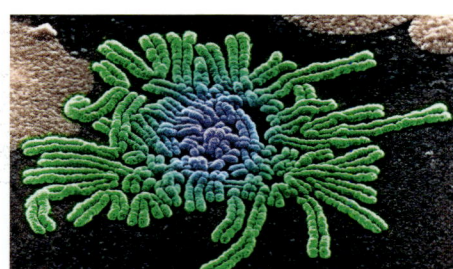

Figure 3 *Colourised scanning electron micrograph of chromosomes during metaphase I of meiosis. The chromosomes are grouped together at the equator of the cell. This is a semi polar view rather that the equatorial view shown in Figure 1*

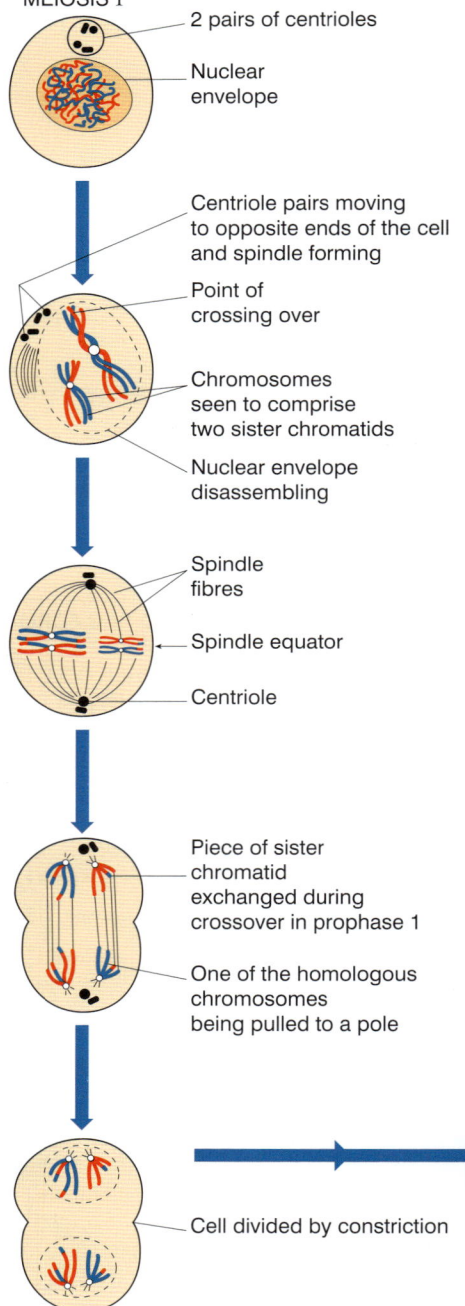

MEIOSIS I

- 2 pairs of centrioles
- Nuclear envelope
- Centriole pairs moving to opposite ends of the cell and spindle forming
- Point of crossing over
- Chromosomes seen to comprise two sister chromatids
- Nuclear envelope disassembling
- Spindle fibres
- Spindle equator
- Centriole
- Piece of sister chromatid exchanged during crossover in prophase 1
- One of the homologous chromosomes being pulled to a pole
- Cell divided by constriction

Figure 1 *Stages of meiosis in an animal cell*

The process of meiosis

Meiosis has two nuclear divisions that normally occur immediately one after the other:

- **The first meiotic division (meiosis I)** is separated for convenience into four stages – prophase I, metaphase I, anaphase I and telophase I. Unlike prophase of mitosis, in prophase I of meiosis homologous pairs come together to form a **bivalent** in a process called **synapsis**. Chromatids of homologous chromosomes wrap around each other and attach at points called **chiasmata**

(singular = **chiasma**) and swap equivalent portions of non-sister chromatids in a process called **crossing over**.

- **The second meiotic division (meiosis II)** is basically a mitotic division: no chiasmata are formed, individual chromosomes line at up the spindle equator, centromeres split and sister chromatids separate and move to opposite poles. The four stages – prophase II, metaphase II, anaphase II and telophase II – occur simultaneously in the two daughter cells formed in meiosis I. At the end of this second meiotic division four daughter cells are formed.

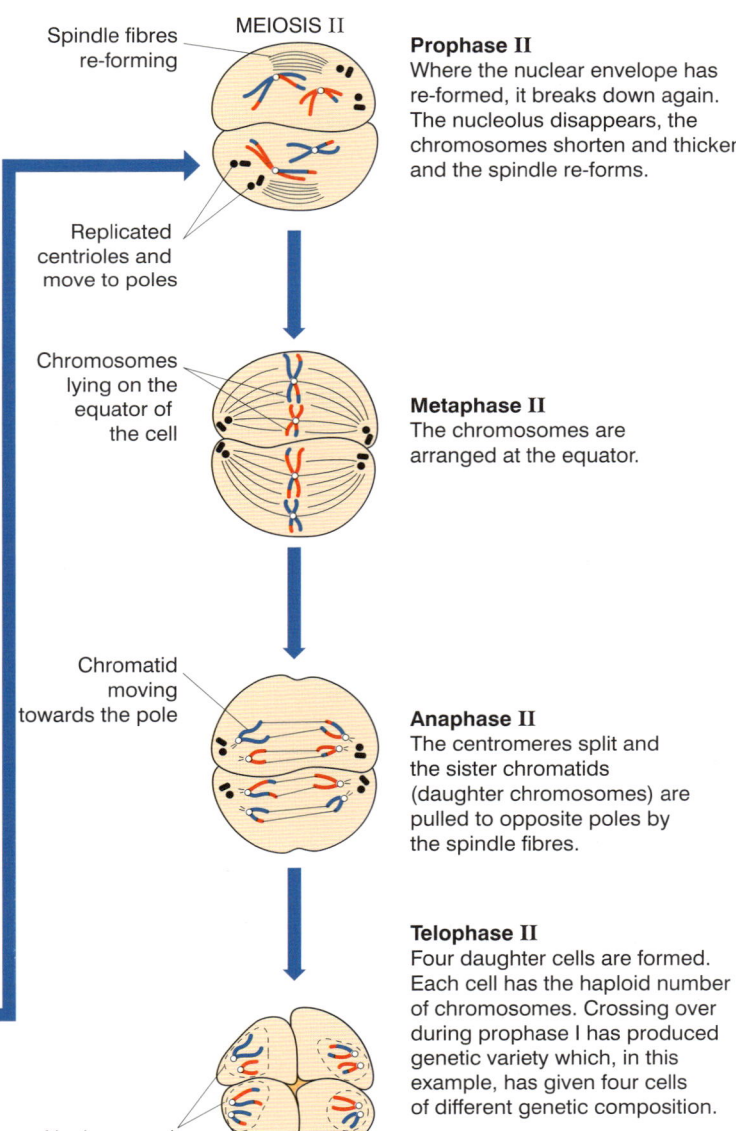

MEIOSIS II

Spindle fibres re-forming

Prophase II
Where the nuclear envelope has re-formed, it breaks down again. The nucleolus disappears, the chromosomes shorten and thicken and the spindle re-forms.

Replicated centrioles and move to poles

Chromosomes lying on the equator of the cell

Metaphase II
The chromosomes are arranged at the equator.

Chromatid moving towards the pole

Anaphase II
The centromeres split and the sister chromatids (daughter chromosomes) are pulled to opposite poles by the spindle fibres.

Telophase II
Four daughter cells are formed. Each cell has the haploid number of chromosomes. Crossing over during prophase I has produced genetic variety which, in this example, has given four cells of different genetic composition.

Nuclear envelope re-forming

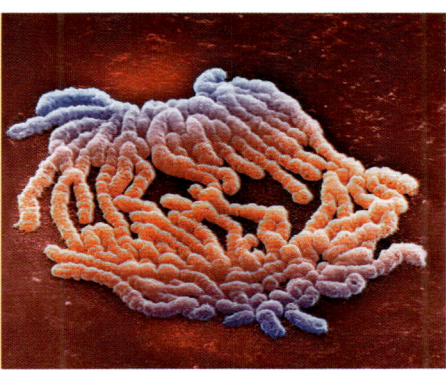

Figure 4 *Colourised scanning electron micrograph of chromosomes during anaphase I of meiosis. The chromosomes are seen pulling apart towards the poles of the cell. The view is from the same angle as in Figure 3*

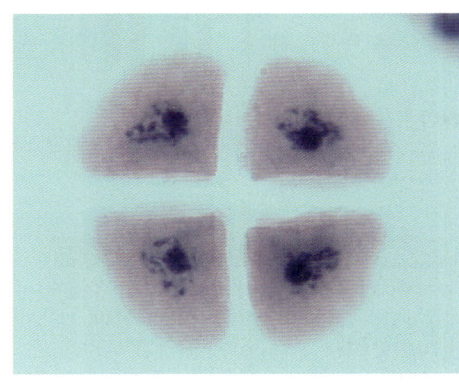

Figure 5 *Photomicrograph of pollen of Zea mays (maize) showing a tetrad of four cells formed at telophase II of meiosis*

On these pages you will learn to:

- Explain how crossing over and random assortment of homologous chromosomes during meiosis and random fusion of gametes at fertilisation lead to genetic variation including the expression of rare, recessive alleles

Meiosis halves the number of chromosomes to form cells such as gametes and ensures that the diploid number is maintained when haploid gametes fuse. Meiosis also produces genetic variation among offspring, allowing an organism to adapt and survive in a changing world. Meiosis brings about this variation in three main ways: genetic recombination by crossing over, random assortment of homologous chromosomes at metaphase I, and production of haploid gametes that fuse randomly at fertilisation.

Genetic recombination by crossing over

We saw in Topic 16.2 that during meiosis 1 each chromosome lines up alongside its homologous partner. The following events then take place.

- The chromatids of each pair become twisted around one another.
- During this twisting process tensions are created and portions of the chromatids break off.
- These broken portions then rejoin with the non-sister chromatids of the homologous partner. The points where crossing over occurs are called chiasmata (singular = chiasma).
- It is usually the equivalent portions of homologous chromosomes that are exchanged.
- In this way new genetic combinations of maternal and paternal alleles (forms or versions of a gene – see Topic 16.4) are produced.

These events are shown in Figure 1.

The effect of this recombination by crossing over on the cells produced at the end of meiosis is shown in Figure 2. Where there is no recombination by crossing over only two different types of cell are produced. However, where recombination

Chromatids of homologous chromosomes twist around one another, crossing over many times. The points where they cross over are called chiasmata

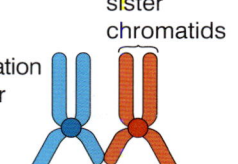

sister chromatids

Simplified representation of a single cross over

Point of breakage (chiasma)

Result of a single cross over showing that equivalent portions of non-sister chromatids have been exchanged

Figure 1 *Crossing over*

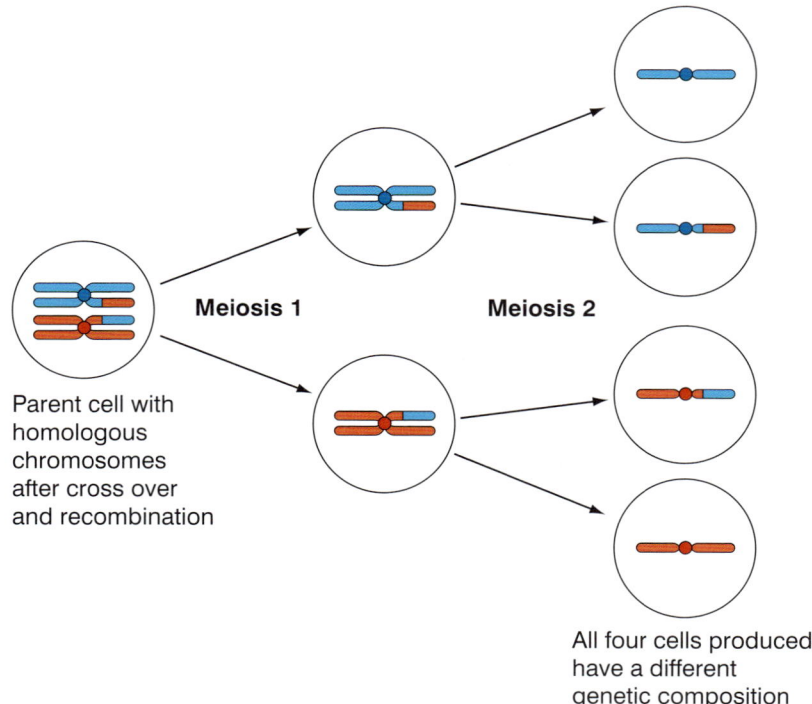

Parent cell with homologous chromosomes after cross over and recombination

Meiosis 1

Meiosis 2

All four cells produced have a different genetic composition

Figure 2 *Genetic variation as a result of recombination by crossing over*

does occur there are four different cell types. Recombination by crossing over therefore increases genetic variety.

Random assortment of homologous chromosomes at metaphase I

When the pairs of homologous chromosomes arrange themselves on the equator of the spindle during metaphase I of meiosis, they do so independently of each other and randomly. This random distribution of homologous pairs at metaphase I is known as random assortment. Although each one of a pair has the same genes, they may differ in the alleles that they possess. Random assortment produces new genetic combinations. A simple example is shown in Figure 3.

Production of haploid gametes that fuse randomly at fertilisation

The haploid gametes produced by meiosis must fuse to restore the diploid state. Each gamete has a different

genetic make-up and their random fusion therefore produces variety in the offspring. Where the gametes are from different parents (as is usually the case) two different genetic make-ups are combined and even more variety results.

Some individuals may carry recessive alleles that are rare, that is, they have resulted from a mutation and only a few individuals in the population may have the allele. The effect of these alleles is not seen as the individual will have another, dominant allele that masks its effect (Topic 16.4). Half the gametes produced as a result of meiosis will contain this rare allele. If random fusion of these gametes occurs (that is, two different individuals, each carrying the recessive allele), then the offspring formed will now have two copies of the recessive allele so that the allele can be expressed and its effects seen.

In **arrangement 1**, the two pairs of homologous chromosomes orientate themselves on the equator in such a way that the chromosome carrying the allele for tall stems and the one carrying the allele for terminal flowers migrate to the same pole. The alleles for short stems and axial flowers migrate to the opposite pole. Cell ❶ therefore carries the alleles for tall stems and terminal flowers while cell ❷ carries the ones for short stems and axial flowers.

In **arrangement 2**, the left-hand homologous pair of chromosomes is shown orientated the opposite way around. As this orientation is random, this arrangement is equally as likely as the first one. The result of this different arrangement is that cell ❸ carries the alleles for short stems and terminal flowers, whereas cell ❹ carries ones for tall stems and axial flowers.

All four resultant cells are different from one another. With more homologous pairs the number of possible combinations becomes enormous. In pea plants with 7 pairs, this is 2^7 combinations; a human, with 23 such pairs, has the potential for $2^{23} = 8\ 388\ 608$ combinations.

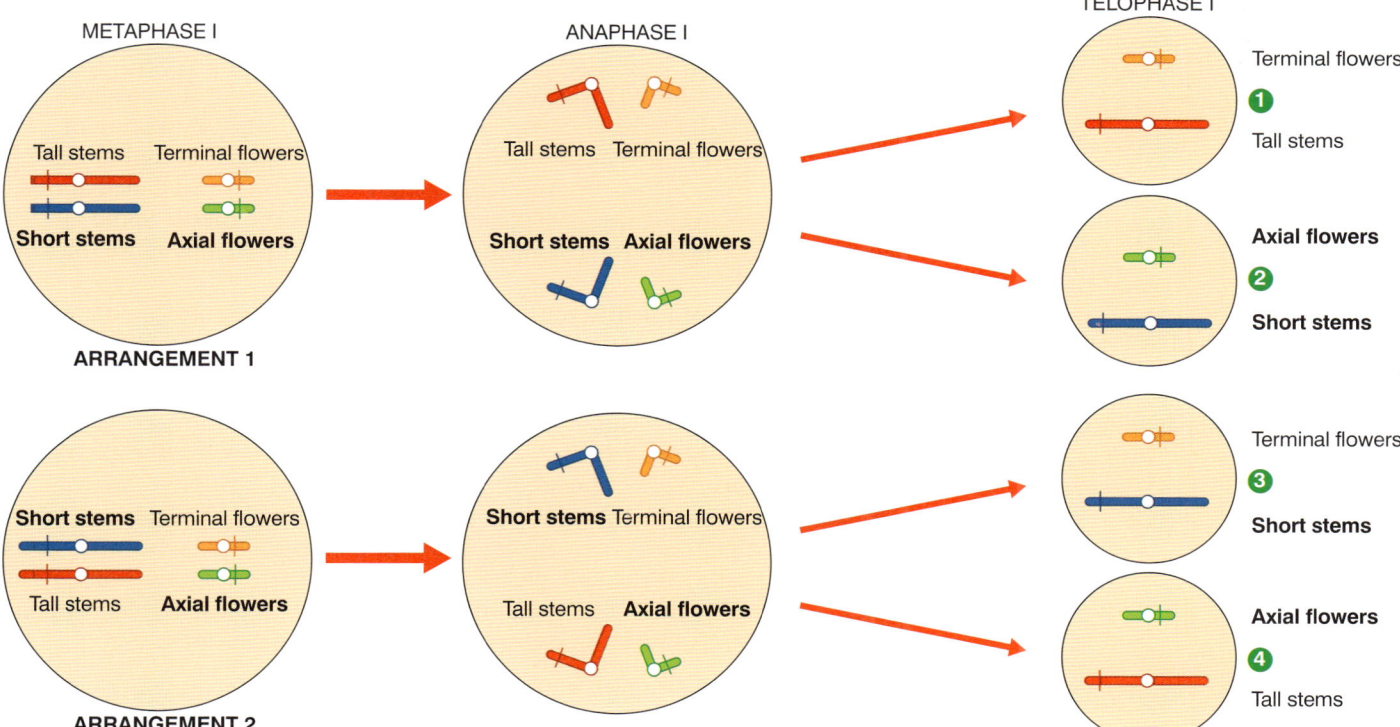

Figure 3 *How random assortment of chromosomes during metaphase I contributes to genetic variation in gametes and hence the offspring*

REMEMBER

Strictly speaking, the terms 'recessive' and 'dominant' should be applied to genetic conditions, traits or inheritance patterns rather than alleles. However, for simplicity and ease of understanding, the practice of applying these terms to alleles is used in this and subsequent topics.

HOMOZYGOUS DOMINANT

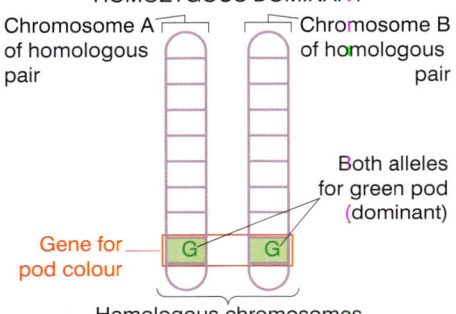

HETEROZYGOUS

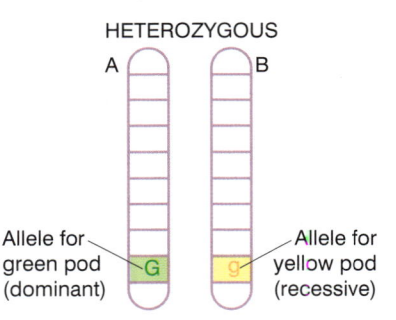

HOMOZYGOUS RECESSIVE

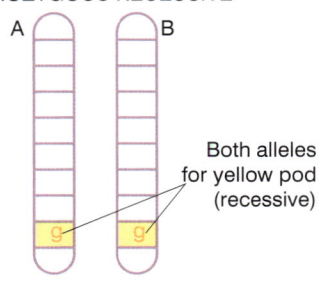

Figure 1 *Pair of homologous chromosomes showing different possible pairings of dominant and recessive alleles*

The fact that children resemble, to a greater or lesser degree, both parents and yet are identical to neither has long been recognised. However, it took the re-discovery, at the beginning of the last century, of the work of a scientist and monk (living in what was at the time part of Austria) called Gregor Mendel (1822–1884) to establish the basic laws by which characteristics are inherited.

Genotype and phenotype

- **Genotype** is the genetic make-up of an organism. It describes all the alleles that an organism contains. The genotype sets the limits within which the characteristics of an individual may vary. It may determine that a human baby could grow to be 1.8 m tall, but the actual height this individual reaches is affected by other factors such as diet. A lack of an element such as calcium (for bone development) or iodine (for production of the hormone thyroxine) at a particular stage of development could mean that the individual never reaches his/her potential maximum height. Any change to the genotype as a result of a change to the DNA is called a **mutation** and may be inherited if it occurs in the gametes.
- **Phenotype** is the observable characteristics of an organism. It is the result of the interaction between the genotype and the environment, which can modify an organism's appearance. Any change to the phenotype that does not affect the genotype is not inherited and is called a **modification**.
- The term genotype is also used when referring to a particular gene, to describe the combination of alleles possessed by an individual for that gene. Here, the stated phenotype would just describe the observable effect of that specific gene.

Genes and alleles

- **A gene** is a length of DNA, i.e. a sequence of nucleotide bases that usually determines a single characteristic of an organism (e.g. blood group). It does this by coding for particular polypeptides that make up the enzymes that are needed in the biochemical pathway leading to the production of the characteristic (e.g. codes for the production of antigen A on red blood cells). Genes exist in two, or occasionally more, different forms called alleles. The position of a gene on a chromosome is known as the **locus**. Some characteristics, such as eye colour in humans, are controlled by more than one gene (Topic 17.1).
- **An allele** is one of the different forms of a gene. In pea plants, for example, there is a gene for the colour of the seed pod. This gene has two different forms, or alleles – an allele for a green pod and another allele for a yellow pod.

Only one allele of a gene can occur at the locus of any one chromosome. However, in sexually reproducing organisms the chromosomes occur in pairs called homologous pairs. There are therefore two loci that can each carry one allele of a gene. If the allele on each of the chromosomes is the same (e.g. both alleles for green pods are present) then the organism is said to be **homozygous** for the characteristic. If the two alleles are different (e.g. one chromosome has an allele for green pods and the other chromosome has an allele for yellow pods) then the organism is said to be **heterozygous** for the characteristic.

In most cases where two different alleles are present in the genotype (heterozygous state) only one of them shows itself in the phenotype. For instance in our example, where the alleles for green and yellow pods are present in the genotype, the phenotype is always green pods. The allele of the heterozygote that expresses itself in the phenotype is said to be **dominant**, the one that is not expressed is said to be **recessive**. A homozygous organism with two dominant alleles is called **homozygous dominant**, whereas one with two recessive alleles is called **homozygous recessive**. The effect of a recessive allele is apparent in the phenotype of a diploid organism only when it occurs in the presence of another identical allele, i.e. when it is in the homozygous state. These different genetic types are shown in Figure 1.

In some cases, two alleles both contribute to the phenotype, in which case they are referred to as **codominant**. In this situation when both alleles occur together, the phenotype is either a blend of both features, e.g. flower petal colour in snapdragons (*Antirrhinum*) where pink flowers result from an allele for red colour and an allele for white colour or both features are represented (e.g. the presence of both A and B antigens in blood group AB).

Sometimes a gene of an organism has more than two possible alleles. The organism is said to have **multiple alleles** for that gene. However, as there are always only two chromosomes in a homologous pair, it follows that only two of the three or more alleles in existence can be present in a single organism. Multiple alleles occur in the human ABO blood grouping system (Topic 16.7).

Figure 2 summarises the different terms used in genetics.

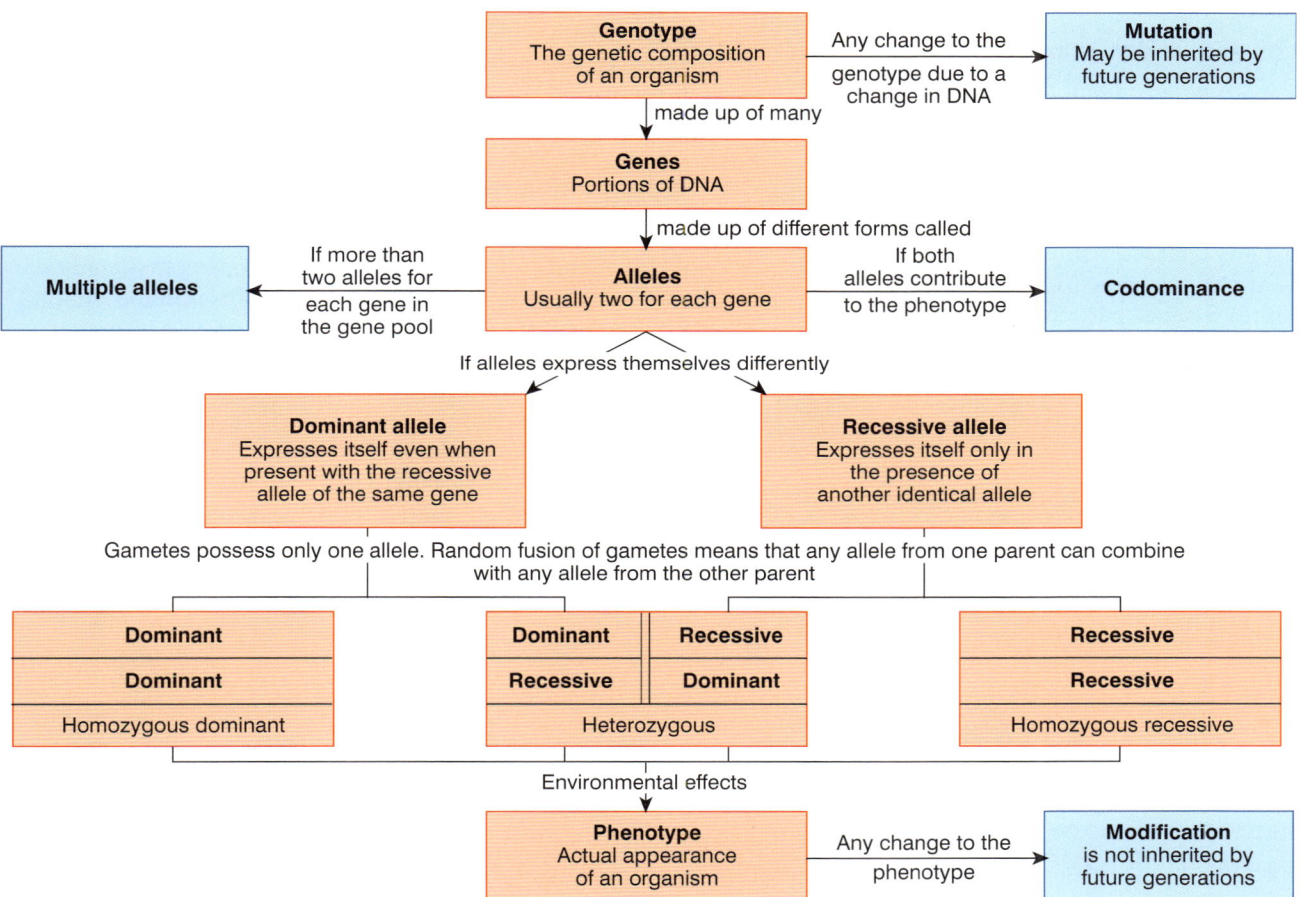

Figure 2 *Summary of genetic terms*

SUMMARY TEST 16.4

The genetic composition of an organism is called the **(1)** and any change to it is called a **(2)** and may be inherited by future generations. The actual appearance of an organism is called the **(3)** and any change to this that is not inherited is called a **(4)**. A gene is a sequence of **(5)** along a piece of DNA that codes for a particular **(6)** that makes the enzyme needed in a biochemical pathway. Each gene has two or more different forms called alleles. If the two alleles on a homologous pair of chromosomes are the same they are said to be **(7)**, but if they are different, they are said to be **(8)**. An allele that is not apparent in the phenotype when paired with a dominant allele is said to be **(9)**.

On these pages you will learn to:

- Use genetic diagrams to solve problems involving monohybrid crosses
- Explain the term F1 and F2

Monohybrid inheritance is the inheritance of a single gene. A characteristic, or trait, is any feature that is measured, for example, blood group type or flower colour.

Inheritance of pod colour in peas

Pod colour in peas is known to be controlled by a single gene. If pea plants with green pods are bred repeatedly with each other so that they always produce plants with green pods, they are said to be **pure breeding** for the characteristic of green pods. Pure breeding strains can be bred for almost any characteristic. What it means is that the organisms are homozygous (i.e. they have two **alleles** that are the same) for that particular gene.

If these pure breeding green-pod plants are then crossed with pure breeding yellow-pod plants, all the offspring, known as the **first filial** or **F_1 generation**, turn out to

> **REMEMBER**
>
> The term F_1 should be used only for the offspring of crosses in which the original parents are homozygous, whereas the term F_2 should be used only for the offspring resulting from crossing the F_1 individuals.

produce green pods. This means that the allele for green pods is dominant to the allele for yellow pods, which is therefore recessive. This cross is shown in Figure 1.

When the heterozygous plants (Gg) of the F_1 generation are crossed with one another (= F_1 intercross), the offspring (known as the **second filial or F_2 generation**) are always in an approximate ratio of 3 plants with green pods to each 1 plant with yellow pods. This cross is shown in Figure 2.

These observed facts led to the formation of the **law of segregation**, sometimes called Mendel's first law, which states: **In diploid organisms, characteristics are determined by alleles that occur in pairs. Only one of each pair of alleles can be present in a single gamete.**

Representing genetic crosses

Genetic crosses are usually represented in a standard form of shorthand (Table 1). Although you may occasionally come across some minor variations to this scheme, that outlined in Table 1 is the one normally used. Always carry out the procedures completely. Once you have practised a number of crosses, you may be tempted to miss out stages or explanations. Not only is this likely to lead to errors, it often makes your explanations difficult for others to follow. **You** may understand what you are doing, but if your teacher or examiner cannot follow it, you are not likely to get full credit for your efforts.

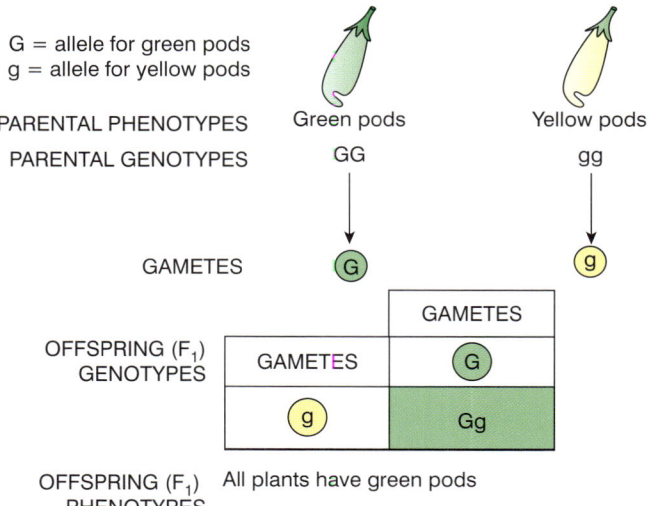

Figure 1 *Cross between a pea plant that is pure breeding for green pods and one that is pure breeding for yellow pods*

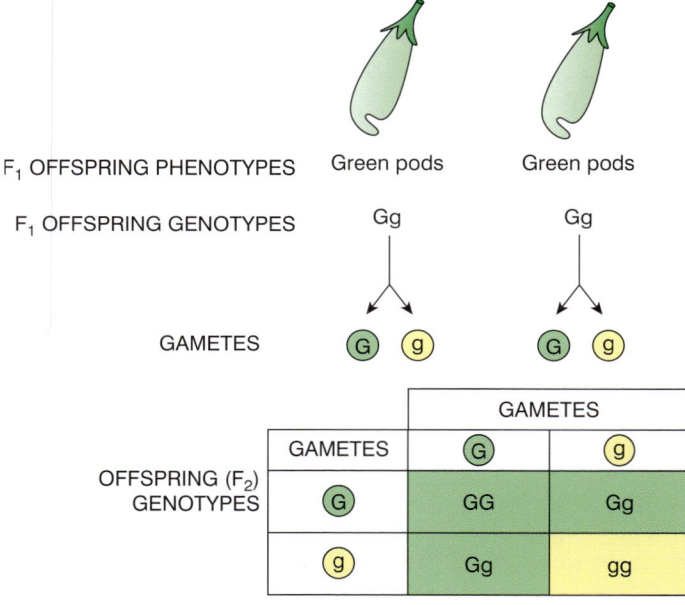

Figure 2 *F_1 intercross between pea plants that are heterozygous for green pods*

Table 1 Representing genetic crosses

Instruction	Reason/notes	Example [green pod and yellow pod]
Choose a single letter to represent each characteristic	An easy form of shorthand. In some genetic crosses, e.g. in *Drosophila*, there are set symbols, some of which use two letters	–
Choose the first letter of one of the contrasting features	When more than one characteristic is considered at one time such a logical choice means it is easy to identify which letter refers to which characteristic	Choose G (green) or Y (yellow)
If possible, choose the letter in which the upper and lower case forms differ in shape as well as size. It is often helpful to choose the letter of the alphabet that represents the dominant allele	If the upper and lower case forms differ it is almost impossible to confuse them, regardless of their size	Choose G because green is the dominant allele and the upper case form (G) differs in shape from the lower case form (g). If yellow is a dominant allele, make sure that the two letters are clearly identified as upper (Y) or lower (y) case, for example the lower case y could have a curly tail.
Let the upper case letter represent the dominant feature and the lower case letter the recessive one. Never use two different letters where one characteristic is dominant	The dominant and recessive feature can easily be identified. Do not use two different letters as this indicates codominance	Let G = green and g = yellow Do *not* use G for green and Y for yellow
Where a feature is compared to 'normal', as in human genetic conditions, then the letter of the alphabet can be used to represent the condition.	Otherwise N for normal would be used, which gives no information about the feature and would be a problem when handling crosses with two genes, each with a dominant normal allele.	Let A = normal dominant allele a = recessive allele for albinism
Represent the parents with the appropriate pairs of letters. Label them clearly as 'parents' and state their phenotypes	This makes it clear to the reader what the symbols refer to	Green pod Yellow pod Parents GG × gg
State the gamete types produced by each parent. Label them clearly, and encircle them.	This explains why the gametes only possess one of the two parental factors. Encircling them reinforces the idea that they are separate	Gametes (G) (g) (g)
Use a type of chequerboard or matrix, called a **Punnett square**, to show the results of the random crossing of the gametes. Only label which parents are male and which are female where the question tells you which is which	This method is less liable to error than drawing lines between the gametes and the offspring. Labelling the sexes is only important when considering sex-linked crosses (see Topic 16.9)	GAMETES (G) (G) (g) Gg Gg (g) Gg Gg
State the phenotypes of each different genotype and indicate the numbers of each type. Always put the higher case (dominant) letter first when writing out the genotype	Always putting the dominant feature first can reduce errors in cases where it is not possible to avoid using symbols with the upper and lower case letters of the same shape	All offspring are plants producing green pods (Gg)

SUMMARY TEST 16.5

When representing a genetic cross, there is a standard form of shorthand that is normally used. Where the allele for one characteristic is dominant to the allele for another, it is normal to use an **(1)** case letter to represent the dominant allele and a **(2)** case letter to represent the **(3)** allele. Where gametes are involved, the letter for the allele should be **(4)**. To show all the possible outcomes of different gametes fusing, a type of chequerboard called a **(5)** is used. The inheritance of a single gene is known as **(6)** inheritance. One example is a cross between pea plants with yellow seed pods and ones with green pods. The first generation of plants produced is called the **(7)** generation and all have pods coloured **(8)**. When these offspring are crossed with one another the next generation, called the **(9)** generation, has a ratio of three **(10)** coloured pods to one **(11)** pod.

Although Gregor Mendel was a monk, he had some training in science, especially in physics, chemistry and mathematics. His experiments show the benefits of thorough preparation, careful experimental procedure and logical interpretation of results. His choice of pea plants was no accident. He chose them because they were:

- easy to grow and

- possessed many contrasting features that could be easily observed.

He carefully controlled pollination, and hence fertilisation, by accurately transferring pollen from one plant to another with a paint brush. He ensured the plants he used were pure breeding for each feature by self pollinating them for many generations. He was the first person to produce quantitative and not just qualitative results. Finally, he counted many offspring to ensure that his results were reliable and relatively free from statistical error. In short, over a nine-year period, Mendel formed a well-organised programme of research, performed it accurately, recorded masses of data and analysed his results with precision and insight. He had worked out the basic principles of inheritance by 1865, when the nature of genetics was unknown; he effectively predicted the existence of genes and meiosis long before they were discovered. Although he circulated his work to many libraries and distinguished scholars (academics) of his day, his theories were not accepted – indeed, they were ignored. Even Darwin, whose theory of evolution was based on genetic variation, failed to understand the significance of Mendel's work. It took some 35 years before other geneticists rediscovered his work and the significance of his remarkable experiments was at last appreciated.

On these pages you will learn to:

- Use genetic diagrams to solve problems involving test crosses
- Explain the term test cross

In a monohybrid cross, an organism whose phenotype displays a dominant characteristic may possess either of two **genotypes**:

- two dominant **alleles** (homozygous dominant)
- one dominant allele and one recessive allele (heterozygous).

A **test cross** is a cross to determine the genotype of a individual. It is carried out between an individual with the unknown genotype and an individual having the homozygous recessive genotype.

Carrying out a test cross

To look at how we carry out a test cross, let us use the example in Topic 16.5 of pea plants with different seed pod colours. Suppose we have a plant from the F$_2$ generation that produces green seed pods. This plant has two possible genotypes with respect to pod colour:

- homozygous dominant (GG)
- heterozygous (Gg).

To discover its actual genotype, we cross the plant with an organism displaying the recessive **phenotype** of the same characteristic, i.e. in our case with a pea plant producing yellow pods (gg). Figure 1 shows that:

- if the organism is homozygous dominant (GG), then all the offspring will be heterozygous (Gg) and will show the dominant feature (green pods)
- if the organism is heterozygous (Gg), then it would be expected that half the offspring would be heterozygous (Gg) and show the dominant feature (green pods) while half the offspring would be homozygous recessive (gg) and show the recessive feature (yellow pods).

The test cross is so called because it 'tests' the unknown genotype of a dominant characteristic. Is the test cross foolproof? The answer is 'yes and no':

- **Yes**, if any single offspring displays the recessive characteristic (in our case yellow pods). This plant is homozygous recessive and must have obtained one recessive allele from each parent. Our unknown parental genotype must therefore have a recessive allele and be heterozygous (in our case Gg). Assuming no mutations, we can say with absolute certainty that our unknown genotype is heterozygous (Gg).
- **No**, if all the offspring display the dominant characteristic (in our case green pods). While the likelihood is that the unknown genotype is homozygous dominant (in our case GG), we cannot completely discount the possibility that it could be heterozygous (Gg). This is because the gametes produced from our parent of unknown genotype contain alleles of two types, either dominant (G) or recessive (g). It is a matter of chance which of these gametes fuses with those from our recessive parent – all these gametes have

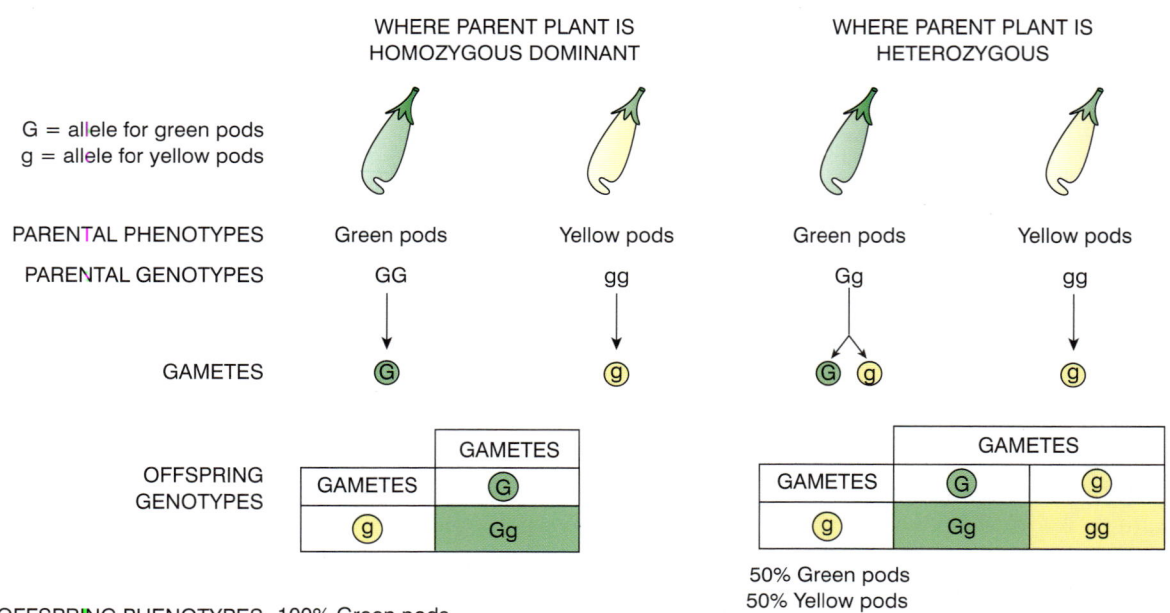

	WHERE PARENT PLANT IS HOMOZYGOUS DOMINANT		WHERE PARENT PLANT IS HETEROZYGOUS	
G = allele for green pods g = allele for yellow pods				
PARENTAL PHENOTYPES	Green pods	Yellow pods	Green pods	Yellow pods
PARENTAL GENOTYPES	GG	gg	Gg	gg
GAMETES	G	g	G g	g

OFFSPRING GENOTYPES

GAMETES	G
g	Gg

OFFSPRING PHENOTYPES 100% Green pods

GAMETES	G	g
g	Gg	gg

50% Green pods
50% Yellow pods

Figure 1 *The test cross*

a recessive allele (g). It is just possible that in every case it is the gametes with the dominant allele that fuse and so all the offspring show the dominant characteristic. Provided the sample of offspring is large enough, however, we can be reasonably sure that the unknown genotype is homozygous dominant.

Why actual results of genetic crosses are rarely the same as the predicted results

If you look at Table 1, you will see the results that Gregor Mendel actually obtained in his experiments. Our knowledge of genetics tells us that for each cross we would expect that, in the F_2 generation, there would be three offspring showing the dominant feature to every one showing the recessive feature. However, in no case did Mendel obtain an exact 3:1 ratio. The same is true of almost any genetic cross. These differences are due to statistical error.

Imagine tossing a coin 20 times. In theory you would expect it to come down heads on 10 occasions and tails on 10 occasions. In practice it rarely does – try it. This is because each toss of the coin is an independent event that is not affected by what went before. If the coin has come down heads 9 times out of 19 tosses, there is still a 50% chance it will come down tails, rather than the head needed to complete the 1:1 ratio. The coin does not 'know' it is expected to come down heads.

The same is true of gametes. It is chance that determines which ones fuse with which. In our cross between the heterozygote (Gg) and the homozygous recessive (gg), all

the gametes of the homozygous parent are recessive (g), whereas the heterozygote parent produces gametes of which half are dominant (G) and half are recessive (g). If it is the dominant gamete that meets the recessive one, plants with green pods are produced (Gg). If it is the recessive gamete, the plants have yellow pods. The larger the sample, the more likely are the actual results to match the theoretical ones. It is therefore important to use large numbers of organisms in genetic crosses if representative results are to be obtained. It is no coincidence that the two ratios nearest to the theoretical value of 3:1 in Mendel's experiments were those with the largest sample size, whereas the ratio furthest from the theoretical value had the smallest sample size (Table 1).

SUMMARY TEST 16.6

Suppose the appearance, otherwise known as the (1), of an organism displays a feature that is determined by a dominant allele. This organism may have a genotype that is either (2) or (3) for this feature. To determine which genotype it has we can carry out a (4) cross. To do this we cross our organism of unknown genotype with an organism of the same species that has a genotype that is (5) for the feature. The gametes of this organism will all contain recessive (6) for this feature. If some of the resultant offspring from the cross show the recessive feature, then the unknown genotype must have been (7). If, however, all the resultant offspring show only the dominant feature, then it is most probable that the unknown genotype was (8). This becomes increasingly probable when the number of offspring is (9).

Table 1 Actual results of Mendel's crosses in pea plants

Characteristic	F_2 results		Ratio
Cotyledon colour	6020 yellow	2001 green	3.01:1
Seed type	5474 smooth	1850 wrinkled	2.96:1
Pod type	882 inflated	299 constricted	2.95:1
Flower position	651 axial	207 terminal	3.14:1
Petal colour	705 purple	224 white	3.15:1
Stem height	787 long	277 short	2.84:1
Pod colour	428 green	152 yellow	2.82:1

On these pages you will learn to:

- Use genetic diagrams to solve problems with crosses, including those involving codominance, multiple alleles and gene interactions

We have looked at the straightforward situations in which there were two possible **alleles** at each **locus** on a chromosome, one of which was dominant and the other recessive. We shall now look at situations in which both alleles contribute to the phenotype, called **codominance**, and where there are more than two alleles, of which only two may be present at the loci of an individual's **homologous chromosomes** – **multiple alleles**.

Codominance

Codominance occurs when both alleles of a gene are expressed in the phenotype. One example occurs in the snapdragon plant, in which one allele codes for an enzyme that catalyses the formation of a red pigment in flowers. The other allele codes for an altered enzyme that lacks this catalytic activity and so does not produce the pigment. In plants that are homozygous for this second allele, no pigment is made and the flowers are white. Heterozygous plants, with their single allele for the functional enzyme, produce just enough red pigment to produce pink flowers. If a snapdragon with red flowers is crossed with one with white flowers, the resulting seeds give rise to plants with pink flowers (Figure 1).

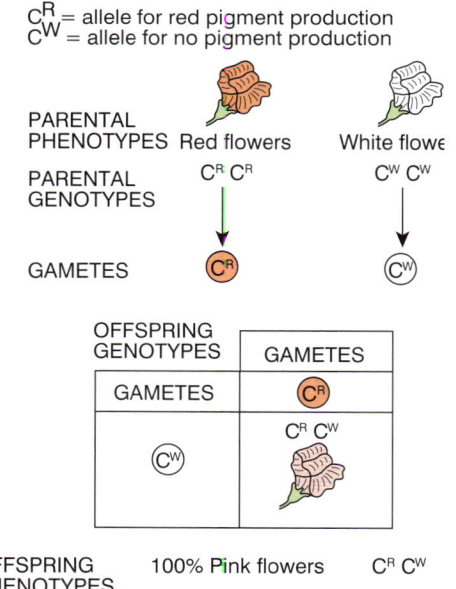

Figure 1 *Cross between snapdragons with red flowers and ones with white flowers*

Note that we cannot use upper and lower case letters for the alleles, as this would imply that one (the upper case) was dominant to the other (the lower case). We therefore use different letters – R for red and W for white – and use these as superscripts on a letter that represents the gene, in this case C for colour. Hence the allele for red pigment is written as C^R and the allele for no pigment as C^W. Figure 1 shows a cross between a red and a white snapdragon and Figure 2 a cross between the pink-flowered plants that result.

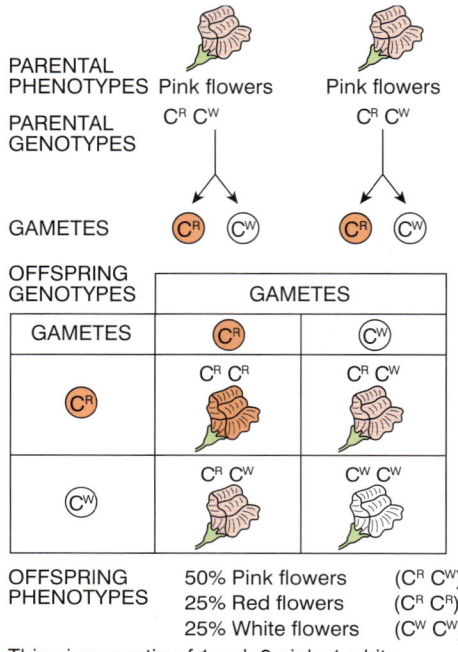

Figure 2 *Cross between two snapdragons with pink flowers*

Multiple alleles

Sometimes a gene has more than two alleles, i.e. it has **multiple alleles**. The inheritance of the human ABO blood groups is an example. There are three alleles associated with the *ABO* gene (I), which lead to the production of different **antigens** on the surface membrane of red blood cells:

- **allele I^A** leads to the production of antigen A
- **allele I^B** leads to the production of antigen B
- **allele I^o** does not lead to the production of any antigens.

Although there are three alleles, only two can be present in an individual at any one time, as there are only two homologous chromosomes and therefore only two gene loci. The alleles I^A and I^B are codominant, whereas the allele I^o is recessive to both. The possible **genotypes** for the four blood groups are shown in Table 1. There are obviously many different possible crosses between different blood groups, but two of the most interesting are:

- A cross between an individual of blood group O and one of blood group AB, rather than producing individuals of either of the parental blood groups, produces only individuals of the other two groups, A and B (Figure 3).
- When certain individuals of blood group A are crossed with certain individuals of blood group B, their children may have any of the four blood groups (Figure 4).

Table 1 *Possible genotypes of blood groups in the ABO system*

Blood group	Possible genotype
A	$I^A I^A$ or $I^A I^O$
B	$I^B I^B$ or $I^B I^O$
AB	$I^A I^B$
O	$I^O I^O$

Multiple alleles and a dominance hierarchy

In blood groups alleles I^A and I^B are codominant and I^o is recessive to both. Sometimes, however, there may be more than three alleles, each of which is arranged in a hierarchy with each allele being dominant to those below it and recessive to those above it. One example is coat colour in rabbits. The gene for coat colour (C) has five alleles. In order of dominance they are:

Agouti coat	C^A
Chinchilla coat	C^{Ch}
Himalayan coat	C^H
Platinum	C^P
Albino coat	C^a

Table 2 shows the possible genotypes of rabbits with each of these coat colours.

Table 2 *Possible genotypes of various coat colours in rabbits*

Coat colour	Possible genotypes
Full colour (Agouti)	$C^A C^A$, $C^A C^{Ch}$, $C^A C^H$, $C^A C^P$, $C^A C^a$
Chinchilla	$C^{Ch} C^{Ch}$, $C^{Ch} C^H$, $C^{Ch} C^P$, $C^{Ch} C^a$
Himalayan	$C^H C^H$, $C^H C^P$, $C^H C^a$
Platinum	$C^P C^P$, $C^P C^a$
Albino	$C^a C^a$

Gene interaction

Genes may interact in different ways. For example, when genes act in sequence as part of a biochemical pathway, an allele that produces a defective enzyme early in the pathway blocks the reactions through the remainder of the pathway. Although the other genes involved in the pathway may result in the production of functioning enzymes, they will not be expressed in the phenotype.

A gene interaction, where one gene interferes with expression of another is an example of a situation known as **epistasis**. In leghorn fowl, there are white and coloured birds. Colour is due to a coloured pigment produced by a dominant allele C. Normally birds are only white when

there are two recessive alleles cc of the gene. However, another dominant allele I of a gene on a different chromosome prevents the action of allele C. When the dominant allele C is present with the dominant allele I, no pigment is produced and the bird is white. The result of a dihybrid cross (Topic 16.8) between one fowl with the genotype CCII and another with genotype ccii produces an F_2 generation with the following possible genotypes:

CCII, CCIi, CcII, CcIi, ccII, ccIi, ccii, which are white and CCii, Ccii, which are coloured. In other words any bird possessing both a dominant C and a dominant I allele will be white as well as genotype ccii.

Without gene interaction only the genotypes ccII, ccIi and ccii would be white. We look at gene interaction in more detail in Topic 16.10.

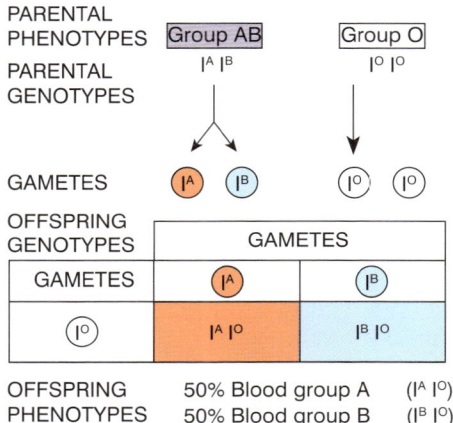

Figure 3 *Cross between an individual of blood group AB and one of blood group O*

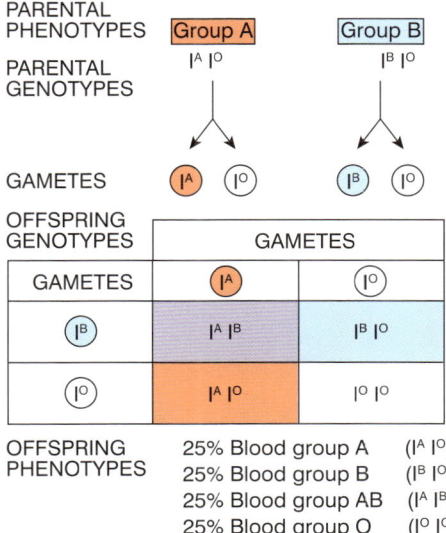

Figure 4 *Cross between an individual of blood group A and one of blood group B*

On these pages you will learn to:

- Use genetic diagrams to solve problems involving dihybrid crosses

In Topic 16.5 we saw how a single characteristic controlled by one gene is passed on from one generation to the next (monohybrid inheritance). In practice, many thousands of characteristics are inherited together. In this topic we shall look at how two characteristics, determined by two **genes** located on non-homologous chromosomes, are inherited. This is referred to as **dihybrid inheritance**.

An example of dihybrid inheritance

In one of his experiments, Gregor Mendel investigated the inheritance of two characteristics of a pea plant at the same time. These were:

Table 1 *Results obtained by Gregor Mendel when he crossed F₁ generation plants with round shaped, yellow coloured seeds*

Appearance of seeds	Condition	Number produced
Round Yellow	Dominant Dominant	315
Round Green	Dominant Recessive	108
Wrinkled Yellow	Recessive Dominant	101
Wrinkled Green	Recessive Recessive	32

- **seed shape** – where round shape is dominant to wrinkled shape
- **seed colour** – where yellow-coloured seeds are dominant to green-coloured ones.

He carried out a cross between the following two pure breeding types of plants:

- one producing round-shaped, yellow-coloured seeds (both dominant features)
- one producing wrinkled-shaped, green-coloured seeds (both recessive features).

In the F₁ generation he obtained plants all of which produced round-shaped, yellow-coloured seeds, i.e. both dominant features. He then raised the plants from these seeds and crossed them with one another to obtain the results shown in Table 1.

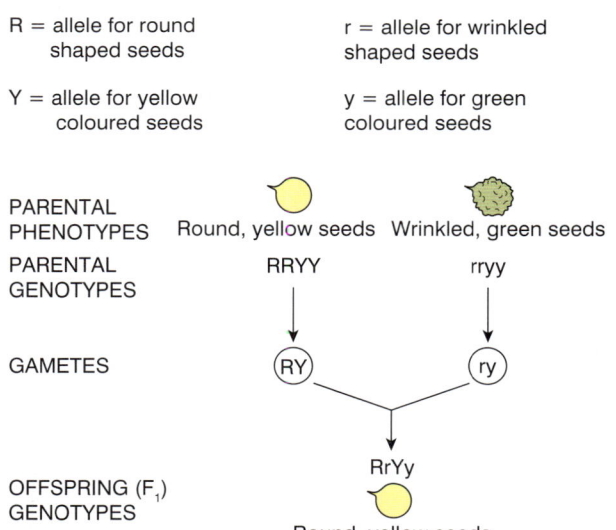

Figure 1 *Genetic explanation of Mendel's cross between a pure breeding plant for round, yellow seeds and a pure breeding one for wrinkled, green seeds*

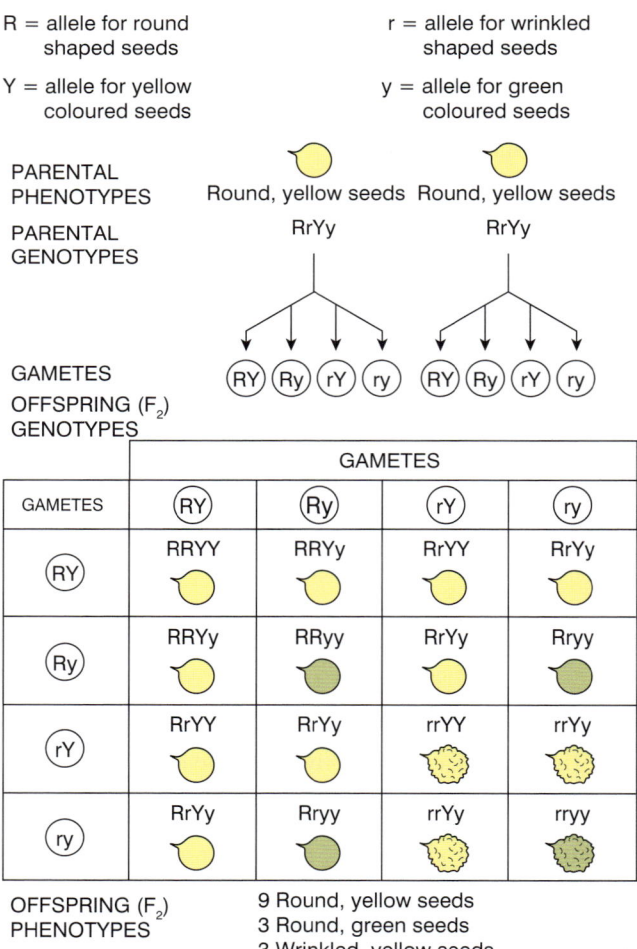

Figure 2 *Genetic explanation of Mendel's intercross between plants of the F₁ generation*

The explanation for these results is given in Figures 1 and 2.

From Figure 2 it can be seen that the plants of the F_1 generation produce four types of gamete (RY, Ry, rY, ry). This is because the gene for seed colour and the gene for seed shape are on non-homologous chromosomes. As the chromosomes arrange themselves randomly on the equator during meiosis (Topic 16.2), any one of the two **alleles** of the gene for seed colour (Y and y) can combine with any one of the alleles for seed shape (R and r). Fertilisation is also random, so that any of the four types of gamete (with respect to seed colour and seed shape) of one plant can combine with any of the four types from the other plant.

The theoretical ratio produced of 9:3:3:1 is close enough, allowing for statistical error (Topic 16.6), to Mendel's observed results of 315:108:101:32. Mendel's observations led him to formulate his **law of independent assortment** which, written in today's biological language states: **For genes that are on separate, non-homologous chromosomes, each member of a pair of alleles may combine randomly with either of another pair**.

Dihybrid test cross

We saw in Topic 16.6 that it is possible to determine whether the **genotype** of an organism that exhibits a dominant characteristic is either homozygous dominant or **heterozygous**. This involved crossing the organism of unknown genotype with one that is homozygous recessive for the same feature. In the same way it is possible to perform a dihybrid test cross. Here the unknown genotype is crossed with an individual with a double homozygous recessive genotype.

In our example, a plant that produces round, yellow seeds (i.e. has both dominant alleles) has four possible genotypes:

RRYY
RrYY
RRYy
RrYy.

To find out the actual genotype of this plant, it can be crossed with one displaying the recessive characteristic for both seed colour and seed shape, i.e. a plant producing wrinkled, green seeds. This plant has only one possible genotype – rryy – and therefore all its gametes are ry with respect to these features. The outcome of each of the four crosses is shown in Table 2. From the table it can be seen that the unknown genotypes may be identified from examining the seeds produced by the offspring of each cross. The presence of even one plant producing wrinkled green seeds (rryy) indicates that the unknown genotype can only have been RrYy. To identify the other possible genotypes it is necessary to count a large sample of seeds – say in excess of 100 – and check the ratios of each type against the table of theoretical ratios. There is a statistical test called the **chi-squared (χ^2) test** that enables us to test the statistical validity of our results. Its use is explained in Topic 16.11.

Table 2 Dihybrid test cross

Possible genotypes of plant producing round, yellow seeds	Possible gametes	Genotypes of offspring crossed with plant producing wrinkled, green seeds (gamete = (ry))	Phenotype (type of seeds produced)
RRYY	(RY)	RrYy	All round and yellow
RrYY	(RY)	RrYy	$\frac{1}{2}$ round and yellow
	(rY)	rrYy	$\frac{1}{2}$ wrinkled and yellow
RRYy	(RY)	RrYy	$\frac{1}{2}$ round and yellow
	(Ry)	Rryy	$\frac{1}{2}$ round and green
RrYy	(RY)	RrYy	$\frac{1}{4}$ round and yellow
	(Ry)	Rryy	$\frac{1}{4}$ round and green
	(rY)	rrYy	$\frac{1}{4}$ wrinkled and yellow
	(ry)	rryy	$\frac{1}{4}$ wrinkled and green

SUMMARY TEST 16.8

The inheritance of two characteristics, determined by two genes located on non-homologous chromosomes is referred to as **(1)** inheritance. Mendel investigated two such characteristics, seed shape, where round shape is **(2)** to wrinkled shape, which is therefore **(3)**, and seed colour where green is **(4)** and yellow is **(5)**. In a cross between a plant producing round-shaped, yellow-coloured seeds (genotype RRYY) and one producing wrinkled-shaped, green-coloured seeds (genotype rryy), all the offspring produced seeds that were **(6)** in shape and **(7)** in colour. Each of these plants could produce four types of gametes depending on the **(8)** of the two genes that they possess. These four types are represented as **(9)**, **(10)**, **(11)** and **(12)**. When these gametes fuse randomly with one another, the offspring produce seeds, of which nine in every 16 are **(13)** in shape and **(14)** in colour, whereas only one in every 16 is **(15)** in shape and **(16)** in colour. If a plant that is heterozygous for seed shape and homozygous dominant for seed colour is crossed with a plant producing green, wrinkled seeds, the offspring produce either **(17)**-shaped and **(18)**-coloured seeds or **(19)**-shaped and **(20)**-coloured seeds.

16.9 Sex determination, sex linkage and autosomal linkage

On these pages you will learn to:

- Use genetic diagrams to solve problems with crosses, including those involving sex linkage and autosomal linkage
- Explain the term linkage

Humans have 23 pairs of chromosomes. 22 of these pairs have homologues that are identical in appearance whether in a male or a female. These are known as **autosomes**. The remaining pair are referred to as **heterosomes** or the **sex chromosomes**. In females, the two sex chromosomes appear the same and are called the **X chromosomes**. In males there is a single X chromosome like those in females, but the second one of the pair is smaller in size and shaped differently. This is the **Y chromosome**.

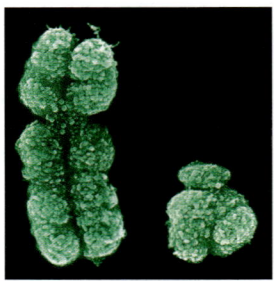

Figure 1 Scanning electron micrograph of human X and Y chromosomes as found in a male. There are some organisms where the sex chromosomes are known by different letters and where the same two sex chromosomes are for the male

Sex determination

Unlike other features of an organism, sex is determined by chromosomes, rather than genes. In humans:

- Females have two X chromosomes and the gametes are all the same in that they contain a single X chromosome. Females are therefore the **homogametic sex**.
- Males have one X chromosome and one Y chromosome and they produce two different types of gamete – half with an X chromosome and half with a Y chromosome. Males are therefore the **heterogametic sex**.

Figure 2 shows sex determination in humans.

Sex linkage – haemophilia

Any gene that is carried on either the X or Y chromosome is said to be **sex-linked**. In practice, very few genes are carried on the Y chromosome in humans. Those that are, such as the SRY gene (sex-determining region on the Y chromosome), are only ever found in males. They also always show themselves, even if **recessive**, because there is no **homologous chromosome** that might have the dominant **allele**.

By contrast the X chromosome carries many genes. One example in humans is the condition called **haemophilia**, in which the blood does not clot, leading to slow and continued bleeding, especially in the joints. As such it is potentially lethal if not treated. This has resulted in some selective removal of

the gene from the population, making its occurrence relatively rare (about 1 person in 20 000 in Europe). Compared to males, there are very few females with haemophilia.

One of a number of causes of haemophilia is the result of a recessive allele with an altered sequence of DNA bases that leads to the production of a non-functioning protein and the inability to produce a clotting factor known as **anti-haemophiliac globulin (AHG)** or **factor VIII**. The extraction of factor VIII from blood donations or produced using recombinant gene technology means that it can now be given to people with haemophilia, allowing them to lead near normal lives. Figure 3 shows the inheritance of haemophilia. Note that the alleles are shown using H for the normal, dominant allele for production of factor VIII and h = recessive allele for non-production of factor VIII (see Topic 16.5). As they are linked to the X chromosome, they are not shown separately, but always attached to the X chromosome, i.e. as X^H and X^h respectively. There is no equivalent allele on the Y chromosome.

As males can only obtain their Y chromosome from their father, it follows that their X chromosome comes from their mother. As the allele for non-production of factor VIII, along with other sex-linked characteristics, is linked to the X chromosome, males always inherit the disease from their mother. As the mothers do not show symptoms of the disease, they must be heterozygous for the characteristic ($X^H X^h$). Such females are called **carriers** because they carry the allele without showing any signs of the condition in their **phenotype**.

As males pass the Y chromosome on to their sons, they cannot pass haemophilia to them. However they can pass the allele to their daughters, via the X chromosome, who would then become carriers of the disease (Figure 4).

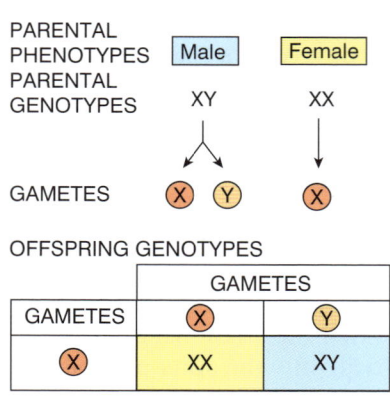

Figure 2 Sex determination in humans

Notice that in Figure 4, the outcome of the cross appears to be the same as a standard example of monohybrid inheritance, with the normal allele being dominant. To confirm sex-linked inheritance, the results of a reciprocal cross would be required. A female with haemophilia (rare) and a normal male, would result in a very different ratio in the first generation, as all the female offspring would be normal (carriers) but all the male offspring would have haemophilia.

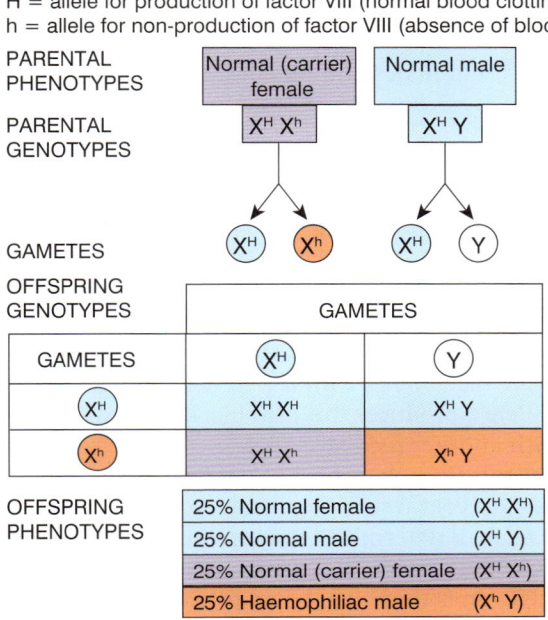

Figure 3 *Inheritance of haemophilia from a carrier female*

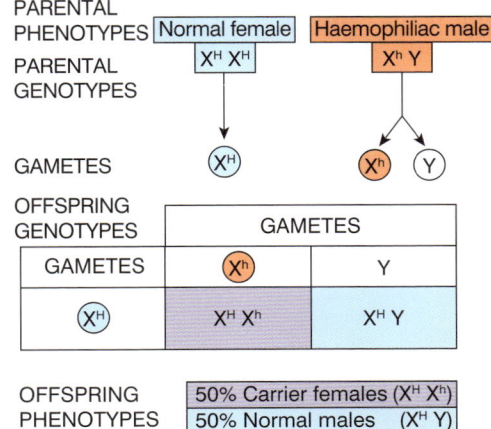

Figure 4 *Inheritance of the haemophiliac allele from a haemophiliac male*

Pedigree charts

One useful way to trace the inheritance of sex-linked characteristics such as haemophilia is to use a pedigree chart. In these a male is represented by a square and a female by a circle. Shading within either shape indicates the phenotypic presence of a characteristic such as haemophilia. Half-shading represents a normal phenotype who carries the allele for non-production of factor VIII.

Autosomal linkage

We have looked at how genes may be linked to the sex chromosomes. We now turn our attention to where genes are linked to the autosomes. For just 23 pairs of chromosomes to determine the many thousands of different human characteristics, it follows that each chromosome must possess many different genes. Any two genes that occur on the same chromosome are said to be **linked**. All the genes on a single chromosome form a **linkage group**.

Under normal circumstances, all linked genes remain together during cell division and so pass into the gamete, and hence the offspring, together. There is no random assortment and they do not segregate (separate) according to the law of independent assortment (Topic 16.8). Figure 5 shows the gametes produced if a pair of genes A and B are linked rather than on separate chromosomes.

Let us take the case in which an organism is heterozygous for both genes A and B. It will have the alleles AaBb.

- If genes A and B occur on separate chromosomes, i.e. they are not linked, the four alleles will be on two homologous pairs of chromosomes. According to Mendel's Law of independent assortment, any one of a pair of contrasted characteristics may combine with any of another pair. Therefore there will be four different possible combinations of these alleles in the gametes: AB Ab aB ab
- If genes A and B occur on the same chromosome, i.e. they are linked, the four alleles will be on just one pair of homologous chromosomes. When these chromosomes segregate during meiosis, the alleles AB will remain together, as will the alleles ab. There will only be two different combinations of alleles in the gametes produced: AB and ab.

Linkage affects the ratios of offspring produced. For example, if the double heterozygote (AaBb) discussed in our example were crossed with a homozygous recessive (aabb), which produces gametes with only ab alleles, the offspring ratios would be:

- not linked – 1 AaBb: 1 Aabb; 1 aaBb; 1 aabb
- linked – 1 AaBb; 1 aabb.

On these pages you will learn to:

• Use genetic diagrams to solve problems with crosses, including those involving autosomal linkage

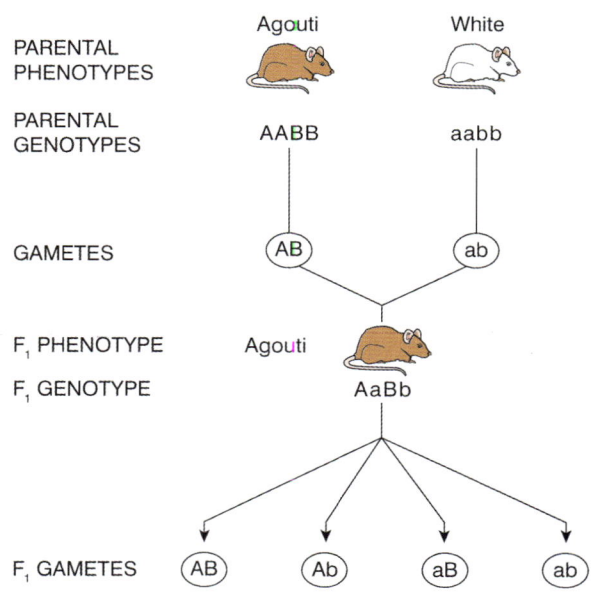

F₂ GENOTYPES

	GAMETES			
GAMETES	AB	Ab	aB	ab
AB	AABB	AABb	AaBB	AaBb
Ab	AABb	AAbb	AaBb	Aabb
aB	AaBB	AaBb	aaBB	aaBb
ab	AaBb	Aabb	aaBb	aabb

F₂ PHENOTYPES

9 Agouti   4 Albino   3 Black

Figure 1 *Gene interaction in mice*

Inheritance involving a single gene with codominant alleles gives a 1:2:1 ratio in the F2, compared to the standard 3:1 Mendelian ratio. Similarly, there are deviations from the 9:3:3:1 ratio of dihybrid crosses when autosomal linkage is involved, for example, when linkage is complete, a 3:1 ratio is obtained. Sex linkage gives different ratios in F1 and F2 ratios when a reciprocal cross is performed, and as the genes are on the X chromosome, the ratios of the phenotypes differ in males and females. Let us consider some examples involving gene interaction.

One form of gene interaction arises when the allele of one gene suppresses or masks the action of another.

An example occurs in mice, where several genes determine coat colour (Figure 2). Let us look at two such genes.

• Gene A controls the distribution of the black pigment melanin in hairs and therefore whether they are banded or not. The dominant allele A of this gene determines that there are black hairs with a yellow band whereas the recessive allele a produces uniform black hairs when it is present with another recessive allele a.

• Gene B controls the colour of the coat by determining pigment to give coat colour is produced or not. The dominant allele B determines the production of pigment and so black hairs are produced whereas the recessive allele b determines there will be no pigment and the hair will therefore be white when it is present with another recessive allele b.

The usual (wild-type) mouse has a grey-brown coat known as agouti. This is the result of having black hairs with yellow banding. If a mouse has uniform black hairs its coat is black, and if the hairs lack melanin altogether its coat is white (albino).

If an agouti mouse with the genotype AABB is crossed with an albino mouse with the genotype aabb, then the offspring are all agouti. If individuals from the F₁ generation are crossed to produce the F₂ generation the following ratio is produced:

• 9 agouti mice
• 4 albino mice
• 3 black mice.

The crosses are shown in Figure 1.

The explanation of the results is that the expression of gene A (black with banding) is affected by the expression of gene B (production of melanin). If gene B is in the homozygous recessive state (bb), then no melanin is produced and the coat is white. In the absence of melanin, gene A cannot be expressed. It makes no difference which alleles are present (AA, Aa or aa); if there is no pigment, then the hairs can be

 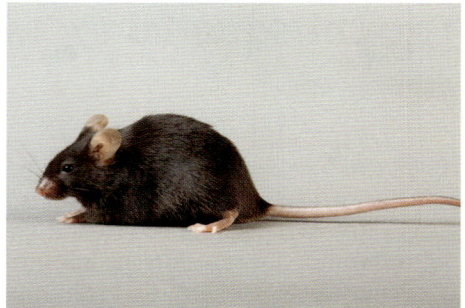 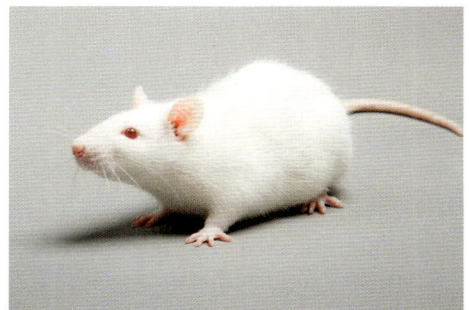

Figure 2 *Agouti, black and albino mice*

neither coloured nor banded. Where a dominant allele B is present, melanin is produced. If this allele is present with a dominant allele A, then banding occurs and an agouti coat results. Where allele B is present with two recessive alleles a, the hairs, and hence the coat, are uniform black.

There are other forms of gene interaction, such as when genes act in sequence by determining the enzymes in a biochemical pathway. An example occurs in maize (corn), *Zea mays*, where some varieties have purple seeds due to the presence of a pigment called anthocyanin in their seed coat. In the absence of the pigment, the seeds are white. The production of anthocyanin is controlled by two genes A and B.

If one pure breeding variety of white-seeded corn with the genotype AAbb is crossed with another pure breeding variety of white seeded corn with the genotype aaBB, all the offspring have purple seeds. A cross between two of the F_1 generation (AaBb) produces a ratio of 9 purple-seeded plants to 7 white-seeded plants.

Figure 3 shows the results of this F_1 intercross.

The production of anthocyanin involves a two-stage process that can be summarised as:

$$\text{starting molecule} \xrightarrow{\text{enzyme A}} \text{intermediate molecule} \xrightarrow{\text{enzyme B}} \text{anthocyanin}$$

The production of enzymes A and B is coded for by genes A and B, respectively. Dominant alleles of each gene code for a functional enzyme, whereas recessive alleles code for a non-functional enzyme. It follows that if the alleles of either gene are both recessive then that enzyme will be non-functional and the pathway cannot be completed. This affects the other gene in that, even if it is functional and produces its enzyme, its effects cannot be expressed. This is because both enzymes need to function to make anthocyanin. If any one enzyme is non-functional then anthocyanin cannot be produced regardless of whether the other enzyme is functional or not.

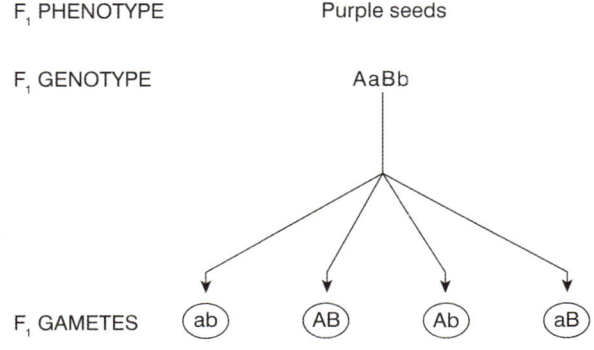

Figure 3 *Results of the F_1 intercross*

SUMMARY TEST 16.10

Using the example of gene interaction in mice in this topic, consider a cross between mouse X that is heterozygous for gene A (colour distribution) and homozygous recessive for gene B (melanin production) with mouse Y that is heterozygous for both genes. The colour of mouse X is **(1)**, whereas the colour of mouse Y is **(2)**. Of the eight genotypes produced from this cross the two genotypes that appear twice are **(3)** and **(4)**. The remaining four genotypes are **(5)**, **(6)**, **(7)** and **(8)**. The number of agouti mice in the F_2 generation is **(9)** and the number of albino mice is **(10)**.

The chi-squared (χ^2) test

On these pages you will learn to:

- Use the chi-squared test to test the significance of differences between observed and expected results

Figure 1 *The single characteristic of comb shape in domestic fowl is controlled by two unlinked genes*

If you toss a coin 100 times it would be reasonable to expect it to land heads on 50 occasions and tails on 50 occasions. In practice, it would be unusual if these exact results were obtained (try it if you like!). If it lands heads 55 times and tails only 45 times, does this mean that the coin is weighted or biased in some way, or is it purely a chance deviation from the expected result? How can we test which of these two options is correct?

What is the chi-squared test?

The chi-squared (χ^2) test is used to test the null hypothesis. The null hypothesis is used to examine the results of scientific investigations and is based on the assumption that there will be no statistically significant difference between sets of observations, any difference being due to chance alone. In our coin tossing example, the null hypothesis would be that there is no significant difference between the number of times it lands heads and the number of times it lands tails. The chi-squared test is a means of testing whether any deviation between the observed and the expected numbers in an investigation is significant or not. It is a simple test that can be used only if certain criteria are met:

- the sample size must be relatively large, i.e. over 20
- the data must fall into discrete categories – i.e. there is discontinuous variation (Topic 17.1)
- only raw counts and not percentages, rates, etc. can be used
- it is used to compare experimental results with theoretical ones, e.g. in genetic crosses with expected Mendelian ratios

The formula is:

$$\text{chi squared} = \text{sum of } \frac{[\text{observed numbers (O)} - \text{expected numbers (E)}]^2}{\text{expected numbers (E)}}$$

summarised as:

$$\chi^2 = \sum \frac{[(O - E)^2]}{E}$$

The value obtained is then read off on a chi-squared distribution table (Table 1) to determine whether any deviation from the expected results is significant or not. To do this we need to know the number of **degrees of freedom**. This is simply the number of classes (categories) minus one, i.e. if a human can have blood group A or B or AB or O, there are 4 classes and 3 degrees of freedom in this case.

Table 1 *Part of a χ^2 table (based on Fisher)*

Degrees of freedom	Number of classes	χ^2							
1	2	0.00	0.10	0.45	1.32	2.71	3.84	5.41	6.64
2	3	0.02	0.58	1.39	2.77	4.61	5.99	7.82	9.21
3	4	0.12	1.21	2.37	4.11	6.25	7.82	9.84	11.34
4	5	0.30	1.92	3.36	5.39	7.78	9.49	11.67	13.28
5	6	0.55	2.67	4.35	6.63	9.24	11.07	13.39	15.09
Probability that deviation is due to chance alone		0.99 (99%)	0.75 (75%)	0.50 (50%)	0.25 (25%)	0.10 (10%)	0.05 (5%)	0.02 (2%)	0.01 (1%)

←———— Accept null hypothesis
(Any difference is due to chance and not significant)

CRITICAL VALUE

Reject null hypothesis and therefore accept experimental hypothesis that 'any difference is not due to chance and is significant'———→

of χ^2 at 0.05p level as this is the smallest value accepted by statisticians for results being due to chance

Calculating chi squared

Using our example of the coin tossed 100 times, we can calculate the chi-squared value:

Class (category)	Observed (O)	Expected (E)	O − E	(O − E)²	$\frac{(O - E)^2}{E}$
Heads	55	50	+5	25	0.5
Tails	45	50	−5	25	0.5
					$\Sigma = 1.0$

Therefore the value of $\chi^2 = 1.0$.

Using the chi-squared table

To find out whether this value of 1.0 is significant or not we use a chi-squared table, part of which is given in Table 1. Before trying to read this table it is necessary to decide how many **classes of results** there are. In our case there are two classes of results, 'heads' and 'tails'. This corresponds to one **degree of freedom**, as the degrees of freedom are the number of classes minus one. We now look along the row showing two classes (i.e. one degree of freedom) for our calculated value of 1.0. This lies between the values of 0.50 (50%) and 0.25 (25%). This means that the probability that chance alone could have produced the deviation is between 0.50 (50%) and 0.25 (25%). In the chi-squared test the critical value is p = 0.05. This is the value accepted by statisticians, i.e. 5% due to chance. If the probability that the deviation is due to chance is greater than 0.05 (5%), the deviation is said to be **not significant** and the null hypothesis would be accepted. If the probability is less than 0.05 (5%), the deviation is said to be **significant**. In other words, some factor other than chance is affecting the results and the null hypothesis must be rejected. In our example the value is greater than 0.05 (5%) and so we assume the deviation is due to chance and accept the null hypothesis. Had we obtained 60 heads and 40 tails, a chi-squared value of slightly less than 0.05 (5%) would be obtained, in which case the null hypothesis would be rejected and we would assume that the coin might be weighted or biased in some way.

Chi-squared test in genetics

The chi-squared test is especially useful in genetics. To take the example of the genetic cross described in Topic 16.8. If we cross F_1 plants producing round, yellow seeds that we know are **heterozygous** we could expect an F_2 ratio of:

9 round, yellow seeds (186)   3 wrinkled, yellow seeds (72)
3 round, green seeds (48)   1 wrinkled, green seeds (14)

Suppose we obtained 320 plants in the ratio 186:48:72:14 as shown in the brackets. Could this variation be due to statistical chance or could some other factor be the reason for the differences? Our null hypothesis states that there is no significant difference between the observed and the expected results. Applying the chi-squared test:

Class (category)	Observed (O)	Expected (E)	O − E	(O − E)²	$\frac{(O - E)^2}{E}$
Round, yellow seeds	186	180	+6	36	0.2
Round, green seeds	48	60	−12	144	2.4
Wrinkled, yellow seeds	72	60	+12	144	2.4
Wrinkled, green seeds	14	20	−6	36	1.8
					$\Sigma = 6.8$

In this example there are four classes and therefore three degrees of freedom. Using the chi-squared table (Table 1) we see that our value falls between 6.25 and 7.82 shown on the row for three degrees of freedom and that these values correspond to between 0.1 (10%) and 0.05 (5%) probability that the deviation is due to chance alone. The chi-squared value of 6.8 is less than 7.82 at the 0.05 level, so that any deviation from the expected can be accepted to be due to chance. Therefore we accept the null hypothesis and accept that the results are a 9:3:3:1 ratio.

In another experiment domestic fowl with walnut combs were crossed with each other. The expected offspring ratio of comb types was 9 walnut, 3 rose, 3 pea and 1 single. In the event, the 160 offspring produced 103 walnut combs, 20 rose combs, 33 pea combs and 4 single combs (see Figure 1). The null hypothesis states that there is no significant difference between the observed and the expected results. Applying the chi-squared test:

Class (category)	Observed (O)	Expected (E)	O − E	(O − E)²	$\frac{(O - E)^2}{E}$
Walnut comb	103	90	+13	169	1.88
Rose comb	20	30	−10	100	3.33
Pea comb	33	30	+3	9	0.30
Single comb	4	10	−6	36	3.60
					$\Sigma = 9.11$

These results give us a chi-squared value of 9.11. In this instance there are four classes of results (walnut, rose, pea and single comb) and this is equivalent to three degrees of freedom. We must therefore use this row to determine whether the deviations are significant. The value of 9.11 lies between 7.82 and 9.84, which is equivalent to a probability of 0.05 (5%) and 0.02 (2%) that the deviation is due to chance alone. This deviation is significant and we must reject the null hypothesis. Instead we must look for some other genetic explanation than the parents being heterozygous for the two alleles involved.

On these pages you will learn to:

- Explain that gene mutation occurs by substitution, deletion and insertion of base pairs in DNA and outline how such mutations may affect the phenotype

Any change to the sequence of nucleotides in DNA of an organism is known as a **mutation**. Mutations occuring in body cells will not be passed on to the next generation. Those occurring during the formation of gametes may, however, be inherited, often producing sudden and distinct differences between individuals. Changes in the structure or number of whole chromosomes are called **chromosome mutations**, whereas changes to DNA that affect a single locus and therefore produce a different allele of a gene are called **gene mutations** or **point mutations**.

Insertions

A gene mutation by insertion occurs when an extra nucleotide is added into the normal sequence of nucleotides in a DNA molecule. In transcription, this sequence is copied to form mRNA. The nucleotides are 'read' in threes (triplet code) when they are translated into a sequence of amino acids (Topic 6.5). The insertion of an extra base can completely alter the sequence of amino acids. This means that the primary structure is altered and this will affect the secondary and tertiary protein structure. As a result of the inserted nucleotide, every triplet of bases that follows the insertion is altered by one nucleotide. This is called a **frame shift**. The frame shift may alter the sequence so that a STOP codon is created and the polypeptide chain that results will be shorter (premature chain termination). In many cases, the cell will not recognise the polypeptide chain produced and so there is no further processing to produce a functioning protein. As polypeptides are components of enzymes, this often means that the enzyme cannot be made. The complete biochemical pathway that involved the missing enzyme stops, with potentially fatal consequences for the organism. Insertions of more than one nucleotide are also possible. Where there is an insertion of three nucleotides there will be a change at the site of insertion but the sequence of amino acids following this will be the same. Even so, a change such as this can greatly alter the tertiary structure of the protein formed.

Deletions

A gene mutation by deletion occurs when a nucleotide is lost from the normal DNA sequence. Again, this causes a frame shift that results in a completely different sequence of amino acids in the polypeptide from that originally coded for. Enzymes are often not produced, or do not function properly, with major consequences for the organism.

One of the most common mutations associated with the genetic disorder cystic fibrosis is a deletion of three nucleotides. In cystic fibrosis the individual produces mucus that is stickier than usual. As a result, the lungs become congested with this stickier mucus, leading to reduced gaseous exchange. The mucus often blocks the pancreatic ducts as well.

Figure 1 shows the effect of an insertion and deletion mutation on an amino acid sequence.

Substitutions

If a nucleotide in a DNA molecule is replaced by another that has a different base, the type of mutation is known as a substitution. Depending on which new base substitutes for the original one, there are three possible consequences. For example, if we take the triplet of bases in the DNA template strand, guanine-thymine-cytosine (GTC) that codes for the amino acid glutamine, a change to a single base could result in one of the following:

- **A nonsense mutation** occurs if the base change results in the formation of one of the three STOP codons that mark the end of a polypeptide chain (Topic 6.5). For example if the first base, guanine, is substituted by adenine, then GTC becomes ATC. The triplet ATC gives rise to one of the three STOP codons on mRNA. As a result the production of the polypeptide would be stopped prematurely. The final protein would almost certainly be significantly shortened and the protein could not perform its usual function.

- **A mis-sense mutation** occurs when the base change results in a different amino acid being coded for. In our example, if the final base, cytosine, is substituted by guanine, then GTC becomes GTG. The amino acid histidine is coded for by GTG and this then replaces the original amino acid, glutamine. The polypeptide produced as a consequence will be different. How important this change proves to be will depend upon the precise role of the

Table 1 *Different types of substitution mutation*

	Usual triplet of DNA bases	Nonsense mutation	Mis-sense mutation	Silent mutation
Sequence of bases in DNA non-template strand	CAG	TAG	CAC	CAA
Sequence of bases in DNA template strand	GTC	ATC	GTG	GTT
Sequence of bases in mRNA	CAG	UAG	CAC	CAA
Amino acid in polypeptide	GLUTAMINE	STOP CODON	HISTIDINE	GLUTAMINE

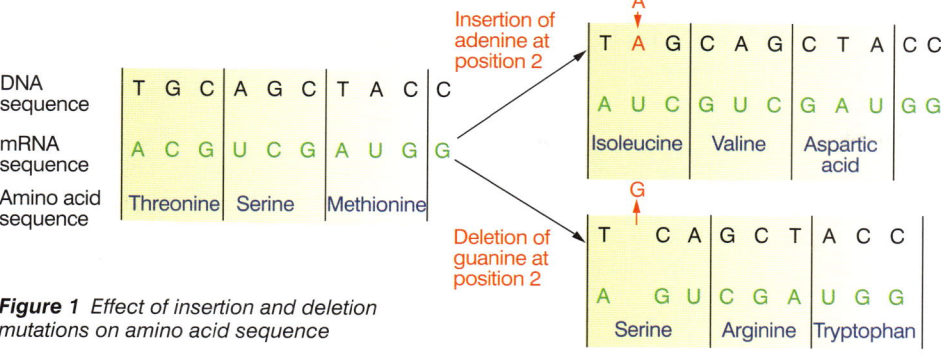

Figure 1 *Effect of insertion and deletion mutations on amino acid sequence*

EXTENSION
Mutagens

Mutagens are agents that increase the natural mutation rate (typically around one to two mutations per 100 000 genes). These agents include:

- Chemicals such as dinitrogen oxide and mustard gas that may directly alter the structure of DNA or interfere with transcription. Hydroxylamine, for example, causes cytosine to pair with adenine rather than guanine. Benzopyrene, a constituent of tobacco smoke, is a powerful mutagen that inactivates the tumour-suppressor gene p53 and so leads to cancer (Topic 8.6).

- High energy radiation, e.g. α particles, β particles and neutrons as well as short wavelength ionising radiation such as X-rays and ultraviolet light. These forms of radiation can disrupt the structure of DNA.

original amino acid. If it is important in forming the bonds that determine the three-dimensional shape of the final protein, then the protein may not function. For example, if the protein is an enzyme, it may no longer have an active site that is complementary to its substrate molecule and so will not catalyse a reaction. Sickle cell anaemia is an example of how a mis-sense substitution mutation may affect the function of a protein.

- **A silent mutation** is one in which the substitution results in a different base occurring in a DNA triplet but one that still codes for the same amino acid. The final polypeptide produced is identical to the original and no effects are apparent. For instance, if the final base in our example is replaced by thymine then GTC becomes GTT. However, as both these triplets code for glutamine, there is no change to the polypeptide produced. Examples of all three types of substitution mutation are given in Table 1.

EXTENSION
Chromosome mutations

Chromosome mutations can take a number of forms:

- **Changes in whole sets of chromosomes** occur when organisms have three or more sets of chromosomes rather than the usual two. This condition is called **polyploidy** and occurs mostly in plants.

- **Changes in the number of individual chromosomes**. Sometimes individual chromosomes fail to segregate during the anaphase stage of meiosis. This is known as **non-disjunction** and results in an organism having one or more additional chromosomes. This is called polysomy. Where **polysomy** in humans occurs with the larger chromosomes, the developmental defects are so severe that infants with them die within a few months. The smaller autosomes, such as numbers 13, 15, 18, 21 and 22, can however be present as three copies and the individual may survive for some time; in the case of chromosomes 21 and 22 they usually survive to adulthood. An example of a non-disjunction in humans is **Down syndrome** in which the effect on the phenotype is that individuals have a broad, flat face, learning difficulties, an increased risk of infections and a shorter life expectancy.

SUMMARY TEST 16.12

Genes may exist in two or more different forms called (**1**). Gene mutations arise through a change to a sequence of (**2**) that make up a gene. They take a number of forms. If the original base sequence was GGCTAGATC, then the types of mutation in the following cases would be: GGCTACGATC = (**3**), GCCTAGATC = (**4**), GGCTAGTTC = (**5**), GGCTAGAC = (**6**), GGCTAGATCC = (**7**). The type of mutation known as substitution takes three forms depending on the consequences of the change. If the mutation results in a different amino acid being coded for, it is called a (**8**) mutation, if the same amino acid is coded for it is a (**9**) mutation. If one of the three STOP codons is formed, it is known as a (**10**) mutation.

On these pages you will learn to:

- Describe the way in which the nucleotide sequence codes for the amino acid sequence in a polypeptide with reference to the nucleotide sequence for *HbA* (normal) and *HbS* (sickle cell) alleles of the gene for the β-globin polypeptide
- Outline the effects of mutant alleles on the phenotype in the following human conditions: albinism, sickle cell anaemia, haemophilia and Huntington's disease
- Explain the relationship between genes, enzymes and phenotype with respect to the gene for tyrosinase that is involved with the production of melanin

A = allele for melanin production

a = allele causing absence of melanin

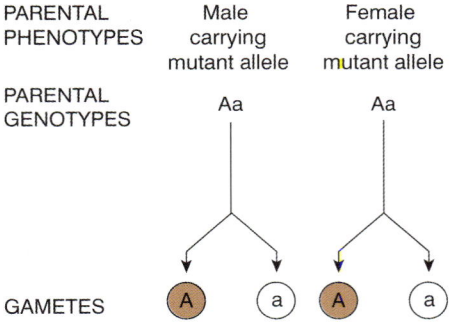

PARENTAL PHENOTYPES — Male carrying mutant allele / Female carrying mutant allele

PARENTAL GENOTYPES — Aa / Aa

GAMETES — A a A a

	GAMETES	
GAMETES	A	a
A	AA	Aa
a	Aa	aa

75% individuals with melanin (AA, Aa)

25% individuals without melanin (aa)

Figure 1 *Cross between two parents heterozygous for albinism*

We have looked at different types of mutation and how they occur. Let us now consider specific examples of the effects of a mutation on the phenotype in humans. One example, haemophilia, was examined in Topic 16.9; others, such as albinism, sickle cell anaemia and Huntington's disease, are considered below.

Albinism

Albinism is an example of the relationship between genes, enzymes and phenotype. A gene (*TYR* gene) codes for the enzyme tyrosinase. This enzyme catalyses the hydroxylation of the amino acid tyrosine to form a chemical known as dopa. It also then catalyses the oxidation of dopa to form dopaquinone, from which the dark pigment melanin is synthesised. Individuals that produce melanin in this way have a phenotype that includes coloured irises of the eyes, as well as hair and skin that are varying degrees of brown and black. The environmental temperature can affect the functioning of tyrosinase, as we shall see in Topic 17.2.

A number of different mutations of the *TYR* gene, located on chromosome 11, have been identified. Most of these result in the production of a non-functioning enzyme. In the absence of the enzyme; no melanin is produced and the individual's phenotype is pale skin, white hair and eyes that appear pink because the red retina is visible through the almost transparent iris.

The mutated allele is recessive and so the condition only occurs when each parent has the recessive mutant allele and the individual inherits a recessive allele from both parents. This may be either when the two parents are heterozygous for the condition, or one is heterozygous and the other has albinism (is homozygous recessive). Figure 1 shows how an albino results from a cross between parents who are each heterozygous for the condition.

Sickle cell anaemia

Sickle cell anaemia was the first human disease to be successfully understood at the molecular level, its cause being suggested by Linus Pauling in 1949. Sickle cell anaemia shows how the smallest of mutations can have a very large effect on the phenotype. It is the result of a gene mutation in the gene producing one of the polypeptides of the haemoglobin molecule and causes the following sequence of events:

- In the gene that codes for the β-globin polypeptide, a single base substitution results in thymine replacing adenine.
- The triplet code on the DNA template strand is CAC rather than CTC.
- As a result, the mRNA transcript has the codon GUG rather than GAG on codon 6, and codes for the amino acid valine (GUG) rather than glutamic acid (GAG) on the sixth amino acid of the polypeptide chain.
- Valine has a side chain that is hydrophobic, whereas glutamic acid has a polar side chain. This minor change produces a molecule of haemoglobin (called haemoglobin-S) that has a 'sticky patch' which results from a change in the surface properties of the molecule owing to the presence of the valine amino acid.
- When the haemoglobin molecules are not carrying oxygen (i.e. at low oxygen concentrations) they tend to adhere to one another by their sticky patches and become insoluble, forming long fibres within the red blood cells.

- These fibres distort the red blood cells, making them inflexible and sickle (crescent) shaped.
- These sickle cells are unable to carry oxygen and may block small capillaries because their diameter is greater than that of capillaries.
- People with sickle cell anaemia have two alleles coding for HbS and so are homozygotes.

Figure 2 shows this sequence of events.

As sickle cell anaemia disables and kills individuals, it might be expected that it would have been eliminated by the process of natural selection. However, for reasons that we shall discuss in Topic 17.8, the condition is relatively common in some parts of Africa and amongst populations of African origin.

Huntington's disease

Although most hereditary diseases are caused by recessive alleles, not all of them are. Huntington's disease is caused by a dominant allele. Everyone who inherits this allele will be affected by the disease. Dominant mutant alleles do not normally last in human populations because the disadvantages caused often prevent individuals surviving long enough to breed and so pass on the allele to their offspring. The effects of Huntington's disease, a progressive deterioration of brain cells, is not usually apparent until individuals are over 30 years old. Some people with the dominant allele will have children before the symptoms develop, so the allele can therefore be passed on before this lethal condition develops.

The huntingtin gene, *HTT*, on chromosome 4 codes for a protein known as huntingtin. Mutations in this gene result in many repeats of the sequence CAG, which results in a protein with many repeats of the amino acid glutamine. Unaffected people have 9 to 36 repeats but people with the mutation have more and with 40 to 100 symptoms will develop; the onset of the disease is earlier and the severity of the symptoms greater the more repeats there are.

As Figure 3 shows, a person who is heterozygous for Huntington's disease has a 50% chance of passing the dominant allele to his or her children, even if the other parent does not have the disease. All people with Huntington's disease have the heterozygous genotype.

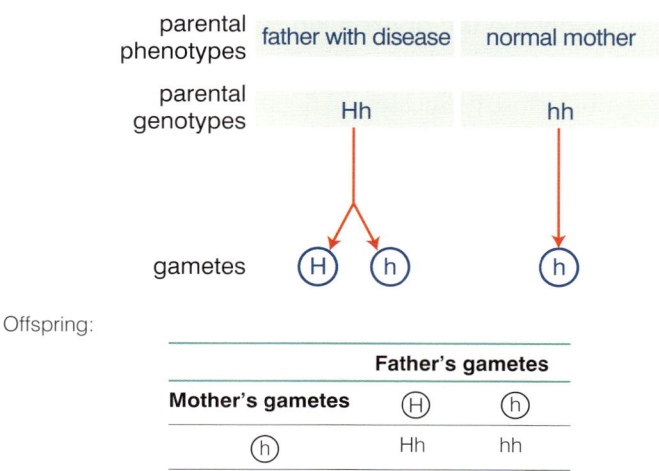

Let allele for Huntington's disease = H
Let allele for normal condition = h

parental phenotypes	father with disease	normal mother
parental genotypes	Hh	hh

gametes (H) (h)   (h)

Offspring:

Mother's gametes	Father's gametes	
	(H)	(h)
(h)	Hh	hh

Half (50%) of offspring will have Huntington's disease (Hh).
Half (50%) of offspring will be normal (hh).

Figure 3 *Cross between one parent with Huntington's disease and the other without the disease*

1. The DNA molecule which codes for the β-globin polypeptide chain in has a mutation whereby the base thymine replaces adenine on its non-template strand

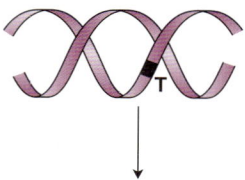

2. The mRNA produced has the triplet codon GUG (for amino acid valine) rather than GAG (for amino acid glutamic acid)

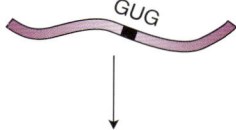

3. The β-globin polypeptide chain produced has one glutamic acid molecule replaced by a valine molecule

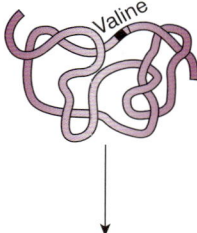

4. The haemoglobin molecule containing the abnormal β chains becomes sticky and forms long fibres when the oxygen level of the blood is low. This haemoglobin is called haemoglobin-S

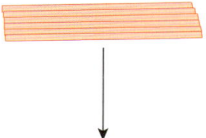

5. Haemoglobin-S causes the shape of the red blood cell to become crescent (sickle) shaped

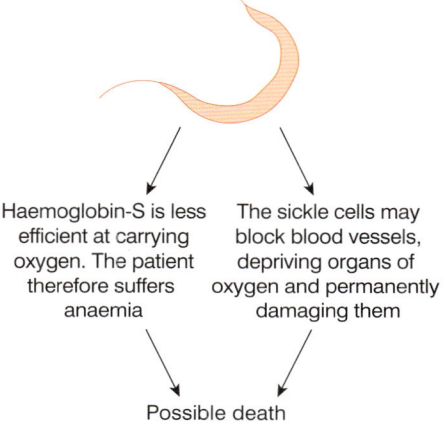

Haemoglobin-S is less efficient at carrying oxygen. The patient therefore suffers anaemia

The sickle cells may block blood vessels, depriving organs of oxygen and permanently damaging them

Possible death

Figure 2 *How a single substitution mutation results in sickle cell anaemia*

On these pages you will learn to:

- Distinguish between structural and regulatory genes and between repressible and inducible enzymes
- Explain genetic control of protein production in a prokaryote using the *lac* operon

All organisms synthesise proteins by the mechanisms described in Topics 6.6 and 6.7. Some of these proteins are needed continually by the cell and so the polypeptides of which they are made are produced continuously. Other proteins are only required in certain circumstances and therefore need not be produced all the time – indeed, to do so would be wasteful. As the proteins produced are usually enzymes, we recognise two types of enzyme depending on whether they are produced continually or only when required.

- **Repressible enzymes** are normally produced continuously because the gene that codes for them is always expressed (switched on). However, their production can be decreased as the concentration of certain substances, such as the products of the reaction they catalyse, increases. The product may bind to an allosteric site of an inactive repressor protein (see **regulatory genes** on this page). The activated repressor then binds to the operator (see **lac operon**) and prevents RNA polymerase binding or progression along the DNA. As this binding is temporary, when the product is used up in the cell, the situation reverses and transcription proceeds.
- **Inducible enzymes** are produced only when the gene that codes for them is switched on as a result of the presence of a specific molecule such as the substrate of the enzyme. Examples of inducible enzymes are those produced in the lac operon.

Control of protein synthesis in bacteria

Under normal circumstances the bacterium *Escherichia coli* (Figure 2) absorbs and respires glucose as its main respiratory substrate. If, however, it is grown on a medium in which lactose is present it is able to produce two enzymes:

- β-galactosidase – the enzyme that hydrolyses lactose to glucose and galactose. It also catalyses the conversion of lactose to allolactose, which is the molecule that binds to the repressor protein: coded for by gene *Z*
- lactose permease – a membrane transport protein for the uptake of lactose: coded for by gene *Y*
- transacetylase – an enzyme thought to have a role in sugar metabolism and detoxification: coded for by

gene *A*. As this enzyme is not directly involved in lactose uptake and metabolism it is left out of the following description of the lac operon. Gene *A* occurs at the end of the operon.

As lactose is not always available, it would be wasteful of material and energy for *E. coli* to produce these two enzymes continuously. It is far better to express the genes that code for these enzymes only when they are needed, i.e. when lactose is available. How then is *E. coli* able to do this?

The lac operon

The lac operon is a length of DNA that is responsible for the production of β-galactosidase and lactose permease. As shown in Figure 1, it is made up of a number of parts.

- **The structural genes** comprise the length of DNA with the bases needed to produce the mRNA that codes for the polypeptides that make up the two enzymes.
- **The operator** is a portion of DNA lying next to the structural genes that effectively switches them on and off.
- **The promoter** is a portion of DNA to which the enzyme RNA polymerase becomes attached in order to begin the process of **transcription** of DNA to form mRNA from the structural genes (Topic 6.6). The promoter is adjacent to the operator – the importance of this will become clear later.
- **The regulatory gene** is a portion of the DNA that is not part of the lac operon and is situated some distance from it (Figure 1). As its name suggests, it regulates the process of protein synthesis through the production of a **repressor protein**.

How the lac operon works

The following account of how *E. coli* controls the production of lactose permease and β-galactosidase is shown in Figure 1.

- The regulatory gene codes for a protein called the repressor protein.
- The repressor protein has two different binding sites:
  - one that binds to lactose (allolactose)
  - one that binds to the operator.
- If there is no lactose present in the medium on which *E. coli* is growing, the repressor protein will bind to the operator.
- Because the operator and promoter are close together, the repressor protein covers part of the promoter when it binds to the operator. This blocks the site on the promoter to which RNA polymerase normally attaches.

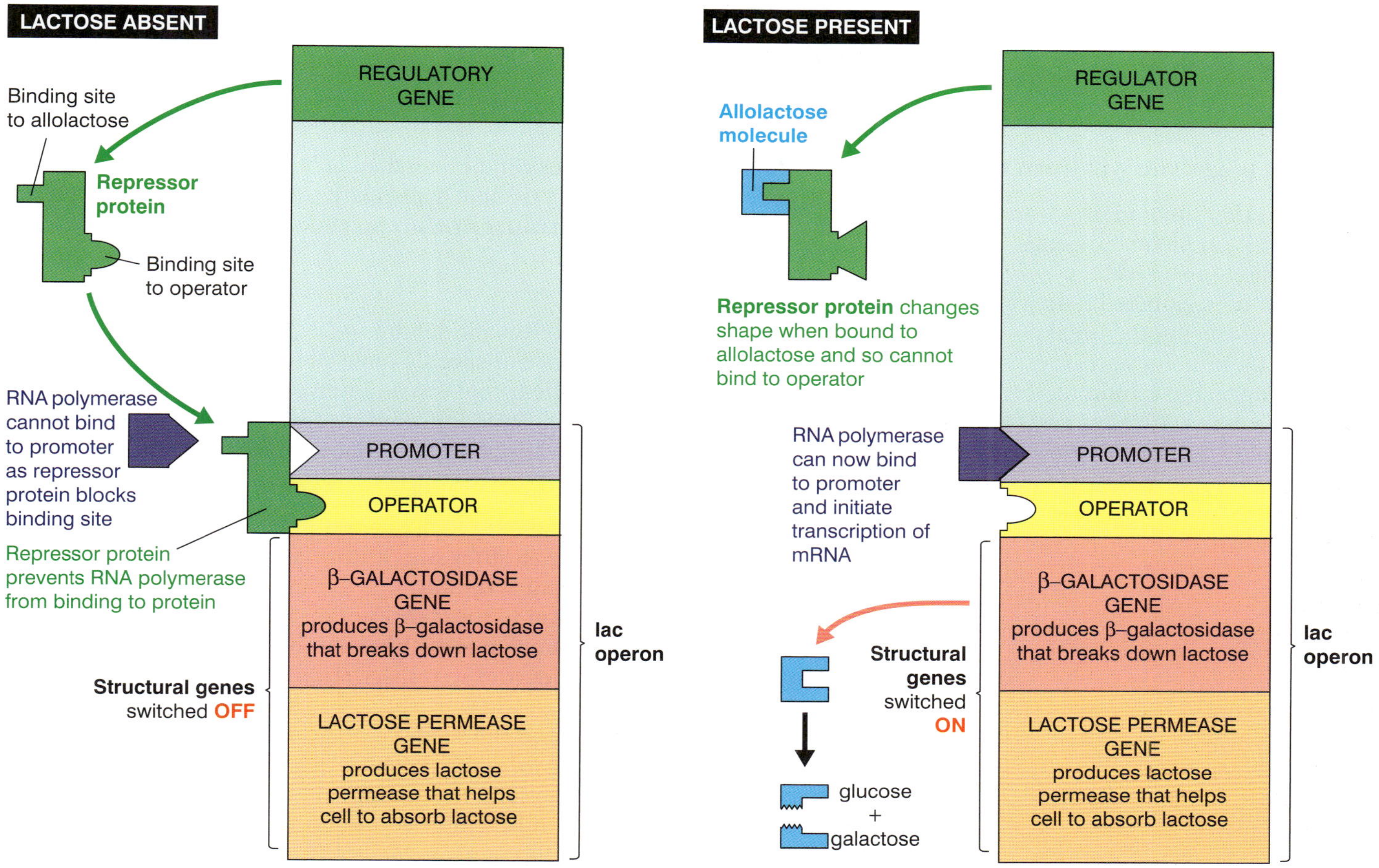

LACTOSE ABSENT

Binding site to allolactose

Repressor protein

Binding site to operator

RNA polymerase cannot bind to promoter as repressor protein blocks binding site

Repressor protein prevents RNA polymerase from binding to protein

Structural genes switched **OFF**

REGULATORY GENE

PROMOTER

OPERATOR

β–GALACTOSIDASE GENE
produces β–galactosidase that breaks down lactose

LACTOSE PERMEASE GENE
produces lactose permease that helps cell to absorb lactose

lac operon

LACTOSE PRESENT

Allolactose molecule

Repressor protein changes shape when bound to allolactose and so cannot bind to operator

RNA polymerase can now bind to promoter and initiate transcription of mRNA

Structural genes switched **ON**

glucose + galactose

REGULATOR GENE

PROMOTER

OPERATOR

β–GALACTOSIDASE GENE
produces β–galactosidase that breaks down lactose

LACTOSE PERMEASE GENE
produces lactose permease that helps cell to absorb lactose

lac operon

Figure 1 *Functioning of the lac operon in* E. coli

- RNA polymerase cannot attach to the promoter and therefore cannot express (switch on) the structural genes that code for the two enzymes.
- mRNA cannot be made on the structural genes and therefore neither β-galactosidase nor lactose permease can be synthesised, i.e. their production stops.
- If lactose is added to the medium on which *E. coli* is growing, it is converted to allolactose, which binds to its site on the repressor protein (there is a very small quantitiy of β-galactosidase present to convert lactose to allolactose).
- As a result of this binding of allolactose, the repressor protein changes shape in such a way that it can no longer use its other binding site to attach to the operator.
- With no repressor protein to block the promoter, RNA polymerase can now attach to the promoter.
- The RNA polymerase is able to begin the transcription of the structural genes and form mRNA.
- At the ribosomes, this mRNA acts as a template for the assembly of the amino acids needed to synthesise β-galactosidase and lactose permease.
- Lactose permease helps *E. coli* to absorb lactose rapidly from the medium and this is hydrolysed by β-galactosidase to glucose and galactose – the glucose being used as a respiratory substrate.

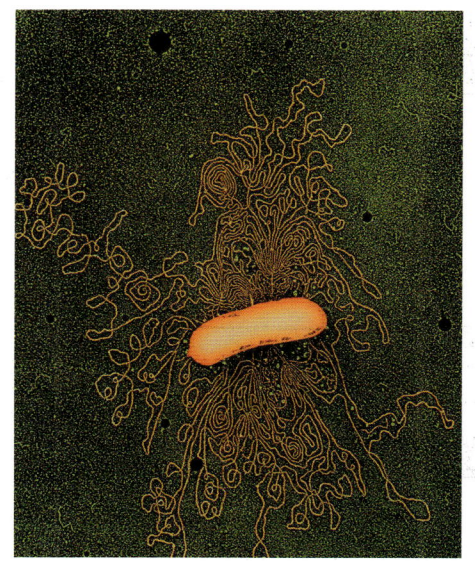

Figure 2 Escherichia coli *treated so that its DNA is ejected from the cell. This DNA is 1.5 mm long – some 1000 times the length of the cell – and part of it forms the lac operon*

Gene control – regulation of transcription

On these pages you will learn to:

- Explain the function of transcription factors in gene expression in eukaryotes
- Explain how gibberellin activates genes by causing the breakdown of DELLA protein repressors, which normally inhibit factors that promote transcription

We saw in Topic 16.14 how genes can be switched on and off to control protein, especially enzyme, synthesis. Let us now investigate some ways in which cells regulate gene expression using **transcription factors**.

Transcription factors

RNA polymerase in eukaryotic cells cannot bind to DNA and initiate (begin) transcription without the presence of specific molecules, known as transcription factors. These are produced in the cytoplasm and move into the nucleus. Most transcription factors are proteins. Different transcription factors have different roles but they are all involved in the control of gene expression (switching genes on or off).

- Some transcription factors bind to specific regions of DNA. These sites may be within the promoter region or at other sites known as enhancer regions that also have a role in initiating transcription.
- Other transcription factors bind to the transcription factors attached to the DNA to form a complex. RNA polymerase binds to this complex to form a transcription initiation complex and so transcription of the structural genes can begin.
- Messenger RNA is produced and the sequences of nucleotides it carries are translated at the ribosomes to form the polypeptide.
- The formation of many of these complexes, especially in higher eukaryotes, requires the presence of many different molecules. If a specific factor is missing to complete the formation of the complex, then transcription will not proceed.
- Some transcription factor complexes may act to prevent RNA polymerase binding or progression along the DNA, so preventing gene expression.

Extracellular factors can be involved in the control of gene expression by transcription factors. One example is the hormone oestrogen (see Figure 1):

- Oestrogen is a lipid-soluble molecule and so diffuses easily through the **phospholipid** portion of membranes, including the cell surface membrane.
- Once inside the cytoplasm of a cell, oestrogen enters the nucleus and binds to a complementary site in an oestrogen receptor molecule (ER). The ER is a transcription factor. The shape of this site and the shape of the oestrogen molecule complement one another.
- Binding of oestrogen to ER causes attached proteins (heat-shock proteins) to dissociate and the shape of the receptor to change.
- This change of shape and the removal of the heat-shock proteins allows the oestrogen-ER complex to bind to DNA.
- Together with other factors, an transcription initiation complex is formed and transcription begins.
- The genes that are expressed as a result of the activation of the transcription factor ER are those associated with proteins involved in growth.

Activation of genes by gibberellins

In Topic 15.13 we looked at some of the effects of plant growth regulators called gibberellins on germination. Let us now examine how gibberellins have their effect by activating genes.

We saw that gibberellins promote the germination of seeds such as barley. Present in plant cells are substances called **DELLA** proteins. DELLA proteins bind to transcription factors, preventing them from initiating the expression of genes associated with proteins involved in growth. In this way, DELLA proteins act as inhibitors of cell growth and therefore maintain dormancy by inhibiting seed germination. By causing the breakdown these DELLA proteins, gibberellins are able to stimulate germination. The process by which gibberellins activate genes is as follows.

- Transcription factors normally promote transcription of a portion of DNA that codes for enzymes involved in the process of germination.
- DELLA proteins inhibit these factors and therefore act as repressors of this transcription and prevent germination.
- Gibberellin binds to a specific receptor (GID1) within the cell and this complex binds to DELLA proteins. This leads to the activation of a large complex of enzymes that are able to break down of the DELLA proteins.
- This allows the transcription factors to operate and so transcription of DNA can proceed. This results in the production of enzymes such as amylase.
- Amylase converts maltose to glucose, which is respired by the embryo during germination.

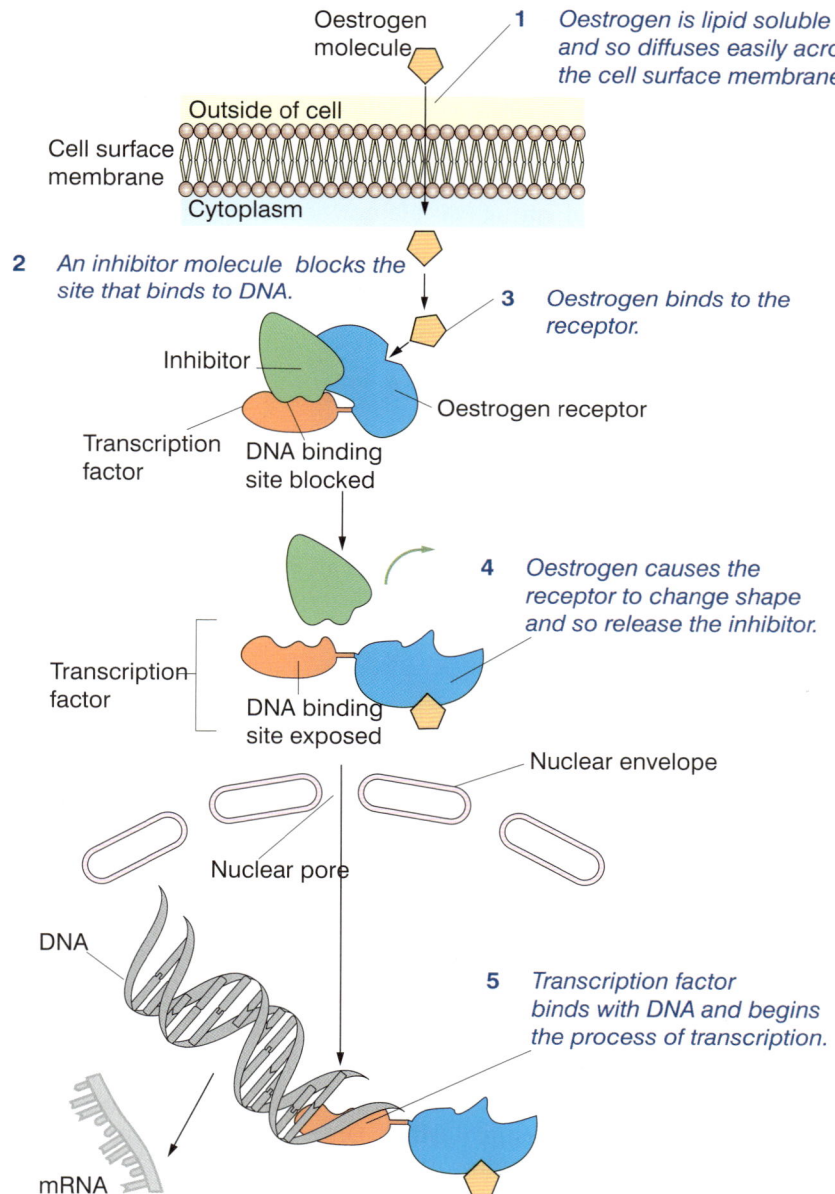

Figure 1 *The effect of oestrogen on gene transcription*

SUMMARY TEST 16.15

For a gene to produce mRNA it needs to be stimulated by specific molecules that move into the **(1)**. These molecules are called transcription factors which bind to a specific region of **(2)**. If one or more factors that make up a transcription initiation complex is missing then RNA polymerase cannot proceed and the gene is switched **(3)**. Hormones like **(4)** can combine with a **(5)** portion of the transcription factor because their shapes are **(6)**. This causes the transcription factor to change shape and release the attached protein molecules, allowing **(7)** to take place. In plants, genes may be activated by plant growth factors called **(8)** that combine with, and cause the breakdown of, **(9)** proteins. These proteins **(10)** transcription factors and so their breakdown allows transcription to take place.

16 Examination Questions

1 Colour blindness is a condition characterised by the inability of the brain to perceive certain colours accurately.

- The most common form is termed red-green colour blindness (RGC).

- RGC results from a recessive allele.

- 0.6% of females worldwide have RGC.

- 8.0% of males worldwide have RGC.

Figure 1 shows the occurrence of RGC in one family.

Figure 1

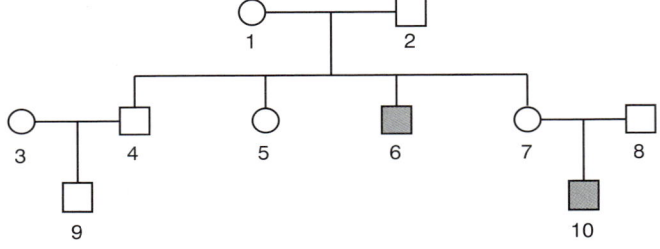

a Explain the meaning of the terms *allele* and *recessive*.
(2 marks)

b Explain why females are less likely than males to have RGC.
(2 marks)

c With reference to Figure 1, and using the symbols **R** for the dominant allele and **r** for the recessive allele, state the genotypes of the individuals **1**, **4**, **6** and **7**.
(4 marks)
(Total 8 marks)

Cambridge International AS and A Level Biology 9700 Paper 4 Q6 June 2008

2 In mice there are several alleles of the gene that controls the intensity of pigmentation of the fur.

The alleles are listed below in order of dominance with **C** as the most dominant.

    C = full colour

    C^ch = chinchilla

    C^h = himalayan

    C^p = platinum

    C^a = albino

The gene for eye colour has two alleles. The allele for black eyes, **B**, is dominant, while the allele for red eyes, **b**, is recessive.

A mouse with full colour and black eyes was crossed with a himalayan mouse with black eyes. One of the offspring was albino with red eyes.

Using the symbols above, draw a genetic diagram to show the genotypes and phenotypes of the offspring of this cross.
(6 marks)
(Total 6 marks)

Cambridge International AS and A Level Biology 9700 Paper 4 Q8 November 2008

3 a The fruit fly, *Drosophila melanogaster*, feeds on sugars found in damaged fruits. A fly with normal features is called a wild type. It has a striped body and its wings are longer than its abdomen. There are mutant variations such as an ebony coloured body or vestigial wings. These three types are shown in Figure 2.

Figure 2

Wild type features are coded for by dominant alleles. **A** for wild type body and **B** for wild type wings.

Explain what is meant by the terms *allele* and *dominant*.
(2 marks)

b Two wild type fruit flies were crossed. Each had alleles **A** and **B** and carried alleles for ebony body and vestigial wings. Draw a genetic diagram to show the possible offspring of this cross.
(6 marks)

c When the two heterozygous fruit flies in (b) were crossed, 384 eggs hatched and developed into adult flies.

A chi-squared (χ^2) test was carried out to test the significance of the differences between observed and expected results.

$$\chi^2 = \Sigma \frac{(O - E)^2}{E}$$

where Σ = sum of
          O = observed value
          E = expected value

i Copy Table 1 and complete the missing values.

(3 marks)

Table 1

	phenotypes of *Drosophila melanogaster*			
	grey body long wing	grey body vestigial wing	ebony body long wing	ebony body vestigial wing
observed number (O)	207	79	68	30
expected ratio	9	3	3	1
expected number (E)	216	72	72	24
O − E	−9	....	−4	6
$(O - E)^2$	81	....	16	36
$\frac{(O - E)^2}{E}$	0.38	....	0.22	1.50

ii Calculate the value for χ^2.   *(1 mark)*
   Table 2 relates χ^2 values to probability values. As four classes of data were counted the number of degrees of freedom was 4−1 = 3. Table 2 gives values of χ^2 where there are three degrees of freedom.

Table 2

probability greater than	0.50	0.20	0.10	0.05	0.01	0.001
values for χ^2	2.37	4.64	6.25	7.82	11.34	16.27

iii Using your value for χ^2, and Table 2, explain whether or not the observed results were significantly different from the expected results.

(2 marks)
(Total 14 marks)

Cambridge International AS and A Level Biology 9700 Paper 41 Q7 November 2009

16 Practice Questions

4 In humans, Huntington's disease is caused by a dominant, mutant allele. Draw a genetic diagram to show the possible genotypes and phenotypes of the offspring produced by a man with one allele for the disease and a woman who does not suffer from the disease.

5 In cocker spaniels, black coat colour is the result of a dominant allele and red coat colour is the result of a corresponding recessive allele.

  a Draw a genetic diagram to show a cross between a pure breeding bitch (female) with a black coat and a pure breeding dog (male) with a red coat.

  b If a dog and a bitch from this first cross are mated, what is the probability that any one of the offspring will have a red coat? Use a genetic diagram to show your working.

6 In shorthorn cattle there is a gene C that determines coat colour. The gene has two alleles:

  • the allele C^W produces a white coat when homozygous

  • the allele C^R produces a red coat when homozygous.

  In the heterozygous state the coat is light red, a colour also known as roan. The roan coat is a mixture of all white hairs and all red hairs. As each hair is either all red or all white, the C^W and C^R alleles are codominant.

  The gene for coat colour is **not** sex-linked.

  a Draw a genetic diagram to show the possible genotypes and phenotypes of a cross between a bull with a white coat and a cow with a roan coat.

  b In each of the following crosses between shorthorn cattle what is the percentage of offspring with a roan coat?
    i   red coat × white coat
    ii red coat × roan coat
    iii white coat × roan coat
    iv roan coat × roan coat.

7 A man claims not to be the father of a child. The man is blood group O while the mother of the child is blood group A and the child is blood group AB. State, with your reasons, whether you think the man could be the father of the child.

8 In some breeds of domestic fowl, the gene controlling feather shape has two alleles that are codominant. The allele A^S when homozygous produces straight feathers. The allele A^F when homozygous produces frizzled feathers. The heterozygote for feather shape gives mildly frizzled feathers. Draw a genetic diagram to show the genotypes and phenotypes resulting from a cross between a mildly frizzled cockerel and a frizzled hen. The gene for feather shape is **not** sex-linked.

Selection and evolution

Variation

On these pages you will learn to:

- Describe the differences between continuous and discontinuous variation and explain the genetic basis of continuous (many, additive genes control a characteristic) and discontinuous variation (one or few genes control a characteristic)

Every one of the billions of organisms on planet Earth is unique. Even monozygotic twins, although genetically identical, vary as a result of their different environmental experiences. The phenotypic variation shown within a species for a particular characteristic can be quantified. There are two main types of variation: **continuous variation** and **discontinuous variation**.

Continuous variation

Some characteristics of organisms appear to have a graded effect and the phenotypes do not appear to fall into distinct classes. In humans, two examples are height and mass. Characteristics that display this type of variation are not controlled by a single gene, but by many genes (polygenes). These genes have an additive effect, each contributing in some way, and to a different extent, to the overall phenotype produced.

Environmental factors play a major role in determining where on the continuum an organism lies. For example, individuals who are genetically predetermined to be the same height grow to different heights as a result of variations in environmental factors, such as diet.

This type of variation is the product of polygenes and the environment. Table 1 shows the number of people in a particular sample (frequency) with various heights. If we take these data and plot them on a graph, we obtain a bell-shaped curve known as a normal distribution curve (Figure 1).

Table 1 *Frequency of heights in a sample of humans (measured to the nearest 2 cm)*

Height/cm	Frequency
140	0
144	1
148	23
152	90
156	261
160	393
164	440
168	413
172	177
176	63
180	17
184	4
188	1
190	0
192	0

Discontinuous variation

Some characteristics in organisms fit into a few distinct forms; there are no intermediate types, or at least there is very little overlap between groups. This is called discontinuous variation. In the ABO blood grouping system (Topic 16.7), for example, there are four distinct groups: A, B, AB and O (Figure 2). A characteristic displaying discontinuous variation is usually controlled by a few genes or often just a single gene – in the case of blood groups the *ABO* gene. Discontinuous variation can be represented on bar charts or pie graphs. Environmental factors have little influence on discontinuous variation.

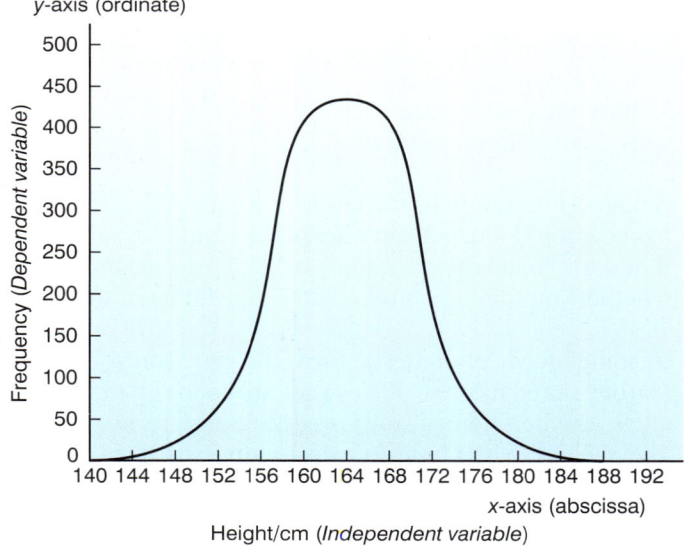

Figure 1 *Graph of frequency against height for a sample of humans*

As features showing discontinuous variation are often controlled by a single gene and gene mutations normally affect a single gene, it follows that many human diseases that result from gene mutations show discontinuous variation. Examples include albinism, sickle cell anaemia and Huntington's disease, which were discussed in Topic 16.3, and haemophilia (Topic 16.9).

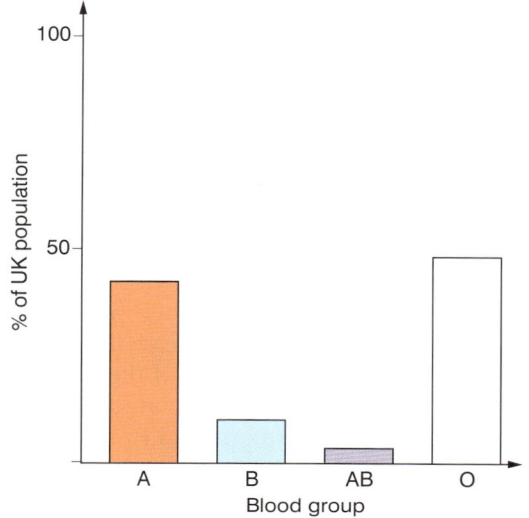

Figure 2 *Discontinuous variation illustrated by the percentage of the UK population with blood groups A, B, AB and O*

Causes of variation – genetic differences

Genetic variation is the result of the genotype of each individual, in other words the genes they possess and the forms of allele of each gene that are present. This genetic make-up varies from generation to generation as a result of:

- **Mutations** – a change in sequence of bases in DNA is called a **gene mutation** (Topic 16.12). While mutations in body cells are not passed on to the next generation, those in reproductive tissues (testes and ovaries in animals and anthers and ovaries in flowering plants) may be inherited. In organisms that are asexually reproducing, mutations are the only source of genetic variation.
- **Crossing over** and the consequent formation of recombinants during prophase I of meiosis (Topic 16.2) leads to equivalent sections of non-sister **chromatids** being exchanged and therefore separates alleles of linked genes that would otherwise be inherited together.

- **Random assortment of pairs of homologous chromosomes** at metaphase I of meiosis (Topic 16.2) results in daughter cells that are genetically different. This is because the chromosome set they have received is the result of a random alignment of homologous pairs at the spindle equator. The greater the haploid number, the greater the number of different possible combinations.
- **Random mating between individuals within a species** – which pair of organisms of a species that mate, and hence which two genotypes combine, is largely a matter of chance, although in some animals (especially humans) there may be an element of choice involved.
- **Random fusion of gametes at fertilisation**. When mating takes place, which gamete fuses with which at fertilisation is random.

Causes of variation – environmental influences

Environmental influences affect the way a genotype is expressed and result in different phenotypes (Topic 17.2). Climatic factors (e.g. temperature, availability of water and sunlight) lead to variation, as do pH and quantity and type of nutrients. The environment may also influence genetic variation by, for example, affecting the rate of mutation. It may also switch genes on and off. The contribution of the environment to variation can be studied when organisms that are genetically identical are used. In the case of humans, monozygotic (identical) twin studies can be carried out.

SUMMARY TEST 17.1

There are two main types of variation in organisms. In continuous variation, the various forms of a characteristic grade into one another. Examples in humans include **(1)** and **(2)**. In a large sample of humans, if we plot the number of people at each point within the range of either of these characteristics, we obtain a bell-shaped graph called a **(3)**. Continuous variation results from the interaction of many genes rather than a single gene. These genes are collectively called **(4)**. Continuous variation is not only the result of a group of genes, but is also affected by **(5)**. In discontinuous variation, organisms fit into a number of distinct groups with no intermediate types. It is usually the result of **(6)** gene and an example in humans is **(7)**. Discontinuous variation is hardly, if ever, influenced by the environment. Variation has many causes. A change in the sequence of bases of DNA, known as a **(8)**, can cause variation, as can crossing over during **(9)** stage of meiosis and the consequent formation of **(10)**. Further variation results from the **(11)** of pairs of homologous chromosomes during the **(12)** stage of meiosis as well as random **(13)** between individuals of a **(14)** and random **(15)**.

Environment and phenotype

Figure 1 *Siamese cats have dark fur at the extremities of their bodies as a result of an enzyme involved in melanin production that only functions at lower temperatures*

Figure 2 *The Arctic fox produces darker pigmentation when the temperature is warmer, giving rise to the grey-brown summer coat. In colder conditions, no pigmentation occurs, giving rise to the white winter coat that camouflages it better against its snowy background*

You may recall that in Topic 16.4 we saw how the final appearance of an organism – its phenotype – is the result of the genotype and the effect of the environment upon it. If organisms of identical genotype are exposed to different environmental influences, they show considerable variety. Because environmental influences, e.g. temperature and light intensity, are themselves very various and because they form gradations, they are largely responsible for continuous variation within a population.

How the environment may affect the phenotype

The **alleles** that make up the genotype of an organism provide a blueprint that determines the limits within which the organism will develop. The degree to which an allele is expressed often depends on the environment. Examples include:

- The **recessive** c^s allele in Siamese cats and the equivalent c^h alelle in Himalayan rabbits code for a heat sensitive form of the enzyme tyrosinase (Topic 16.7). This enzyme is involved in the production of the dark pigment, melanin. The c^s/c^h form of the enzyme does not function at temperatures above 33 °C. Over much of the body surface of Siamese cats and Himalayan rabbits the temperature is above 33 °C and so the enzyme is inactive and no melanin is produced during development. The fur in these regions is therefore light in colour. At the extremities such as the tips of the tail, ears, feet and nose, the temperature is usually below 33 °C and so the heat sensitive form of tyrosinase is active and melanin is produced. These regions are therefore much darker in colour (Figure 1).

- A small Californian plant, *Potentilla glandulosa*, has a number of genetic forms, each adapted to growing at different altitudes. Experiments were carried out as follows:
  - plants of *Potentilla* were collected from three altitudes – high, medium and low
  - one plant from each location was split into three cuttings, each of which therefore had an identical genotype
  - one of the cuttings from each location was grown at each altitude (high, medium and low)
  - three separate sets of genetically identical plants were therefore grown in three different environments.

  The results are illustrated in Figure 3 and show that plants with identical genotypes differ in phenotype (height, number of leaves, overall size and shape) and even survival rate, according to the environment in which they live.

- Arctic foxes (Figure 2) have the alleles to make fur pigments and so produce dark coats. These pigments are, however, produced only in warm temperatures. They are therefore not produced as the colder temperatures of winter approach and the surface hairs are slowly replaced by white ones. By the time the winter snows cover the ground, the Arctic fox is completely white and better camouflaged and therefore more able to capture its prey.

- The height of humans is determined by the range of alleles for height that each of us inherits from our parents. However, even if our alleles allow us to grow tall, our diet will influence whether we do so. For example, a lack of calcium, phosphate or poor overall nutrition especially at critical growth periods (early years and adolescence) may prevent maximum bone and body growth and so we fail to realise our full potential height.

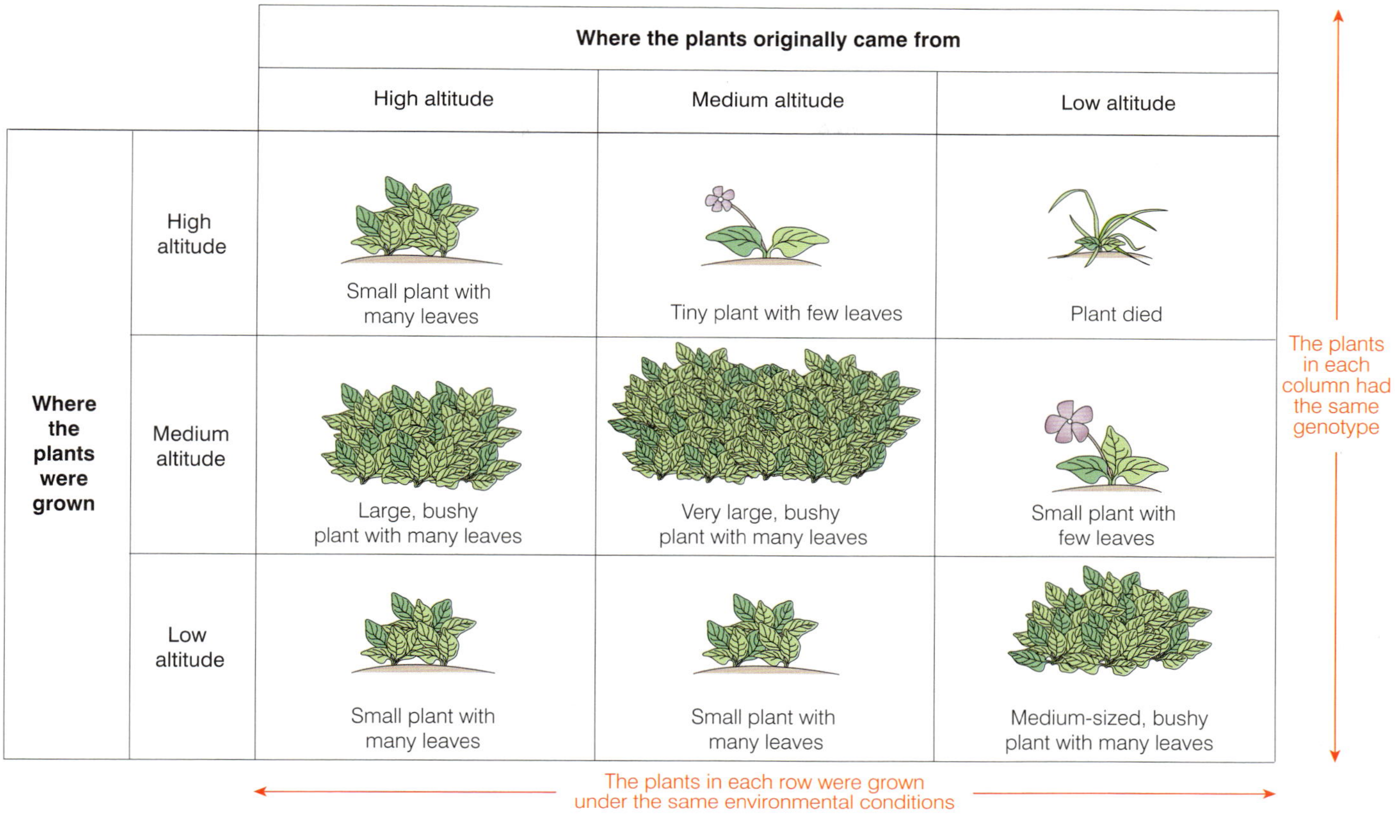

		Where the plants originally came from		
		High altitude	Medium altitude	Low altitude
Where the plants were grown	High altitude	Small plant with many leaves	Tiny plant with few leaves	Plant died
	Medium altitude	Large, bushy plant with many leaves	Very large, bushy plant with many leaves	Small plant with few leaves
	Low altitude	Small plant with many leaves	Small plant with many leaves	Medium-sized, bushy plant with many leaves

The plants in each column had the same genotype

The plants in each row were grown under the same environmental conditions

Figure 3 *Effect of environment on phenotype – growing genetically identical* Potentilla glandulosa *at different altitudes*

- A set of plants grown in a soil deficient in nitrogen will develop far less **biomass** than another genetically identical set grown in soils with a plentiful supply of nitrogen.
- The environment may induce a mutation which affects the phenotype. For example, ultraviolet radiation from the sun or tanning lamps can disrupt DNA replication and lead to the production of melanomas on the skin. Melanomas are a form of skin cancer which cause changes to the appearance and patterning of moles on the skin.

The effect of the environment on the phenotype is greater for those characteristics that are determined by more than one gene – polygenes (Topic 17.1). The genotype determines the range of possible phenotypes, but the environment often determines where within that range the actual appearance of an organism lies.

SUMMARY TEST 17.2

The phenotype of an organism is the result of the effect of the **(1)** and the organism's **(2)**. For example, in Siamese cats there is a heat sensitive form of the **(3)** called tyrosinase that is involved in the production of the pigment known as **(4)**. The production of this heat sensitive form of tyrosinase is controlled by an **(5)** represented by c^s. This form of the enzyme is inactive at temperatures **(6)** 33 °C. Body temperatures below this normally occur at the **(7)** of the body and so in these regions the pigment is produced and the fur is coloured **(8)**. If genetically identical forms of the plant *Potentilla glandulosa* are grown at different **(9)** their phenotype differs in **(10)** and **(11)** due to the different environments they experience. It is clear therefore that the **(12)** of an organism determines the **(13)** of possible phenotypes, but it is the **(14)** that influences its final appearance.

EXTENSION

Further proof!

How can we be sure that the dark extremities of Siamese cats are the result of the temperature at which fur develops, rather than simply being genetically determined? A couple of simple experiments prove the case:

- Some dark fur is removed from the tail of a Siamese cat which is then kept in a warmer than usual environment. The new fur that develops is light in colour.
- Some light fur is removed from the back of a Siamese cat and the shaved area kept at a lower than normal temperature. The new growth of fur is black.

If fur colour is determined only by genes, with no environmental influence, then in both cases the new fur would have matched the original colour.

On these pages you will learn to:

- Use the *t*-test to compare the variation of two different populations

Remember

A large standard deviation means a lot of variety; a small standard deviation means little variety.

The *t*-test is used to find out if the difference between the mean of two sets of continuous data is significant or if it is purely due to chance. The *t*-test is used if:

- the data that have been collected are continuous
- the data are from a population that is normally distributed
- the standard deviations are approximately the same
- each of the two samples has fewer than 30 values.

The equation for the t-test is in two parts that are expressed as:

$$t = \frac{\bar{x}_1 - \bar{x}_2}{\sqrt{\dfrac{s_1^2}{n_1} + \dfrac{s_2^2}{n_2}}}$$

and

$$v = n_1 + n_2 - 2$$

Where:

$\bar{x}$ = mean value
s = standard deviation
n = sample size (number of observations)
v = degrees of freedom

The *t*-test makes use of the **mean** and **standard deviation**, so let us begin by looking at these.

Mean and standard deviation

A normal distribution curve always has the same basic shape (Figure 1). It differs in two measurements: its maximum height and its width.

- The **mean** is the measurement at the maximum height of the curve. The mean of a sample of data provides an average value and useful information when comparing one sample with another. It does not, however, provide any information about the range of values within that sample. For example, the mean number of children in a sample of eight families may be 2. However, this could be made up of eight families each with two children or six families with no children and two families with eight children each.
- The **standard deviation** (s) is a measure of the width of the curve. It gives an indication of the range of values either side of the mean. A standard deviation is the distance from the mean to the point where the curve changes from being convex to concave (the point of inflexion). Of all the measurements, 68% lie within this range. Increasing this width to almost two (actually 1.96) standard deviations takes in 95% of all measurements. These measurements are shown in Figure 1.

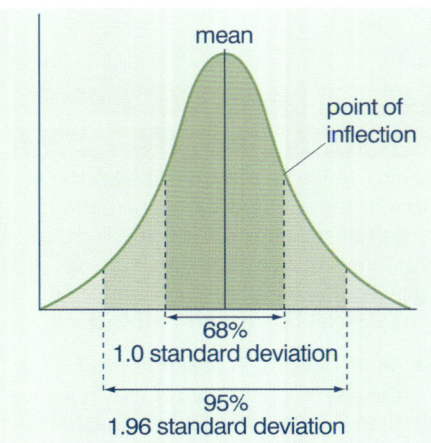

Figure 1 *The normal distribution curve, showing values for standard deviation*

Calculating the standard deviation

At first sight, the formula for standard deviation can look complex:

$$\text{standard deviation} = \sqrt{\frac{\sum(x - \bar{x})^2}{n - 1}}$$

Where:

$\sum$ = the sum of

x = measured value (from the sample)

$\bar{x}$ = mean value

n = total number of values in the sample.

However, it is straightforward to calculate and less daunting if you take it step by step. The following very simple example, using the six measured values (x) 4, 1, 2, 3, 5 and 0, illustrates each step in the process.

Step 1 Calculate the mean value ($\bar{x}$), i.e. $4 + 1 + 2 + 3 + 5 + 0 = 15$

$$15 \div 6 = 2.5.$$

Step 2 Subtract the mean value (2.5) from each of the measured values ($x - \bar{x}$). This gives: $+1.5, -1.5, -0.5, +0.5, +2.5, -2.5$.

Step 3 As some of these numbers are negative, we need to make them positive. To do this, square **all** the numbers ($x - \bar{x}$)^2. Remember to square all the numbers and not just the negative ones. This gives: 2.25, 2.25, 0.25, 0.25, 6.25, 6.25.

Step 4 Add all these squared numbers together:

$$\sum(x - \bar{x})^2 = 17.5$$

Step 5 Divide this number by the original number of measurements less one, i.e. 5:

$$\frac{\sum(x - \bar{x})^2}{n - 1} = \frac{17.5}{5} = 3.5$$

Step 6 As all the numbers have been squared, the final step is to take the square root in order to get back to the same units as the mean:

$$\sqrt{\frac{\sum(x - \bar{x})^2}{n - 1}} = \sqrt{3.5} = 1.87$$

Significant figures

You will need to use a calculator to find the value of standard deviations, as it will considerably speed up your calculation. In doing so you will often find the calculator gives a long figure running to many decimal places. In our calculation, for example, the calculation $\sqrt{3.5}$ produces the answer 1.870828693. Clearly the significance of the latter digits is less than the earlier ones. It is normal to reduce these figures to a certain number of significant figures. In our case we have rounded down the answer to three significant figures, namely 1.87. We did this because we calculated our square values to be 2.25, 0.25, 6.25, etc. As these had three significant figures, we used the same number in our final calculation.

To demonstrate how the *t*-test works, let us consider an imaginary example. A farmer wants to find out if one fertiliser is better than another at improving her yield of wheat. To do so she divides her fields into 16 plots, eight of which she treats with fertiliser A and eight with fertiliser B. The number of tonnes of wheat obtained from each plot is given in Table 1.

The first stage of the *t*-test is to calculate the standard deviation for each sample. To do this we must calculate:

- the mean of each sample
- the deviation of each reading from the mean
- the square of this deviation and the sum of the squares.

All these values are calculated and shown in Table 2.

Table 2 *Calculating the standard deviation for each sample*

Table 1 *Yield of wheat for fertilisers A and B*

Number of tonnes of wheat per plot		
Fertiliser A	Fertiliser B	
5	4	
9	3	
11	6	
9	7	
10	5	
7	3	
5	3	
8	5	
64	36	Total
8	8	Number of plots
8	4.5	Mean

Fertilizer 1			Fertilizer 2		
Observation (x)	Deviation from the mean $(x - \bar{x}_1)$	Square of the deviation $(x - \bar{x}_1)^2$	Observation (x)	Deviation from the mean $(x - \bar{x}_2)$	Square of the deviation $(x - \bar{x}_2)^2$
5	−3	9	4	−0.5	0.25
9	+1	1	3	−1.5	2.25
11	+3	9	6	+1.5	2.25
9	+1	1	7	+2.5	6.25
10	+2	4	5	+0.5	0.25
7	−1	1	3	−1.5	2.25
5	−3	9	3	−1.5	2.25
8	0	0	5	+0.5	0.25
Sum of squares of deviation:		34	Sum of squares of deviation:		16
Standard deviation $\sqrt{\dfrac{\Sigma(x - \bar{x}_1)^2}{n - 1}}$		$\sqrt{\dfrac{34}{7}} = 2.20$	Standard deviation $\sqrt{\dfrac{\Sigma(x - \bar{x}_2)^2}{n - 1}}$		$\sqrt{\dfrac{16}{7}} = 1.51$

We can now substitute the calculated values in the *t*-test equation as follows:

$$t = \frac{\bar{x}_1 - \bar{x}_2}{\sqrt{\dfrac{s_1^2}{n_1} + \dfrac{s_2^2}{n_2}}}$$

$$t = \frac{8 - 4.5}{\sqrt{\dfrac{2.20^2}{8} + \dfrac{1.51^2}{8}}}$$

$$= \frac{3.5}{\sqrt{\dfrac{4.84}{8} + \dfrac{2.28}{8}}}$$

$$= \frac{3.5}{\sqrt{0.61 + 0.29}}$$

$$= \frac{3.5}{\sqrt{0.9}}$$

$$= \frac{3.5}{0.95}$$

$$= \mathbf{3.68}$$

Finally, to discover if our value of 3.68 indicates whether the different readings are significant, or merely due to chance, we need to look up 3.68 on a statistical table called the t-table, part of which is reproduced as Table 3. To do this we need to know the degrees of freedom. This is calculated according to the formula:

Degrees of freedom $(v) = (n_1 + n_2) - 2$

In our example $v = (8 + 8) - 2$

$$= 14$$

We now find that looking along the row for 14 degrees of freedom, our value of 3.68 lies between 2.98 and 4.14 which corresponds to a probability value of between 0.01 and 0.001. These values refer to the probability that chance alone is the reason for the difference between our two sets of data. In our example the probability that the different wheat yields when using our fertilisers was pure chance was between one in a hundred ($\rho = 0.01$) and one in a thousand (p = 0.001). Assuming all other factors in the experiment were constant, it can be stated that, at the 1% confidence level, there is a difference in the yields from the use of the two different fertilisers, and that the differences are not due to chance effects.

Table 3 *The t-table*

Degrees of freedom	Value of *t*			
1	6.31	12.71	63.66	636.62
2	2.92	4.30	9.93	31.60
3	2.35	3.18	5.84	12.94
4	2.13	2.78	4.60	8.61
5	2.02	2.57	4.03	6.86
6	1.94	2.45	3.71	5.96
7	1.90	2.37	3.50	5.41
8	1.86	2.31	3.36	5.04
9	1.83	2.26	3.25	4.78
10	1.81	2.23	3.17	4.59
11	1.80	2.20	3.11	4.44
12	1.78	2.18	3.06	4.32
13	1.77	2.16	3.01	4.22
14	1.76	2.15	2.98	4.14
15	1.75	2.13	2.95	4.07
16	1.75	2.12	2.92	4.02
17	1.74	2.11	2.90	3.97
18	1.73	2.10	2.88	3.92
19	1.73	2.09	2.86	3.88
20	1.73	2.09	2.85	3.85
21	1.72	2.08	2.83	3.82
22	1.72	2.07	2.82	3.79
23	1.71	2.07	2.81	3.77
24	1.71	2.06	2.80	3.75
25	1.71	2.06	2.79	3.73
26	1.71	2.06	2.78	3.71
27	1.70	2.05	2.77	3.69
28	1.70	2.05	2.76	3.67
29	1.70	2.05	2.76	3.66
30	1.70	2.04	2.75	3.65
40	1.68	2.02	2.70	3.55
60	1.67	2.00	2.66	3.46
Probability that chance produced this value of t	**0.1**	**0.05**	**0.01**	**0.001**
Confidence level	**10%**	**5%**	**1%**	**0.1%**

- Explain that natural selection occurs as populations have the capacity to produce many offspring that compete for resources; in the 'struggle for existence' only the individuals that are best adapted survive to breed and pass on their alleles to the next generation
- State the general theory of evolution that organisms have changed over time
- Discuss the molecular evidence that reveals similarities between closely related organisms with reference to mitochondrial DNA and protein sequence data

Evidence from fossils shows that organisms have changed over time. Natural selection is the process by which organisms that are better adapted to their environment tend to survive and reproduce while those less well adapted tend not to. Those that are adapted and so survive to reproductive age will be the ones that pass on their favourable alleles to the next generation.

Survival of the fittest

Charles Darwin and Alfred Wallace in 1865 independently developed the theory of evolution by natural selection based on the following principles:

- All organisms produce more offspring than can be supported by the supply of food, light, space, etc.
- Despite the over-production of offspring, most **populations** remain relatively constant in size.
- There must hence be competition between members of a species to be the ones that survive = intraspecific competition, with individuals competing for resources such as food, breeding sites, space, light and water.
- Within any population of a species there will be a wide variety of genetically different organisms (Topic 17.1).
- Some of these individuals will possess **alleles** that make them better adapted to survive (fitter) and so more likely to breed.
- Only those individuals that do survive and breed will pass on their alleles to the next generation.
- The advantageous alleles that gave these individuals the edge in the struggle to survive and breed are therefore likely to be passed on to the next generation.
- Over many generations, the individuals with beneficial alleles are more likely to survive to breed and therefore increase in number at the expense of the individuals with less favourable alleles.
- The frequency of favourable alleles in the population will increase over time.

Specific examples of how natural selection produces changes within a species include:

- antibiotic resistance in bacteria
- **industrial melanism**.

Antibiotic resistance in bacteria

It was not long after the discovery of antibiotics that it was realised after treating people with bacterial infections that some antibiotics no longer killed bacteria as effectively as before. It was found that these populations of bacteria had acquired resistance, as the result not of a cumulative tolerance to the antibiotic, but rather a chance **mutation** within the bacteria. The bacteria with the mutation could produce an enzyme, penicillinase, which broke down the antibiotic penicillin before it was able to kill them. When penicillin is used to treat the same disease, only the susceptible (non-resistant) forms of the bacteria are killed. There is therefore a selection pressure favouring the resistant form when exposed to penicillin. These penicillin-resistant bacteria therefore gradually form the greatest proportion of the population. The frequency of the allele for penicillin resistance increases in the population. This type of selection is called directional selection and is described in Topic 17.6.

It is also important that the allele for antibiotic resistance is carried on **plasmids** and these circular DNA molecules can be transferred from cell to cell by natural as well as artificial means. Resistance can therefore find its way into other bacterial species. Overuse of antibiotics, e.g. for minor infections that present no danger, increases the likelihood of selection of resistant strains over ones that are more susceptible to the antibiotic. More details of how antibiotics work and antibiotic resistance are given in Topic 10.9.

Industrial melanism

Some species of organisms have two or more distinct forms or morphs. These different forms are genetically distinct but exist within the same interbreeding population. This situation is called **polymorphism** ('poly' = many; 'morph' = form). One example is the peppered moth (*Biston betularia* in England). It existed only in its natural light form until the middle of the nineteenth century. Around this time a melanic (black) form occurred as the result of a mutation. These mutants had undoubtedly occurred before (one existed in a collection made before 1819) but they were highly conspicuous (very easily seen) against the light background of lichen-covered trees and rocks on which they normally rest. Insect-eating birds such as robins and hedge sparrows are predators of the peppered moth. The melanic form of moths were more easily seen and eaten by these birds than the better camouflaged, normal light forms.

When in 1848 a melanic form of the peppered moth was captured in Manchester, a large city, most buildings, walls and trees were blackened by the soot of 50 years of industrial development. The sulfur dioxide in smoke emissions killed the lichens that previously covered trees and walls. Against this black background the melanic form was less, not more, conspicuous than the light natural form. As a result, the light form was taken by birds more frequently than the melanic form and, by 1895, 98% of Manchester's population of the moth was of the melanic type (Figure 1).

This is an example of how a change in the environment can rapidly alter the allele frequency in populations. Here the variation that exists is the light and dark forms and the selection pressure exerted is predation by birds. The moths are still members of the same species and can interbreed. To become two distinct species, the two populations would need to become reproductively isolated from one another (Topic 17.9).

Molecular evidence for similarities between organisms

We have seen how natural selection leads to changes in a species. We shall see in Topic 17.9 how this can lead to new species. As these changes occurred over millions of years, how can we determine which organisms are closely related? One method is to compare the mitochondrial DNA of different species.

The DNA found in mitochondria is made up of relatively few genes. The nucleotide bases of these genes can be sequenced to reveal patterns that are recognisable in different species. Mitochondrial DNA is only inherited along the female line and so remains relatively unchanged from generation to generation, as no meiosis occurs to introduce variety. The only change to mitochondrial DNA is by mutations, which are very rare. However, assuming very occasional mutations, the patterns of nucleotide bases will change over time, although very slowly. If we compare the mitochondrial DNA of two species, then the more similar their nucleotide base patterns are, the more closely they are related.

Another molecular device to reveal the evolutionary relationships between species is to look at the sequences of amino acids in their proteins. The sequence of amino acids in proteins is determined by DNA. The degree of similarity in the amino acid sequence of the same protein in two species will therefore reflect how closely related the two species are. Once the amino acid sequence for a chosen protein has been determined for two species, the two sequences are compared. This can be done by counting either the number of similarities or the number of differences in each sequence. Figure 2 shows a short sequence of seven amino acids of the same protein in six different species, highlighting the number of differences and the number of similarities.

Figure 1 *Industrial melanism in the peppered moth* (Biston betularia). *Against a natural background (above) the dark melanic form is far more visible and more readily predated on by birds. This natural selection leads to a predominance of the light form in rural areas. In polluted areas, however, the melanic form is better camouflaged and this selective advantage leads to this form predominating*

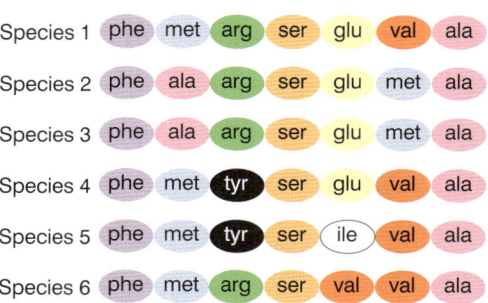

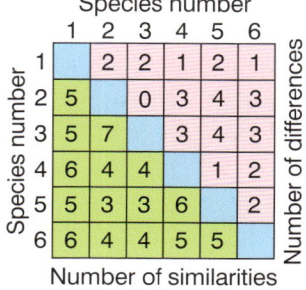

Figure 2 *Comparison of amino acid sequence in part of the same protein in six species*

SUMMARY TEST 17.4

The theory of natural selection by survival of the fittest was developed independently by Charles Darwin and **(1)**. It is based on the principles that all organisms produce **(2)** offspring than can be supported by the food, light and space available for them. However the size of most populations is **(3)** as a result of competition between members of each species for the limited resources available. This type of competition is called **(4)**. Within any population there will be many types of **(5)** different individuals. Those individuals with **(6)** that better suit them to the prevailing conditions are more likely to survive and hence more likely to produce **(7)**. As only those that survive can pass their characteristics to the next generation, there is an increased chance that the next generation will have a greater proportion of these advantageous characteristics. An example of natural selection in practice is **(8)** resistance in bacteria. Another example of natural selection has occurred in the peppered moth. This moth has two forms, a natural light coloured one and a dark coloured one called the **(9)** form. The general term for the situation where a species has two or more distinct forms is **(10)**.

313

The roles of over-production and variation in natural selection

On these pages you will learn to:

- Explain why genetic variation is important in selection
- Explain that natural selection occurs as populations have the capacity to produce many offspring that compete for resources; in the 'struggle for existence' only the individuals that are best adapted survive to breed and pass on their alleles to the next generation
- Explain why organisms become extinct, with reference to climate change, competition, habitat loss and killing by humans

Remember

If an environmental change is great enough, there may be no phenotype suited to the new conditions, in which case the population will die out.

The process of evolution by means of natural selection depends upon a number of factors. Two of the most important are that:

- organisms produce more offspring than can be supported by the available supply of food, light, space, etc.
- there is genetic variety within the **populations** of all species.

Over-production of offspring

Darwin appreciated that all species have the potential to increase their numbers exponentially. He realised that, in nature, populations rarely, if ever, increased in size at such a rate (Figure 1). He rightly concluded that the death rate of even the most slow-breeding species must be extremely high. For most species, the rate of reproduction and the production of offspring is high, but only a very small proportion survive. The reason why reproductive rates are high is because a species cannot control the climate, rate of predation, availability of food, etc. Therefore to ensure a sufficiently large population survives to breed and produce the next generation, each species must produce vast numbers of offspring. This is to compensate for considerable death rates from predation, lack of food (including light in plants) and water, extremes of temperature, natural disasters such as earthquake and fire, disease, etc.

How organisms over-produce depends on the species in question and its means of reproduction. Some examples include:

- A bacterium can divide by binary fission about every 20 minutes when conditions are favourable. A single bacterium could theoretically give rise to 4×10^{21} cells in just 24 hours.
- Some fungi can produce over 500 000 spores each minute at the peak of production. Each spore has the potential to develop a new fungal **mycelium**.
- Higher plants can spread rapidly by **vegetative propagation**, e.g. the production of bulbs, rhizomes, runners, etc.
- Flowering plants produce vast amounts of pollen from their anthers. These can fertilise the many ovules in plants of the same species, leading to the production, in some cases, of millions of seeds from a single plant.
- Animals produce vast numbers of sperm, and sometimes large numbers of eggs also. A female oyster, for example, can produce 100 million eggs in a year and the male oyster produces many more times this number of sperm.
- Many organisms, e.g. birds such as blue tits and mammals like the rabbit, produce several clutches/litters every year, each of which comprises several offspring.

The importance of over-production to natural selection lies in the fact that, where there are too many offspring for the available resources, there is competition amongst individuals (**intraspecific competition**) for the limited resources available. The greater the numbers, the greater this competition and the more individuals will die in the struggle to survive. These deaths are, however, not random. Those individuals best suited to prevailing conditions (have adaptations that make them better able to hide from or escape predators, better able to obtain light or catch prey or better able to resist disease) will be more likely to survive than those less well adapted.

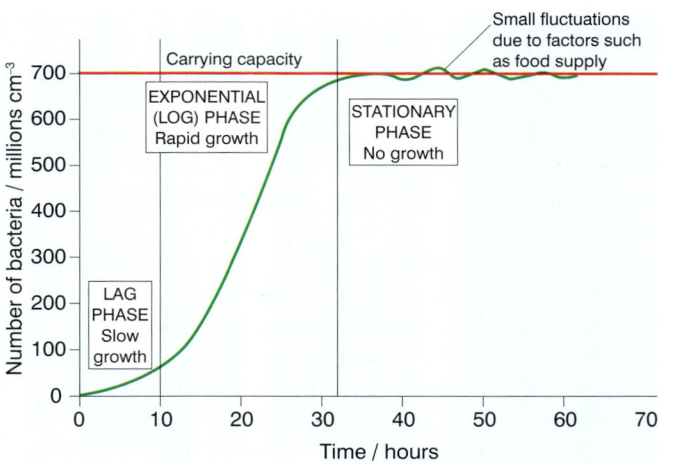

Figure 1 *Typical population growth curve starting with a small initial population. A point is reached where factors such as lack of food cause the population to plateau (level out). The population is said to have reached its carrying capacity*

These individuals will be more likely to breed and so pass on these favourable characteristics, via their alleles, to the next generation, which will therefore be slightly different from the previous one – i.e. the species will have changed over time to be better adapted to the prevailing conditions. This selection process, however, depends on individuals of a species being genetically different from one another.

Variation and natural selection

Figure 3 *Variation within a species. Despite the immense variety they show, all dogs, including this Pug dog and Great Dane, all belong to the same species – Canis familiaris*

If an organism can survive in the conditions in which it lives, you may wonder why it doesn't produce offspring that are identical to itself. These will, after all, be equally capable of survival in these conditions, whereas variation may produce individuals that are less suited. However, conditions change over time and having a wide range of different individuals in the population means that some will have the combination of alleles needed to survive in almost any set of new circumstances. Populations showing little individual variation are vulnerable to new diseases and climate changes. It is also important that a species adapts to changes resulting from changes to the allele frequencies within other species. If, for example, rabbits in a particular region have alleles as a result of mutation that allow them to run faster, then foxes and other predators will be less able to catch them. The foxes will therefore have less food, unless they in turn develop greater speed as a result of new adaptations occurring from mutation. Mutations occur in populations that are neutral in their effect. These may become beneficial in a changing environment. A species cannot predict future changes; it does not know whether the climate will become wetter/drier, warmer/colder or how its prey or predator populations will change, or what new disease agent may occur. However, the larger a population is, and the more genetically varied the organisms within it, the greater the chance that one or more individuals will have the genetic characteristics that give them an advantage in the struggle for survival. These individuals will therefore be more likely to breed and pass their more advantageous alleles on to future generations. Variation therefore provides the potential for a species to evolve and so adapt to new circumstances.

The influence of variation on natural selection is best summarised by Darwin himself who, nearly a hundred and fifty years ago, wrote:

'How can it be doubted, from the struggle each individual has to obtain subsistence, that any minute variation in structure, habits or instinct, adapting that individual better to the new conditions, would tell upon its vigour and health? In the struggle it would have a better chance of surviving; and those of its offspring which inherited the variation, be it ever so slight, would have a better chance'.

Why organisms become extinct

Extinction is a normal and natural process; indeed it is essential to the process of evolution. More than 99% of species known from fossil records are extinct. However, the current rate of extinction is alarmingly high, with some scientists estimating that up to 20% of current species could become extinct in the next 30 years. There is a number of possible reasons for this increase.

- **Climate change**: Global warming is changing vegetation patterns throughout the world, leading to a redistribution of species. In some cases, the members of a species are unable to migrate, because of geographical barriers, and the pace of the warming is too rapid for them to adapt. The death of the last golden toad in Central America in 1999 is an example of a species probably made extinct by climate change. Increased tropical sea temperatures are causing the extinction of certain algal species on coral reefs. Rises in sea level as a result of global warming could result in the loss of nesting sites for turtles and their possible extinction.
- **Habitat loss**: In exploiting natural habitats, humans often destroy them and so cause the extinction of species in those habitats. Examples include timber extraction, which destroys forests and endangers species such as the orang-utan, as well as industrial and agricultural developments that threaten many plant species of the rainforests. In addition, modern farming methods drain ponds as well as remove trees and hedgerows. These practices endanger the species that live and breed there. Pollution has led to the loss of some pond and river habitats and overgrazing has led to the loss of some terrestrial ones. More information on habitat loss is provided in Topic 18.10.
- **Competition**: Interspecific competition has always led to extinctions. Species compete with each other for food, territory and nesting sites, but the competition other species face from humans has accelerated the process. As more and more land is taken up for building, industry and farming, there is more intense competition among wildlife for food, shelter and breeding sites in the remaining natural habitats. Humans often introduce new species that outcompete natural ones. Many marsupials in Australia have been exterminated by competition from introduced rabbits and predation by introduced foxes and cats.
- **Hunting and fishing**: Humans hunt for sport, for animal skins, for trophies, for the pet trade and for research purposes. These are in addition to the numerous species hunted purely as food or for use in traditional medicine. These activities have led to the extinction of species such as the passenger pigeon. Other species, such as the black rhinoceros, have been brought to the brink (edge) of extinction. More details are given in Topic 18.10.

315

17.6

How environmental factors act as forces of natural selection

On these pages you will learn to:

- Explain, with examples, how environmental factors can act as stabilising, disruptive and directional forces of natural selection
- Explain how selection may affect allele frequencies in populations

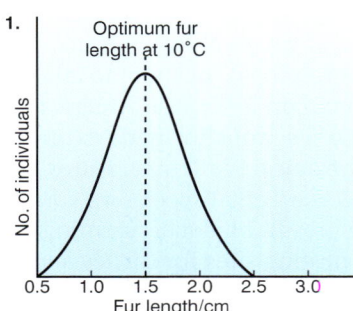

1.

Optimum fur length at 10°C

No. of individuals

0.5 1.0 1.5 2.0 2.5 3.0
Fur length/cm

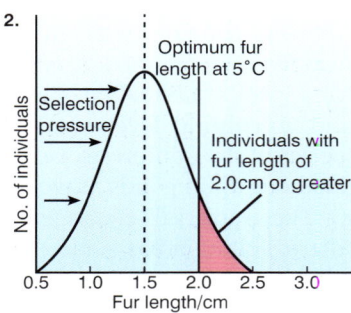

2.

Optimum fur length at 5°C

Selection pressure

Individuals with fur length of 2.0cm or greater

No. of individuals

0.5 1.0 1.5 2.0 2.5 3.0
Fur length/cm

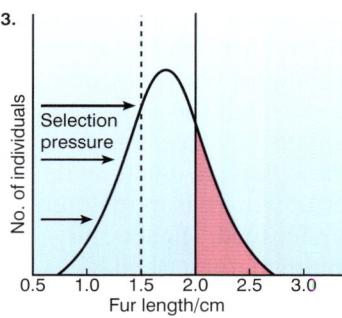

3.

Selection pressure

No. of individuals

0.5 1.0 1.5 2.0 2.5 3.0
Fur length/cm

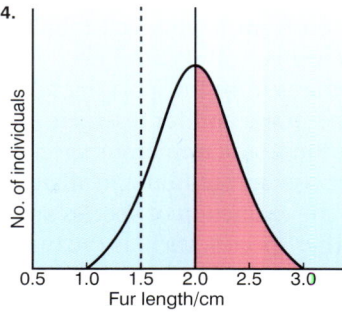

4.

No. of individuals

0.5 1.0 1.5 2.0 2.5 3.0
Fur length/cm

Environmental factors help to contribute to variation within a population. These environmental factors may be an agent for constancy or an agent for change according to the type of selection pressure they exert.

Selection pressure

Every organism faces a process of selection, based upon the organism's suitability for survival under the conditions that exist at the time. The environmental factors that act on and limit a population of a species are called **selection pressures** or **environmental resistances**. These selection pressures include:

- competition for food
- competition for a space in which to live, breed and rear young
- need for light, water, oxygen, etc.
- climate changes, e.g. temperature, rainfall, wind/water currents
- predation
- disease.

The extent and direction of selection pressures varies from time to time and place to place. These selection pressures determine the frequency of an allele within the gene pool. A **gene pool** is the total of all the alleles of all the genes of all individuals within a particular population at a given time.

There are three main types of selection:

- Selection that preserves the characteristics of a population by favouring average individuals (those at or near the mean of the population) = **stabilising selection**.
- Selection that changes the characteristics of a population by favouring individuals that vary in one direction from the mean of the population = **directional selection**.
- Selection that changes the characteristics of a population by favouring individuals at the extremes rather than those around the mean of the population = **disruptive selection**.

Directional selection

Within a population there will be a range of individuals in respect of any one characteristic. The continuous variation amongst these individuals forms a normal distribution curve which has a mean that represents the optimum value for the characteristic under the existing conditions. If the environmental conditions change, so will the optimum value needed for survival. Some individuals, either to the left or the right of the mean, will possess a phenotype

Figure 1 Directional selection

1. When the average environmental temperature is 10°C, the optimum fur length is 1.5cm. This then represents the mean fur length of the population.

2. A few individuals in the population already have a fur length of 2.0cm or greater. If the average environmental temperature falls to 5°C, these individuals are better insulated and so are more likely to survive to breed. There is a selection pressure favouring individuals with longer fur.

3. The selection pressure causes a shift in the mean fur length towards longer fur over a number of generations. The selection pressure continues.

4. Over further generations the shift in the mean fur length continues until it reaches 2.0cm – the optimum length for the average environmental temperature of 5°C. The selection pressure now stops provided the environmental temperature stays the same.

with the new optimum for the characteristic and so there will be a selection pressure moving the mean to either the left or the right of its original position. Directional selection therefore results in one extreme of a range of variation being selected against in favour of the other extreme. Figure 1 illustrates a theoretical example of directional selection. A specific example is antibiotic resistance in bacteria (Topic 17.4).

Stabilising selection

Stabilising selection tends to eliminate the extremes of the **phenotype** range within a population and with it the opportunity for evolutionary change. It arises where the environmental conditions are constant. One example is fur length in a particular mammalian species. In years when the environmental temperatures are hotter than usual, the individuals with shorter fur length will be at an advantage because they can lose body heat more rapidly. In colder years the opposite is true and those with longer fur length will survive better as they are better insulated. Therefore, if the environment fluctuates from year to year, both extremes will survive because each will have some years when it can thrive at the expense of the other. If, however, the environmental temperature is constantly 10°C, individuals at the extremes will never be at an advantage and will therefore be selected against in favour of those with average fur length. The mean will remain the same, but there will be fewer individuals at either extreme (Figure 2). An actual example of stabilising selection is the body mass of human children at birth. Babies born with a body mass greater or less than the optimum of 3.2 kg have a higher mortality rate.

Disruptive selection

Disruptive selection is the opposite of stabilising selection. It favours the two extreme phenotypes at the expense of the intermediate phenotype. Although the least common form of selection, it is the most important in bringing about evolutionary change. Disruptive selection occurs when an environmental factor, such as temperature, takes two or more distinct forms. In our example this might occur if the temperature alternated between 5°C in winter (favouring long fur length) and 15°C in summer (favouring short fur length). This could lead at some time in the future to two separate species of the mammal – one with long fur and active in winter, the other with short fur and active in summer (Figure 3). An example is coho salmon, where large males and small males have a selective advantage over intermediate-sized males in passing on their alleles to the next generation. The small males are able to sneak up to the females in the spawning grounds. The large males are fierce competitors. This leaves intermediate-sized males at a disadvantage.

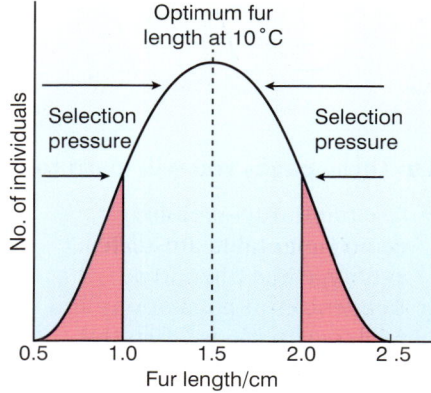

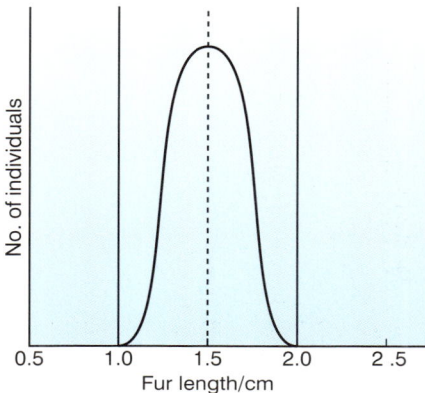

1. Initially there is a wide range of fur length about the mean of 1.5 cm. The fur lengths of less than 1.0 cm or greater than 2.0 cm in individuals are maintained by rapid breeding in years when the average temperature is much warmer or colder than normal.

2. When the average environmental temperature is consistently around 10°C with little annual variation, individuals with very long or very short hair are eliminated from the population over a number of generations.

Figure 2 *Stabilising selection*

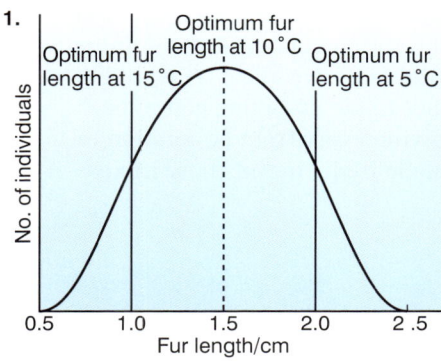

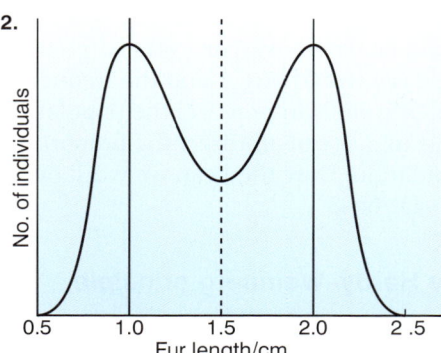

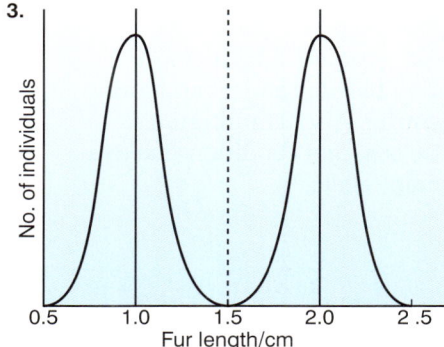

1. When there is a wide range of temperatures throughout the year, there is continuous variation in fur length around a mean of 1.5 cm.

2. Where the summer temperature is static around 15°C and the winter temperature is static around 5°C, individuals with two distinct fur lengths predominate: 1.0 cm types which are active in summer and 2.0 cm types which are active in winter.

3. After many generations two distinct sub-populations are formed.

Figure 3 *Disruptive selection*

Allelic frequencies

On these pages you will learn to:

- Use the Hardy–Weinberg principle to calculate allele, genotype and phenotype frequencies in populations and explain situations when this principle does not apply

In theory, any sexually mature individual in a **population** is capable of breeding with any other. This means that the **alleles** of any individual organism may combine with the alleles of any other individual in the population.

Before looking at allelic frequencies, we need to understand what is meant by a 'population'. A population is a group of organisms of the same species that occupies a particular space at a particular time and may potentially interbreed. Any species may exist as one or more populations. More information on populations is given in Topic 18.4.

All the alleles of all the genes of all the individuals in a population at any one time is known as the **gene pool**. Sometimes the term is used to refer to all the alleles of one particular gene in a population, rather than all the genes. The number of times an allele occurs within the gene pool is referred to as the **allelic frequency**.

Let us look at this more closely by considering just one gene that has two alleles, one of which is dominant and the other recessive. An example is the gene responsible for cystic fibrosis, a disease in humans in which the mucus produced by affected individuals is thicker than usual. The gene has a dominant allele (F) that leads to normal mucus production and a recessive allele (f) that leads to the production of thicker mucus and hence cystic fibrosis. Any individual human has two of these alleles in every one of their cells, one on each of the pair of homologous chromosomes on which the gene is found. As these alleles are the same in every cell, we only count one pair of alleles, per gene, per individual when considering a gene pool. If there are 10 000 people in a population, there will be twice as many (20 000) alleles in the gene pool of this gene.

The pair of alleles of the cystic fibrosis gene has three different possible combinations, namely homozygous dominant (FF), homozygous recessive (ff) and heterozygous (Ff). When we look at genotype frequencies, however, it is important to appreciate that the heterozygous combination can exist in two different arrangements, namely Ff and fF (it is just convention that we put the dominant allele first in all cases). This is because the male parent may contribute either F or f as gametes and the female parent also F or f.

In any population the total number of alleles is taken to be 1.0. In our population of 10 000 people, if everyone had the genotype FF, then the frequency of the dominant allele (F) would be 1.0 and the frequency of the recessive allele (f) would be 0.0. If everyone was heterozygous Ff, the frequency of the dominant allele (F) would be 0.5 and the frequency of the recessive allele (f) would be 0.5. Of course, in practice, the population is not made up of one genotype, but a mixture of all three, the proportions of which vary from population to population. How then can we work out the allele frequency of these mixed populations?

The Hardy–Weinberg principle

The Hardy–Weinberg principle provides a mathematical equation that can be used to calculate the frequencies of the alleles of a particular gene in a population. The principle predicts that the proportion of dominant and recessive alleles of any gene in a population remains the same from one generation to the next provided that five conditions are met:

- no mutations arise
- the population is isolated, i.e. there is no flow of alleles into or out of the population
- there is no selection, i.e. all alleles are equally likely to passed to the next generation
- the population is large
- mating within the population is random.

Although these conditions are probably never totally met in a natural population, the Hardy–Weinberg principle is still useful when studying allele and genotype frequencies.

To help us understand the principle let us consider a gene that has two alleles, a dominant allele A and a recessive allele a.

Let the frequency of allele A = p.
Let the frequency of allele a = q.

The first equation we can write is:

$$p + q = 1.0$$

because there are only two alleles and so the frequency of one plus the other must be 1.0 (100%).

A genotype results from the fertilisation of the male and female gamete.
The probability of 'taking' an A gamete from the pool = p
The probability of 'taking' an a gamete from the pool = q

(male A × female A) = p × p = p^2
(male a × female a) = q × q = q^2
(male A × female a) = p × q = pq
(male a × female A) = q × p = pq

Therefore we can state that:

AA + Aa + aA + aa = 1.0   or, expressing this as genotype frequencies:

p^2 +   2pq   + q^2 = 1.0 (Hardy–Weinberg equation)

We can now use these equations to determine the frequency of any allele in a population and to calculate the frequency of the heterozygous genotype in a population. For example, suppose that a particular characteristic is the result of a recessive allele a, and we know that one person in 25 000 displays the characteristic.

- The characteristic, being recessive, will only be observed in individuals who have two recessive alleles aa.
- The frequency of aa must be 1/25 000 or 0.00004.
- The frequency of aa is q^2.
- If q^2 = 0.00004, then q = $\sqrt{0.00004}$ or 0.0063 approx.
- We know that the frequency of both alleles A and a is p + q and is equal to 1.0.
- If p + q = 1.0, and q = 0.0063 then, p = 1.0 − 0.0063 = 0.9937, i.e. the frequency of allele A = 0.9937.
- We can now calculate the frequency of the heterozygous individuals in the population.
- From the Hardy–Weinberg equation we know that the frequency of the heterozygotes is 2pq.
- In this case 2pq = (2 × 0.9937 × 0.0063) = 0.0125.
- In other words, 125 individuals in 10 000 carry the recessive allele for the characteristic. This is the equivalent of 313 in our population of 25 000.
- These individuals act as a reservoir of recessive alleles in the population, although they do not express the allele in their phenotype.

Remember

Unless an allele leads to a phenotype with an advantage or a disadvantage compared with other phenotypes, its allele frequency in a population will stay the same from one generation to the next.

SUMMARY TEST 17.7

All the alleles of all the genes of all the individuals in a population at any one time is known as the (1), and the number of times an allele occurs within it is referred to as the (2). A mathematical equation that can be used to calculate the frequencies of the alleles of a particular gene in a population is known as the (3) principle. It requires that five conditions are met: the population must be both (4) and (5), no (6) or (7) should occur, and mating within the population should be (8). If the frequency p of a dominant allele is 0.942, then the frequency of the heterozygous genotype in the population will be (9). You will need to a calculator to help with this last question.

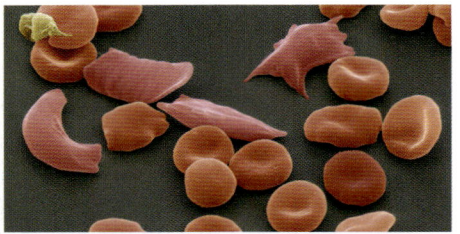

Figure 1 *Scanning electron micrograph of red blood cells in sickle cell anaemia. The sickle shaped cells (pink) are the result of a single base substitution in the DNA that codes for the beta-globin chain of haemoglobin.*

Environmental changes affect the probability of an allele surviving in a population and therefore the number of times it occurs within the gene pool. It must be emphasised that environmental factors do **not** affect the probability of a **particular** mutant allele occurring, they simply affect the frequency of a mutant allele that is already present in the gene pool. It should also be remembered that some environmental factors may influence the overall **mutation** rate (Topic 16.12), but that this is a **general** and **random** process rather than one that affects a specific allele in a specific way. To illustrate the effect of an environmental factor on the frequency of an allele, we shall look at the condition in humans known as sickle cell anaemia.

Sickle cell anaemia

We saw in Topic 16.12 that sickle cell anaemia is the result of a gene mutation in which a single base substitution in DNA causes the wrong amino acid to be incorporated into two polypeptides in **haemoglobin** molecules. The result is red blood cells with a sickle (crescent) shape. Sickle cell anaemia is the result of a single **gene** for the β-globin polypeptide chain of haemoglobin which has two codominant (Topic 16.7) alleles, Hb^A (normal) and Hb^S (sickled). The malarial parasite, *Plasmodium*, is unable to exist in red blood cells with Hb^S. Table 1 shows the three possible genotype combinations of these two alleles and the corresponding phenotypes. The selection pressures on each genotype differ as follows:

- **Homozygous for haemoglobin-S (Hb^SHb^S)** – individuals with sickle cell anaemia and are considerably disadvantaged without medical attention. They rarely live long enough to pass their alleles on to the next generation. Their anaemia is so severe it outweighs being resistant to one form of malaria and so individuals are always selected against.
- **Homozygous for haemoglobin-A (Hb^AHb^A)** – individuals lead normal healthy lives, but are susceptible to malaria in areas of the world where the disease is **endemic** and they are therefore selected against in these regions only.
- **Heterozygous for haemoglobin (Hb^AHb^S)** – individuals are said to have **sickle cell trait**, but are not badly affected except when the oxygen concentration of their blood is low, e.g. in exercising muscles, when the abnormal haemoglobin makes the red blood cells sickle shaped and less able to carry oxygen. In these cases the person may therefore become tired more easily, but in general the condition is symptomless. They do, however, have protection against the serious forms of malaria and this advantage outweighs the disadvantage of tiredness in areas of the world where malaria occurs.

(a) Non-malarial region

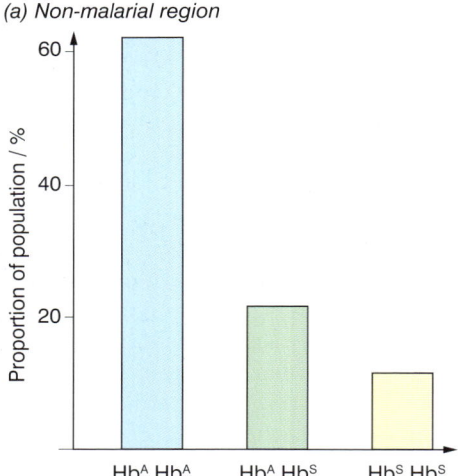

(b) Malarial region

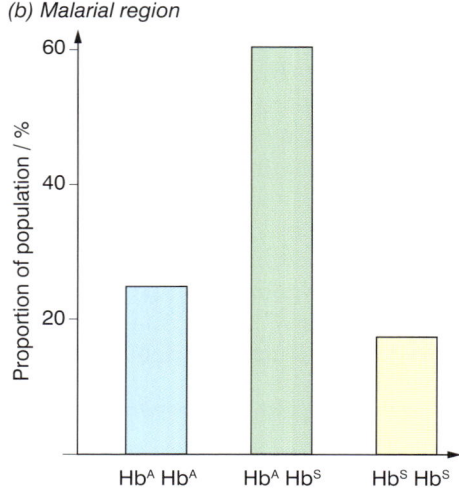

Figure 2 *Distribution of the sickle cell allele in a non-malarial and a malarial population*

Table 1 *Comparison of different genotypes of the gene for haemoglobin-A*

GENOTYPE	Hb^AHb^A	Hb^AHb^S	Hb^SHb^S
PHENOTYPE	Normal	Sickle cell trait	Sickle cell anaemia
Type of haemoglobin	Normal	50% normal and 50% mutant	Mutant
Type of red blood cell	Normal	Usually normal, but sickle shaped at low oxygen concentrations	Sickle shaped
Oxygen carrying capacity	Normal	Reduced (mild anaemia)	Poor (severe anaemia)
Protection against malaria	None	Moderate	High

This situation is called **heterozygote superiority** or **heterozygote advantage**.

To summarise, in parts of the world where malaria is prevalent, the heterozygous state (Hb^{A}Hb^S) will be selected for at the expense of both homozygous states. This is a form of stabilising selection (Topic 17.6). In areas where malaria does not occur, the homozygous state for haemoglobin-A (Hb^{A}Hb^A) (and the heterozygous state, to a large extent) is selected for – a form of directional selection (Topic 17.6). The homozygous state for haemoglobin-S (Hb^{S}Hb^S) is so debilitating that it is always selected against. (Figure 2 shows the proportion of these three genotypes in malarial and non-malarial regions.) Individuals homozygous for *HbS* remain in the population because an *HbS* allele is always present in each heterozygous individual of the population. When two heterozygous individuals produce offspring, there is a one in four chance that one will be homozygous Hb^{S}Hb^S. Because a greater proportion of the population in a malarial region are heterozygous than in a non-malarial region, it follows that the frequency of the sickle cell allele (Hb^S) is greater in areas where malaria is present than in ones where it is absent (Figure 3).

The founder effect

The founder effect occurs when just a few individuals from a population colonise a new region. These few individuals will carry with them only a small fraction of the alleles of the population as a whole. This means that the gene pool of the founder population may not be representative of the gene pool of the larger population. The new population that develops from the few colonisers will therefore show less genetic diversity than the population from which they came. The founder effect often takes place when new volcanic islands rise up out of the sea. The few individuals that colonise these barren islands give rise to populations that are genetically distinct from the populations they left behind. The new population may, in time, develop into a separate species. As these species have lower genetic diversity they are less able to adapt to changing conditions.

SUMMARY TEST 17.8

The total number of all individual alleles of all the genes in a population is called the **(1)** and the number of times an allele occurs within a population is called the **(2)**. The occurrence of an allele in a population is affected by environmental factors through the process of **(3)**. An example of such an environmental factor occurs in the condition known as sickle cell anaemia that results from a single base in DNA being **(4)** by another base. As a result, affected individuals have **(5)** cells that are sickle (crescent) shaped. The gene for the beta-globin chain of haemoglobin has two distinct alleles, Hb^A and Hb^S, that are **(6)**. Individuals with sickle cell anaemia have the genotype **(7)** and are so disadvantaged that they rarely survive long enough to breed. Individuals with normal haemoglobin lead normal lives but are susceptible to malaria. Heterozygous individuals are affected when the **(8)** concentration of their blood is low. These individuals are, however, more protected against malaria, which puts them at an advantage over both homozygous states in areas where this disease is common. This is an example of **(9)** selection. In areas where malaria is absent, individuals with normal haemoglobin are selected in favour of those with the HbS HbS genotype – an example of **(10)** selection.

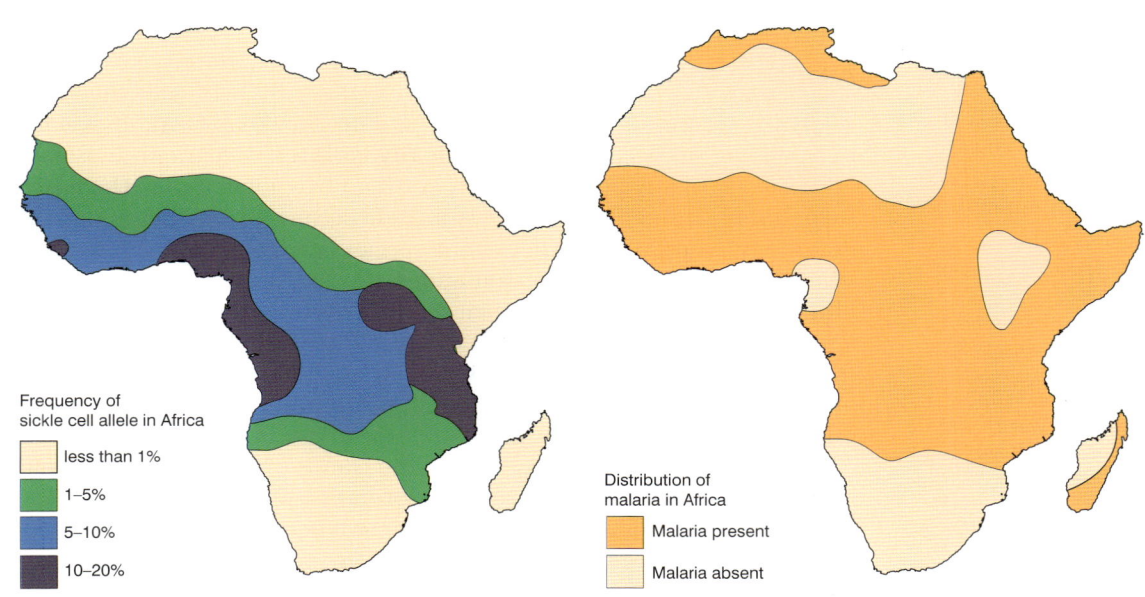

Frequency of sickle cell allele in Africa

less than 1%

1–5%

5–10%

10–20%

Distribution of malaria in Africa

Malaria present

Malaria absent

Figure 3 Comparison of the frequency of the sickle cell allele and the distribution of malaria in Africa

On these pages you will learn to:

- Explain how genetic drift may affect allele frequencies in populations
- Explain how speciation may occur as a result of geographical separation (allopatric speciation), and ecological and behavioural separation (sympatric speciation)
- Explain the role of pre-zygotic and post-zygotic isolating mechanisms in the evolution of new species

1. Species X occupies a forest area. Individuals within the forest form a single gene pool and freely interbreed.

Species X lives and breeds in the forest
Forest

2. Climatic changes to drier conditions reduce the size of the forest to two isolated regions. The distance between the two regions is too great for the two groups of species X to cross to each other.

Forest A
Group X₁
Arid grassland
Forest B
Group X₂

3. Further climatic changes result in the one region (Forest A) becoming colder and wetter. Group X₁ adapts to these new conditions. Physiological and anatomical changes occur in this group.

COLDER AND WETTER   WARMER AND DRIER

Forest A
Group X₁
Arid grassland
Forest B
Group X₂

4. Continued adaptation leads to evolution of a new form group Y in forest A.

COLDER AND WETTER   WARMER AND DRIER

Forest A
Group Y
Forest B
Group X

5. A return to the original climatic conditions results in regrowth of forest. Forests A and B are merged and groups X and Y are reunited. The two groups are no longer capable of interbreeding. They are now two species, X and Y, each with its own gene pool.

Species Y
Forest
Species X

Speciation is the evolution of new species from existing ones. A **species** is a group of individuals that have a common ancestry and so share the same set of **genes** and are capable of breeding with one another to produce fertile offspring. In other words, members of a species are **reproductively isolated**.

It is through the process of speciation that evolutionary change has taken place over millions of years. This has resulted in great diversity of forms amongst organisms, past and present.

How new species are formed

The formation of new species can occur in two different ways:

- **Cross fertilisation between individuals of two different species** that leads to the formation of a hybrid. This is thought to have occurred in the production of modern wheat plants from a chance hybridisation of emmer wheat and goat grass followed by chromosome doubling (**polyploidy**) (Topic 17.11).
- **Reproductive isolation followed by genetic change due to natural selection**. Within a population of any species there are groups of individuals that breed with one another. These breeding subpopulations are called **demes**. Although individuals tend to breed only with others in the same deme, they are capable of breeding with individuals in other demes. In other words, the population has a single **gene pool**. Suppose, however, that the demes become isolated in some way and each undergoes different mutations and becomes genetically different. Each deme will then adapt to the different environmental influences it is subjected to. This is known as adaptive radiation and results in changes to the **allele** frequencies known as **genetic drift** in each population and in the various **phenotypes** present. As a result of these genetic differences it may be that, even if the species were no longer physically isolated from one another, they would be unable to interbreed successfully. Each group would now be a different species, each with its own gene pool. This type of speciation has two main forms, **allopatric speciation** and **sympatric speciation**.

Allopatric speciation

Allopatric means 'different countries' and describes the form of speciation where two populations become **geographically isolated**. Geographical isolation may be the result of any physical barrier between two populations which prevents them interbreeding. These barriers include oceans, rivers, mountain ranges and deserts. What proves a barrier to one species may be no problem to another. While an ocean may isolate populations of goats, it can be crossed by many birds and for marine fish it is their mode of getting from place to place. A tiny stream may be a barrier to snails, whereas the whole of the Pacific Ocean fails to separate populations of certain birds. If environmental conditions either side of the barrier vary, then natural selection will influence the two populations differently and each will adapt in order to survive in their local conditions. These changes take many hundreds or even thousands of generations, but in the end lead to reproductive isolation and the formation of separate species. Figure 1 shows how speciation might occur when two populations of a forest-living species become geographically isolated by a region of arid grassland.

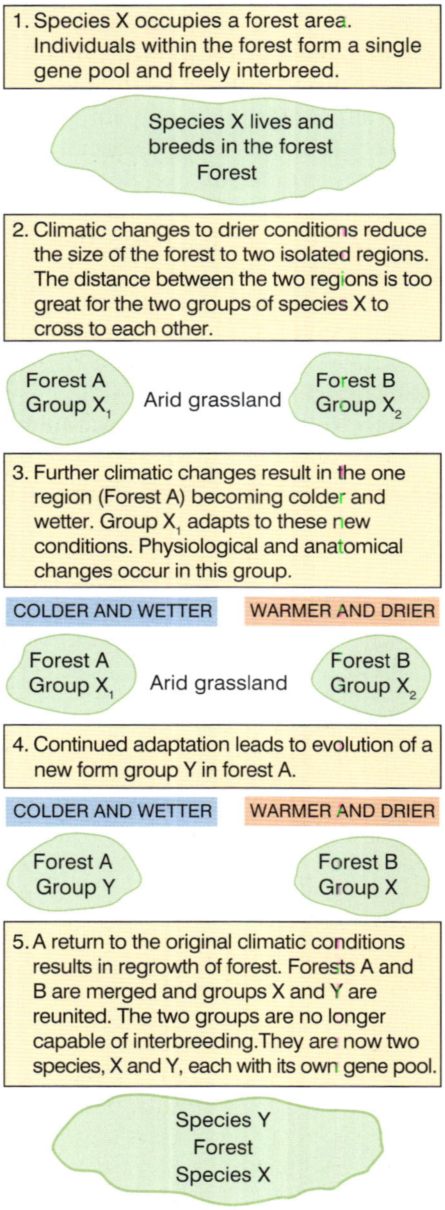

Figure 1 *Speciation as a result of geographical isolation*

Sympatric speciation

Sympatric means 'same country' and describes the form of speciation that results from populations living together becoming reproductively isolated. There are various forms of isolation that lead to sympatric speciation and two examples of these are ecological isolation and behavioural isolation.

In **ecological isolation**, members of the population living in one area form subpopulations that may live in different microhabitats, experiencing different microclimates and differences in available food resources. Individuals rarely move out of their small area and over time changes occur so that interbreeding becomes no longer possible.

In **behavioural isolation**, changes to behaviour, usually in mating rituals and in courtship behaviour, can bring about reproductive isolation. Mutations occurring in subpopulations may result in differences in morphology, for example changes in colours that are associated with attracting mates, or differences in courtship behaviour. Only members within the subpopulation respond to these changes during courtship and mating, so that eventually groups are so different that interbreeding no longer occurs.

The role of pre-zygotic and post-zygotic isolation mechanisms in the evolution of new species

The formation of new species requires time for the gene pools of the reproductively isolated populations to become so different that interbreeding is no longer possible (or if it is possible, no fertile offspring are produced). Two types of mechanism may operate to ensure that groups remain reproductively isolated: prezygotic mechanisms and postzygotic mechanisms. These are summarised in Table 1. Three types of prezygotic mechanism, geographic, ecological and behavioural have already been described.

- **Prezygotic mechanisms** – occur before mating takes place and prevent the exchange of gametes. They are more efficient than post-mating mechanisms in bringing about speciation.
- **Postzygotic mechanisms** – occur after mating has taken place and in some way prevent the development of zygotes into offspring. As a result of different mutations occurring within the subpopulations, viable offspring can no longer be produced when individuals from the two groups mate.

Table 1 summarises all forms of reproductive isolating mechanisms.

Table 1 *Summary of the forms of reproductive isolating mechanisms*

Time of isolation	Type of variation	Explanation of isolation
Prezygotic	Geographical	Populations are isolated by physical barriers such as oceans, mountain ranges, rivers, etc
	Ecological	Populations inhabit different habitats within the same area and so individuals rarely meet
	Temporal	The breeding seasons of each population do not coincide and so they do not interbreed. Figure 2 shows this in relation to four types of frog
	Behavioural	Mating is often preceded by courtship, which is stimulated by the colour or markings of the opposite sex, the call or particular actions of a mate. Any variations in these patterns may prevent mating, e.g. if a female stickleback does not respond in the right way to the actions of the male, he ceases to court her
	Mechanical	Anatomical differences may prevent mating occurring, e.g. it may be physically impossible for the penis to enter the vagina in mammals
	Gametic	The gametes may be prevented from meeting due to genetic or biochemical incompatibility. For instance, some pollen grains fail to germinate or grow when they land on a stigma of different genetic makeup. Some sperm are destroyed by chemicals in the female reproductive tract
Postzygotic	Hybrid sterility	Hybrids formed from the fusion of gametes from individuals of the different populations are often sterile because they cannot produce gametes. For example, alterations to chromosome morphology may make pairing up at meiosis impossible and so no gametes are formed by the hybrid, making the hybrid sterile
	Hybrid inviability	Despite fertilisation taking place, further development does not occur or fatal abnormalities arise in early growth. As the offspring do not reach sexual maturity, breeding does not occur
	Hybrid breakdown	The first generation of hybrids is fertile but the second generation fails to develop or, if it does, it is sterile

SUMMARY TEST 17.9

The evolution of new species from existing ones is known as **(1)**. A species is a population of organisms that is **(2)** isolated from other populations. New species may arise when a breeding sub-unit, called a **(3)**, of a population becomes isolated in some way. The isolated sub-unit may become genetically different as it adapts to different environmental conditions. This process is called **(4)** and results in changes to **(5)** and phenotypes of the two populations. In time the two populations may become so changed that they are unable to interbreed, at which point they have become different **(6)**. There are two main types of speciation: **(7)** speciation occurs when two populations become geographically isolated, e.g. by rivers, oceans or mountain ranges. Where the two populations become reproductively isolated but are not geographically isolated, this is known as **(8)** speciation. Mechanisms to maintain reproductive isolation in the formation of new species can be divided into **(9)** and **(10)** mechanisms.

On these pages you will learn to:

- Describe how selective breeding (artificial selection) has been used to improve the milk yield of dairy cattle

Selective breeding, also known as artificial selection, involves identifying individuals with the desired characteristics and using them to parent the next generation. Offspring that do not exhibit (show) the desired characteristic are killed, or at least prevented from breeding. In this way the **gene pool** is deliberately restricted to a number of desired alleles and the diversity of individuals within the **population** is reduced. Over many generations, this leads to a population all of which possess the desired qualities. Some differences between selective breeding and the evolutionary process (natural selection) are given in Table 1.

Table 1 *Differences between selective breeding and the evolutionary process (natural selection)*

Selective breeding	Natural selection
Selection pressure exerted by humans	Selection pressure exerted by environmental factors
Genetic diversity is lowered	Genetic diversity remains high
Does not lead to new species forming	May lead to new species forming
Inbreeding is common, leading to loss of vigour in offspring	Outbreeding is common, leading to hybrid vigour
Proportion of heterozygotes in the population is reduced	Proportion of heterozygotes in the population remains high
Genetic isolation mechanisms do not operate	Genetic isolation mechanisms operate

Carrying out selective breeding

There are two main methods of carrying out selective breeding:

- **Inbreeding** is used to keep, as far as possible, a desirable characteristic that has arisen by chance **mutation**. By breeding the individual with close relatives, the chances of the offspring showing the desired characteristic are greater. There are harmful effects with inbreeding. For example, there is loss of vigour, with the population being weakened by a lack of diversity and reduced fertility. There is also an increased danger of a harmful recessive allele expressing itself, because there is a greater risk of a **homozygous recessive** individual arising. As a result, inbreeding is not carried out indefinitely and outbreeding may need to occur from time to time to make the population stronger and healthier.
- **Outbreeding** involves the breeding of unrelated individuals. This may be used to try to combine two different desirable characteristics each possessed by separate individuals, for example by crossing a crop plant

that gives an excellent yield with one resistant to disease in the expectation of a plant with a high yield and disease resistance. It frequently produces tougher individuals with a better chance of survival. This is called **hybrid vigour**.

Selective breeding in cattle

Selection of modern-day cattle for milk production (dairy herds) will normally include consideration of the following factors:

- volume of milk produced each day
- length of milking (lactation) period
- protein and fat content of milk
- type of udder, e.g. degree of support, length, shape and angle of teats (important for use of milking machine)
- quantity and type of feed required
- disease resistance, e.g. to mastitis
- temperament (e.g. to cope with length of time attached to milking machine).

The selective breeding process is outlined as follows:

- Selecting a suitable cow and bull by consulting the pedigree records of each and through progeny testing.
- Collection of sperm from the selected bull and storing by freezing.
- Detection of when the cow is in **oestrus** by observing changes in her behaviour, e.g. increasing restlessness, mounting other cows and being mounted by them, feeding less.
- Artificially inseminating the defrosted semen into the cow.
- Checking that fertilisation has occurred and that the cow is in calf.

Figure 1 *Domesticated milking cow*

Figure 2 *Wild cow (Bos sp.)*

An alternative to artificial insemination is the use of embryo transplantation (see extension).

Progeny testing

Progeny testing involves the maintenance of detailed data on all the offspring (progeny) of a particular organism. In the case of dairy cattle this might include growth rate of the calves, body size, amount of body fat, milk yield, quality of milk, length of lactation period, lifespan and veterinary history. The data produced are extremely useful in selecting the correct animal for any particular set of desired characteristics, based upon the offspring it has already produced.

Figure 3 *Artificial insemination of cattle*

Artificial insemination (AI) is the collection of semen and its introduction into the vagina or uterus by artificial means. It has a number of advantages over natural insemination:

- Specific characteristics required in the offspring can be selected.
- Semen can be used from a bull with certain desired characteristics even though it is hundreds, even thousands, of kilometres away.
- Semen can be frozen and transported easily as well as be kept for years, often long after the death of the donor bull.
- The costs of keeping bulls are reduced or eliminated altogether.
- The rate of conception is greater.
- The risk of contracting a sexually transmitted disease is reduced.

One disadvantage of artificial insemination is that the semen of one male can be used to inseminate many hundreds of females and there is therefore less genetic diversity amongst offspring than where natural processes are used.

Artificial insemination is used in cattle as follows:

- A suitable bull is stimulated to ejaculate into the vagina of a 'false' cow, using suitable smells to attract the bull.
- The semen is collected, checked in a laboratory and, depending on the concentration of sperm, may be diluted.
- The semen is stored in liquid nitrogen at $-196\,°C$.
- When needed, the semen is defrosted and inserted into a receptive cow's vagina around 6–8 hours before oestrus, using a plastic pipette.

Embryo transplantation involves fertilising ova in the laboratory before implanting the developing embryos into the uterus of the natural or a surrogate mother. There are two basic methods of embryo transplantation in cattle:

- Ova are removed from cows at the abattoir, grown for five days and then fertilised, *in vitro*, with semen from a bull with the desired characteristics. After 5–6 days of growth in the laboratory, the embryos are frozen and stored ready for transplantation into a suitable cow.

- A cow is treated with follicle stimulating hormone (FSH), causing her to release a large number of ova. These are fertilised by artificial insemination and allowed to implant in the uterus. After about six days they are 'flushed' out through the vagina and transferred to a culture medium before being implanted into receptive cows, either through the vagina or surgically.

SUMMARY TEST 17.10

Selective breeding is also known as **(1)** and differs from natural selection in that genetic diversity is **(2)** and the proportion of **(3)** in the population is reduced. The two main methods of carrying out selective breeding are **(4)** and **(5)**. The process can be used with dairy herds to increase milk production and begins with the selection of a suitable bull and cow by consulting the **(6)** records of each and the checking of data on the offspring each have produced – a process called **(7)**. Semen is then collected from the chosen bull and inserted into the **(8)** of the chosen cow by a technique known as **(9)**. Alternatively, ova may be removed from the cow and fertilised in the laboratory before being returned to the donor cow. This process is known as **(10)**.

Crop improvement by selective breeding

On these pages you will learn to:

- Outline the following examples of crop improvement by selective breeding:
  - the introduction of disease resistance to varieties of wheat and rice
  - the incorporation of mutant alleles for gibberellin synthesis into dwarf varieties so increasing yield by having a greater proportion of energy put into grain
  - inbreeding and hybridisation to produce vigorous, uniform varieties of maize

To supply the food required to satisfy the increasing demands of an expanding world population, crops need to be constantly improved so they yield more. One major loss of yield is due to crop diseases. Toxins (mycotoxins) produced as a result of some fungal diseases may be harmful to humans when they eat crop products. Although many of these diseases can be managed by pesticides, selectively breeding resistant varieties is a more ecologically sound and sustainable way of achieving control of disease. The main stages in selective breeding are:

- to decide which characteristics are desirable
- to select the parents that show these characteristics
- to choose the offspring that most clearly display these characteristics and use them to breed the next generation
- to repeat the process continuously.

Let us now consider some examples of how selective breeding is used to improve crops.

Introduction of disease resistance to varieties of wheat

Diseases of wheat, *Triticum aestivum*, such as rusts and mildews caused by fungal pathogens, are a major factor in reducing crop yields. The more resistant a variety of wheat is to disease, the greater the potential yield. Farmers have always selected those plants that were most productive and, in doing so, will have selected disease-resistant varieties along with other favourable features.

More recently, scientists have deliberately crossed wheat plants known to be disease resistant with others that were known to be high yielding. Initially they targeted resistance to particular diseases but soon chance mutations led to the **pathogen** developing resistance to the disease, leaving the wheat vulnerable once again. Attention has since turned to selecting varieties of wheat with features that give more general resistance to most diseases rather than just to a specific one.

Modern breeders now scan the genome of wheat plants and detect regions that differ between disease-resistant ones and those that are not resistant. They can then use gene markers (Topic 19.4) to test new varieties of wheat for those genes that give disease resistance. This allows them to select plants more quickly than growing large numbers to maturity to see whether they have inherited the desired resistance.

Introduction of disease resistance to varieties of rice

Blast is among the most widespread and damaging diseases of Asian rice, *Oryza sativa*, and can cause more than 50% losses in yield. Selection of varieties of rice that are resistant to blast has followed the same pattern as selecting disease-resistant varieties of wheat. Breeders screen and breed for resistance to blast, along with other desired traits that will suit various ecosystems. Selected plants showing resistance to blast are grown and F_1 seeds are increased to produce the F_2 generation. Starting from the F_2, breeders evaluate the lines for blast resistance. Selection for blast resistance, along with other desired traits, such as high yield, begins at the F_2 generation and continues for a number of generations.

Gene markers are also used to speed the selection process. 'Tetep', a rice cultivar from Vietnam, has several genes that give resistance to blast. These have been tagged with markers linked to the genes. These markers are then used to look for the same genes in plants used in breeding experiments, so that they can be selected.

Another disease affecting Asian rice is bacterial blight disease. African rice, *Oryza glaberrima*, is known for its resistance to blast and to bacterial blight disease. Crop improvement by carrying out interspecific crosses (*O. sativa* × *O. glaberrima*) is another way of producing disease-resistant Asian rice. Here, varieties can be screened for the best resistance to disease and the ability to grow in a wide range of environments. Desirable features of disease resistance include long-term resistance and resistance in seeds.

Incorporation of mutant alleles for gibberellin synthesis

Dwarf varieties of cereals were developed in the 1960s. They produce higher yields for two reasons:

- less of the products of photosynthesis are needed to grow tall, so more goes into the grain
- dwarf plants are less likely to fall, or be blown over, so it is easier to collect more of the grain during harvesting.

In Topic 15.13 we saw that gibberellins have a role in stem elongation. Therefore it is often desirable to produce varieties of cereals that do not produce the normal quantity of gibberellins. The synthesis of gibberellins involves a number of enzyme-controlled steps. A mutation in a gene, known as semi-dwarf allele *sd-1*, results in a non-functioning oxidase enzyme that is required as part of the synthesis pathway. As the plants have very low concentrations of gibberellin, DELLA proteins are not broken down (Topic 16.15). Dwarf and semi-dwarf cereal varieties that have been bred with the semi-dwarf allele include varieties of rice and barley.

Inbreeding and hybridisation in maize

For generations humans have selectively bred organisms in order to obtain varieties that produce characteristics they find beneficial. They largely do this by cross fertilising two closely related individuals displaying the desired characteristic. This is called **inbreeding** and was discussed in Topic 17.10. We also saw that crossing two individuals from different inbred strains (outbreeding) can produce organisms that are genetically superior to their parents. This is known as **hybrid vigour**.

Hybridisation is combining the genes of different varieties or species of organism to produce a **hybrid**.

Up to the beginning of the twentieth century, maize, *Zea mays*, was bred to produce greater yields. The yield from these inbred varieties gradually declined, probably due to an increase in the frequency of harmful recessive alleles. Then scientists crossed two different inbred varieties (Southern

dent and Northern flint) to produce a hybrid that increased yields up to four times. Since that time many hybrid varieties of maize have been developed. These produce the taller, more resistant and higher yielding plants that are grown today.

The new genetic variety within hybrids is clearly beneficial to farmers. It has a disadvantage, however, in that crosses between numbers of different strains produces variety between individual plants. Farmers prefer uniformity (Figure 1) because similar plants are easier to harvest and produce a more uniform crop which is easier to sell.

To overcome this, companies now produce different strains of maize where each one is the result of a single cross. The farmer can then choose a strain knowing that all the seed purchased will produce F_1 plants that have the same genotype. These will provide uniformity while still showing hybrid vigour. With numerous F_1 strains available, different farmers can choose the one most suitable

Figure 1 *Field of hybrid maize plants showing the uniformity preferred by farmers*

for their particular circumstances. Some may select drought-resistant varieties, others, ones that thrive (grow) best on a clay soil and yet others ones that are resistant to a particular pest found in their region. It follows that farmers no longer keep their own seed to sow because the plants produced would not breed true for the characteristic they want. Instead they purchase fresh F_1 seeds from commercial suppliers every year.

SUMMARY TEST 17.11

Selective breeding to create disease resistance in wheat is used to increase (**1**). To improve varieties, breeders scan the (**2**) of wheat plants to detect differences between resistant and non-resistant varieties. Gene markers can then be used to test new varieties for genes that give resistance. A similar technique is used to select varieties of (**3**) that are resistant to a disease called blast. Another type of selective breeding is used to produce dwarf varieties of cereals. These produce higher yields because less of the products of (**4**) are needed to make the plant tall, leaving more to go into the (**5**). They are also less likely to be blown over and therefore easier to (**6**). The dwarf varieties have a mutant (**7**) that makes the plants able to synthesise a growth regulator called (**8**). Selective breeding can involve combining the genes of different varieties or species, a process known as (**9**).

Adaptation of organisms to their environment

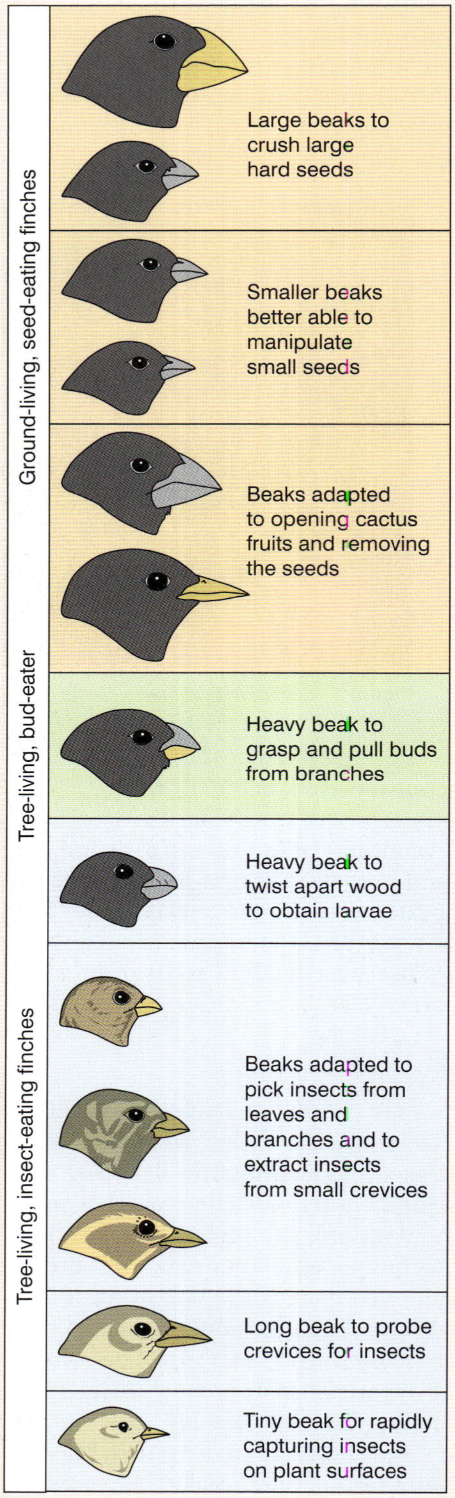

Figure 1 *Adaptive radiation amongst the thirteen species of Galapagos (Darwin's) finches*

An organism is considered to be adapted to a particular environment if it survives and reproduces better than other organisms in that environment. Adaptation is a relative term that compares performance both within and between species.

Adaptation and speciation

Adaptation and speciation are often related. As species adapt to different environments, they will develop differences that may lead to reproductive isolation and therefore speciation. For example, the male *Anolis* lizards of the Caribbean court females by extending a colourful flap of skin under their throat called a dewlap. There is considerable variation in the colour of these dewlaps. Some are easier to see in open **habitats**, others in shaded areas. As lizards occupy new habitats, there is selection pressure favouring the dewlaps that are most conspicuous, e.g. light coloured ones in a dark shaded forest. This adaptive change in dewlap colour means that the male may only be attractive to females stimulated by lighter dewlaps and not to ones attracted by darker dewlaps, i.e. they have become reproductively isolated and hence form a separate species.

Adaptive radiation – Darwin's finches

While visiting the Galapagos Islands during his voyage on HMS Beagle, Darwin was greatly interested by the range of different beaks displayed by the 13 species of finches found there and in particular how each was adapted to obtaining different food. It is generally accepted that this variety of beaks arose as a result of what is called **adaptive radiation**. This process occurred as follows:

- The Galapagos Islands are geographically isolated, being some 1000 km from Ecuador, the nearest country on the South American mainland.
- By some means, e.g. blown by gales, or carried on a boat or vegetation, some seed-eating finches made the improbable journey from the mainland to one of the volcanic Galapagos Islands.
- This single ancestral species found few competitors on this sparsely colonised (very few other organisms) island, and so flourished.
- As with all species, **mutations** occurred, leading to natural selection favouring those individuals that were better suited to some of the many **ecological niches** available on the island. In particular, changes in beak shape allowed them to exploit (make use of) new food sources.
- The adaptations to these niches meant that these finches were now different from the ancestral ones on the mainland.
- Some finches spread to other islands in the Galapagos, although the island groups were some distance apart and so were geographically isolated.
- These new arrivals survived well because there was little competition.
- Mutations again led to increased variety and natural selection favoured those changes of beak that allowed some types to use new and different varieties of food.
- The geographical isolation of the islands from the mainland, and from each other, meant that the finches on each island group were reproductively isolated and so, in time, became separate species. Even where they returned to former islands or the mainland, they could not breed successfully with populations that had remained there.
- Even on a single island group, the adaptations to different niches led to reproductive isolation and speciation.

The range of beak adaptations to suit different food sources among Galapagos finches is shown in Figure 1.

Other examples of structural adaptations to different environments

As heat is lost and gained through the body surface it follows that, the larger the body surface area is compared with the body volume, the faster will be the rate of exchange. In general, animals in cold environments have a smaller surface area to volume ratio than their relatives living in warmer climates. Mammals are **endotherms**; they gain their heat from the metabolic activities taking place inside their bodies in order to maintain a more or less constant body temperature. Those living in cold climates, such as an Arctic fox, have a smaller body surface area to volume ratio than their relations, such as the fennec fox, that live in warm conditions. This is largely achieved through the size of their ears – Arctic foxes have very short external ears that reduce heat loss. Fennec foxes, by contrast, have very large external ears, enabling them to lose heat, especially during and after exertion (Figure 2).

For plants, the ability to withstand a shortage of water is very important. Plants transpire and so lose water continuously. If water is in short supply they must reduce **transpiration** if they are to survive. Structural adaptations to reduce transpiration are known as **xeromorphic** features and include:

- a thick waxy **cuticle** on leaves
- leaves that roll and trap moist air within the leaf
- hairy leaves that trap moist air next to the surface
- **stomata** in grooves or pits that reduce the diffusion gradient
- leaves reduced in size that give a small surface area to volume ratio
- leaves absent and photosynthesis carried out by stems as these have fewer stomata
- succulent leaves and stems that store water
- extensive root systems that collect water quickly when it is available.

Examples of physiological adaptations to different environments

Desert-living animals need to survive water shortage. In the case of the kangaroo rat, a desert rodent, the kidney shows a range of physiological adaptations designed to conserve water by producing concentrated urine. It oxidises fat rather than carbohydrate, which yields almost twice as much water, but it is still crucial to keep water losses to a minimum. Kangaroo rats reduce evaporation from the lungs, do not sweat and produce very dry faeces, as well as reducing water loss when removing waste products through the kidney. They produce urine that is four times more concentrated than that of humans (24% urea as opposed to a maximum of 6% in humans) and seventeen times more concentrated than their own blood. This is achieved by having an extremely long loop of Henlé (Topic 14.5) that is important in creating a **counter-current system** involved in the reabsorption of water back into the blood stream.

Reabsorption of water is aided by the unusually high levels of ADH (Topic 14.6) in the blood.

Figure 2 *Arctic fox (top), living in a colder environment, has much shorter ears than the fennec fox (below) that lives in a warmer climate*

SUMMARY TEST 17.12

When visiting the Galapagos Islands, Darwin observed the variety of beaks of finches. A single ancestral species is thought to have initially inhabited the Islands. As a result of **(1)** in the genes of this species, new varieties of finches occurred. Some of the new individuals were better suited to the different **(2)** available on the Islands. Over time these varieties became **(3)** leading to differences in their gene pools and each became a separate **(4)**. Adaptive radiation produces organisms with characteristics that suit them to different environments. For example, plants in dry regions develop **(5)** features to reduce transpiration. These include a thick waxy **(6)** on leaves, extensive **(7)** systems and **(8)** that are reduced in number located only in pits or grooves. Mammals like the kangaroo rat that live in dry regions have extremely long **(9)** in their kidneys that help to produce very concentrated **(10)**.

17 Examination Questions

1 *Asellus aquaticus* is a small freshwater crustacean.

200 *A. aquaticus* were released into a pond where there had previously been none. The pond was favourable for their growth and reproduction.

a Describe **and** explain the expected changes in the population size of *A. aquaticus* over the following few months. *(5 marks)*

b In order for natural selection to occur a population must show phenotypic variation. Explain why variation is important in natural selection. *(2 marks)*
(Total 7 marks)
Cambridge International AS and A Level Biology 9700
Paper 41 Q2 November 2009

2 The Atlantic herring, *Clupea harengus*, lives in large populations called shoals and may grow up to 40 cm long.

Figure 1 shows the appearance of *C.harengus*.

Figure 1

The length of *C.harengus* shows wide variation.

Figure 2 shows the numbers of fish of different lengths in a population of *C. harengus*.

The arrows show the selection pressures, **P** and **S**.

Figure 2

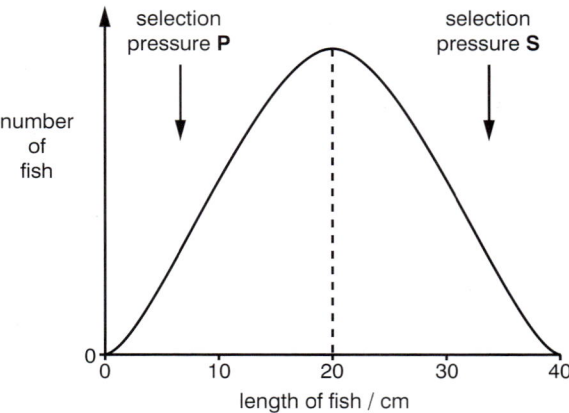

a i On a copy of the axes below (Figure 3), sketch a graph to show the distribution of length of *C. harengus*, when selection pressures **P** and **S** operate for a few years. *(2 marks)*

Figure 3

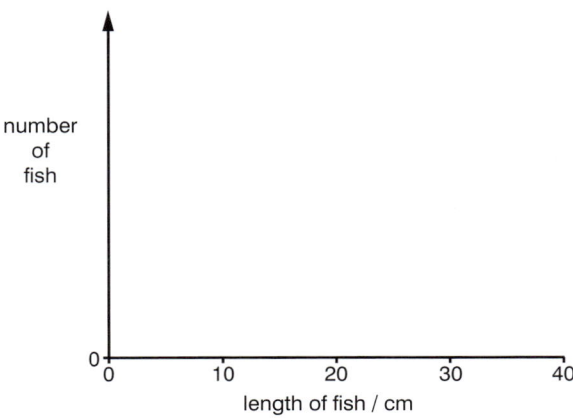

ii Name this type of natural selection. *(1 mark)*

b i On a copy of the axes below (Figure 4), sketch a graph to show the distribution of length of *C. harengus*, when selection pressure **S** alone operates for a few years. *(2 marks)*

Figure 4

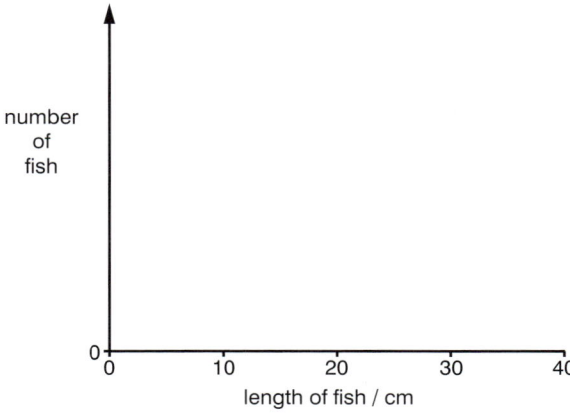

ii Name this type of natural selection. *(1 mark)*
iii Suggest two examples of selection pressure **S**.
(2 marks)
(Total 8 marks)
Cambridge International AS and A Level Biology 9700 Paper 4 Q8 June 2008

3 Anole lizards are found throughout the Caribbean and the surrounding mainland. There are many species. Each species is found only on one island or a small group of islands, apart from *Anolis carolinensis* which is found in mainland Florida.

Figure 5 shows the distribution of four species of anole lizards.

Figure 5

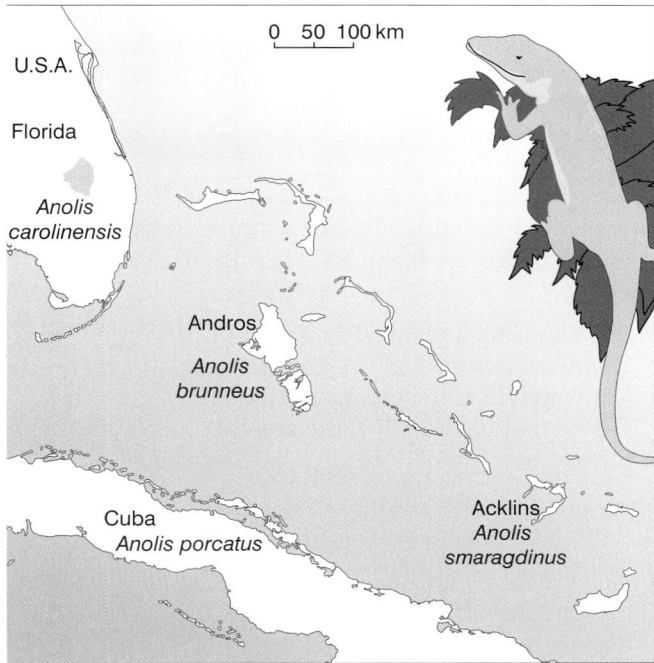

An investigation was carried out into the relationships between these four lizard species, using DNA analysis. The base sequences of a region of mitochondrial DNA from the four species were compared. The results are shown in Table 1.

The smaller the number, the smaller the differences between the base sequences of the two species.

Table 1

	A.brunneus		
A.brunneus		A.smaragdinus	
A.smaragdinus	12.1		A.carolinensis
A.carolinensis	16.7	15.0		A.porcatus
A.porcatus	11.3	8.9	13.2

a With reference to Table 1, state the species to which *A. brunneus* appears to be most closely related. *(1 mark)*

b The researchers put forward the hypothesis that the three species, *A. brunneus*, *A. smaragdinus* and *A. carolinensis*, have originated from three **separate** events in which a few individuals of *A. porcatus* spread directly from Cuba to three different places.

Explain how the results in Table 1 support the researchers' hypothesis. *(3 marks)*

c Explain how a population of *A. porcatus* that became isolated on an island could evolve into a new species.
(4 marks)
(Total 8 marks)

Cambridge International AS and A Level Biology 9700 Paper 41 Q5 June 2010

17 Practice Questions

4 a What is natural selection?

b State the differences between directional and stabilising selection.

c A severe cold period in 1996 killed over 50% of swallows (a type of bird) living on cliffs in Nebraska. Biologists collected nearly 2000 dead swallows from beneath the cliffs and captured around 1000 living ones. By measuring the body mass of the birds, they found that birds with larger than average body mass survived the cold spell better than ones with smaller than average body mass. State, giving your reasons, which type of selection was taking place here.

5 A type of bird known as a cuckoo lays its eggs in the nests of other birds. The host birds will often raise these parasite chicks alongside their own. In many valleys in southern Spain, great cuckoos and common magpies have lived together for hundreds of years. In some valleys, however, magpies have been around for centuries but cuckoos have only recently arrived.

Scientists placed artificial cuckoo eggs into magpie nests in both types of valley. Where cuckoos and magpies had lived together for a long period, 78% of the magpies removed the cuckoo eggs from their nests. Where cuckoos had only recently colonised the valleys, only 14% of the magpies removed the cuckoo eggs.

It would appear that, in the valleys where cuckoos are well established, selection has favoured those magpies that removed the cuckoo eggs.

a Suggest one advantage to the magpies of removing cuckoo eggs from their nest.

b Explain how removing cuckoo eggs increases the probability of the alleles for this type of behaviour in magpies being passed on to subsequent generations.

c Suggest why this form of behaviour is not shown by magpies in those valleys where cuckoos have only recently arrived.

d State, with your reasons, which type of selection is taking place here.

18.1 Classification

On these pages you will learn to:

- Define the term 'species'
- Describe the classification of species into the taxonomic hierarchy of domain, kingdom, phylum, class, order, family, genus and species

The concept of a species

A species is the basic unit of classification. A definition of a species is not easy, but members of a single species have certain things in common:

- **They are capable of breeding** to produce living offspring **which themselves are fertile** and so can successfully produce more offspring.
- **They have a common ancestry**. In other words they are descendents by adaptation of pre-existing or existing species.
- **They have the same genes** and share the same gene pool.
- **They share very similar biochemical, morphological behavioural, and physiological features**. These features, together with the DNA they possess, are more similar to each other than individuals of other species. If animals, they will share the same immunological features.
- **They occupy the same ecological niche** (Topic 18.4), to the exclusion of other species.

Species are not fixed forever, but change and evolve over time. Within an individual species there can be considerable variation amongst individuals. All dogs, for example, belong to the same species, but artificial selection has led to a variety of different breeds.

Organisms are identified by two names and hence the system is called the **binomial system**. Its features include:

- It is a universal system based upon Latin/Greek names.
- The first name, called the **generic** name, indicates the genus to which the organism belongs.
- The second name, called the **specific** name, indicates the species to which the organism belongs.

There are a number of rules that are applied to the use of the binomial system in scientific writing:

- The names are printed in italics or, if hand written, they are underlined to indicate that they are scientific names.
- The first letter of the generic name is in upper case (capitals), but the specific name is not.
- If the specific name is not known, it can be written as *sp.*, e.g. *Felix sp.*
- When referring to all members of a genus, the specific name is written as the plural, *spp.*, e.g. all members of the genus *Amoeba* is written as *Amoeba spp.*
- Once the generic name has been used initially, it can be abbreviated in later text to the first letter, e.g. the creeping buttercup *Ranunculus repens* can be written as *R. repens*.

A horse and a donkey are capable of mating and producing offspring – known as mules. A horse and a donkey are, however, different species and the resulting mules are infertile hybrids, i.e. they can almost never produce offspring when mated with each other. There is some evidence that a few female mules are fertile, although this is exceedingly rare. So rare that the Romans had a saying 'Cum mula peperit' which means 'when a mule foals' and is the equivalent of our modern phrases that mean 'it almost never happens' such as 'once in a blue moon'. Why then are mules infertile? It is all down to the number of chromosomes and the first prophase of meiosis. You may remember from Topic 15.1 that, during prophase I of meiosis, chromosomes line up in their pairs across the equator of the cell. These pairs are homologous, i.e. both of the chromosomes are exactly the same in structure and have genes that code for the same characteristics. Now a horse has 64 chromosomes (32 pairs) and a donkey has 62 chromosomes (31 pairs). The gametes of a horse and a donkey therefore have 32 and 31 respectively. On fusion of the gametes of a horse and a donkey, the offspring (the mule) has 63 chromosomes and you cannot exactly match up 63 chromosomes into pairs. However, mitosis can take place and therefore a mule grows and develops normally, but because the chromosomes cannot form homologous pairs at prophase I, meiosis cannot occur. As this is how gametes are formed, a mule cannot produce gametes and is therefore infertile.

Grouping species together – the principles of classification

Around 1.9 million different **eukaryotic** species have been identified and named, and yet this represents only a small proportion of the total thought to exist on Earth. There are probably at least another 10 million eukaryotic species either undiscovered or yet to be named. It makes sense to organise these species into manageable groups, if only to allow better communication between scientists. The grouping of organisms is known as **classification**, while the study of biological classification is called **taxonomy**. There are two basic types of biological classification:

- **Artificial classification** divides organisms according to differences that are useful at the time. Such features may include colour, size, number of legs, leaf shape, etc. These are described as **analogous** features where they have the same function but do not have the same evolutionary origins. For example, the wings of butterflies and birds are both used for flight but they originated in different ways.
- **Phylogenetic (natural) classification** is more widely used in biology and
  - is based upon the evolutionary relationships between organisms and their evolutionary descent (**phylogeny**)
  - classifies species into groups using shared features derived from their ancestors
  - is arranged in a **hierarchy** in which groups are contained within larger groups with no overlap.

Relationships in a phylogenetic classification are partly based upon **homologous** characteristics rather than analogous ones. Homologous characteristics have similar evolutionary origins regardless of their functions in the adult of a species. For example, the wing of a bird, the arm of a human and the front leg of a horse all have the same basic structure and similar evolutionary origins and are therefore homologous.

Organising the groups of species – taxonomy

Each group within a phylogenetic biological classification is called a **taxon** (plural = **taxa**). Taxonomy is the study of these groups which allows them to be placed in a hierarchical order, known as **taxonomic ranks,** based upon the evolutionary line of descent of the group members. Organisms are placed into one of the three domains, within which the largest groups are known as kingdoms, which are then divided into phyla, and organisms in each phylum have a body plan radically different from organisms in any other phylum. Diversity within each phylum allows it to be divided into classes. Each class is divided into orders of organisms that have additional features in common. Each order is divided into families and at this level the differences are less obvious. Each family is divided into genera and each genus (singular) into species. The classification of four organisms belonging to the domain Eukarya is given in Table 1.

> ### Remember
> A useful mnemonic for remembering the order of these taxonomic ranks is 'Delicious King Prawn Curry Or Fat Greasy Sausages' (Domain, Kingdom, Phylum, Class, Order, Family, Genus, Species).

Figure 1 Lathyrus odoratus *(sweet pea)*

Figure 2 Felix tigris *(Indian tiger)*

Table 1 *Classification of four organisms*

Common name / Rank	Spirogyra	Pin mould	Sweet pea	Tiger
Kingdom	Protoctista	Fungi	Plantae	Animalia
Phylum	Chlorophyta	Zygomycota	Angiospermophyta	Chordata
Class	Gamophyceae	Zygomycetes	Dicotyledonae	Mammalia
Order	Zygnematales	Mucorales	Rosales	Carnivora
Family	Zygnemataceae	Mucoraceae	Fabaceae	Felidae
Genus	*Spirogyra*	*Mucor*	*Lathyrus*	*Felis*
Species	*ellipsospora*	*mucedo*	*odoratus*	*tigris*

On these pages you will learn to:

- Outline the characteristic features of the three domains Archaea, Bacteria and Eukarya
- Outline the characteristic features of the kingdoms Protoctista, Fungi, Plantae and Animalia
- Explain why viruses are not included in the three-domain classification and outline how they are classified, limited to type of nucleic acid (RNA or DNA) and whether these are single stranded or double stranded

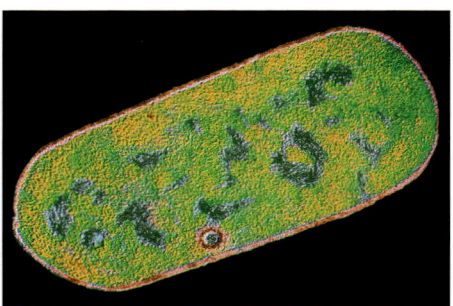

Figure 1 *An example of a bacterium is* Clostridium perfringens, *which causes blood poisoning and gas gangrene in humans*

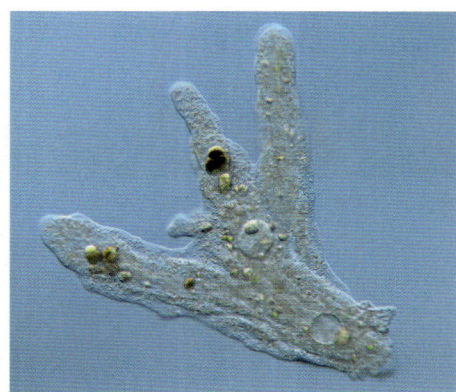

Figure 2 *The unicellular organism* Amoeba proteus *belongs to the kingdom Protoctista*

The main characteristics of the three domains of living organisms are listed below.

Bacteria are a group of single-celled prokaryotes with the following features:

- the absence of double membrane-bounded organelles such as nuclei or mitochondria
- unicellular, although cells may occur in chains or clusters
- ribosomes are smaller (70S) than in eukaryotic cells
- cell walls are present and made of peptidoglycan (murein) but never **chitin** or cellulose
- single loop (circular) of naked DNA made up of nucleic acids but no **histones**
- small in size – typically 0.5–5.0 µm in length.

Archaea are a group of single-celled prokaryotes that were originally classified as bacteria, which they resemble in appearance. Now they form their own group as they have some unique features and also share some features with either bacteria or eukarya:

- they have a wide range of morphologies, varying in shape and can be single celled, form filaments or group together
- they are found in a wide range of habitats and are common in extremes of temperature and salinity
- they do not have membrane-bound organelles and have circular DNA, although often much smaller than in bacteria, and they also have plasmids
- the promoter region (Topic 19.4) of the DNA and the RNA polymerase is more similar to those of eukarya than those of bacteria
- they have 70S ribosomes but the ribosome structure is different to that of bacteria and shows a very similar response to eukaryotic 80S ribosomes to antibiotics
- their genes and protein synthesis are more similar to eukaryotes
- their membranes contain very different lipids to bacteria and eukarya. Although archaeans may have some phospholipids as in bacteria and eukarya, the highest proportion of phosphate-containing lipids are ether lipids; here, glycerol is attached by highly resistant ether bonds to branched chains called alkyl chains.
- there is no peptidoglycan in their cell walls and there is more variety in composition than in bacteria.

Eukarya are a group of organisms made up of one or more eukaryotic cells. Their features are:

- their cells have chromosomes composed of linear DNA complexed with histone proteins and their chromosomes are enclosed in a double-membrane (nuclear envelope)
- their cells possess both single (e.g. ER, Golgi body) and double (e.g. mitochondria, chloroplasts) membrane-bound organelles
- not all possess cells with a cell wall, but where they do it contains no peptidoglycan, e.g. cellulose in plant cells and photosynthetic prokarotyes, and chitin in most fungi
- ribosomes are larger (80S) than in bacteria and archaea (with 70S ribosomes in mitochondria and chloroplasts).

Viruses are simple structures comprising a nucleic acid surrounded by a protein coat. They are not included in the three-domain system because they are not usually considered to be living organisms. They do not fit into the key concept that cells are the units of life and that living organisms are composed of one or more cells. They do not have their own metabolism and therefore need a host cell to make new products and to reproduce. Viruses can, however, be classified into different types, for example on the basis of their nucleic acid as they have only DNA or only RNA as their genetic material. A simple system groups them into four categories depending whether they possess DNA or RNA and whether this is single or double stranded.

Let us now look more closely at the four kingdoms of the Eukarya: **Protoctista**, **Fungi**, **Plantae** and **Animalia**.

Protoctista

The Protoctista are an extremely varied group of organisms with little in common except that they are unicellular or made up of groups of similar cells and they are **eukaryotic**. The group includes algae (e.g. *Spirogyra* spp.), *Amoeba* spp. and *Plasmodium* spp. Many protoctists share features with other kingdoms and were at one time classified in those kingdoms. However as each kingdom became more clearly defined it meant that some organisms were reclassified into the Protoctista.

The distinguishing features of the Protoctista are:

- eukaryotic cells, i.e. they have linear DNA complexed with histone protein and they possess membrane-bounded organelles such as a nucleus and mitochondria
- the majority are unicellular or groups of similar cells
- some are multicellular, but less complex than plants and animals; most form groups of similar cells although the macroscopic marine algae, known commonly as seaweeds, can have relatively complex morphologies with tissues specialised for particular functions.

Fungi

The Fungi are a large group of organisms that were once classified as plants but are now allocated to a kingdom in their own right. The group includes moulds, yeasts, mushrooms and toadstools.

The distinguishing features of the Fungi are:

- they are eukaryotic organisms
- absence of chlorophyll and therefore they do not photosynthesise but feed **heterotrophically** by absorbing their food, either as saprophytes (decomposers) or as parasites
- cell walls usually made of chitin and never of cellulose
- some are unicellular (e.g. yeasts) but most are made up of thread-like hyphae that collectively form a **mycelium**
- carbohydrate is stored as glycogen
- they reproduce sexually and/or asexually by means of spores that lack a flagellum.

Figure 3 The mould growing on this lemon is Penicillium sp *which belongs to the kingdom Fungi*

Plantae

The Plantae is a diverse group that ranges in size from liverworts a few millimetres across to giant redwood trees over 120 metres high. The group includes liverworts, mosses, ferns, coniferous trees and flowering plants.

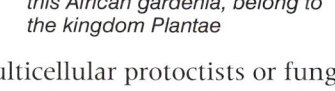

Figure 4 All flowering plants, like this African gardenia, belong to the kingdom Plantae

The distinguishing features of the Plantae are:

- they are eukaryotic organisms
- they are multicellular organisms with a higher level of complexity than multicellular protoctists or fungi – for example, specialised vascular tissues (xylem and phloem) and organs for photosynthesis (leaf)
- they possess chlorophyll and other pigments and therefore feed **autotrophically** by photosynthesis
- they possess cells with cell walls that are composed of cellulose
- carbohydrate is stored as starch
- there is an alternation between a multicellular gamete-producing **haploid** generation and a multicellular spore-producing **diploid** generation (Topic 5.3).

Animalia

The Animalia show the greatest diversity of form of any of the kingdoms. Much of this diversity is a result of their ability to move from place to place, which has led to the evolution of a wide range of different methods of locomotion.

The distinguishing features of the Animalia are:

- they are eukaryotic organisms
- they are multicellular organisms that have a high level of cellular organisation (specialised tissues and organs)
- their cells do not possess chlorophyll and therefore feed heterotrophically
- their cells do not possess cell walls
- carbohydrate is stored as glycogen
- can respond to stimuli in a coordinated way, with most having a nervous system.

SUMMARY TEST 18.2

The Protoctista are a very varied group of organisms that possess membrane-bounded organelles and are therefore known as (1) cells. The Fungi comprise organisms that feed (2). Most are made up of thread-like (3) that collectively form a (4) and have cell walls made up of (5). Plants are multicellular organisms that possess (6) and other pigments. They feed (7) by photosynthesis, most have cell walls made of (8) and store carbohydrate as (9). Animals store carbohydrate as (10) and most possess a (11) that allows them to coordinate activities.

18.3 Biodiversity

On these pages you will learn to:

- Explain that biodiversity is considered at three different levels:
  - variation in ecosystems or habitats
  - the number of species and their relative abundance
  - genetic variation within each species
- Discuss the reasons for the need to maintain biodiversity

Biodiversity is the general term used to describe variety in the living world. It refers to the number and variety of living organisms in a particular area and can be considered at a number of different levels: genetic diversity, species diversity and ecosystem diversity.

Genetic diversity

Organisms of the same species can show considerable genetic diversity; although they share the same genes, they will each have their own combinations of alleles. Even within a species, different populations may show differences in their genetic diversity. This may be because of the different selection pressures acting on the populations or the effects of humans (which often leads to a loss of genetic diversity).

Species diversity

One measure of biodiversity is species diversity. It may be measured within a community (a group of interacting populations in a defined area), or it may be measured in a larger area such as a desert, which has a number of different communities. Species diversity has also been compared between different countries and not surprisingly countries with tropical rainforests (Figure 1), such as Brazil, come out at the top. Species diversity has two components:

- the number of different species in a given area, which is known as **species richness**
- the proportion of the community that any individual species makes up, which is known as **relative abundance** or species evenness.

Two communities may have the same number of species but differ markedly in the proportion of the community that each species makes up. A natural grassland and a field of wheat may both have 25 species. However, in the grassland all 25 might be equally abundant whereas in the wheat field over 95% of the plants may be a single species of wheat. The community with one or two dominant species is considered to have a lower species diversity.

Figure 1 In a tropical rainforest there is high species diversity

Ecosystem and habitat diversity

Biodiversity is a measure of how well an ecosystem functions. The higher the species diversity the more stable an ecosystem usually is and the less it is affected by climate change. For example, if there is a drought, a community with a high species diversity is much more likely to have at least one species able to tolerate drought than a community with a low species diversity. At least some members are therefore likely to survive the drought and maintain the community. High diversity within an ecosystem means a high diversity of habitats (the place where an organism lives, see Extension in Topic 18.4).

In extreme environments such as the sub-arctic tundra (Figure 2) and hot deserts, only a few species have the necessary adaptations to survive the harsh conditions. The species diversity is therefore normally low. This usually results in an unstable ecosystem in which communities are dominated by climatic factors rather than by the organisms within the community. In less hostile environments like a tropical rainforest, the species diversity is normally high.

Figure 2 In the sub-arctic tundra there is low species diversity

This usually results in a stable ecosystem in which communities are dominated by living organisms rather than climate.

The biodiversity of a tropical rainforest is high because there is:

- the largest productivity (see Extension in Topic 18.5) of all terrestrial communities, which allows it to support a large variety of organisms
- a large number of species – up to 500 different species of tree in each square kilometre
- high genetic diversity between these species
- many different habitats and niches (see Extension in Topic 18.4)
- constant biological activity throughout the year, which means that the leaves, flowers and fruits that provide animals with food and shelter are always available.

Ecosystems with a high diversity are likely to have complex food webs and a high level of nutrient recycling.

Importance of maintaining biodiversity

The world possesses a rich diversity of species and ecosystems that are of great potential value to humankind. It is important to maintain this rich biodiversity for a number of reasons:

- **Ecological importance** – high biodiversity helps with the recycling of nutrients, in the formation and protection of soils and in the maintenance of ecosystems. By retaining water and minerals the soil, and the various plants growing in it, can help prevent natural disasters such as floods and **desertification.** High biodiversity also increases the stability of ecosystems and therefore allows them to more readily recover from unpredictable events, e.g. fires and drought.
- **Economic importance** – a high biodiversity means that there is a larger variety of species which can provide a larger variety of useful materials and a larger genetic diversity which is essential to adapt to a changing world. These include food, fibres for clothing, medicines, and timber for construction. Most of the world's food comes from a small number of plant species, many of which have been inbred to improve yields. Wild strains possess a **gene pool** which may be required in the future to further improve yields or introduce disease resistance. Around half of all medicines, e.g. aspirin, used today have active ingredients extracted from living organisms. Conserving biodiversity helps safeguard our future supply of food and medicines. In addition, areas of high biodiversity have considerable potential for tourism. In Costa Rica, for example, 12% of its land area, including much of its tropical rain forests and cloud forests, has been placed under protection. These regions attract two million visitors a year and produce an income of around two billion dollars. The prevention of natural disasters mentioned in ecological importance also has an economic benefit.
- **Ethical and social importance** – A high biodiversity is necessary to safeguard future resources and to maintain a diversity of genes. It allows the indigenous (native) population to maintain its own cultural values and way of life as well as providing the opportunity for ecotourism and recreation. It also permits research to be carried out and for others to learn about the vast range of organisms on the planet. Many people feel that humans have a responsibility to conserve biodiversity, not just for ecological and economic reasons, but also for its aesthetic value. Imagine how less fulfilling life would be without the beauty of flowers and other wildlife.

Figure 3 *In harsh environments, like this hot desert in Jordan, only a few species are adapted to survive the extreme conditions and therefore species diversity is low*

SUMMARY TEST 18.3

Biodiversity has three different components. Species diversity refers to both the number of different species and the number of **(1)** of each species within a **(2)**. Ecosystem diversity includes the range of **(3)** within a particular area. The variety of different alleles possessed by individuals of any one species is called **(4)** diversity. An ecosystem with a high species diversity is more **(5)** than one with a low species diversity and is therefore less affected if there is a **(6)** change. An example of an ecosystem with a high species diversity is a **(7)** whereas one with a low species diversity is a **(8)**. It is important to maintain biodiversity for ecological reasons because it helps to recycle **(9)** and promotes the formation and protection of **(10)**. Biodiversity also has an **(11)** and a **(12)** importance.

18.4 Ecosystems

On these pages you will learn to:

- Define the terms 'ecosystem' and 'niche'

Ecology is the study of the inter-relationships between organisms and their environment. The term **environment** refers to the conditions that surround an organism. These include both non-living (**abiotic**) components and living (**biotic**) components. Ecology is therefore a complex area of study which incorporates not only most aspects of biology but also elements of physics, chemistry, geography and geology. It is, in effect, the study of the life-supporting layer of land, air and water that surrounds the Earth, called the **biosphere**.

Ecosystems

An ecosystem is made up of all the interacting biotic (living) and abiotic (non-living) elements in a specific area. Ecosystems are more or less self-contained functional units. Within an ecosystem there are two major processes to consider:

- the flow of energy through the system
- the cycling of nutrients within the system.

In theory, the biosphere can be considered as a single ecosystem because energy flows through it and nutrients are recycled within it. In practice, there are much smaller units that are more or less self-contained in terms of energy and nutrients. A fresh-water pond, for example, has its own community of plants to collect the necessary sunlight energy to supply the organisms within it. Nutrients such as nitrates and phosphates are recycled within the pond, with little or no loss or gain between it and other ecosystems. An example of a much larger ecosystem is a tropical rainforest (Figure 1). Within each ecosystem, there are a number of interacting populations known as a community.

Figure 1 *Tropical rainforest habitat*

EXTENSION

Some ecological terms

Community

A community is defined as **all the populations** of different organisms living and interacting in a **particular place** at the **same time**. Within a tropical rainforest, a community might include a large range of organisms such as banana trees, insects, macaws, monkeys, iguanas, jaguars, woodlice, fungi and bacteria. Each species is made up of many groups of individuals, called populations.

Populations

A population is a group of organisms of the **same species** that occupy the **same place** at the **same time** and that have the chance to **interbreed** with one another. The boundaries of a population are difficult to define, except perhaps within a small pond. In our tropical rainforest, for example, all the mature jaguars can breed with one another and so form a single population. However, the woodlice on a decaying log can, in theory, breed with

those on a log a kilometre or more away. In practice, the long distance makes interbreeding unlikely and therefore they can be considered as subpopulations. Where exactly the boundary lies between these two subpopulations is, however, unclear.

Habitat

A habitat is the place where an organism lives. Within an ecosystem there are many habitats. For example, in our tropical rain forest, the leaf canopy of the trees may be a habitat for macaws while a decaying log is the habitat for woodlice. A river flowing through the forest provides a very different habitat, within which aquatic plants and fish live. For a caiman (a member of the Crocodile order), the river and banks are its habitat. Within each habitat there are smaller units, each with its own microclimate. These are called microhabitats. The mud at the bottom of the stream may be the microhabitat for worms while a crevice on the bark of an oak tree may be the microhabitat for a lichen.

Ecological niche

An ecological niche is all of the ranges of environmental conditions and resources required for an organism to survive, reproduce and maintain a viable population. It is also sometimes referred to as the ecological role of a species within its community. Some species may appear very similar, but their nesting habits or other aspects of their behaviour will be different, or they may show different levels of tolerance to, e.g. a pollutant or a shortage of oxygen or nitrates. Any differences in niche, however small, limit competition between species. No two species occupy exactly the same niche.

EXTENSION

Tropical rain forests

Tropical rain forests have been estimated to contain 50% of the world's standing timber. They represent a huge store for carbon and sink for carbon dioxide and their destruction may increase atmospheric concentrations of carbon dioxide by 50%. They are important in conserving soil nutrients and preventing large-scale erosion in regions of high rainfall. They contain a large gene pool of plant resources and their potential for the production of food, fibre and pharmaceutical products is not known. At present rates of destruction, all tropical rain forest, except that in reserves, will have disappeared by the middle of this century.

The traditional small-scale slash-and-burn agriculture of the hunter-gatherer cultures caused few environmental problems but attempts to exploit the forest on a large scale has had a very damaging effect. The problems have been most apparent in the rain forests of Brazil and other countries of the Amazon basin. The forest has been cleared for rubber plantations, timber extraction, planting of cash crops like cocoa, coffee and oil palm, extraction of minerals, especially bauxite, and for pulp and paper manufacture. The consequences of deforestation have been soil infertility, floods, soil erosion and increased sediment in rivers. The scale of the problem is huge. However, some areas are being set aside as reserves and some attempts have been made at replanting.

SUMMARY TEST 18.4

The study of the inter-relationships between organisms and their environment is called **(1)**. The layer of land, air and water that surrounds the Earth is called the **(2)**. An ecosystem is a more or less self-contained functional unit made up of all the living or **(3)** elements and non-living or **(4)** elements in a specific area. Within each ecosystem are groups of organisms, called a **(5)**, which live and interact in a particular place at the same time. A group of interbreeding organisms occupying the same place at the same time is called a **(6)**, and the place where they live is known as a **(7)**.

18.5

Food chains and food webs

Table 1 *Net primary production in different ecosystems*

Ecosystem	Mean NPP / kJ m^{-2} yr^{-1}
Desert	260
Ocean	4700
Temperate grassland	15 000
Intensive agriculture	30 000
Tropical rainforest	40 000

The organisms found in a tropical rainforest, or any other **ecosystem**, rely on a source of energy to carry out all their activities. The initial source of this energy is sunlight, which is converted to chemical energy by plants (producers) and then passed as food from one animal (consumer) to another.

Producers

Most producers are photosynthetic organisms that manufacture organic substances using light energy, water and carbon dioxide:

$$6CO_2 + 6H_2O \quad \xrightarrow[\text{chlorophyll}]{\text{light energy}} \quad C_6H_{12}O_6 + 6O_2$$

The rate at which they produce this organic food is referred to as their **productivity**.

- **Gross primary productivity (GPP)** is the total production of organic food in a given area and in a given time. It depends on the types of plant growing there, their density and the climate.
- **Net primary productivity (NPP)** is the rate of production of organic food after allowing for that lost via respiration by the plant – in other words, the production of material that is available to be eaten by consumers. Table 1 shows the mean NPP for a number of different ecosystems.

Consumers

Animals can only gain energy and nutrients by eating other organisms. Those that eat plants are **herbivores**, those eating animals are **carnivores**, and those eating both are **omnivores**. In terms of level within a food chain or web, herbivores are also referred to as **primary consumers** and they are eaten by **secondary consumers**. Although most secondary, tertiary and quaternary consumers are predators, there are some animals that feed on dead the bodies of dead organisms. Parasites can be found feeding in or on the living bodies of plants or animals. The carnivore at the end of the food chain is known as a top carnivore (Figure 1).

Decomposers and detritivores

When producers and consumers die, some energy is 'locked up' in the complex organic molecules of which they are made. This energy is used by a group of organisms that break down these complex materials into simple components again. In doing so, they also release valuable minerals and elements in a form that can be absorbed by plants and so contribute to recycling. The majority of this work is carried out by saprobiontic fungi and bacteria called **decomposers** and, to a lesser extent, breakdown is carried out by certain animals, such as earthworms, called **detritivores**.

Figure 1 *The jaguar is the top carnivore in the rain forests of South and Central America*

Food chains

The term **food chain** describes a feeding relationship in which plants are eaten by herbivores, which are in turn eaten by carnivores. Each stage in this chain is referred to as a **trophic level**. The first trophic level is represented by producers, the second by herbivores, and all subsequent ones by carnivores. The shortest food chains usually have three levels:

$$\text{grass} \rightarrow \text{sheep} \rightarrow \text{human}$$

and the longest usually no more than four or five:

banana tree	→	herbivorous insect	→	spider	→	tree frog	→	jaguar
producer		primary consumer		secondary consumer		tertiary consumer		quaternary consumer

The arrows on food chain diagrams represent the **direction of energy flow**.

Food chains in aquatic ecosystems are often longer than in terrestrial ecosystems as less energy is lost at each trophic level (see Topic 18.6).

Food webs

In reality, most animals do not rely upon a single food source and, within a single **ecosystem**, many food chains will be linked together to form a **food web**. For example, in a tropical rainforest food chains can be linked to form the web shown in Figure 2.

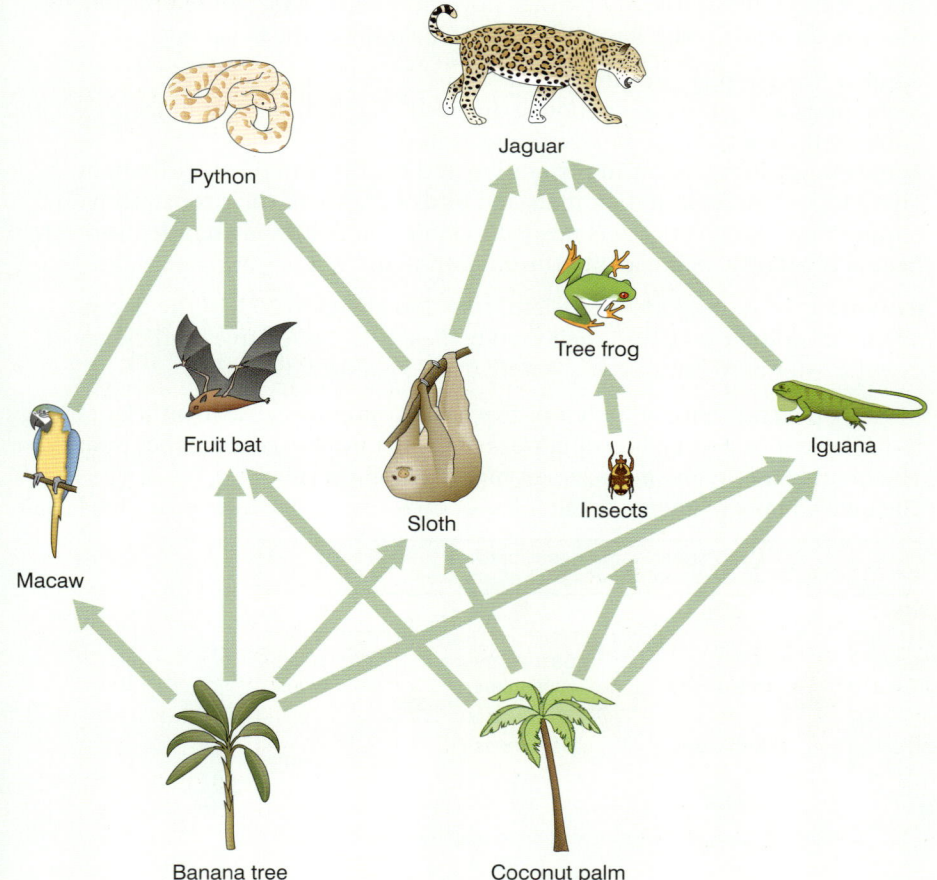

Figure 2 *Part of a simplified tropical rain forest food web (some arrows have been omitted for simplicity)*

SUMMARY TEST 18.5

The sun's energy is passed from one feeding level to another through the ecosystem. Each feeding level in this chain is called a **(1)**. As green plants manufacture complex organic molecules from simple ones during the process of **(2)**, they are known as **(3)**. The total amount of organic food made in a given time for a given area is the **(4)**. Some of this food is used by the plant for the process of **(5)**. The remainder, known as the **(6)**, is available for consumption by other organisms called consumers. **(7)** are consumers that feed directly off green plants, whereas **(8)** feed off other animals. In this way energy is passed along a hierarchy of feeding levels. This is referred to as a **(9)**. When organisms die, some energy remains locked up in the chemicals of which they are made. This energy may be used by a group of organisms called detritivores and **(10)**.

Energy transfer in ecosystems

The sun is the source of energy for **ecosystems**. Only around 1% of this light energy is captured by producers such as green plants and made available to successive organisms in the food chain. These in turn pass on only a fraction of the available energy at each stage.

Energy losses in food chains

Plants normally convert between 1% and 3% of the Sun's energy available to them into organic matter. Losses occur in a number of ways.

- Over 90% of solar energy is reflected back into space by clouds and dust or absorbed by the atmosphere and re-radiated.
- Not all wavelengths of light can be absorbed and used for photosynthesis.
- Light may not fall on a chlorophyll molecule.
- Low carbon dioxide levels may limit the rate of photosynthesis.

Plants then lose 20–50% of their gross primary production (Topic 18.5) via respiration, leaving little to be stored as potential food for primary consumers (herbivores). Even then, only about 10% of the **net primary production** of plants is used by primary consumers for growth. This low percentage is the result of the following:

- Some of the plant is not eaten.
- Some parts are eaten but cannot be digested (e.g. they are lost in faeces).
- Some of the energy is lost in excretory materials (e.g. urine).
- Some energy losses occur in respiration and heat loss to the environment. These losses are high in mammals and birds because of their constant body temperature. Much energy is needed to maintain their body temperature when heat is constantly being lost to the environment.

Carnivores are slightly more efficient, transferring up to 20% of the energy available from their prey into their own bodies. It is the relative inefficiency of energy transfer between **trophic levels** that explains why:

- Most food chains have only four or five trophic levels because insufficient energy is available to support a breeding **population** at trophic levels higher than these.
- The biomass of organisms is less at higher trophic levels.
- The total amount of energy stored is less at each level as one moves up a food chain.

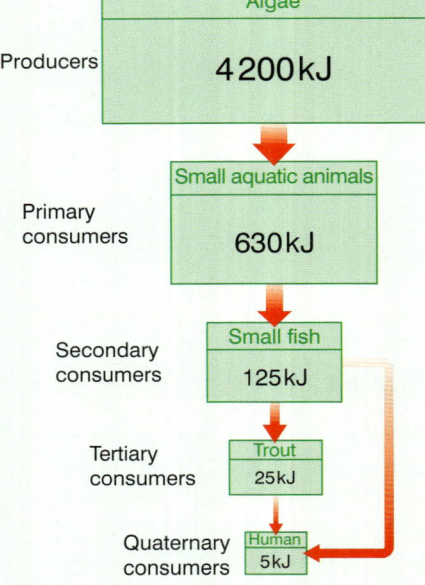

Figure 1 *Food chain in Cayuga Lake, New York State. Figures illustrate the relative amount of energy available at each stage in the food chain.*

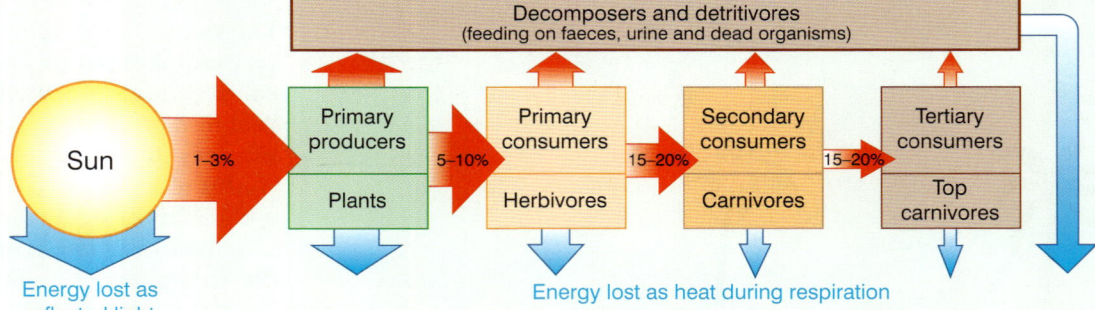

Figure 2 *Energy flow through different trophic levels of a food chain. The arrows are not to scale and give only an idea of the proportion of energy transferred at each stage. Likewise the figures for % energy transfer between trophic levels are only a rough average as they vary considerably between different plants, animals and habitats. Some of the decomposers and detritivores will be eaten and enter other food webs.*

It is therefore more energy-efficient for humans to eat food from lower trophic levels, but tastes often prevent this happening. This is shown in Figure 1, where we can see that the eating of small fish, called smelt, would yield five times the biomass of the more popular trout.

Energy flow along food chains is summarised in Figure 2. It is possible to construct **ecological pyramids** representing the numbers, biomass or stored energy of organisms at different trophic levels in a food chain.

Pyramids of number

Usually the numbers of organisms at lower trophic levels are greater than the numbers at higher levels. This can be shown by drawing bars with lengths proportional to the numbers present at each trophic level (Figure 3). However, Figure 3 b) and c) indicate that there can be disadvantages to using these pyramids to describe a food chain, as the following examples show.

- No account is taken of size – one tree is equated to one aphid and each parasite has the same numerical value as its larger host.
- The number of individuals can be so great that it is impossible to represent them accurately on the same scale as other species in the food chain. For example, one tree may have millions of greenfly living off it.
- No account is taken of juveniles (young) and other immature forms of a species, whose diet and energy requirements may differ from those of the adult.

Pyramids of biomass

Biomass is the total mass of the plants and/or animals in a particular place. It is normally measured over a fixed period of time. The term is sometimes used to refer to all living organisms on Earth, or a major part of the Earth, such as the oceans. It may also refer to plant or animal material that is used as fuel or raw material for industry. A more reliable, quantitative description of a food chain is provided when, instead of counting the organisms at each level, their biomass is measured. The fresh mass is quite easy to assess, but the presence of varying amounts of water makes it unreliable. The use of dry mass measurement overcomes this problem but, because the organisms must be killed, it is usually only made on a small sample, and this sample may not be representative. In both pyramids of numbers and pyramids of biomass, only the organisms present at a particular time are shown; seasonal differences are not apparent. This is particularly noticeable when the biomass of some marine ecosystems is measured: over the course of a whole year, the mass of **phytoplankton** (plants) must exceed that of **zooplankton** (animals), but at certain times of the year this is not seen. For example, in early spring around the British Isles, the standing crop (biomass at one particular time) of zooplankton is greater than that of phytoplankton (Figure 4).

Pyramids of energy

Collecting the data for pyramids of energy (Figure 5) can be difficult and complex, but the result is a true representation of the energy flow through a food web, with no anomalies. Data are collected in a given area (e.g. one square metre) for a set period of time, usually a year. The results are much more reliable than those for biomass, because two organisms of the same dry mass may store different amounts of energy. For example, 1 gram of fat stores twice as much energy as 1 gram of carbohydrate. The energy flow in these pyramids is usually shown as kJ m^{-2} yr^{-1}.

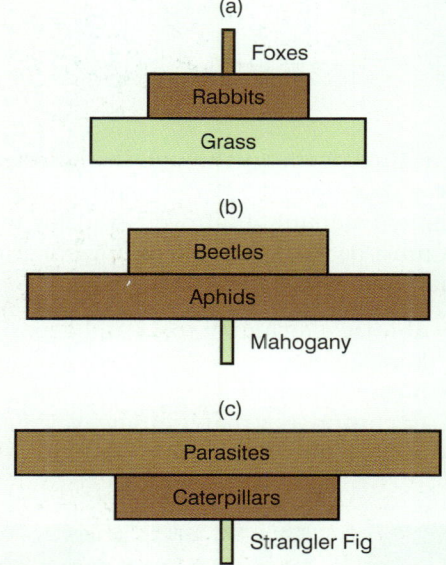

Figure 3 *Pyramids of numbers*

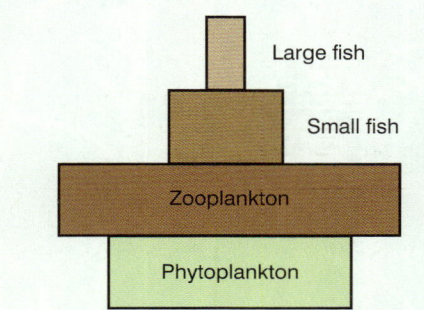

Figure 4 *Pyramid of biomass for a marine system*

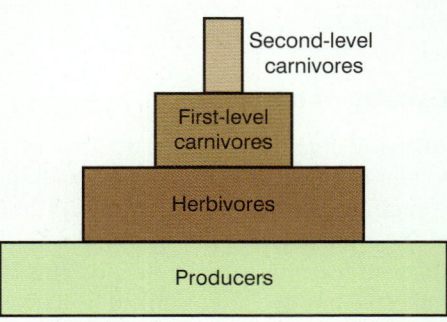

Figure 5 *Pyramid of energy, based on oak trees*

SUMMARY TEST 18.6

The sun is the initial source of energy for all living organisms on Earth. Between 1% and **(1)**% of this solar energy is converted to carbohydrate by plants. Of this amount, around **(2)**% is used by plants during respiration. Of the remainder, known as **(3)**, only around 10% is consumed by **(4)**. Carnivores convert around 20% of the energy in their prey, for their own use. This inefficiency in transferring energy from one **(5)** to the next along a food chain is the reason why each food chain is short and the numbers of organisms at each level usually reduce, producing a pyramid of numbers. It is more reliable, however, to use the total mass of organisms, rather than numbers, to give a pyramid of **(6)**. More reliable still is a pyramid of **(7)**.

18.7

The nitrogen cycle

The flow of energy through living systems is **linear** (one-way flow), but the flow of matter is **cyclical**. There is a limited amount of nitrogen available to living organisms so it must be used over and over again. Most nutrient cycles have two components:

- **abiotic** – including rocks and deposits in oceans and the atmosphere
- **biotic** – including **producers**, **consumers** and **decomposers** that, in some way, help to convert one form of the nutrient into another.

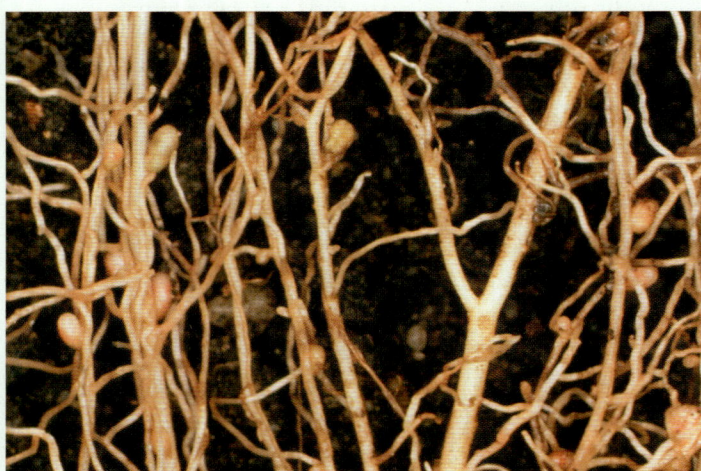

Figure 1 *Legume with root nodules containing the nitrogen-fixing bacterium,* Rhizobium

Cycling of nitrogen

All living organisms require a source of nitrogen from which to manufacture proteins, nucleic acids and other nitrogen-containing compounds, such as **ATP** and NAD. Although 78% of the atmosphere is nitrogen, there are very few organisms that can use nitrogen gas directly. This is because the nitrogen molecule has two atoms linked by a triple **covalent bond**, which makes it extremely stable and unreactive. Plants take up most of the nitrogen they need in the form of nitrate ions (NO_3^-), from the soil. These are absorbed, using **active transport**, by the root hairs (Topic 7.5). Animals obtain nitrogen-containing compounds by eating and digesting plants.

Nitrate ions are very soluble, and easily leach through the soil, beyond the reach of plant roots. One way of re-building the nitrate levels is to add fertilisers, but this is also achieved through the natural recycling of nitrogen-containing compounds. When plants and animals die, the process of decomposition begins a series of steps by which microorganisms replenish (restore) the nitrate levels in the

soil. Within the nitrogen cycle (Figure 2), five main stages can be recognised:

- nitrogen fixation
- assimilation
- denitrification.
- ammonification
- nitrification

Nitrogen fixation

This is the process by which nitrogen gas is converted into nitrogen-containing compounds. There are three ways of doing this, all of them requiring energy.

- **Lightning** allows nitrogen and oxygen to combine, producing oxides of nitrogen. These are washed into the soil by rain and absorbed by plant roots in the form of nitrates.
- **Industrial processes**, such as the Haber process, use high temperatures and pressures to combine nitrogen and hydrogen to produce ammonia. Much of this is added to the soil as nitrogen-containing fertilisers.
- **Fixation by microorganisms** is carried out by many bacteria and cyanobacteria living freely in the soil, and by some that live in nodules on the roots of leguminous plants.
  - **Free-living nitrogen-fixers** include bacteria that reduce gaseous nitrogen to ammonia, which they then use to manufacture amino acids. Nitrogen-rich compounds are released from them when they die and decay.
  - **Mutualistic** nitrogen-fixers include *Rhizobium*, which lives in nodules on the roots of leguminous plants such as peas and beans. Nodules are swellings on the roots of these plants (see Figure 1) in which *Rhizobium* uses an enzyme called **nitrogenase** to convert nitrogen gas (N_2) into ammonium ions (NH_4^+) using hydrogen ions (H^+) and ATP. The process requires **anaerobic** conditions and so the plant produces a pigment similar to **haemoglobin**, called leghaemoglobin, which absorbs any oxygen in the root nodule. The plant combines the ammonium ions produced with carbohydrate to form amino acids for making proteins. In return, *Rhizobium* receives energy and a place to live. Because both partners to this relationship benefit, it is called **mutualism**.

Assimilation

In plants which have a mutualistic relationship with *Rhizobium*, some nitrogen is assimilated in the form of ammonium ions from the nodules. All plants, however, can absorb nitrates from the soil via their root hairs. These are then reduced to nitrite ions (NO_2^-) and then ammonium ions (NH_4^+) for incorporation into amino acids, and then proteins. Animals assimilate their nitrogen in the form of protein, which forms part of the plants or animals that they eat.

Ammonification

Ammonification is the production of ammonia from organic nitrogen-containing compounds. In nature, these include urea (from the breakdown of excess amino acids) and proteins, nucleic acids and vitamins (found in faeces and dead organisms). Decomposers, mainly fungi and bacteria, feed on these materials, releasing ammonia which forms ammonium compounds in the soil.

Nitrification

The conversion of ammonium ions to nitrates involves oxidation reactions, which release energy for the nitrifying bacteria which carry out this process.

This conversion occurs in two stages:

- **oxidation of ammonium ions to nitrites (NO_2^-)** by nitrifying bacteria, e.g. *Nitrosomonas*, which live freely in well-aerated soil:

$$2NH_3 + 3O_2 \rightarrow 2NO_2^- + 2H^+ + 2H_2O$$
ammonia   oxygen   nitrite ions   hydrogen ions   water

- **oxidation of nitrites to nitrates (NO_3^-)** by other free-living nitrifying bacteria, e.g. *Nitrobacter*.

$$2NO_2^- + O_2 \rightarrow 2NO_3^-$$
nitrite ions   oxygen   nitrate ions

The oxygen requirements of nitrifying bacteria mean that it is important for farmers to keep soil structure light and well aerated by ploughing. Good drainage also prevents the air spaces from being filled with water, which would displace air, and hence oxygen, from the soil.

Denitrification

When soils become waterlogged, and therefore short of oxygen, a different type of microbial flora flourishes. Fewer nitrifying and free nitrogen-fixing bacteria are found, and there is an increase in anaerobic denitrifying bacteria, which reduce soil nitrates into gaseous nitrogen. The stages in reduction are as follows:

$$NO_3^- \rightarrow NO_2^- \rightarrow N_2O \rightarrow N_2$$
nitrate   nitrite   dinitrogen oxide   nitrogen

This reduces the availability of nitrogen-containing compounds for plants.

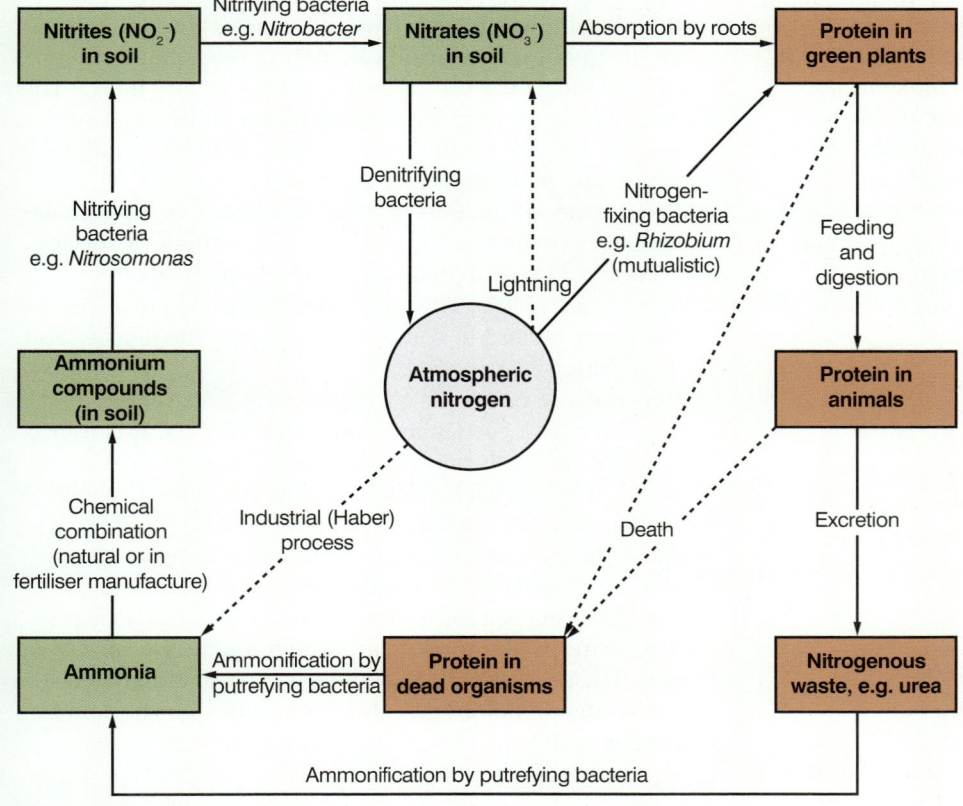

Figure 2 The nitrogen cycle

SUMMARY TEST 18.7

Nitrogen gas makes up **(1)**% of the atmosphere. A few organisms can convert this gas into compounds useful to other organisms in a process known as **(2)**. These organisms can be free-living or live in a **(3)** relationship with other organisms. One example is the bacterium called **(4)**, which lives in **(5)** on the roots of plants such as **(6)**. Most plants obtain their nitrogen by absorbing **(7)** from the soil through their root hairs. Animals obtain this nitrate when they eat the plants and then convert it into **(8)** in their bodies. On death, **(9)** break down these organisms, releasing **(10)**, which can then be oxidised to form **(11)** by nitrifying bacteria such as **(12)**. Further oxidation by other nitrifying bacteria like **(13)** will form nitrate ions. These nitrate ions may be converted back to atmospheric nitrogen by the activities of **(14)** bacteria.

On these pages you will learn to:

- Explain the importance of random sampling in determining the biodiversity of an area
- Use suitable methods, such as frame quadrats, line transects, belt transects and mark–release–recapture, to assess the distribution and abundance of organisms in a local area

It is often necessary to measure the abundance of organisms in particular habitats, and as it is virtually impossible to identify and count every organism, only small samples of the **ecosystem** are usually studied in detail. Provided these are representative of an area as a whole, any conclusion drawn from the findings will be valid. There are a number of sampling techniques including:

- random sampling using frame quadrats (Figures 1 and 2)
- systematic sampling along transects.

In addition to giving information about the numbers of individuals of each species (abundance), these techniques can also be used to assess the distribution of each species, that is, where a species occurs within an area.

Random sampling using frame quadrats

A quadrat is a sturdily built square frame (Figure 1) divided by string or wire into equally sized subdivisions. It is often designed so that it can be folded to make it more compact for storage and transport. The size of quadrat used will depend on the size of the plants or animals being counted. Suppose we wish to investigate the effects of grazing animals on the species of plants growing in a field. We would begin by choosing two fields as close together as possible to minimise soil, climatic and other **abiotic** differences. One field should be regularly grazed by animals

such as sheep, whereas the other should not have been grazed for many years. We would take random samples at many sites on each field by placing the quadrat on the ground and recording the name and abundance of each species found within the area of the quadrat. The problem is: how do we get a truly random sample? We could simply stand in one of our fields and throw the quadrat over our shoulder. Even with the best of intentions, it is difficult not to introduce an element of personal bias using this method. Are we as likely to stand in a muddy wet area as in a dry one? Will we deliberately try to avoid the area covered in sheep droppings or rich in nettles (stinging plants)?

A better form of random sampling is to:

- lay out two long tape measures at right angles, along two sides of the study area
- using random numbers from a table or generated on a computer or certain types of calculator, obtain a series of coordinates
- place a quadrat at the intersection of each pair of coordinates and record the species within it.

Abundance can be measured in a number of ways, depending on the size of the species being counted and the habitat.

- **Species density**: calculated by counting the number of times an individual of a particular species occurs within all the quadrats used and calculating the mean number of individuals per unit area, e.g. 12 per square metre. This method can be time-consuming where individuals of a species are very small, and it is often difficult to judge where one plant ends and the next begins.
- **Frequency of occurrence**: the likelihood of a particular species occurring in a quadrat. If, for example, a species occurs in 15 out of 30 quadrats, the frequency of occurrence is 50%. This method is useful where a species, e.g. grass, is hard to count, but it ignores the density and distribution of a species.
- **Percentage cover**: an estimate of the area within a quadrat that a particular plant species covers. It is useful where a species is particularly abundant or is difficult to count. It is less useful where organisms, more probably plants, occur in several overlapping layers.
- **Abundance scales**: specific measures that give the relative abundance of a particular species. The scales vary from one species to the next. One such scale lists organisms as abundant, common, frequent, occasional or rare. The process is simple and easy to use but subjective – investigators often give different values for the same sample quadrat.

To obtain reliable results, it is necessary to ensure that the sample size is large, i.e. many quadrats are used and the

0.5 metre (internal dimension)

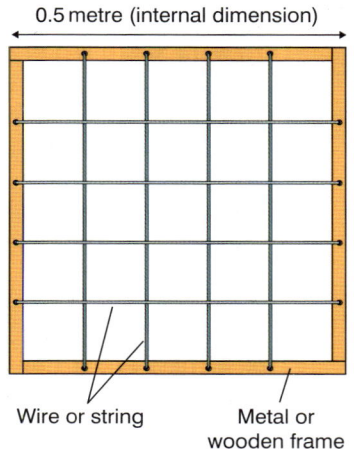

Wire or string

Metal or wooden frame

Figure 1 A frame quadrat

mean of all the samples is obtained. The larger the number of samples, the more representative of the field as a whole will be the results obtained.

Systematic sampling along transects

It is sometimes more informative to measure the abundance and distribution of a species in a systematic rather than a random manner. For example, the distribution of organisms on a tidal seashore is determined by the relative periods of time they spend under water and time they spend exposed to the air, i.e. by their vertical height up the shore. In these circumstances more useful data are obtained using a transect, of which there are two main types:

- **line transect**: a string or tape stretched across the ground in a straight line; any organism over which the line passes is recorded
- **belt transect**: a strip, usually a metre wide, marked by putting a second line transect parallel to the first; the species occurring within the belt between the lines are recorded. Alternatively, a metre-square frame quadrat can be laid alongside a single line transect and the species within it recorded. The quadrat is then moved its own length along the line and the process repeated.

Both line and belt transects are of two types:

- **continuous transect**: in which sampling takes place over a relatively short distance and therefore takes place from one end of the line/belt to the other
- **interrupted (ladder) transect**: in which a much larger distance is involved and therefore it is only practical to take samples at intervals, e.g. every 10 metres, along the line/belt.

Mark–release–recapture techniques

The methods of measuring abundance described above work well with plant communities but not with most animals. Most animals are mobile and move away when approached. They are often hidden and therefore difficult to find and identify. To estimate the abundance of most animals requires an altogether different technique.

A known number of animals are caught, marked in some way, and then released back into the community. Some time later, a given number of individuals are collected randomly and the number of marked individuals is recorded. The size of the population is then calculated as follows:

Estimated size of population =

$$\frac{\text{TNI in 1st sample } \times \text{ TNI in 2nd sample}}{\text{Number of marked individuals recaptured}}$$

where TNI = Total number of individuals

This technique makes a number of assumptions:

- the proportion of marked to unmarked individuals in the second sample is the same as the proportion of marked to unmarked individuals in the population as a whole
- the marked individuals released from the first sample distribute themselves evenly among the remainder of the population and have sufficient time to do so
- the population has a definite boundary so that there is no immigration into or emigration out of it
- there are few, if any, deaths and 'births' within the population
- the method of marking used is not toxic to the individual, nor does it make the individual more conspicuous (easily seen) and therefore more liable to predation
- the mark or label is not lost or rubbed off during the investigation.

Figure 2 *Ecology student using a quadrat*

SUMMARY TEST 18.8

We can measure the abundance of a species within a frame quadrat by counting the number of times individuals belonging to that species occur – known as the **(1)** of a species. Where a species is very **(2)** it is preferable to measure its percentage **(3)** within the quadrat. Systematic, rather than random, sampling is preferable in places such as a **(4)**. Here a transect is used. Sampling along a transect may be continuous or at intervals, in which case it is known as an **(5)** transect. To estimate the abundance of animals we use the **(6)** technique. Using this technique an ecologist collected and marked 100 individuals of a species. A week later she collected 80 individuals of which five were marked. The estimated size of this population is therefore **(7)**.

On these pages you will learn to:

- Use Spearman's rank correlation and Pearson's linear correlation to analyse the relationships between the distribution and abundance of species and abiotic or biotic factors
- Use Simpson's index of diversity (D) to calculate the biodiversity of a habitat, using the formula

$$D = 1 - \left(\sum \left(\frac{n}{N} \right)^2 \right)$$

and state the significance of different values of D

To make sense of the data that we collect from ecological studies we often need to analyse it. There are a number of statistical tests that can be used to analyse data. Some of these enable us to determine whether or not there is a relationship between two sets of data. We have already looked at one such test, the t-test (Topic 17.3). Let us look at others.

Spearman's rank correlation

Spearman's rank correlation measures the relationship between two sets of **ranked** data. There are some criteria that must be met for the test to be valid.

- Ordinal data are used or any data that have been collected can be converted to an ordinal scale using ranking. Ordinal data are data that have two or more categories that can be ordered or ranked.
- The data points within samples must be independent of one another.
- A scatter diagram indicates that there is a relationship whereby as the value of one increases, so does the value of the other variable **or** as the value of one increases, the value of the other variable decreases.
- All individuals must be selected at random from a population.
- Each individual must have an equal chance of being selected.
- More than five paired observations are needed, but 10 to 30 are ideal.

The Spearman's correlation coefficient is represented by the Greek letter rho (ρ) or r_s and the equation is:

$$r_s = 1 - \left(\frac{6 \times \sum D^2}{n^3 - n} \right)$$

$\sum$ = sum of
D = difference between each pair of ranked measurements
n = number of pairs of items in the sample

The best way to explain how this equation operates is to work through a particular example.

An ecologist is trying to determine whether the number of plants found at different sites in a woodland is related to the light intensity at each site. He uses random quadrats to count the number of plants and a light meter to measure light intensity. His results are shown in Table 1.

Table 1

Quadrat number	Number of plants	Light intensity/lux
1	40	9000
2	12	7000
3	8	2000
4	27	8000
5	24	7000
6	21	6000
7	25	5000
8	60	15000
9	18	3000
10	64	19000
11	19	4000
12	70	22000

The next stage is to rank the two sets of data from highest to lowest. For the number of plants this is straightforward, but for light intensity there are two identical values. For example, the value of 7000 is sixth in rank but there are two of these values. In other words rank 6 and rank 7 are both 7000. We therefore average the two ranks $(6 + 7 = 13 \div 2 = 6.5)$ and give each the value of 6.5. **Remember that as ranks 6 and 7 have both been used, the next rank value is 8**. Table 2 shows the rank values for our data.

Table 2

Quadrat number	Number of plants	Rank order of plant numbers	Light intensity/ lux	Rank order of light intensity
1	40	4	9000	4
2	12	11	7000	6.5
3	8	12	2000	12
4	27	5	8000	5
5	24	7	7000	6.5
6	21	8	6000	8
7	25	6	5000	9
8	60	3	15000	3
9	18	10	3000	11
10	64	2	19000	2
11	19	9	4000	10
12	70	1	22000	1

The next stage is to calculate the difference (D) between the two sets of rank values by taking the second value away from the first. Some values of D will be negative, but as the next stage is to square these values (D^2) then they all become positive. Finally we add all the values of D^2 to give a total of 31.50. See Table 3.

Table 3

Quadrat number	Number of plants	Rank order of plant numbers	Light intensity/ lux	Rank order of light intensity	Difference (D)	D^2
1	40	4	9000	4	0.0	0.00
2	12	11	7000	6.5	4.5	20.25
3	8	12	2000	12	0.0	0.00
4	27	5	8000	5	0.0	0.00
5	24	7	7000	6.5	0.5	0.25
6	21	8	6000	8	0.0	0.00
7	25	6	5000	9	−3.0	9.00
8	60	3	15000	3	0.0	0.00
9	18	10	3000	11	−1.0	1.00
10	64	2	19000	2	0.0	0.00
11	19	9	4000	10	−1.0	1.00
12	70	1	22000	1	0.0	0.00
					Total	**31.50**

We now substitute the values of n and D^2 in the Spearman's rank correlation equation as follows:

$$r_s = 1 - \left(\frac{6 \times 31.5}{1728 - 12} \right)$$

$$= 1 - 0.11013986$$

$$= 0.88986014$$

Table 4

n	$p = 0.05$ level	$p = 0.01$ level
12	0.591	0.777
14	0.544	0.715
16	0.506	0.665
18	0.475	0.625
20	0.450	0.591

The correlation coefficient will range in value from -1.0 to $+1.0$. The value of 0.890 (to three decimal places), which is close to $+1$, gives an indication that there is a relationship between light intensity and the number of plants at the selected sites. To obtain a more accurate measure of the strength of the relationship, it is necessary to look up the coefficient in critical value tables. This will provide a level of significance for the relationship. Table 4 shows a small, but relevant, part of the critical values table for the Spearman's rank correlation coefficient. In this table, with $n = 12$, (the number of quadrats used) the calculated value of r exceeds the critical value of 0.777 at the 0.01 probability level. As a probability level of 0.05 or lower is considered acceptable, it can be stated that there is a significant correlation between the two factors: as light intensity increases, so does the number of plants.

Pearson's correlation coefficient

Pearson's correlation coefficient is used to investigate the relationship between two quantitative, continuous variables, for example, age and blood pressure. Pearson's correlation coefficient (r) is a measure of the strength of the association between the two variables. It measures continuous variables rather than the categorical ones needed for Spearman's rank correlation. Again we need to satisfy some criteria.

- The data collected must be continuous.
- A scatter diagram indicates a linear relationship.
- The data collected must be normally distributed.

There are a number of different ways of expressing the Pearson's correlation coefficient equation. One is:

$$r = \frac{\sum xy - n\bar{x}\bar{y}}{n\, s_x s_y}$$

Where Σ = sum of
n = number of observations
x and y = observations
$\bar{x}$ and $\bar{y}$ = mean of observations
s = standard deviation

The values obtained for the Pearson's correlation coefficient lie between -1 and $+1$.

For both Spearman's rank correlation and Pearson's correlation, the coefficient tells us two things about the linear relationship between our two variables.

- **Strength** – the larger the value the stronger the relationship; 0.0 indicates the absence of a relationship, whereas 1.0 is a perfect relationship.
- **Direction** – The sign ($+$ or $-$) indicates the direction of the relationship; the coefficient is positive if both variables increase or decrease together; the coefficient is negative if one variable increases while the other decreases.

These relationships are shown in Figure 1.

As the names suggest, the coefficient only indicates a correlation. It does not imply a causative relationship.

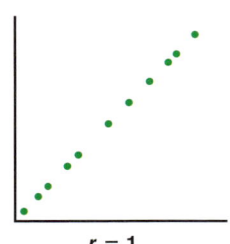

r = 1
Data lie on a perfect straight line with a positive slope

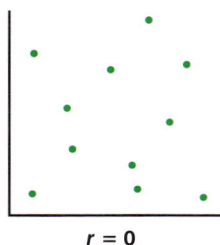

r = 0
No linear relationship between the variables

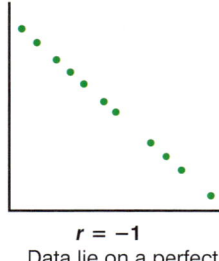

r = −1
Data lie on a perfect straight line with a negative slope

Figure 1 *Values of Pearson's and Spearman's correlations coefficients*

Measuring species diversity

Consider the data shown in Table 1 about two habitats. The final two rows don't tell us much about the differences between the two habitats because in both cases the total number of species and the total number of individuals are identical. However, if we measure the species diversity we get a different picture.

Table 5 Number and types of species found in two habitats within the same ecosystem

Species found	Numbers (n) found in habitat X	Numbers (n) found in habitat Y
A	10	3
B	10	5
C	10	2
D	10	36
E	10	4
No. species	5	5
Total no. all individuals (N)	50	50

One way of measuring species diversity is to use the Simpson's index of diversity using the formula:

$$D = 1 - \left(\Sigma \left(\frac{n}{N} \right)^2 \right)$$

Where D = species diversity index

N = total number of all individuals of all species

n = total number of individuals of each species

Σ = the sum of

Calculation of the diversity index for habitat X:

$$D = 1 - \left(\left(\frac{10}{50} \right)^2 + \left(\frac{10}{50} \right)^2 + \left(\frac{10}{50} \right)^2 + \left(\frac{10}{50} \right)^2 + \left(\frac{10}{50} \right)^2 \right)$$

$$D = 1 - (0.2^2 + 0.2^2 + 0.2^2 + 0.2^2 + 0.2^2)$$

$$D = 1 - (0.04 + 0.04 + 0.04 + 0.04 + 0.04)$$

$$D = 1 - 0.2$$

$$D = \mathbf{0.8}$$

Calculation of the diversity index for habitat Y:

$$D = 1 - \left(\left(\frac{3}{50} \right)^2 + \left(\frac{5}{50} \right)^2 + \left(\frac{2}{50} \right)^2 + \left(\frac{36}{50} \right)^2 + \left(\frac{4}{50} \right)^2 \right)$$

$$D = 1 - (0.06^2 + 0.10^2 + 0.04^2 + 0.72^2 + 0.08^2)$$

$$D = 1 - (0.0036 + 0.0100 + 0.0016 + 0.5184 + 0.0064)$$

$$D = 1 - 0.54$$

$$D = \mathbf{0.46}$$

The higher the value D the greater the species diversity. So in this case, although both habitats have the same total number of species and the same total number of individuals, the species diversity of habitat X is greater (0.8) than that of habitat Y (0.46).

Remember

Calculating a species diversity index provides a number that makes it easier to compare the variety in different habitats. It would be much harder, and less precise, if we had to rely on descriptions of different habitats to make these comparisons.

18.10

Threats to biodiversity

On these pages you will learn to:

- Discuss the threats to the biodiversity of aquatic and terrestrial ecosystems

There is often a conflict between the needs of a country to produce sufficient food to feed its inhabitants and the need to conserve habitats and wildlife. As a result, many habitats have been lost, and many species have become extinct. Species extinction has always been a normal part of natural selection. What is different now is the accelerating pace of species extinctions and habitat losses as a consequence of human activities. Let us look at some human activities that endanger wildlife.

Habitat loss and fragmentation

Humans exploit many natural habitats, destroying them in the process. Examples include:

- Timber cutting (Figure 1) destroys forests and endangers species such as the orang-utan.
- Industrial and agricultural developments threaten many plant species of the Amazon forest.
- Clearing of river banks destroys the natural habitat of the otter and beaver.
- Modern farming methods remove trees and hedgerows as well as drain wetlands. These practices endanger the species that live and breed there. In Central America, for example, a traditional coffee plantation has coffee bushes grown in the shade beneath large trees. The canopy of these trees supports a diverse variety of species and the crop requires few pesticides. To increase productivity, modern coffee plantations use varieties of coffee that require full sun. To accommodate these, the trees are removed, with consequent loss of biodiversity, and large applications of pesticides are needed to control pests.

Figure 1 *Deforestation in Indonesian Borneo. Ten thousand square kilometres of primary rainforest are destroyed or altered each year in Indonesia alone, affecting biodiversity, soils and drainage basins.*

Maintaining the diversity of organisms in a habitat is important as species may have economic importance outside their habitat, e.g. pollinating insects. As natural habitats are destroyed, those remaining become increasingly **fragmented** and isolated. These patches often cannot support a species. For example, when the Central American tropical rainforest becomes **fragmented**, species such as the spider monkey and the tufted capuchin, which have large ranges, are the first to disappear.

Competition from humans and their animals

The human population now exceeds 7 billion, putting ever-increasing pressure on land-use. As more and more land is taken up for building, industry and farming, there is more intense competition amongst wildlife for food, shelter and breeding sites in the remaining natural habitats. Where a species is restricted to a small area, e.g. the giant tortoises in the Galapagos Islands, they are often unable to compete with the influx of humans and their animals. Because their habitat is restricted, in this case by water, they cannot escape. Humans often introduce new species that out-compete native ones. Almost half of all small- to medium-sized Australian marsupials have been exterminated by competition with introduced rabbits and predation by introduced foxes and cats.

Hunting, poaching and fishing

Humans hunt tigers for sport, crocodiles and foxes for their skins, oryx as trophies, elephants for ivory, whales for oil and rhinoceros for their horn. Other organisms are collected for the pet trade, e.g. tamarins and parrots; and for research purposes, e.g. frogs. These are in addition to the numerous species hunted purely as food. These activities have led to the loss of species such as the passenger pigeon. Thought to have once been the most abundant bird on the planet, the passenger pigeon (Figure 2) was hunted for food and sport. In addition, its woodland habitat was destroyed to make room for agriculture. These combined effects led to its extinction, the last survivor dying in a zoo in 1914.

Other species, such as the black rhinoceros, have been brought to the brink (edge) of extinction.

Climate change

The emission of greenhouse gases when fossil fuels are burned has led to global warming. This has caused changes in the weather patterns with consequent changes to habitats. The abundance and distribution of individual species around the globe has been affected as a result of areas becoming wetter or drier, warmer or colder. Rising sea levels are beginning to create problems for many coastal ecosystems. Throughout the tropics, an effect known as mass coral bleaching has occurred. This is a stress response to rising ocean temperatures and has caused the breakdown of the very close beneficial relationship between coral and a group of unicellular algae (zooxanthellae). This has meant that the coral have lost up to 90% of their energy source and many have not survived. The climate is also becoming more unpredictable, and extreme – often devastating – events are becoming more frequent. All these changes threaten biodiversity.

Elimination of organisms considered dangerous

Many species are persecuted because they carry diseases of domesticated species, e.g. badgers (tuberculosis of cattle) and eland (various cattle diseases). Others, such as crocodiles and the big cats are seen as a direct threat to human life and are removed to make people feel more secure.

Pollution

Oil pollution, such as the 2010 Deepwater Horizon oil spill in the Gulf of Mexico (Figure 3), threatens some rare species of sea birds as well as turtles and the dwarf seahorse. The build-up of certain insecticides along food chains endangers predatory birds like the peregrine falcon and the golden eagle.

Wars

Conflicts around the globe lead to the destruction of habitats and the death of other organisms as well as humans. The use of defoliants such as Agent Orange in the Vietnam war of the 1960s destroyed forests and the other wildlife within them.

Figure 2 *This photo shows Martha, believed to have been the last passenger pigeon. She died in the Cincinnati Zoological Park in 1914.*

Figure 3 *A pelican coated in oil from the Deepwater Horizon oil spill in the Gulf of Mexico in 2010. The survival of species like these can be threatened by pollution of this kind.*

SUMMARY TEST 18.10

Many human activities threaten species. Some human activities destroy habitats. These include the clearing of **(1)** which threatens otters, timber cutting which threatens the **(2)** and fragmentation of forests which threatens **(3)** in Central America. The introduction of **(4)** to Australia led to the extinction of many native marsupials as they competed for similar food. Hunting and loss of **(5)** led to the extinction of the **(6)** which was thought to once have been the most numerous species of its type. Predatory birds such as the peregrine falcon are at risk from man-made **(7)** that accumulate along food chains.

On these pages you will learn to:

- Discuss methods of protecting endangered species, including the roles of zoos, botanic gardens, conserved areas (national parks and marine parks), 'frozen zoos' and seed banks
- Discuss the roles of non-governmental organisations, such as the World Wide Fund for Nature (WWF) and the Convention on International Trade in Endangered Species of Wild Fauna and Flora (CITES), in local and global conservation

Figure 1 Giant panda at Chengdu Research Base of Giant Panda Breeding in China. The base is carrying out a captive breeding programme with a view to reintroducing this endangered species back into the wild.

Figure 2 African elephant herd in the Amboseli National Park, Kenya. National reserves like these help in the conservation of endangered species.

A species is considered endangered when it is facing a very high risk of extinction in the near future. Some non-governmental organisations (NGOs) play a major role in protecting and conserving endangered species. These organisations complement the work of governments. They can often work more quickly than governments to bring about cooperation between countries for the conservation of endangered species (see Topic 18.12 for a definition of conservation) and for the preservation of habitats (leaving habitats undisturbed by humans). Their role also includes trying to find a balance between the needs of local people and the need to protect endangered species.

Development of national parks and nature reserves

National parks and nature reserves are habitats legally safeguarded and patrolled by wardens. They may preserve a vulnerable food source, e.g. in China areas of bamboo forest are protected to help conserve the giant panda (Figure 1). In Africa, game parks help to protect endangered species

such as the African elephant (Figure 2). Efforts are being made to conserve the dwindling areas of tropical rainforest in Central America and elsewhere. Planning authorities have greater powers to control developments and activities within these areas. Nest sites and the young can be monitored and protected to help ensure they mature to adulthood and have the opportunity to breed.

The importance of conserving the biodiversity of the seas is being increasingly appreciated. As a result, more and more **marine parks** (Figure 3) are being designated by governments throughout the world. Areas of the sea are set aside to preserve a specific habitat and ensure that the ecosystem is sustained for the organisms that live there. Coral reefs are one of the most biodiverse habitats on the planet, and yet they are especially vulnerable because:

- climate change is raising the temperature of the seas and increasing their carbon dioxide concentration
- more dissolved carbon dioxide decreases the pH of the seas, which causes the skeletons of corals to dissolve and the reef to break up
- rising sea levels and pollution by oil and sewage are killing species that live on them
- intensive fishing and disturbance from tourists diving and snorkelling on them also threaten their fragile ecology.

The status of a marine park allows fishing and diving to be restricted and special measures taken to avoid pollution. Two of the largest marine parks are the Great Barrier Reef Marine Park in Australia and the Chagos Marine Park in the Indian Ocean.

Legal protection for endangered species

In many countries it is illegal to collect or kill certain species, e.g. the koala in Australia. The **Convention on International Trade in Endangered Species of Wild Fauna and Flora (CITES)** came into force in 1975. Its aim is to prevent international trade from threatening the survival of certain endangered species. The freely available CITES appendices contains a list of these species and notes about products obtained from some species. Some 5000 species of animals and 28 000 species of plants are protected against overexploitation through international trade. Although participation is voluntary, it is legally binding on those countries that choose to sign up. CITES provides a framework for each country to adopt its own legislation to implement it. Most of these countries have very strict laws banning the import or export of endangered plants and animals and their products.

Non-governmental organisations such as the **World Wide Fund for Nature (WWF)** also have a role to play.

WWF works on conserving, preserving and restoring the environment. It is the world's largest independent conservation organisation, supporting around 1300 conservation and environmental projects. Its aim is to stop the degradation of the planet's natural environment and to build a future in which humans live in harmony with nature. Its work concentrates on the conservation of the oceans and coasts, forests, and freshwater ecosystems, as these contain most of the world's biodiversity. It is also concerned with endangered species, pollution and climate change.

Captive breeding in zoos and botanical gardens

Endangered species may be bred in the protected environment of a zoo and when numbers have been sufficiently increased they may be reintroduced into the wild. One species conserved in this way has been the Hawaiian Goose or Ne-Ne. Its population in the wild fell to around 20 pairs before being supplemented by thousands of birds bred in captivity and released in Hawaii. Captive breeding programmes have the following advantages:

- There is less need to capture wild animals and plants to supply zoos and botanical gardens.
- Natural populations can be maintained or increased by reintroduction of individuals bred in captivity.
- Breeding success is improved by techniques such as in-vitro fertilisation (IVF) and monitoring mothers during the gestation period (antenatal and postnatal care).

Captive breeding is not without its problems however. These include:

- Inbreeding leading to a gene pool that is small and offspring that may be weak with the potential for genetic defects.
- When reintroduced into the wild, captively bred individuals may not adapt to their new circumstances. They may have problems feeding, mating and be more prone to disease and more vulnerable to predators.
- Introducing pathogens and diseases that have been acquired in captivity, into the wild population.
- Captivity is an unnatural state for wild animals. They often become stressed which makes them unable to mate and makes them more vulnerable to disease.

Establishing seed, embryo and sperm banks

In addition to captive-breeding programmes, zoos often 'freeze' (cryopreservation) ova, sperm and embryos ('**frozen zoos**') for later use when natural habitats become available and finances permit. Long-term storage is also useful as an insurance against epidemics and natural disasters. Reproductive tissue from deceased animals can be frozen to preserve genetic diversity. The use of frozen sperm has the advantage of being less stressful to animals than transporting them over large distances for breeding. Using embryo transfer and surrogacy (see Topic 18.12), these 'frozen genes' can be later used to yield additional individuals of endangered

species. In the same way, plant species may be protected in botanical gardens, either as adult individuals or their genetic material temporarily preserved in **seed banks**. The Millennium Seed Bank partnership at Kew Gardens in England has already banked seeds from 10% of the world's wild plant species and aims to have banked 25% by 2020. Seeds from plants and regions most at risk from climate change and other human activities have been targeted.

Seeds, embryos and tissues are stored at –196°C (cryopreservation). The advantages of seed banks are that the samples are small so less space is required, maintenance is simple and relatively inexpensive and the samples can last for a very long time.

EXTENSION
Other methods of protection

- **Education** – It is important to educate people in ways of preventing habitat destruction and encouraging conservation.
- **A ban on hunting and fishing coupled with commercial farming** – Legislation to control hunting and fishing of endangered species helps maintain their numbers.
- **Removal of animals from threatened areas** – Organisms in habitats threatened by humans, or by natural disasters such as floods, may be removed and resettled in more secure habitats.
- **Control of introduced species** – Organisms introduced into a country by humans often require strict control if they are not to out-compete the indigenous species.
- **Ecological study of threatened habitats** – Careful analysis of all natural habitats is essential if they are to be managed in a way that permits conservation of a maximum number of species.
- **Pollution control** – Measures to control pollution such as smoke emissions, oil spillage, over-use of pesticides, fertiliser run-off, etc. help to prevent habitat and species destruction.
- **Recycling** – The more material that is recycled, the less need there is to obtain that material from natural sources, e.g. through mining.

Figure 3 *A ranger collecting freshly laid eggs from a hawksbill turtle at Turtle Island Marine Park, Sabah, Malaysia. The eggs are relocated to an artificial nest in a hatchery as soon as they are laid to protect them from being dug up by other turtles or monitor lizards. The loss of nesting beaches and the effects of hunting and pollution have endangered some turtle species.*

Conservation

On these pages you will learn to:

- Discuss methods of assisted reproduction, including IVF, embryo transfer and surrogacy, used in the conservation of endangered mammals
- Discuss the use of culling and contraceptive methods to prevent overpopulation of protected and non-protected species
- Use examples to explain the reasons for controlling alien species
- Outline how degraded habitats may be restored, with reference to local or regional examples

Conservation is the protection, maintenance and management of the Earth's natural resources in such a way that maximum use can be made of them in the future. This involves active intervention by humans to maintain **ecosystems** and **biodiversity**. It is a dynamic process that involves careful management of existing resources and reclamation of those already damaged by humans. The ecological, economic, ethical and social reasons for conserving biodiversity were covered in Topic 18.3.

Assisted reproduction

When the population of a species becomes so low that it is threatened with extinction, it is sometimes necessary to assist the natural reproductive process in order to build up its numbers again. One problem is that a small population has a limited **gene pool** and so a priority is to maintain or increase genetic diversity and to avoid inbreeding, which causes poor fertility and increased susceptibility to disease.

To be successful, information on an animal's reproductive cycle is required. As this is controlled by hormones, it is important to know when each hormone is being produced. At one time this would have required taking blood samples from anaesthetised animals, but new techniques allow us to measure hormone levels in faeces, removing the need to ever touch the animal. Useful information can be obtained about the time of ovulation, whether an animal is pregnant and, if so, the likely time of birth. Animals whose reproduction is assisted in this way include the cheetah, snow leopard, ocelot and black rhinoceros (Figure 1). The techniques employed are often the ones used to improve human fertility, including:

- **In vitro fertilisation** – involves female animals being treated with gonadotrophins to stimulate the maturation of oocytes. These are then collected by using ultrasound to guide a needle to the ovaries and to collect the oocytes by 'sucking' them into the needle (aspiration). The oocytes are then fertilised in the laboratory by sperm, either frozen or fresh, from a genetically different male donor. The embryos can then be transplanted back into the donor female or into a surrogate mother, or they can be frozen until a suitable recipient becomes available. This technique has been successfully used to produce offspring of the endangered black rhinoceros.
- **Embryo transfer** – involves the collection of pre-implantation embryos from a donor female, freezing and storing them for long periods of time before thawing them and transferring them into the reproductive tract of a surrogate mother (recipient female of the same or closely related species). The process may involve treating the donor female with gonadotrophins to stimulate the production of a number of oocytes, followed by artificial insemination (see below) to allow fertilisation and the resulting embryos to implant. The embryos are usually removed by flushing the uterus with a solution designed to provide the best possible conditions for the survival of the embryos.
- **Surrogacy** – where the embryo of one individual is implanted into the uterus of another, where it develops and is carried by the surrogate mother until birth. It is used to simultaneously raise many embryos produced by artificial insemination.
- **Artificial insemination** – where sperm are introduced into the vagina of a female by artificial means. Sperm from a male that is genetically different

Figure 1 *The numbers of black rhinoceros* (Diceros bicornis) *have been increased using assisted reproduction*

from the female is usually chosen to increase genetic diversity. Essential to this is the large genetic database that has been assembled, detailing the genome of a large number individuals. To improve the chances of success, semen can be screened to select the most fertile, disease-free sperm, and insemination can be carried out around the time of female ovulation as determined by hormone analysis. Artificial insemination has been successful in many animals, including koalas, wallabies and giant pandas.

Preventing over-population by culling and contraception

It may seem odd, but sometimes the best way of conserving a species is to limit its population. There is a limit on how many individuals of a species a particular habitat can support. If this is exceeded, then a shortage of food and breeding territories may threaten the welfare of the species as a whole. Over-population of one species might threaten the welfare of another. In these circumstances it is better to cull some of the population to allow the remainder the best conditions in which to thrive. Culling is usually selective and involves killing the physically weaker, or genetically damaged, individuals in order to improve the quality of the remaining stock.

It has been estimated that the Kruger National Park can sustain a population of about 7500 elephants. It is, in fact, home to around twice that many. This over-population is causing damage to trees and the species dependent on these trees. Some elephants have been relocated to other parks, but a more sustainable solution is needed. Culling is controversial, because it may create a trade in ivory. Instead of culling, female elephants are being administered contraceptives delivered by darts fired from a vehicle or helicopter. Contraception takes time to reduce the population and at least 75% of female elephants must be made infertile for it to be effective.

Culling and the use of contraceptives are also a means of controlling non-protected species that compete with protected ones for food, water and territories or spread disease. By controlling the populations of these competitors, there is less disease and more resources to allow the populations of protected species to recover.

Control of alien species

Some non-native, or alien, species have been introduced into new areas unintentionally (by accident), for example, arriving in ship containers or brought back unknowingly by tourists. Others, however have been introduced for beneficial purposes or for pleasure, (e.g. for food, for biological pest control, as pets, for zoos, as attractive garden plants).

The introduction of non-native species into a new area can create major problems. There are often no natural predators, or grazing animals, to control their populations, which therefore expand rapidly. They can aggressively outcompete the native species and so present a serious threat to biodiversity and a major cause of extinction globally. In Europe the red squirrel populations are on the brink of extinction in the UK and Italy, following the introduction of the larger American grey squirrel. The American grey squirrel damages trees, with huge consequences for the timber industry. Dutch elm disease is caused by an introduced fungus and has vastly reduced elm trees in the forests of central Europe. Japanese knotweed is a particularly invasive non-native plant in some parts of the world. It competitively eliminates native species but is found to support 40% fewer insect species. This damages habitats and reduces food supply for insect-eating animals. It is essential to control alien species to prevent them from disrupting habitats and causing the individual problems mentioned above.

Restoring degraded habitats

Conservation is not only about maintaining natural habitats but also restoring ones that have been damaged, degraded or destroyed for whatever reason. There are many examples, including the restoration of areas damaged naturally by fire or volcanic ash. More commonly, projects are involved in repairing damage caused by human activities. Examples include stabilising hills created by mining and industrial waste and replanting them with tolerant species, cleaning and restocking ponds, rivers and streams damaged by pollutants, and the use of marine parks to help restore sea beds damaged by trawling. There are specific projects in almost all areas around the world, and it would be worthwhile to investigate those in your locality.

Summary Test 18.12

Conservation involves active management to maintain ecosystems and (**1**). One method of conservation is assisted reproduction. This takes a number of forms, including introducing sperm into the vagina of a female, in a process called (**2**), implanting the embryo of one individual into the uterus of another, in a process called (**3**) and freezing sperm and embryos of endangered species for future use, in a technique called (**4**). Sometimes it is necessary to prevent overpopulation of endangered species, in which case two methods that can be employed are (**5**) and (**6**). The introduction of non-native animals to an area can create problems as they (**7**) the native species because there are no natural (**8**) to control their numbers.

18 Examination Questions

1 The element nitrogen is present in many biological molecules, such as amino acids, proteins and nucleotides.

Figure 1 shows part of the nitrogen cycle.

Figure 1

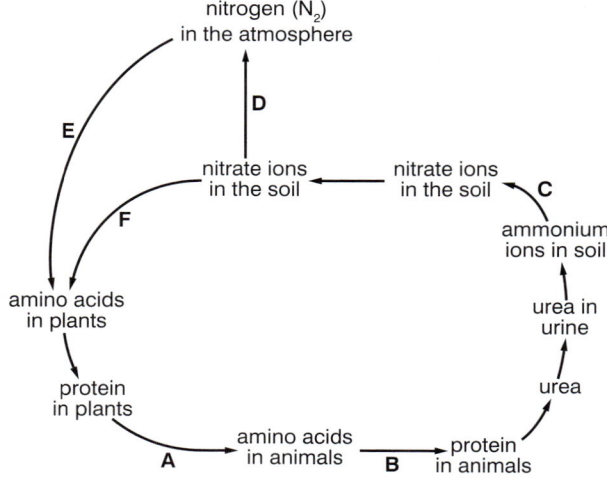

The statements 1 to 10 are processes that occur during the nitrogen cycle.

For each of the stages **B** to **F** shown on Figure 1, select the appropriate description from the list of statements and write it in a copy of the box provided.

Write only **one** number in each box.

The first one (**A**) has been selected and completed for you.

1 digestion by primary consumers

2 amino acid synthesis in plants

3 protein synthesis in primary consumers

4 nitrification

5 decomposition

6 nitrogen fixation

7 excretion

8 deamination in primary consumers

9 denitrification

10 deamination by bacteria and fungi

A	1
B	
C	
D	
E	
F	

(Total 5 marks)
Cambridge International AS and A Level Biology 9700
Paper 2 Q6 November 2008

2 The flatback turtle, *Natator depressus*, is an endangered species that nests on northern Australian beaches.

Figure 2 shows a flatback turtle.

Figure 2

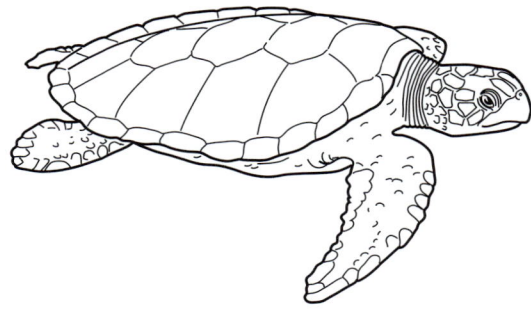

Each female lays approximately 50 eggs per nest, which is a smaller number than all other species of marine turtle. The eggs are buried in the sand and when the hatchlings emerge each has a mass of approximately 43 g. Unlike most marine turtles, flatback turtles spend most of their time in coastal waters. This is where they feed and mate.

Figure 3 shows the numbers of female flatback turtles nesting on a beach in northern Australia between 1993 and 2002.

Figure 3

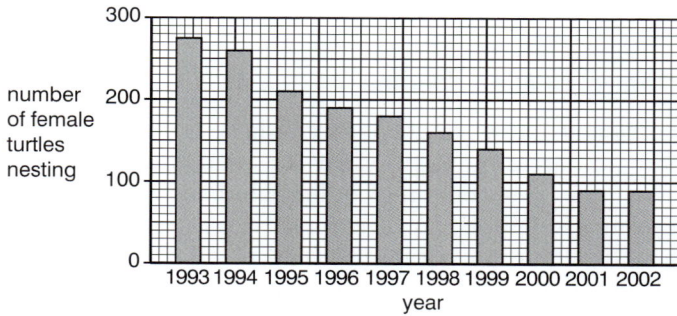

a Calculate the mean **rate** of decrease in the numbers of females nesting between 1993 and 2002. Show all the steps in your calculation. *(2 marks)*

b Suggest ways in which the flatback turtle could be protected. *(5 marks)*
(Total 7 marks)

Cambridge International AS and A Level Biology 9700 Paper 41 Q1 June 2010

3 Lancaster Sound in the Canadian Arctic is a very productive marine environment and supports large populations of sea birds and marine mammals.

Studies of the area have shown the importance of Arctic cod, *Boreogadus saida*, in the flow of energy to marine birds, such as guillemots and fulmars, and marine mammals, such as narwhals and belugas. Arctic cod forms the main, or only, source of food for many such animals.

The flow of energy through the food web in Lancaster Sound is shown in Figure 4

Figure 4

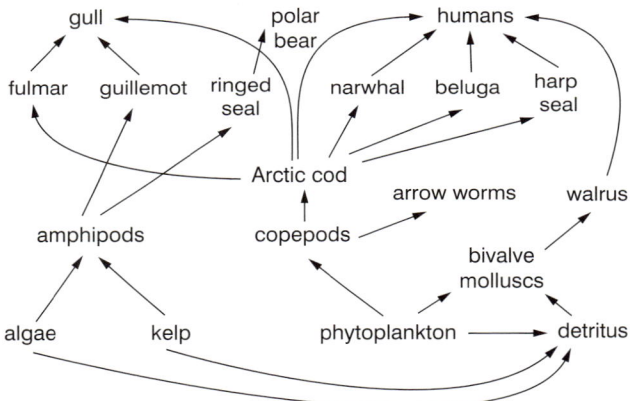

Note: detritus is dead and decaying matter

a Name the trophic levels occupied by the following organisms in the food web in Figure 4:

kelp

arrow worms.

narwhals *(3 marks)*

b The population of polar bears in the Lancaster Sound area is quite small in comparison to populations of animals that feed on Arctic cod.

Using only the information shown in Figure 4, explain why the population of polar bears is small. *(4 marks)*

c Populations of many fish species are under threat of extinction as a result of over-fishing.

Explain the likely consequences of over-fishing of Arctic cod. *(2 marks)*
(Total 10 marks)

18 Practice Questions

4 a State four features that are shared by members of the same species.

b State the main features of a phylogenetic (natural) system of classification?

c *Rana temporaria* is the frog commonly found in Britain. The incomplete table below shows some of its classification. Give the name for each of the blanks in the table represented by the numbers 1–7.

Kingdom	Animalia
1	Chordata
2	Amphibia
3	Anura
4	Ranidae
Genus	5
6	7

5 Scientists believe that the production of greenhouse gases by human activities is contributing to climate change.

a Explain why an increase in greenhouse gases is more likely to result in damage to communities with a low species diversity than ones with a high species diversity.

b The figure below shows the effect of environmental change on the stability and the functioning of ecosystems.

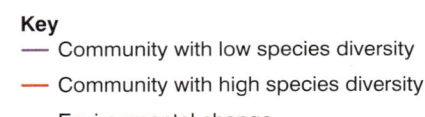

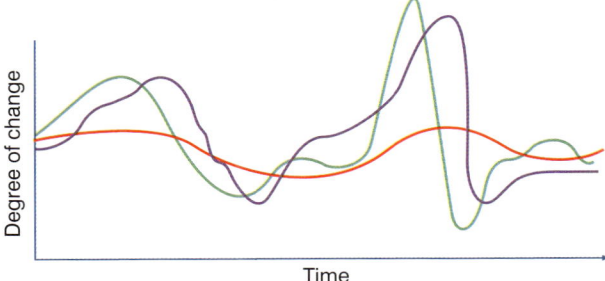

i Describe the relationship between environmental change and the community with low species diversity.
ii Explain the different responses to environmental change between communities with low and high species diversity.

Genetic technology

Identifying and isolating genes

On these pages you will learn to:

- Define the term 'recombinant DNA'
- Explain that genetic engineering involves the extraction of genes from one organism, or the synthesis of genes, in order to place them in another organism (of the same or another species) such that the receiving organism expresses the gene product
- Explain the roles of restriction endonucleases and reverse transcriptase in genetic engineering

One of the most important scientific advances since the discovery of the structure of DNA in 1953 has been the development of gene technology. In a branch of biology known as genetic engineering (also known as genetic, or recombinant DNA, technology), techniques have been devised that allow the synthesis of genes, the removal and alteration of genes, and the transfer of genes from organism to organism. An organism with introduced DNA has altered characteristics: this may be from a direct alteration of its genes, or by transferring into it genes from another organism. As a result, the affected cells of the organism make a different product during protein synthesis.

When DNA is formed (novel DNA) by joining together DNA from two different sources, it is known as **recombinant DNA**. An organism containing recombinant DNA is known as a **genetically modified organism (GMO)**. Where the DNA has been transferred into another organism for a protein product to be synthesised and harvested, the organism may be termed the **recombinant host**. Sometimes the introduced DNA is termed 'foreign DNA' or 'foreign gene'.

A number of human diseases result from individuals being unable to produce for themselves various metabolic chemicals. Many of these chemicals are proteins e.g. insulin. They are therefore the product of a specific portion of DNA i.e. the product of a gene. Techniques have been developed to produce large quantities of 'pure' proteins by isolating genes, cloning them and transferring them into microorganisms. These microorganisms are then grown to provide a 'factory' for the continuous production of a desired protein. The process of making a protein using genetic engineering has a number of stages:

- **synthesis of a gene** for the desired protein
- **cloning** (making many genetically identical copies) of the DNA using DNA polymerase
- **insertion** of DNA fragment into a vector (carrier)

- **transformation** – the transfer of DNA into suitable host cells
- **identification** of host cells that have successfully taken up the gene by use of **gene markers**
- **large-scale production** of the population of host cells.

The required gene may consist of a sequence of a few hundred bases among, for example, the many million in human DNA. It must be identified and extracted for use, or must be synthesised. Genes can be synthesised by using mRNA as a template or by using a combination of chemical and molecular biology methods to build up short nucleotide sequences that can then be joined to form the gene. The fragment of DNA transferred to the host may contain additional nucleotide sequences necessary for transcription.

Reverse transcriptase

Retroviruses are a group of viruses, the best known of which is the Human Immunodeficiency Virus (HIV). Retroviruses have their genetic information in the form of RNA. However, they are able to synthesise DNA from their RNA using an enzyme called reverse transcriptase. It is so named because it catalyses the production of DNA from RNA which is the reverse of the more usual transcription of RNA from DNA. This enzyme can be used in genetic engineering to synthesise the DNA required. The process of using reverse transcriptase to synthesise a gene is described below and shown in Figure 2.

- A cell that readily produces the protein is selected: e.g. for insulin, β cells of the islets of Langerhans from the pancreas are used. These cells have large quantities of the relevant mRNA, which is extracted.
- Reverse transcriptase is then used to make DNA from RNA. This DNA is known as **complementary DNA (cDNA)** because it is made up of the **nucleotides** that are complementary to the mRNA.
- To make the other strand of DNA, the enzyme DNA polymerase is used to build up the complementary nucleotides on the cDNA template. This double strand of DNA is the required gene.

The advantage of synthesising genes in this way is that it is much easier to extract the specific mRNA and the mRNA is present in many copies (unlike the gene). Also the host cells used may be bacteria and they may not have the metabolic capability of processing RNA transcripts that contain non-coding portions of DNA (introns).

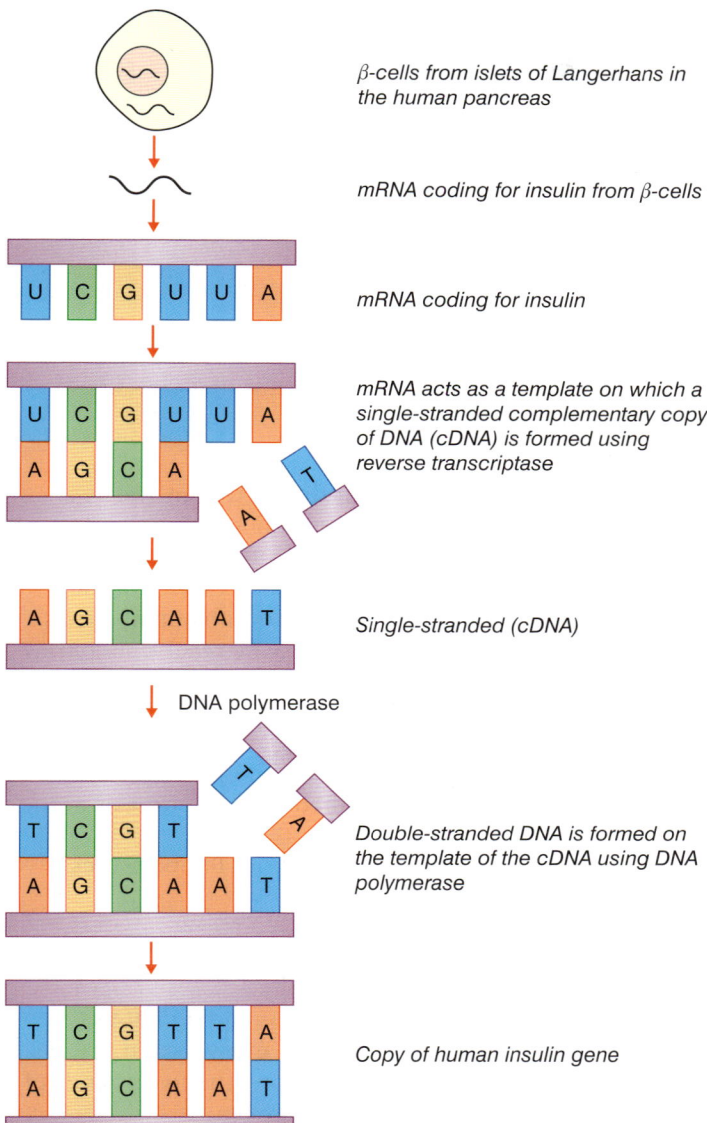

Figure 2 *Using reverse transcriptase to synthesise the gene that codes for insulin*

β-cells from islets of Langerhans in the human pancreas

mRNA coding for insulin from β-cells

mRNA coding for insulin

mRNA acts as a template on which a single-stranded complementary copy of DNA (cDNA) is formed using reverse transcriptase

Single-stranded (cDNA)

DNA polymerase

Double-stranded DNA is formed on the template of the cDNA using DNA polymerase

Copy of human insulin gene

REMEMBER

Each restriction endonuclease recognises and cuts DNA at a specific sequence of bases. These sequences occur in the DNA of all species of organisms – but not in the same places!

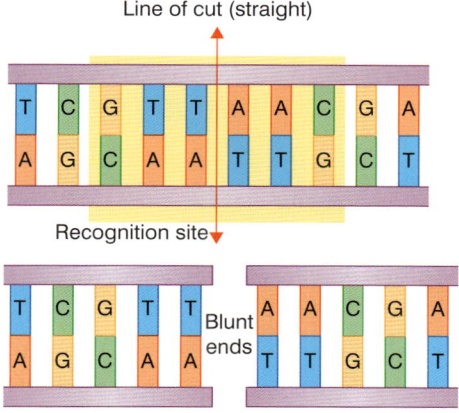

a HpaI *restriction endonuclease has a recognition site GTTAAC, which produces a straight cut and therefore blunt ends:*

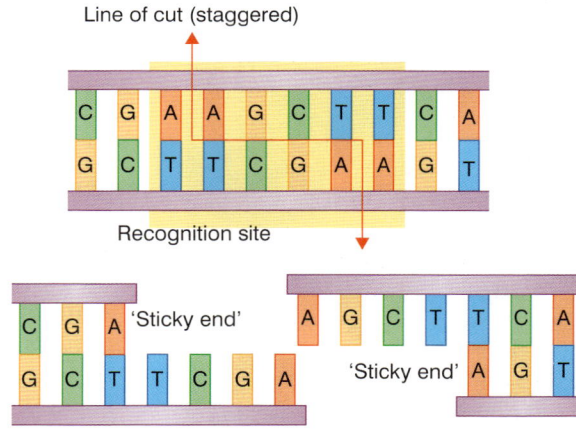

b HindIII *restriction endonuclease has the recognition site AAGCTT, which produces a staggered cut and therefore 'sticky ends':*

Figure 3 *Action of restriction endonucleases*

Restriction endonucleases

Bacteria are frequently invaded by viruses that inject their DNA into them in order to take over the cell. Some bacteria defend themselves by producing enzymes that cut up the viral DNA. These enzymes are called restriction endonucleases.

Many types of restriction endonucleases exist. Each one cuts a DNA double strand at a specific sequence of bases called a recognition sequence. Sometimes this cut occurs between two opposite base pairs. This leaves two straight edges known as blunt ends. For example, one restriction endonuclease cuts in the middle of a base recognition sequence GTTAAC (Figure 3a). Other restriction endonucleases cut DNA in a staggered fashion. This leaves an uneven cut in which each strand of the DNA has exposed, unpaired bases. An example is a restriction

endonuclease that recognises a six-base pair (six bp) AAGCTT, as shown in Figure 3b. In this figure, look at the sequence of unpaired bases that remain. If you read both the four unpaired bases at each end from left to right, the two sequences are opposites of one another i.e. they are a **palindrome**. The recognition sequence is referred to as a six bp palindromic sequence. This feature is typical of the way restriction endonucleases cut DNA to leave 'sticky ends'. We shall look at the importance of these 'sticky ends' later.

DNA cloning – the polymerase chain reaction

On these pages you will learn to:

- Describe the principles of the polymerase chain reaction (PCR) to clone and amplify DNA

REMEMBER

DNA polymerase causes nucleotides to join together as a strand. It does **not** cause complementary bases to join together.

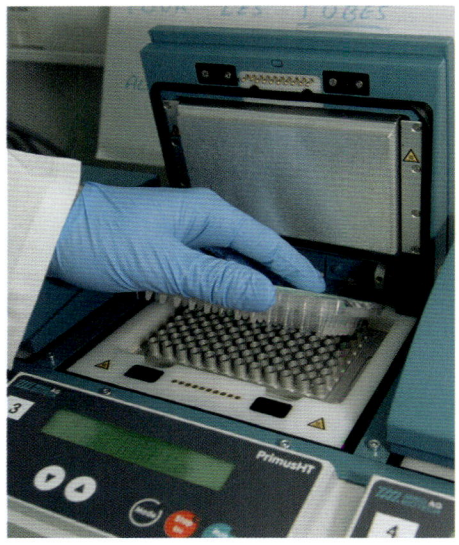

Figure 1 *Thermocycler – a machine that carries out the polymerase chain reaction (PCR)*

REMEMBER

The polymerase chain reaction is **not** the same as semi-conservative replication of DNA in cells.

Once the required fragment containing the gene is extracted and synthesised, the next stage is to create many identical DNA fragments for use in the next step. This is known as DNA cloning and one means of carrying it out is to use the polymerase chain reaction.

Polymerase chain reaction

The polymerase chain reaction (PCR) is an automated process, making it both rapid and efficient. It requires the following:

- **The DNA fragment** to be copied.
- *Taq* **polymerase** – DNA polymerase obtained from the bacterium *Thermus aquaticus*, after which it is named. The bacterium lives in hot springs, and so the remarkable feature of *Taq* polymerase is that it is very tolerant to heat (it is thermostable) and does not denature at the high temperatures of the polymerase chain reaction so that it can be used throughout the many cycles of replication that take place. DNA polymerase is an enzyme capable of joining together tens of thousands of **nucleotides** in a matter of minutes.
- **Primers** – short sequences of nucleotides that have a set of bases complementary to those at one end of each of the two DNA fragments.
- **Nucleotides** – which contain each of the four bases found in DNA. They are nucleotide triphosphates (dNTPs) as energy is required for the synthesis of the phosphodiester bonds.
- **Thermocycler** – a computer-controlled machine that varies temperatures precisely over a period of time (Figure 1).

The polymerase chain reaction is illustrated in Figure 2 and is carried out in a three stages:

- **Separation of the DNA double helix** – the mixture containing DNA fragments, primers, dNTPs and *Taq* polymerase is placed in a vessel in the thermocycler. The temperature is increased to 95 °C causing the two strands of the DNA fragments to separate as hydrogen bonds are broken.
- **Annealing of the primers** – the mixture is cooled to 55 °C causing the primers to join (anneal) to their complementary bases at the end of the DNA fragment. The primers provide the starting sequences for *Taq* polymerase to begin DNA copying because *Taq* polymerase can only attach nucleotides to the end of an existing chain. Primers also prevent the two separate strands from simply rejoining.
- **Synthesis of DNA** – the temperature is increased to 72 °C. This is the optimum temperature for the *Taq* polymerase to add complementary nucleotides along each of the separated DNA strands. It begins at the primer on both strands and adds the nucleotides in sequence until it reaches the end of the chain.

Because both separated strands are copied simultaneously (at the same time) there are now two copies of the original fragment. Once the two DNA strands are completed, the process is repeated by carrying out the temperature cycle again. This gives four molecules, and so on until millions of copies have been made. This is known as **DNA amplification**. The complete cycle takes around two minutes. After only 25 cycles over a million copies of the DNA can be made and 100 billion copies can be manufactured in just a few hours. The polymerase chain reaction has revolutionised many aspects of science and medicine. Even the most minute sample of DNA from a single hair or a speck of blood can now be multiplied to allow forensic examination and accurate cross-matching.

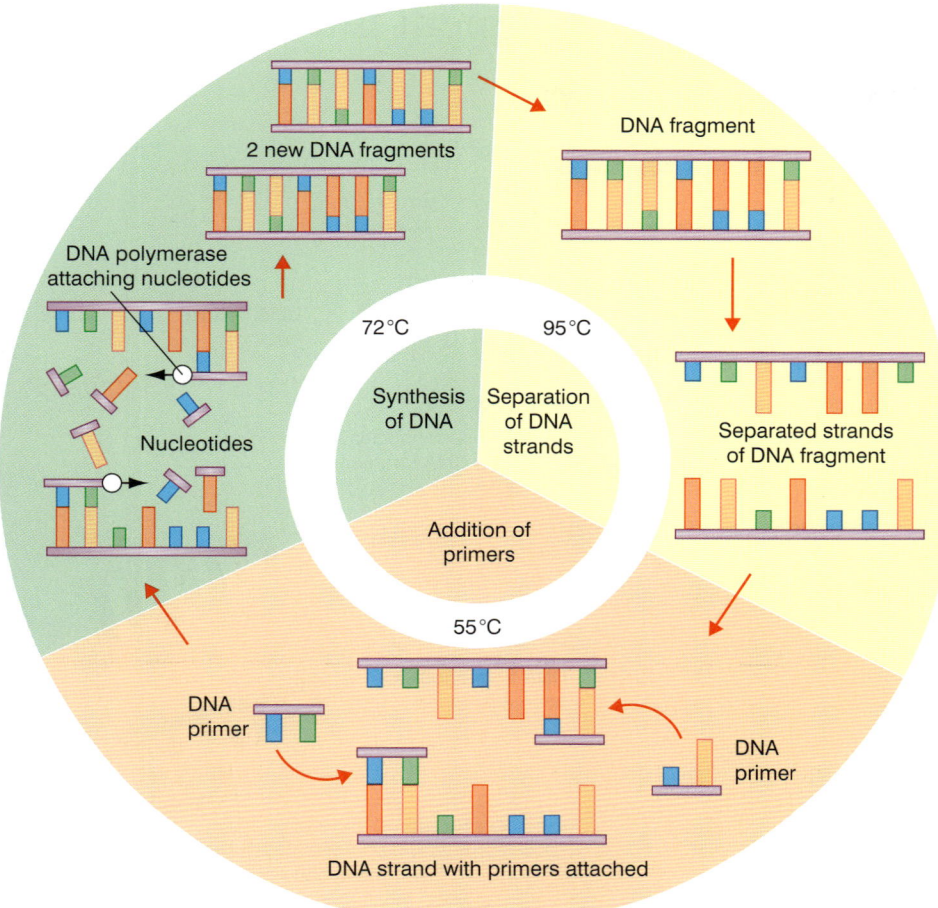

Figure 2 *The polymerase chain reaction showing a single cycle*

Advantages of gene cloning

The advantages of this type of gene cloning are that it:

- **is extremely rapid.** Within a matter of hours 100 billion copies of a gene can be made. This is particularly valuable where only a minute amount of DNA is available, for example at the scene of a crime. This can quickly be increased using the polymerase chain reaction and so valuable time is not lost before forensic analysis and matching can take place.
- **does not require living cells.** All that is required is a base sequence of DNA that needs amplification. No complex culturing techniques are needed which require time and effort.

SUMMARY TEST 19.2

The polymerase chain reaction (PCR) is both rapid and efficient because it is an **(1)** process. PCR begins with DNA fragments, *Taq* polymerase, dNTPs and primers being placed in a machine called a **(2)**. The temperature of the mixture is then increased to **(3)** which cause the strands of the two DNA fragments to **(4)**. The temperature of the mixture is then changed to **(5)** which makes the primers join to their **(6)** at the end of the DNA fragment. The primers also prevent the strands **(7)**. Finally the temperature is changed to **(8)** which allows *Taq* polymerase to add **(9)** to each of the separated DNA strands. The complete cycle takes around **(10)** minutes.

Once the required fragment of DNA containing the gene has been cut from the rest of the DNA or has been synthesised, the next task is to transfer it to the host cell. One method for doing this is to join it into a carrying unit, known as a **vector**. Before we look at how this is carried out, let us first consider the importance of the 'sticky ends' left when DNA is cut by **restriction endonucleases** (restriction enzymes).

Importance of 'sticky ends'

The sequences of DNA that are cut by restriction endonucleases are called recognition sites. Where the recognition site is cut in a staggered fashion, the cut ends of the DNA double strand are left with a single strand that is a few **nucleotide** bases long (Topic 19.1). The nucleotides on the single strand at one side of the cut are obviously complementary to those at the other side, because they were previously paired together.

If the same restriction endonuclease is used to cut DNA, then all the fragments produced will have bases that are complementary to one another. This means that the single-stranded end of any one fragment can be joined (stuck) to the single-stranded end of any other. In other words their ends are 'sticky'. Once the complementary bases of two 'sticky ends' have paired up and hydrogen bonds have formed, an enzyme called **DNA ligase** is used to join the phosphate-sugar framework of the two sections of DNA and so unite them as one.

'Sticky ends' have considerable importance because, provided the same restriction endonuclease is used, we can combine the DNA by phosphodiester bonds of any organism with that of any other organism (Figure 1).

Inserting DNA into a plasmid vector

There are different types of vector but the most commonly used is the **plasmid**. Plasmids are circular lengths of DNA, found in bacteria, which are separate from the main bacterial DNA. In addition to carrying the required gene, an additional 'marker' gene can be added to the plasmid (Figure 3). In the past, marker genes were actually those carried naturally by the plasmid, genes for antibiotic resistance. Concerns about the spread of

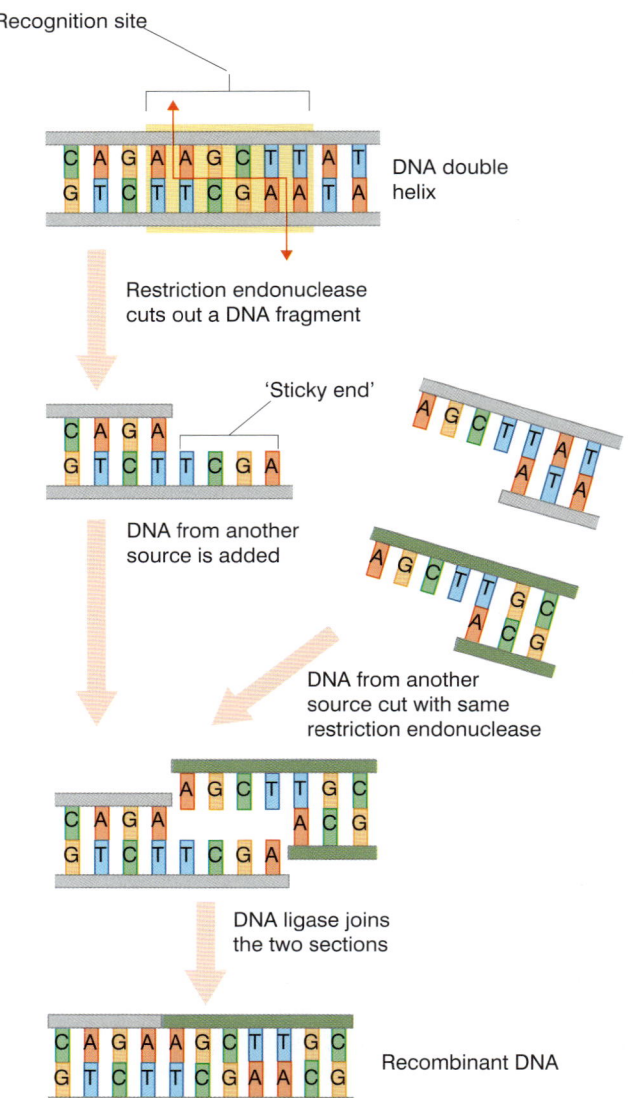

Figure 1 *The use of 'sticky ends' to combine DNA from different sources*

antibiotic resistance from GMOs and their products has led to a decline in the use use of antibiotic resistance marker genes. These have been replaced by the use of genes such as the gene coding for GFP, green fluorescent protein (see Topic 19.4).

The same restriction endonuclease is used as the one that cut out the DNA fragment. This ensures that the 'sticky ends' of the opened-up plasmid are complementary to the 'sticky ends' of the DNA fragment. When the DNA fragments are mixed with the opened-up plasmids, they may become incorporated into them. Where they are incorporated, the join is made permanent using the enzyme DNA ligase. These plasmids now have recombinant DNA and are termed recombinant plasmids. These events are summarised in Figure 2.

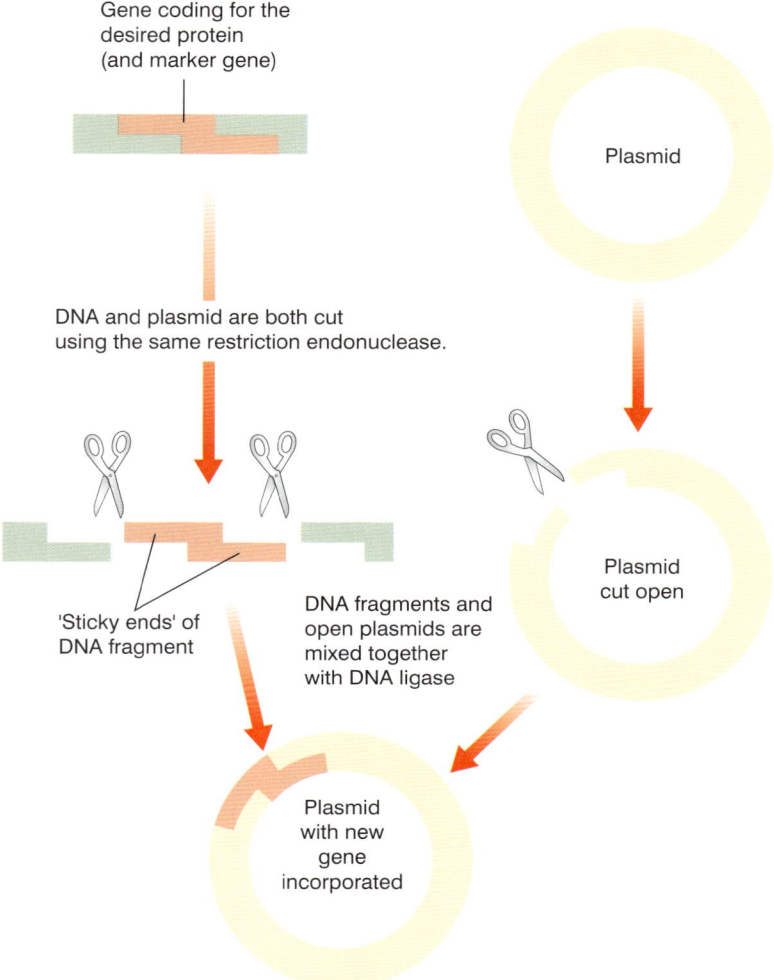

Gene coding for the desired protein (and marker gene)

DNA and plasmid are both cut using the same restriction endonuclease.

Plasmid

Plasmid cut open

'Sticky ends' of DNA fragment

DNA fragments and open plasmids are mixed together with DNA ligase

Plasmid with new gene incorporated

Figure 2 *Inserting a gene into a plasmid vector*

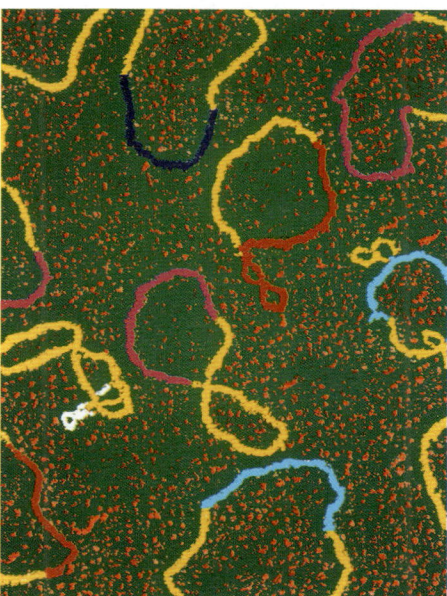

Figure 3 *Colourised TEM of genetically engineered DNA plasmids from the bacterium Escherichia coli. The plasmids (yellow) have had different gene sequences (various colours) inserted into them.*

Properties of plasmids that make them suitable for use in gene cloning

- They are small and so do not contribute much additional DNA to the host cell. It is also easier to get small pieces of DNA into a bacterium.
- They replicate rapidly in the host cell and independently of the host cell. Replication produces genetically identical copies (gene cloning occurs).
- They are easily taken up by bacterial cells.
- It is easy to determine their DNA sequence, making it easier to use recombinant DNA techniques.
- They have a number of recognition sites so that different restriction endonucleases may be used.
- They possess marker genes such as antibiotic resistance that allow them to be readily recognised. In particular they often carry at least two such marker genes (Topic 19.4).
- They are easily manipulated. For example, as they do not carry genes essential to survival, unnecessary genetic material can be removed to decrease the size of the plasmid for easier uptake by the host.
- If there are no other marker genes that can be used, then some plasmids possess natural marker genes for antibiotic resistance.

SUMMARY TEST 19.3

The sequences of DNA that are cut by restriction endonucleases are called (**1**). Where the cut is in a staggered fashion, the ends of the DNA double strand are left with a single strand that has a few (**2**). Once the complementary bases of two 'sticky ends' have paired up, an enzyme called (**3**) is used to join the (**4**) framework of the two sections of DNA. To insert a DNA fragment into another cell a carrying unit called a (**5**) is used. One commonly used type is a plasmid. These are (**6**) lengths of DNA. An additional gene, known as a (**7**) gene may be added together with the desired gene. When the DNA fragments are mixed with the opened-up plasmids, they may become incorporated into them. These plasmids now have (**8**) DNA.

On these pages you will learn to:

- Explain why promoters and other control sequences may have to be transferred as well as the desired gene
- Explain the use of genes for fluorescent or easily stained substances as markers in gene technology

With the DNA fragment incorporated into at least some of the plasmids, these must then be re-introduced into bacterial cells along with a region of DNA called a **promoter** and other control sequences. The promoter is located close to the nucleotide sequence to be transcribed and is required so that RNA polymerase can bind and transcription can begin. Some promoters, called inducible promoters, can be triggered to start the process in response to a stimulus such as heat or a specific chemical, e.g. lactose (Topic 19.5). Without the promoter, the gene would have to be inserted near an existing promoter, which would be difficult and could disrupt expression of the gene. Other control sequences may be required, such as those coding for transcription factors that are not present in the host cell or those responsible for the termination of transcription. The promoter and the other control sequences are therefore used to:

- control where the gene is located
- begin the process of transcription of the gene
- determine the level of gene expression
- terminate transcription.

This process of adding DNA to a bacterial cell is called **transformation**, and involves the plasmids and bacterial cells being mixed in a medium containing calcium ions. The temperature of the medium is increased. These conditions appear to increase the uptake of plasmids through the bacterial cell wall and cell surface membrane. However, not all the bacterial cells will possess the DNA fragments. There are two main reasons for this.

- Only a few bacterial cells (as few as 1%) take up the plasmids when the two are mixed together.
- Some plasmids will have closed up again without incorporating the DNA fragment.

The task now is to identify which bacterial cells contain the DNA fragment (gene). This can be done using **gene markers**. There are a number of different ways in which this can be achieved. They all involve using a separate gene on the plasmid that contains the gene that we want. This second gene is easily identifiable for one reason or another. For example:

- It is resistant to an antibiotic.
- It makes a fluorescent protein that is easily seen.
- It produces an enzyme whose action we can identify.

Antibiotic resistance markers

Bacteria have, over the years, evolved resistance to the effects of antibiotics, typically by producing an enzyme that breaks down the antibiotic before it can destroy the bacterium (Topic 10.9). The genes for the production of these enzymes are found in plasmids. Some plasmids carry genes for resistance to more than one antibiotic. An example is the R-plasmid, which carries genes for resistance to the two antibiotics, ampicillin and tetracycline.

The use of antibiotic resistance markers risks spreading resistance to other bacteria, for example, by uptake of plasmids released when bacterial cells die, or by conjugation. This is undesirable because it makes antibiotics less effective at treating diseases. For these reasons, their use is now rare and they have been replaced by other types such as fluorescent markers and enzyme markers.

Fluorescent markers and enzyme markers

A more recent and more rapid method of identifying the bacterial cells which contain the DNA fragment (gene) is to use a gene that is transferred into the plasmid from a jellyfish. This gene produces a green fluorescent protein (GFP). When this protein is present, it fluoresces green under ultraviolet (UV) light, so it is easily detected. GFP does not require any other molecules for fluorescence, which makes the use of the gene for GFP a ideal marker gene. In addition, the gene can be transferred into prokaryotic and eukaryotic hosts and the protein will be expressed. It appears to have no harmful effects on the host organism.

The gene can be inserted downstream from the promoter region, so that both the marker and the required genes are transcribed together. This produces the desired protein produce with a short 'marker' protein attached. In this way, recombinant host cells are easily identified. The proteins produced in this way do not have their function affected by the presence of GFP.

The gene we are trying to clone is transplanted into the centre of the GFP gene. Any bacterium that has taken up the plasmid with the gene we want to clone will not be able to produce GFP. It will therefore not be fluorescent green while the ones that have not taken up our gene will be. As the bacterial cells with our gene are not killed, there is no need for replica plating. Results can be obtained by simply viewing cells under a microscope and retaining those that do not fluoresce. This makes the process more rapid.

Plasmids can be used that contain the gene marker *lacZ*. The gene insertion site (where the plasmid is cut open) of this plasmid is within the *lacZ* gene, which codes for the enzyme β-galactosidase (see Section 16.14). If a recombinant plasmid is formed, *lacZ* is destroyed. When bacteria are grown on a medium containing a synthetic lactose substrate known as X-gal, only those bacteria that do not contain a recombinant plasmid are able to break down the colourless X-gal to form a blue product. With recombinant bacteria, X-gal is not broken down and bacterial colonies appear white. This is known as blue white screening. It is commonly used with another marker for resistance to ampicillin antibiotic.

The gene coding for the enzyme ß-glucuronidase (GUS), originally from *E. coli*, can also be used as a gene marker. If the enzyme is present with a suitable substrate (e.g. a chemical known as 4-MUG), using UV light a blue fluorescence results from the product (4-MU). This marker is only used to transform host cells or organisms that do not produce their own GUS (or very low quantities), so that when a fluorescence is obtained it indicates success in producing a recombinant host. This procedure works well with plants as they do not produce GUS naturally and the production of the enzyme with the introduced gene marker has no harmful effects.

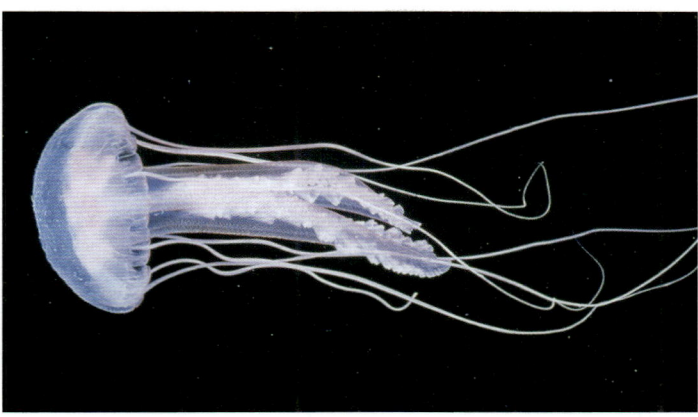

Figure 1 *The gene in this jellyfish, which produces a green fluorescent protein, can be transferred together with the required gene into other organisms and used as a fluorescent marker*

On these pages you will learn to:

- Explain the advantages of producing human proteins by recombinant DNA techniques:
  - can be cultured using cheaper feedstock and have simple nutrient requirements for growth
  - easier downstream processing for purification of the product
  - using microorganisms to obtain protein products overcomes ethical problems and moral / religious dilemmas
  - less space needed as can be cultured in fermenters
  - can be cultured anywhere in the world, with any climate and in labs under controlled conditions
  - high multiplication/population growth rate increases yield of protein product
  - overall cheaper to produce

In the past few topics we have seen how we can use recombinant DNA (genetic) technology to introduce into bacteria a gene that codes for a particular protein. The technique can be used to produce human proteins necessary to treat a variety of disorders.

Treatment of diabetes mellitus

There are two forms of diabetes mellitus:

- **Type I (insulin dependent)** is due to the body being unable to produce insulin. It normally begins in childhood and is therefore also called juvenile-onset diabetes. It may be the result of an **autoimmune response** whereby the body's immune system attacks its own cells – in this case the β cells of the **islets of Langerhans**.
- **Type II (insulin independent)** is normally due to the **glycoprotein** receptors on the body cells losing their responsiveness to insulin.

Type I diabetes is treated by injections of insulin (Figure 1). It cannot be taken by mouth because, being a protein, it would be digested in the alimentary canal. It is therefore injected.

Through the use of recombinant DNA technology, people with Type I diabetes can use insulin produced by genetically engineered bacteria. Before this technology, the only source of insulin extraction from the pancreases of animals such as cows and pig.

Genetically engineered insulin has a number of advantages over that extracted from animals when it comes to treating diabetes:

- It is more effective because it is an **exact** copy of human insulin, whereas animal insulin has slight differences.

- It is more rapid in its action because it is identical to human insulin.
- There is no immune response, whereas animal insulin, with its slight differences, can sometimes stimulate an immune reaction.
- There is no risk of infection being transferred with the insulin. Animal insulin can transfer certain diseases.
- Some patients develop tolerance to animal insulin and become less sensitive to it, requiring them to have increasingly large doses to achieve the same effect. This is less of a problem with genetically engineered insulin.
- It is cheaper to produce in large volumes than extracting and purifying animal insulin. For example, the costs of feed, large space required and heating are much greater when rearing animals compared with culturing bacteria in a fermenter.
- It has fewer ethical and moral objections because animals are not involved in its production. While the animals from which insulin was extracted were killed primarily for food (insulin being purely a by-product), many vegetarians and others were unhappy using an animal product in this way.

For some people, genetically engineered insulin has not proved to better than animal insulin. Some have found that their response is better and they have experienced fewer side effects with animal insulin.

Factor VIII production

Factor VIII is also known as anti-haemophilic factor (AHF). It is essential in the process of blood clotting. A defect in the gene producing factor VIII is a recessive X chromosome-linked disorder known as haemophilia (Topic 16.9). Factor VIII is a protein that circulates in the blood stream in an inactive form. When injury occurs it is activated and leads, by a series of reactions, to clotting of blood. Haemophiliacs required factor VIII extracted from blood donations to enable them to lead normal lives. In 1984 the genes that control the production of factor VIII were isolated and six years later it was produced using DNA recombinant techniques. Factor VIII is produced by recombinant mammalian cells in large-scale fermenter cultures, then purified and freeze dried for storage.

The advantages of producing factor VIII by recombinant DNA techniques is that it overcomes most of the problems that come with obtaining it from blood donations. For example extracting factor VIII from blood is difficult as it needs many donations to obtain just a small amount of factor VIII. There is also a risk that the factor VIII extracted could be contaminated of transfer disease from the donor. Recombinant DNA techniques produce far more factor VIII, more easily and without the risk of disease or

EXTENSION
Producing insulin

Genetically engineered human insulin can be produced as follows:

- Using the techniques described in the earlier topics of this chapter, a tiny quantity of bacteria that contain the gene for human insulin is isolated.
- Bacterial cells, like all cells, have many genes each producing a different protein. Many of these genes are only expressed when the protein it codes for is required. The expression (switching on) of each gene is controlled by a region of DNA known as a **promoter**. It is therefore essential to insert promoter DNA into these bacteria together with the gene that codes for insulin. One method was to insert the insulin gene into a plasmid with a lac operon, which includes a promoter and the *lacZ* gene, which breaks down lactose. When lactose is present, *lacZ* turns on the β-galactosidase gene and, because it is linked to it, the insulin gene as well. If lactose is included in the medium on which we are growing our genetically engineered bacteria, they will produce insulin at the same time as β-galactosidase.
- The final task is now to produce and purify insulin in large enough quantities to satisfy the needs of around 177 million diabetics worldwide. This is achieved by growing our tiny sample of genetically engineered bacteria in a nutrient medium that includes lactose using an industrial fermenter.

Figure 1 A person with diabetes injecting insulin

SUMMARY TEST 19.5

There are two forms of diabetes mellitus. Type I is also called **(1)** diabetes and may result from the body's white blood cells attacking the **(2)** cells found in the **(3)** of the pancreas. This form of self-attack is known as an **(4)** response. Type I diabetes can be treated by injections of insulin. Insulin cannot be taken by mouth because it is a **(5)** and would therefore be digested. Insulin can be produced using genetic technology. To switch on the insulin gene requires a portion of DNA known as a **(6)** which must be present in the region close to the insulin gene. The insulin gene is inserted so that it is transcribed together with the *LacZ* gene coding for **(7)**, that breaks down lactose. Genetically engineered insulin is preferable to that extracted from animals because it is an **(8)** copy of human insulin and so is more **(9)** in its action and does not induce an **(10)** response. There is no risk of **(11)** being transferred, it is **(12)** to produce in large quantities and there is less risk of **(13)** to insulin developing. As animals are not involved in its production, it is **(14)** more acceptable. Two other proteins produced using gene technology are Factor VIII, used to treat **(15)**, and **(16)**, used to treat severe combined immunodeficiency.

contamination. They also overcome any ethical objections that people may have to extracting and transferring human material from one person to another.

Adenosine deaminase production

The enzyme adenosine deaminase (ADA) is produced by the *ADA* gene. It is made in all cells, but the highest levels of adenosine deaminase are found in lymphocytes. The function of the adenosine deaminase enzyme is to break down a molecule called deoxyadenosine, which is formed when DNA is broken down. Deoxyadenosine is toxic and so without the enzyme adenosine deaminase it builds up and interferes with the development and maintenance of lymphocytes. There are more than 70 mutations of the *ADA* gene, most involving a single amino acid change in the enzyme. This leads to the condition called **severe combined immunodeficiency (SCID)** where the patient does not show a cell-mediated response (Topic 11.5) nor is able to produce antibodies (Topic 11.2). DNA recombinant techniques can be used to introduce the normal *ADA* gene into bacterium which is then cultured to produce adenosine deaminase which can be administered to SCID patients. The use of gene therapy as an alternative treatment is considered in Topic 19.9.

The advantage of using recombinant DNA techniques to treat SCID is that it provides an effective treatment for the first time. Previously there was no means of extracting or manufacturing ADA and so SCID was untreatable.

Ethical and social implications of genetic technology

On these pages you will learn to:

- Discuss the ethical and social implications of using genetically modified organisms (GMOs) in food production

REMEMBER

Ethics are a set of standards that are followed by a particular group of individuals and are designed to regulate their behaviour. They determine what is acceptable and legitimate in pursuing the aims of the group.

REMEMBER

Social issues relate to human society and its organisation. They concern the mutual relationships of human beings, their interdependence and their cooperation for the benefit of all.

Figure 1 *This maize has been genetically modified to be herbicide resistant. When the crop is treated with herbicide the maize is unaffected, but weeds are killed. This aims to increase the yield of such crops by reducing competition from weeds.*

Gene technology undoubtedly brings many benefits to mankind, but it is not without its risks. Let us consider these benefits and risks and evaluate the ethical and social issues associated with the use of gene technology.

Genetically modified crops can be engineered to have economic and environmental advantages. These include making plants more tolerant to environmental extremes, e.g. able to survive drought, cold, heat, salt or polluted soils, etc. This permits crops to be grown commercially in places where they are not at present. Growing of genetically modified plants, such as salt-tolerant tomatoes, could bring land back into productivity. Other examples include producing genetically modified forms of crop plants that are resistant to herbicides. This allows weeds competing for light, water and nutrients to be killed by the herbicide but with minimum effect on the crop. This gives greater yields. An example of such a crop treated in this way is oilseed rape, more information on which is given in Topic 19.11. In a world where millions lack a basic nutritious diet, and with a predicted 90 million more mouths to feed by 2025, can we ethically oppose the use of such crop plants?

Genetically modified crops can help prevent certain diseases. Rice can have a gene for vitamin A production added. This rice, when fed to people with vitamin A deficiency, can prevent one form of blindness (Topic 19.11). Can we ethically justify not developing more vitamin A enriched crops when 250 million children worldwide are at risk from vitamin A deficiency leading to 500 000 cases of irreversible blindness each year?

Genetically modified animals that have resistance to disease or produce a greater yield have also been developed. An example is genetically modified salmon that grow at a much faster rate and so provide more food more quickly as described in Topic 19.11.

Genetically modified organisms are able to produce expensive drugs, antibiotics and hormones, e.g. insulin, relatively cheaply. Many enzymes used in the food industry are manufactured by genetically modified bacteria. These include amylases used to break down starch during beer production, lipases used to improve the flavour of cheeses and proteases used to tenderise meat.

Gene therapy might be used to cure certain genetic disorders, such as cystic fibrosis. Details are given in Topic 19.9.

Genetic fingerprinting can be used in forensic science. Details are given in Topic 19.8.

The ethical and social implications of genetically modified organisms

Against the benefits of gene technology, must be weighed the risks – both real and potential. These include the following.

- It is impossible to predict with complete accuracy what the ecological consequences will be of releasing genetically-engineered organisms into the environment. The delicate balance that exists in any habitat may be irreversibly damaged by the introduction of organisms with engineered genes. This could have social implications should a genetically modified organism out-compete other crop plants to the point that the overall production of food was reduced. This might lead to widespread famine.

- A modified gene may pass from the organism it was placed in, to a completely different one. We know, for example, that viruses can transfer genes from one organism to another. What if a virus were to transfer genes for herbicide resistance and vigorous growth from a crop plant to a weed that competed with the crop plant? What if the same gene were transferred by pollen to other plants? How would we then be able to control this weed?

- Any manipulation of the DNA of a cell will have consequences for the metabolic pathways within that cell. We cannot be sure until after the event what unpredictable by-products of the change might be produced. Could these lead to metabolic malfunctions, cause cancer, or create a new form of disease?
- Genetically modified bacteria often have antibiotic resistance marker genes that have been added. These bacteria can spread antibiotic resistance to harmful bacteria, making certain diseases harder to treat.
- All genes mutate. What then, might be the consequences of our engineered gene mutating? Could it turn the organism into a **pathogen**, which we have no means of controlling? This pathogenic form might kill plants and animals that are part of our food chain.
- An important ethical issue is what might be the long-term consequences of introducing new gene combinations? We cannot be certain of the effects on the future evolution of organisms. Will the artificial selection of 'desired' genes reduce the genetic variety that is so essential to evolution?
- What might be the social and economic consequences of developing plants and animals to grow in new regions? Developing bananas that can grow in Europe could have disastrous consequences for the Caribbean economies that rely heavily on this crop for their income.
- What will be the consequences of the ability to manipulate genes getting into the wrong hands? Will some individuals, groups or governments use this power to achieve political goals, control opposition or gain overall power?
- Is the cost of genetic engineering justified, or would the money be better used fighting hunger and poverty, that are the cause of much human misery? Will advanced treatments, with their more high-profile images, be put before the everyday treatment of rheumatoid arthritis or haemorrhoids? Will such treatments only be within the financial reach of wealthier people?
- Is it ethically right to interfere with genes at all? Should we let nature take its own course in its own time?
- Is it right that an individual or company can patent, and therefore effectively own, a gene? What are the ethical implications of individuals owning a gene that can prevent famine by increasing crop yields? Could they hold large populations to ransom?

It is inevitable that we remain inquisitive about the world in which we live, and that we will seek to try to improve the conditions in which we live. Genetic research is bound to continue, but the challenge will be to develop the safeguards and ethical guidelines that will allow genetic engineering to be used in a safe and effective manner.

EXTENSION

The benefits of gene technology in food production

- Microorganisms can be modified to produce a range of substances, e.g. antibiotics, hormones and enzymes, which are used to treat diseases and disorders.
- Microorganisms can be used to control pollution, e.g. to break up and digest oil slicks or destroy harmful gases released from factories. Care needs to be taken to ensure that such bacteria do not destroy oil in places where it is required, e.g. car engines. To do this, a 'suicide gene' can be incorporated that causes the bacteria to destroy themselves once the oil slick has been digested.
- Genetically modified plants can be transformed to produce a specific substance in a particular organ of the plant. These organs can then be harvested and the desired substance extracted. If a drug is involved, the process is called plant pharming. One promising application of this technique is in the production of plants that manufacture antibodies to a particular disease, or manufacture **antigens** which, when injected into humans, induce natural **antibody** production.

Bioinformatics

- Explain, in outline, how microarrays are used in the analysis of genomes and in detecting mRNA in studies of gene expression
- Define the term 'bioinformatics'
- Outline the role of bioinformatics following the sequencing of genomes, such as those of humans and parasites, e.g. *Plasmodium*

Bioinformatics is the science of collecting and analysing complex biological data using complex computers. It uses complex computers to read, store and organise biological data at a much faster rate than previously. It also uses algorithms (mathematical formulae) to analyse and interpret biological data. The data is freely available and includes gene (nucleotide) sequences, genome data (total of all the genetic material of an organism), amino acid sequences, proteins and phenotypic data, information about pharmaceutical products (drugs). Using computers and the internet allows information to be shared and databases to be linked. Researchers continually add information to the databases so that information is accumulating (building up) at an enormous rate. This information can be retrieved and analysed or compared to new information to look for patterns.

Role of bioinformatics

One particular use of bioinformatics is in the sequencing of DNA (Figures 1 and 2). When you consider that the human genome consists of over 3 billion base pairs organised into 21 000 genes, sequencing every one of those bases is a huge task. Without the use of bioinformatics it is unlikely that we would have achieved success for many years to come. The sequencing of the human genome (the Human Genome Project) took just 13 years. The medical advances that have been made as a result of sequencing the human genome are many. For example, over 1.4 million single nucleotide polymorphisms (SNPs) have been found in the human genome. SNPs are single-base variations in the genome that may be associated with disease and other disorders. Medical screening of individuals has allowed quick identification of potential medical problems and for treatment to be given at a much earlier stage. Sequencing the DNA of different organisms has also made it possible to establish the evolutionary links between species.

The sequencing of the DNA of parasites such as *Plasmodium falciparum*, which causes malaria (Topic 10.5), has given us an insight into its metabolism. Knowledge of the proteins and other substances it produces will be invaluable in helping us to develop a cure for this disease. As a eukaryote, *Plasmodium* is known to have many antigens, and that different antigens are produced at different stages of it's life cycle. Knowledge of the protein structure of these antigens, the amino acid sequences and the genes that code for them, will help researchers develop vaccines and therapeutic drugs. There is also the potential to develop genetically engineered *Plasmodium* for use in vaccines. Again, only the power of bioinformatics has allowed us to sequence all 5300 genes on *Plasmodium's* 14 chromosomes in such a short period.

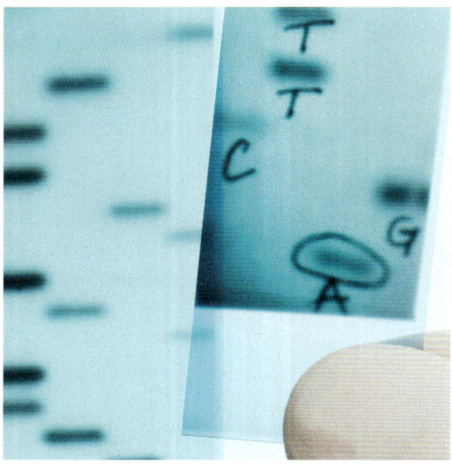

Figure 1 *Scientist holding an X-ray image of a gel with DNA nucleotide bases (A, C, T, G) sequenced using the Sanger method*

Microarrays

DNA microarrays have the following features:

- Thousands of spots are arranged in rows and columns on a solid surface of glass, plastic or silicon.
- Microarrays can be the size of a microscope slide, or even smaller.
- Each spot on a microarray contains multiple copies of single-stranded DNA.
- The DNA sequence on each spot is unique and known as a probe.
- Each spot represents a single gene or a sequence from part of a gene.
- The exact location and sequence of each spot is recorded in a computer database.

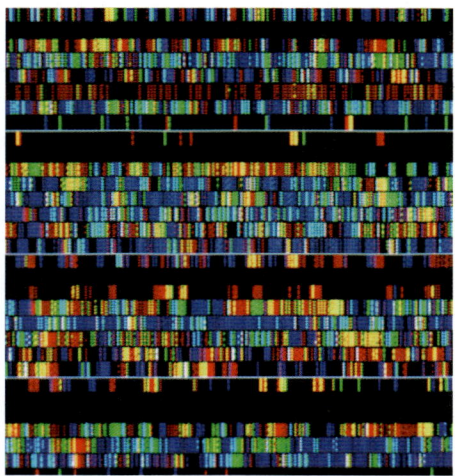

Figure 2 *Computer screen display of a DNA sequence. Each coloured band represents one of the four nucleotide bases.*

Using DNA microarrays, scientists can carry out an analysis on thousands of genes simultaneously. This has a number of uses including:

- **Gene expression profiling** – At any given moment a human cell has some combination of its 21 000 genes being expressed (switched on) and others are not expressed (switched off). The pattern of gene expression at any one time is called the **gene profile**. One way to determine which genes are being expressed in the cells of an organism and which are not is to start with a DNA microarray that is spotted with DNA fragments that represent specific gene coding regions. The DNA sequences that are to be used as probes can be obtained commercially. All the mRNA in the cell to be tested is extracted and converted into complementary DNA (cDNA) using reverse transcriptase. The cDNA is labelled using fluorescent tags of a particular colour, for example red, and the mixture of mRNA and cDNA is treated so that any mRNA is broken down. The purified cDNA is then added to the microarray. Where a gene is being expressed, the cDNA with complementary bases to that gene will attach (hybridise). Where the gene is not being expressed, no cDNA will attach. The microarray can be scanned by lasers to produce fluorescent spots, in this case red fluorescence, that are analysed to show exactly which genes have attached to cDNA and are therefore being expressed. The intensity of the fluorescence can also be recorded to show the level of gene expression that is occurring. The higher the intensity, the more mRNA was produced and the more active the gene. This technique is particularly useful in comparing gene expression in healthy and diseased tissue. Here, mRNA from diseased cells is converted into cDNA that will have a different fluorescence colour (e.g. red) than the cDNA produced from mRNA extracted from healthy cells (e.g. green). Spots that only fluoresce red show that the diseased state causes particular genes to switch on. If there are some spots that are only green then the diseased state has led to those genes being switched off. Yellow spots indicate that genes are active for both the diseased and normal states.

- **Genome analysis** – Microarrays can allow the quick and efficient comparison of two closely related genomes. To do this, the DNA from the two sources to be compared is isolated and each labelled with a different coloured fluorescent tag. The DNA is denatured to make it single stranded and it is then added to the microarray. Hybridisation of the two samples with the DNA on the microarray takes place and fluorescence patterns of each sample are compared. Differences may indicate gains or losses of genetic material, for instance due to mutation. This method can be used to determine abnormal DNA sequences, which may indicate disease. Also, differences in the number of repeat sequences (such as SNPs) between the test individual and the prepared microarray can be analysed.

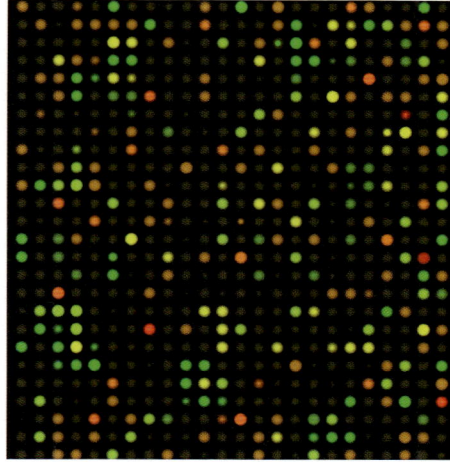

Figure 3 *A microarray*

EXTENSION
Sequencing the nucleotides of a gene

Before we look at the role of bioinformatics in DNA sequencing, let us start by considering how the process is carried out. Firstly DNA is cut into fragments using restriction endonucleases. There are a number of different methods to sequence the exact order of nucleotides in each of these fragments. One is called the Sanger method. In this process, a single strand of DNA is synthesised using normal nucleotides and special ones called terminator nucleotides. There are four types of terminator nucleotide, each with one of the four bases adenine, guanine, cytosine and thymine. Terminator nucleotides cannot attach to the next base in the sequence when they are being joined together and so terminate DNA synthesis at that point. Sometimes DNA synthesis is terminated after just a few nucleotides, and sometimes after many nucleotides. As a consequence, many fragments of varying length are produced, each ending in one of the four types of terminator nucleotide. The next stage is to separate out these DNA fragments of different lengths using a technique called gel electrophoresis (Topic 19.8).

EXTENSION
Automation of DNA sequencing

DNA sequencing is an example of the use of bioinformatics. It is routinely done by automatic machines and the data produced analysed by computers. In these computerised systems, instead of radioactively labelling the DNA primer, the four types of terminators are labelled with fluorescent dye. Each type uses a different colour – adenine (green), thymine (red), cytosine (blue) and guanine (yellow). Electrophoresis is carried out in a single narrow capillary gel. The results are scanned by lasers and interpreted by computer software to give the DNA sequence in a fraction of the time taken by conventional methods. Further automation is the use of a polymerase chain reaction (PCR) machine to produce the DNA fragments required in these techniques (Topic 19.2). With new innovations in the fields of DNA technology and computer software, these methods are continually being updated.

On these pages you will learn to:

- Describe and explain how gel electrophoresis is used to analyse proteins and nucleic acids, and to distinguish between the alleles of a gene
- Outline the use of PCR and DNA testing in forensic medicine and criminal investigations

Genetic fingerprinting uses the fact that any organism's genome contains many repetitive, non-coding sequences of nucleotides of DNA. The process in humans is based on the fact that 95% of human DNA does not code for any characteristic. These non-coding DNA sequences are known as **introns** and they contain repetitive sequences of DNA called **variable number tandem repeats (VNTRs)**. Each VNTR is at a particular chromosomal locus. For every individual, the number and length of VNTRs has a unique pattern. They are different in all individuals except identical twins, and the probability that two individuals have identical VNTRs is extremely small. These VNTRs are more similar, the more closely related any two individuals are. The making of a genetic fingerprint involves analysing VNTRs and uses the technique of gel electrophoresis. The steps in genetic fingerprinting are summarised in Figure 3.

Gel electrophoresis

In step 3 of Figure 3, the DNA fragments are placed onto an agar gel and a voltage is applied across it (Figure 1) to create an electric field. The fragments move through the gel toward the anode as they are negatively charged (owing to the presence of phosphate groups). Due to the resistance of the gel, they move at different rates depending on their mass and, to a lesser extent, their shape. The larger the fragments, the more slowly they move. Over a fixed period, therefore, the smaller pieces move further than the larger ones. In this way fragments of DNA of different lengths are separated. Electrophoresis of DNA is commonly only on the basis of size as a buffer can be used so that all fragments have the same charge.

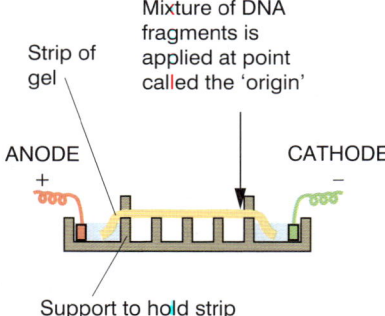

Figure 1 *Apparatus for carrying out electrophoresis*

Interpreting the results

DNA fingerprints from two samples, e.g. from a suspect and from blood found at the scene of a crime, are visually checked. If there appears to be a match, the pattern of bars of each fingerprint is passed through an automated scanning machine, which calculates the length of the DNA fragments from the bands. It achieves this using data obtained by measuring the distance travelled during electrophoresis by known lengths of DNA. Finally, the odds are calculated of someone else having an identical fingerprint. The more the two patterns match, the greater the probability that the two sets of DNA have come from the same person.

The use of PCR in forensic medicine and criminal investigations

The DNA to be analysed in forensic medicine and from crime scenes can come from very small original samples. Using the polymerase chain reaction, millions of copies of the DNA can be produced so that there is enough for DNA fingerprinting to be carried out and for samples to be kept for future testing. The procedure must be carried out to very strict guidelines to avoid contamination of the samples.

Uses of DNA fingerprinting

DNA fingerprinting is used extensively in forensic science, to indicate whether or not an individual is connected with a crime, e.g. from blood or semen samples found at the scene.

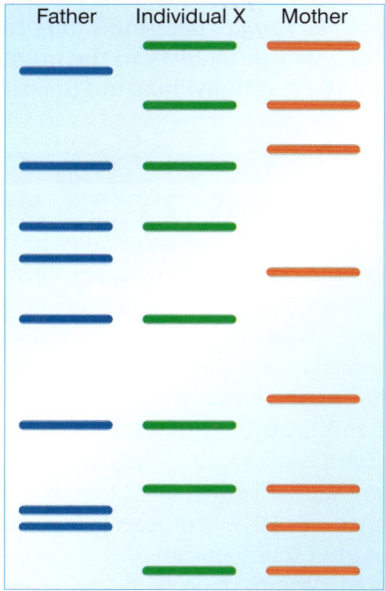

Figure 2 *DNA fingerprints of two parents and Individual X, who is their child. Note that each band on the child's fingerprint corresponds to a band on one or other parent's fingerprint*

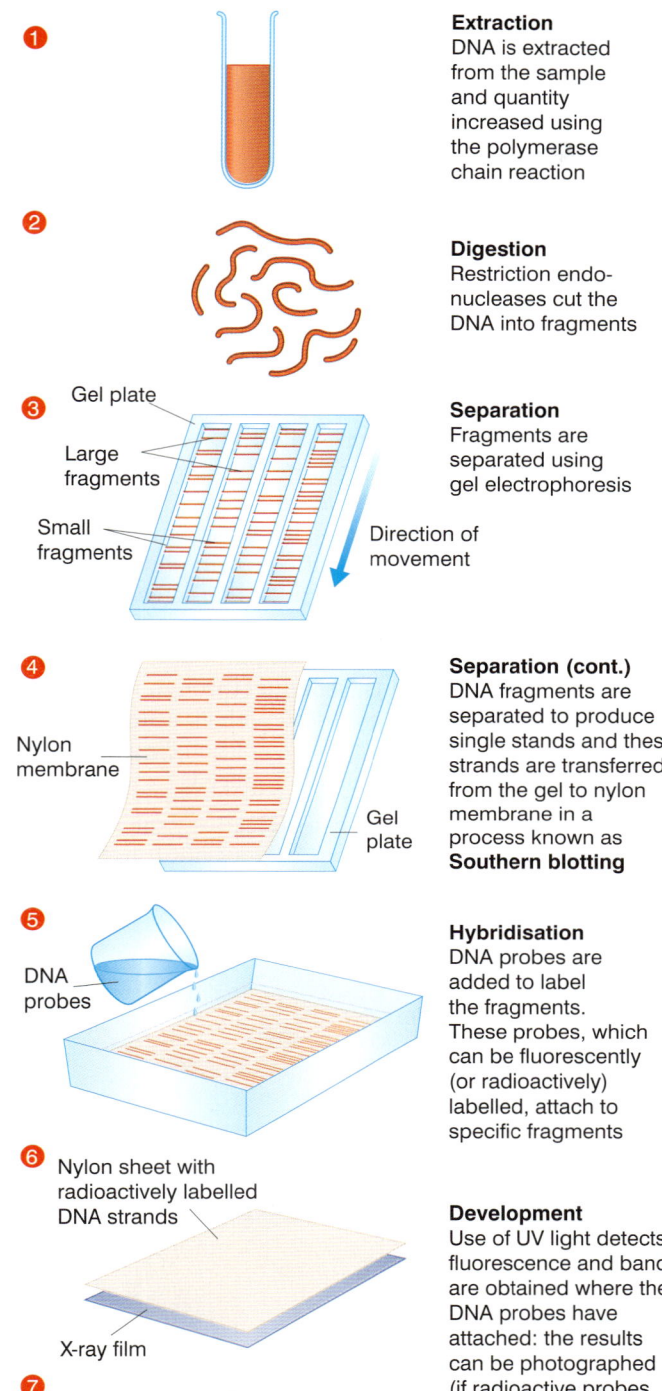

Extraction
DNA is extracted from the sample and quantity increased using the polymerase chain reaction

Digestion
Restriction endo-nucleases cut the DNA into fragments

Separation
Fragments are separated using gel electrophoresis

Gel plate
Large fragments
Small fragments
Direction of movement

Separation (cont.)
DNA fragments are separated to produce single stands and these strands are transferred from the gel to nylon membrane in a process known as **Southern blotting**

Nylon membrane
Gel plate

Hybridisation
DNA probes are added to label the fragments. These probes, which can be fluorescently (or radioactively) labelled, attach to specific fragments

DNA probes

Nylon sheet with radioactively labelled DNA strands

X-ray film

Development
Use of UV light detects fluorescence and bands are obtained where the DNA probes have attached: the results can be photographed (if radioactive probes are used, a sheet of X-ray film is placed over the nylon sheet and then developed to reveal dark bands)

Figure 3 *Summary of genetic fingerprinting*

Sometimes the DNA fingerprints of close relatives can be near matches to a DNA fingerprint from a crime scene. This sometimes helps police narrow down their search for suspects or people who were present at the crime scene. For example, Figure 2 shows that individual X is the child of the two parents. Every individual inherits half their genetic material from their mother and half from their father. Each band on a DNA fingerprint of any individual should, therefore, have a corresponding band in the parents' DNA fingerprint (Figure 3).

Genetic fingerprinting is also useful in determining genetic variability within a population. The more closely two individuals are related, the more closely their genetic fingerprints will resemble one another. A population whose genetic fingerprints are very similar shows little genetic diversity. A population with more varied genetic fingerprints has greater genetic diversity.

Uses of gel electrophoresis

By altering the composition of the gel (givies different pore sizes), the time the electrophoresis is run for and the voltage applied, gel electophoresis can be adapted for the separation of many different mixtures of macromolecules.

DNA fingerprinting is one use of gel electrophoresis. The technique is very useful for separating DNA fragments on the basis of size, for example for analysing restriction enzyme action or for identifying mutations. RNA can also be separated by electrophoresis, for example different mRNA molecules will be of different lengths as the genes from which they were transcribed are of different lengths.

Gel electrophoresis is used to separate not just DNA fragments, but also proteins. Proteins can be separated on the basis of size and charge. They can be treated so that they lost their tertiary structure and form linear molecules. They can then be separated on the basis of overall charge (on the basis of their different R groups). The beta-globin polypeptides for normal and for sickle cell haemoglobin can be identified by carrying out electrophoresis against known marker molecules.

We can also use gel electrophoresis to distinguish between different alleles of a gene. As the two alleles differ slightly in DNA sequence, it is also likely that they will differ in one or more restriction sites. If so, each will produce different sized fragments when cut by the same restriction enzyme. Gel electrophoresis can then be used to separate the different sized fragments from the two alleles. Different band patterns will therefore be produced, allowing us to distinguish the two alleles.

On these pages you will learn to:

- Outline how genetic diseases can be treated with gene therapy and discuss the challenges in choosing appropriate vectors, such as viruses, liposomes and naked DNA
- Discuss the social and ethical considerations of using gene testing and gene therapy in medicine

Up to 2% of the human population is affected by one of the 4000 diseases caused by a chromosomal mutation where one or more genes may be missing or by a gene mutation where the gene does not express itself properly. One example is cystic fibrosis. A partial cure is to replace defective alleles with ones cloned from healthy individuals in a technique called **gene therapy**. Gene therapy has many potential uses, including the treatment of cancer. In gene therapy, the length of DNA which has the sequence of nucleotides corresponding to the normal allele, is frequently called the healthy gene or the normal gene. Where the treatment is for a disease caused by a recessive allele, all that is required is the successful delivery of one copy of the gene, the normal allele.

Cystic fibrosis

Cystic fibrosis is the most common genetic disorder among the white population of Europe and North America, with around 1 in every 20 000 people having the disease. Of these, 70% of cases are caused by a mutant **recessive allele** in which three DNA bases, adenine–adenine–adenine are missing. This form is therefore an example of a deletion **mutation** (Topic 16.12). Other types of mutation can lead to the same condition. The normal allele of the *CFTR* gene (cystic fibrosis transmembrane conductance regulator) gene produces a protein of some 1480 amino acids. The deletion results in a single amino acid being left out of the protein. This, however, is enough to make the protein unable to perform its role of transporting chloride ions across epithelial membranes (Figure 1). CFTR is a unique transport protein that requires ATP to transport chloride ions out of epithelial cells. Part of the protein forms a channel through which chloride ions pass (not to be confused with channel proteins and facilitated diffusion). In this way, epithelial membranes are kept moist as a watery mucus is produced.

In a person with cystic fibrosis, the defective gene means that the protein is either not made or does not function normally depending on which type of mutation the person has. The epithelial membranes are therefore dry, and the mucus they produce remains viscous and sticky.

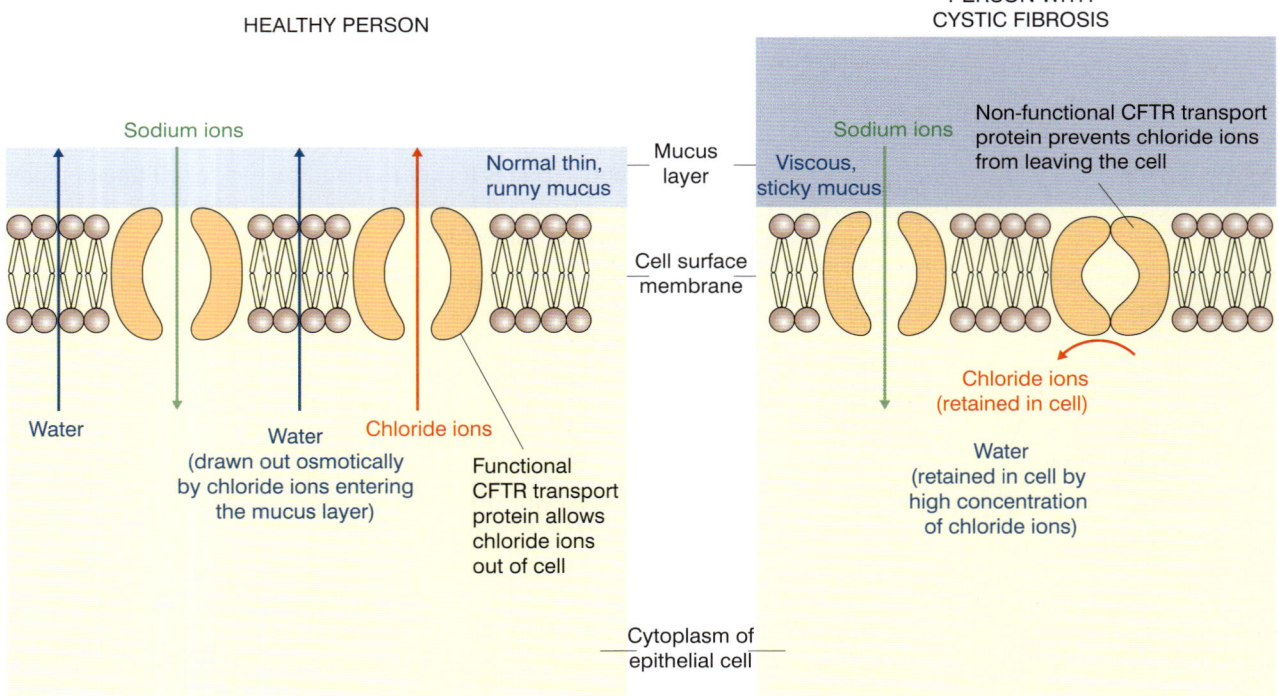

Figure 1 *Transport of ions by CFTR transport protein in a healthy person and someone with cystic fibrosis.*

As cystic fibrosis is caused by a recessive allele, it is possible that two normal parents could have a child who has the disease. Where there is a history of the disease in both families, the parents may choose to be genetically screened to see whether they carry the allele. More details on genetic screening are given in Topic 19.10.

Treatment of cystic fibrosis using gene therapy

There are two ways in which gene therapy may be used to treat cystic fibrosis:

- **Gene replacement** – in which the defective alleles are replaced by normal alleles.
- **Gene supplementation** – in which one or more copies of the normal allele of the gene are added so that they are present in addition to the defective alleles. As the added alleles are dominant, the effects of the defective recessive alleles are masked.

In addition, two different techniques of gene therapy can be used, according to which type of cell is being treated:

- **Germ-line gene therapy** – in which the defective gene is replaced or supplemented in the fertilised egg. This ensures that all cells of the organism will develop normally, as will all the cells of their offspring. This is therefore a much more permanent solution, affecting future generations. However, the moral and ethical issues of manipulating such a long-term genetic change mean that the process is currently prohibited.
- **Somatic-cell gene therapy** targets just the affected tissues, e.g. lungs, and the additional gene is therefore not present in sperm or eggs, and so not passed on to future generations. As the cells of the lungs are continually dying and being replaced, the treatment needs to be repeated periodically – as often as every few days. At present, the treatment has had limited success. The long-term aim is therefore to target undifferentiated **stem cells** that give rise to mature tissues. The treatment would then be effective for the life span of the individual.

Delivering the cloned *CFTR* alleles

The aim of somatic-cell gene therapy is to introduce cloned normal alleles into the DNA of epithelial cells of the lungs. This can be carried out in three ways: using a harmless virus, wrapping the gene in lipid molecules and using naked DNA.

Using a harmless virus

Viruses, called adenoviruses, cause colds and other respiratory diseases by injecting their DNA into the epithelial cells of the lungs. They therefore make useful vectors to transfer the normal *CFTR* allele into the host cells. Lentiviruses have also been researched as potential vectors. These are retroviruses that are known to be able to deliver DNA into non-dividing cells. Adeno associated viruses have also been trialled for gene therapy. These viruses are able to infect dividing and non dividing cells and as they do not cause disease in humans are become increasingly common in gene therapy trials. The process with adenoviruses works as follows:

- The adenoviruses are made harmless by interfering with a gene involved in their replication.
- These adenoviruses are then grown in epithelial cells in the laboratory together with plasmids that have had the normal *CFTR* allele inserted.
- The *CFTR* allele becomes incorporated into the DNA of the adenoviruses.
- These adenoviruses are isolated from the tissue culture of epithelial cells and purified.

REMEMBER

Gene supplementation **does not involve replacing** a gene, it simply introduces a dominant allele so that it is present with the existing recessive ones. The dominant allele therefore masks the effect of the recessive ones.

REMEMBER

Inserting a functional allele does not remove the defective alleles – it just means that cells produce both functional and non-functional proteins at the same time.

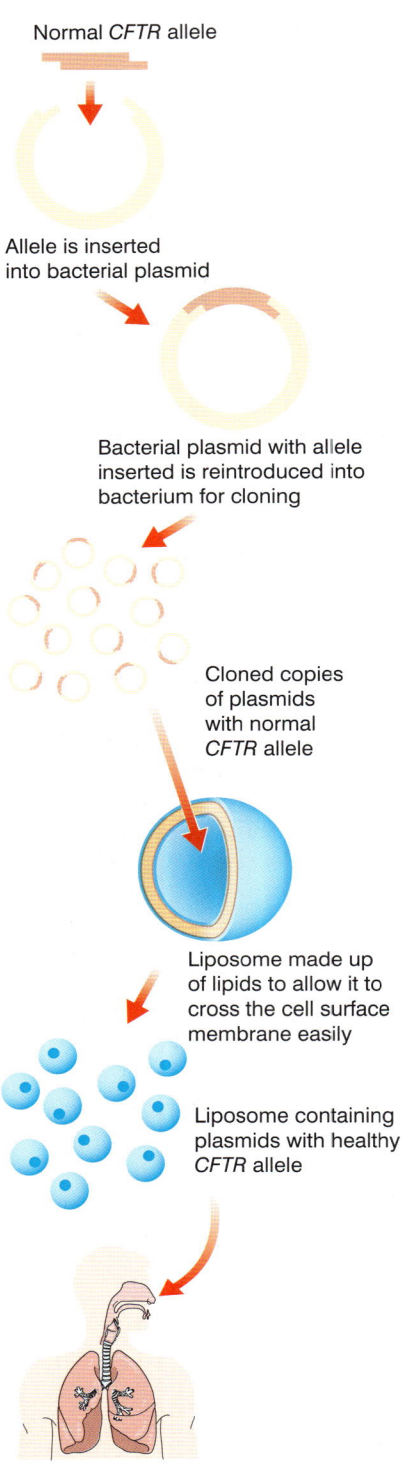

Normal *CFTR* allele

Allele is inserted
into bacterial plasmid

Bacterial plasmid with allele
inserted is reintroduced into
bacterium for cloning

Cloned copies
of plasmids
with normal
CFTR allele

Liposome made up
of lipids to allow it to
cross the cell surface
membrane easily

Liposome containing
plasmids with healthy
CFTR allele

Liposome introduced into lungs of a
person with cystic fibrosis via a nasal spray

Figure 2 *Summary of treatment of cystic
fibrosis by gene therapy*

- The adenoviruses with the *CFTR* allele are introduced into the person with cystic fibrosis via the nostrils or directly placed into the lungs, or delivered via an aerosol.
- The adenoviruses inject their DNA, which includes the normal *CFTR* gene, into the epithelial cells of the lungs.

Wrapping the normal allele in lipid molecules

The reason for wrapping the genes in lipid molecules is because lipid molecules can relatively easily pass through the phospholipid portion of the cell surface membranes. The process of delivering the *CFTR* alleles to their target cells, the epithelial cells of the lungs, is as follows:

- *CTFR* alleles are isolated from healthy human tissue and inserted into bacterial plasmid vectors (Topic 19.4).
- The plasmid vectors are reintroduced into their bacterial host cells and gene markers used to detect which bacteria have successfully taken up the plasmids with the *CFTR* allele (Topic 19.4).
- These bacteria are cloned to produce multiple copies of the plasmids with the *CFTR* allele.
- The plasmids are extracted from the bacteria and wrapped in lipid molecules to form a liposome.
- The liposomes containing the *CFTR* allele are sprayed into the nostrils of the patient as an aerosol and are drawn down into the lungs during inhalation.
- The liposomes pass across the phospholipid portion of the cell surface membrane of the lung epithelial cells.

These events are summarised in Figure 2.

Using naked DNA

Naked DNA is **histone**-free DNA. Naked DNA that includes the required gene can be transferred from one organism to another. This is done through the cell surface membrane, which has been made porous by treatment with an electrical current or UV lasers. It is relatively easy to administer as it can be injected directly into cells.

Challenges in choosing appropriate vectors

There is no perfect method for delivering cloned genes; each has its advantages and disadvantages. The overall success rate using naked DNA is much lower than with either viral or liposomal vectors. It is also unsuitable for systemic use and so is limited to only a few applications involving tissues that are easily accessible to direct injection such as skin and muscle cells.

The advantages of viral vectors are that they are very good at targeting and entering cells and can be targeted at specific types of cell. They can also be modified so that they do not replicate and destroy cells. They have the disadvantage that they can carry a limited amount of genetic material, and therefore some genes may be too big to fit into some viruses. They can also cause immune responses in patients, making them ill. In addition the immune system may block the virus from delivering the gene to the patient's cells, or it may kill the cells once the gene has been delivered.

The advantage of using liposomes is that they can carry larger genes, and rarely trigger an immune response. The disadvantage is that they are much less efficient than viruses at getting genes into cells.

Treatment of severe combined immunodeficiency disorder using gene therapy

Severe combined immunodeficiency (SCID) is a rare inherited disorder. Individuals with SCID do not show a cell-mediated immune response, nor are they able to produce antibodies. Until recently SCID was found only in young children because individuals with the disorder soon died of an infection. The disorder occurs when individuals inherit a defective allele to the gene that codes for the enzyme adenosine deaminase (ADA). This enzyme destroys toxins that otherwise kill white blood cells.

Survival has depended upon children being raised in the strictly sterile environment of an isolation tent or 'bubble' and giving them bone marrow transplants and/or injections of ADA. In recent years there has been some success in treating children with the condition. One type of *ADA* gene therapy includes the following:

- The normal *ADA* allele is isolated from healthy human tissue using restriction endonucleases (Topic 19.1).
- The *ADA* allele is inserted into a retrovirus.
- The retroviruses are grown with host cells in the laboratory to increase their number and hence the number of copies of the *ADA* allele.
- The retroviruses are mixed with the patient's T lymphocytes.
- The retroviruses inject their DNA, which includes a copy of the normal *ADA* allele, into the T lymphocytes.
- The T lymphocytes are reintroduced into the patient's blood to provide the genetic code needed to make ADA.

The effectiveness of this treatment is limited because T lymphocytes live only 6–12 months and so the treatment has to be repeated at intervals. Successful treatment for some children has involved harvesting stem cells from the bone marrow of the individual and transforming these rather than T lymphocytes. These are placed back into the children where the stem cells divide to produce T lymphocytes. Sadly in one trial, two of the children who were cured of the disease developed leukaemia. There have been a number of reports since where children appear to have been cured of the disease by gene therapy, with healthy production of T lymphocytes. More recent treatment involves transforming bone marrow stem cells rather than T lymphocytes. As bone marrow stem cells divide to produce T lymphocytes, there is a constant supply of the *ADA* allele and hence the enzyme ADA. Although not totally effective as there is an increased risk of leukaemia, the results have been promising.

Treatment of Leber's congenital amaurosis

Leber's congenital amaurosis (LCA) is a rare inherited eye disease that affects around 1 in 80 000 of the population. It is the result of an autosomal recessive mutation thought to be caused by abnormal development of photoreceptor cells in the eye. In 2008, a young man with this form of genetic blindness had his sight partially restored using gene therapy. In this case, the vector used was adeno associated virus.

Social and ethical considerations of genetic testing and gene therapy

As with many scientific applications, the use of genetic testing (Topic 19.10) and gene therapy brings risks and benefits and raises ethical and social issues. These include:

- Do disabilities need to be cured or prevented, or are they just part of the genetic variety that makes up all species?
- Who decides what is normal and what is a disability? While serious diseases are clearly disabling, there are many other conditions that some individuals consider disabling. These might include being very tall or short, having a particular hair colour, skin colour, body shape, birthmark, etc. Are these candidates for gene therapy?
- Gene therapy research and treatment is very expensive. Could the money be better spent on more proven treatments, where success is guaranteed?
- Germ-line therapy would be very effective, but what might be the long-term consequences of introducing heritable genes into the population? Could it lead to selection of one race in favour of another (eugenics)?
- Could the use of gene therapy make people less accepting of those who are different?
- Should gene therapy be used to enhance basic human traits such as height, intelligence, or athletic ability? How long before there are genetic tests for these traits, so that we can choose the features of our offspring? Genetic testing involves similar selection when parents choose whether or not to have a child with a disorder.
- Should people undergoing in-vitro fertilisation (IVF) have the right to genetically screen the eggs and sperm? Should they be able to pre-select the ones they use? Should a biopsy be carried out on the in-vitro embryo to screen for genetic disorders before it is implanted in the uterus?
- Should genetic testing of embryos in the uterus be carried out routinely? Should parents have the choice of a therapeutic abortion if there is some genetic abnormality?
- Who should know about the results of a genetic test? Would employers choose not employ someone with a genetic disorder? Would insurance premiums rise for that individual?
- Is it reasonable to carry out a genetic test for a disease for which there is no known cure?

Summary Test 19.9

Cystic fibrosis is caused by a mutant (**1**) allele of the CFTR gene. The normal variety of this gene produces a protein that transports (**2**) ions out of (**3**) cells. This in turn causes water to move out by (**4**). Treatment for cystic fibrosis using gene therapy involves either (**5**) or (**6**) of the defective gene. Where this is carried out by targeting the affected tissue it is called (**7**) gene therapy. In addition to using naked DNA, vectors that can be used to deliver the gene include (**8**) and (**9**).

On these pages you will learn to:

- Outline the advantages of screening for genetic conditions (reference may be made to tests for specific genes such as those for breast cancer, *BRCA1* and *BRCA2*, and genes for haemophilia, sickle cell anaemia, Huntington's disease and cystic fibrosis)

Genetic techniques can determine whether a person has a genetic condition, is likely to develop a disease that has a genetic basis, or is a carrier of a genetic condition. They can also determine for some genetic conditions whether an embryo or fetus will have a genetic disorder. The process by which this is done is called **genetic screening** or **genetic testing**. It is also carried out on children and adults for the reasons discussed in this section.

Genetic screening

Many genetic disorders, such as cystic fibrosis, are the result of **gene mutations**. We saw in Topic 16.12 that gene mutations may occur if the sequence of **nucleotides** in DNA are changed, either by substitution or deletion or insertion. If the mutation results in a **dominant allele**, all individuals will have the genetic disorder. If the allele is **recessive**, it will only appear in those individuals that have two recessive alleles, i.e. are **homozygous** recessive. Individuals that are **heterozygous** will not display symptoms of the disease but will carry one copy of the mutant allele. They have the capacity to pass the disease to their offspring if the other parent is also heterozygous or homozygous recessive.

It is important to screen individuals who may carry a mutant allele. These individuals will often have a family history of the disease. Screening can determine the probabilities of a couple having offspring with a genetic disorder. Using family history and the results of screening tests, potential parents can obtain advice from a genetic counsellor (see below) about the implications if they choose to have children.

Genetic screening for a gene mutation has a number of advantages:

- Test results, even where a mutation is found, can provide relief from uncertainty.
- On the basis of test results, individuals and families can make informed decisions about what to do.
- A negative result may remove the need for unnecessary checkups and screening tests.
- A positive result can lead to individuals taking preventative measures and/or continuing to monitor the situation and/or seeking treatment.

- Some test results may assist people to make decisions about whether or not to have children.
- Screening of newborn babies can identify genetic disorders early so treatment can be started as soon as possible.

Genetic screening can be used for a wide variety of genetic disorders including:

- **Breast cancer** The risk of developing breast cancer is greatly increased if a woman or man (men can get breast cancer too although it is much more rare) inherits a mutation in the *BRCA1* gene or the *BRCA2* gene. They may also be at increased risk of other forms of cancer. *BRCA1* and *BRCA2* genes produce tumour suppressor proteins, which help to repair damaged DNA. Mutation of either of these genes can mean that the tumour-suppressor proteins are not produced or do not function correctly. As a result, damage to DNA may not be properly repaired. Cells are therefore more likely to develop genetic changes that can lead to cancer.

  Genetic tests, involving taking DNA from a blood or saliva sample, can check for mutations in the *BRCA1* and *BRCA2* genes. These are targeted at people whose family history suggests the likely presence of a harmful mutation in one of these genes. If a harmful *BRCA1* or *BRCA2* mutation is found, several options are available to help a person manage their cancer risk. Some women have chosen to have a total mastectomy (both breasts removed): this is an example of preventive action. Others are aware that more frequent checks for abnormalities may help to spot early signs of cancer.

- **Huntington's disease** is a progressive disorder of the central nervous system caused by a dominant allele (Topic 16.12). A child who has one unaffected parent and one who has a dominant allele for Huntington's (heterozygote) has a 50% chance of inheriting the dominant allele and in these cases will develop the disease. However, as the symptoms only arise later in life, it is normal to wait until a child is 18 years old before carrying out a genetic test. If they have inherited the mutant gene, they will develop Huntington's, although it is not possible to know at what age. As there is no cure for the condition, some people choose not to have the test. Of those asking to be tested, some use the results to help them decide whether or not to have children.

- **Sickle cell anaemia** is a blood disorder caused by a recessive allele (Topic 17.8). In England, all pregnant women are offered screening to find out if they are carriers. If they are found to be carriers then the baby's father is also offered a screening blood test. It is then possible for a prenatal test to be carried out on the embryo so that the parents can make very difficult choices about continuing with the pregnancy. In addition, all newborn babies are screened for the condition by taking a sample of blood from the baby's heel.

- **Haemophilia** is the inability of the blood to clot effectively as a result of a recessive mutation on the X chromosome (Topic 16.9). Genetic tests can be carried out if there is a family history of the disease. This is especially important if pregnant because of the risk of excessive bleeding of the baby at birth. Genetic screening of the embryo can be carried out by taking a sample from the placenta or from the amniotic fluid. Any problems that occur during pregnancy for the mother or the fetus can then be managed.

- **Cystic fibrosis** causes mucus congestion and breathing difficulties (Topic 19.9) and is the result of a recessive allele. A genetic test checks for the mutant allele by either analysing a saliva sample taken from inside the cheek or a blood sample. A test can be done on a woman when she is pregnant to see if her embryo has cystic fibrosis. The test uses chorionic villus sampling. Antenatal testing for cystic fibrosis is usually only offered to mothers who are thought to be at high risk of having a child with the disease, such as women with a family history of the condition. As with sickle cell anaemia, parents are then faced with difficult choices about whether to continue with a pregnancy. Testing couples to determine if they are carriers of cystic fibrosis helps them to make decisions about whether to have children or not. In some cases couples may opt to go through IVF and have embryos checked before implantation.

Genetic screening goes hand in hand with genetic counselling during which expert advice is provided to enable individuals to understand the results and implications of the screening and so make appropriate decisions.

Genetic counselling

Genetic counselling is normally recommended before and after any genetic test to:

- explain how the test is carried out and its technical accuracy
- discuss whether genetic testing is appropriate
- explain the medical implications of a positive or a negative test result
- point out the possibility that a test result might not be conclusive
- discuss psychological risks and benefits of genetic test results
- explain the risk of passing a mutation to children.

Genetic counselling is rather like genetic social work. It involves the giving of advice and information to allow others to make personal decisions about themselves or their offspring. An important aspect of genetic counselling is to research the family history of inherited disease and to advise parents on the likelihood of it arising in their children.

Imagine a mother whose family has a history of cystic fibrosis. If she is unaffected but carries the gene, she can only be heterozygous for the condition. Suppose she wishes

to produce children by a man with no history of cystic fibrosis. It must be assumed that he does not carry the allele for the disease and therefore none of the children have the disease, although they may be carriers. If on the other hand, the potential father's family has a history of the disease, it is possible that he too carries the allele. In this case, genetic counselling can make the couple aware that there is a one in four chance of the children being affected and a two in four chance that their children will be carriers.

Counselling can also inform them of the symptoms of cystic fibrosis and any emotional, psychological, medical, social and economic consequences of disease. This may include the likely cost of any care that an affected child might need and the life expectancy of the child and how this might be extended using drugs or gene therapy. On the basis of this advice the couple can choose whether or not to have children or whether to use a sperm donation from an unaffected male. If conception has already occurred, the couple can decide whether to have a termination. Counselling can also make them aware of any further medical tests that might produce a more accurate prediction of whether their children will have the condition.

Genetic counselling is closely linked to genetic screening and the results provide the genetic counsellor with the basis for an informed discussion. For example, screening for cancer can help to detect:

- Oncogene mutations that can determine the type of cancer the patient has and hence the most effective drug or radiotherapy to use.
- Gene changes that predict which patients are more likely to benefit from certain treatments and have the best chance of survival. For example, the drug herceptin is most effective at treating certain types of breast cancer.
- A single cancer cell amongst millions of normal cells and so identify patients at risk of relapse from certain forms of leukaemia.

The information provided by the screening can help a counsellor to discuss with the patient the best course of treatment and their prospects of survival.

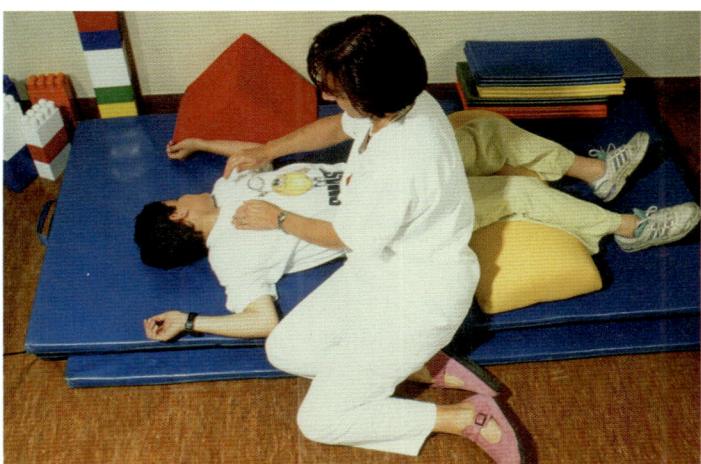

Figure 1 *Physiotherapist treating a young cystic fibrosis patient*

On these pages you will learn to:

- Explain the significance of genetic engineering in improving the quality and yield of crop plants and livestock in solving the demand for food in the world, e.g. Bt maize, vitamin A enhanced rice (Golden rice™) and GM salmon
- Outline the way in which the production of crops such as maize, cotton, tobacco and oil seed rape may be increased by using varieties that are genetically modified for herbicide resistance and insect resistance

Recombinant DNA technology produces genetically modified (GM) organisms that have some added feature which makes them of increased benefit to humans. The vector suitable for modifying plant cells is the bacterium *Agrobacterium tumefaciens*.

Herbicide resistance in oilseed rape and tobacco

Oilseed rape is a crop grown both as an animal feed and for its oil, which is used in the manufacture of vegetable oils and margarine. As with all crops, its yield will be less when it competes with weeds for sunlight, water and nutrients from the soil. Tobacco is cultivated in more than 100 countries. The leaves of the tobacco plant are used commercially and are dried, fermented, cut and processed into cigars and cigarettes or made into chewing tobacco or snuff.

Genetically modified varieties of oilseed rape and tobacco have been developed in which a gene is transferred into them from a soil bacterium. This gene makes them resistant to the non-selective herbicide glufosinate. Ordinarily, if this herbicide were applied to fields of oilseed rape and tobacco it would kill not only the competing weeds, but also the crop. The genetically modified crops are, however, not affected by the herbicide and so continue to grow while only the weeds are destroyed. Clearly this is of considerable benefit to the farmer, who can spray whole fields with the herbicide rather than having to specifically target the weeds – a difficult and expensive task. With the weeds eliminated, the yield of the crops is substantially improved. The herbicide can also be used shortly before harvesting to clear away other foliage and make gathering the crop easier.

The use of genetically modified plants alongside the herbicide glufosinate may potentially have detrimental effects. These include:

- The development of glufosinate tolerance in weeds could lead to increased doses of the herbicide being necessary to produce the same effect. This may present environmental or human health risks.
- Spread of glufosinate resistance from oilseed rape and tobacco to weeds. As we have seen in Topic 17.11, hybridisation between plants can occur naturally. This could lead to genes for herbicide resistance being spread to other species. If the gene were to be transferred to weeds, this would make glufosinate ineffective as a weed killer.
- The removal of the weeds by the herbicide reduces biodiversity. These weeds form part of the food chain for animals such as insects and birds. The population of these other species could be reduced or they might disappear altogether.
- The herbicide-resistant varieties of plants are usually patented by companies. These companies often also produce the particular herbicide to which the new variety is resistant. The two are therefore sold as a pair. The company has a monopoly on both products leaving no room for commercial competition. This leads to higher charges for the farmer, which leads to high charges for the consumer.

Figure 1 *Spraying oilseed rape with a herbicide. Oilseed rape can be genetically modified to resist the herbicide which therefore only kills weeds that compete with it.*

Insect resistance in maize and cotton

Maize and cotton are both important crops worldwide. Maize is used as a staple food for humans and as animal feed. Cotton is grown for its fibre, which is used in textiles which are made into furnishings and clothes. Other products from cotton include oil and livestock feed.

Both crops can be attacked by insects therefore reducing yields. Rootworm larvae and stem-boring caterpillars are major pests of maize. Cotton boll weevils and caterpillars feed within the fruits of cotton plants with devastating effects. Indeed cotton has such a range of insect pests that more insecticides are sprayed on it per hectare, than on any other crop – often 10–15 sprayings a season and, in extreme cases, up to 30 sprayings.

At the beginning of the twentieth century a soil bacterium, *Bacillus thuringiensis* (Bt) was found to produce a toxin with insecticidal properties that killed certain caterpillars. The bacteria were used to produce a range of different insecticides, which were then sprayed on crops to control a variety of insect larvae.

Figure 2 *Cotton boll severely damaged by a boll weevil. Damage like this can be prevented by sowing insect-resistant varieties of cotton plants.*

In the 1990s, the gene for the Bt toxin was isolated and transferred into potato, maize and cotton plants. The genetically modified plants produce their own Bt toxin which kills insect larvae that feed on it. The advantages of Bt maize and Bt cotton plants include:

- There is less environmental impact because only insects feeding on the plant are killed. Spraying crops with insecticides kill most insects, including beneficial ones such as pollinating butterflies and bees.
- The plant produces sufficient Bt toxin to be effective in killing insect larval pests that eat it.
- It is more economic as the use of conventional insecticides is much reduced.
- Control of pests is easier, especially for smaller farmers, because less specialist knowledge and equipment is required than when spraying insecticides.

Some potentially detrimental effects include:

- The possible transfer of the Bt toxin gene to related species through hybridisation cannot be ruled out.
- There have been concerns that the Bt toxin could be exuded from the roots of Bt plants or remain in the parts of the plants not harvested. The Bt toxin could then adversely affect soil organisms.
- There are similar concerns that the Bt toxin may be harmful to humans or animals that eat the GM maize.
- Genetically modified seeds cost more than non-GM seeds. These extra costs need to be weighed against the considerable costs of spraying insecticides.
- Insects may become resistant to the Bt toxin, making Bt maize and Bt cotton vulnerable to insect attack. Again this needs to be weighed against the use of less insecticide which reduces the chance of resistance developing in the first place.

Vitamin A enhanced rice

Rice contains vitamin A which is present in the aleurone layer towards the outside of the grain. This aleurone layer is often removed in a process called polishing. This is done because it allows the rice to be stored for longer periods. The white rice that results is therefore deficient in vitamin A.

Vitamin A is important in the diet as it is necessary to make rhodopsin, an important chemical in the functioning

of the retina of the eye. Serious vitamin A deficiency can lead to blindness and this is a problem in countries where white rice forms a major part of the diet. It is estimated that around half a million children develop blindness and millions more have impaired sight in these countries every year as a result of vitamin A deficiency.

Vitamin A is made in the human body using β-carotene, a yellow-orange pigment, found in many plants, which is involved in photosynthesis. Although rice can produce β-carotene, it does not do so in the endosperm that makes up the white portion of rice grains.

Genetically modified varieties of rice have been developed by transferring two genes for the synthesis of β-carotene into it. One comes from the daffodil plant and the other from a soil bacterium, *Erwinia uredovora*. Both genes were necessary to enable rice to produce β-carotene in the endosperm of its seeds. The resultant rice is pale yellow in colour giving it its name, golden rice. The presence of β-carotene in the grains of the golden rice mean that those eating it can synthesise vitamin A and so avoid the blindness associated with its deficiency. This is especially beneficial to developing nations where poor health, as a result of a diet deficient in certain nutrients, is a handicap to their development.

Genetically engineered salmon

GM salmon are being developed for their ability to grow two to four times faster than other farmed salmon. The GM Atlantic salmon, called AquAdvantage, include a growth hormone-regulating gene from the Chinook salmon as well as a promoter gene from an ocean pout fish. The GM salmon have the potential to grow to market size in half the time (18 months rather than 3 years), because the genes increase the amount of growth hormone in the blood and allow the fish to grow throughout the year rather than just in spring and summer as with conventional salmon.

Table 1 *Summary of the benefits and concerns surrounding GM crops*

BENEFITS

- Crop yields can be improved reducing hunger and starvation
- Production is cheaper leading to less expensive food
- Unproductive land can be farmed using GM varieties engineered to suit extreme conditions
- Some deficiency diseases can be prevented leading to better health
- Produce can be stored and transported without spoiling
- Less insecticide is needed, reducing the ecological damage they cause

CONCERNS

- Cost of GM seeds is high
- Global companies become more powerful
- Engineered genes may spread to other species through hybridisation
- GM crops may be difficult to sell due to consumer concerns
- GM varieties may be genetically unstable
- There are no long-term studies on the effects to human health
- Biodiversity may be reduced

19 Examination Questions

1 **a** Describe the role of insulin in the regulation of blood glucose concentration. *(3 marks)*

b State two advantages of treating diabetes with insulin produced by gene technology. *(2 marks)*

c One of the steps in the production of bacteria capable of producing human insulin is the insertion of the gene coding for human insulin into a plasmid vector.

Figure 1 shows one of the artificial plasmids constructed to act as a vector.

Figure 1

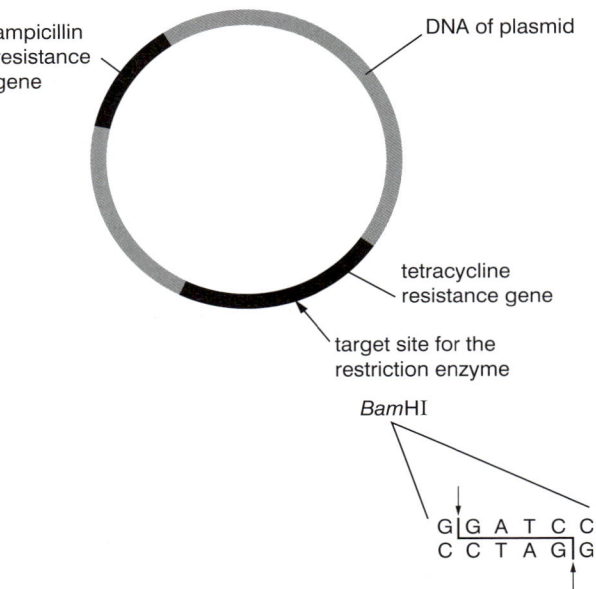

i With reference to Figure 1, explain the importance of the plasmid having a single target site for a particular restriction enzyme, such as *Bam*HI. *(2 marks)*

ii The genes for ampicillin resistance and tetracycline resistance on the plasmid allow the genetic engineer to distinguish between bacteria that have taken up different circles of DNA.
Copy and complete the table to show whether bacteria which have taken up each different circle of DNA are, or are not, resistant to ampicillin, to tetracycline or to both. Show presence of resistance with a tick (✓) and absence of resistance with a cross (✗). *(3 marks)*

Circle of DNA taken up by bacteria	Bacteria resistant to ampicillin	Bacteria resistant to tetracycline
Unaltered plasmids		
Recombinant plasmids that have taken up the wanted gene		
Circles of the wanted gene		

d i Explain why genes for antibiotic resistance are now rarely used as markers in gene technology. *(3 marks)*

ii Describe the use of **one** alternative marker gene that can be used instead of an antibiotic gene. *(2 marks)*

(Total 15 marks)

Cambridge International AS and A Level Biology 9700 Paper 4 Q6 June 2009

2 Figure 2 shows the CFTR (cystic fibrosis transmembrane conductance regulator) protein in a plasma (cell surface) membrane.

Figure 2

a i Describe the normal function of the CFTR protein.
(2 marks)

ii On a copy of Figure 2, use the letter **E** to indicate the external face of the membrane. State how you identified this face. *(1 mark)*

b Cystic fibrosis is caused by a recessive allele of the *CFTR* gene.

i Explain the meaning of the term *recessive allele*. *(2 marks)*

ii Explain how cystic fibrosis affects the function of the lungs. *(3 marks)*

c As cystic fibrosis is caused by a recessive allele of a single gene, it is a good candidate for gene therapy. Trials were undertaken in the 1990s, attempting to deliver the normal allele of the *CFTR* gene into cells of the respiratory tract, using viruses or liposomes as vectors.

Explain how viruses deliver the allele into cells.
(2 marks)

d In some people with cystic fibrosis, the allele has a single-base mutation which produces a 'nonsense' (STOP) codon within the gene.

i Explain how this mutation would prevent normal CFTR protein being produced. *(2 marks)*

ii A new type of drug, PTC124, enables translation to continue through the nonsense codon. Trials in mice homozygous for a *CFTR* allele containing the nonsense codon have found that animals treated with PTC124 produce normal CFTR protein in their cells. The drug is taken orally, and is readily taken up into cells all over the body.
Using your knowledge of the progress towards successful gene therapy for cystic fibrosis, suggest why PTC124 could be a simpler and more reliable treatment for this disease. *(3 marks)*
(Total 15 marks)

Cambridge International AS and A Level Biology 9700 Paper 4 Q2 November 2008

3 a A husband and wife who already have a child with cystic fibrosis (CF) elected to have their second child tested for the condition while still a fetus in very early pregnancy. The results of the test, a DNA banding pattern, were discussed with a genetic counsellor.

The relevant DNA banding pattern produced by electrophoresis is shown in Figure 3.

Figure 3

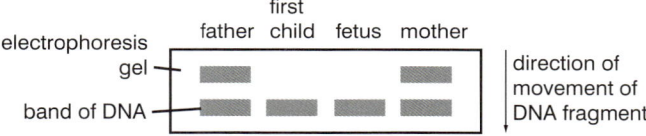

With reference to Figure 3, explain why,
i the fetus will develop CF, *(1 mark)*

ii the positions of the bands of DNA of the first child and of the fetus indicate that the mutant allele for CF has a deletion in comparison with the normal allele. *(2 marks)*

b Explain briefly the need to discuss the result of the test with a genetic counsellor. *(4 marks)*
(Total 7 marks)

Cambridge International AS and A Level Biology 9700 Paper 41 Q6 November 2009

19 Practice Questions

4 a What is the role of a vector during gene cloning?

b Why are gene markers necessary during gene cloning?

The figure below shows the results of an experiment using antibiotic resistance gene markers to find which bacterial cells have taken up a gene X. The circles within each plate represent a colony of growing bacteria.

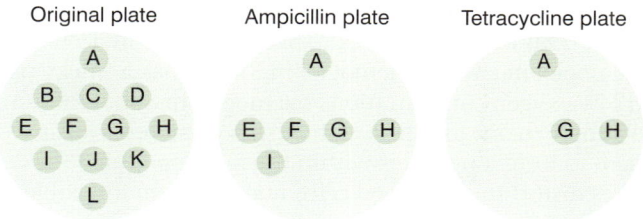

c Which colonies on the original plate:
i did not take up any plasmids with gene X
ii contained plasmids possessing gene X?
Give reasons for your answers.

5 a Explain why the polymerase chain reaction is often necessary before producing a genetic fingerprint.

b The figure below shows the genetic fingerprints of four DNA samples collected following a crime.

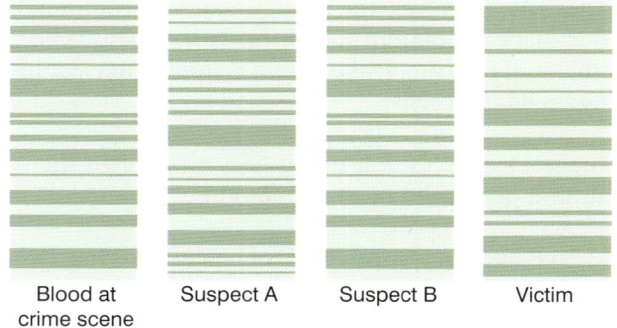

i Which of the two suspects do you think was present at the scene of the crime? Give a reason for your answer.
ii Suggest why a genetic fingerprint of a DNA sample from the victim was made.

Question A

Sickle-cell anaemia is a disease caused by a gene mutation in the gene that codes for haemoglobin.

In the DNA molecule that produces one of the amino acid chains in haemoglobin, the normal DNA triplet on the template strand is changed from CTC to CAC. As a result, the mRNA produced has a different code.

1 a Identify the type of gene mutation that causes sickle cell anaemia.

  b Deduce the:
    i   normal mRNA codon produced from the DNA
    ii mRNA codon produced as a result of the mutation.

The changed mRNA codes for the amino acid valine rather than for glutamic acid. This produces a molecule of haemoglobin (called haemoglobin S) that has a 'sticky patch'. At low oxygen concentrations haemoglobin S molecules tend to adhere to one another by their sticky patches, causing them to form long fibres within the red blood cells. These fibres distort the red blood cells, making them inflexible and sickle (crescent) shaped. These sickle cells are unable to carry oxygen and may block small capillaries.

2 Suggest:

  a how a change in a single amino acid might lead to the change in protein structure described

  b why sufferers of sickle-cell anaemia easily become tired.

Sickle-cell anaemia disables and kills individuals, and so the gene causing it has been eliminated from most populations. However, the gene is relatively common among black populations of African origin. This is because the malarial parasite, *Plasmodium*, is unable to exist in sickled red blood cells.

3 Suggest a process that might have eliminated the mutant gene from most populations.

Sickle cell anaemia is the result of a gene that has two codominant alleles, Hb^A (normal) and Hb^S (sickled).

4 What is meant by the term 'codominant'?

The three possible genotype combinations of these two codominant alleles and their corresponding phenotypes are as follows.

• Homozygous for haemoglobin S (Hb^SHb^S). Individuals suffer from sickle-cell anaemia and are considerably disadvantaged if they do not receive medical attention. They rarely live long enough to pass their genes on to the next generation.
• Homozygous for haemoglobin A (Hb^AHb^A). Individuals lead normal healthy lives, but are susceptible to malaria in areas of the world where the disease is endemic.
• Heterozygous for haemoglobin (Hb^AHb^S). Individuals are said to have sickle-cell trait, but are not badly affected. Sufferers may become tired more easily but, in general, the condition is symptomless. They do, however, have resistance to malaria.

5 a In parts of the world where malaria is prevalent, the heterozygous state (Hb^AHb^S) is selected for at the expense of both homozygous states. Consider the information above and suggest why this is the case.

  b Name the type of selection taking place. Explain your answer.

6 a If two heterozygous individuals produce offspring, calculate the chance of any one of them having sickle-cell anaemia.

  b Genetic screening for sickle-cell anaemia can be carried out. Explain why the advice given by a genetic counsellor to individuals with the same genotypes might differ depending on where they live.

7 In a population of 175 individuals the frequency of the Hb^A allele is 0.6 and the frequency of the Hb^S allele is 0.4.

  a Calculate the frequencies of the Hb^A and Hb^S alleles that would be expected in the next generation if the individuals mated randomly.

  b Using the Hardy–Weinberg equation, calculate the number of individuals with each phenotype. Show your working.

Question B

Figure 1 shows the energy flow through a freshwater ecosystem.

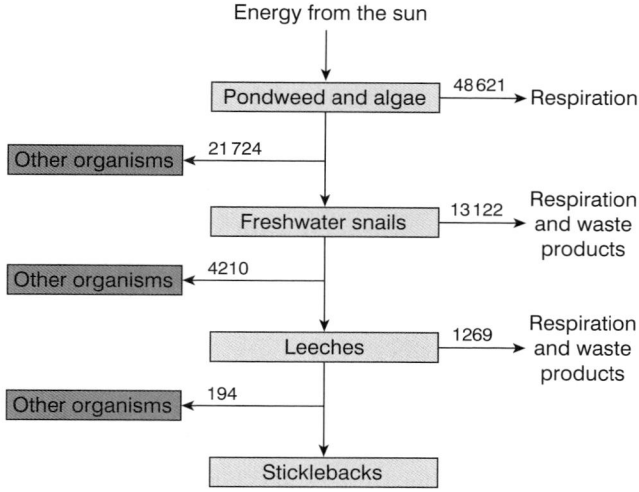

The units shown are kilojoules per metre squared per year (kJ m^{-2} year^{-1})

Figure 1

1 a i State which organisms in this food chain are the primary consumers.
  ii Calculate the energy lost in respiration and waste products by the freshwater snails. Show your working.
  iii Calculate the percentage efficiency of energy transfer from leeches to sticklebacks. Show your working and give your answer to three significant figures.

The sun's energy for food chains like the one shown in Figure 1 is converted to chemical energy by the process of photosynthesis. Photosynthesis has two distinct stages, the light independent stage and the light dependent stage. Figure 2 is a simplified sequence of the light dependent stage.

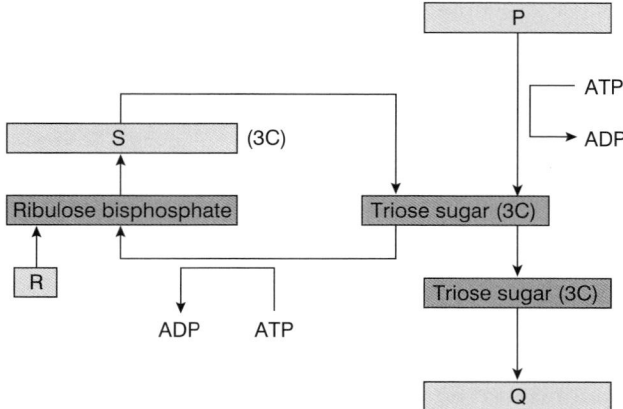

Figure 2

2 a Name the substances P, Q, R and S.
  b State the number of carbon atoms in ribulose bisphosphate.

The products of photosynthesis are transported from the leaves to the regions of the plant using or storing them through the tissue shown in Figure 3.

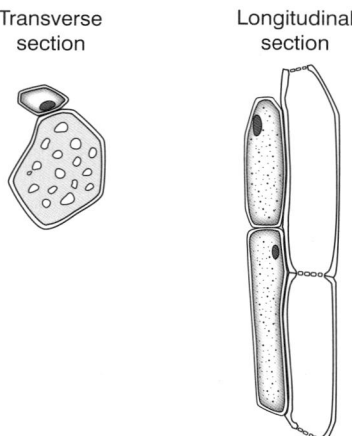

Figure 3

3 a i Name the tissue shown in Figure 3.
  ii Figure 3 shows two types of cells that are typical of this tissue. Name the two types of cell.

  b i Name two organic substances that are transported in this tissue.
  ii Outline four pieces of evidence that support the view that organic substances are transported in the phloem.

  c Explain the advantages of studying cells such as these with an electron microscope (EM) rather than a light microscope.

387

Question C

Figure 1 shows a nephron from a mammalian kidney.

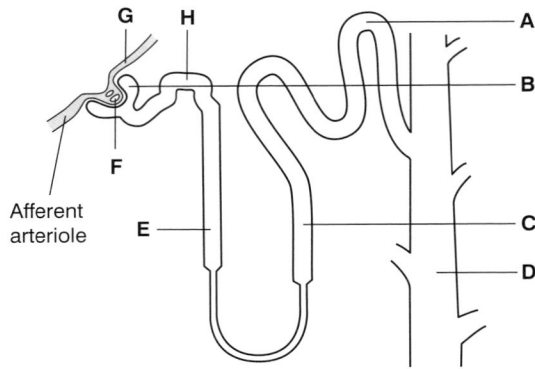

Figure 1

1 **a** Name the parts labelled A–H.

 b i Name a substance found in the blood plasma in the afferent arteriole but **not** present in the structure labelled H.

 ii Explain why this substance is absent from structure H.

 c i Name a substance found in the structure labelled B but absent from the fluid in the structure labelled E.

 ii Explain why this substance is absent from structure E.

2 Name the two blood vessels through which blood passes on its journey from the heart to the afferent arteriole.

Figure 2 represents a cell from the wall of structure H.

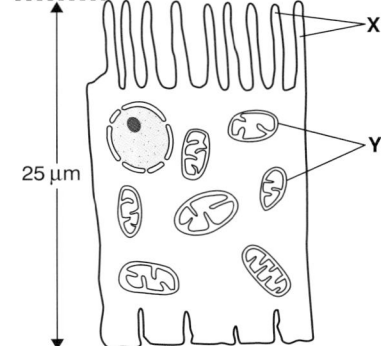

Figure 2

3 **a** Calculate the magnification of the cell in Figure 2. Show your working.

 b i Name the structures labelled X.
 ii Describe the function of the structures labelled X.

 c i Name the structures labelled Y.
 ii Explain why these cells have a large number of structure Y.

Goodpasture's disease is a very rare condition in which antibodies attack an antigen that is found in cells of the glomerulus of the kidney. One symptom of the disease is blood in the urine.

4 **a** Suggest an explanation for the sufferers of the disease having blood in the urine.

 b i Name the type of cell that produces antibodies in the body.
 ii Antibodies do not directly destroy cells with the antigen they are complementary to, but rather prepare these cells for destruction. Explain the role of antibodies in preparing cells with the antigen for destruction.

Question D

Cystic fibrosis is caused by a mutant recessive allele in which three DNA bases, adenine–adenine–adenine, are missing from the cystic fibrosis trans-membrane-conductance regulator (CFTR) gene. The deletion results in a single amino acid being left out of the protein produced by this gene. As a result the protein is unable to perform its role of transporting chloride ions across epithelial membranes. CFTR is a chloride-ion channel protein that transports chloride ions out of epithelial cells, and water naturally follows keeping epithelial membranes moist.

1 Explain how it is possible for parents without cystic fibrosis to have a child who suffers from the disease.

2 Name the process by which water will follow chloride ions across epithelial membranes.

In a patient with cystic fibrosis, the epithelial membranes are dry and the mucus they produce remains viscous and sticky. The symptoms this causes include:

- mucus congestion in the lungs, leading to a much higher risk of infection, breathing difficulties and less efficient gaseous exchange
- accumulation of thick mucus in the pancreatic ducts, preventing pancreatic enzymes, such as lipases, from reaching the duodenum.

3 **a** Explain why mucus congestion in the lungs can lead to a higher risk of infection.

 b Describe precisely the action of lipases.

Research is taking place to treat cystic fibrosis using gene therapy. This involves replacing or supplementing the defective gene with a healthy one. One method is to introduce cloned normal genes into the epithelial cells of the lungs, but the treatment needs to be repeated as often as every few days.

4 a State the meaning of the word 'cloned'.

b Explain why the treatment of lung epithelial tissue has to be repeated frequently.

Viruses make useful vectors for the transfer of the normal CFTR gene into the epithelial cells. They are grown in epithelial cells in the laboratory along with plasmids that have had the normal CFTR gene inserted. The CFTR gene becomes incorporated into the DNA of the viruses, which are isolated and purified before being introduced into the nostrils of the patients.

5 a Define each of the following terms:
  i virus
  ii vector
  iii plasmid.

b From your knowledge of viruses, suggest a reason why they are used to introduce the healthy CFTR gene into lung epithelial cells.

c Suggest a possible disadvantage of using viruses in this way.

Glossary

abiotic an ecological factor that makes up part of the non-biological environment of an organism. Examples include temperature, pH, rainfall and humidity. See also *biotic*.

abscisic acid *plant growth regulator* that has a role when a plant is under stress, such as in the closing of stomata during times of drought.

absorption spectrum a graph that results from plotting the degree of absorption of light of different wavelengths by a pigment such as chlorophyll.

acetylcholine one of a group of chemicals, called *neurotransmitters*, released an *axon*. It diffuses across the gap (*synapse*) between adjacent neurones and so passes an impulse from one neurone to the next.

action potential change that occurs in the electrical charge across the membrane of an *axon* when it is stimulated and a nerve impulse passes.

action spectrum graph displaying the proportion of each wavelength of light that is used in a process such as photosynthesis.

actin filamentous protein which is involved in contraction within cells, especially muscle cells. See also *myosin*.

activation energy energy required to bring about a reaction. The activation energy is lowered by the presence of enzymes.

active immunity protection from infection, so that the person does not become ill: it results from an immune response involving the production of memory cells to give long-term protection.

active site a group of amino acids that makes up the region of an enzyme into which the substrate fits in order to catalyse a reaction.

active transport movement of a substance from a region where it is in a low concentration to a region where it is in a high concentration. The process requires the expenditure of energy.

acute a sudden, but short-term, illness from which the patient recovers quickly.

adenosine triphosphate (ATP) an activated *nucleotide* found in all living cells that acts as an energy carrier. The *hydrolysis* of ATP leads to the formation of adenosine diphosphate (ADP) and inorganic phosphate, with the release of energy.

adhesion attraction between the molecules of different types such as between water molecules and cellulose molecules. See also *cohesion*.

adipose tissue a form of connective tissue that is made up of cells storing large amounts of fat.

adrenaline a hormone produced by the adrenal glands in times of stress that prepares the body for an emergency.

aerobic connected with the presence of free oxygen. Aerobic respiration requires free oxygen to release energy from glucose and other foods. See also *anaerobic*.

aerobic exercise forms of physical exertion, such as swimming and cycling, that involve the lungs and heart in providing oxygen to the muscles for aerobic respiration.

aerobic fitness a measure of the ability of an individual to take in and use oxygen. It depends upon the effective ventilation of the lungs, transport of oxygen by the blood and its use by cells.

aleurone layer protein-rich layer beneath the testa of a cereal seed that makes amylase during germination.

allele one of a number of alternative *nucleotide* sequences at a single gene *locus*. For example, the gene for the shape of pea seeds has two alleles, one for 'round' and one for 'wrinkled'.

allergen a normally harmless substance that causes the immune system to produce an immune response. See also *allergy*.

allergy the response of the immune system to an *allergen*. Examples include hay-fever and asthma.

allopatric speciation the formation of a new species as a result of *populations* of a parent species becoming geographically isolated. See also *sympatric speciation*.

anabolism an energy-requiring process of metabolism in which small molecules are combined to make larger ones.

anaerobic connected with the absence of free oxygen. Anaerobic respiration releases energy from glucose or other foods without the presence of free oxygen. See also *aerobic*.

anion negatively charged *ion* that is attracted to the anode during electrolysis. See also *cation*.

antibiotic a substance produced by one kind of microorganism that inhibits the growth of, or kills, another.

antibiotic resistance the development in microorganisms of mechanisms that prevent *antibiotics* from killing them.

antibody a *glycoprotein* produced by a plasma cell in response to the presence of the appropriate *antigen*.

antidiuretic hormone (ADH) a hormone produced by the *hypothalamus* that passes to the posterior *pituitary gland* from where it is secreted. ADH reduces the amount of water in urine by increasing water reabsorption in the kidneys.

antigen a molecule that is recognised by the body as *non-self* and triggers an *immune response*.

antioxidant chemical which reduces or prevents *oxidation*. Often used as an additive to prolong the shelf-life of certain foods.

apoplast pathway route through the cell walls and intercellular spaces of plants by which water and dissolved substances are transported. See also *symplast pathway*.

aquaporin integral protein that forms a water channel in cell surface membranes. It selectively conducts water molecules through the membrane while preventing ions and other solutes from doing so.

arteriosclerosis *atherosclerosis* that occurs in the arteries.

artificial immunity protection against infection, where the person does not become ill, acquired as a result of the deliberate exposure of the body to *antibodies* or *antigens* in non-natural circumstances, e.g. *vaccination*.

asthma a *chronic illness* in which there is resistance to airflow to the alveoli of the lungs as a result of the airways becoming inflamed due to an allergic response to an *allergen*.

atheroma fatty deposits on the walls of arteries, often associated with high *cholesterol* levels in the blood.

atherosclerosis disease of blood vessels in which the lumen is narrowed due to thickening of the arterial wall caused by fat, fibrous tissue and salts being deposited on it and over time, a gradually hardening so that the blood vessel is less flexible.

ATP see *adenosine triphosphate*.

atrioventricular node (AV node) area of muscle between the atria and ventricles of the heart that plays an important role in coordinating the heartbeat.

autosome a *chromosome* that is not a sex chromosome.

autotrophic nutrition form of feeding in which an organism uses energy from light or chemicals to build up complex organic molecules from simple inorganic substances. See also *heterotrophic nutrition*.

auxin a *plant growth regulator* that affects plant growth by stimulating cell division and enlargement.

axon a process extending from a *neurone* that conducts *action potentials* away from the cell body.

Benedict's test a simple biochemical reaction to detect the presence of reducing sugars.

biodiversity the range and variety of living organisms within a particular region. The term can also be applied to ecosystems, habitats, and genetic variation within species.

biomass the total mass of living material, normally measured in a specific area over a given period of time.

biosensor a device that uses biological molecules to measure the levels of certain chemicals.

biotic an ecological factor that makes up part of the living environment of an organism. Examples include food availability, competition and predation. See also *abiotic*.

Biuret test a simple biochemical reaction to detect the presence of protein.

B lymphocyte type of white blood cell that is produced and matures within the bone marrow. B lymphocytes produce *antibodies* as part of their role in *immunity*. See also *T lymphocyte*.

Bohr effect the reduced affinity of *haemoglobin* for oxygen in the presence of carbon dioxide.

Bowman's capsule see *renal capsule*.

bronchitis a respiratory *disease* in which the bronchi of the lungs become inflamed and congested with mucus. *Acute* bronchitis is the result of a bacterial or viral infection while *chronic* bronchitis is frequently associated with smoking.

buffer solution with the ability to absorb hydrogen *ions* and which therefore does not significantly change its pH when moderate amounts of acid or alkali are added.

Calvin cycle a biochemical pathway that forms part of the *light independent reaction* of photosynthesis during which carbon dioxide is reduced to form carbohydrate.

cambium dividing layer of cells in higher plants, parallel to the surface of stems and roots, which produces new cells leading to an increase in their diameter.

cancer a disease resulting from *mutations* that leads to uncontrolled cell division and the eventual formation of a group of abnormal cells called a tumour, from which cells may break away (*metastasis*) and form secondary tumours elsewhere in the body.

carcinogen a chemical, a form of radiation or other agent that causes *cancer*.

cardiac cycle a continuous series of events which make up a single heartbeat.

cardiac muscle type of muscle found only in the heart that can contract and relax continuously throughout life without stimulation by nerve impulses. See also *smooth muscle*.

cardiac output the total volume of blood which the heart can pump each minute. It is calculated as the volume of blood pumped at each beat (*stroke volume*) multiplied by the number of heart beats per minute (heart rate).

carrier protein (carrier molecule) a protein that helps to transport molecules across the plasma membrane.

cartilage flexible supporting tissue found at the end of bones and between vertebrae, where it cushions the shocks and jolts that occur during movement.

Casparian strip a distinctive band of suberin around the endodermal cells of a plant root which prevents water passing into *xylem vessels* via the cell walls. The water is forced through the living part (protoplast) of the endodermal cells.

catabolism chemical reactions of metabolism involving the breakdown of large molecules.

cation positively charged *ion* which is attracted to the cathode during electrolysis. See also *anion*.

cell signalling the means by which cells interact with the environment or with the cells around them.

centrifugation process of separating out particles of different sizes and densities by spinning them at high speed in a centrifuge.

channel protein water-filled channel in protein molecules of the cell surface membrane for the diffusion of water-soluble ions.

chemiosmosis the synthesis of *ATP* in mitochondria and chloroplasts using energy that is stored as a hydrogen *ion* concentration gradient across a membrane.

chitin tough, nitrogen-containing polysaccharide that forms the walls of fungi and the exoskeleton of insects.

chloride shift the movement of negatively charged chloride *ions* from the plasma into red blood cells to replace the loss of negatively charged hydrogencarbonate ions during the transport of carbon dioxide. In this way the overall electrochemical neutrality of the red blood cells is maintained.

cholesterol lipid occurring in large quantities in the brain, spinal cord and liver. It is an intermediate in the formation of vitamin D and steroid hormones. Excess in the blood can lead to *atherosclerosis*.

chromatid one of the two identical structures formed after DNA replication in late interphase, joined at the centromere to make the chromosome that becomes visible during prophase of mitosis and meiosis.

chromatin the diffuse material that comprises chromosomes in their decondensed state. It consists of DNA and proteins, especially *histones*.

chromatography technique by which substances in a mixture are separated according to their different solubilities in a solvent.

chromosome a cell structure made of linear DNA complexed with histone proteins by which hereditary information is physically passed from one generation to the next.

chronic a continuing ailment that arises gradually and may recur over a number of years.

chronic obstructive pulmonary disease name given to the lung condition where *emphysema* and chronic *bronchitis* occur together.

cilium (plural **cilia**) short projection from the surface of a *eukaryotic cell* that has an internal 9 + 2 arrangement of microtubules.

clone a group of genetically identical organisms formed from a single parent as the result of asexual reproduction or by artificial means.

codominance condition in which both alleles for one gene in a *heterozygous* organism are expressed and so contribute to the *phenotype*.

codon a sequence of three adjacent *nucleotides* in mRNA that codes for one amino acid.

cohesion attraction between molecules of the same type. It is important in the movement of water up the plant. See also *adhesion*.

collagen fibrous protein that is the main constituent of connective tissues such as *tendons, cartilage* and bone.

collenchyma plant tissue which has cell walls thickened by cellulose especially in the corners. It provides mechanical support, especially in young stems and leaves.

colloidal describes a mixture consisting of microscopic particles evenly dispersed throughout another substance.

community all the *populations* of different species within a particular area at a given time.

complementary DNA DNA which is made from messenger RNA using the enzyme *reverse transcriptase* in a process which is the reverse of normal *transcription*.

condensation reaction chemical process in which two molecules combine to form a more complex one with the elimination of a simple substance, usually water. Many biological *polymers*, such as polysaccharides and polypeptides, are formed by *condensation reactions*. See also *hydrolysis*.

conjugation the transfer of DNA from one cell to another by means of a thin tube between the two.

conservation method of maintaining *ecosystems* and the living organisms that occupy them. It requires planning and organisation to make best use of resources while preserving the natural landscape and wildlife.

consumer *heterotrophic* organism that obtains energy by eating or decomposing other organisms. Herbivores feed on plants and are known as primary consumers and carnivores feeding on herbivores are known as secondary consumers. See also *producer*.

continuous variation variation in which measurements of any one characteristic do not fall into distinct categories, but rather there are gradations from one extreme to the other, e.g. height in humans. See also *discontinuous variation*.

coronary heart disease any condition, e.g. *atherosclerosis* and *thrombosis,* affecting the coronary arteries that supply *cardiac muscle*.

corpus luteum a group of yellow cells in a mammalian ovary which produces progesterone and is formed from a Graafian follicle after ovulation.

cotyledon an embryonic leaf found in the seed of a plant.

counter-current system a mechanism by which the efficiency of exchange between two substances is increased by having them flowing in opposite directions.

covalent bond type of chemical bond in which two atoms share a pair of *electrons*, one from each atom.

crossing over the process whereby a *chromatid* breaks during *meiosis* I and rejoins to the chromatid of its *homologous chromosome* so that their *alleles* are exchanged.

cuticle exposed non-cellular outer layer of certain animals and the leaves of plants. It is waxy and therefore helps to reduce water loss.

cystic fibrosis inherited *disease* in which the body produces abnormally thick mucus that obstructs breathing passages and prevents secretion of pancreatic enzymes. It is a *recessive* condition that leads to production of a non-functioning membrane protein needed to transport chloride *ions*.

cytokinins group of *plant growth regulators* that stimulate cell division.

cytokines chemicals secreted by T helper cells that stimulate an *immune response* in other white blood cells.

deamination removal of an amino group (-NH$_2$) from a compound, particularly an amino acid.

decomposer a saprobiontic organism that breaks down the organic matter of dead organisms and waste products to form water, carbon dioxide and inorganic *ions*.

denaturation permanent changes due to the breaking of one or more types of bond that maintain the three-dimensional structure of a protein as a result of factors such as changes in temperature or pH.

dendrite a process, usually branched, extending from the cell body of a *neurone* that conducts impulses toward the cell body.

denitrifying bacteria bacteria that convert nitrates to nitrogen gas as part of the nitrogen cycle.

depolarisation temporary reversal of charges on the axon membrane of a *neurone* that takes place when a nerve impulse is transmitted.

desertification process by which a desert slowly spreads into neighbouring areas of semi desert.

diabetes metabolic disorder in which there is abnormal thirst and the production of large amounts of urine. Diabetes mellitus is caused by a reduction or absence of insulin production by the pancreas or insensitivity of insulin receptors on cells, leading to changes in the blood glucose level.

diastole the stage in the *cardiac cycle* when the heart muscle relaxes. See also *systole*.

dicotyledonous plants any member of the class of flowering plants called Dicotyledonae. Their features include: having two seed leaves (cotyledons), broad leaves, flower parts in rings of four or five, and vascular tissue arranged in a ring in stems. See also *monocotyledonous plants*.

diffusion the net movement of molecules or ions from a region where they are in high concentration to one where their concentration is lower.

diploid a term applied to *eukaryotic* cells in which the nucleus contains two sets of *chromosomes*. See also *haploid*.

directional selection selection that operates towards one extreme in a range of variation.

disease an abnormal condition affecting an organism and reducing the effectiveness of its functions.

discontinuous variation variation shown when different forms or phenotypes for any one characteristic fall into distinct categories, e.g. blood groups in humans. See also *continuous variation*.

disulfide bridges S–S bonds between two cysteine amino acids that are important in maintaining the structure of proteins.

diuretic drug used to increase the output of urine especially in the treatment of *hypertension*.

DNA probe a single strand of DNA used to identify a particular *gene*.

DNA helicase enzyme that acts on a specific region of the DNA molecule to break the *hydrogen bonds* between the bases causing the two strands to separate and expose the *nucleotide* bases in that region.

DNA replication the process in which the double helix of a DNA molecule unwinds and each strand acts as a template on which a new strand is constructed.

dominant a term applied to an *allele* that is always expressed in the *phenotype* of an organism even when present with an allele that is *recessive* to it.

ecological niche role of a species within its *community*. It includes what the species is like, where it occurs, how it behaves, its interactions with other species and how it responds to the *abiotic* environment.

ecology study of the interrelationships of organisms with each other and their environment.

ecosystem all the living (*biotic*) and non-living (*abiotic*) components of a particular area interacting together to make one functional unit.

ectotherm an organism that uses the environment to control body temperature interacting together to make one functional unit.

electron negatively charged sub-atomic particle that orbits the positively charged nucleus of all atoms. See also *proton*.

element one of just over 100 substances that cannot be split into simpler substances by chemical means.

emphysema a *disease* in which the walls of the alveoli break down, reducing the surface area for gaseous exchange, thereby causing breathlessness in the sufferer. See also *chronic obstructive pulmonary disease*.

endemic describes any *disease* that occurs regularly in a particular region or amongst a particular *population*. See also *epidemic* and *pandemic*.

endocrine gland a gland with cells which secrete a hormone into the blood at a distance from the hormone's target organ.

endocytosis the inward transport of large molecules through the cell surface membrane. See also *exocytosis*.

endergonic a chemical reaction in which the products contain more energy than the reactants so that, if the reaction is to proceed, free energy must be provided from outside. See also *exergonic*.

endosperm a storage tissue found in the seeds of plants such as cereals.

endotherm an animal which uses physiological processes to maintain its body temperature at a more or less constant level. Birds and mammals are endotherms.

environment the external conditions, resources and stimuli with which organisms interact.

epidemic describes any disease that rapidly spreads through a population to affect a large number of individuals. See also *endemic* and *pandemic*.

epidemiology the study of the spread of disease and the factors that affect this spread.

epidermis the outermost layer of cells in a multicellular organism.

essential amino acid an amino acid that cannot be synthesised by the human body and which must therefore be included in the diet.

eugenics human improvement by the controlling the production of offspring.

eukaryote an organism whose cells have a membrane-bound nucleus that contains *chromosomes*. The cells also possess a variety of other membranous organelles such as mitochondria and endoplasmic reticulum. See also *prokaryote*.

excretion the removal of metabolic waste products from the body.

exergonic a chemical reaction in which the products contain less free energy than the reactants and so free energy is released during the reaction. See also *endergonic*.

exocytosis the bulk transport of materials out of cells by cytoplasmic vesicles fusing with the cell surface membrane and releasing their contents to the outside. See also *endocytosis*.

facilitated diffusion diffusion down a concentration gradient through cell membranes using specific protein carrier channels.

flagellum (plural **flagella**) long, whip-like extension of a cell used as a means of locomotion.

gel electrophoresis technique for separating a mixture of charged particles in a gel, by applying a voltage across the fluid. It is used in the analysis of mixtures of substances, especially proteins.

gene sequence of *nucleotides* on a DNA molecule coding for a specific polypeptide. It is now known that some genes may code for more than one polypeptide.

gene marker a gene used as a label in genetic engineering because its effects can be easily recognised.

gene pool total number of *alleles* in a particular *population* at a specific time.

gene technology general term that covers the processes by which *genes* are manipulated, altered or transferred from organism to organism. Also known as *genetic engineering*.

gene therapy a mechanism by which genetic *diseases* such as *cystic fibrosis* may be cured or treated by masking the effect of the defective *gene* by inserting a functional gene.

generator potential *depolarisation* of the membrane of a receptor cell as a result of a stimulus.

genetically modified organism (GMO) organism that has had its DNA altered as a result of *gene technology*.

genetic engineering see *gene technology*.

genome all of the genetic material of an organism.

genotype the genetic composition of an organism.

global warming the recent increase in average temperatures at the Earth's surface, thought to be the result of the increased production of *greenhouse gases* such as carbon dioxide and methane. These gases help to trap solar radiation at or near the Earth's surface.

gibberellins *plant growth regulators* that stimulate cell division, stem elongation and germination.

glomerulus a cluster of blood capillaries enclosed by the *renal (Bowman's) capsule* in the kidney.

glucagon a hormone produced by α cells of the *islets of Langerhans* in the pancreas that increases blood glucose levels by initiating the breakdown of glycogen to glucose.

glycolysis first part of cellular respiration in which glucose is broken down anaerobically in the cytoplasm to two molecules of pyruvate.

glycoprotein substance made up of a carbohydrate molecule and a protein molecule. Part of the cell surface membrane and certain hormones are glycoproteins.

goblet cell mucus-producing cell found in the epithelium of the intestines and bronchi, so called because its shape resembles a wine glass or goblet.

granum (plural **grana**) a stack of *thylakoids* in a chloroplast that resembles a pile of coins.

greenhouse gases gases such as methane and carbon dioxide which in the atmosphere cause more heat energy to be trapped, so raising the temperature at the Earth's surface.

gross primary production the total energy production in the form of *biomass* made by *producers* (plants) during photosynthesis. It is normally expressed as the biomass per unit area in unit time. See also *net primary production*.

guard cell one of a pair of cells that surround a *stoma* in plant leaves and control its opening and closing.

habitat the place where an organism normally lives and which is characterised by physical conditions and the types of other organisms present.

haemoglobin globular protein in mammalian blood that readily combines with oxygen to transport it around the body. It comprises four polypeptide chains around an iron-containing haem group. See also *myoglobin*.

haploid term referring to *eukaryotic* cells that contain only a single copy of each *homologous chromosome* e.g. the sex cells or gametes.

herbaceous term applied to non-woody plants.

heterotrophic nutrition form of feeding in which the organism consumes complex organic material. See also *autotrophic nutrition*.

heterozygous condition in which the *alleles* at a particular *gene locus* of a *diploid* cell are different.

histones proteins associated with the DNA in *chromosomes*. Their function is to condense *chromatin* and coil the chromosomes during cell division.

homeostasis the maintenance of a more or less constant internal environment despite fluctuations in the external environment and involving negative feedback mechanisms.

homologous chromosomes a pair of *chromosomes,* one maternal and one paternal, that have the same gene loci and therefore determine the same features. They are not necessarily identical, however, as individual *alleles* of the same *gene* may vary, e.g. one chromosome may carry the allele for normal haemoglobin (*HbA*), the other the allele for sickle cell haemoglobin (*HbS*). Homologous chromosomes are capable of pairing during *meiosis* I.

homozygous condition in which the *alleles* at a particular *gene locus* of a *diploid* cell are identical.

human genome all of the DNA sequences on the *chromosomes* of a single human cell.

hydrogen bond chemical bond formed between the positive charge on a hydrogen atom and the negative charge on another atom of an adjacent molecule, e.g. between the hydrogen atom of one water molecule and the oxygen atom of an adjacent water molecule.

hydrolysis the breaking down of large molecules into smaller ones by the addition of water molecules. See also *condensation reaction*.

hydrophyte plant that is adapted to live in water or wet habitats. See also *xerophyte*.

hypertension persistent high blood pressure when a person is at rest.

hypertonic a solution that has a higher solute concentration, and therefore a lower water potential, than another solution. See also *hypotonic*.

hyperventilation breathing deeper and faster than normal resulting in excessive removal of carbon dioxide from the blood.

hyphae thread-like structures that make up the body mass of a fungus.

hypothalamus region of the brain adjoining the *pituitary gland* that acts as the control centre for the *autonomic nervous system* and regulates body temperature, fluid balance, thirst, hunger and sexual activity.

hypotonic a solution that has a lower solute concentration, and therefore a higher water potential, than another solution. See also *hypertonic*.

immune response the complex series of reactions of the body to an *antigen* by which the body protects itself from infection.

immunisation an artificial means of producing *immunity,* either by injection of *antibodies* (passive immunity) or by inducing the body to produce its own antibodies (active immunity).

immunity the means by which the body protects itself from infection.

incidence the number of new cases of a *disease* in a *population* in a given time, e.g. in one month or one year.

industrial melanism the evolutionary process in which the frequency of organisms that are initially light coloured becomes less, and the frequency of those that are dark increases, as a result of natural selection in areas blackened by pollution.

infectious disease *disease* caused by a *pathogen* that can be transmitted from one organism to another.

insulin a hormone, produced by the β cells of the *islets of Langerhans* in the pancreas, that decreases blood glucose levels by, amongst other things, increasing the conversion of glucose to glycogen.

intraspecific competition competition between organisms of the same species.

intrinsic proteins proteins of the cell surface membrane that completely or partially span the *phospholipid* bilayer from one side to the other.

introns portions of DNA within a *gene* that do not code for a polypeptide. The introns are removed from messenger RNA after *transcription*.

ion an atom or group of atoms that have lost or gained one or more *electrons*. Ions therefore have either a positive or negative charge. See also *anion* and *cation*.

ion channel a passage across a cell surface membrane made up of a protein that spans the membranes: some open and close to allow *ions* to pass in and out of the cell.

islets of Langerhans groups of cells in the pancreas comprising large α cells that produce the hormone *glucagon*, and small β cells that produce the hormone *insulin*.

isotonic solutions that possess the same concentration of solutes and therefore have the same *water potential*.

isotope variations of a chemical element that have the same number of *protons* and *electrons* but different numbers of neutrons. While their chemical properties are similar they differ in mass. One example is carbon which has a relative atomic mass of 12 and an isotope with a relative atomic mass of 14.

kinetic energy energy that an object possesses due to its motion.

Krebs cycle series of biochemical reactions in most *eukaryotic* cells by which energy is obtained through the oxidation of acetyl coenzyme A produced from the breakdown of glucose.

leach process in which chemicals are removed from soil by being dissolved in rainwater and washed away.

light dependent reaction stage of photosynthesis in which light energy is required to produce ATP and reduced NADP.

light independent reaction stage of photosynthesis which does not require light energy directly but does need the products of the *light dependent reaction* to reduce carbon dioxide and so form carbohydrate.

lignin a complex, non-carbohydrate polymer associated with cellulose in plant cell walls. Lignin makes the cell walls stronger, allowing them to resist tension and compression. It also makes them more waterproof.

link reaction the process linking *glycolysis* with *Krebs cycle* in which pyruvate is dehydrogenated and decarboxylated to form acetyl coenzyme A in the matrix of the mitochondria.

locus the position of a *gene* on a *chromosome*.

loop of Henlé the portion of a *nephron* that forms a hairpin loop which extends into the medulla of the kidney. It has a role in the reabsorption of water.

low density lipoprotein (LDL) a compound containing both protein and lipid molecules that occurs in blood plasma and *lymph*. The proportion of protein is lower than in high density lipoproteins.

lumen the hollow cavity inside a tubular structure such as the gut or a *xylem vessel*.

lymph a slightly milky fluid found in lymph vessels and made up of *tissue fluid*, fats and *lymphocytes*.

lymphocytes type of white blood cell responsible for the *immune response*. They become activated in the presence of *antigens*. There are two types: *B lymphocytes* and *T lymphocytes*.

lysis the breakdown of a cell or compound.

magnification the size of the image of an object compared to the actual size.

meiosis the type of nuclear division in which the number of *chromosomes* is halved. In a cell with two sets of chromosomes (diploid), meiosis results in daughter cells, each with one set of chromosomes (haploid).

memory cells type of *lymphocyte* that circulates in the blood and *tissue fluid* long after the original *antigen* that caused them to develop has gone. They are reactivated when the same antigen returns, triggering an immediate secondary *immune response*.

menstrual cycle cycle in female humans, apes and monkeys of reproductive age during which the body is prepared for pregnancy. It differs from the *oestrous* cycle in other mammals in that, where fertilisation does not occur, the lining of uterus is shed along with a little blood in a process called menstruation.

mesophyll tissue found between the two layers of *epidermis* in a plant leaf comprising an upper layer of *palisade* cells and a lower layer of *spongy* cells.

messenger RNA form of ribonucleic acid that carries the information held by DNA, in the form of a nucleotide sequence, from the nucleus to the ribosomes in the cytoplasm, where it acts as a template on which polypeptides are assembled.

metabolism all the chemical processes that take place in living organisms.

metastasis process in which cancer cells break free from the original (primary) tumour and are carried by blood or *lymph* system to other parts of the body, where they may start another (secondary) tumour.

mitosis the type of nuclear division in which the daughter cells have the same number of *chromosomes* as the parent cell.

monocotyledonous plant any member of the class of flowering plants called Monocotyledonae. Their features include having a single seed leaf (cotyledon) and leaves that are parallel veined. See also *dicotyledonous plants*.

motor neurone *neurone* that transmits *action potentials* from the central nervous system to an effector such as a muscle or gland.

multiple alleles term used to describe a *gene* that has more than two possible *alleles*.

mutagen any agent that induces a *mutation*.

mutation a permanent change in the amount or arrangement of a cell's DNA. A gene mutation is a change in the sequence of nucleotides that may result in an altered polypeptide.

mutualism a nutritional relationship between two species in which both gain some advantage.

mycelium a mass of fungal hyphae.

myelin a fatty substance that surrounds *axons* and *dendrites* in certain *neurones*.

myocardial infarction otherwise known as a heart attack, results from the interruption of the blood supply to the heart muscle causing damage to an area of the heart with consequent disruption to its function.

myoglobin red-coloured pigment found in muscle and used to store oxygen. See also *haemoglobin*.

myosin the thick filamentous protein found in skeletal muscle.

NAD (nicotinamide adenine dinucleotide) a molecule that carries high energy *electrons* and hydrogen ions from oxidised molecules to pathways that produce *ATP* during *aerobic* respiration.

natural immunity protection against infection so that the person does not become ill: it is provided by the transfer of antibodies from a mother to her fetus or through breastfeeding or is acquired as a result of having an infection.

negative feedback a series of changes, important in *homeostasis*, that result in a substance being restored to its normal level. See also *positive feedback*.

nephron basic functional unit of the mammalian kidney responsible for the formation of urine.

net primary production the rate at which material produced during photosynthesis is built up in a plant. Also known as the net assimilation rate, it is the *gross primary production* less the 20% or so used by the plant in processes such as respiration.

neuromuscular junction a *synapse* that occurs between a *neurone* and a muscle.

neurone a nerve cell, comprising a cell body, *axon* and *dendrites*, that is adapted to conduct *action potentials*.

neurotransmitter one of a number of chemicals that are involved in communication between adjacent nerve cells or between nerve cells and muscles. Two important examples are *acetylcholine* and noradrenaline.

neutron uncharged sub-atomic particle that occurs in the nucleus of an atom.

niche the functional role or place of a *species* within an *ecosystem*. It includes all of the ranges of environmental conditions and resources required for an organism to survive, reproduce and maintain a viable *population*.

nitrifying bacteria microorganisms that convert ammonium compounds to nitrites and nitrates or convert nitrites to nitrates.

nitrogen-fixing bacteria group of microorganisms that incorporate atmospheric nitrogen into nitrogen-containing compounds using the enzyme nitrogenase. They may be either free-living or act in conjunction with leguminous plants.

node of Ranvier a gap in the *myelin* sheath that surrounds the *axon* of a *neurone*.

non-infectious disease disease that is not caused by a *pathogen* e.g. genetic diseases.

non-self substances (usually proteins) that are not recognised by the immune system and can therefore trigger an *immune response* in the body.

nucleotides complex chemicals made up of an organic base, a sugar and a phosphate. They are the basic units of which the nucleic acids DNA and RNA are made.

oedema swelling due to the accumulation of fluid beneath the skin or in body cavities.

oestrus the period in the oestrous cycle immediately after ovulation when the female is most fertile.

operator portion of DNA next to structural *genes* of an operon that switches them on or off.

operon a group of *genes* that codes for one or more proteins and controls its/their production.

oral rehydration therapy (ORT) means of treating dehydration, involving giving, by mouth, a balanced solution of salts and glucose that stimulates the gut to re-absorb water.

osmosis the passage of water from a region where it is at a higher water potential to a region where its water potential is lower, through a partially permeable membrane.

osteoarthritis degeneration of the *cartilage* of the joints causing pain and stiffness of these joints.

osteoporosis condition in which bones lose some of their bony substance and so become brittle and fragile. Causes include a lack of calcium or vitamin D in the diet, old age, and a reduction in oestrogen levels in women, notably post-menopausal ones.

oxidation chemical reaction involving the loss of *electrons*.

oxidative phosphorylation the formation of *ATP* in the electron transport system of *aerobic* respiration.

oxygen debt the quantity of oxygen needed to oxidise the lactate that accumulates during *anaerobic* respiration.

palisade mesophyll cells long, narrow cells, packed with chloroplasts, that are found in the upper region of a leaf and which carry out photosynthesis. See also *spongy mesophyll*.

pandemic describes any *disease* that spreads over vast areas of the world, e.g. HIV/AIDS. See also *endemic* and *epidemic*.

parasite an organism that lives on or in a host organism. The parasite gains a nutritional advantage and the host is harmed in some way.

passive immunity protection from infection so that the person does not become ill: it is acquired from the introduction of *antibodies* from another individual, rather than an individual's own immune system, e.g. across the placenta or in the mother's milk. It is usually short-lived.

pathogen any biological agent that causes disease.

peristaltic describes a wave of contraction and relaxation of muscles that move contents along a muscular tube.

phagocytosis form of endocytosis by which cells transport large particles across the cell surface membrane into the cell.

phenotype the physical, detectable expression of the particular *alleles* of an organism resulting from both its *genotype* and the effects of the environment.

phospholipid lipid molecule in which one of the three fatty acid molecules is replaced by a phosphate group, attached to a simple organic molecule such as choline. Phospholipids are important in the structure and functioning of the plasma membranes.

photolysis splitting of a water molecule by light such as occurs during the *light dependent reaction* of photosynthesis.

photon 'particle' of light with a quantum of energy.

photosystem an organised group of chlorophyll and other pigment molecules situated in the *thylakoids* of chloroplasts that traps *photons* of light in a process called light harvesting.

phytoplankton small, often microscopic, photosynthesising organisms that live suspended in large bodies of water such as oceans and lakes. See also *zooplankton*.

pinocytosis form of *endocytosis* by which cells take up liquids from their environment.

pituitary gland master gland of the endocrine (hormone) system situated at the base of the brain.

plaque deposits of fatty material, fibrous tissue and dead cells that form large irregular patches on the lining of the inner walls of arteries causing a condition called *atherosclerosis*.

plant growth regulator (plant hormone) chemicals produced by plants in tiny quantities that affect their growth or development. Examples include *auxins*, *gibberellins*, *abscisic acid* and *cytokinins* and *ethene*.

plasmid a small circular piece of DNA found in bacterial cells and often used as a *vector* in *gene technology*.

plasmodesmata fine strands of cytoplasm that extend through pores in adjacent cell walls and connect the cytoplasm of one cell with another.

platelets cells found in blood which play an important role in blood clotting.

podocyte cell from the inner lining of the *renal capsule* that has many processes and is adapted to help *ultrafiltration*.

polymer large molecule made up of repeating sub-units.

polymerase chain reaction (PCR) process of making many copies of a specific sequence of DNA or part of a *gene*. It is used extensively in *gene technology* and genetic fingerprinting.

polymerases group of enzymes that catalyse the formation of long-chain molecules (*polymers*) from similar basic units (monomers).

polymerisation production of large molecules called polymers that are made of numerous similar sub-units.

polyploidy the possession of three or more sets of *chromosomes*.

polyunsaturated fatty acid (PUFA) fatty acid that possesses carbon chains with many double bonds.

population all the organisms of a particular *species* that occur in the same place at the same time.

positive feedback process which results in a substance that departs from its normal level becoming further from its norm. See also *negative feedback*.

prevalence the number of people in a *population* who have a particular *disease* at a particular time.

producer an *autotrophic* organism that synthesises organic molecules from simple inorganic ones such as carbon dioxide and water. Most producers are photosynthetic and form the first *trophic level* of a food chain. See also *consumer*.

prokaryote an organism belonging to the kingdom Prokaryotae that is characterised by having cells less than 5 μm in diameter which lack a nucleus and double membrane-bound organelles. Examples include bacteria and blue-green bacteria. See also *eukaryote*.

promoter the portion of an *operon* to which *RNA polymerase* attaches to begin the *transcription* of mRNA from the structural *genes*.

protoctist an organism belonging to the kingdom Protoctista, many of which are single celled *eukaryotes* such as certain algae and *protozoa*.

proton positively charged sub-atomic particle found in the nucleus of the atom. See also *electron*.

protoplast the living portion of a plant cell, i.e. the nucleus and cytoplasm along with the organelles it contains.

protozoa a sub-group of the kingdom *Protoctista* made up of single-celled organisms such as *Amoeba* and *Plasmodium*.

Purkyne tissue a region of specialist heart muscle that conducts a wave of excitation that causes contraction of the ventricles.

reaction centre a molecule of chlorophyll *a* that collects light energy that has been absorbed from the surrounding accessory pigments in the *photosystem*.

recessive the condition in which the effect of an *allele* is apparent in the *phenotype* of a *diploid* organism only in the presence of another identical allele. See also *dominant*.

reduction division see *meiosis*.

reflex arc the nerve pathway in the body taken by an *action potential* that leads to a rapid, involuntary response to a stimulus.

refractory period period during which the membrane of a *neurone* cannot be *depolarised* and no new *action potential* can be initiated.

renal capsule the cup-shaped portion at the start of a *nephron* that encloses the *glomerulus*.

repolarisation return of the *resting potential* in a *neurone* after an *action potential*.

resolution ability of a microscope to distinguish two objects as separate from each other.

respiratory quotient (RQ) a measure of the ratio of carbon dioxide given out by an organism to the oxygen taken in over a certain period.

resting potential the difference in electrical charge maintained across the cell membrane of an *axon* when not stimulated.

restriction endonucleases a group of enzymes that are able to cut DNA into shorter lengths at specific points. Found naturally in certain bacteria, they are important in *gene technology*.

reverse transcriptase enzyme capable of producing a DNA molecule from the corresponding *messenger RNA*. Found in many viruses, reverse transcriptase is used in *gene technology*.

rheumatoid arthritis autoimmune *disease* in which there is *chronic* inflammation of the joints leading to disability and disfigurement.

rickets a deficiency *disease* in children, caused by a lack of vitamin D, in which the bones are weakened due to a lack of calcium, leading to bowing of the legs.

RNA polymerase enzyme that joins together RNA *nucleotides* to form messenger RNA during *transcription*.

saltatory conduction propagation of an nerve impulse along a *myelinated dendron* or *axon* in which the *action potential* jumps from *node of Ranvier* to node of Ranvier.

sarcomere a section of myofibril between two Z-lines that forms the basic unit of striated muscle.

Schwann cell cell around a *neurone* whose cell surface membrane wraps around the *dendron* or *axon* to form the *myelin* sheath.

sclerenchyma plant tissue whose cells have become rigid due to the presence of cell walls thickened with *lignin*. The cells are dead and function to provide support to the plant.

selection process that results in the best adapted individuals in a *population* surviving to breed and so pass their favourable *alleles* to the next generation.

self substances that are the product of the body's own *genes* and therefore do not trigger an *immune response* in that body.

semi-conservative replication the means by which DNA makes exact copies of itself by unwinding the double helix so that each strand acts as a template for the next. The new copies therefore possess one original and one new strand of DNA.

sensory neurone a *neurone* that transmits an *action potential* from a sensory receptor to the central nervous system.

serum clear liquid that is left after blood has clotted and the clot has been removed. It is therefore blood plasma without the clotting factors.

sickle cell anaemia inherited blood disorder in which abnormal *haemoglobin* leads to red cells becoming sickle-shaped and less able to carry oxygen.

sieve plate the perforated end wall of the phloem component called the sieve tube element.

sieve tube elements part of phloem tissue, elongate cells that join end to end to form tube-like structures for the transport of dissolved organic substances.

sinoatrial node an area of *cardiac muscle* in the right atrium that controls and coordinates the contraction of the heart. Also known as the pacemaker.

smooth muscle also known as involuntary or unstriated muscle, smooth muscle is not under conscious control.

sodium–potassium pump carrier proteins across cell surface membranes that use *ATP* to move sodium *ions* out of the cell in exchange for potassium ions that move in.

speciation the evolution of two or more *species* from existing ones.

species a group of reproductively-isolated organisms which interbreed to produce fertile offspring.

spongy mesophyll cells irregularly shaped photosynthetic cells in the lower section of the leaf, below the palisade mesophyll layer. They have many air-spaces between them and are important in exchanging gases between the atmosphere and the rest of the leaf. See also *palisade mesophyll cells*.

stabilising selection selection that tends to eliminate the extremes of the *phenotype* range within a *population*. It occurs when environmental conditions are constant.

stem cells undifferentiated cells that can divide by mitosis to form new stem cells and can form specialised cells during development.

steroid a lipid, of which cholesterol is an example, that does not contain fatty acids.

stoma (plural stomata) pore, mostly in the lower epidermis of a leaf, through which gases diffuse in and out of the leaf.

stroke volume the volume of blood pumped at each ventricular contraction of the heart.

stroma fluid matrix of a chloroplast where the *light independent reaction* of photosynthesis takes place.

suberin a waxy, waterproof substance found in certain plant cell walls like endodermal cells of a root.

supernatant liquid the liquid portion of a mixture left at the top of the tube when suspended particles have been separated out at the bottom during *centrifugation*.

sympatric speciation the formation of new *species* that occurs when organisms that are living together become reproductively isolated, e.g. different *populations* may have different breeding seasons. See also *allopatric speciation*.

symplast pathway route through the cytoplasm, vacuoles and *plasmodesmata* of plant cells by which water and dissolved substances are transported. See also *apoplast pathway*.

synapse a junction between *neurones* in which they do not touch but have a narrow gap, the synaptic cleft, across which a *neurotransmitter* can pass.

systole the stage in the *cardiac cycle* in which the heart muscle contracts. It occurs in two stages: atrial systole when the atria contract and ventricular systole when the ventricles contact. See also *diastole*.

telomere ends of chromosomes (and *chromatids*) where a sequence of *nucleotides* is repeated many times.

tendons tough, flexible connective tissue that joins muscle to bone.

threshold level/value the minimum intensity that a stimulus must reach in order to trigger an *action potential* in a *neurone*.

thrombosis formation of a blood clot within a blood vessel that may lead to a blockage.

thylakoid flattened membranous sac in a chloroplast that contains chlorophyll and the associated molecules needed for the *light dependent reaction* of photosynthesis.

tidal volume the volume of air breathed in and out during a single breath when at rest.

tissue of cells of one (e.g. squamous epithelium) or a few types (e.g. phloem), specialised to perform a particular function.

tissue fluid fluid that surrounds the cells of the body. Its composition is similar to that of blood plasma except that it lacks some of the larger proteins, in particular those that cause the blood to clot. It supplies nutrients to the cells and removes waste products.

T lymphocyte type of white blood cell that is produced in the bone marrow but matures in the thymus gland. T lymphocytes coordinate the immune response and kill infected cells. See also *B lymphocyte*.

totipotent having the ability to differentiate into all cell types.

transcription the formation of *messenger RNA* molecules from the DNA that makes up a particular *gene*. It is the first part of protein synthesis.

transduction the process by which one form of energy is converted into another. In microbiology, the natural process by which genetic material is transferred between one host cell and another by a virus.

translation process whereby the code on a section of messenger RNA is converted to a particular sequence of amino acids that will go on to make a polypeptide and ultimately a protein.

translocation process where the information in the nucleotide sequence of a messenger RNA molecule is converted to a particular sequence of amino acids on a polypeptide chain: processing after translation forms the functioning protein.

transpiration the loss of water vapour from the aerial parts of a plant, usually the leaves: evaporation from the surfaces of mesophyll cells is followed by diffusion out of water vapour to the atmosphere down the water potential gradient.

triglyceride an individual fat or oil molecule made up of a glycerol molecule and three fatty acids.

trophic level the position of an organism in a food chain. See also *producer* and *consumer*.

tumour a swelling in an organism that is made up of cells which continue to divide in an abnormal way.

tumour suppressor genes *genes* that code for proteins that repress the cell cycle.

ultrafiltration filtration under pressure. A term applied to the first stage of urine formation in the kidney.

urea organic molecule, $CO(NH_2)_2$, that is the main nitrogenous excretory product in mammals.

vaccination the introduction of a vaccine containing appropriate *disease antigens* or *antibodies* into the body, by injection or mouth, in order to induce *artificial immunity*.

vascular tissue any tissue that forms a network of vessels through which fluids are transported in organisms. Blood vessels are the usual vascular tissue of vertebrates, while in plants it is phloem and xylem.

vasoconstriction narrowing of the internal diameter of blood vessels. See also *vasodilation*.

vasodilation widening of the internal diameter of blood vessels. See also *vasoconstriction*.

vector a carrier. The term may refer to something such as a *plasmid,* which carries DNA into a cell, or to an organism that carries a parasite to its primary host.

vegetative propagation form of asexual reproduction in higher plants involving the separation of a piece of the original plant (stem, root or leaf), which then develops into a separate plant.

virulent able to cause disease or cause harm. The term is applied to a *disease* that spreads rapidly through a population.

vital capacity the total volume of air that can be forced out of the lungs after a maximum inspiration.

voltage-gated channel channel protein across a cell surface membrane that opens and closes according to changes in the electrical potential across the membrane.

water potential the tendency of water to move from one area to another. It is a measure of potential energy and is expressed in units of pressure. The greater the number of water molecules present, the higher (less negative) the water potential. Pure water has a water potential of zero.

xerophyte a plant adapted to living in dry conditions. See also *hydrophyte*.

zooplankton small animals, often microscopic, that live suspended in large bodies of water such as oceans and lakes. See also *phytoplankton*.

Answers to Summary Tests

Summary Test 1.1

1 100
2 100 nm
3 separate / distinct
4 μm / microns
5 electron
6 clear / precise

Summary Test 1.2

1 phospholipids
2 chromatin
3 mitochondria
4 glycogen
5 starch
6 cell wall
7 cellulose
8 cell sap
9 tonoplast

Summary Test 1.3

1 0.1 nm
2 electromagnets
3 air
4 transmission
5 scanning

Summary Test 1.4

1 light
2 eyepiece
3 stage
4 calibrated
5 20
6 3000

Summary Test 1.6

1 nuclear envelope
2 40–100 nm
3 nucleolus
4 ribosomal
5 grana / thylakoids
6 chlorophyll
7 }
8 } Krebs cycle; oxidative phoshorylation
9 matrix
10 stalked / elementary
11 cristae

Summary Test 1.8

1 protein
2 } glycoproteins;
3 } cholesterol
4 2
5 nuclear division
6 spindle
7 chromosomes

Summary Test 1.10

1 nucleus
2 bacteria
3 circular
4 eukaryotic
5 70S
6 peptidoglycan/murein

Summary Test 2.3

1 amylose
2 glycogen
3 amylopectin
4 α-glucose
5 amylose, amylopectin, glycogen
6 amylopectin
7 amylose, amylopectin, glycogen
8 amylopectin, glycogen
9 α-glucose, β-glucose

Summary Test 2.4

1 β-glucose
2 1,4 glycosidic
3 hydrogen bonds
4 microfibrils
5 structural / support
6 bursting
7 osmosis
8 turgid / semi-rigid
9 photosynthesis

Summary Test 2.5

1 triglycerides
2 glycerol
3 polyunsaturated
4 2
5 hydrophobic
6 hydrogen
7 energy

Summary Test 2.6

1 —COOH
2 amino
3 amphoteric
4 condensation
5 peptide
6 hydrogen
7 α-helices
8 β-pleated sheets

Summary Test 2.7

1 amino acids
2 polypeptide
3 hydrogen bonds
4 —CO
5 amino acid
6 sulfur
7 disulfide bridges
8 ionic bonds
9 hydrophobic interactions
10 prosthetic

Summary Test 2.8

1 globular
2 fibrous
3 glycine – proline – alanine
4 tendons
5
6 } strength; flexibility
7 4
8 haem
9 Fe^{2+} / ferrous ion

Summary Test 2.9

1 dipolar
2 electrons
3 hydrogen bonds
4 specific heat capacity
5 latent heat of vaporisation

Summary Test 3.1

1 catalysts
2 globular
3 active site
4 activation energy
5 substrate
6 products
7 lock and key
8 induced fit

Summary Test 3.2

1 time course
2 disappearance
3 product
4 maltose
5 starch
6 increase
7 active site
8 denatured
9 40 °C
10 2
11 salivary amylase

Summary Test 3.3

1 initial rate of reaction
2 increases
3 active sites
4 halved
5 be constant

Summary Test 4.1

1 bilayer
2 hydrophilic
3 hydrophobic
4 intrinsic / integral
5 extrinsic / peripheral
6
7 } cholesterol; glycolipids; glycoproteins
8

Summary Test 4.2

1 higher / greater
2 lower / less
3 kinetic
4 passive
5 less / reduced / lower
6 less / reduced / lower
7 more / greater / higher
8 carrier proteins/channel proteins

Summary Test 4.3

1 water
2 selectively permeable
3 pressure
4 zero
5 lower / more negative
6 B
7 passive

Summary Test 4.4

1 higher / less negative
2 (solute) concentration / water potential
3 osmosis
4 protoplast
5 (cellulose) cell wall
6 higher / less negative
7 plasmolysis

Summary Test 4.5

1 against
2 ATP
3 mitochondria
4 respiratory
5 cytosis
6 endocytosis
7 exocytosis
8 phagocytosis
9 pinocytosis

Summary Test 5.2

1 interphase
2 prophase
3 centrioles
4 poles
5 spindle apparatus
6 nuclear envelope
7 nucleolus
8 metaphase
9 equator
10 anaphase
11 centromere
12 chromatids / daughter chromosomes
13 telophase

Summary Test 5.3

1 mutation
2 red blood
3
4 } growth; repair; asexual reproduction
5
6 stem cells
7 tumour

Summary Test 6.1

1
2 } nitrogen; phosphorus
3 pentose
4 5
5
6 } ribose; deoxyribose
7 pyrimidines
8
9 } cytosine; uracil
10 purines
11 5
12
13 } guanine; adenine
14 thymine
15 ribosomal RNA
16 transfer RNA
17 messenger RNA

403

Summary Test 6.2

1 pentose
2 deoxyribose
3 phosphate
4 adenine

5 cytosine
6 antiparallel
7 double helix

Summary Test 6.3

1 polynucleotide
2 helicase
3 replication forks
4 DNA polymerase

5 complementary
6 DNA ligase
7 semi-conservative

Summary Test 6.4

1 ^{14}N
2 ^{15}N
3 isotope
4 control
5 centrifuged

6 } light; hybrid
7 }
8 light
9 hybrid
10 semi-conservative

Summary Test 6.5

1 gene
2 3
3 codon
4 3.2 billion (3.2×10^9)
5 degenerate
6 non-overlapping

7 stop
8 Tyr (tyrosine)
9 } ATA; ATG (complementary
10 } to UAU and UAC on mRNA)
11 mutation

Summary Test 6.6

1 messenger RNA
2 cistron
3 helicase
4 template
5 RNA polymerase

6 uracil
7 guanine
8 adenine
9 introns
10 nucleus

Summary Test 6.7

1 translation
2 ATP
3 amino-acyl tRNA

4 ribosome
5 codon

Summary Test 7.2

1 0.6
2 0.3 / halved
3 increase
4 2
5 double

6 osmosis
7 less negative / higher
8 active transport

Summary Test 7.4

1 vessels
2 lignin
3 } annular; reticulate;
4 } spiral
5 }

6 lumen
7 pits
8 (sclerenchyma) fibres

Summary Test 7.5

1 root hairs
2 lower (more negative)
3 apoplast
4 symplast
5 plasmodesmata
6 endodermis

7 Casparian strip
8 suberin
9 protoplast
10 mineral
11 decreases/lowers
12 root pressure

Summary Test 7.6

1 evaporates
2 stomata
3 guard cells
4 lower / more negative

5 water potential
6 cohesion
7 cellulose
8 adhesion

Summary Test 7.7

1 evaporation
2 stomata
3 cuticle
4 potometer
5 watertight
6 decrease

7 decrease
8 decrease
9 increase
10 decrease
11 increase

Summary Test 7.8

1 } sand dunes; salt marshes
2 }
3 cuticle
4 } hairy leaves; stomata in pits or grooves;
5 } leaves that roll up
6 }

Summary Test 7.9

1 } sugars; amino acids
2 }
3 storage
4 sieve plates
5 phloem-protein
6 }
7 } nucleus; Golgi body; ribosome
8 }
9 companion cells
10 plasmodesmata

Summary Test 7.10

1 phloem
2 sources
3 sinks
4 mass flow
5 companion cell
6 apoplast / cell wall

7 cotransport
8 lower / more negative
9 xylem
10 higher / less negative
11 osmosis
12 hydrostatic

Summary Test 8.1

1 closed
2 away from
3 } muscle; elastic fibres
4 }
5 capillaries
6 arteries
7 lumen
8 arteries
9 tunica intima/endothelium
10 left ventricle
11 aorta
12 carotid artery
13 liver
14 hepatic portal vein
15 hepatic vein
16 vena cava
17 right atrium

Summary Test 8.2

1 plasma
2 red (blood) cells
3 7–8 μm
4 120 days
5 haemoglobin
6 oxygen
7 white (blood) cells
8 phagocytes
9 monocyte / neutrophil
10 antibodies
11 lymphocytes

Summary Test 8.3

1 plasma
2 hydrostatic
3 arterial
4 ultrafiltration
5 } fatty substances; lymphocytes
6 }

Summary Test 8.4

1 respiratory pigment
2 68 000
3 4
4 iron (Fe^{2+})
5 polypeptides
6 4
7 tension / partial pressure
8 oxygen dissociation curve

Summary Test 8.5

1 respiration
2 carbaminohaemoglobin;
3 hydrogen
4 10
5 5
6 plasma
7 85
8 hydrogencarbonate ions
9 carbonic acid
10 carbonic-anhydrase
11 haemoglobinic acid
12 buffer
13 Bohr effect

Summary Test 8.6

1 21%
2 one-third (or less)
3 acclimatise
4 hypoxia
5 erythopoietin
6 red blood cell production

Summary Test 8.7

1 atria
2 ventricles
3 left atrioventricular / bicuspid / mitral
4 right atrioventricular / tricuspid
5 pulmonary veins
6 left atrium
7 aorta
8 right atrium

Summary Test 8.8

1 70
2 diastole
3 atrial systole
4 atrioventricular
5 ventricles
6 semilunar
7 pulmonary artery
8 aorta

Summary Test 8.9

1 myogenic
2 sinoatrial node
3 right atrium
4 atria
5 atrioventricular node
6 Purkyne tissue

Summary Test 8.10

1 16–18.5 kPa
2 10.5–12.0 kPa
3 rises / increases
4 systemic
5 aorta
6 arterioles
7 veins
8 fall / decrease
9 rise / increase
10 hypertension

Summary Test 9.1

1 C
2 goblet
3 } bacteria; pollen;
4 } dust (any 2)
5 cilia
6 bronchioles
7 smooth muscle
8 elastic fibres

Summary Test 9.2

1	100–300 μm	5	surface area
2	squamous epithelium	6	pulmonary
3	0.1–0.5 μm	7	} slowed; flattened
4	thin	8	

Summary Test 9.3

1	nicotine	5	mucus
2	} bronchi; bronchioles	6	pathogens
3		7	carcinogens
4	cilia	8	mitosis

Summary Test 9.4

1 chronic obstructive pulmonary disease
2 breathlessness
3 elastase
4 elastin
5 epithelial
6 goblet
7 cilia
8 carcinogens / benzopyrene *(as an example)*
9 tumour
10 metastasis
11 ⎫
12 ⎬ persistent cough; blood in sputum; shortness of breath
13 ⎭

Summary Test 9.5

1	cholesterol	8	cerebrovascular accident
2	plaques	9	aneurysm
3	atheromas	10	haemorrhage
4	lumen	11	carbon monoxide
5	thrombus	12	haemoglobin
6	oxygen	13	adrenaline
7	myocardial infarction	14	blood pressure

Summary Test 9.6

1	mental	6	cirrhosis
2	ethanol	7	10
3	liver	8	21
4	} hepatitis; fatty liver	9	14
5			

Summary Test 9.7

1 mass (in kilograms)
2 height (in metres)
3 30
4 20
5 blood pressure / hypertension
6 cholesterol
7 type II / mature onset
8 ⎫
9 ⎬ osteoarthritis; rheumatoid arthritis
10 salt
11 saturated

Summary Test 10.1

1 ⎫
2 ⎬ social; mental; physical
3 ⎭
4 multifactorial
5 acute
6 chronic
7 parasite
8 pathogens
9 infectious / communicable

Summary Test 10.2

1 arthritis
2 mental
3 acquired immune deficiency syndrome
4 kwashiorkor
5 degenerative
6 Alzheimer's disease
7 inherited
8 sickle-cell anaemia
9 self-inflicted
10 deficiency
11 rickets

Summary Test 10.3

1	epidemiology	5	tuberculosis
2	prevalence	6	influenza / HIV/AIDS
3	mortality	7	pandemic
4	endemic	8	100 000

Summary Test 10.4

1	pathogens	6	sanitation
2	virulence	7	vaccination
3	*Vibrio cholerae*	8	antibiotics
4	water	9	oral rehydration therapy
5	food		

Summary Test 10.5

1 protoctist (protoctistan)
2 *Anopheles*
3 vector
4 200
5 (sub-Saharan) Africa
6 larva / pupa
7
8 } insecticide; oil
9
10 } fish; bacteria *(Bacillus thuringiensis)*
11 nets
12 insect repellent / clothes
13
14 } chloroquine; proguanil

Summary Test 10.6

1 *Mycobacterium tuberculosis*
2 *Mycobacterium bovis*
3 air
4 30
5 5–10
6 over-crowded
7 HIV positive
8 Bacille-Calmette-Guerin / BCG
9 6–9
10 (multi-drug) resistant / MDR

Summary Test 10.8

1 respiratory
2 *Morbillivirus* virus
3
4 } sub-Saharan Africa; South East Asia
5 MMR
6 18 months
7 antibodies
8 *Variola*

Summary Test 10.9

1 autolysins
2 stretch
3 peptidoglycan (murein)
4 enzymes
5 peptide
6 osmotic lysis
7 penicillinase
8 mutation
9 plasmids
10 vertical gene transmission
11 horizontal gene transmission

Summary Test 11.1

1 epithelial
2 mucus
3 hydrochloric acid
4 phagocytosis
5
6 } macrophages; neutrophils
7 marrow
8 long
9 opsonins
10 lysosomes

Summary Test 11.2

1 foreign / non-self
2 protein
3 immunoglobulins
4 B lymphocytes
5 heavy
6 light
7 disulfide bridges
8 binding sites
9 toxins
10 lysis

Summary Test 11.3

1 B
2 hybridoma
3 clone
4 diagnosing
5 herceptin
6 cytotoxic
7 human chorionic gonadotrophin (hCG)
8 placenta

Summary Test 11.5

1 viruses
2 cell-mediated
3 T helper
4 cytokines (lymphokines)
5 macrophage
6 phagocytosis
7 B
8 plasma
9 T cytotoxic (T killer)

Summary Test 11.6

1 natural
2 passive
3 vaccination
4
5 } living attenuated; dead;
6 } genetically engineered
7 natural passive
8 artificial active
9 artificial passive

Summary Test 12.1

1 work
2 kinetic
3 potential
4 anabolic
5 active transport
6 (high) body temperature
7 activation

Summary Test 12.3

1 cytoplasm
2 glucose
3 phosphate
4 ATP
5 fructose 1,6–bisphosphate
6 triose phosphate
7 hydrogen
8 NAD (nicotinamide adenine dinucleotide)
9 pyruvate
10 ATP

Summary Test 12.4

1 glycolysis
2 matrix
3 active transport
4 decarboxylation
5 dehydrogenation
6 acetyl coenzyme A
7 fatty acids
8 oxaloacetate
9 citrate
10 nicotinamide adenine dinucleotide (NAD)

Summary Test 13.2

1 palisade (mesophyll)
2 stomata
3 xylem
4 phloem
5 sugars / sucrose
6 envelope
7 } DNA; ribosomes
8 }
9 stroma
10 light independent
11 starch
12 thylakoids
13 grana
14 light dependent

Summary Test 13.5

1 Calvin
2 ribulose bisphosphate (RuBP)
3 glycerate 3-phosphate (GP)
4 } ATP; reduced NADP (NADPH + H$^+$)
5 }
6 light dependent
7 triose phosphate
8 hexose / six-carbon
9 polymerisation

Summary Test 13.6

1 limiting factor
2 carbon dioxide
3 oxygen
4 light compensation point
5 0.04
6 0.1

Summary Test 13.7

1 carbon dioxide
2 photorespiration
3 } maize; sorghum
4 }
5 phosphoenol pyruvate / PEP
6 bundle sheath
7 Calvin cycle
8 pyruvate

Summary Test 14.2

1 elimination /egestion
2 water
3 carbon dioxide
4 deamination
5 carbon dioxide
6 ornithine
7 (fibrous) capsule
8 cortex
9 medulla
10 nephrons
11 renal artery
12 ureter

Summary Test 14.3

1 renal (Bowman's) capsule
2 glomerulus
3 afferent
4 podocytes
5 proximal
6 cuboid
7 microvilli
8 loop of Henlé
9 distal
10 collecting duct
11 peritubular

Summary Test 14.5

1 hairpin
2 medulla
3 proximal (first) convoluted tubule
4 descending
5 descending
6 ascending
7 low
8 osmosis
9 lowest
10 counter-current multiplier
11 pH
12 } mitochondria; microvilli
13 }

Summary Test 14.6

1 2500
2 2300
3 respiration
4 urine
5 sweat
6 homeostatic
7 distal (second) convoluted tubule
8 sweating
9 fall / reduce / decrease
10 hypothalamus
11 posterior pituitary
12 permeability
13 urea
14 rises / increases
15 negative feedback

Summary Test 14.7

1 blood
2 target
3 permanent
4 steroids
5 exocrine
6 β
7 islets of Langerhans
8 α
9 glucagon
10 blood sugar (level)

Summary Test 14.8

1	respiratory substrate	11	glycogen
2	90	12	respiration
3	homeostasis	13	endocrine
4	brain	14	islets of Langerhans
5	osmotic	15	smaller
6	carbohydrates	16	insulin
7	glycogen	17	lower / reduce / fall
8	muscle	18	glucagon
9	amino acids	19	antagonistically
10	gluconeogenesis	20	adrenaline

Summary Test 14.10

1
2 } carbon dioxide, water vapour

3 ATP synthase
4 chloroplasts
5 active transport
6 potassium (K^+)
7 solute
8 reduces/decreases/becomes more negative
9 turgid
10 stress hormone
11 calcium

Summary Test 15.2

1	perception	5	generator
2	dissolved chemicals	6	action potential
3	receptor	7	all or nothing
4	transducers		

Summary Test 15.4

1	electrical	5	positively
2	50–90	6	40
3	65	7	depolarised
4	polarised		

Summary Test 15.5

1	outside	6	sodium ions
2	polarised	7	potassium
3	action potential	8	electrochemical
4	depolarised	9	repolarised
5	voltage-gated		

Summary Test 15.6

1	0.5	7	Ranvier
2	100	8	saltatory conduction
3	slower	9	(voltage-gated) channels
4	lower / less	10	absolute
5	diffusion	11	relative
6	speeds up / increases	12	threshold value

Summary Test 15.7

1	dendrite	6	synaptic vesicles
2	cleft	7	postsynaptic
3	20–30	8	unidirectional
4	presynaptic	9	junctions
5	neurotransmitter		

Summary Test 15.8

1	calcium	5	sodium
2	synaptic vesicles	6	diffusion
3	exocytosis	7	excitatory postsynaptic
4	synaptic cleft		

Summary test 15.10

1 Z-lines
2
3 } I-band, H-zone

4 action potential
5 neuromuscular junction
6 transverse system/T
7 sarcoplasmic reticulum
8 troponin
9 tropomyosin
10 myosin
11 ATP

Summary Test 15.11

1 28
2 menstruation
3 5
4 follicular
5 follicle stimulating hormone (FSH)
6 (anterior lobe of) pituitary
7 gonadotrophin releasing hormone (GnRH)
8 14
9 (secondary) oocyte
10 luteinising hormone (LH)
11 luteal
12 corpus luteum
13 fertilisation
14 endometrium

Summary Test 15.12

1	zygote	5	ovulation
2	progesterone	6	cervix
3	combined	7	endometrial
4	oestrogen	8	implantation

Summary Test 16.4

1	genotype	6	polypeptide
2	mutation	7	homozygous
3	phenotype	8	heterozygous
4	modification	9	recessive
5	nucleotides		

Summary Test 16.5

1	upper	6	monohybrid
2	lower	7	first filial / F_1
3	recessive	8	green
4	encircled / put in a circle	9	second filial / F_2
5	Punnett square	10	green
		11	yellow

Summary Test 16.6

1 phenotype
2
3 } heterozygous; homozygous dominant
4 test
5 homozygous recessive
6 alleles
7 heterozygous
8 homozygous dominant
9 large / greater

Summary Test 16.8

1	dihybrid	13	round
2	dominant	14	yellow
3	recessive	15	wrinkled
4	recessive	16	green
5	dominant	17+18 } round and yellow;	
6	round	19+20 } wrinkled and yellow	
7	yellow		
8	alleles		
9			
10			
11			
12			

9
10
11
12 } 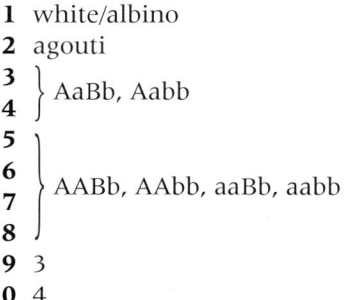 (RY); (Ry); (rY); (ry)

Summary test 16.10

1 white/albino
2 agouti
3
4 } AaBb, Aabb
5
6
7 } AABb, AAbb, aaBb, aabb
8
9 3
10 4

Summary Test 16.12

1	alleles	6	deletion
2	nucleotides	7	insertion
3	insertion	8	mis-sense
4	substitution	9	silent
5	substitution	10	nonsense

Summary Test 16.14

1	glucose	9	mRNA
2	respiratory	10	regulator
3	} glucose, galactose	11	operator
4		12	RNA polymerase
5	lac operon	13	repressor protein
6	structural	14	lactose permease
7	operator	15	β-galactosidase
8	promoter		

Summary Test 16.15

1	nucleus	6	complementary
2	DNA	7	transcription
3	off	8	gibberellins
4	oestrogen	9	DELLA
5	receptor	10	inhibit

Summary Test 17.1

1
2 } height; mass
3 normal distribution curve
4 polygenes
5 environment
6 one / a single
7 blood groups / albinism / sickle cell anaemia / Huntington's disease / haemophilia
8 recombinants
9 prophase I
10 recombinants
11 random assortment
12 metaphase I
13 mating
14 species
15 fusion of gametes

Summary Test 17.2

1 environment
2 genotype
3 enzyme
4 melanin
5 (recessive) allele
6 above
7 extremities / tips / *any suitable example*
8 black / dark
9 altitude
10 } height / number of leaves / overall size / shape /
11 } survival rate *(any two)*
12 genotype
13 range / extent / limit
14 environment

Summary Test 17.4

1 Alfred Wallace
2 more
3 constant / stable
4 intraspecific (competition)
5 genetically
6 alleles / genes
7 offspring
8 antibiotic
9 melanic
10 polymorphism

Summary Test 17.7

1 gene pool
2 allelic frequency
3 Hardy–Weinberg
4 } large, isolated
5
6 } selection, mutations
7
8 random
9 10.9% / 0.109
   $(p + q = 1.0$ and $p = 0.942$
   Therefore $q = 1.0 - 0.942 = 0.058$
   Frequency of the heterozygous
   genotype $= 2pq$
   $\qquad = 2 \times 0.942 \times 0.058$
   $\qquad = 0.109$
   As a %, $0.109 \times 100 = 10.9\%)$

Summary Test 17.8

1 gene pool
2 allelic frequency
3 selection
4 substituted / replaced
5 red (blood) cells
6 codominant
7 Hb^SHb^S
8 oxygen
9 stabilising
10 directional

Summary Test 17.9

1 speciation
2 reproductively
3 deme
4 adaptive radiation
5 allele frequencies
6 species
7 allopatric
8 sympatric
9 } prezygotic; postzygotic
10

Summary Test 17.10

1 artificial selection
2 lowered / less / reduced
3 heterozygotes
4 } inbreeding; outbreeding
5
6 pedigree
7 progeny testing
8 vagina/uterus
9 artificial insemination
10 embryo transplantation

Summary Test 17.11

1 crop yield
2 genome
3 rice
4 photosynthesis
5 grain / seed
6 harvest / collect
7 allele
8 gibberellin
9 hybridisation

Summary Test 17.12

1 mutations
2 (ecological) niches
3 reproductively isolated
4 species
5 xeromorphic
6 cuticle
7 root
8 stomata
9 loops of Henlé
10 urine

Summary Test 18.2

1 eukaryotic
2 heterotrophically
3 hyphae
4 mycelium
5 chitin
6 chlorophyll
7 autotrophically
8 cellulose
9 starch
10 glycogen
11 nervous system

Summary Test 18.3

1 individuals
2 community
3 habitats
4 genetic
5 stable
6 climatic
7 tropical rain forest
8 arctic tundra / hot desert
9 nutrients
10 soils
11 } economic; ethical; social
12 } (any two)

Summary Test 18.4

1 ecology
2 biosphere
3 biotic
4 abiotic
5 community
6 population
7 habitat

Summary Test 18.5

1 trophic level
2 photosynthesis
3 producers
4 gross primary production
5 respiration
6 net primary production
7 herbivores
8 carnivores
9 food chain
10 decomposers

Summary Test 18.6

1 3
2 20–50
3 net primary production
4 herbivores / primary consumers
5 trophic level
6 biomass
7 energy

Summary Test 18.7

1 78
2 nitrogen fixation
3 mutualistic
4 *Rhizobium*
5 nodules
6 peas / beans
7 nitrates
8 amino acids / proteins
9 decomposers
10 ammonia
11 nitrites
12 *Nitrosomonas*
13 *Nitrobacter*
14 denitrifying

Summary Test 18.8

1 density
2 abundant/common/frequent
3 cover
4 seashore
5 interrupted/ladder
6 mark–release–recapture
7 1600 (100 × 80 ÷ 5)

Summary Test 18.10

1 river banks
2 orang-utan
3 spider monkeys / tufted capuchin
4 rabbits
5 habitat / woodland
6 passenger pigeon
7 pesticides

Summary Test 18.12

1 biodiversity
2 artificial insemination
3 surrogacy
4 cryopreservation
5 }
6 } culling, contraception
7 outcompete
8 predators

Summary Test 19.2

1 automated
2 thermocycler
3 $95\,°C$
4 separate
5 $55\,°C$
6 complementary bases
7 rejoining
8 $72\,°C$
9 (complementary) nucleotides
10 2

Summary Test 19.3

1 recognition sites
2 nucleotide bases
3 DNA ligase
4 phosphate – sugar
5 vector
6 circular
7 marker
8 recombinant

Summary Test 19.5

1 insulin dependent
2 β
3 islets of Langerhans
4 autoimmune
5 protein
6 promoter
7 β-galactosidase
8 exact / identical
9 rapid
10 immune
11 infections / diseases
12 cheaper
13 tolerance
14 morally / ethically
15 haemophilia
16 adenosine deaminase

Summary Test 19.9

1 recessive
2 chloride
3 epithelial
4 osmosis
5 }
6 } replacement, supplementation
7 somatic cell
8 }
9 } liposomes, harmless viruses

Answers to Exam and Practice Questions

Chapter 1

1 a A = nuclear membrane; B = (crista of) mitochondrion; C = (Golgi) vesicle.

b Centrioles replicate during interphase before each nuclear division and during the process one moves to each pole. Here they form spindle fibres which are important in drawing chromosomes to the poles. Cilia and flagella possess a form of centriole.

c Membranes are 7 nm wide. A light microscope can only resolve objects which are at least 200 nm apart. We are therefore unable to see membranes through a LM. An electron microscope uses beams of electrons that have a shorter wavelength than light beams and so have a higher resolving power i.e. they can distinguish objects that are as little as 0.5 nm apart. Internal membranes are therefore visible using an EM.

d i As the temperature increases, the percentage transmission decreases. There is a gradual decrease initially – between 20 °C and 60 °C the percentage transmission only decreases from 96% to 70%. Further decreases in temperature produce a much greater decline in percentage transmission – between 60 °C and 70 °C it falls from 70% to 19%. Between 70 °C and 80 °C the fall is reduced slightly – from 19% to 6%.

ii At low temperatures, the vacuolar membrane retains the betalain pigment within the vacuole. At temperatures above 60 °C, the proteins in the vacuolar membrane are denatured and the membrane becomes more fluid. This allows the betalain to diffuse out. Higher temperatures also increase the rate of diffusion and this too increases the rate of outward movement of betalain.

2 a i H = nucleolus; J = Golgi body; K = cell wall; L = vacuolar membrane.

ii *Any two from the following.* A prokaryotic cell lacks a nucleus / lacks mitochondria / lacks large vacuoles / lacks a Golgi body / lacks endoplasmic reticulum (rough and smooth) / has circular rather than linear DNA / its DNA is not associated with histones / its internal organelles are not bounded by a double membrane / its ribosomes are smaller / its cell wall is made of murein/peptidoglycan.

b The gene is made up of DNA which acts as a template on which complementary nucleotide bases become

attached and form messenger RNA. This mRNA leaves the nucleus via a nuclear pore and associates itself with ribosomes on the rough endoplasmic reticulum (RER) in the cytoplasm. Transfer RNA molecules, each attached to an amino acid, associate with complementary triplets of bases on the mRNA. The amino acids become linked by peptide bonds to form a polypeptide which is combined with others to form a protein. The energy for this assembly comes from ATP provided by mitochondria. This protein is transported by the RER to the Golgi body where sugars are added and it is enclosed within a membrane to form a vesicle. The vesicle fuses with the cell surface membrane and the TMPI protein is released to the outside.

3 A = absent; B = present; C = present; D = sometimes; E = sometimes; F = sometimes; G = present; H = present; I = sometimes; J = absent; K = present; L = present; M = absent; N = present

4 a Magnification is how many times bigger the image is compared to the original object. Resolution is the minimum distance apart that two objects can be in order for them to appear as separate items.

b 200 times

c 10 mm

d 500 nm (0.5 μm)

e The EM uses a beam of electrons that has a much shorter wavelength than light.

f Electrons are absorbed by the molecules in air and, if present, this would prevent the electrons reaching the specimen.

g i plant cell and bacteria; **ii** all of them; **iii** plant cell, bacterium and virus

h The preparation of the specimens may not be good enough.

5 a protein synthesis

b i mitochondrion; **ii** nucleus; **iii** Golgi body; **iv** lysosome

c i mitochondria, nucleus; **ii** Golgi body, lysosomes; **iii** microvilli, mitochondria; **iv** rough endoplasmic reticulum / ribosomes, mitochondria

6 a A Golgi body; B nuclear envelope; C nucleoplasm; D mitochondrion; E nuclear pore; F rough endoplasmic reticulum; G nucleolus; H lysosome

b i D **ii** G **iii** D **iv** F

Chapter 2

1 a i Glycogen is a branched molecule, which is not coiled and has 1,6 glycosidic links as well as 1,4 glycosidic links. Amylose is an unbranched, coiled molecule with only 1,4 glycosidic links.

   ii Glycogen is insoluble so does not easily diffuse out of cells and it has no osmotic effect. Glucose is soluble and lowers the water potential of cells causing water to be drawn in by osmosis. The cell would therefore swell and burst (animal cells) or require a thicker cell wall (plant cells) to prevent it doing so. Glycogen is also a compact molecule allowing much to be stored in a small space.

b

Addition of water (H_2O) breaks this glycosidic bond. (= hydrolysis)

glucose molecule

Each carbon that was linked via the oxygen atom has now got an hydroxyl (OH) group attached.

2 a i glycosidic bond

   ii hydrolysis

   iii *Any two answers from*: Maintains the turgidity of cells / is a raw material for photosynthesis / is a solvent in which reactions take place / is a medium for the transport of substances / maintains a water potential gradient between cells / is involved in cell expansion and therefore growth / is necessary for the opening and closing of stomata.

b Globular means that the enzyme has a tertiary structure that forms a spherical-shaped molecule which is water soluble.

c i active site

   ii The substrate has a complementary shape to the active site (it is specific to it). It therefore fits into the active site in the way that a key fits a lock and forms an enzyme-substrate complex. This causes stress in the substrate molecule which lowers the activation energy of the reaction and so it can take place at a lower temperature than would otherwise be the case. After the reaction, the products leave the active site and so the enzyme can be used repeatedly.

3 a A = glycerol B = fatty acid / hydrocarbon chain of fatty acid

b part A (glycerol) attracts water (hydrophilic) while part B (fatty acids) repels water

c carbon, hydrogen and oxygen

d the hydrocarbon chain has no double bonds

e chip shop portion contains 7.6 (19.2 − 11.6) g more fat. Total energy produced from this = 288.8 (7.6 × 38) kJ

4 a COOH

b peptide

c condensation

d biuret

e i peptide
   ii peptide and hydrogen
   iii peptide, hydrogen, ionic, disulfide bridges and hydrophobic interactions
   iv peptide and disulfide bridges

5 a G   **b** H   **c** B and C   **d** E   **e** G   **f** B   **g** G   **h** D and H

Chapter 3

1 a Record when the brown colour first appears / use colour charts or colorimetry.

b Initially there is no reaction because there is no catechol present. At catechol concentrations lower than 5.0 mM the concentration of catechol is limiting the reaction, as there are few collisions between the enzyme and catechol. The rate increases as catechol concentration increases, and so more active sites can be occupied – for example, at 2.0 mM concentration of catechol the rate has increased from 0 to 60 units. The maximum rate is reached at 4.5–5.0 mM catechol concentration. The rate then remains constant at catechol concentrations greater than 5.0 mM because all active sites of the enzyme are occupied and so enzyme concentration becomes the limiting factor and further increases in catechol concentration do not increase the rate.

c i The curve will be a similar shape but always lower than the curve without the inhibitor. The curve will eventually reach the maximum.
   ii As PHBA has a similar shape to catechol, it will be able to bind to the enzyme's active site and block catechol from binding to the enzyme. There will be fewer enzyme–substrate collisions that form an enzyme–substrate complex and so the rate of reaction decreases.

d Enzymes only operate in a limited pH range, either side of which the rate of reaction decreases. Acids, like citric acid, have a high concentration of H^+ ions, which can denature enzymes by breaking ionic or hydrogen bonds and thereby changing the shape of the active site, preventing the substrate from fitting it.

2 a To denature the sucrase and stop its action in all tubes at the same time so that a reliable comparison of the results could be made. In addition, a high temperature is required for the Benedict's test that is used to detect any reducing sugars produced.

b

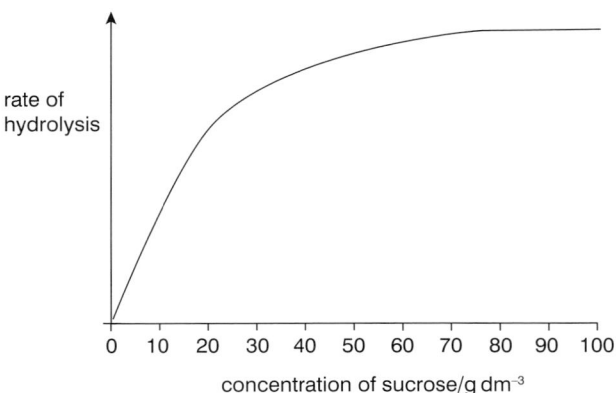

concentration of sucrose/g dm^{-3}

c *Description.* Question says 'rate' of hydrolysis so this is calculated by using the reciprocal of the time taken. The results are as follows:

Conc. g/dm^3	rate
5	0.0036
10	0.0069
15	0.0105
20	0.0133
50	0.0213
100	0.0222

The rate of hydrolysis increases up to around a concentration of 50 g dm^3 after which it levels off and remains constant from around 50 g dm^3 to 100 g dm^3.

Explanation. The sucrase hydrolyses the glycosidic bond of the sucrose molecule (a non-reducing sugar) to form a glucose and a fructose molecule (both reducing sugars). At low concentrations of sucrose, the quantity of sucrose is limiting the reaction as some of the active sites on the sucrase molecule are not filled. At higher concentrations of sucrose, there is excess substrate and all the active sites are filled so the enzyme is working at its maximum rate. The factor limiting the reaction at this stage is likely to be the concentration of the enzyme.

3 a i Enzyme X is not very specific, as it acts on a number of different proteins. Enzyme Y is highly specific, as it acts on a single protein.
ii Enzyme X could be used in biological washing powders to digest/remove stains from clothes. Enzyme Y could be used in making yoghurt/cheese from milk.

iii Milk is the only food in the diet of young mammals. The enzyme coagulates the milk, causing it to remain in the stomach for longer. This gives time for enzymes there to act on it so that it can be broken down into products that can then be absorbed. If it had remained liquid, it would pass through the stomach more quickly and only be partially digested.

iv Enzyme X functions at much higher temperatures than enzyme Y and so must have a much more stable tertiary structure to prevent it becoming denatured. The bonds holding the polypeptide chains in its precise 3D arrangement that makes up the active site must therefore be less easily broken than those of enzyme Y. It is therefore likely that the bonds of enzyme X are mostly, or entirely, disulfide bonds, as these are not easily broken by heat. Enzyme Y is likely to have more of the heat-sensitive ionic and hydrogen bonds.

b i

Temperature / °C	Time / min for hydrolysis of protein	Rate of reaction / ^{1}/time
15	5.8	0.17
25	3.4	0.29
35	1.7	0.59
45	0.7	1.43
55	0.6	1.67
65	0.9	1.11
75	7.1	0.14

ii

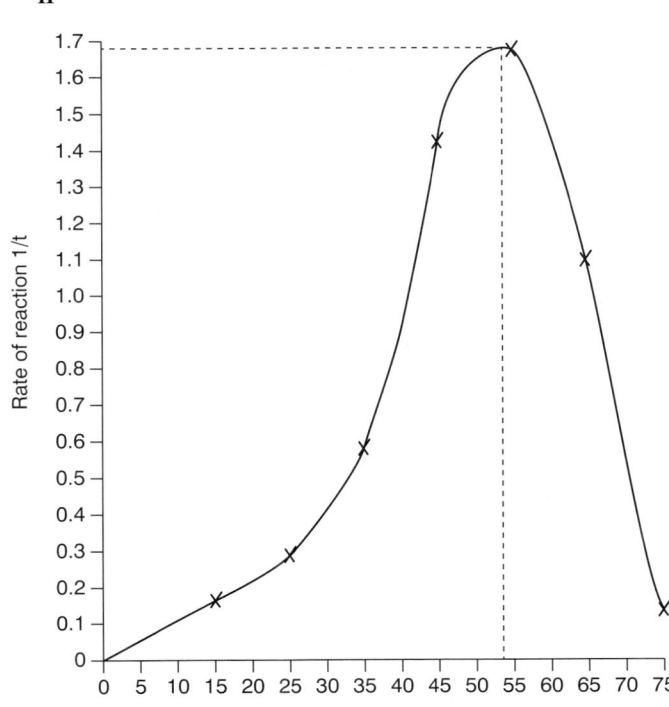

iii The optimum temperature is found by dropping a vertical line from the highest point on the curve and reading the temperature where it transects the temperature (*x*) axis. The value is in the range 50–55 °C.

   iv Carry out the experiment in exactly the same way but use narrower temperature intervals (e.g. 1 °C) over the range 45–60 °C.

4 a To function, enzymes must physically collide with their substrate. Lower temperature decreases the kinetic energy of both enzyme and substrate molecules which move around less quickly. They hence collide less often and therefore react less frequently.

  b The heat causes hydrogen and other bonds in the enzyme molecule to break. The shape of the enzyme molecule changes as does the active site. The substrate no longer fits the active site.

  c i High temperatures denature the enzymes and so they cannot spoil the food.

   ii Vinegar is very acidic. The low pH will denature the enzymes and so preserve the food

5 a 2

  b 4

Chapter 4

1 a i *Any two from the following:* calcium ions are: water soluble rather than lipid soluble / are charged ions / are hydrophilic while the phospholipid layer is hydrophobic.

   ii Calcium ions are moved against their concentration gradient by active transport. The ions bind with receptors on the channels of carrier proteins which span the membrane. On the inside of the cell, ATP binds to the carrier protein causing it to change shape and open to the opposite side of the membrane. The calcium ions are then released to the other side of the membrane.

  b Receptors on the membrane recognise antibodies that have attached themselves to antigens on the bacterial surface. As a result of a series of reactions, the surface of the bacterium becomes coated with proteins called opsonins. This facilitates the process where the membrane invaginates to enclose the bacteria and then fuses to form a vesicle called a phagosome.

  c Lysosomes move towards the vesicle and fuse with it. Enzymes within the lysosomes firstly break down the cell wall of the bacteria by hydrolysing the murein of which it is made. Other enzymes from the lysosome hydrolyse the peptide bonds of proteins, the ester bonds of lipids and the glycosidic bonds of carbohydrates to produce smaller, soluble material. These soluble products are absorbed into the cytoplasm of the phagocyte.

2 a 7.0 nm

  b K allows the passage of ions, small water-soluble molecules and polar (charged) ones across the membrane by facilitated diffusion or active transport.

    L acts as a recognition site for specific chemicals. It also stabilises the membrane by forming hydrogen bonds with water.

    M permits the movement of lipid-soluble substances across the membrane while acting as a barrier to water soluble ones. The fatty acid tails of this layer also help to keep the membrane fluid.

    N the cholesterol regulates the fluidity of the membrane depending on the temperature and influences its permeability. It also restricts the lateral movement of phospholipids and is a storage material.

  c Glucose is a relatively large polar molecule and, being water soluble, cannot pass through the phospholipid bilayer.

  d The graph shows that, at low glucose concentrations, the rate of uptake is dependent on glucose concentration and increases as the concentration of glucose increases up to a point after which the rate becomes constant. The levelling off is due to all the carrier proteins involved in facilitated diffusion being fully occupied in moving glucose. If it were a passive process the rate would not level off but continue to rise. If active transport were responsible the concentration of glucose would not affect the rate except at very low glucose concentrations.

  e Active transport is an active process using ATP to move substances against a concentration gradient while facilitated diffusion is passive, does not use ATP and occurs down a concentration gradient.

3 a Transport mechanism – box 2 = diffusion, box 3 = endocytosis / phagocytosis.

    Example – box 5 = glucose / amino acids / ions / polar molecules, box 6 = water.

  b Box 1 = facilitated diffusion, box 2 = active transport / sodium-potassium pump,

    box 3 = diffusion / osmosis, box 4 = endocytosis (pinocytosis / phagocytosis) / exocytosis.

4 a Because only lipid-soluble substances diffuse across the phospholipid bilayer easily, water-soluble ones such as glucose diffuse only very slowly.

  b It could increase its surface area with microvilli or have more proteins with pores that span the phospholipid bilayer (N.B. the thickness of the plasma membranes does not vary to any degree).

c i increases two times / doubles; **ii** no change; **iii** decreases four times / it is one quarter; **iv** increases two times / doubles (the CO_2 concentration is irrelevant)

5 a A membrane that is permeable to water molecules (and a few other small molecules) but not to larger molecules.

b Zero

c C, D, A, B

6 a Both cells have a lower water potential than the pure water and so water enters them by osmosis. The animal cell is surrounded only by a thin cell surface membrane and so it swells until it bursts. The plant cell is surrounded by a rigid cellulose cell wall. Assuming it is turgid water cannot enter as the cellulose cell wall prevents any expansion of the cell and hence it bursting.

b A – turgid, B – incipient plasmolysis, C – plasmolysed, D – turgid.

c Solutions A, B and D. (All except C).

Chapter 5

1 a It increases cell numbers and so allows growth, produces genetically identical cells and therefore replaces dead and damaged cells, making repair possible and allowing asexual reproduction / cloning.

b There should be a tick in all the boxes.

c The chromosomes first become visible (due to condensation of chromatin) as two chromatids joined by a centromere. Spindle fibres form and centrioles move to opposite poles, the nucleolus disappears and the nuclear envelope breaks down.

d The timing of the cell cycle will be disrupted, with prophase starting again immediately and repeated mitotic divisions taking place.

2 a i 6
ii centromere; the site of attachment to spindle fibres.
iii The shaded pair must be identical (there are three possible pairings).
iv

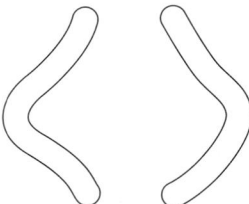

b The chromosomes unravel and become indistinct being visible only as chromatin. The DNA is then used as a template to transcribe mRNA which is used to form new proteins from which new cell organelles are constructed. The DNA then replicates by splitting into two separate strands each of which forms a complementary strand (semi-conservative replication) to produce new DNA.

c The chromosome number is halved from six to three during the process of meiosis. This is necessary to maintain the diploid number of chromosomes (six) after fertilisation of an egg by a sperm. Without this reduction the number of chromosomes would double at each generation.

3 a i Metaphase
ii The chromosomes / chromatids are lined up at the equator. The centromeres are attached to the spindle fibres. The centrioles have reached the poles of the cell.

b It increases cell numbers and so allows growth to occur. It produces genetically identical cells and so is used in asexual reproduction and cloning. It replaces dead and damaged cells, thereby replacing and repairing tissues. It maintains the number of chromosomes at each cell division.

c Different length cell cycles allow each tissue to produce cells when required, i.e. skin cells need to be replaced more often than muscle cells. Being carefully controlled means that growth is coordinated and tumours are prevented from forming, which might otherwise invade other tissues.

4 a In prokaryotic cells the DNA is circular and is not associated with proteins (i.e. does not have chromosomes).

b It fixes the DNA into position

c It is looped and coiled a number of times.

d i 50 mm (46 chromosomes in every cell)
ii 2.3 metres (all cells have same quantity of DNA)

5 a 24 minutes. Number of cells in metaphase ÷ total number of cells observed × time for one cycle in minutes ($20 \div 1000 \times 1200 = 24$).

b 11%. Chromosomes are visible in all stages except interphase. Total of these stages is 110 (73 + 20 + 9 + 8). This number is then divided by the total number of cells observed (1000) and multiplied by 100 to give a percentage. $110 \div 1000 \times 100 = 11\%$.

6 a 0.2 million (200 000)

b 50%

c 8.33 times ($0.5 \div 0.06$)

d More cancer cells are killed because they divide more rapidly than healthy cells and so are more susceptible to the drug.

e Cancer cells take longer to recover. Cancer cells divide more slowly / rate of mitosis is reduced.

f One dose of the drug does not kill all the cancer cells – those that remain continue to divide and build up the number of cancer cells again.

Chapter 6

1 a interphase

b i hydrogen bonds

ii M = adenine; O = cytosine.

c A molecule of DNA has two polynucleotide strands. Both of these DNA strands act as a template for the formation of a complementary DNA strand. As the newly formed double helix retains one parent strand and has one new strand, the process is called 'semi-conservative'.

d Any error is, in effect, a mutation. The altered DNA could lead to a different protein, or no protein at all, being made. This could lead to the loss of a function especially where the altered protein is an enzyme. The body's immune system may not recognise the new protein and so reject the cell that possesses it. The altered protein could result in conditions such as sickle cell anaemia or cancer.

2 a i DNA, because thymine is present and RNA has uracil rather than thymine.
ii Phosphodiester
iii Deoxyribose

b tRNA attaches to a specific amino acid and takes it to a ribosome on mRNA, where the anticodon on tRNA binds with the complementary codon on mRNA. The ribosome brings together two tRNA molecules, with amino acids attached, and a peptide bond forms, joining them together. The process continues until a polypeptide is formed.

c A carboxyl (–COOH) group from one amino acid combines with an amino (–NH$_2$) group from another amino acid and a water molecule is lost = a condensation reaction.

3 a pentose (sugar), phosphate group and organic (nitrogenous) base

b Adenine and guanine are longer molecules than thymine and cytosine. The distance between the two phosphate/deoxyribose 'uprights' is constant in the DNA molecule. Pairing adenine and guanine would produce a long 'rung' while pairing thymine and cytosine produces a short 'rung'.

c The bases are linked by hydrogen bonds. The molecular structures might be such that hydrogen bonds do not form between adenine and cytosine and between guanine and thymine.

d ACCTCTGA

e 30.1%. If 19.9% is guanine then, as guanine always pairs with cytosine, it also makes up 19.9% and together they make up 39.8% of the bases in DNA. This means the remaining 60.2% of DNA must be adenine and thymine and as these also pair, each must make up half, i.e. 30.1%.

4 a 5

b The first and last (5th) / the two coded for by the bases TAC

c Because some amino acids have up to six different codes.

d A different base might code for a different amino acid. The sequence of amino acids in the polypeptide produced will be different. This change to the primary structure of the protein might result in a different shaped tertiary structure. The enzyme shape will be different and may not fit the substrate. The enzyme-substrate complex cannot be formed and so the enzyme is non-functional.

5 a The enzyme RNA polymerase moves along the template DNA strand causing the already paired nucleotides on this strand to join together. The RNA polymerase adds the nucleotides one at a time, to build a strand of mRNA until it reaches a particular sequence of nucleotide bases on the DNA which it recognises as a 'stop' code.

b DNA helicase – it acts on a specific region of the DNA molecule to break the hydrogen bonds between the bases, causing the two strands to separate and expose the nucleotide bases in that region.

c Splicing is necessary because pre-mRNA has nucleotide sequences derived from introns in DNA. These introns are non-functional and if left on the mRNA would lead to the production of non-functional polypeptides or no polypeptide at all. Splicing removes these non functional introns from pre-mRNA.

d i UACGUUCAGGUC
ii 4 amino acids (one amino acid is coded for by 3 bases – 12 bases code for 4 amino acids)

e Some of the base pairs in the genes are introns / non-functional DNA. These introns are spliced from pre-mRNA so that the resulting mRNA has fewer nucleotides.

6 a i Universal – because a triplet codes for the same amino acid in all organisms.
ii Degenerate – because most amino acids have more than one codon.
iii Non-overlapping – because each base in the sequence is read only once.

b *Any three from*: RNA is smaller than DNA; RNA is a single helix, DNA is a double helix; the sugar in RNA is ribose, in DNA it is deoxyribose; in RNA the base uracil replaces the thymine found in DNA.

c A codon is the triplet of bases on messenger RNA that codes for an amino acid. An anticodon is the triplet of bases on a transfer RNA molecule that is complementary to the codon.

d i DNA needs to be stable to allow it to be passed from generation to generation unchanged and thereby allow offspring to be very similar to their parents. Any change to the DNA is a mutation and is normally harmful.
  ii mRNA is produced to help manufacture a protein e.g. enzymes. It would be wasteful to produce these continuously when they are only needed periodically. mRNA therefore breaks down once it has been used and is produced again only when next required.

7 a i GUA
  ii TCA

b Leucine

c Glutamine; alanine; proline; tyrosine; alanine.

d GTAGGACTGGAT.

8 a Ribosome

b i UAG on tRNA; **ii** TAG on DNA.

c A tRNA molecule attaches an amino acid at one end and has a sequence of three bases, called an anticodon, at the other end. The tRNA molecule is transferred to a ribosome on a mRNA molecule. The anticodon on tRNA pairs with the complementary codon sequence on mRNA. Further tRNA molecules with amino acids attached line up along the mRNA in the sequence determined by the mRNA bases. The amino acids are joined by peptide bonds, therefore the tRNA has helped to ensure the correct sequence of amino acids in the polypeptide.

d One of the CODONS is a STOP codon that indicates the end of polypeptide synthesis. STOP codons do not code for any amino acid hence there is one less amino acid than there are codons.

Chapter 7

1 a i As the cell wall is freely permeable, water passes across it by diffusion until it reaches the cell surface membrane, which is only selectively permeable. Water moves across the membrane by osmosis down a water potential gradient and into the cell. This movement is made faster by specialised protein channels called aquaporins, which span the membrane.

  ii The phospholipid layer of cell surface membranes makes them relatively impermeable to water, and so osmosis is too slow to supply water quickly enough for the plants' needs. The aquaporins span this layer and provide a channel for the rapid entry of water.

b Water passes across cortical cells either through their cell walls (apoplast pathway) or through plasmodesmata and the cytoplasm (symplast pathway) until it reaches the endodermis. Here, the suberin in the walls of the endodermal cells (Casparian strip) block the apoplast pathway, forcing water across them via the symplast pathway into the xylem.

c i During the day stomata are open to allow diffusion of carbon dioxide for photosynthesis.
  ii In the mutant plants (A) the rate of transpiration is always higher. As the enzyme does not affect stomata, the rate of stomatal transpiration in the light is the same for both A and B. In the dark, transpiration only occurs through the cuticle. The observed differences in transpiration are the result of differences in cuticular transpiration and show that the mutant plants (A) have cuticles that are less effective at preventing water loss.

2 a Hydrogen ions are actively pumped from companion cells into the apoplast (spaces within cell walls) using ATP. These hydrogen ions then flow down a concentration gradient through carrier proteins into the sieve tube elements. Sucrose molecules are transported along with the hydrogen ions in a process known as co-transport. The sucrose moves by facilitated diffusion down a concentration gradient through plasmodesmata.

b The ingrowths increase the surface area of the membrane so that many co-transporter proteins can be accommodated.

c i An electron microscope (EM) uses a beam of electrons that has a shorter wavelength than a beam of light. This means an EM has a higher resolving power and can reveal much greater detail in the specimens being observed. Structures not visible in a light microscope are revealed using an EM.
  ii The sieve tubes are long with plates across them (sieve plates), which have pores (sieve pores) through them. The walls are thin and there is some peripheral cytoplasm inside the walls.

3 a The rain rapidly drains through the sand out of reach of the roots / Sand dunes are usually in windy situations; this reduces humidity and so increases the water potential gradient, leading to increased water loss by transpiration.

b The soil solution is very salty, i.e. it has a very low water potential, making it difficult for root hairs to draw water in by osmosis.

c Because in winter the water in the soil is frozen and therefore cannot be absorbed by osmosis.

d Being enzyme-controlled, photosynthesis is influenced by temperature. In cold climates enzymes work slowly, and this limits the rate of photosynthesis. Therefore there is a reduced need for light as photosynthesis is taking place only slowly. In warm climates, photosynthesis occurs rapidly and therefore requires more light.

4 a At 12.00 hours, because this is when water flow is at its maximum. As transpiration creates most of the water flow, they are both at a maximum at the same time.

b Rate of flow increases from a minimum at 00.00 hours to a maximum at 12.00 hours and then decreases to a minimum again at 24.00 hours.

c As evaporation / transpiration from the cell walls of spongy mesophyll cells increases during the morning (due to higher temperature / higher light intensity) it pulls water molecules through the xylem because water molecules are cohesive / stick together. This transpiration pull creates a negative pressure / tension. The greater the rate of transpiration, the greater the water flow. The reverse occurs as transpiration rate decreases during the afternoon and evening.

d As transpiration increases up to 12.00 hours, so there is a higher tension (negative pressure) in the xylem. This reduces the diameter of the trunk. As transpiration rate decreases, from 12.00 hours to 24.00 hours, the tension in the xylem reduces and the trunk diameter increases again.

e Transpiration pull is a passive process / does not require metabolic energy. Xylem is non-living and so cannot provide energy. Although root cortex and leaf mesophyll cells are living – the movement of water across them uses passive processes, e.g. osmosis, and so continues at least for a while, even though the cells have been killed.

5 a i As xylem is under tension, cutting the shoot in air would lead to air being drawn into the stem, which would stop transport of water up the shoot. Cutting under water means water, rather than air, is drawn in and a continuous column of water is maintained.
ii Sealing prevents air being drawn into the xylem and stopping water flow up it / Sealing prevents water leaking out which would produce an inaccurate result.

b that all water taken up is transpired

c Volume of water taken up in 1 minute:
$3.142 \times (0.5 \times 0.5) \times 15.28 = 12.00\,mm^3$.
Volume of water taken up in 1 hour:
$12.00 \times 60 = 720\,mm^3$

d their surface area / surface area of the leaves

e An isolated shoot is much smaller than the whole plant / may not be representative of the whole plant / may be damaged when cut.
Conditions in the lab may be different from those in the wild, e.g. less air movement / greater humidity / more light(artificial lighting when dark).

Chapter 8

1 a i Red blood cells / erythrocytes
ii Myoglobin 78%, haemoglobin 21%
iii At low partial pressures of oxygen myoglobin has higher affinity for oxygen and so binds oxygen while haemoglobin releases it. In this way myoglobin acts as a store of oxygen and only releases it at low oxygen partial pressures, e.g. when oxygen demand exceeds supply, such as during exercise.

b i Fetal haemoglobin has higher affinity for oxygen than adult haemoglobin at all partial pressures of oxygen. The fetus will therefore always be able to take up oxygen from the mother's blood. This is important because the mother provides the only source of oxygen for the fetus.
ii At low partial pressures of oxygen both adult and fetal haemoglobin unload their oxygen, and at higher ones the adult haemoglobin will take up sufficient oxygen – therefore no symptoms are apparent.

c The overall shape of the new line should be the same, but all of it should be to the right of the adult haemoglobin curve; the line should begin at 0.2 kPa and end at 97%.

2 a i In a closed system, blood moves within blood vessels; and in a double system, blood passes through the heart twice during a complete circuit of the body.

b

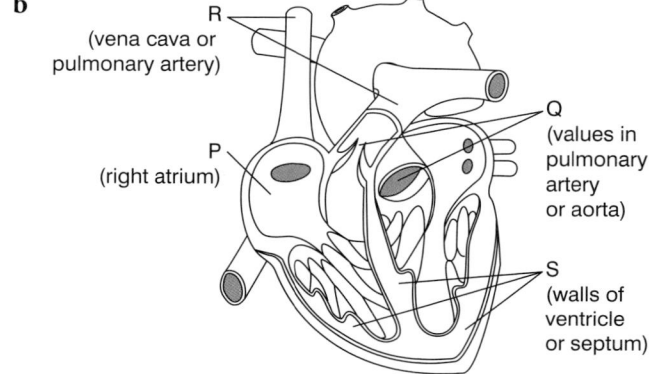

R
(vena cava or pulmonary artery)

Q
(values in pulmonary artery or aorta)

P
(right atrium)

S
(walls of ventricle or septum)

c i In the 8 seconds shown on the *x*-axis, there are 10 peaks, indicating that there are 10 beats in this period. This means that in a minute there are $10 \div 8 \times 60 = 75$ beats (per minute).

ii The pressure in the aorta never drops below 10.8 kPa, whereas in the left ventricle the lowest pressure is always 0.0 kPa. The range of pressure (difference between lowest and highest) is less in the aorta (4.8 − 5.2 kPa) than it is in the left ventricle, where it is always 16.0 kPa. The explanation for these differences is that pressure in the aorta is kept relatively high at all times due to the elastic recoil of the aorta. In the left ventricle, however, there is a high pressure when its muscular walls are contracting (ventricular systole). There is a low pressure when the walls are relaxed (ventricular diastole), because the semilunar valves prevent blood flowing back into the ventricle, which is almost empty of blood as a result.

d i coronary arteries

ii A heart attack or angina as a result of inadequate oxygen/glucose to the cardiac muscle.

e A coronary by-pass operation / insertion of a coronary stent / angioplasty.

3 a At this partial pressure of oxygen, lugworm haemoglobin is 90% saturated – more than enough to supply sufficient oxygen to the tissues of a relatively inactive organism. Human haemoglobin, by contrast, is only 10% saturated – insufficient to supply enough oxygen to keep tissues alive.

b The dissociation curve of the lugworm is shifted far to the left. This means that it is fully loaded with oxygen even when there is very little available in its environment.

c Respiration produces carbon dioxide. This builds up in the burrow when the tide is out. If lugworm haemoglobin exhibited the Bohr effect, it would not be able to absorb oxygen when it was present in only very low concentrations in the burrow.

d The higher part of the beach is uncovered by the tide for a much longer period of time than the lower part. During this longer period all the oxygen in the burrow would be used up and the lugworm might die before the next tide.

e It is shifted to the left.

Chapter 9

1 a

	Cartilage	Ciliated epithelium	Elastic fibres	Goblet cells	Smooth muscle
A	✓	✓	✓	✓	✓
B	✓	✓	✓	✓	✓
C	✗	✓	✓	✓	✓
D	✗	✗	✓	✗	✗

b Goblet cells produce mucus, which is sticky and traps pathogens/particles/dust. If a person is infected, the goblet cells produce more mucus. This increases the distance between the pathogen and the epithelium of the alveoli and so acts as a barrier and helps to prevent infections.

Cilia move in a synchronous rhythm that moves mucus up from the lungs to the back of the throat, from where it is swallowed. In this way pathogens/dust/particles are moved out of the lungs, and so the gas exchange surface is kept free of them and therefore healthy.

2 a 8.57 μm (X–Y = 33 mm = 33 000 μm. Actual size = 33 000 ÷ 3500 = 9.43 μm)

b i Elastic fibres stretch when breathing in to allow the lungs to expand and therefore increase the volume of air inspired and consequently increase the rate of gas exchange. They also prevent the alveoli from bursting. When breathing out, the elastic fibres cause the alveoli to spring back (recoil) and therefore help to expel air from them.

ii Alveoli provide a large surface area for the diffusion of oxygen into, and carbon dioxide out of, the blood. Alveoli are surrounded by numerous capillaries and there is a very short diffusion pathway between the blood in them and the air in the alveoli because the alveoli walls are only one cell thick. Movement of air in the lungs, and blood in the capillaries, helps to maintain a concentration gradient of oxygen and carbon dioxide between the two.

c i *Any two from*: The lung tissue of someone with emphysema has: less elastin / alveoli which have burst / alveoli with a reduced surface area / fewer capillaries / inflammation of bronchial tissue and so narrow lumen.

ii *Any two from*: breathlessness / wheezing noise when breathing / chronic cough / blue colour to the skin / tiredness.

3 a The chromosomes are still at the equator / the nuclear envelope is still breaking down / the chromosomes are still together rather than separated into two groups.

b Tobacco smoke contains a number of carcinogens which cause damage to the genes of the bronchial epithelium. Among these mutant genes are ones that control normal cell division and give rise to lung cancer. One such carcinogen is benzopyrene (BP) found in the tar of tobacco smoke. A derivative of BP binds directly to the tumour-suppressor gene p53, which normally stops cell division and destroys mutated cells. With this gene inactivated, epithelial cells divide by mitosis in an uncontrolled manner leading to the formation of a tumour.

c There is a long time lag (20 years or more) between smoking taking place and the development of the symptoms of lung cancer and death from the disease. The peak mortality rate around 1980 (Figure 4) is the result of the high percentage of people who smoked in the 1950s and 1960s (Figure 3). The increasing number of deaths from 1950 to 1980 was probably the result of the large numbers of people smoking in the 1930s, 1940s and 1950s. The decrease in lung cancer deaths from 1980 onwards was probably due to better and earlier diagnosis of lung cancer preventing many deaths and better treatment extending life expectancy of those with the disease.

4 a 17.14 breaths min^{-1}. Measure the time interval between any two corresponding points on either graph that are at the same phase of the breathing cycle (e.g. two corresponding peaks on the volume graph or two corresponding troughs on the pressure graph). The interval is always 3.5 s. This is the time for one breath. The number of breaths in a minute (60 seconds) is therefore 60 s ÷ 3.5 s = 17.14.

b It is essential to first convert all figures to the same units. For example 3000 cm^3 is equal to 3.0 dm^3. From the graph you can calculate that the exhaled volume is 0.48 dm^3 less than the maximum inhaled volume. The exhaled volume is therefore 3.0 − 0.48 = 2.52 dm^3. If working in cm^3, the answer is 2520 cm^3.

5 a *Any 4 from*: smoking, air pollution, genetic makeup, infections and occupation.

b Any figures in the range 50–55%.

c Two times. Around 80% of non-smokers live to age 70 compared to around 40% of people who smoke more than 25 cigarettes a day.

d In general terms, she will live longer. More specifically she has a 50% chance of living to be 65 years if she carries on smoking but a 50% chance of living to 80 years if she gives up. Her life expectancy will increase by 10–15 years.

Chapter 10

1 a Tropical climates provide the best breeding and living conditions for the *Anopheles* mosquito, which transmits malaria. *Plasmodium* needs temperatures in excess of 20 °C for it to complete its life cycle within the mosquito. The mosquito life cycle requires areas of still water and these are more common in the wetter tropics. The disease has largely been eradicated in areas outside of the tropics.

b i A = 28; B = 14
 ii The chromosome number is halved during the process of meiosis. This is necessary to maintain the diploid number of chromosomes after fertilisation of an egg by a sperm. Without this reduction the number of chromosomes would double at each generation. Meiosis also introduces some genetic variety in the offspring.

c *Plasmodium* has the ability to change the proteins that make up its surface antigens (antigenic variability) because it is a eukaryotic organism and therefore has many genes that can code for a wide variety of surface antigens. In addition it has many stages in its life cycle within humans, each one having different surface antigens. It divides by meiosis which increases variety and it has haploid stages that allow recessive alleles to be expressed. This constant changing of antigens (antigen shifting) is the main reason why a vaccine is difficult to develop. No sooner has one been developed against one set of antigens, before they change to another form and a new vaccine is required. *Plasmodium* also lives inside red blood cells where it is hidden away from the body's immune system and any antibodies present in the blood that might kill the pathogen.

2 a *Vibrio cholerae*

b *Two possible answers*:

If the inhibitor has a complementary shape to the active site it will compete with the substrate for the active site and so fewer enzyme–substrate complexes can be formed, there will be less enzyme activity, a lower respiratory rate and less ATP produced. To have maximum effect a high concentration of the inhibitor would need to be used.

or

If the inhibitor binds to the enzyme at a site other than the active site it may change the shape of the active site so the shape of substrate will no longer be complementary to the active site. The substrate will be unable to bind to the active site, there will be no enzyme activity, no respiration and no ATP production.

c i 2.70 / 2.71.

ii In regions such as Africa and Asia compared to Europe and North America there might be:
A greater risk of consuming contaminated water or food / the drinking water may be less safe, as there is little access to clean bottled water nor are there water treatment plants / lack of hygiene, e.g. hands not washed after defaecation / poor sanitation so faeces may mix with drinking water / vaccines may be less effective / little access to vaccines / less education about how to avoid contracting cholera / antibiotic resistance may be more common / a lack of drugs and medical care for treatment of cholera.

3 a A female *Anopheles* mosquito takes a blood meal from someone infected with *Plasmodium* and takes in the parasite. The mosquito later takes another blood meal from an uninfected person and injects the parasite with its saliva which is used to prevent the blood clotting while sucking it up.

b i Each protein on the surface membrane of *P.falciparum* acts as an antigen. The B cell in the body that has antibodies on its surface that exactly fit this antigen attaches to the antigen and begin to divide by mitosis to form a clone of plasma cells. These plasma cells produce antibodies that correspond to the antigen. Some plasma cells develop into memory cells which can live for long periods. When the person is later infected again by *P.falciparum,* these memory cells divide and some develop into plasma cells. These plasma cells produce antibodies that attach to the antigen on the surface of the pathogen and so prevent it attaching to red blood cells. This secondary response is very rapid.

ii *Plasmodium* has the ability to change the proteins that make up its surface antigens (antigenic variability) because it is a eukaryotic organism and therefore has many genes which can code for a wide variety of surface antigens. In addition it has many stages in its life cycle within humans, each one having different surface antigens. It divides by meiosis which increases variety and it has haploid stages that allow recessive alleles to be expressed. This constant changing of antigens (antigen shifting**)** is the main reason why a vaccine is difficult to develop. No sooner has one been developed against one set of antigens, before they change to another form and a new vaccine is required. *Plasmodium* also lives inside red blood cells where it is hidden away from the body's immune system and any antibodies present in the blood that might kill the pathogen.

c *One of the following alternative explanations*:

If the drug is a competitive inhibitor of the enzyme, it will have a similar shape to the surface protein on *P.falciparum*. The drug fits into the active site of the enzyme and prevents the formation of the enzyme-substrate complex.

If the drug is a non-competitive inhibitor it will fit into a site other than the active site. Its attachment causes the enzyme to change shape in a way that alters the shape of the active site and so it can no longer fit onto the surface protein.

The drug may attach to the active site by covalent bonding and block access by the surface protein and so prevent formation of the enzyme-substrate complex.

4 a *Any three of the following*: clean, uncontaminated water supplies / water treatment / chlorination / proper sanitation and sewage treatment / personal hygiene (washing hands after using the toilet) / proper food hygiene.

b Breast milk does not contain the bacterium that causes cholera (and contains antibodies that provide some resistance to it). Bottle-fed babies have milk made up with water and this may be contaminated with bacteria that cause cholera.

c It could prevent the bacterium penetrating the mucus barrier of the intestinal wall and so stop it reaching the epithelial cells.

d The antibiotic may be digested by enzymes and therefore not function / the antibiotic may be too large to diffuse across the intestinal epithelium. Severe diarrhoea is a symptom of cholera. Therefore any antibiotic taken orally may pass through the intestines so rapidly that it is passed out of the body before it can come in contact with the cholera bacterium. Taken via the blood the antibiotic is not digested / does not have to diffuse into the body / can reach the bacterium and kill it.

5 a A substance produced by a living organism that kills, or prevents the growth of other living organisms.

b Cross linkages hold the bacterial cell wall together. Without them the wall is weakened. Water enters bacteria by osmosis and the bacteria swell. Normally the wall prevents swelling and hence osmosis. Where wall is weakened, it breaks, water continues to enter and the cell bursts – osmotic lysis.

c When antibiotics are used they only kill the non-resistant bacteria. This reduces competition and makes it easier for the resistant bacteria to survive and so pass on the resistance to subsequent generations and to other species.

d Because they feel better and think they are cured.

e The course of treatment is very long (6–9 months).

f Because many different types of antibiotics are used, often in relatively large amounts. This creates a selection pressure favouring multiple resistant strains.

6 a R and S **g** P
b Q **h** S
c R **i** R
d R and S **j** P, Q, R and S
e S **k** P
f R and S

Chapter 11

1 a A mouse is exposed to an antigen against which an antibody is required. B cells within the mouse produce a mixture of antibodies and these are extracted from the spleen of the mouse. To enable these B cells to divide outside the body, they are mixed with cells that divide readily outside the body e.g. cells from a cancer tumour. These fused cells are called hybridoma cells and are separated out under a microscope and each single cell is grown into a group (clone). Each clone is tested to see if it is producing the required antibody. Any group producing the required antibody is grown on a large scale and the antibodies extracted from the growing medium.

b i Herceptin induces only slightly more cell death (0.6 units) than no treatment (0.5 units)) with X-ray treatment producing slightly more cell death (0.75 units). The two treatments together produce considerably more cell death (2.0 units) than either by itself.
ii $\dfrac{2.0 - 0.6}{0.6} \times 100 = 233\%$

c i Increasing the dose of X-rays in the presence of herceptin decreases the number of breast cancer cells that survive more than using X-rays in the absence of herceptin. The difference is most apparent above a dose of $2\,\mathrm{J\,kg^{-1}}$ and greatest at $8\,\mathrm{J\,kg^{-1}}$.
ii Herceptin may increase the effect of the X-rays by some means, e.g. may identify cancer cells by binding to a receptor site on them and then induce an immune response against these cells or may enter cancer cells and kill them in some way.

2 a i *Bracket should extend across the bilayer of phospholipids at the bottom of the figure.*
ii The membrane is fluid because its shape can change as the phospholipids and proteins can move. It is a mosaic because the proteins of the membrane are scattered within the phospholipid layer.
iii *Any two of:* composed of polypeptides / have a variable region / have a non-variable region / have an antigen binding site / are held together by disulfide bonds

b T helper cells produce cytokines which stimulate B cells to divide by mitosis and develop into plasma cells which produce antibodies. Cytokines also stimulate macrophage cells to engulf pathogens by phagocytosis. T cytotoxic cells attach to any body cell presenting the viral antigen and produce perforins to make holes in the cell surface membrane and so destroy the cell, along with the viruses it contains.

c *Any three from:* acts as a recognition site for substances other than antigens e.g. hormones / controls movement of substances in to and out of the cell / allows movement of substances across it by diffusion/ facilitated diffusion/active transport / exocytosis and endocytosis / retains the cell contents.

3 a Specific mechanisms distinguish between different pathogens. Non-specific mechanisms treat all pathogens in the same way. Non-specific mechanisms respond more rapidly than specific ones.

b Self refers to the body's own chemicals and cells while non-self refers to material that is foreign to the body.

c The lymphocytes that will finally control the pathogen need to build up their numbers and this takes time.

d The body responds immediately by 'recognising' the pathogen (and by phagocytosis), the delay is in building up numbers of lymphocytes and therefore controlling the pathogen.

4 a In the primary response the antigens of the pathogen have to be ingested, processed and presented by B cells. T helper cells need to link with the B cells that then clone some of the cells developing into the plasma cells that produce antibodies. These processes occur consecutively and therefore take time. In the secondary response memory cells are already present and the only processes are cloning and development into plasma cells that produce antibodies. Fewer processes – quicker.

b Examples include:

Cell-mediated	Humoral
Involves T cells	Involves mostly B cells
No antibodies	Antibodies produced
First stage of immune response	Second stage of immune response after cell mediated stage
Cytokines produced	No cytokine production
Effective through cells	Effective through body fluids

c Rough endoplasmic reticulum to make and transport the proteins of the antibodies. Golgi body to sort, process and compile the proteins. Mitochondria to release the energy needed for such massive antibody production.

d There must be a massive variety of antibodies as each responds to a different antigen, of which there are millions. Only proteins have the diversity of molecular structure to produce millions of different types.

5 a An organism or substance, usually a protein, which is recognised as foreign by the immune system and therefore stimulates an immune response.

b Both are types of white blood cell / have a role in immunity and both are produced from stem cells.

c T lymphocytes mature in the thymus gland while B lymphocytes mature in the bone marrow. T lymphocytes are involved in cell-mediated immunity while B lymphocytes are involved in humoral immunity.

6 a H5N1 infects the lungs leading to a massive production of T cells. Accumulation of these cells may block the airways / fill the alveoli and cause suffocation.

b Birds carry H5N1 virus. Birds can fly vast distances across the world in a very short space of time.

7 a An antigen.

b A pathogen or material engulfed by the phagocyte.

c Macrophage or antigen-presenting cell

d T helper cell

e Memory cells

f *Any two from:*

stimulates macrophage cells / phagocytes to engulf pathogens by phagocytosis.

stimulates B cells to divide and develop into antibody-producing plasma cells.

activates T cytotoxic (T killer) cells.

8 a Passive immunity – antibodies are introduced from outside rather than being produced by the individual. Immunity is normally only short-lived. Active immunity – individuals are stimulated to produce their own antibodies. Immunity is normally long-lasting.

b Living microorganisms that have been weakened so that they do not cause disease symptoms while retaining the ability to stimulate an immune response.

c The influenza virus displays antigen variability. Its antigens change frequently and so antibodies no longer recognise the virus. New vaccines are required to stimulate the antibodies that complement the new antigens.

d *Any three from:* HIV infections mean that more people have impaired immune systems / more refugees carrying TB move frequently and live in overcrowded temporary conditions / more mobile population / more elderly people – they often have less effective immune systems.

Chapter 12

1 a i B = adenine.
ii S = ribose.

b ATP has a 'universal role' because it is found in most cells of all organisms. It is the 'energy currency' because it releases much energy when it is hydrolysed. This hydrolysis occurs easily and the energy released is used in reactions within cells. ATP links catabolic and anabolic reactions and is rapidly interconverted with ADP + P. Being soluble makes it particularly useful as it can easily be moved from cell to cell as required.

c *Any two of*: the stalked particles of the cristae of the inner mitochondrial membrane / the thylakoids in the grana of chloroplasts / the cytoplasm.

2 a i 18
ii 7.2 [18(CO_2) ÷ 25(O_2)]

b The RQ value falls steeply (40–80 min) and then falls less steeply (80–160 min) before rising slightly (160–200 min). The change in RQ is because sugar is metabolised initially and then, as the supply of sugar runs out, fat is metabolised causing the overall fall in RQ.

c A rise in temperature causes an increase in the rate of respiration because the kinetic energy of the enzyme and substrate molecules increases. This means that more collisions occur between these molecules and so more enzyme-substrate complexes are formed. If the temperature increases too much, the enzymes become denatured and the rate of respiration falls.

3 a Homogenate is spun at slow speed. Heavier particles (e.g. nuclei) form a sediment. Supernatant is removed, transferred to another tube and spun at a greater speed. Next heaviest particle is removed – process repeated.

b Nuclei and ribosomes because neither carbon dioxide nor lactate (products of respiration) are formed in any of the samples.

c i Mitochondria
ii Krebs cycle converts carbon dioxide and results show that carbon dioxide is produced when mitochondria only are incubated with pyruvate.

d (Remaining) cytoplasm (N.B. complete homogenate is not a 'portion' of the homogenate).

e Cyanide prevents electrons passing down transport chain. Reduced NAD therefore accumulates and blocks Krebs cycle where carbon dioxide is produced. Glycolysis can still occur because the reduced NAD it produces is used to make lactate. Glucose can therefore be converted to lactate, but not carbon dioxide, in the presence of cyanide.

f The conversion of glucose to carbon dioxide involves glycolysis (occurs in cytoplasm) and Krebs cycle (occurs in mitochondria). Only the complete homogenate contains both cytoplasm and mitochondria.

g Ethanol and carbon dioxide.

h Liver cell, epithelial cell and muscle cell.

4 a To show that the yeast suspension was responsible for any changes that occurred and the glucose did not change DCPIP nor did DCPIP change by itself.

b i Yeast uses glucose as a respiratory substrate producing hydrogen atoms that are taken up by DCPIP causing it to become reduced and changing from blue to colourless.

ii As in 4a except that production of hydrogen atoms is slower.

c Contents of tube might have remained blue because at 60 °C the enzymes that catalyse the production of the hydrogen that reduces DCPIP would be denatured. The DCPIP would therefore not turn colourless.

d Air contains oxygen which would reoxidise DCPIP turning it blue.

e This is a single experiment. The same results would need to be obtained on many occasions to increase reliability.

Chapter 13

1 a i Both chlorophyll a and b have two peaks, one around wavelengths 400–500 nm (blue light), the other around wavelengths 600–700 nm (red light). The peaks for chlorophyll a are at 430 nm and 660 nm while those for chlorophyll b are at 450 nm and 635 nm. Chlorophyll a has a small peak at 410 nm which chlorophyll b does not. Chlorophyll a absorbs little light at wavelengths 450–600 nm while chlorophyll b absorbs little light at wavelengths 500–600 nm. In both chlorophylls the peak in blue light is higher than in red light, with the peak for chlorophyll a being higher than chlorophyll b in red light and chlorophyll b peak being higher than the chlorophyll a peak in blue light.

ii The light being absorbed by the chlorophylls is used for photosynthesis. The action spectrum shows that the rate of photosynthesis is higher in blue and red light and lower at intermediate wavelengths (green light). The action spectrum curve generally follows the shapes of the absorption curves for the

two chlorophylls although this is less obvious in green light (middle section). This is because other pigments, such as carotenoids, absorb some light at these wavelengths.

iii Plants contain chlorophyll, which does not absorb light in the green region of the spectrum. Green light is therefore reflected into the eyes giving the plant a green appearance.

b

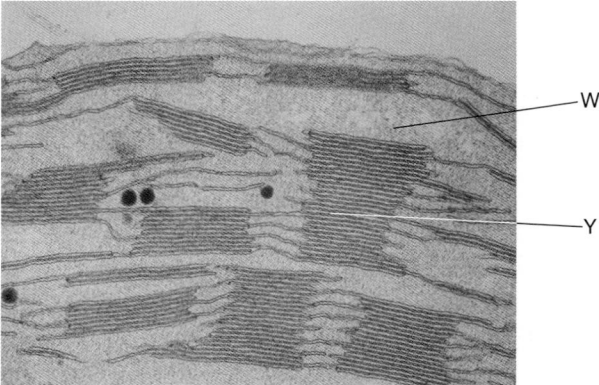

c At high light intensities, light does not limit the rate of photosynthesis and so a lot of ATP and reduced NADP will be available from the light dependent reaction. As there is very little CO_2 in the atmosphere (0.04%), CO_2 is likely to be the factor which limits the rate of photosynthesis. Adding more CO_2 will therefore mean that more CO_2 will combine with ribulose bisphosphate using the enzyme rubisco. As a result, the Calvin cycle can proceed at a greater rate and more glycerate-3-phosphate will be converted to triose phosphate which in turn is made into more carbohydrate (hexose sugar).

2 a i **J** = epidermal cell; **K** = mesophyll cell; **L** = bundle sheath cell.

ii Around the vascular bundle is a tight ring of bundle sheath cells which itself is surrounded by a ring of tightly fitting mesophyll cells. This arrangement ensures that the bundle sheath cells, where the light independent stage of photosynthesis takes place, are isolated from the air inside the leaf. This prevents photorespiration by preventing oxygen reaching the bundle sheath cells. It also prevents carbon dioxide being lost, which therefore accumulates within them and can be used when supplies from outside the leaf are in short supply. Plants which use C4 photosynthesis, add a carbon dioxide molecule to a three carbon molecule called phosphoenol pyruvate (PEP). The enzyme which catalyses this reaction is called PEP carboxylase which does not carry out oxidation and so there is no photorespiration. In addition, PEP carboxylase has a greater affinity for carbon dioxide than rubisco does and operates at a higher optimum temperature without being denatured.

b Wax reduces water loss and its shiny surface reflects sunlight, helping sorghum withstand dry conditions. Its higher melting point means it does not melt and so continues to function at the high temperatures of the tropics.

c i There was a greater reduction in CO_2 uptake in sorghum (from 5.5 to 1.2 mg CO_2 g^{-1}) than in soybean (5.2 to 1.6 mg CO_2 g^{-1}).

ii CO_2 is used in photosynthesis and so anything that reduces the rate of photosynthesis will reduce CO_2 uptake. The changes will reduce the surface area of the thylakoids, there will be less absorption of light, less light dependent reaction and therefore less reduced NADP and ATP produced. As these are necessary for the light independent stage, it too slows down and so less CO_2 is required to combine with PEP and therefore its uptake is reduced. As the spaces between thylakoids are reduced, there will be less chemiosmosis and this too will reduce the amount of NADP and ATP available for the light independent reaction.

3 a i At high light intensity, as the temperature increased, the volume of oxygen released increased from 1.0 mm^3 h^{-1} at 10 °C up to a peak of 4.0 mm^3 h^{-1} at 30 °C and then declined to 0.4 mm^3 h^{-1} at 50 °C. At low light intensity, as temperature increased, the volume of oxygen released remained constant at 0.4 mm^3 h^{-1} and then fell to 0.2 mm^3 h^{-1} at 50 °C.

ii Light is no longer the limiting factor but temperature is and the higher temperature has caused some hydrogen bonds that hold enzyme molecules to break, altering the shape of the active site (denaturation). Fewer enzyme-substrate complexes are formed and so there is less photolysis and less oxygen is produced.

b i Photolysis
ii Photosystem II (P680)
iii Respiration uses oxygen

4 a Chloroplasts vary in shape and size but are typically biconvex discs, 3–10 μm in diameter. They are surrounded by a double membrane called the chloroplast envelope. The inner membrane is folded into a series of lamellae. Inside the chloroplast envelope are two distinct regions. The stroma is a fluid-filled matrix within which are a number of structures such as starch grains and oil droplets. The stroma also contains ribosomes and DNA as well as the enzymes involved in the Calvin cycle. The grana are stacks of up to 100 disc-like structures called thylakoids. Within the thylakoids are the chloroplast pigments, which are arranged in a structured way and form a complex called photosystem II. Some thylakoids have tubular extensions that join up with thylakoids in adjacent grana. These are called inter-granal lamellae and are the sites of photosystem I.

b Palisade mesophyll cells are adapted to carry out photosynthesis because they are: closely packed and thin-walled to absorb maximum light; arranged vertically so there are fewer cross walls that could filter out the light; are packed with numerous chloroplasts that can move within the cells and so arrange themselves in the best positions to collect the maximum amount of light. Palisade mesophyll cells also have a large vacuole that pushes the cytoplasm and chloroplasts to the edge of the cell allowing them to absorb maximum light and leave a short diffusion pathway for carbon dioxide. In addition they have a large surface area and moist, thin walls for rapid diffusion of gases. Air spaces between the cells act as a reservoir of carbon dioxide for photosynthesis.

5 a The volume of oxygen produced / carbon dioxide absorbed.

b Light intensity because an increase in light intensity produces an increase in photosynthesis over this region of the graph.

c Raising the carbon dioxide level to 0.1% because this almost doubles the rate of photosynthesis while increasing the temperature to 35 °C produces only around a 20% increase.

d Because light is limiting photosynthesis and so an increase in temperature will not increase the rate of photosynthesis.

e More carbon dioxide is available to combine with RuBP to form more GP, then more TP and ultimately more hexose sugar and starch.

Chapter 14

1 a The label line from G should lead to the cells in the centre of the circular structure in the centre of the photo. This is the glomerulus. The label line from R should lead to any part of the white area surrounding the central cells. This is the renal capsule.

b Antidiuretic hormone

c i Because protein molecules are too large to pass through the basement membrane.
ii Glucose is reabsorbed into the blood from the filtrate in the proximal convoluted tubule.
iii There is a higher concentration of urea in the urine than in the filtrate because water is reabsorbed from the filtrate in the distal convoluted tubule leaving most of the urea to pass out of the body in the urine.

2 a i 17.9%
ii Fluid is able to pass through glomerular capillaries because there are pores in the capillary endothelium, so material only has to cross the basement membrane that acts as a filter. So no substances with a molecular mass greater than around 70 000 can cross, and

certainly no cells. Fluid can pass through podocytes because they have foot-like processes, which leave gaps between these processes.

b i Microvilli

ii To produce ATP for the active transport of sodium ions (Na^+) out of the epithelial cell

iii *Any two from the following*: glucose / amino acids / vitamins / any named mineral ion, e.g. chloride ions / some urea.

3 a As sweating involves a loss of water from the blood, its water potential will decrease / be lower / more negative.

b i Osmotic cells (in the hypothalamus)

ii Kidney

c As it is a hormone, it is transported in the blood.

d Absorption / taking in / consumption / drinking / of water. Because water has been lost during sweating. As the water potential of the blood returns to normal, the lost water must have been replaced. However, the kidney only excretes less water, it does not replace it. Therefore process X must be the way that water is replaced.

e Negative feedback.

4 a Adrenaline/cortisol

b The rise in insulin level is greater in group Y than in group X. The rise in insulin level is more rapid in group Y than in group X.

c Glucose is removed from blood by cells using it for respiration.

d Glucose level rises at first because the glucose that is drunk is absorbed into the blood (glucose line on the graph rises). This rise in blood glucose causes insulin to be secreted from cells (β cells) in the pancreas (insulin line rises steeply). Insulin causes increased uptake of glucose into liver and muscle cells, converts glucose into glycogen and fat and increases cellular respiration. All these actions have the effect of reducing glucose levels (glucose line falls from 2.5 hours onwards). As the glucose level rises after one hour, so the glucagon level falls. The reduction in glucagon level decreases glucose production from other sources (from glycogen, amino acids and glycerol) and so also helps to reduce blood glucose levels. As the blood glucose level falls (after 2.5 hours) so the glucagon level increases to help maintain the blood glucose at its normal level.

e Group X has diabetes and therefore the intake of glucose does not stimulate insulin production (insulin level as shown by the graph is low). The glucose level in the blood therefore continues to rise (glucose line rises steeply) as there is no insulin to reduce its level. Blood glucose level remains high, falling only slightly as it is respired by cells.

f As it was respired by cells, the glucose level would decrease steadily until it fell below the normal level.

5 a chloroplasts

b The thicker inner walls are less elastic than the thinner outer ones. As the volume of the guard cells increases, this difference in thickness causes the guard cells to bow and so widen the stomatal pore.

c i ATP synthase

ii potassium

iii active transport

iv The water potential of guard cells is lower than that of the epidermal cells.

v The volume of the guard cells increases.

d i to specific receptors on the cell surface membrane of the guard cells

ii calcium

iii potassium, which moves out by diffusion

Chapter 15

1 a At C, depolarisation of the axon membrane is occurring with sodium ions flowing into the axon making the inside of the membrane more positive. At D, repolarisation of the axon membrane is occurring with potassium ions flowing out of the axon making the inside of the membrane more negative. At E, hyperpolarisation is occurring causing a refractory period where the potential difference across the membrane is more negative than at resting potential.

b Stimulus A did not have a generator potential that was above the threshold value needed to produce an action potential.

2 a Oestrogen = follicle cells; Progesterone = corpus luteum.

b The progesterone and oestrogen cause a decrease in the secretion of both follicle stimulating hormone (FSH) and luteinising hormone (LH) from the anterior pituitary and so prevent the development of follicles in the ovary and ovulation. They also thicken secretions around the cervix and so form a barrier to sperm reaching the uterus as well as making the endometrium of the uterus thinner, so that implantation of an ovum is less likely.

c *Any two from the following*: The world population can be limited so that there is sufficient food and other resources for mankind and therefore less poverty / people can choose if and when to have children to suit their social and economic conditions / the children born are therefore more likely to be better cared for / women have greater control over their fertility and therefore greater freedom to go to work / there are health risks such as an increased risk of contracting a sexually transmitted disease and cervical cancer as a result of greater sexual freedom and promiscuity.

3 a

hormone	site of secretion	target tissue(s)	action during human menstrual cycle
FSH	anterior pituitary gland	follicles within ovaries	stimulates growth of follicle / stimulates follicle to produce oestrogen
progesterone	corpus luteum	anterior pituitary / endometrium	increases thickness of endometrium / inhibits LH secretion / inhibits FSH secretion

b The oestrogen and progesterone in the pill cause a decrease in the secretion of both follicle stimulating hormone (FSH) and luteinising hormone (LH) from the anterior pituitary and so prevent the development of follicles in the ovary and ovulation. They also thicken secretions around the cervix and so form a barrier to sperm reaching the uterus as well as making the endometrium of the uterus thinner, so that implantation of an ovum is less likely.

4 a A spinal reflex arc, such as withdrawing the hand from a hot object, begins with a stimulus of sufficient strength to exceed the threshold value of the (temperature) receptor in which case a generator potential is established. The generator potential leads to an action potential passing along the sensory neurone into the spinal cord via the dorsal root. The action potential then passes to a relay (intermediate) neurone that links the sensory neurone via synapses to the motor neurone within the grey matter of the spinal cord. The motor neurone carries an action potential away from the spinal cord to the effector (muscle of the forearm) which is stimulated to contract so that there is a response – the hand is raised away from the hot object. Reflex arcs are important because they are involuntary and so do not need the decision-making powers of the brain, leaving it free to carry out more complex responses. In this way the brain is not overloaded with situations in which the response is always the same. Some impulses are nevertheless sent to the brain, so that it is aware of what is happening. Reflex arcs protect the body from dangerous stimuli and they are effective from birth as they do not have to be learned. They are fast, because the neurone pathway is short with very few, typically one or two, synapses (which are the slowest link in a neurone pathway). This is important in withdrawal reflexes.

b Schwann cells wrap themselves around the axon many times, so that layers of their membranes build up around the axon. These membranes are rich in a lipid known as myelin and so form a covering called the myelin sheath which insulates the axon. The space between adjacent Schwann cells lacks myelin, forming gaps 2–3 μm long, called nodes of Ranvier. Sodium and potassium ions can only pass through the axon membrane at these points and so depolarisation cannot occur where there is a myelin sheath, but only at these nodes. Neurones with a myelin sheath transmit nerve impulses faster than neurones without the myelin sheath because local circuits are established between adjacent nodes. The action potential therefore 'jumps' from one node to the next in a process called saltatory conduction which increases the speed of transmission from $0.5\,\mathrm{m\,s^{-1}}$ in a unmyelinated neurone up to $100\,\mathrm{m\,s^{-1}}$ in a similar myelinated one.

5 a When an action potential arrives at the end of a presynaptic neurone it causes calcium channels to open and calcium ions enter the synaptic knob. This causes vesicles to fuse with the membrane of the presynaptic neurone leading to the release of acetylcholine into the synaptic cleft by the process of exocytosis. Acetylcholine diffuses across the synaptic cleft and fuses with receptor proteins on the membrane of the postsynaptic neurone. The receptor proteins change shape causing sodium channels to open allowing sodium ions to rapidly diffuse down a concentration gradient. This influx of sodium ions depolarises the postsynaptic membrane producing an action potential. The acetylcholine is then broken down by the enzyme cholinesterase.

b Synapses convey impulses from one neurone to the next and it is from this basic function that all the others arise. Synapses can only pass impulses in one direction, because only the presynaptic neurone has vesicles containing the neurotransmitter and only the postsynaptic neurone has receptors for the neurotransmitter. Synapses act as junctions, allowing nerve impulses to diverge and converge. This allows a single stimulus to create a number of simultaneous responses or combine many impulses into a single response. They also filter out low level stimuli. Low frequency impulses that produce insufficient amounts of neurotransmitter to trigger a new action potential in the postsynaptic neurone, can be made to do so by a process called summation. This entails a build-up of neurotransmitter in the synapse until the total amount exceeds the threshold value of the postsynaptic neurone, then a new action potential is triggered. Some synapses are inhibitory whereby they make it less likely that they will create a new action potential in a postsynaptic neurone. It is thought that synapses have a role in the brain where they are involved in memory and learning.

6 a The greater the diameter of an axon the faster the speed of conductance. From the table, compare data for the two myelinated axons – 20 μm conducts at $120\,\mathrm{m\,s^{-1}}$ while the smaller 10 μm diameter conducts at only $50\,\mathrm{m\,s^{-1}}$. Compare also data for unmyelinated axons – 500 μm diameter conducts at $25\,\mathrm{m\,s^{-1}}$ while 1 μm diameter conducts at $2\,\mathrm{m\,s^{-1}}$.

b In myelinated axons, the myelin acts as an electrical insulator. Action potentials can only form where there is no myelin (at nodes of Ranvier). The action potential therefore jumps from node to node (= saltatory conduction) which makes its conductance faster.

c Schwann cells.

d The presence of myelin has the greater effect because a myelinated human sensory axon conducts an action potential at twice the speed of the squid giant axon despite being only 1/50th of its diameter. (Similar comparisons can be made between other types e.g. squid and human motor neurone.)

e Temperature affects the speed of conductance of action potentials. The higher the temperature, the faster the conductance. Squid's conductance of action potentials will therefore change as the environmental temperature changes. It will react more slowly at lower temperatures.

7 a Muscles require much energy for contraction. Most of this energy is released during the Krebs cycle and oxidative phosphorylation, which take place in mitochondria.

b The actin and myosin filaments lie side by side in a myofibril and overlap at the edges where they meet. Where they overlap, both filaments can be seen. Where they do not, we see one or other filament only.

c Myosin is made of two proteins. The fibrous one is long and thin in shape, allowing it to combine with others to form a long filament, along which the actin filament can move. The globular protein forms two spherical structures (the head) at the end of a filament (the tail). This shape allows it to exactly fit recesses in the actin molecule, where it can become attached. The shape also means it can be moved at an angle. This allows it to change its angle when attached to actin and so move it along, causing the muscle to contract.

d Single ATP molecule is enough to move actin a distance of 40 nm. Total distance moved by actin = 0.8 μm (= 800 nm). Number of ATP molecules required = 800/40 = 20.

e One role of ATP in muscle contraction is to attach to the myosin heads, thereby causing them to detach from the actin filament and making the muscle relax. As no ATP is produced after death, there is none to attach to the myosin, which therefore remains attached to actin and leaves the muscle in a contracted state, i.e. rigor mortis.

8 a The glass plate prevents any lateral movement of IAA while allowing light to pass through.

b Experiment 2 shows that light causes IAA to accumulate more on the side away from the light source than on the side nearest to the light source when IAA is free to move within the tip. The glass plate in the agar block shows that the movement of IAA takes place in the tip and not in the agar block. Experiment 3 shows that when lateral movement of IAA in the tip is prevented, there is no redistribution of IAA.

c The total amount of IAA in the experiments in the dark (1 and 3) is approximately the same as the experiments with light from one side (2 and 4). If light destroyed IAA, one would expect there to be less when it was exposed to light.

9 The stimulus of the insect touching hairs generates an action potential that passes to the hinge cells. This causes auxin levels in the hinge cells to increase, leading to a proton pump moving protons out of the hinge cells into the cell walls. The increase in acidity makes the cell walls more plastic. The loss of protons from the hinge cells makes them more negative and so positively charged ions, such as calcium ions, are attracted into the cells, increasing their water potential. Water therefore enters the cells by osmosis. As the cell walls are very plastic, the cells rapidly expand, causing the lobes of the leaf to become concave, closing together and trapping the insect.

Chapter 16

1 a An allele is one of the different forms of a gene. Recessive refers to the condition in which the effect of an allele is apparent in the phenotype of a diploid organism only in the presence of another identical allele.

b The allele for RGC is on the X chromosome. As a male has only one X chromosome he requires just one recessive allele to be colour blind, a female has two X chromosomes and so requires two recessive alleles, which is less likely.

c 1 = $X^R X^r$; 4 = $X^R Y$; 6 = $X^r Y$; 7 = $X^R X^r$;

2

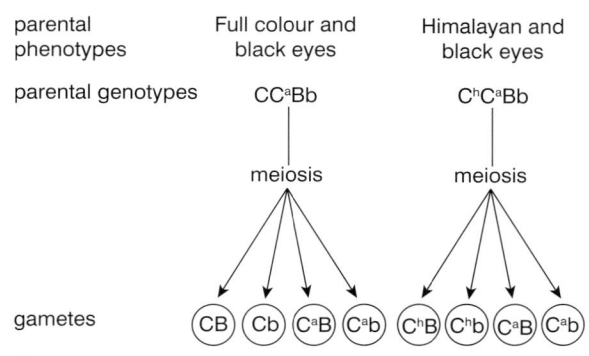

parental phenotypes — Full colour and black eyes — Himalayan and black eyes

parental genotypes — CC^aBb — C^hC^aBb

meiosis — meiosis

gametes — CB Cb C^aB C^ab — C^hB C^hb C^aB C^ab

Offspring

	CB	Cb	C^aB	C^ab
C^hB	CC^hBB	CC^hBb	C^hC^aBB	C^hC^aBb
C^hb	CC^h Bb	CC^hbb	C^hC^a Bb	C^hC^a bb
C^aB	CC^aBB	CC^aBb	C^aC^aBB	C^aC^aBb
C^ab	CC^a Bb	CC^abb	C^aC^a Bb	C^aC^abb

6 offspring – full colour and black eyes (C–B–)

2 offspring – full colour and red eyes (C– bb)

3 offspring – Himalayan and black eyes (C^hC^aB–)

1 offspring – Himalayan and red eyes (C^hC^a– bb)

3 offspring – albino and black eyes (C^aC^aB–)

1 offspring – albino and red eyes (C^aC^abb)

3 a An allele is one of the different forms of a gene. Dominant is a term applied to an allele that is always expressed in the phenotype of an organism even when present with an allele that is recessive to it.

b A = allele for striped body a = allele for ebony body
B = allele for long wings b = allele for vestigial wings

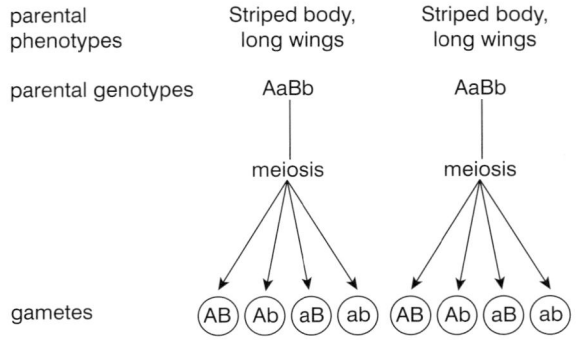

parental phenotypes — Striped body, long wings — Striped body, long wings

parental genotypes — AaBb — AaBb

meiosis — meiosis

gametes — AB Ab aB ab — AB Ab aB ab

Offspring

	AB	Ab	aB	ab
AB	AABB	AbBb	AaBB	AaBb
Ab	AABb	AAbb	AaBb	Aa bb
aB	AaBB	AaBb	aaBB	aaBb
ab	AaBb	Aabb	aaBb	aabb

9 offspring – striped body, long wings (A–B–)

3 offspring – striped body, vestigial wings (A–bb)

3 offspring – ebony body, long wings (aaB–)

1 offspring – ebony body, vestigial wings (aabb)

c i Missing values for grey body, vestigial wings from top down: 7; 49; 0.68.

ii 2.78

iii If the probability that the deviation is due to chance is greater than 0.05, any deviation is said to be not significant. Our value for chi squared of 2.78 represents a probability of between 0.2 and 0.5. This is greater than 0.05 and so the difference is not significant.

4 Let allele for Huntington's disease = H

let allele for normal condition = h

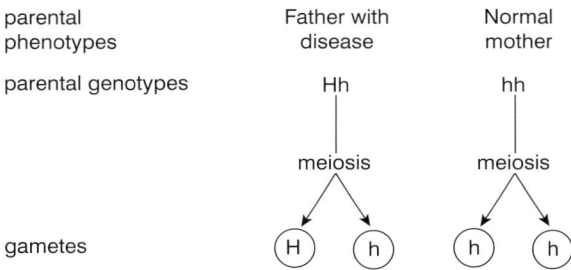

parental phenotypes — Father with disease — Normal mother

parental genotypes — Hh — hh

meiosis — meiosis

gametes — H h — h h

Offspring

	Father's gametes	
Mother's gametes	(H)	(h)
(h)	Hh	hh
(h)	Hh	hh

50% of offspring will have Huntington's disease (Hh)

50% of offspring will be normal (hh)

5 a Let allele for black coat = B

let allele for red coat = b

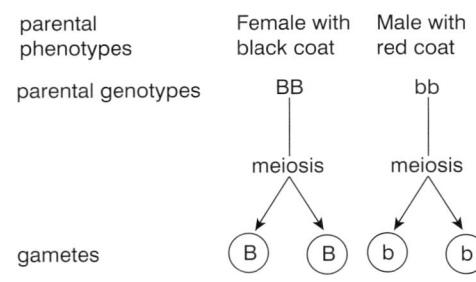

parental phenotypes — Female with black coat — Male with red coat

parental genotypes — BB — bb

meiosis — meiosis

gametes — B B b b

Offspring

	male gametes	
female gametes	ⓑ	ⓑ
Ⓑ	Bb	Bb
Ⓑ	Bb	Bb

All (100%) offspring will have black coats

b Allele for black coat = B

allele for red coat = b

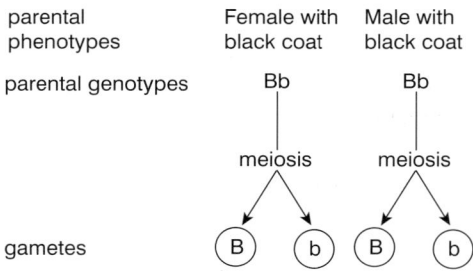

| parental phenotypes | Female with black coat | Male with black coat |
| parental genotypes | Bb | Bb |

Offspring

	male gametes	
female gametes	Ⓑ	ⓑ
Ⓑ	BB	Bb
ⓑ	Bb	bb

3 offspring (75%) with black coat (BB, Bb and Bb)

1 offspring (25%) with red coat (bb)

Probability of offspring having red coat = 1 in 4 (25%)

6 a

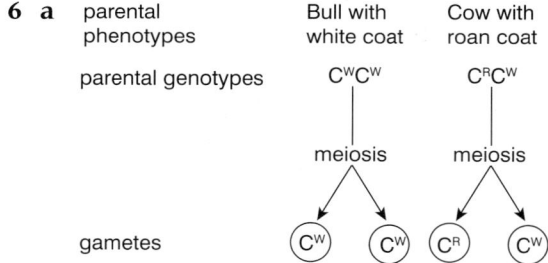

| parental phenotypes | Bull with white coat | Cow with roan coat |
| parental genotypes | C^WC^W | C^RC^W |

Offspring

	Male gametes	
Female gametes	Ⓒ^W	Ⓒ^W
Ⓒ^R	C^RC^W	C^RC^W
Ⓒ^W	C^WC^W	C^WC^W

b Half (50%) roan coat (C^RC^W)

Half (50%) white coat (C^WC^W)

   i 100%   **ii** 50%   **iii** 50%   **iv** 50%

7 The man is not the father.

Reasons: child has blood group AB and therefore has alleles I^AI^B. The mother is blood group A and therefore either I^AI^O or I^AI^A. In either case she could have provided the I^A alleles to the child but not the I^B allele. The I^B allele must have come from the real father. The supposed father is blood group O and therefore has alleles I^OI^O. He cannot provide an I^B allele and so cannot be the father.

8

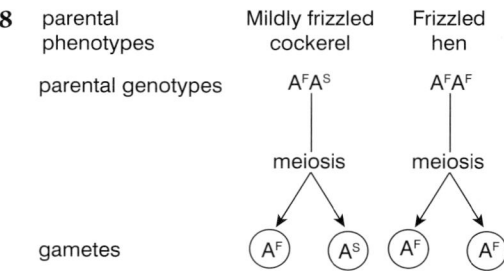

| parental phenotypes | Mildly frizzled cockerel | Frizzled hen |
| parental genotypes | A^FA^S | A^FA^F |

Offspring

	Cockerel gametes	
Hen gametes	Ⓐ^F	Ⓐ^S
Ⓐ^F	A^FA^F	A^FA^S
Ⓐ^F	A^FA^F	A^FA^S

Half (50%) Frizzled fowl (A^FA^F)

Half (50%) Mildly frizzled fowl (A^FA^S)

Chapter 17

1 a At first the population would increase only slowly (lag phase) because there are relatively few individuals to begin with and they are adjusting to the new environment. Later the population would increase rapidly (log phase) because the pond was favourable for their growth and reproduction e.g. there was plenty of food. The population would then stabilise (stationary phase) before decreasing in size (death phase) due to some factor such as predation or a build up of toxic wastes.

  b Phenotypic variation means that within a population there are individuals that possess different characteristics. Depending upon which of these characteristics suit the prevailing conditions, some individuals will be better able to survive and breed than others = natural selection.

2 a i The mean remains the same at 20 cm, but the graph is narrower (*see following sketch*).

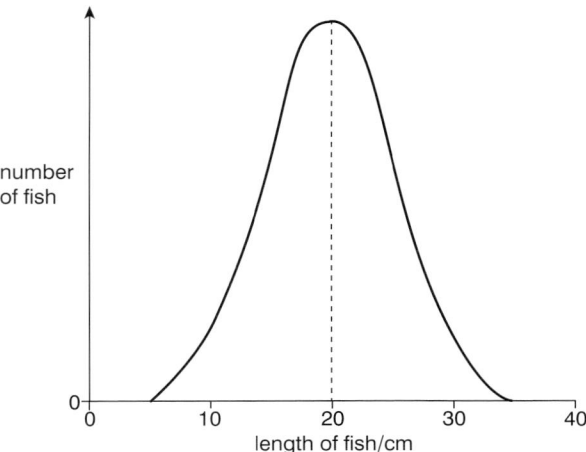

ii Stabilising

b i The graph is narrower and shifted to the left so that the mean is now less than 20 cm e.g. 15 cm (*see following sketch*).

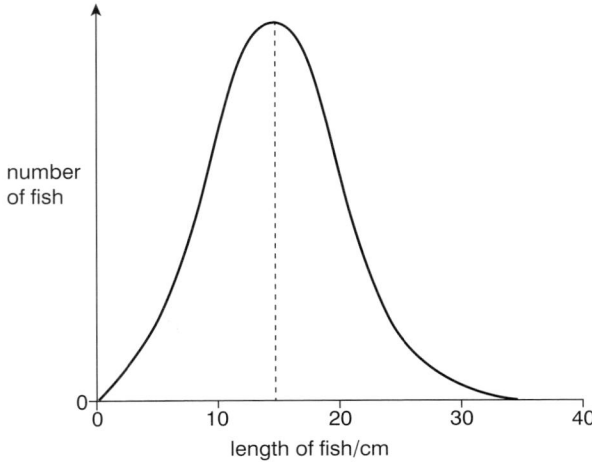

ii Directional
iii Predation and fishing

3 a *A. porcatus*

b The figures for *A. porcatus* and each of the other species are always lower than those for any two of the other species. For example, for *A. brunneus*, the figures are *A. smaragdinus* = 12.1, *A. carolinensis* = 16.7, *A porcatus* = 11.3. This means that they have smaller differences with *A. porcatus* than they have with each other showing that they are more closely related to *A. porcatus* than they are to one another.

c A group of *A. porcatus* on an island is geographically isolated from the rest of its population by the water between them. There is no flow of genes between the two populations. Mutations arise in both populations but each population (deme) is subject to different selection pressures because environmental conditions are different. As a result each deme becomes genetically different as it adapts to the different environmental influences it is subjected to. This is known as adaptive radiation and results in changes to allele frequencies, called genetic drift, in each population and in the various phenotypes present. As a result of these genetic differences it may be that, even if the species were no longer physically isolated from one another, they would be unable to interbreed successfully. Each group would now be a different species, each with its own gene pool. This type of speciation is called allopatric speciation.

4 a Selection is the process by which organisms that are better adapted to their environment survive and breed, while those less well adapted fail to do so.

b

Direction selection	Stabilising selection
Favours/selects phenotypes at one extreme of a population	Favours/selects phenotypes around the mean of a population
Changes the characteristics of a population	Preserves the characteristics of a population
Distribution curve remains the same shape but the mean shifts to the left or right	Distribution curve becomes narrower and higher but mean does not change

c Directional selection – because birds to one side of the mean (heavier birds) were being favourably selected, while those to the other side of the mean (lighter birds) were being selected against. The population's characteristics are being changed, not preserved.

5 a Removing cuckoo eggs means there will be more food for the magpie's own chicks. These chicks have a greater probability of being successfully raised to adulthood.

b Alleles for this type of magpie behaviour are obviously present in the adult birds. There is a high probability that some of the magpie chicks will inherit these alleles. Removing cuckoo eggs increases the probability of more of these magpie chicks surviving to breed and therefore passing on the alleles for this behaviour to subsequent generations.

c Displaying this behaviour has previously been of no advantage to magpies and so no selection for this behaviour has taken place. Although cuckoos have now arrived, it will take many generations for selection to operate and for allele frequencies to change.

d Directional selection – because the population's characteristics are being changed, not preserved.

Chapter 18

1   B = 3; C = 4; D = 9; E = 6; F = 2

2   **a**   Number of females nesting in 1993 = 275 compared to 90 in 2002. The period involved = 10 years. Therefore mean rate of decrease is:

$$\frac{(275 - 90)}{10} = 18.5.$$

   b   *Any five of the following*: Protect nest sites and young from predators / protect nest sites and nesting females from disturbance / be careful not to snare turtles in fishing nets / ban hunting of turtles and the sale of turtle products / prevent pollution of the seas with general rubbish, oil, etc. / establish conservation areas for turtles / breed turtles in captivity for later release into the wild / educate the population on the importance of turtles and how to protect them.

3   **a**   Kelp = (primary) producer

   Arrow worms = secondary consumers

   Narwhals = tertiary consumers

   b   Polar bears are tertiary or quaternary consumers, and as there is around a 90% energy loss at each trophic level (due to heat loss, movement, respiration, excretion, etc.) there is very little energy left to support a large population. In addition, polar bears only feed on ringed seals, and so their food supply is very limited, especially as ringed seals depend on a plentiful supply of their own food (e.g. amphipods, Arctic cod) to sustain a large population.

   c   A decrease in Arctic cod affects both the trophic levels below and above. The species at lower trophic levels (e.g. copepods) are likely to increase in numbers due to lack of predation from Arctic cod. This means the species they feed on (e.g. phytoplankton) will decrease in number as they are consumed by the increased number of copepods. The species at higher trophic levels (e.g. ringed seals, narwhal) are likely to decrease in numbers, as they have less food / have greater competition from other species for their food, which may lead them to migrate to other areas in search of food.

4   **a**   They have similar genes and therefore resemble one another, immunologically, biochemically and anatomically. They are capable of breeding to produce offspring which themselves are fertile. They have common ancestry. They occupy the same ecological niche.

   b   It is based on evolutionary relationships between organisms and their ancestors. It classifies species into groups using shared characteristics derived from their ancestors. It is arranged in a hierarchy in which groups are contained within larger composite groups with no overlap

   c   1. phylum   2. class   3. order   4. family   5. *Rana*   6. species   7. *temporaria*

5   **a**   Greenhouse gases lead to climate change. Communities with a high species diversity are likely to have at least one species adapted to withstand the change and therefore survive. Where the species diversity is low, there is less chance of a species being adapted to withstand the change and the community is therefore at greater risk of being damaged.

   b   i   The community fluctuates in line with environmental change – rising and falling in the same way but a little later in time.

   ii   Communities with high species diversity are more stable because they have greater variety of species and therefore are more likely to have species that are adapted to the changed environment. Those with low species diversity are less stable because they have fewer species and are less likely to have species adapted to the change. If the population of one species is reduced by the environmental change there is more likely to be an alternative species that serves a similar role where the index is high than where it is low.

Chapter 19

1   **a**   Insulin helps to maintain the blood sugar level at around 90 mg glucose per 100 cm^3 of blood by increasing the rate of cellular respiration rate, causing the conversion of glucose into glycogen (glycogenesis) in the cells of the liver and muscles, increasing the rate of conversion of glucose to fat in adipose tissue and increasing the rate of absorption of glucose into cells, especially muscle cells.

   b   *Any two of the following*: It is more effective because it is an exact copy of human insulin / it is more rapid in its action because it is identical to human insulin / there is no immune response, whereas animal insulin, with its slight differences, can sometimes stimulate an immune reaction / there is no risk of infection being transferred with the insulin while animal insulin can transfer certain diseases / some patients develop tolerance to animal insulin and become less sensitive to it / it is cheaper to produce in large volume than extracting and purifying animal insulin / animals are not involved in its production – many vegetarians were unhappy using an animal product in this way.

c i A single target site is important because more cuts would lead to the plasmid becoming fragmented. The single target site will be found in the correct resistance gene. The gene that is to be inserted will have complementary sticky ends to the sticky ends of the target site.

ii

Circle of DNA taken up by bacteria	Bacteria resistant to ampicillin	Bacteria resistant to tetracycline
Unaltered plasmids	✓	✓
Recombinant plasmids that have taken up the wanted gene	✓	✗
Circles of the wanted gene	✗	✗

d i Using genes for antibiotic resistance can lead to resistance being spread to other bacteria by either horizontal gene transmission or vertical gene transmission (conjugation). In these ways a phage can be introduced into bacteria transforming them from a form that is susceptible to the antibiotic to one that is resistant to it making the antibiotic less useful in treating disease. Bacteria can develop multiple resistance to many antibiotics making the treatment of some diseases e.g. tuberculosis, very difficult.

ii Using a gene that is transferred into the plasmid from a jellyfish. This gene produces a green fluorescent protein (GFP). The gene we are trying to clone is transplanted into the centre of the GFP gene. Any bacterium that has taken up the plasmid with the gene we want to clone will not be able to produce GFP. It will therefore not be fluorescent green while the ones that have not taken up our gene will be. Results can be obtained by simply viewing cells under a microscope and retaining those that do not fluoresce.

2 a i The CFTR protein is a chloride ion channel protein that actively transports chloride ions out of epithelial cells, and water naturally follows, by the process of osmosis.

ii *The external face of the membrane to be marked with a letter E is the upper face.* This is identified by the presence of the chains of hexagons that represent a glycoprotein.

b i Recessive refers to the condition in which the effect of an allele is apparent in the phenotype of a diploid organism only in the presence of another identical allele. An allele is one of the different forms of a gene.

ii In someone with CF, the epithelial membranes are dry and the mucus they produce remains viscous and sticky. This causes mucus congestion in the lungs, leading to a much higher risk of infection as the mucus, which traps disease-causing organisms, cannot be removed. It also causes breathing

difficulties and less efficient gaseous exchange. In some cases the lungs may be scarred.

c The viruses have the CFTR gene inserted into them using a plasmid. Viruses with the CFTR gene are then introduced into the nostrils of the patients. The viruses bind with lung cells and inject their DNA, which includes the normal CFTR gene, into the epithelial cells of the lungs.

d i The STOP codon will prevent translation occurring normally because the synthesis of the protein will stop when the STOP codon is reached. The protein will be shorter than normal and so not function properly.

ii *Any three of the following*: PTC124 can be taken by the patient alone whereas gene therapy (GT) requires medical staff / PTC124 is easily taken up by cells whereas take up using GT is poor / PTC124 is taken orally but GT uses a vector to deliver it to the lungs making it a more complex process / there are fewer side effects with PTC124 as no vector is involved but there are more side effects with GT because a vector is used / delivery of PTC124 is easy as it only has to enter the cytoplasm of cells whereas GT involves inserting a gene into host DNA / PTC124 does not require a gene to be switched on but GT does, and this can prove difficult.

3 a i Because the fetus has the same DNA band as the first child who has CF.

ii A deletion makes the allele lighter and so it is carried further during electrophoresis. This accounts for the four bands of DNA at the bottom of the figure. They are the mutant allele for CF. The two bands at the top of the figure show that the parents each also have the normal allele (without a deletion) which are heavier.

b A counsellor can explain the results of the test so that they are understood by the parents. Counselling can also inform them of the likely cost of any care for an additional affected child and the life expectancy of the child and how this might be extended using new, improved drugs or gene therapy. On the basis of this advice the couple can choose whether or not to have children or whether to use a sperm donation from an unaffected male. If conception has already occurred, the couple can decide whether to have a termination.

4 a A vector transfers genes/DNA from one organism into another.

b Gene markers are necessary to show which cells/bacteria have taken up the plasmid/gene.

c i B, C, D, J, K, L. (Those that did not take up the plasmid will not have taken up the gene for ampicillin resistance and so will be the ones that are killed on the ampicillin plate i.e. the colonies that have disappeared.)

ii E, F, I. (Those with the plasmid containing gene X will have lost the gene for tetracycline resistance and therefore the colonies will have been killed on the tetracycline plate i.e. the colonies will have disappeared.)

5 a PCR is used to increase the quantity of DNA because the quantity available, e.g. at a crime scene, is often very small.

b i Suspect B because the bands on this suspect's genetic fingerprint match those of the genetic fingerprint of blood found at the crime scene.
ii To eliminate the victim as the source of the blood sample found at the scene.

Chapter 20

1 a **A** = germinal epithelium; **B** = Graafian follicle.

b i primary oocyte.
ii

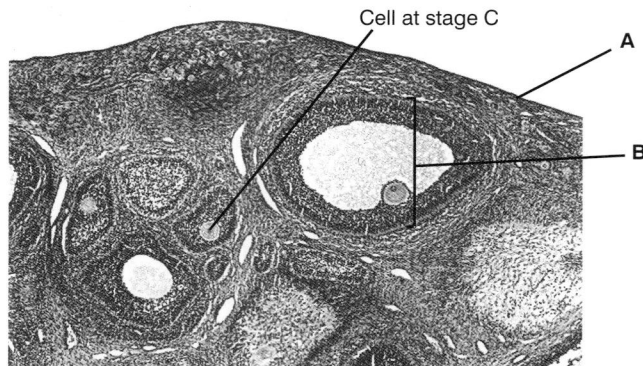

Cell at stage C

A

B

iii P = mitosis; **Q** = meiosis.

c *Either of the following*: Independent assortment of homologous chromosomes on the equator leads to random segregation of chromosomes leading to any combination of maternal and paternal chromosomes in the daughter cells / crossing over between equivalent portions of chromatids on homologous chromosomes leads to the exchange of genetic material and therefore new combinations of alleles.

2 a **A** corpus luteum; **B** primary oocyte; **C** secondary oocyte; **D** Graafian follicle; **E** primary follicle; **F** germinal epithelium.

b The primary follicle (**E**) is present in the ovary at birth. After puberty it may mature into a primary oocyte (**B**) and later a Graafian follicle (**D**). The secondary oocyte (**C**) from this Graafian follicle is released at ovulation and the empty follicle develops into a corpus luteum (**A**). The correct sequence is therefore **E, B, D, C, A**.

c i diploid **ii** haploid **iii** diploid.

3 a X and Y are produced as a result of division by mitosis but A and B are produced as a result of division by meiosis.

b In gametogenesis in females, of the four cells produced at the end of the second meiotic division, only one develops into an ovum. The other three form tiny, non-functional polar bodies. This means there are fewer gametes and so the available energy can be used for the growth of one, rather than four gametes, which are larger.

c The journey to the ovum is long and so to increase the chances of at least one sperm reaching it, many millions are produced. Mitosis is essential in producing these vast quantities.

d i Acrosome
ii It releases an enzyme that softens the plasma membrane covering the oocyte and it inverts to form a needle-like filament that pierces the softened plasma membrane and allows the nucleus of the sperm to enter.

e i Mitochondria
ii They carry out the respiratory processes that release the energy required for the sperm to swim up the oviducts to the oocyte.

Answers to Synoptic Questions

Question A

1 a Substitution

b i GAG

ii GUG

2 a The change in an amino acid may affect how it bonds with other amino acids in the same, or other, polypeptide chains. This can alter the tertiary structure of the protein, changing it from a compact globular shape to a more fibrous structure. Some amino acids may bind to amino acids on different haemoglobin molecules rather than those within their own molecule – hence the haemoglobin molecules stick together.

b Sufferers become tired because the haemoglobin S does not carry oxygen, and capillaries become blocked. Cells, especially those of muscle and the brain, are unable to produce sufficient energy from respiration due to low oxygen availability.

3 Natural selection

4 Codominance is the expression of both alleles in the phenotype.

5 a Individuals with the genotype Hb^SHb^S have anaemia that is so severe it outweighs the advantage of being resistant to malaria and so are always selected against. Individuals with the genotype Hb^AHb^A are susceptible to malaria and so are selected against in malarial regions. Individuals with the genotype Hb^AHb^S suffer a degree of tiredness, but this is more than outweighed by their increased resistance to malaria, and so their genotype is selectively favoured in areas where malaria is common and this genotype is therefore more frequent.

b Stabilising selection – because it favours average individuals rather than those at the extremes (i.e. those that are disadvantaged by being either susceptible to malaria or having sickle-cell anaemia).

6 a One in four (25%)

b The sickle-cell trait confers some resistance to malaria. Therefore, if individuals who are heterozygous for this condition live in a region where malaria is prevalent, the advantage of not developing malaria may more than offset the disadvantage of having sickle-cell trait. Individuals who live in non-malarial regions remain at a disadvantage because they have the sickle-cell trait.

7 a They will be the same (Hb^A allele = 0.6 and Hb^S allele = 0.4).

b The Hardy-Weinberg equation is $p^2 + 2pq + q^2 = 1.0$

H^A frequency = 0.6 = p

H^S frequency = 0.4 = q

$0.6^2 + (2 \times 0.6 \times 0.4) + 0.4^2 = 1.0$

$0.36 + 0.48 + 0.16 = 1.0$

Individuals with $H^AH^A = p^2 = (0.36 \div 1.0 \times 175) = 63$

Individuals with $H^AH^S = 2pq = (0.48 \div 1.0 \times 175) = 84$

Individuals with $H^SH^S = q^2 = (0.16 \div 1.0 \times 175) = 28$.

Question B

1 a i Freshwater snails

ii Energy present in the snails less that consumed by leeches and other organisms, i.e. $13122 - (1529 = 4210) = 7383\,kJ\,m^{-2}\,year^{-1}$

iii Energy passed down to sticklebacks is $1529 - (1269 + 194) = 66\,kJ\,m^{-2}\,year^{-1}$

Efficiency = $\times 100 = 4.32\%$

2 a P = reduced nicotinamide adenine dinucleotide phosphate ($NADPH + H^+$)

Q = starch

R = carbon dioxide

S = glycerate-3-phosphate (GP)

b 5

3 a i Phloem

ii Sieve tube elements and companion cells

b i Sucrose and amino acids

ii When phloem is cut, a solution of organic molecules is exuded. Plants provided with radioactive carbon dioxide can be shown to have radioactively labelled carbon in phloem after a short time.

Aphids that have penetrated the phloem with their needle-like mouthparts can be used to extract the contents of the sieve tubes. These contents show diurnal variations in the organic content of leaves that are mirrored a little later by identical changes in the organic content of the phloem.

The removal of a ring of phloem from around the whole circumference of a stem leads to the accumulation of organic substances above the ring and their disappearance from below it.

c An EM has a greater resolving power because it uses a shorter wavelength. Therefore more detail can be seen/things are clearer at the same magnification/ more structures are visible.

Question C

1 a A = distal convoluted tubule
B = Renal (Bowman's) capsule
C = ascending limb loop of Henle
D = collecting duct
E = descending limb of loop of Henle
F = glomerulus / glomerular capillary
G = efferent arteriole
H = proximal convoluted tubule
b i proteins
ii The pressure created by ultrafiltration in the glomerulus (F) is only sufficient to squeeze out molecules with a relative molecular mass (RMM) of up to 70 000. Protein molecules have a greater RMM.
c i glucose / amino acids
ii They are reabsorbed into the blood from the proximal convoluted tubule (H).
2 The aorta and the renal artery
3 a Magnification = 4.8 cm ÷ 25 μm
$$= \frac{4.8 \times 10^{-2}}{25 \times 10^{-6}}$$
$$= 1.06 \times 10^7$$
b i Microvilli
ii To increase the surface area for the reabsorption of substances such as glucose, amino acids, water.
c i Mitochondria
ii Active transport is involved in the reabsorption of substances by these cells. This is an active process requiring ATP. Mitochondria produce ATP during respiration.
4 a The glomerulus is a group of blood capillaries containing blood under high pressure. Any damage to these capillaries is likely to cause blood to pass into the Bowman's capsule and ultimately into the urine.
b i B cells
ii Antibodies attach to the antigen and act as markers that stimulate phagocytes to engulf cells to which they are attached.

Question D

1 The parents could each be heterozygous and carry the recessive CF allele, which if both passed it on would make the child homozygous recessive and a sufferer of CF.
2 Osmosis
3 a Because the mucus traps pathogens and it is usually removed by cilia lining the air passages. In CF sufferers the thick mucus cannot be removed and the pathogens remain in the lungs to cause infection.
b Lipases hydrolyse the ester bond in triglycerides forming fatty acids and glycerol.
4 a Genetically identical
b The cells constantly die and are replaced by new ones without the healthy gene.
5 a i A virus is an acellular, non-living particle with DNA or RNA as its genetic material. It can only multiply inside living host cells. The nucleic acid is enclosed within a protein coat called the capsid.
ii A vector is a carrier, e.g. a plasmid, which carries DNA into a cell, or to an organism that carries a parasite to its host.
iii A plasmid is a small circular piece of DNA found in bacterial cells.
b Viruses inject their DNA/RNA into host cells and so the incorporated CFTR gene will be introduced into the epithelial cells at the same time.
c They may cause infections.

Index